Advances of Healthy Environment Design in Urban Development

Advances of Healthy Environment Design in Urban Development

Guest Editors

Xia Zhang
Hui He

MDPI

Basel • Beijing • Wuhan • Barcelona • Belgrade • Novi Sad • Cluj • Manchester

Guest Editors
Xia Zhang
School of Urban Design
Wuhan University
Wuhan
China

Hui He
School of Architecture and Urban Planning
Huazhong University of Science and Technology
Wuhan
China

Editorial Office
MDPI AG
Grosspeteranlage 5
4052 Basel, Switzerland

This is a reprint of the Special Issue, published open access by the journal *Buildings* (ISSN 2075-5309), freely accessible at: https://www.mdpi.com/journal/buildings/special_issues/517PHPFOLF.

For citation purposes, cite each article independently as indicated on the article page online and as indicated below:

Lastname, A.A.; Lastname, B.B. Article Title. *Journal Name* **Year**, *Volume Number*, Page Range.

ISBN 978-3-7258-3983-4 (Hbk)
ISBN 978-3-7258-3984-1 (PDF)
https://doi.org/10.3390/books978-3-7258-3984-1

Contents

Preface vii

Zhiyi Zhou, Pengfei Wang, Jun Deng, Cheng Ouyang, Yuxuan Xu, Wanping Jiang and Kai Ma
Numerical Simulation of Street Canyon Morphology and Microclimate in Hot Summer and Cold Winter Zone
Reprinted from: *Buildings* **2023**, *13*, 2433, https://doi.org/10.3390/buildings13102433 1

Yue Hu, Ziyi Lin, Sheng Jiao and Rongpeng Zhang
High-Density Communities and Infectious Disease Vulnerability: A Built Environment Perspective for Sustainable Health Development
Reprinted from: *Buildings* **2024**, *14*, 103, https://doi.org/10.3390/buildings14010103 20

Hui He, Tong Zhang, Qinghao Zhang, Sheng Rong, Yihe Jia and Fengqian Dong
Exploring the Impact of Campus Landscape Visual Elements Combination on Short-Term Stress Relief among College Students: A Case from China
Reprinted from: *Buildings* **2024**, *14*, 1340, https://doi.org/10.3390/buildings14051340 39

Soujanya Mogra and Mohd Faris Khamidi
Applicability of Vegetation to Reduce Traffic-Borne PM2.5 Concentration in Roadside User Zones in Hot Arid Climates: The Case of Central Doha, Qatar
Reprinted from: *Buildings* **2024**, *14*, 1388, https://doi.org/10.3390/buildings14051388 57

Soo In Jee
Enhancing Dementia Nursing Homes in South Korea: Lessons from German Building Standards
Reprinted from: *Buildings* **2024**, *14*, 1427, https://doi.org/10.3390/buildings14051427 78

Yilin Wang, Wenjun Peng, Yanyan Huang, Tiancheng Yang, Xinyu Du, Xinyan Wang, et al.
Research on the Design Strategies of Public Outdoor Spaces in Government Centers from a Health Perspective
Reprinted from: *Buildings* **2024**, *14*, 1505, https://doi.org/10.3390/buildings14061505 113

Yu Wen, Bingbing Liu, Yulan Li and Lin Zhao
A Review of Research Progress on the Impact of Urban Street Environments on Physical Activity: A Comparison between China and Developed Countries
Reprinted from: *Buildings* **2024**, *14*, 1779, https://doi.org/10.3390/buildings14061779 127

Erke Zhang, Wei Zhao, Zihan Mei, Zhexi Yang, Fei Chen, Yuanyuan Xia and Yihan Wang
Experimental Study on the Universal Design of Signage Size and Brightness Contrast for Low Vision Individuals
Reprinted from: *Buildings* **2024**, *14*, 2063, https://doi.org/10.3390/buildings14072063 165

Zhiyi Zhou, Yuxuan Xu, Cheng Ouyang, Mengyao Gui, Wanping Jiang, Chunlei Zhou, et al.
Climate Adaptability Research of Vernacular Dwellings in Jiangxi Based on Numerical Simulation—An Example from Nanfeng County
Reprinted from: *Buildings* **2024**, *14*, 2211, https://doi.org/10.3390/buildings14072211 183

Yichi Zhang, Hui Tang, Kecheng Huo and Jiangfan Tang
Research on Urban Community Street Environment Evaluation and Optimization Strategy under the Concept of a Healthy City: A Case Study of the Dingwangtai Area of Changsha City
Reprinted from: *Buildings* **2024**, *14*, 2449, https://doi.org/10.3390/buildings14082449 216

Yuhua Li and Hui Tang
Health-Oriented Evaluation and Optimization of Urban Square Space Elderly Suitability: A Case Study of Yiyang City Center
Reprinted from: *Buildings* **2024**, *14*, 2482, https://doi.org/10.3390/buildings14082482 . . . . . . **234**

Shichao Du, Chunyu Wang and Longjian Mao
Exploring the Influence of the Perceived Neighborhood Built Environment on the Fall Risks among Older Adults in China
Reprinted from: *Buildings* **2024**, *14*, 2631, https://doi.org/10.3390/buildings14092631 . . . . . . **254**

Xinyue Yan and Tao Geng
Healing Spaces Improve the Well-Being of Older Adults: A Systematic Analysis
Reprinted from: *Buildings* **2024**, *14*, 2701, https://doi.org/10.3390/buildings14092701 . . . . . . **268**

Ying Cao, Danming Li, Xuewen Gao, Shibo Bi, Kexin Yu and Di Zhou
Age-Friendly Environment Design of High-Speed Railway Stations from a Healthy Ageing Perspective: A Case Implementation in Nanjing, China
Reprinted from: *Buildings* **2024**, *14*, 3280, https://doi.org/10.3390/buildings14103280 . . . . . . **299**

Lan Wang, Xia Zhang and Aolin Xie
Exploring the Link Between Social Media Engagement, Social Anxiety, and Environmental Preferences in University Students
Reprinted from: *Buildings* **2024**, *14*, 3284, https://doi.org/10.3390/buildings14103284 . . . . . . **321**

Santiago Quesada-García, Pablo Valero-Flores and María Lozano-Gómez
Residential Care Facilities for Users with Alzheimer's Disease: Characterisation of Their Architectural Typology
Reprinted from: *Buildings* **2024**, *14*, 3307, https://doi.org/10.3390/buildings14103307 . . . . . . **343**

Xiao Wang, Jingping Huang, Zhengyou Qin, Wei Gan, Zhen He and Xiang Li
Is the Children's 15-Minute City an Effective Framework for Enhancing Children's Health and Well-Being? An Empirical Analysis from Western China
Reprinted from: *Buildings* **2025**, *15*, 248, https://doi.org/10.3390/buildings15020248 . . . . **388**

Feifei Liu, Yuzhe Wang and Qi An
Spatio-Temporal Analysis of Beijing Residents' Lifestyles: Data-Driven Insights into Apartment Interior Design
Reprinted from: *Buildings* **2025**, *15*, 576, https://doi.org/10.3390/buildings15040576 . . . . **414**

Preface

The built environment has profoundly shaped human health, influencing physical, psychological, and social well-being. This Special Issue, Advances of Healthy Environment Design in Urban Development, compiles 18 studies that address urgent challenges in urban health through innovative design strategies, interdisciplinary methods, and evidence-based solutions.

Key Themes

1. Healthy Buildings: Climate and Inclusivity

Research highlights the synergy between traditional wisdom and modern technology. Zhou et al. (2023) demonstrate how vernacular Chinese dwellings optimize thermal comfort via natural ventilation, while Cao et al. (2024) propose age-friendly standards for transport hubs, emphasizing tactile navigation and ergonomic seating. For specialized needs, Quesada-García et al. (2024) identify compact layouts and sensory-controlled spaces as vital for Alzheimer's care facilities.

2. Healthy Communities: Resilience for Aging Populations

High-density urban areas require balanced design to mitigate health risks. Hu et al. (2024) link poor ventilation and overcrowding to disease spread, advocating decentralized green infrastructure. Li and Tang (2024) prioritize shaded rest areas and walkability in elderly friendly squares, supported by Du et al. (2024), who quantify how greenery and perceived safety reduce seniors' fall risks.

3. Healing Spaces: Mental Health in the Digital Age

Campus environments emerge as critical settings for mental health interventions. He et al. (2024) validate stress reduction through water features and shaded seating. Yan and Geng (2024) synthesize global evidence on healing spaces for older adults. Wang et al. (2024) address the intersection of digital behaviors and physical environments in campus design.

4. Active Urban Spaces: Generational Equity

Wen et al. (2024) call for "activity-friendly corridors" in Chinese cities, combining cycling lanes and pocket parks to boost physical activity. Wang et al. (2025) prove that 15-minute child-friendly neighborhoods enhance independent mobility through safe play zones and mixed-use pathways. Zhang et al. (2024) further link street vibrancy to holistic well-being in Changsha, prioritizing air quality and noise control.

Interdisciplinary Innovations and Future Directions

This Special Issue transcends traditional disciplinary boundaries, merging methodologies from environmental science, data analytics, and social epidemiology to advance healthy urban design. For instance, Mogra and Khamidi (2024) use computational fluid dynamics (CFD) simulations to quantify how roadside vegetation in Doha reduces PM2.5 concentrations by up to 35%, offering scalable solutions for arid regions. Liu et al. (2025) leverage spatiotemporal behavior data from Beijing residents to create adaptive apartment layouts that respond to evolving lifestyle trends, emphasizing modular furniture and flexible partitions. Such innovations highlight the transformative potential of technology in bridging design theory and practice.

Conclusions

Advances of Healthy Environment Design in Urban Development synthesizes global expertise to chart a path toward healthier, more equitable cities. By redefining the role of the built environment as both a determinant and enabler of health, this collection empowers stakeholders to prioritize human well-being in urban agendas. From adaptive vernacular architecture to AI-driven design tools, the studies herein illustrate that health-centric urbanization is not merely aspirational—but achievable through collaborative innovation, evidence-based policymaking, and community engagement. As cities continue to evolve, this Special Issue serves as a critical reference for transforming urban spaces into engines of holistic health, resilience, and social vitality.

Xia Zhang and Hui He
Guest Editors

buildings

MDPI

Article

Numerical Simulation of Street Canyon Morphology and Microclimate in Hot Summer and Cold Winter Zone

Zhiyi Zhou, Pengfei Wang, Jun Deng, Cheng Ouyang, Yuxuan Xu, Wanping Jiang and Kai Ma *

School of Architecture and Design, Nanchang University, No. 999 Xuefu Avenue, Honggutan District, Nanchang 330031, China; zhouzhiyi@ncu.edu.cn (Z.Z.); 406000210022@email.ncu.edu.cn (P.W.); dj@email.ncu.edu.cn (J.D.); 408800220006@email.ncu.edu.cn (C.O.); 408800220009@email.ncu.edu.cn (Y.X.); jiangwanping@ncu.edu.cn (W.J.)

* Correspondence: makai@ncu.edu.cn

Abstract: The design of street canyons has become a focus of attention under the requirements of high-quality urban modernization, while existing research has gradually broken through the basic norms of aesthetic design to include ecological considerations. However, it is only in recent decades that relevant research has been carried out in super and super-large cities in China. In this article, we take Nanchang, one of the largest cities in China, as an example, and use ENVI-met software (v5.5.1.) to simulate and analyze the street canyon of the city. Certain measurements were made and verified to compare the microclimatic conditions of street canyons at different scales. The relationship between street canyon morphology and outdoor thermal comfort was explored in terms of near-surface air temperature, wind speed, and thermal comfort indicators. The results show that there is a high correlation between the morphology of street canyons, such as orientation, aspect ratio interface continuity, and outdoor thermal comfort. Therefore, starting from adjusting the morphological characteristics of street canyons, practical suggestions can be provided for urban planners to guide the sustainable development of contemporary cities and improve the comfort of urban street spaces.

Keywords: outdoor spaces; pedestrian comfort; sustainable urban development; simulation; urban canyon airflow; urban microclimate

Citation: Zhou, Z.; Wang, P.; Deng, J.; Ouyang, C.; Xu, Y.; Jiang, W.; Ma, K. Numerical Simulation of Street Canyon Morphology and Microclimate in Hot Summer and Cold Winter Zone. *Buildings* **2023**, *13*, 2433. https://doi.org/10.3390/buildings13102433

Academic Editor: Chi-Ming Lai

Received: 25 August 2023
Revised: 23 September 2023
Accepted: 24 September 2023
Published: 25 September 2023

1. Introduction

With the continuing development of cities, their influence on the local climate is stronger than ever. According to the sixth report of the United Nations Intergovernmental Panel on Climate Change (IPCC), the global average temperature will rise by 2.1–3.5 °C if the greenhouse gas emission scenario is not changed [1]. Recent global climate change intensities, temperature anomalies, frequent rainstorms, and other issues have attracted many scholars' attention [2,3]. In response to the problems of rising global carbon dioxide and other greenhouse gas emissions, China proposed "Carbon peak" and "Carbon neutrality"; "Carbon peak" refers to an inflection point of greenhouse gas emissions. "Carbon neutrality" refers to achieving net-zero carbon dioxide emissions. While implementing these two plans, the latest "China Peak Carbon Dioxide Emissions Carbon Neutralisation Progress Report (2022)" emphasizes the need to implement differentiated and detailed plans in different regions and industries [4]. As a result, green and low-carbon transformation in various regions requires the development of regional reform plans based on local basic conditions. The Master Plan of National Land and Space of Nanchang City (2021–2035), published in 2022, also aspires to "promoting urban transformation, building an innovative and green central city" [5]. The development of green cities and the quality of urban human settlements will be the focus of future development.

However, due to high concentrations of population and industry in cities, a series of urban street spaces have emerged, with narrow urban street spaces surrounded by continuous and tall buildings on both sides. Street valleys are not only a fundamental component of a city, but are also an important place for urban residents to live and play. They are typically one of the types of public space with the highest use frequency, the most severe automobile exhaust pollution, and the highest daily population density. As a result, studying the microclimate environment in street valleys can help to optimize the urban environment and have a positive impact on the cities' future sustainable development.

Temperature, wind speed, humidity, and solar radiation are crucial climatic parameters for assessing the thermal conditions of street valleys, and they also serve as significant indicators for evaluating human thermal comfort. Urban meteorologist T.R. Oke was among the pioneers in exploring the connection between street design and the physical climate. They believed that urban design will influence the urban climate and that the geometric shape of street valleys will also influence the outdoor thermal environment [6].

Research that investigates thermal environmental quality using temperature as an evaluation factor concentrates on the block scale and geometry, including the aspect ratio of street valley space and the cooling effect of the sky's visual factors. Busato et al. (2014) conducted extensive measurements and simulations at different time periods, and the types of H/W, SVF, and surface determine very different heat exchanges between urban sites, air, and sky [7]. Bourbia et al. (2004) investigated the effect of street size and proportion on the thermal environment in low-latitude dry and hot areas, such as Algeria. The findings revealed that the surface temperature was significantly influenced by the street shape, and was closely related to the street geometry and the sky visibility factor [8]. Deng et al. (2018) also simulated and studied different forms of urban block schemes, and found that the aspect ratio and openness of street valleys played major roles in the influence of temperature and thermal comfort [9].

Wind as a research point for thermal environmental quality concentrates on the wind-blocking effects of building density and street geometry. Kja et al. (2020) used numerical simulation to investigate the relationship between extreme wind environments and urban morphology on the block level [10]. Wu et al. (2009) studied the relationship between the architectural layout and the wind environment in Shenzhen on the macro scale. They discovered that changing the architectural layout significantly changes the wind speed and direction of the dominant wind in the city [11]. Feng et al. (2019) discovered that a dense road network structure increased wind accessibility and improved ventilation effects by investigating the internal correlation between wind environment and spatial form [12]. By controlling the numerical changes of street morphological characteristics, Ying et al. (2018) studied the changing laws of the influence of shape factors such as line attachment rate, street interface density, and street height–width ratio on the wind environment. The results show that the decrease in street interface density would reduce the comfort in the street [13].

In relation to materials, we concentrate on the regulatory effect of physical properties on heat islands. Han et al. (2015) analyzed the ability of biomimetic building envelopes to reflect solar radiation. The results indicate that biomimetic structures can reduce the reflected heat of solar radiation from buildings in close-range spaces, thereby reducing the heat island effect [14]. Kousis et al. (2023) conducted a retrospective study on CPs (Cool Pavements). The performance of various evaluation scales was analyzed from the perspective of material properties, and corresponding limitations, research gaps, and future development directions were deeply explored [15]. Fabiania et al. (2019) studied the potential of urban roof materials to counteract urban overheating in summer. The results demonstrate that using thermochromic materials produces a smart optical response to local environmental stimuli, and allows for enhanced shortwave solar reflection in summer conditions, reduced reflected solar fraction in winter, and adaptive properties during transition periods [16].

Research on solar radiation concentrates on the obstruction of street trees and buildings to direct sunlight. Li et al. (2018) investigated the street geometry and the spatial distribution of street trees in Boston, and studied the shading effect of street trees with the shadow rate as the evaluation index [17]. Yang et al. (2018) analyzed the thermal environment characteristics of typical streets in winter and summer, and concluded that the openness of streets is the key factor affecting the thermal comfort of street space [18]. In the process of studying the influence of solar radiation on the thermal environment and thermal comfort of streets, Li et al. (2016) divided the street space into latent heat and sensible heat areas. The street space where buildings and pavements are directly exposed to the sun is the sensible heat area, and the street space where physical masks such as street trees block the direct sunlight is the latent heat area [19]. Wang et al. (2008), for example, conducted a simulation study on the street thermal environment in Wuhan. The results show that the best street thermal environment in Wuhan is in the Gezizhuang area, which is at an angle of 30 degrees off the true north direction. The results also show that the sloping roof helps to provide shelter in summer, while wide streets should be oriented east–west, and narrow lanes should be north–south [20].

Limor et al. (2004) summarized the typical geometric shapes of streets and studied the influence of street interface density, street height–width ratio, and building depth on air temperature through numerical simulation software [21]. Sun et al. (2016) established the ideal urban space model and block space model using software, and found that street orientation, height–width ratio, envelope coefficient, shadow coefficient, enclosure, and albedo have great influences on the thermal environment [22]. Yang et al. (2018) simulated six streets and lanes in the old city of Beijing using ENVI-met software, and found that the microclimate environment in the streets is very different from the overall climate of the city, and the thermal environment in the streets is affected by factors such as physical space environment and its use state [23]. Du et al. (2015) believe that the trend makes the greatest contribution to the thermal environment of living streets and valleys in Guangzhou, and the optimal trend is the southeast–northwest trend, based on numerical simulations of the thermal environment of living streets and valleys in Guangzhou. The thermal environment optimization design strategies and methods of Guangzhou living streets and valleys at all levels are obtained through optimization and induction when combined with field investigations and existing codes. The thermal environment of street valleys is influenced by local climatic conditions and geographical latitude, and the research results in different climatic regions cannot be simply analogized [24,25].

Based on the comprehensive analysis of the research results at home and abroad, it can be found that computer numerical simulation has become an important way to quantitatively analyze the impact of the urban thermal environment on the basis of field measurement. However, existing research mostly focuses on the spatial optimization of super cities and super-large cities, as these types of cities typically represent areas with high population concentrations. But the number of medium to large cities is larger, and the living population is not in the minority either. It is quite necessary to improve the quality of living space in such cities. Simultaneously, considering the different climates in different regions, selecting representative medium-sized and large cities in specific climate zones for research may have certain reference significance for the climate adaptation transformation of similar cities in the region. In addition, most of the existing studies focus on a single physical factor, while the thermal comfort of streets is affected by many factors. At the same time, in the simulation study of the microclimate influence mechanism, there are many studies on influencing factors such as urban greening and water bodies. But from the perspective of urban construction land, the area of green space and water bodies that can be used for planning and layout is relatively limited, while the area of streets or buildings accounts for a huge proportion of the city. Hence, the paper selects the microclimate of street canyons in the central urban area of Nanchang in the hot summer and cold winter regions for research to study the impact of urban street or building layouts on the local thermal environment. Due to the main focus of this study being on several influencing factors closely related to morphology, street greening, building materials, interface density, and other content, they are temporarily not within the scope of discussion. This aspect will continue in future research.

2. Case Study

2.1. Study Subjects

Nanchang, the provincial city of Jiangxi Province, is an important central city in China's middle reaches of the Yangtze River and the central city of the State Council-approved Poyang Lake Eco-Economic Zone. Nanchang was designated as a type I big city in China in December 2022 due to its rapid development and expanding city scale, it being a city with a permanent population of more than 3 million but less than 5 million. Nanchang, in the northwestern part of central Jiangxi, has a subtropical monsoon climate with long summers and long winters. It is a typical Chinese city with "hot summers and cold winters" and is one of China's "four stoves". Nanchang's urban construction dates back to the early Han Dynasty and gradually formed a unique urban pattern of "one large river, three little rivers, and ten lakes" in the evolution process of successive dynasties. But the city side is still evolving and expanding. The growth of a city is inextricably linked to the construction of the central city.

This paper investigates the street canyons in downtown Nanchang, extracts typical geometric shapes and builds physical models, and simulates and analyzes the thermal environment of urban streets under different layout forms from the standpoint of urban design. On the basis of meeting the technical regulations of Nanchang urban planning management, the characteristics of the microclimate inside the streets are analyzed, and their regular changes are summarized, in order to provide some reference and basis for street design and optimization in Nanchang urban renewal.

The annual average temperature in downtown Nanchang is 17–17.7 °C, the extreme historical maximum temperature is 40.9 °C, the extreme historical minimum temperature is −15.2 °C, and the annual average relative humidity is 78.5%. According to monthly climate data from China Ground International Exchange Station, the average summer temperature is 29.84 °C, the average maximum temperature is 38.72 °C, and the wind direction in summer is southeast wind. The average winter temperature is 3.12 °C, the average minimum temperature is −3.39 °C, and the wind direction in winter is north wind.

2.2. Description of the Studied Area

The main factors affecting the street layout of the microclimate are street aspect ratio, street orientation, sky visibility factor, and street interface density. This paper mainly analyzes the orientation and aspect ratio of streets in the central urban area of Nanchang, and the target area has a high population density and high development intensity. Through the arrangement of urban planning documents and related papers about Nanchang, it is concluded that the road network structure is regular and complete, and is basically vertically distributed.

With the classification of roads in China, the streets in the central urban area of Nanchang can be divided into expressways, main roads, secondary roads, and branch roads. According to the road network system planning map of the central urban area of Nanchang Land and Space Master Plan (2021–2035), there are 20 main roads in the central urban area, 8 in the north–south direction and 12 in the east–west direction; 174 secondary roads, 91 from north to south and 83 from east to west; and approximately 74 branches, 25 in the north–south direction and 49 in the east–west direction, according to incomplete statistics.

Because there are no pedestrians on the expressways, the expressway study will not be undertaken in this article. Furthermore, plots containing historical blocks of the old city will be excluded, since the buildings along the street in the old city street valley are mostly one to two floors, and can no longer meet the development requirements of modern urban areas. The living street is the most closely related to citizens' daily lives in the city, mainly comprising urban secondary roads and branches. As a result, the research object defined in this paper is the canyon-like space formed by urban secondary roads or branches with living functions and buildings on both sides.

In this paper, 24 streets were chosen from the representative urban streets of Nanchang using Google Earth and Street View, and a sample plot survey was conducted in

March 2023 on Hongguzhong Avenue, Shimao Road, Huizhan Road, and Shajing Road (Table 1). Because the distance between two intersections in downtown Nanchang is typically 100–260 m, the length of the street valley is chosen to be approximately 150 m to eliminate the influence of intersection morphological characteristics on the microclimate of the street valley. As the heights of buildings along the street vary, the research objects were selected to be as symmetrical as possible. The height calculation was based on the average building height of a uniformly continuous 150 m section. The sample plot survey was primarily concerned with street layout form, which had no correlation with survey time, so the actual survey had no impact on the simulation data.

Table 1. Statistical table of road conditions.

Road Name	Road Nature	Road Orientation	Average Building Height (/m)	Road Width (/m)	Aspect Ratio
Jinrong Avenue	minor road	North–south	78	17	4.6
Hongguzhong Avenue			64	23	2.7
Fenghuangnan Avenue			58	18	3.2
Shajing Road			52	18	2.8
Supu Road			19	6	3.1
Taohua Road			46	22	2.1
Fusheng Road			42	22	1.9
Chuanshan Road			25	12	2.0
Yuanming North Road			9	10	0.9
Huizhan Road		East–west	82	18	4.5
Lijing Road			18	16	1.2
Lingkou Road			42	22	1.9
Shimao Road			62	14	4.4
Yongshu Road			22	10	2.2
Zhongshan Road			40	6	6.7
Dieshan Road			28	18	1.5
Minde Road			23	6	3.8
Ziyun Road	branch road	North–south	20	10	2.0
Cuilin Road			38	10	3.8
Jiangbao Road			36	10	3.6
Minde West Road		East–west	19	6	3.1
Jiandeguan Road			22	4	5.5
Fanglin Road			25	6	4.2
Yindu Road			46	10	4.6

2.3. Physical Models

Based on on-site investigations, 2 east–west streets and 2 north–south street valleys were selected for simulation from 24 streets. For operability in the actual measurement process, adjacent roads are preferred (Figure 1). Finally, four representative roads were selected as the study's simulated streets: Hongguzhong Avenue and Shajing Road in the north–south direction, and Huizhan Road and World Trade Road in the east–west direction (Figure 2). Sketchup creates a street model based on the actual width and height of the four selected streets (Figure 3). Due to the limited performance of computers, the experiment appropriately simplified the internal details of the street model while maintaining accuracy. Simultaneously, to ensure the simulation's effectiveness, the simulation scale was set to 540 m. This paper mainly investigated the impact of street layout on microclimate using the aspect ratio and street orientation.

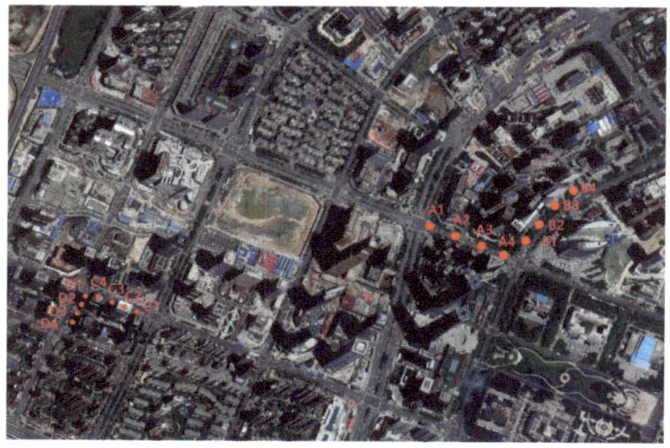

Figure 1. Schematic representation of the road measurement points.

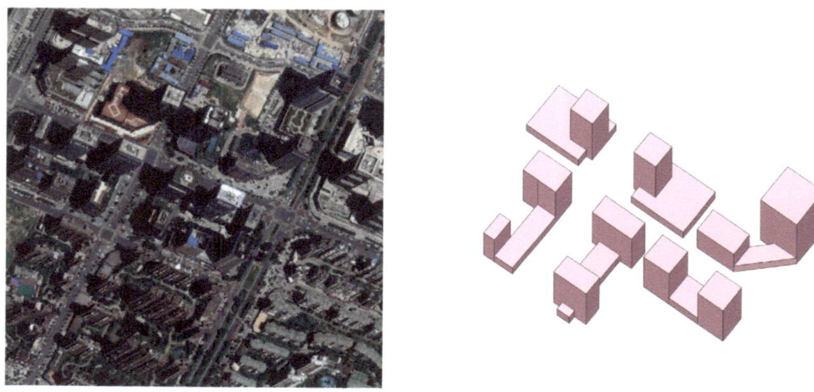

(**a**) (**b**)

Figure 2. (**a**) Satellite aerial view of Shajing–Shimao Road. (**b**) Simplified model of the measured site of Shajing–Shimao Road.

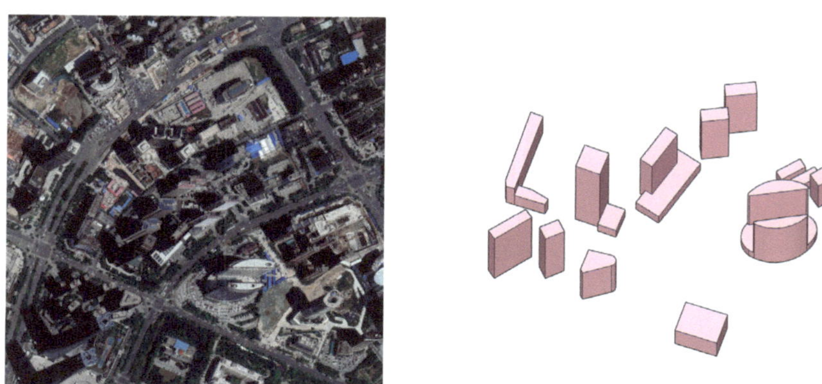

(**a**) (**b**)

Figure 3. (**a**) Satellite aerial view of the actually measured site of Huizhan–Hongguzhong Avenue. (**b**) Schematic simplified model of the actually measured site of Huizhan–Hongguzhong Avenue.

3. Materials and Methods

The geometric data of street morphology (street width, building height, and orientation) were collected through field investigations and used to study the influence of street canyon morphological factors on the urban thermal environment. Simultaneously, the open space of the research site was monitored using data measuring instruments, and the data were compared with those of Jiangxi Meteorological Bureau to ensure the instruments' accuracy. On this basis, the thermal environment model of typical street valleys was established using climate data from field investigations and measurements, such as temperature, humidity, and wind speed, and the original thermal environment of each street valley without greening was obtained by importing ENVI-met software for simulation operation. The simulation's accuracy was then validated by comparing the measured and simulated data. Finally, based on the simulation results, PET was used as a human thermal comfort evaluation index to measure the comfort within each street valley.

3.1. Field Measurement

At the moment, software simulation is used to study the outdoor street thermal environment. During the simulation experiment, the model can be appropriately simplified, such as by simplifying the scale and shape of buildings in street valleys, taking into account the computing capacity of the computer. However, no fixed standard for the model's simplification degree has been given in the past, and different simplification degrees produced different errors in the experimental results, so the accuracy of the simplified model had to be verified. Field measurement and simulation, on the other hand, have a sequential relationship and should be used repeatedly to verify the problem after determining the problem, so field measurement is critical in this study.

To compare the current situation of street microclimates under different street orientations and street aspect ratios, four measuring points were arranged on four streets, with a 40 m interval between each measuring point, for a total of 16 measuring points (Figure 1).

The research thoroughly considered the influence of air temperature, humidity, wind, and solar radiation on the street thermal environment, and adopted data on air temperature, humidity, and wind speed given by a TESTO-410 handheld temperature and humidity recorder, with the measurement ranges of −10 + 50 °C, 0.4–20 m/s, and 0–100% RH, and measurement accuracies of ±0.5 °C, ±0.2 m/s, and ±1.5% RH. HQZY-1 was used to measure black ball temperature, with a measuring range of −20 °C to +80 °C and a measuring accuracy of 0.3 °C.

Because the research focused on pedestrian thermal comfort, the test instruments were mounted on tripods so that the black ball of the temperature acquisition recorder and the sensor of the hand-held temperature and humidity anemometer were 1.5 m above the ground (Figure 4), and the test data recording interval was 1 min. At each measuring point, three groups of data were measured, and the average value of the three groups of data was used to calculate the average data per hour.

On 31 December 2022, the instrument collected 9–15 points of climate data in the open space near the measurement area, including temperature, humidity, and wind speed. The accuracy of instrument measurement was verified by comparing the meteorological bureau average data to the field measurement data.

3.2. Computational Settings and Parameters

3.2.1. Setting of Boundary Conditions

Shimao Road–Hongguzhong Avenue has a calculated horizontal area of 639 m × 639 m. Because the tallest building in the area is a super highrise, the calculation area's height should be double or more, so the vertical area dimension is 330 m, and the overall simulation environment is 639 m × 639 m × 330 m. There are 71 × 71 × 110 modeling grids with grid resolutions of dx, dy, and dz of 9 m, 9 m and 3 m, respectively.

Figure 4. Measurement schematic of instrument placement.

The calculated horizontal area in the Huizhan Road–Shajing Road area is 540 m × 540 m. Because the tallest building in the area is a super-highrise, the calculation area's height should be double or more, so the vertical area height is 210 m, and the modeling grids with the entire simulation environment of 540 m × 540 m × 210 m are 60 × 60 × 70, with grid resolutions dx, dy, and dz of 9 m, 9 m and 3 m, respectively.

Field measurement data were used to simulate the initial temperature, humidity, wind speed, and black ball temperature during the simulation feasibility verification process. The initial temperature, humidity, wind speed, and black ball temperature were substituted for data from typical meteorological days in the subsequent analysis and research. A typical meteorological day, as defined in the Standard for Design Standard for Thermal Environment of Urban Residential Areas (JGJ286-2013), is a day chosen in a typical meteorological year to represent seasonal climate characteristics [26]. A typical meteorological day in summer is the day when the daily average of temperature, daily range, humidity, and solar irradiance in the hottest month of a typical meteorological year is the closest to that of that month, so the data from 26 July 2022 were chosen for research.

When using ENVI-met to calculate PET, men with a height of 175 cm, weight of 70 kg, age of 35 years, clothing thermal resistance of 0.9 clo, and walking speed of 1.34 m/s were used as the human thermal environment perception parameters.

3.2.2. Data Processing Method

(1) Formula of the human thermal balance:

$$M - W = Q_{cv} + Q_r + E_{sk} + Q_{res} + S \tag{1}$$

(2) Thermal comfort equation:

$$\begin{aligned} M - W = {} & f_{cl} h_c (T_{cl} - T_a) + 3.96 \times 10^{-8} f_{cl} \left[(T_{cl} + 273)^4 - (T_{mrt} + 273)^4 \right] + 3.05[5.73 - 0.007(M - W) - P_a] \\ & + 0.42[(M - W) - 58.15] + 0.0173M(5.87 - P_a) + 0.0014M(34 - T_a) \end{aligned} \tag{2}$$

(3) The average radiation temperature, Tmrt, is calculated as:

$$T_{mrt} = \left[\left(T_g + 273.15\right)^4 + \frac{1.10 \times 10^8 {v_a}^{0.6}}{\varepsilon D^{0.4}} \times \left(T_g - T_a\right) \right]^{0.25} - 273.15 \tag{3}$$

Remark 1. *M—Metabolic quantity* (w/m^2). *W—Workload* (w/m^2). Q_{cv}*—Convective heat dissipation on the skin surface* (w/m^2). Q_r*—Radiation heat dissipation on skin surface* (w/m^2). E_{sk}*—Evaporative heat dissipation on skin surface* (w/m^2). Q_{res}*—Breathing heat dissipation* (w/m^2). *S—heat storage capacity of human body* (w/m^2). f_{cl}*—Clothing area coefficient (%).* h_c*—Convective heat transfer coefficient* ($w/(m^2{}^\circ C)$). T_{cl}*—The average temperature of the outer surface of human body or clothing (°C).* P_a*—water vapor partial pressure (kPa).* T_a*—Ambient temperature (c).* T_{mrt} *is the average radiation temperature (°C);* T_g *is the black ball temperature (°C); εIs is emissivity (black sphere takes 0.95); D is the diameter of the black ball (150 mm).*

4. Results and Discussion

4.1. Verification of the Measured Results

To test the consistency of ENVI-met software simulation and actual measurements, the winter parameters of four street valleys in the actual measurement area were simulated hourly from 10 am to 3 pm, and the simulation results of temperature, humidity, and wind speed were compared to the actual measurement values. At the same time, the measuring points near the intersection in the street valley would inevitably be disturbed during the actual measurement. So, the measuring points connected to the middle were taken for data comparison during verification, and the measuring points on Huizhan Road were named A1, A2, A3 and A4. Those on Hongguzhong Avenue were named B1, B2, B3 and B4. Those on Shimao Road were named C1, C2, C3 and C4, and those on Shajing Road were named D1, D2, D3 and D4, with Huizhan Road and Shimao Road running east–west, while Hongguzhong Avenue and Shajing Road run north–south.

4.1.1. Air Temperature

The figure depicts the air temperature distribution 1.5 m above the ground at 10 a.m. and 2 p.m., as well as the statistics of the measured and simulated temperatures at each measuring point (Figure 5). The difference between the simulated temperature and the measured temperature during the day was small, as shown in the figure, and the temperature in the east–west direction was obviously higher than that in the north–south direction, while the temperature on the east–west street was higher than that in the north–south direction. The difference was mostly within 1 °C, but at 2 pm, there was a deviation of about 1.8 °C at the measured temperature (Figure 6).

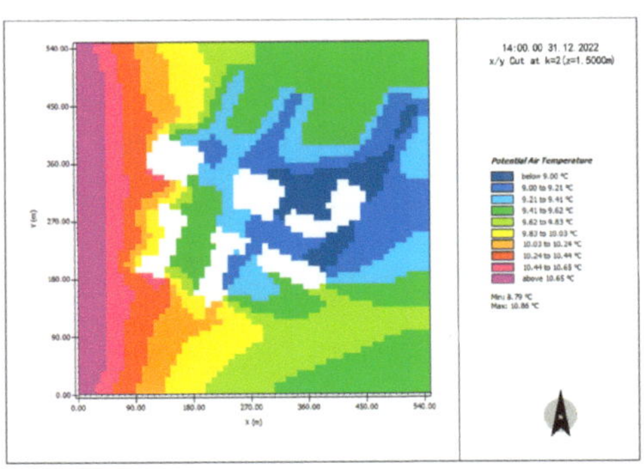

(a)

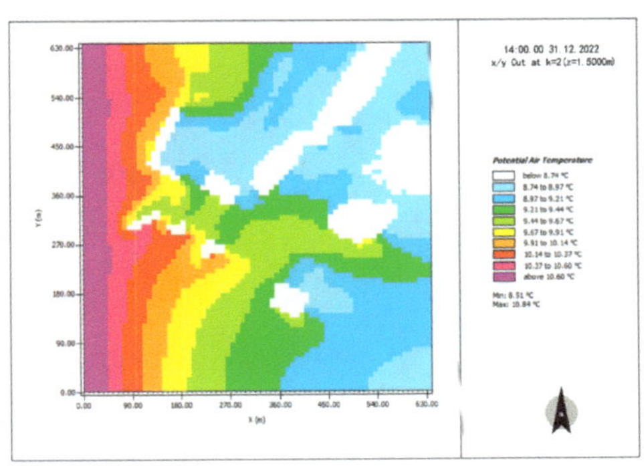

(b)

Figure 5. (**a**) Simulation of the air temperature in Shajing–Shimao Road at 14:00 in winter. (**b**) Simulation of the air temperature at 14:00 in winter on Huizhan–Hongguzhong Avenue.

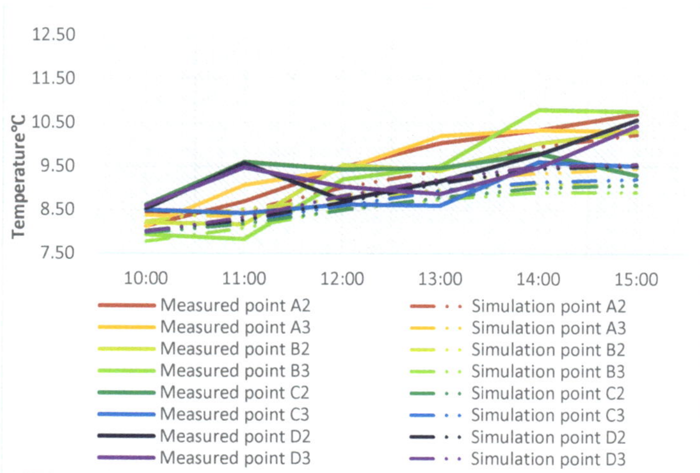

Figure 6. Comparison of the simulated and actual measured hourly temperatures at individual measuring points.

4.1.2. Air Humidity

By comparing the measured and simulated air humidity data of each measuring point at 10:00 and 15:00 in streets (see Figures 7 and 8), we can determine that the measured and simulated air humidity values were not closely coupled, and the errors between the measured and simulated values were about 20%, which may be due to the influence of greening on both sides of the road during actual measurement, but the measured and simulated values of the overall air humidity distribution were roughly consistent.

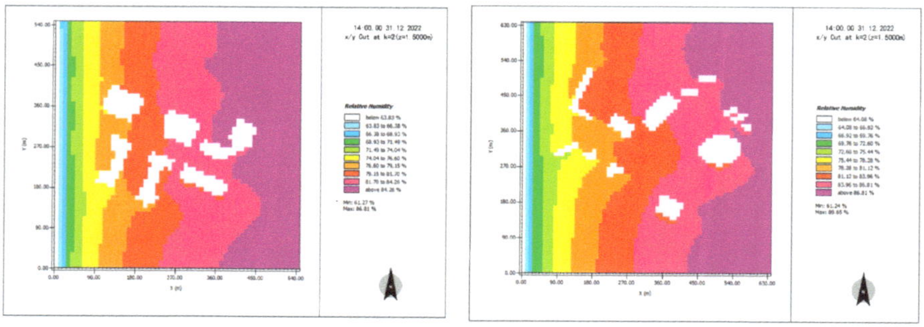

(**a**) (**b**)

Figure 7. (**a**) Simulation of the air humidity in Shajing–Shimao Road at 14:00 in winter. (**b**) Simulation of the air humidity at 14:00 in winter on Huizhan–Hongguzhong Avenue.

4.1.3. Wind Speed

By comparing the measured and simulated wind speed data of each measuring point at 10:00 and 15:00 in the streets, it was found that the difference between the simulated and measured wind speed lay in its stability, and the measured values were more volatile, while the simulated values tended to be more stable. However, the measured values of the change trend of the overall wind speed were roughly consistent with the simulated values (see Figures 9 and 10).

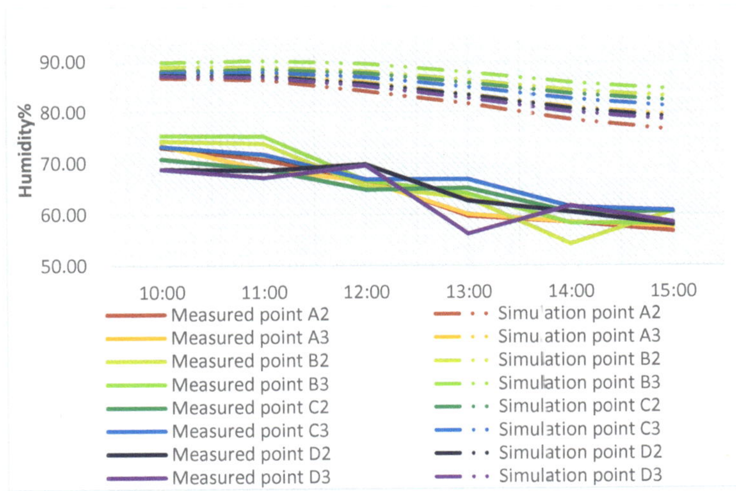

Figure 8. Comparison of the simulated and actual measured hourly humidity at individual measuring points.

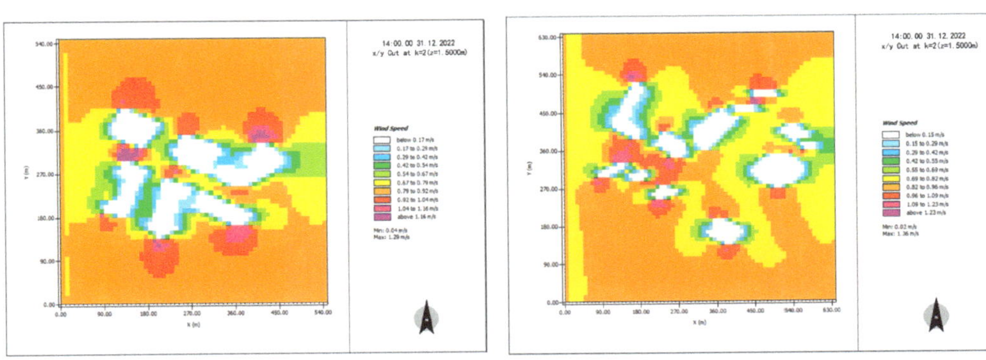

(a) **(b)**

Figure 9. (**a**) Simulation of the air wind speed on Shajing–Shimao Road at 14:00 in winter. (**b**) Simulation of the air wind speed at 14:00 in winter on Huizhan–Hongguzhong Avenue.

In order to verify the accuracy of the average value of the measurement input in the simulation, the root mean square error (MRSE) operation was performed on the measured values of each measuring point. The results show that the maximum MRSE value of the wind speed in each measuring point is 1.78, and the maximum value of the MRSE temperature is 0.78, indicating that the measurement errors of individual measuring tend to be stable for both wind speed and temperature measurements, which proves that the average values of individual measuring points are reliable.

In order to further analyze the errors between the measured values and the simulated values, an analysis was performed using the mean absolute percentage error (MAPE). The results show that the MAPE values of the wind speed and temperature measured in winter and the simulated values when the models were used for simulation analysis were mostly between 1% and 12%; a few of them were above 15%, and many groups were below 1%.

Through simulation verification, it can be concluded that ENVI-met is a climate simulation software with certain accuracy in simulating the microclimate environment in street valleys. So it is of certain significance to use this software to simulate and predict the microclimate environment in different street valleys.

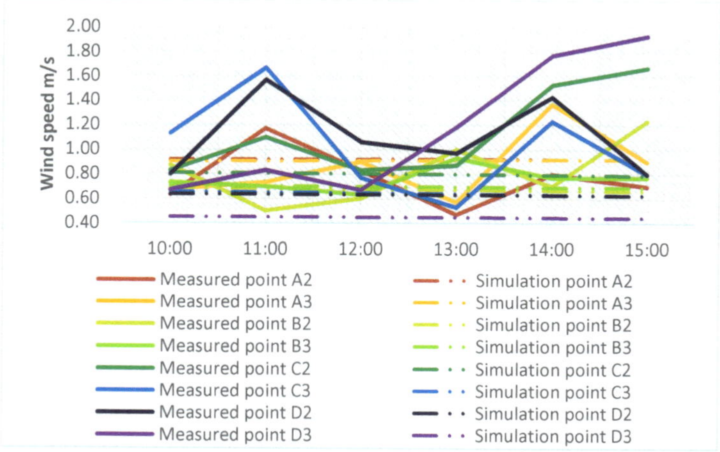

Figure 10. Comparison of the simulated and actual measured hourly wind speed at individual measuring points.

4.2. Analysis of Simulation Results

ENVI-met, a simulation analysis software that can couple the outdoor wind, heat, humidity and sunshine environment of the city, was mainly used to simulate the street and valley environment of the research object in summer, and four main indexes, air temperature, wind speed, average radiation temperature and thermal comfort PET, were compared. Because the large amount of simulation data and the changing trends of individual data were consistent from 10:00 to 15:00, and they reached their maximum values for the whole day at 14:00, the following were selected (see Figures 11–14).

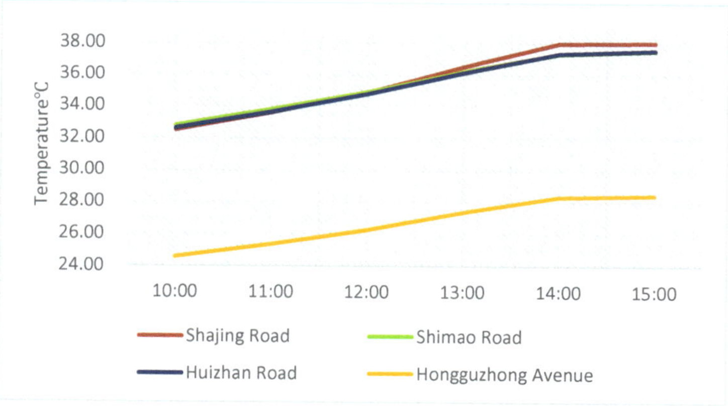

Figure 11. Average temperatures of four streets.

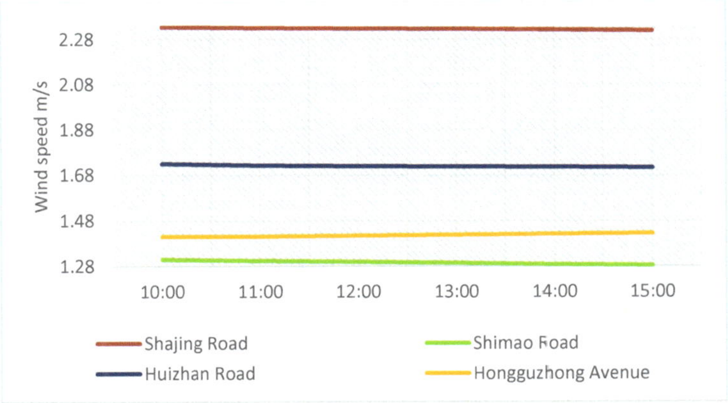

Figure 12. Average wind speeds in four streets.

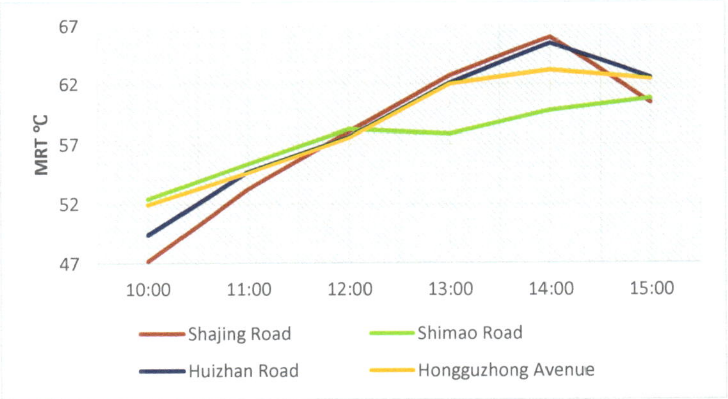

Figure 13. Average MRTs of four streets.

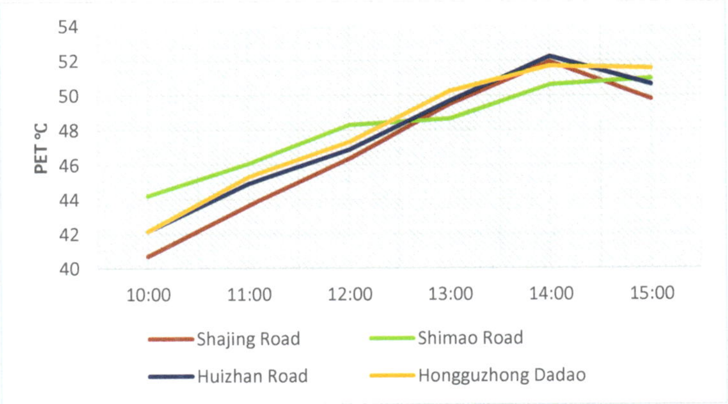

Figure 14. Average PETs of four streets.

4.2.1. Air Temperature

The incoming flow temperature of the Shimao Road–Shajing Road area and Huizhan Road–Hongguzhong Avenue area was around 38.70 °C, according to the simulation results, and their internal temperatures were lower than the incoming flow temperatures (Figure 15). The temperatures of Shajing Road in the Shimao Road–Shajing Road area were 37.13–38.75 °C in the north–south direction and 36.59–37.67 °C in the east–west direction. The temperature of Hongguzhong Dadao in the north–south direction in the Huizhan Road–Hongguzhong Avenue area was 37.05–38.66 °C, while that at Huizhan Road in the east–west direction was 36.52–37.32 °C, with a difference of 0.00–2.14 °C. Overall, the simulation results for the Huizhan Road–Hongguzhong Avenue area are comparable to those for the Shimao Road–Shajing Road area. The east–west street valley had a lower average temperature than the north–south street valley, and the high-temperature area of the east–west street was primarily concentrated in the gap between the buildings along the street. Although the temperature of the north–south street varied, the high-temperature area appeared primarily at the intersection in the inflow direction, and the temperatures of the other areas were within 1.00 °C of the temperature of the east–west street.

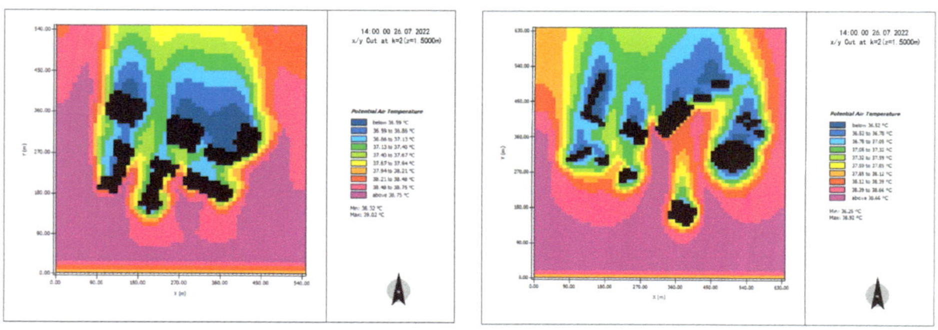

(**a**) (**b**)

Figure 15. (**a**) Simulation of the air temperature in Shajing–Shimao Road at 14:00 in summer. (**b**) Air temperature simulation of Huizhan–Hongguzhong Avenue at 14:00 in summer.

4.2.2. Wind Speed

The simulation results show that the areas of Shimao Road–Shajing Road and Huizhan Road–Hongguzhong Avenue were all southerly, with wind speeds ranging from 1.68 to 2.10 m/s. Inside the street, the wind environment was relatively large, with noticeable differences between regions (Figure 16). Among them, the north–south Shajing Road in the Shimao Road–Shajing Road area conformed to the incoming wind direction, and most of the wind speeds were 1.65–1.95 m/s, but it is easy to form a wind field with wind speeds of 2.32–2.98 m/s with the blocking of buildings on both sides, which had a significant environmental impact. However, the east–west World Trade Road deviated from the incoming wind. The north–south direction of Hongguzhong Avenue in the Huizhan Road–Hongguzhong Avenue area also conformed to the incoming wind direction, with a relatively high wind speed. However, because the road is relatively wide, the blocking effect of buildings was minimal, and there was no wind field with high wind speeds, as there was on Shajing Road. The overall wind speeds were 1.26–2.10 m/s, with a local increase to 2.10–2.52 m/s. However, the east–west layout of Huizhan Road is fundamentally similar to that of the World Trade Road. The wind speeds were relatively high in the gaps between buildings along the street, reaching 2.10–2.52 m/s, while the wind speeds were relatively low in other areas, with overall wind speeds of 0.42–1.26 m/s.

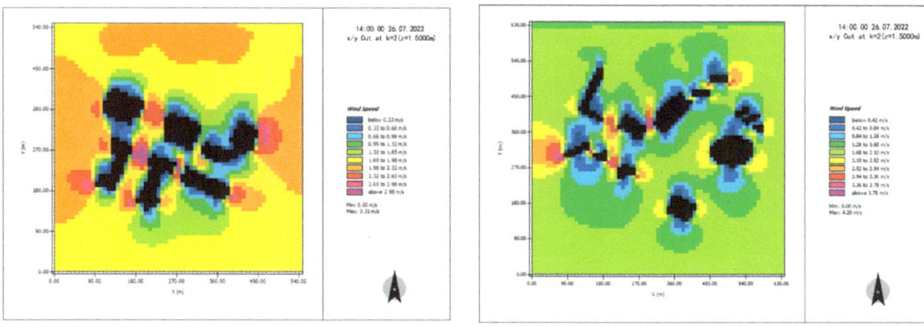

(**a**) (**b**)

Figure 16. (**a**) Simulation of the air wind speed at 14:00 in Shajing–Shimao Road in summer. (**b**) Simulation of the air wind speed at 14:00 in summer on Huizhan–Hongguzhong Avenue.

4.2.3. Mean Radiation Temperature MRT

According to the simulation results, the average radiation temperatures outside the Shimao Road–Shajing Road area and the Huizhan Road–Hongguzhong Avenue area were about 63.75–64.82 °C, while the average radiation temperature inside the street was significantly higher than that in the external environment, and the street temperatures in all directions were similar, with the only relatively low-temperature area being near the north of the building (Figure 17). The average radiation temperatures of Shajing Road in the north–south direction in the area of Shimao Road–Shajing Road were between 65.24 and 66.44 °C. The average radiation temperatures of World Trade Road in the east–west direction were 56.86–59.25 °C in the area near the north of the building, while the rest were between 65.24 and 66.44 °C. The average radiation temperatures of Hongguzhong Avenue in the north–south direction were between 64.67 and 65.82 °C in the area from Huizhan Road to Hongguzhong Avenue, and the average radiation temperatures of Huizhan Road in the east–west direction were 56.65–57.79 °C in the area near the north of the building, while the rest were between 64.67 and 65.82 °C. The area of Huizhan Road–Hongguzhong Avenue was slightly better than that of Shimao Road–Shajing Road, and the average radiation temperatures at the intersection in the middle were between 64.04 and 65.24 °C.

4.2.4. Thermal Comfort PET

The simulation results show that the external PET temperatures were roughly 51.75–52.90 °C, but the average radiation temperatures inside the street were clearly different (Figure 18). The north–south street temperatures were similar to the external environment, and the east–west street temperatures were obviously lower than the external environment, but the entire street was in an extreme thermal stress reaction environment (Table 2). The average radiation temperatures of the east–west Shimao Road in the Shimao Road–Shajing Road area were 48.62–51.86 °C, but there would be a local high-temperature area in the west, and the temperatures could reach 56.18–57.26 °C. Shajing Road's average radiation temperatures in the north–south direction ranged between 50.78 and 51.86 °C. The average radiation temperatures of the east–west Huizhan Road in the Huizhan Road–Hongguzhong Avenue area were 50.44–51.65 °C, with a high-temperature area on the west side ranging from 52.87 °C to 54.09 °C. In the north–south direction, the average radiation temperatures of Hongguzhong Avenue were between 50.44 °C and 52.87 °C. Inside the street, the temperatures dropped as one moved further north of the building, and the average radiation temperatures of World Trade Road's continuous street surface were also lower than those of Huizhan Road's relatively discontinuous street surface. The average radiation temperatures at the intersection of Huizhan Road and Hongguzhong Avenue were clearly higher than 26.71 °C.

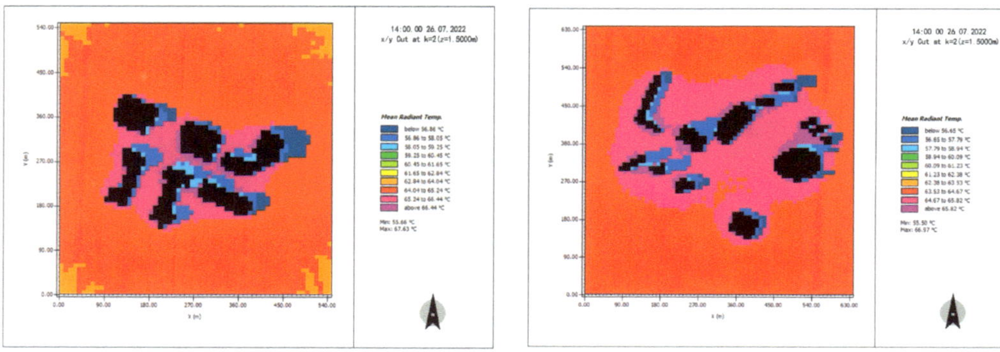

(a) (b)

Figure 17. (**a**) Simulation of the MRTs at 14:00 in Shajing–Shimao Road in the summer. (**b**) MRT simulation of Huizhan–Hongguzhong Avenue at 14:00 in the summer.

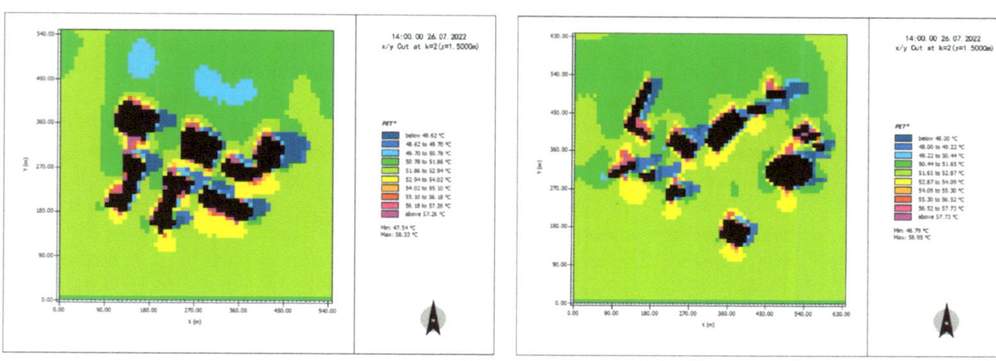

(a) (b)

Figure 18. (**a**) Simulation of the PET at 14:00 in Shajing–Shimao Road in summer. (**b**) Simulation of the PET at 14:00 in summer on Huizhan–Hongguzhong Avenue.

Table 2. Physiological equivalent temperatures and the corresponding human sensations [27].

Physiologically Effective Temperature (°C)	Thermal Sensation and Thermal Comfort	Physiological Stress Level
<4	Quite cold	Extreme cold stress reaction
4–8	Cold	Strong cold stress reaction
8–13	Cool	Moderate cold stress response
13–18	Slightly cool	Mild cold stress reaction
18–23	Comfortable	No cold stress reaction
23–29	Slightly warm	Mild thermal stress reaction
29–35	Warm	Moderate heat stress reaction
35–41	Hot	Strong thermal stress reaction
>41	Very hot	Extreme heat stress reaction

4.2.5. Description of Experimental Results

Through ENVI-met numerical simulation, the following information can be obtained.

(1) As the air temperature changed, the north–south streets conformed to the incoming flow direction, and the ventilation cooling effect improved. The air velocity would increase in the gap between the buildings along the east–west street, causing it to cool down. The roads in the area of Huizhan Road–Hongguzhong Avenue are wider, with an aspect ratio of 2.7. Hongguzhong Avenue has the smallest aspect ratio among the four streets, which can better utilize the cooling effect of natural ventilation.

(2) In response to the wind direction, continuous building surfaces are prone to a narrow tube effect, which leads to localized wind enhancement and is not conducive to normal pedestrian activity. At building breaks, on the other hand, they are relatively empty and the flow tends to level off. On streets facing away from the wind direction, continuous building surfaces can also cause large areas of static wind, which is not conducive to the natural dispersion and reduction of household waste and pollutants.

(3) In terms of changes in average radiant temperature, the buildings, as substantial walls in a street canyon, have a significant impact on thermal radiation. The heating of building surfaces by sunlight leads to a significant increase in the mean radiant temperature in the surrounding area. The average radiant temperature on the north side of the building is relatively low due to the shading effect of the building. On the one hand, it reduces the direct radiation from the sun. On the other hand, hot air is also blown from the south, and the radiation propagation process is asymmetrical between the windward and leeward sides.

(4) According to the change in the value of the thermal comfort indicator PET, it could be seen that the temperatures of the four streets in the order of high to low were Shimao Road > Huizhan Road > Shajing Road > Hongguzhong Avenue. This order of arrangement is the same as the order of the sizes of the aspect ratios of the four streets. The relatively high temperature in the space area indicates that the buildings have some shading effect on the solar radiation. The shading of buildings could effectively reduce the average radiant temperature within the street. The larger the aspect ratio of the building, the better the shading provided by the building.

5. Conclusions

The main contribution of this study is to investigate the thermal comfort of street canyon space in typical medium to large cities in a hot summer and cold winter region from the aspects of neighborhood morphology. Based on the foregoing, we can conclude that street orientation, aspect ratio, and interface continuity had significant effects on the thermal environment. We can make the following conclusions and recommendations in light of these findings:

(1) The street aspect ratio should be maintained between 0.5 and 1.0 to ensure optimal thermal comfort in the summer. Extremely low ratios fail to block direct sunlight, while high ratios create excessive shadow areas, affecting fluid dynamics and reducing comfort;

(2) North–south streets should be positioned between 30 degrees east of south to due south. This helps buildings and streets receive favorable sunlight in Nanchang, a low-latitude area in the southern hemisphere. It also accommodates the local southeast monsoon and enhances ventilation in the summer, while resisting the cold north wind in winter;

(3) Street design should avoid long continuous interfaces and incorporate localized open spaces. Discontinuous interfaces in north–south streets delay the narrow tube effect, while discontinuous interfaces in east–west streets reduce the coverage area of static wind zones. Additionally, expanding the south side of east–west streets accommodates wind patterns in winter and summer;

(4) Adopting a pattern of lower buildings on the south side and taller buildings on the north side for east–west streets aids air flow when the summer monsoon blows from the south to the east. This arrangement also creates a comfort zone on the south side during winter winds from the north, due to the asymmetric heat transfer process.

Author Contributions: Conceptualization, Z.Z.; methodology, K.M.; software, P.W. and J.D.; validation, P.W. and J.D.; formal analysis, C.O.; investigation, Y.X. and J.D.; writing—original draft preparation P.W.; writing—review and editing K.M.; supervision, W.J. All authors have read and agreed to the published version of the manuscript.

Funding: This research received no external funding.

Institutional Review Board Statement: Not applicable.

Informed Consent Statement: Not applicable.

Data Availability Statement: Not applicable.

Conflicts of Interest: The authors declare no conflict of interest.

References

1. Lee, H.; Calvin, K.; Dasgupta, D.; Krinner, G.; Mukherji, A.; Thorne, P.; Trisos, C.; Romero, J.; Aldunce, P.; Barrett, K.; et al. *Climate Change 2023: Sixth Assessment Report Synthesis Report*, 2nd ed.; IPCC: Geneva, Switzerland, 2023; pp. 35–115.
2. Liu, J.N.; Chen, T.; Wang, L.Y. Thermal Health Risk Identification, Asessment and Urban Design Intervention in High-Density Cities from the Perspective of Climate Change: A Case Study of Macao. *Landsc. Urban Plan.* **2022**, *1*, 1–14.
3. Abbass, K.; Qasim, M.Z.; Song, H.M.; Murshed, M.; Mahmood, H.; Younis, I. A Review of the Global Climate Change Impacts, Adaptation, and Sustainable Mitigation Measures. *Environ. Sci. Pollut. Res.* **2022**, *29*, 42539–42559. [CrossRef] [PubMed]
4. Zhang, H. China's Carbon Peak and Carbon Neutrality Progress Report (2021) Released in Beijing. *China Natl. Cond. Natl. Strength* **2022**, *1*, 79.
5. Chen, J.; Yao, S.; Tian, Y. Morphological Typology Theory and the Localisation of its Exploration. *Int. Urban Plan* **2015**, *2*, 57–64.
6. Nunez, M.; Oke, T.R. The Energy Balance of an Urban Canyon. *J. Appl. Meteorol. Climatol.* **1977**, *1*, 11–19. [CrossRef]
7. Busato, F.; Lazzarin, R.M.; Noro, M. Three Years of Study of the Urban Heat Island in Padua: Experimental Results. *Sustain. Cities Soc.* **2014**, *10*, 251–258. [CrossRef]
8. Bourbia, F.; Awbi, H.B. Building Cluster and Shading in Urban Canyon for Hot Dry Climate. *Renew. Energy* **2004**, *2*, 291–301. [CrossRef]
9. Deng, Y.; Zheng, X.; Wong, N.H. The Effect of Street Level Isthmus Morphology on Outdoor Air Temperature and Thermal Comfort in Summer. *J. Zhejiang Univ. Eng.* **2018**, *5*, 873–885.
10. Kja, B.; Vmnab, C. Interactions Between Extreme Climate and Urban Morphology: Investigating the Evolution of Extreme Wind Speeds from Mesoscale to Microscale. *Urban Clim.* **2020**, *31*, 100544.
11. Wu, Z.; Yan, T.; Fu, X. Analysis of Urban Wind Environment in Shenzhen Based on CFD Simulation. *Constr. Qual.* **2009**, *11*, 49–53.
12. Feng, W.; Fei, M.; Zhen, M.; Feng, S. Research on Numerical Simulation of Wind Environment in Typical Neighbourhoods of Xi'an and Urban Optimization Design Strategy. *Mod. Urban Res.* **2019**, *8*, 35–40.
13. Ying, X.; Kan, Q. Impacts of Key Elements of Spatial Layout Patterns of Hangzhou Commercial Streets on the Surrounding Wind Environment. *Sci. Geogr. Sin.* **2018**, *12*, 2093–2099.
14. Han, Y.; Taylor, J.E.; Pisello, A.L. Toward Mitigating Urban Heat Island Effects: Investigating the Thermal-energy Impact of Bio-Inspired Retro-reflective Building Envelopes in Dense Urban Settings. *Energy Build.* **2015**, *102*, 380–389. [CrossRef]
15. Kousis, I.; Pisello, A.L. Evaluating the Performance of Cool Pavements for Urban Heat Island Mitigation under Realistic Conditions: A Systematic Review and Meta-Analysis. *Urban Clim.* **2023**, *49*, 101470. [CrossRef]
16. Fabiani, C.; Pisello, A.L.; Bou-Zeid, E.; Yang, J.; Cotana, F. Adaptive Measures for Mitigating Urban Heat Islands: The Potential of Thermochromic Materials to Control Roofing Energy Balance. *Appl. Energy* **2019**, *247*, 155–170. [CrossRef]
17. Li, X.; Ratti, C. Mapping the Spatial Distribution of Shade Provision of Street Trees in Boston Using Google Street View Panoramas. *Urban For. Urban Gree.* **2018**, *32*, 109–119. [CrossRef]
18. Yang, X. Research on the Typology of Thermal Comfort Shade Regulation of Urban Streets in Xi'an under the Influence of Solar Radiation. Master's Thesis, Xi'an University of Architecture and Technology, Xi'an, China, 2018.
19. Li, D. Research on Microclimate Analysis of Xi'an City Square and Street Space under the Influence of Solar Radiation. Master's Thesis, Xi'an University of Architecture and Technology, Xi'an, China, 2016.
20. Wang, Z. Research on Climate Adaptability Design Strategy of Block Gorge Based on Urban Micro-climate in Hot Summer and Cold Winter Areas. Master's Thesis, Huazhong University of Science and Technology, Wuhan, China, 2008.
21. Limor, S.B.; Tzamir, Y.; Milo, E.H. Thermal Effects of Building Geometry and Spacing on the Urban Canopy Layer Microclimate in a Hot-humid Climate in Summer. *Int. J. Climatol.* **2004**, *13*, 1729–1742.
22. Sun, X.; Yang, J.Y.; Wen, S.S. *Research on Spatial Morphology and Thermal Environment of Urban Center Area Based on ENVI-met Simulation-The Case of Nanjing Xinjiekou*, 2nd ed.; China, ACP. 60 Years of Planning: Achievements and Challenges—Proceedings of the 2016 Annual Conference on Urban Planning in China; China Building Industry Press: Beijing, China, 2016; pp. 124–140.
23. Yang, X.; He, S.; Lu, X. Research on the Evaluation of Thermal Comfort of Public Space in the Old City of Beijing Based on Software Simulation—Taking 6 Hutongs in Baita Temple Area as an Example. *Urban Archit.* **2018**, *6*, 51–56.

24. Du, H.; Shi, Y.; Zhang, Y. The Measured Study on Thermal Environment of Typical Living Street Valleys in Guangzhou. *Build Sci.* **2015**, *12*, 8–13.
25. Du, H. Study on Thermal Environment Design Strategy of Living Street Valleys in Guangzhou. Master's Thesis, South China University of Technology, Guangzhou, China, 2015.
26. Ministry of Housing and Urban-Rural Development of the People's Republic of China. *Design Standard for Thermal Environment of Urban Residential Areas JGJ 286-2013*, 3rd ed.; China Architecture & Building Press: Beijing, China, 2014; pp. 44–52.
27. Hppe, P.R. The Physiological Equivalent Temperature, a Universal Index for the Biometeorological Assessment of the Thermal Environment. *Int. J. Biometeorol.* **1999**, *2*, 71–75. [CrossRef] [PubMed]

buildings

MDPI

Article

High-Density Communities and Infectious Disease Vulnerability: A Built Environment Perspective for Sustainable Health Development

Yue Hu [1], Ziyi Lin [1], Sheng Jiao [1,2,*] and Rongpeng Zhang [1,2,3,*]

1 School of Architecture and Planning, Hunan University, Changsha 410082, China; huyue66@hnu.edu.cn (Y.H.); linziyi@hnu.edu.cn (Z.L.)
2 Hunan Key Laboratory of Sciences of Urban and Rural Human Settlements in Hilly Areas, Hunan University, Changsha 410082, China
3 Hunan International Innovation Cooperation Base on Science and Technology of Local Architecture, Changsha 410082, China
* Correspondence: jiaosheng2008@163.com (S.J.); zhangrongpeng@hnu.edu.cn (R.Z.)

Abstract: High-density communities have proliferated globally during rapid urbanization. They are characterized by a high population density and limited per capita public spaces, making them susceptible to infectious disease risks. The impact of infectious diseases in these communities, as evident during the COVID-19 pandemic, underscores their vulnerabilities. Yet, research on disease prevention in high-density areas remains limited. This study aims to investigate the relationship between the built environment and the transmission of infectious diseases in high-density urban communities, with a particular focus on the lessons learned during the COVID-19 pandemic. Utilizing Shenzhen city as a case study, this study collected data on the built environment and epidemic trends and involved a generalized linear regression analysis, aiming to understand the key built environment factors that affect epidemic spread in high-density areas. The results from the study revealed that high-density communities experience higher rates of infectious disease transmission compared to their medium- to low-density counterparts. The significant factors identified include land use mixture and walkability, with land use mixture showing the most substantial impact on infection rates. Through a combination of qualitative analysis and empirical research, we constructed a conceptual framework linking containment measures, non-pharmaceutical interventions, and the built environment. The findings emphasize the significance to focus on the health development of high-density communities and offer valuable insights for tailored urban planning and built environment design. These insights are crucial for promoting the healthy and sustainable transformation of existing high-density communities.

Keywords: built environment; high-density community; epidemic infection; healthy; sustainable

Citation: Hu, Y.; Lin, Z.; Jiao, S.; Zhang, R. High-Density Communities and Infectious Disease Vulnerability: A Built Environment Perspective for Sustainable Health Development. *Buildings* **2024**, *14*, 103. https://doi.org/10.3390/buildings14010103

Academic Editor: Pablo Pujadas Álvarez

Received: 8 December 2023
Revised: 26 December 2023
Accepted: 28 December 2023
Published: 30 December 2023

1. Introduction

Since the 20th century, cities have been significantly impacted by recurrent outbreaks of infectious diseases [1]. These recurring global outbreaks underscore the fact that cities have become epicenters for disease transmission [2]. Infectious diseases such as AIDS, SARS, HPAI H5N1, and COVID-19 have wrought unprecedented consequences on human health, social stability, and economic progress [3]. The confluence of urbanization and globalization has amplified population mobility and substantial migration to urban centers, creating favorable conditions for the spread of infectious diseases [4,5]. In this context, the community, as a fundamental unit within cities, plays a pivotal role in pandemic preparedness and response. In the post-pandemic era, urban planning aimed at curbing community transmission and fostering the development of resilient and healthy communities has become paramount.

We conducted a literature search in the WOS core database and retrieved a total of 2616 articles in the field of urban planning from 2020 to 2023. Through a keyword analysis, we observed a heightened scholarly interest in epidemic outbreaks, including the impact of community built environment factors on epidemic transmission, particularly in the context of the COVID-19 pandemic (Figure 1). Built environments (BEs), through encouraging close interactions among individuals, the presence of fomites (potential carriers of infectious agents), and airborne viral exchange, emerge as potential mediums for disease propagation [6]. Particularly in high-density BEs, intensified human interactions often lead to increased transmission risks for diseases relying on direct contact and proximity, such as influenza and tuberculosis [7]. In addition, Central European scholars have also discussed BIM (Building Information Modeling) technology and related legislative issues in the context of healthy living environments [8].

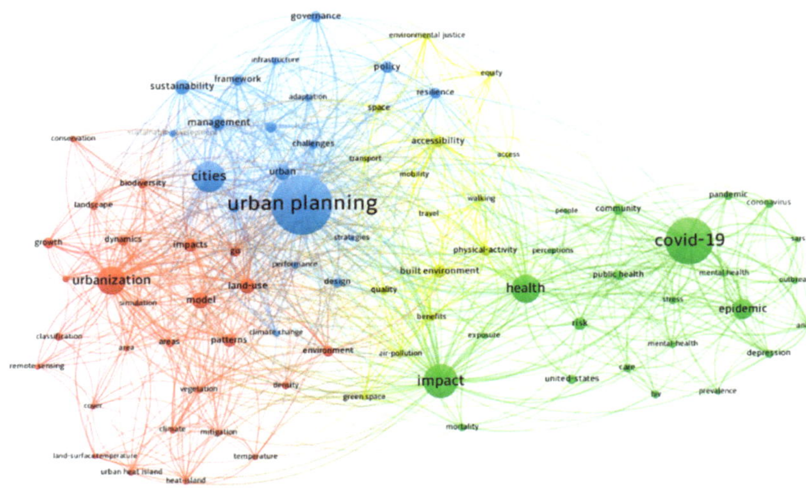

Figure 1. Keyword co-occurrence analysis.

The surge in high-density communities has emerged as an inescapable trend in the rapid and sustainable global urbanization [9]. The concentration of populations in large cities and mega-cities stands as a predominant feature of urbanization across Asia, exemplifying high-density BEs as a pivotal aspect of urban development [10]. Before the epidemic, high-density cities were believed to offer numerous benefits through compact development [11]. However, this pandemic has severely dampened economic growth [12], while highly dense and interconnected urban networks have significantly facilitated the rapid transmission and diffusion of diseases [13]. Notably, areas with high COVID-19 infection rates, both domestically and internationally, such as communal dormitories for foreign workers in Singapore, the "Dayuan F4" community in Chengdu, urban villages in Guangzhou, and the Xiangxi community in Shenzhen, share a common trait of excessive population concentration [14]. Given the recurrent occurrence of epidemics, the enhancement of community planning and built environment (BE) design within high-density communities becomes imperative to effectively curtail epidemic transmission, reduce associated societal and economic costs, and safeguard public health.

The outbreak of the COVID-19 pandemic has notably confined urban residents' living spaces, fostering increased reliance on the surrounding built and natural environments [15]. The community's BEs can exert direct or indirect influences on infectious disease transmission by altering microclimate conditions or influencing behaviors that impact health [1,16]. Scholars from diverse regions have conducted empirical investigations into the BE factors affecting epidemic propagation within communities. Nevertheless, the relationship between the floor area ratio (FAR), building density (BD), and COVID-19 transmission remains contentious. Some researchers assert that high FAR and BD values contribute to higher population densities [17,18] and poor ventilation [13], consequently heightening the virus exposure risk. Conversely, others posit that stringent Non-Pharmaceutical Interventions (NPIs) can curtail COVID-19 transmission within high-density communities [19,20]. For instance, during the COVID-19 pandemic, businesses have adjusted their recruitment strategies and the nature of new positions in response to the challenges posed by remote work and the pandemic [21]. Likewise, the correlation between walkability (WB) and COVID-19 transmission has sparked debate. While certain studies suggest that high community WB promotes physical activity, thereby reducing infection risk [22,23], other research underscores the challenge of maintaining social distancing within highly developed urban areas [24,25]. Various community facilities are closely correlated with resident behavior and influence epidemic spread. Elevated-density commercial facilities may lead to overcrowding [19,26], whereas high-density healthcare establishments may heighten the risk of case clustering during consultations [27–29]. Furthermore, convenient public transportation can enhance community mobility and consequently elevate exposure risk [30–32].

Although the specific directional significance of different BE attributes in COVID-19 infection remains unresolved, the significant role of community BEs in COVID-19 prevention and control is well recognized. Yet, most studies have focused predominantly on general communities, with limited attention paid to an in-depth analysis of high-density communities. In such communities, the high population density poses additional challenges for effective residential management, potentially exacerbated by an excessive FAR, which can compromise living conditions, reduce the per capita green space, and diminish natural lighting [33]. Such unique challenges may impact COVID-19 control strategies within high-density communities.

To realize post-pandemic health-oriented community development, a focused analysis of high-density communities is imperative. Our primary goal is to comprehend the impact of the BE within high-density communities on infectious disease transmission, especially in comparison to low-density communities. Our secondary aim involves exploring the correlation between BE factors, such as land use and BD, and the spread of epidemics, while evaluating their potential implications for public health.

This study, based on a review of the literature on epidemics and BEs, presents a framework for BE elements that are integral to community epidemic prevention. Using Shenzhen city as a study case, we utilize a combination of spatial analysis in various dimensions and quantitative calculations to compare high-density communities with regular communities. Through the application of Generalized Linear Models (GLMs), we quantitatively assess the differences in the BE factors that may affect epidemic transmission between high-density and general communities.

We hypothesize that specific BE factors within high-density communities exhibit a significant correlation with the transmission of infectious diseases, potentially resulting in higher infection rates. In conclusion, this research integrates its findings into an analytical framework that encompasses containment measures, NPI, and various pathways involving the BE. The results of this study offer valuable guidance and technical support for the development of new high-density communities and the improvement of existing ones. This research contributes to enhancing the preparedness and control measures against unforeseen public health events.

2. Methods

In the analysis of COVID-19 infection within communities, it is crucial to recognize the multifaceted nature of the BE and its potential impact. Focusing solely on a single level of BE analysis when assessing its influence on COVID-19 infection in communities may lead to biased and incomplete findings, overlooking localized and specific factors. To address this issue, our study initially conducted a statistical analysis of existing empirical research to identify key indicators of the BE that may influence the transmission of the virus within communities. To further investigate the relationship between these indicators and community infection rates, we conducted a preliminary correlation analysis in the typical high-density city of Shenzhen. However, it is essential to note that this pairwise correlation analysis does not account for potential interactions between variables. To address this limitation, all relevant variables will be further incorporated into a GLM. This will allow us to separately examine the effects of multiple independent variables on the dependent variable, both at the scales of ordinary communities and high-density communities. Furthermore, we will utilize the significance test results (p-values) and standardization coefficients (β) of independent variables to assess the importance of each indicator in influencing the community's epidemic spread.

2.1. Index Selection

Table 1 presents the results of 21 empirical studies examining the relationship between the BE and epidemics in general communities, with investigated indicators classified according to the five "5D" elements of the BE [34].

Table 1. Built environment factors influencing the prevention and control of new crown pneumonia outbreaks in the community.

Authors	Density		Diversity		Design		Destination Accessibility			Distance to Transit
	BD	FAR	LUM	RD	WB	OD	D_cf	D_h	D_gs	D_pt
Behram Wali [35]			-		-	+				
Shakil Bin Kashem et al. [36]			+		-	-			0	
Quynh C. Nguyen [25]					+					
Tianming Zheng [20]	-	-					+	+	+	
Wu Li et al. [32]	+		-	+					-	+
Xin Huang [37]	+						+	+	-	
Tribby and Hartmann [22]					-				0	
Niu et al. [38]	+	+					+		-	+
Credit [28]					-			+		
Guo, Yu and Zhang [23]					-				-	
Asfour [39]						+				
DiMaggio [40]	+									
Rahman [27]				0				+		+
Yong Xu et al. [41]	+			+			+	+		
Bo Li et al. [31]							+	+		+
Emre Tepe [42]							+			
Zerun Liu et al. [43]		-	-				+	-		
Jingwei Wang [19]		-					+	-		-
Eric Gaisie [44]							+			+
Kate H. Choi [45]					-		+			
Dennis Schmiege [46]							-		-	

The direction of association was coded as '+', '-', or '0'; '+' indicates a statistically significant positive association, while '-' represents a statistically significant negative association, and '0' indicates no significant association.

Based on previous reviews, a characteristic of the BE is considered to have strong evidence if the number of positive/negative associations is greater than or equal to the sum of its negative/positive or inconclusive associations [47]. The table shows that there is strong evidence of a positive association between virus infection risk and BD, open design (OD), densities of commercial facilities (D_cf), densities of hospitals (D_h), and densities of public transportation (D_pt). Negative associations with infection risk were found for FAR, land use mixture (LUM), and WB. The evidence for associations of infection risk with other BE factors (e.g., road density and densities of green space) was weak. Notably, scholars have identified that BD, LUM, WB, D_cf, D_h, and D_pt are significantly associated with

COVID-19 infection, and thus, they were selected as the six major BE factors to be further investigated regarding their impacts on COVID-19 infections in high-density communities.

To accurately model and interpret the data, the COVID-19 incidence rate defined as the ratio of COVID-19 cases to the total population of the community was chosen as the dependent variable in this study. This variable was selected due to its ability to consider the population base of each community, and thus, it provides a more precise measure of the probability of infection occurring within a specific population. In terms of independent variables, the study considers various BE indicators, including BD, LUM, WB, D_cf, D_h, and D_pt.

2.2. Theoretical Overview

Within communities, the transmission dynamics of many infectious diseases are intricately linked to human mobility patterns and social interactions, which are factors that are intimately tied to the characteristics of the BE [11]. The BE elements are associated with the features of epidemic transmission, affecting transmission rates through ventilation and congregation dynamics [48,49]. Concurrently, community BE plays a role in controlling infection sources (like home isolation), cutting transmission routes (such as community closures), and offering relief and treatment options [50].

Notably, indoor airborne transmission significantly contributes to COVID-19 infections, especially in congested and poorly ventilated environments [51]. Key elements within the built environment—such as healthcare facilities, retail spaces, cultural hubs, green areas, and transportation networks—serve as pivotal areas for intervening in public spaces, impacting congregation and necessitating focus when devising epidemic prevention strategies [52]. The quantity and scale of these elements can shape congregation dynamics, warranting specific attention to be paid to prevention strategies. Vital preventive measures like physical distancing and population lockdowns remain pivotal [53].

COVID-19 transmission can occur via airborne means or direct/indirect contact within built environments, intricately entangled with human behavior, thus influencing epidemic prevention and control strategies [54]. Our research focuses on analyzing the BE factors influencing epidemic transmission within high-density communities and combining spatial intervention measures to chart the trajectory of health-oriented development in such areas.

2.3. Research Region and Data Sources

Large cities typically pose a higher risk for the transmission of infectious diseases due to the following characteristics: high population density, diversity in residents' health statuses, frequent interpersonal interactions, and a high level of population mobility [55]. Therefore, this study focused on Shenzhen, which was chosen as a representative case due to its high population density and rapid urbanization. Shenzhen is situated in the southeastern coastal region of Guangdong Province, China and is adjacent to Hong Kong. This vibrant city is divided into 9 districts, encompassing 74 sub-district offices and 677 communities (Figure 2), covering a total area of 1997.47 km^2 with a resident population of 17,681,600 [56]. The city's extensive urbanization has led to the creation of numerous high-density community units, as evidenced by a growing trend in this direction [57]. Notably, Shenzhen's communities exhibit a distinct dual status characterized by a juxtaposition of urban villages and modern residential buildings, exemplifying typical traits of high population density, a substantial influx of migrants, and a youthful demographic profile.

From 26 February to 16 September 2022, data collected from the Shenzhen Government Data Open Platform revealed 2128 confirmed indigenous COVID-19 cases [58]. Demographic information on local communities was retrieved from official sub-district office websites [59], facilitating the computation of the incidence of COVID-19 per community. The BE data used were gathered from the collection and processing of multivariate big data and vectorized grid maps. Our data extraction strategy included obtaining various Points of Interest (POIs) and Areas of Interest (AOIs) data from Baidu Maps, building outline and height data from Gaud map data, road data from the OpenStreetMap database, and community boundary data from the Shenzhen Municipal Planning and Natural Resources Bureau's Shenzhen City Map (Outline Version III).

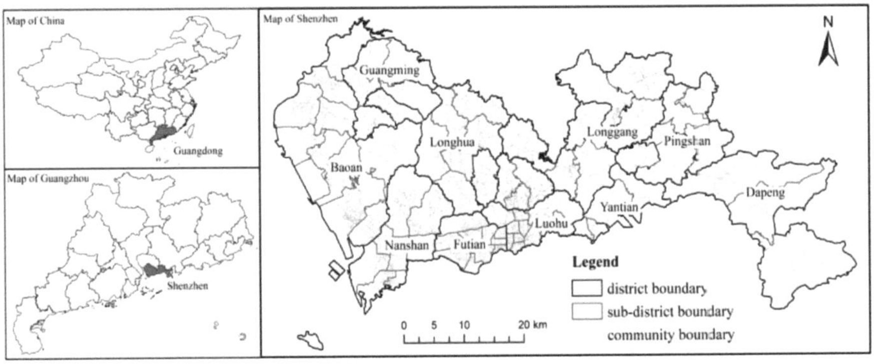

Figure 2. Context map of Shenzhen.

2.4. Statistical Analyzing

To investigate the impacts of these BE factors on COVID-19 infection rates, we employed a three-step statistical analysis (Figure 3). First, we cleaned the case data and obtained data for 311 infected communities in Shenzhen, including the incidence rates, through spatial processing using ArcGIS. Next, we processed the collected BE data, resulting in 6 BE indicators. In the third step, we used the qcc.overdispersion.test function in R language to perform an Equi dispersion test on the dependent variable [60], and the results indicate that the distribution of COVID-19 infections within the community is excessively dispersed, and multiple factors within the BE can potentially influence their spread. For our initial analysis, we performed Pearson and Spearman correlation analyses to identify the built environmental influences on epidemic spread in communities. Subsequently, we assessed multicollinearity using variance inflation factors (VIFs). Finally, we utilized Generalized Linear Models (GLMs), including the Poisson model (PM), negative binomial model (NBM), and zero-truncated models (ZTM), as they are better suited for modeling discrete data.

2.4.1. Calculation of Indicators

Six indicators were selected as independent variables in the present study: BD, LUM, WB, D_cf, D_h, and D_pt. The measurement methods for these BE indicators are outlined in Table 2.

In terms of BD, we obtained Shenzhen's building outline data from Baidu Maps and calculated the BD for each community by dividing the base area of community buildings by the community's land area.

In terms of LUM, it can be measured using the entropy index, with values ranging from 0 to 1. When the proportions of different land uses within the study unit are equal, the index is 1 [61]. We obtained AOI (Area of Interest) data for Shenzhen from Baidu Maps to examine the degree of mixture of eight land uses within each community's boundaries: residential; commercial and office; commercial shopping; scientific, educational, cultural and health; public green space; industrial; transportation; and other land uses.

In terms of WB, we obtained data from OpenStreetMap, including 1024 pedestrian and bicycle lanes within the confirmed communities. These data provided information on road categories, road lengths, and geographic locations. WB was calculated as the length of pedestrian and bicycle lanes within the community divided by the community's area.

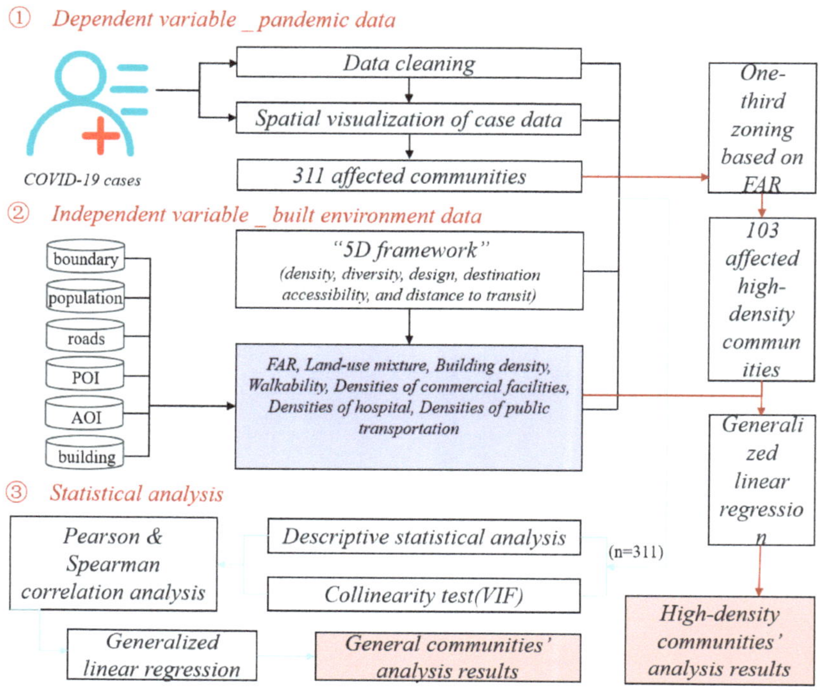

Figure 3. Data analysis workflow diagram.

In terms of facility accessibility, research has shown that the more public service facilities per unit area, the closer the facilities, indicating higher accessibility [17,36]. In this study, we obtained POI data for confirmed communities in Shenzhen from Baidu Maps, including 150,514 commercial facilities, 3773 medical facilities, and 3529 public transportation stops. Accessibility was measured as the number of facilities per square kilometer within each community's boundaries.

Table 2. Built environment metrics used in this study.

Indicator Name	Formula	Description	Data Source
BD	$BD = \frac{BA}{CA}$	BD represents the building density, BA represents the base area of the community building, and CA represents the footprint of the community.	BA: Shenzhen's building outline data from Baidu Maps; CA: Shenzhen Municipal Planning and Natural Resources Bureau's Shenzhen City Map (Out-line Version III)
LUM	$LUM = -\sum \text{pk} ln(pk)\, ln(N)$	pk represents the percentage of sites within K in each community, and N is the number of site types in the community.	AOI: Area of Interest data for Shenzhen from Baidu Maps; N: the number of AOI types
WB	$WB = \frac{Lfw+Lcw}{CA}$	The width of sidewalk/bicycle lane is assumed to be normal. Lfw represents the length of the sidewalk in the community, Lcw represents the length of the bike lane in the community, and CA represents the footprint of the community.	The width of sidewalk/bicycle lane: obtained data from OpenStreetMap

Table 2. *Cont.*

Indicator Name	Formula	Description	Data Source
D_cf	D_cf = $\frac{Ncf}{CA}$	Ncf represents the number of commercial facilities in the community, and CA represents the footprint of the community.	POI data obtained from Baidu Maps, including 150,514 commercial facilities, 3773 medical facilities, and 3529 public transportation stops
D_h	D_h = $\frac{Ns}{CA}$	Nh represents the number of medical facilities in the community, and CA represents the footprint of the community.	
D_pt	D_pt = $\frac{Nps}{CA}$	Nps represents the number of public transportation stops in the community, and CA represents the footprint of the community.	

2.4.2. Correlation Analysis

The Pearson correlation coefficient, also known as the Pearson product moment correlation coefficient, is the most widely used measure of correlation in statistical analysis. It assesses the degree of linear association between two variables, X and Y, assuming that both variables are normally distributed. For binary or other non-normally distributed data, Spearman's correlation analysis is a preferred method, as it employs monotonic functions to evaluate the correlation without relying on the overall distribution or sample size of the data and is less sensitive to outliers. However, the disadvantage of Spearman's method is its lower efficiency compared to Pearson's method. Therefore, it is recommended to consider both Pearson's and Spearman's coefficients, considering the nature and distribution of the data.

The formula for the Spearman correlation coefficient is as follows (1):

$$r = \frac{\Sigma(X-\overline{X})(Y-\overline{Y})}{\sqrt{\Sigma(X-\overline{X})^2\Sigma(Y-\overline{Y})^2}} \tag{1}$$

Here, X and Y are the values of the two variables, and $\overline{X}$ and $\overline{Y}$ represent their respective means.

The formula for the Spearman correlation coefficient is as follows (2):

$$\rho = 1 - \frac{6\sum d^2}{n(n^2-1)} \tag{2}$$

where d represents the differences between ranks, and n is the number of data points.

2.4.3. Regression Analysis

The GLM is an extension of the general linear model that allows us to create regression models for various types of response variables, including count, binary, proportions, and positive-valued continuous distributions. The GLM equation encompasses this flexibility and can be expressed as follows (3):

$$g\{E(y)\} = LP = a + b_1x_1 + b_2x_2 + \cdots b_px_p \tag{3}$$

where g{} represents a linear function of the regressors, $E(Y)$ represents the expected value, LP indicates the linear predictive value, a is the constant of the regression equation, $x_1, \ldots,$ x_p represent the p environmental variables, and $b_1, \ldots, b_p$ are the p regression coefficients.

The Poisson distribution is the most commonly used approach for modeling count data, and it represents a special case of the GLM framework that requires the mean and variance to be equal [23]. However, in real-world scenarios, data can be overly discrete. For this reason, NBM is better suited for such cases and can result in improved fitting [62]. In the NBM, the variance is assumed to be a quadratic function of the mean [63]. Since we collected communities with COVID-19 infections, where the IR is not zero, the ZTM was used to model count data where the value of zero cannot be observed and if there is evidence of overdispersion [64]. The effectiveness of our model was assessed using four indicators: deviance, AIC value, AICc value, and BIC. A model with lower AIC and AICc values is considered to be better fitted when the difference between the AIC and AICc values of two models is more than or equal to 3 [65]. The statistical analyses were performed using R software (V.4.0.4), with the MASS and pscl packages employed.

3. Results

3.1. Spatial Patterns of COVID-19 Infections and BEs in Shenzhen Communities

We performed a spatial analysis of the 2128 COVID-19 cases, using ArcGIS software to visualize their distribution across 341 communities in Shenzhen (Figure 4). Among these, the Futian District recorded the highest cumulative number of cases, with 1091 cases, representing 51.2% of the total. Following closely were the Luohu District and Nanshan District, with 241 (11.3%) and 292 cases (13.7%), respectively, primarily concentrated in the southwestern region of Shenzhen. On the other hand, the Dapeng New District and Guangming District each reported only three cumulative cases, while Pingshan District had just two cases, ranking them as the top three districts with the most effective epidemic control. These districts are situated in the eastern and northwestern areas of Shenzhen. To gain a more comprehensive insight into the distribution of infection cases, we conducted COVID-19 kernel density analyses based on both population and area in the ArcGIS software (refer to Figure 4b,c). The distribution patterns revealed overall consistency with the spatial distribution of case numbers. Hotspots remained concentrated in the Futian District, Luohu District, and Nanshan District. However, when considering the population density, the Nanshan District exhibited a notably higher infection rate. Conversely, coldspots were identified in the western areas, particularly in the Dapeng New District and Pingshan District.

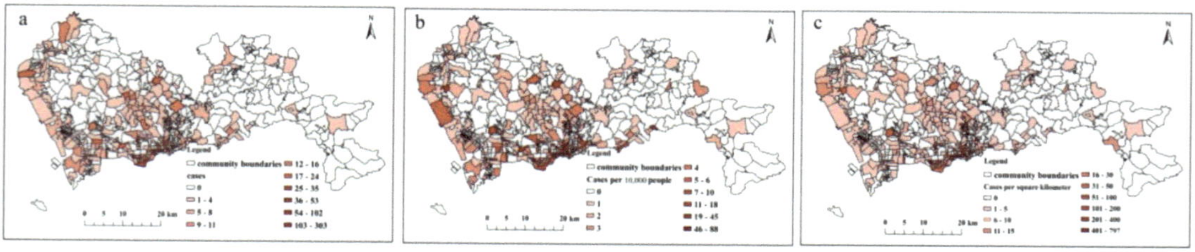

Figure 4. Distribution of locally confirmed cases in Shenzhen communities. (**a**) COVID-19 cases in communities; (**b**) COVID-19 cases per 10,000 people in communities; (**c**) COVID-19 cases per square kilometer in communities.

Furthermore, we conducted an in-depth exploration of the spatial distribution of BEs in the infected communities of Shenzhen, encompassing parameters such as population density, FAR, BD, LUM, WB, D_h, D_cf, and D_pt (Figure 5). The spatial distribution analysis of the BE factors yielded several significant findings. Firstly, all hotspots for the distribution of BEs were concentrated in the Futian District, Luohu District, and Nanshan District, whereas coldspots were evident in the Dapeng New District and Pingshan District. Secondly, we observed significant variations in the kernel density distribution of medical facilities, commercial facilities, and transportation hubs. These factors exhibited distinct spatial patterns, reflecting their varied concentrations and accessibility across the city. Thirdly, when considering LUM and WB, we found them to exhibit intermediate patterns in their spatial distributions. These factors were less concentrated in the central districts

(Futian, Luohu, and Nanshan) compared to medical facilities, commercial establishments, and transportation hubs. Lastly, the population density, FAR, and BD demonstrated a relatively smaller presence and a more uniform distribution across the city.

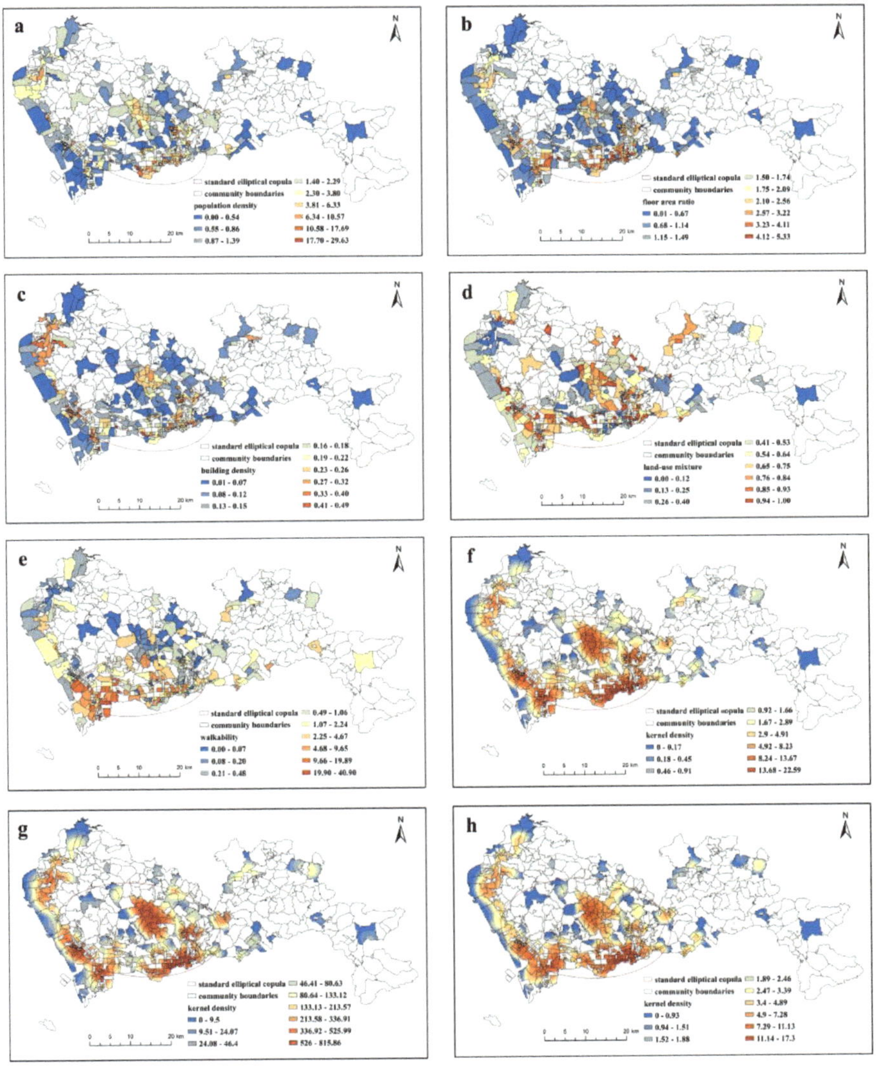

Figure 5. Spatial distribution of BEs in eight types of communities (**(a)** population density; (**b**) floor area ratio; (**c**) building density; (**d**) land use mixture; (**e**) walkability; (**f**) medical facilities; (**g**) commercial facilities; (**h**) public transportation).

It is imperative to acknowledge that while our spatial analysis has unveiled these distinctive patterns, there remains a critical need for additional quantitative investigations to elucidate the intricate relationships between infection rates and these specific BEs. This essential next step will yield a more refined comprehension of how the community BEs may influence the transmission of infectious diseases, including COVID-19. Subsequently, the insights gained from such analyses can serve as valuable guidance for the formulation of precise urban planning strategies and informed public health interventions.

3.2. Relationship between the Community Epidemic Incidence and Built Environment

Following the exclusion of communities with incomplete data, our analysis focused on 311 case communities. This paper used a correlation analysis to determine the BEs that influence community COVID-19 infection. Table 3 shows the descriptive statistical results of the indicators of the community's BEs. The correlations between community COVID-19 incidence and six explanatory variables were determined using Pearson and Spearman correlation coefficients before conducting specific regression modeling. In general, we found large differences in the associations between the variables and COVID-19 in the community, with a range of 0.03–0.26, indicating that different BEs differed significantly in the importance of the impact of the epidemic (Figure 6).

Table 3. Descriptive statistical analysis of built environment factors and correlation coefficient statistics.

Variable	Descriptive Statistics		Pearson		Spearman		Collinearity	
	Mean	Std. Dev.	Coef.	*p*-Value	Coef.	*p*-Value	VIF	1/VIF
BD	0.199	0.097	0.1048 *	0.0650	0.1412 **	0.0127	1.24	0.807
LUM	0.570	0.293	−0.1033 *	0.0689	−0.0318	0.5758	1.01	0.989
WB	2.860	4.594	0.1512 ***	0.0076	0.2430 ***	<0.001	1.05	0.948
D_cf	529.794	641.605	0.0658	0.2476	0.1375 **	0.0152	1.95	0.513
D_h	14.350	16.053	0.0968 *	0.0885	0.1934 ***	<0.001	1.99	0.503
D_pt	10.607	9.940	0.0916	0.1067	0.1009 *	0.0755	1.48	0.675

* $p < 0.1$; ** $p < 0.05$; *** $p < 0.01$.

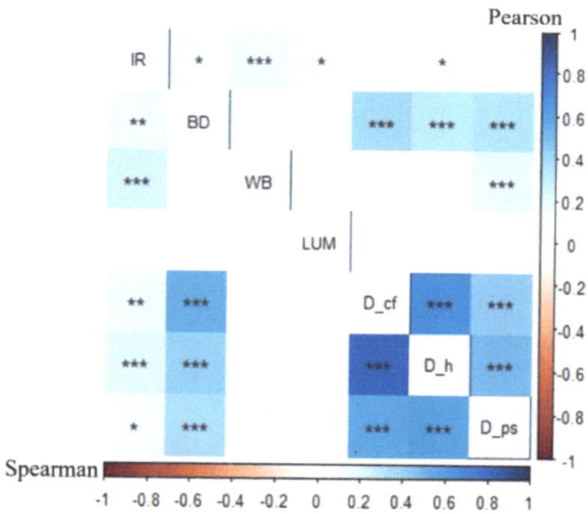

* $p < 0.1$; ** $p < 0.05$; *** $p < 0.01$.

Figure 6. Pearson and Spearman correlation matrix plot.

The selection of the α critical value in hypothesis testing should be based on the specific context. Normally, values such as α = 0.01, 0.05, or 0.1 are used, corresponding to probabilities of correctly accepting the null hypothesis at 99%, 95%, or 90%, respectively [66]. Our Pearson and Spearman correlation analyses identified three environmental factors that were significantly associated with community morbidity: WB ($p < 0.01$), D_h ($p < 0.1$), and BD ($p < 0.1$). However, three other factors, specifically LUM (Ppearson < 0.1, Pspearman = 0.5758), D_cf (Ppearson = 0.2476, Pspearman < 0.05), and D_pt (Ppearson = 0.1067, Pspearman < 0.1), displayed different levels of association with community morbidity between the two correlation analyses. To avoid any issues of excessive multicollinearity among the six selected explanatory variables, we employed the variance inflation factor (VIF) test, which did not reveal concerns of collinearity (VIF < 10) [67].

The results of our correlation analysis indicated that the significance of the BEs we examined with regard to COVID-19 infection varied, which may be attributed to the distribution characteristics of the data. To enhance the accuracy of our analysis, we employed various methods and compared the fitting effects of different models.

3.3. GLM Regression Results for General Communities

In this study, we conducted PM, NBM, and ZTNBM regression analyses on a sample of 311 communities. Table 4 presents the results of our regression analyses regarding BEs that influence community epidemic transmission. The Negative Binomial_F model demonstrated a higher level of explanation, as evidenced by the lower values of the Deviance, AIC, AICc, and BIC, respectively. Our findings suggest that three indicators were significantly and positively associated with community epidemic transmission, which were BD, WB, and D_h, while LUM was negatively associated. The regression coefficients revealed that the impact of BD (1.685) on community epidemic transmission was the most significant, while the impact of D_h (0.012) was found to be the least significant. WB (0.055) and LUM (−0.738) instead exhibited moderate impacts.

Table 4. GLM regression results for general communities.

Variable	Poisson_F			Negative Binomial_F			Zero Truncated nb_F		
	Estimate	S.E.	*p*	Estimate	S.E.	*p*	Estimate	S.E.	*p*
Intercept	3.100 ***	0.033	<0.001	0.701 ***	0.204	<0.001	2.781 ***	0.252	<0.001
BD	1.671 ***	0.122	<0.001	1.685 **	0.766	0.027	1.846 **	0.916	0.044
WB	0.044 ***	0.002	<0.001	0.055 ***	0.014	<0.001	0.064 ***	0.021	0.002
LUM	−0.739 ***	0.036	<0.001	−0.738 ***	0.229	0.001	−0.767 ***	0.264	0.004
D_h	0.012 ***	0.001	<0.001	0.012 **	0.006	0.029	0.013 *	0.007	0.052
D_cf	−0.000 **	0.000	0.032	−0.000	0.000	0.664	−0.000	0.000	0.533
D_pt	−0.005 ***	0.001	<0.001	0.000	0.008	0.970	0.002	0.009	0.817
Deviance		13,972.62			370.603			2573.4	
AIC		15,343.665			1323.309			2589.388	
AICc		15,344.035			1323.786			2589.865	
BIC		15,369.844			1353.227			2619.306	

* $p < 0.1$; ** $p < 0.05$; *** $p < 0.01$.

3.4. GLM Regression Results for High-Density Communities

We initiated a group analysis on the processed data, dividing communities with COVID-19 cases into three categories based on the FAR (Figure 7): high-density, medium-density, and low-density communities. The results indicate that high-density communities exhibit significantly higher values for average case numbers, average incidence rates, and cumulative diagnosis densities compared to medium- and low-density communities, with the disparities being twice as large. These findings strongly emphasize the substantially elevated risk of epidemic infection in high-density communities, highlighting the critical need for focused attention in this regard.

We conducted a regression analysis on the group of 103 high-density communities obtained after grouping, and the results of this analysis are presented in Table 5. Among the three models, PM_H has the highest deviance, AIC, AICc, and BIC values. Therefore, PM_H is initially excluded. The AIC, AICc, and BIC are designed to balance the goodness of fit and complexity of the models, penalizing model complexity to prevent overfitting. In the comparison between NBM_H and ZTNBM_H, NBM_H exhibits significantly lower deviance than ZTNBM_H, indicating its superiority in fitting the data. AIC, AICc, and BIC are slightly higher for NBM_H compared to ZTNBM_H, suggesting that NBM_H may be more complex; hence, NBM_H is selected. This led us to conclude that LUM (p = 0.009) and WB (p = 0.047) displayed high levels of significance. Specifically, LUM demonstrates a significant negative correlation, while WB shows a positive correlation with the absolute value of the correlation coefficient, as the LUM value is greater than that for WB.

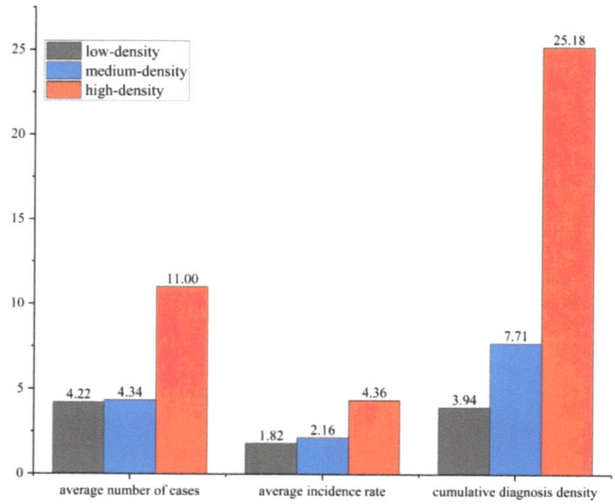

Figure 7. Epidemic situation in low-, medium-, and high-density communities in Shenzhen.

Table 5. GLM regression results for high-density communities.

Variable	Poisson_H			Negative Binomial_H			Zero Truncated nb_H		
	Estimate	S.E.	p	Estimate	S.E.	p	Estimate	S.E.	p
Intercept	4.250 ***	0.069	<0.001	3.873 ***	0.506	<0.001	3.797 ***	0.492	<0.001
BD	0.530 **	0.225	0.018	1.072	1.511	0.489	1.123	1.784	0.529
WB	0.039 ***	0.002	<0.001	0.039 **	0.020	0.047	0.041	0.025	0.105
LUM	−1.140 ***	0.049	<0.001	−1.010 ***	0.347	0.004	−1.066 ***	0.412	0.009
D_h	0.000	0.000	0.904	0.001	0.001	0.466	0.006	0.008	0.478
D_cf	0.000	0.000	0.330	0.000	0.000	0.878	0.000	0.000	0.835
D_pt	−0.016 ***	0.001	<0.001	−0.014	0.010	0.167	−0.014	0.011	0.197
Deviance		7582.157			118.576			950	
AIC		8091.276			977.801			965.963	
AICc		8092.455			979.333			967.495	
BIC		8109.719			998.879			987.041	

* $p < 0.1$; ** $p < 0.05$; *** $p < 0.01$.

4. Discussion

4.1. Analysis of Shenzhen's Epidemic Distribution Patterns and BE Characteristics

By combining the epidemic distribution characteristics in Shenzhen (Figure 4) with the spatial distribution of FAR (Figure 5b), we observed that regions with severe outbreaks often corresponded to areas with a high FAR. Furthermore, by quantifying the infection rates in communities of different densities, we found that high-density communities had infection rates more than twice as high as those in medium- and low-density communities (Figure 7). This underscores the significant risks associated with high-density communities during epidemic outbreaks, highlighting the need for focused attention in epidemic prevention and control management as well as in health planning.

The spatial distribution of BE factors in Shenzhen also reveals the distinctiveness of its high-density areas (Figure 5). Specifically, the analysis reveals that central urban areas, characterized by a high population density and intensified urban development, exhibit a more substantial presence of various BE elements. These encompass critical components such as medical facilities, commercial establishments, and transportation hubs. What is particularly noteworthy is the pronounced disparities observed between hotspots and coldspots across these categories. This disparity underscores a key point: specific elements of the BE, including healthcare infrastructure, commercial amenities, and transportation networks, tend to be disproportionately concentrated in high-density communities, making high-density communities more conducive to prevention and control efforts to some extent.

4.2. Built Environment Factors Affecting COVID-19 Cases in High-Density Communities

High-density BEs are characterized by several key features, including spatial high-intensity development, three-dimensional transportation organization, high-density multi-functionality, frequent population interactions, interwoven underground spatial networks, and challenging outdoor physical conditions [10]. These attributes have substantially exacerbated the spread and diffusion of diseases due to the highly concentrated population and interconnected urban networks associated with high-density BEs.

Our study identified LUM and WB as the most significant BEs impacting COVID-19 infection rates in high-density communities, which is consistent with the sign of the correlation analysis results. However, several other factors, including BD, D_cf, D_h, and D_pt, which exhibited significant correlations in the correlation analysis, did not demonstrate substantial effects in the regression analysis. This inconsistency can be attributed to the difference between the correlation analysis, which evaluates the relationship between two variables, and the regression model, which considers the interplay among independent variables.

Interestingly, our analysis revealed a positive correlation between WB and COVID-19 infection rates at the high-density community level. This finding is consistent with a study conducted on COVID-19 infection rates in communities across 20 states in the United States [25]. This suggests that enforcing social distancing measures may be more challenging in densely developed urban areas. Conversely, other research has indicated that pedestrian-oriented streets are associated with lower mortality and hospitalization rates [22,68]. This could be because high walkability encourages physical activity, enhances residents' physical fitness, and thereby reduces the likelihood of infection.

In contrast, LUM exhibited a negative correlation with COVID-19 infection rates, aligning with the findings of the majority of the existing research [38,69]. A higher degree of LUM implies a better spatial proximity of urban services and diverse functions that cater to residents' needs. This results in reduced travel distances and less frequent travel, ultimately lowering the risk of infection due to reduced exposure to traffic-related contagion. The LUM may exert a more significant influence than walkability, as it impacts both travel mode and distance [70]. It increases the probability of the origin and the destination of travelers being within the same vicinity, meeting residents' daily needs during the epidemic, and minimizing non-commuting interactions across communities. In comparison to walkability, it may also decrease residents' likelihood of visiting high-risk areas.

4.3. Comparative Analysis between High-Density Communities and General Communities

The emphasis on maintaining social distancing has emerged as a crucial intervention measure in the prevention of infectious diseases. In China, high-density communities characterized by high floor area ratios have gained prominence as land-saving urban layouts. However, they may not always facilitate effective ventilation and can present certain transmission risks, making them susceptible to disease spread [13]. Our analysis revealed that high-density communities exhibited the highest risk of COVID-19 infection when compared to general communities. In general communities, the key BEs influencing COVID-19 cases included BD, WB, LUM, and D_h. Remarkably, the LUM and WB were significantly associated with COVID-19 cases in both high-density and general communities.

Furthermore, our regression analyses, conducted separately for high-density and general communities, highlighted significant disparities in the impact of various indicators. Notably, BD and D_h lost their significance in high-density communities. These indicators exhibited consistent high values in high-density communities, with minimal variations and no discernible impact on epidemic transmission. These findings underscore the necessity for tailored planning and design strategies based on the density level of the communities in question. To further extend this discussion, it is essential to consider the broader implications of these findings. High-density communities, which are prevalent in rapidly urbanizing regions, present a unique challenge in disease transmission management. While they often offer benefits in terms of efficient land use, reduced commute times, and proximity to urban amenities, they also entail greater population density and complexity. Therefore, mitigating the transmission risks within high-density communities requires multifaceted strategies.

4.4. Guidance on the Built Environment Design of High-Density Communities

In the context of high-density communities, the discussion around the design of the BE takes on critical significance. Spatial interventions have been a cornerstone in achieving effective epidemic containment, encompassing strategies aimed at isolating infection sources, disrupting transmission pathways, and safeguarding vulnerable populations [71,72]. The conceptual framework, as depicted in Figure 8, illustrates the intricate interplay between containment measures, NPIs, and the BEs of communities. An analyzation of the epidemic prevention process reveals that infection sources primarily originate from residents with transmission capabilities, who subsequently construct transmission chains within their living areas [52].

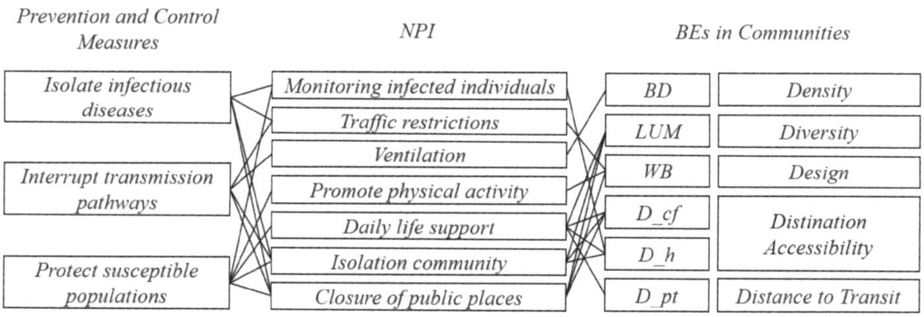

Figure 8. The conceptual framework of various pathways between containment measures, NPIs, and community BEs.

When considering the isolation of infection sources, it becomes evident that grassroots health service institutions within communities play a pivotal role. These institutions are responsible for swift infection testing within the community and the implementation of isolation measures for confirmed cases. The adequacy of medical facility infrastructure, to a certain extent, influences a community's ability to quarantine and isolate individuals affected by novel viruses [73]. Concurrently, NPIs like traffic restrictions, community lockdowns, and the closure of public spaces emerge as indispensable measures. These restrictions limit population movement, both within and beyond the community, effectively reducing the risk of epidemic transmission. In the realm of safeguarding susceptible populations, several measures have proven vital. These include promoting proper ventilation to mitigate aerosol-based infections, encouraging physical activity to enhance overall physical fitness, and providing a diverse array of public facilities within the community to meet residents' daily needs. In high-density communities, optimizing the layout and accessibility of these facilities becomes particularly critical.

Furthermore, the consideration of the LUM and WB takes center stage when designing high-density communities. In essence, communities characterized by a higher degree of LUM are strategically designed to minimize travel distances for their inhabitants. This reduced need for travel results in fewer interactions, ultimately decreasing the risk of infection associated with traffic exposure. Conversely, WB, although generally associated with higher COVID-19 infection rates in high-density communities, requires nuanced consideration. While promoting physical activity through walkability to enhance residents' physical fitness, planners should consider widening sidewalks or establishing multiple square nodes to increase space and avoid crowding.

Our findings provide new insights and guidance for designing high-density communities and controlling COVID-19 in existing ones. These insights offer valuable guidance for the future development of high-density urban areas, emphasizing the importance of creating resilient, health-conscious environments that are prepared to confront and mitigate the challenges posed by infectious diseases.

5. Conclusions

As urbanization continues to accelerate, with high-density urban areas proliferating rapidly, it becomes increasingly imperative for urban planners and policymakers to acknowledge and address the vulnerabilities that are inherent in the built environment. This study underscores the significant influence of the BE on infectious disease transmission, particularly in high-density communities. Our research, focusing on the emblematic high-density city of Shenzhen, selected six key BE indicators based on the statistical analyses of existing research conclusions. By examining 311 infected communities in Shenzhen and categorizing 103 high-density communities, we combined spatial and quantitative analyses of epidemic infections and BE factors to unravel the extent of the outbreak and the characteristics of the BE elements in high-density communities.

Our findings unequivocally demonstrate a stark contrast between high-density areas and medium- to low-density communities. High-density areas exhibited significantly higher case numbers, incidence rates, and diagnosis density, indicating a heightened risk of epidemic infections. Through a correlation analysis and GLM models, specific BE factors affecting high-density and general communities were identified. In general communities, BD, WB, and D_h were positively correlated, with BD being the most influential. In high-density communities, LUM showed a negative correlation, while WB exhibited a positive correlation, with LUM being the most influential factor. To address the unique challenges of epidemic prevention in high-density areas, our study proposes a comprehensive conceptual framework integrating containment measures, NPIs, and the BE. This framework emphasizes isolating infectious cases, disrupting transmission pathways, and safeguarding vulnerable populations. Our regression analysis emphasizes the critical role of prioritizing land use and walkability in the planning of new or refurbished high-density communities.

This research offers practical implications for enhancing epidemic prevention and control in high-density urban areas, facilitating tailored urban planning and BE design and the healthy and sustainable transformation of existing high-density communities. It is essential to note that the findings and recommendations of this study are specific to the context of the studied region, and their applicability to other geographical locations may vary due to differing socio-economic, cultural, and environmental factors. Future research directions may include delving into the mechanisms of epidemic transmission, considering occupant behavior in diverse climate regions, and examining the resident population composition within communities to achieve a more comprehensive understanding of the factors affecting epidemic transmission.

Author Contributions: Y.H.: Conceptualization, methodology, software, validation, formal analysis, investigation, data curation, writing, and visualization. Z.L.: Validation, formal analysis, resources, data curation, and writing. S.J.: Conceptualization, supervision, project administration, and resources. R.Z.: Conceptualization, methodology, writing, supervision, project administration, and funding. All authors have read and agreed to the published version of the manuscript.

Funding: This research was funded by the Hunan Province Innovation Development Program, grant number 2020TP1009.

Data Availability Statement: Data available in a publicly accessible repository that does not issue DOIs. Publicly available datasets were analyzed in this study. This data can be found via [https://pan.baidu.com/s/1WJGJt7OVnw644FchmA3rHQ?pwd=9mzq].

Conflicts of Interest: Author Rongpeng Zhang was employed by the Hunan International Innovation Cooperation Base on Science and Technology of Local Architecture. The remaining authors declare that the research was conducted in the absence of any commercial or financial relationships that could be construed as a potential conflict of interest.

References

1. Pinter-Wollman, N.; Jelić, A.; Wells, N.M. The Impact of the Built Environment on Health Behaviours and Disease Transmission in Social Systems. *Philos. Trans. R. Soc. B Biol. Sci.* **2018**, *373*, 20170245. [CrossRef] [PubMed]
2. Ma, J.; Xu, J.; Zhao, X.; Huo, S.; Duan, X.; Mu, Y.; Wang, Y.; Wei, Y.; Chang, J.; Jin, X.; et al. Several major issues concern-ing the environmental transmission and risk prevention of SARS-CoV-2. *Sci. China Earth Sci.* **2022**, *65*, 1047–1056. [CrossRef] [PubMed]
3. Wu, X.; Tian, H.; Zhou, S.; Chen, L.; Xu, B. Impact of global change on transmission of human infectious diseases. *Sci. China Earth Sci.* **2014**, *57*, 189–203. [CrossRef] [PubMed]
4. Neiderud, C.-J. How Urbanization Affects the Epidemiology of Emerging Infectious Diseases. *Infect. Ecol. Epidemiol.* **2015**, *5*, 27060. [CrossRef] [PubMed]
5. Woolhouse, M.E.J.; Gowtage-Sequeria, S. Host range and emerging and reemerging pathogens. *Emerg. Infect. Dis.* **2005**, *11*, 1842–1847. [CrossRef]
6. Adams, R.I.; Bhangar, S.; Dannemiller, K.C.; Eisen, J.A.; Fierer, N.; Gilbert, J.A.; Green, J.L.; Marr, L.C.; Miller, S.L.; Siegel, J.A.; et al. Ten Questions Concerning the Microbiomes of Buildings. *Build. Environ.* **2016**, *109*, 224–234. [CrossRef]
7. Glaeser, E. Cities, Productivity, and Quality of Life. *Science* **2011**, *333*, 592–594. [CrossRef]
8. Bodea, C.N.; Purnus, A. Legal implications of adopting Building Information Modeling (BIM). *Jurid. Trib.-Trib. Jurid.* **2018**, *8*, 63–72.
9. Li, H.; Liu, Z. Temporal and Spatial Evolution of Urban Density in China and Analysis of Urban High Density Development: From 1981 to 2014. *Urban Dev. Studies.* **2019**, *26*, 46–54.
10. Shi, B.; Marvin, S.; Yang, J. Study on Quality Improvement of Built Environment in High-density Urban Areas Under the Background of Post Smart City Transformation. *Urban Plan. Int.* **2021**, *36*, 16–21. [CrossRef]
11. Yao, Y.; Pan, H.; Cui, X.; Wang, Z. Do Compact Cities Have Higher Efficiencies of Agglomeration Economies? A Dynamic Panel Model with Compactness Indicators. *Land Use Policy* **2022**, *115*, 106005. [CrossRef]
12. Correia, S.; Luck, S.; Verner, E. Pandemics Depress the Economy, Public Health Interventions Do Not: Evidence from the 1918 Flu. *J. Econ. Hist.* **2022**, *82*, 917–957. [CrossRef]
13. Wang, Z.; Zhang, Y. Review on the Relationship between High-density Built Environment and Public Health. *South Archit.* **2023**, *6*, 21–31. Available online: https://kns.cnki.net/kcms/detail/44.1263.TU.20230406.1325.008.html (accessed on 20 April 2022).
14. Duan, J.; Zhang, J.; Chai, Y.; Zhai, G.; Li, Z.; Tan, Z.; Yang, D.; Yang, J.; Yuan, Y.; Liu, W.; et al. "What lesson does COVID-19 teach planning?" Aca-demic writing. *Urban Plan. Forum.* **2022**, *3*, 1–10. [CrossRef]
15. Allam, Z.; Jones, D.S. Pandemic Stricken Cities on Lockdown. Where Are Our Planning and Design Professionals [Now, Then and into the Future]? *Land Use Policy* **2020**, *97*, 104805. [CrossRef] [PubMed]
16. Deng, L.; Wang, L. The Impact of Urban Expansion and High Density on Infectious Diseases and Planning Strategies. *New Archit.* **2021**, *6*, 88–93. [CrossRef]
17. Li, X.; Zhou, L.; Jia, T.; Peng, R.; Fu, X.; Zou, Y. Associating COVID-19 Severity with Urban Factors: A Case Study of Wuhan. *Int. J. Environ. Res. Public. Health* **2020**, *17*, 6712. [CrossRef] [PubMed]
18. Kan, Z.; Kwan, M.-P.; Wong, M.S.; Huang, J.; Liu, D. Identifying the Space-Time Patterns of COVID-19 Risk and Their Associations with Different Built Environment Features in Hong Kong. *Sci. Total Environ.* **2021**, *772*, 145379. [CrossRef]
19. Wang, J.; Zeng, F.; Tang, H.; Wang, J.; Xing, L. Correlations between the Urban Built Environmental Factors and the Spatial Distribution at the Community Level in the Reported COVID-19 Samples: A Case Study of Wuhan. *Cities* **2022**, *129*, 103932. [CrossRef]
20. Zheng, T.; Liu, H. Exploration of the Built-Environmental Elements that Influence the Spread of COVID19 Pandemic on Community Scale: A Case Study of Wuhan, China. *Mod. Urban Res.* **2020**, *10*, 20–29. [CrossRef]
21. Campello, M.; Kankanhalli, G.; Muthukrishnan, P. Corporate Hiring Under COVID-19: Financial Constraints and the Nature of New Jobs. *J. Financ. Quant. Anal.* **2023**, *Published online*, 1–45. [CrossRef]
22. Tribby, C.P.; Hartmann, C. COVID-19 Cases and the Built Environment: Initial Evidence from New York City. *Prof. Geogr.* **2021**, *73*, 365–376. [CrossRef]
23. Guo, Y.; Yu, H.; Zhang, G.; Ma, D.T. Exploring the Impacts of Travel-Implied Policy Factors on COVID-19 Spread within Communities Based on Multi-Source Data Interpretations. *Health Place* **2021**, *69*, 102538. [CrossRef] [PubMed]
24. Guan, C.; Tan, J.; Li, Y.; Cheng, T.; Yang, J.; Liu, C.; Keith, M. How do density, employment and transit affect the prevalence of COVID-19 pandemic? A study of 3,141 counties across the United States. *Health Place* **2023**, *84*, 103117. [CrossRef] [PubMed]
25. Nguyen, Q.C.; Huang, Y.; Kumar, A.; Duan, H.; Keralis, J.M.; Dwivedi, P.; Meng, H.-W.; Brunisholz, K.D.; Jay, J.; Javanmardi, M.; et al. Using 164 Million Google Street View Images to Derive Built Environment Predictors of COVID-19 Cases. *Int. J. Environ. Res. Public. Health* **2020**, *17*, 6359. [CrossRef] [PubMed]
26. Zhou, W. Research on the Correlation between Urban Built Environment and High-risk Communities under Epidemic Situation Based on the Sample of Seven Urban Communities in Hankou. *New Archit.* **2020**, *6*, 52–55. [CrossRef]

27. Rahman, M.H.; Zafri, N.M.; Ashik, F.R.; Waliullah, M.; Khan, A. Identification of Risk Factors Contributing to COVID-19 Incidence Rates in Bangladesh: A GIS-Based Spatial Modeling Approach. *Heliyon* **2021**, *7*, e06260. [CrossRef]
28. Credit, K. Neighbourhood Inequity: Exploring the Factors Underlying Racial and Ethnic Disparities in COVID-19 Testing and Infection Rates Using ZIP Code Data in Chicago and New York. *Reg. Sci. Policy Pract.* **2020**, *12*, 1249–1271. [CrossRef]
29. Wang, L.; Hu, Z.; Zhou, K.; Kwan, M.-P. Identifying spatial heterogeneity of COVID-19 transmission clusters and their built-environment features at the neighbourhood scale. *Health Place* **2023**, *84*, 103130. [CrossRef]
30. Li, B.; Liu, Q.; He, H.; Peng, Y. The Influence of Urban Built Environment on COVID-19 Epidemic and Its Prevention and Control Measures. *Chin. Overseas Archit.* **2021**, *12*, 2–7. [CrossRef]
31. Li, B.; Peng, Y.; He, H.; Wang, M.; Feng, T. Built Environment and Early Infection of COVID-19 in Urban Districts: A Case Study of Huangzhou. *Sustain. Cities Soc.* **2021**, *66*, 102685. [CrossRef] [PubMed]
32. Li, W.; Zhao, S.-C.; Ji, X.-F.; Ma, J.-W. Impact of Traffic Exposure and Land Use Patterns on the Rick of COVID-19 Spread at the Community Level. *China J. Highw. Transp.* **2020**, *33*, 43–54. [CrossRef]
33. Duan, J.; Yang, B.; Zhou, L.; Zhang, J. Planning Improves City's Immunity: A Written Conversa-tion on COVID-19 Breakout. *City Plan. Rev.* **2020**, *44*, 115–136.
34. Ewing, R.; Cervero, R. Travel and the Built Environment: A Meta-Analysis. *J. Am. Plann. Assoc.* **2010**, *76*, 265–294. [CrossRef]
35. Wali, B.; Frank, L.D. Neighborhood-Level COVID-19 Hospitalizations and Mortality Relationships with Built Environment, Active and Sedentary Travel. *Health Place* **2021**, *71*, 102659. [CrossRef] [PubMed]
36. Bin Kashem, S.; Baker, D.M.; González, S.R.; Lee, C.A. Exploring the Nexus between Social Vulnerability, Built Environment, and the Prevalence of COVID-19: A Case Study of Chicago. *Sustain. Cities Soc.* **2021**, *75*, 103261. [CrossRef] [PubMed]
37. Huang, X.; Yang, Q.; Yang, J. Importance of Community Containment Measures in Combating the COVID-19 Epidemic: From the Perspective of Urban Planning. *Geo-Spat. Inf. Sci.* **2021**, *24*, 363–371. [CrossRef]
38. Niu, Q.; Wu, W.; Shen, J.; Huang, J.; Zhou, Q. Relationship between Built Environment and COVID-19 Dispersal Based on Age Stratification: A Case Study of Wuhan. *Int. J. Environ. Res. Public. Health* **2021**, *18*, 7563. [CrossRef]
39. Asfour, O.S. Housing Experience in Gated Communities in the Time of Pandemics: Lessons Learned from COVID-19. *Int. J. Environ. Res. Public. Health* **2022**, *19*, 1925. [CrossRef]
40. DiMaggio, C.; Klein, M.; Berry, C.; Frangos, S. Black/African American Communities Are at Highest Risk of COVID-19: Spatial Modeling of New York City ZIP Code-Level Testing Results. *Ann. Epidemiol.* **2020**, *51*, 7–13. [CrossRef]
41. Xu, Y.; Guo, C.; Yang, J.; Yuan, Z.; Ho, H.C. Modelling Impact of High-Rise, High-Density Built Environment on COVID-19 Risks: Empirical Results from a Case Study of Two Chinese Cities. *Int. J. Environ. Res. Public. Health* **2023**, *20*, 1422. [CrossRef] [PubMed]
42. Tepe, E. The Impact of Built and Socio-Economic Environment Factors on Covid-19 Transmission at the ZIP-Code Level in Florida. *J. Environ. Manag.* **2023**, *326*, 116806. [CrossRef] [PubMed]
43. Liu, C.; Liu, Z.; Guan, C. The Impacts of the Built Environment on the Incidence Rate of COVID-19: A Case Study of King County, Washington. *Sustain. Cities Soc.* **2021**, *74*, 103144. [CrossRef] [PubMed]
44. Gaisie, E.; Oppong-Yeboaha, N.Y.; Cobbina, P.B. Geographies of infections: Built environment and COVID-19 pandemic in metropolitan Melbourne. *Sustain. Cities Soc.* **2022**, *81*, 103838. [CrossRef] [PubMed]
45. Choi, K.H.; Denice, P. Socioeconomic Variation in the Relationship Between Neighbourhoods' Built Environments and the Spread of COVID-19 in Toronto, Canada. *Can. Stud. Popul.* **2022**, *49*, 149–181. [CrossRef]
46. Schmiege, D.; Haselhoff, T.; Ahmed, S.; Anastasiou, O.E.; Moebus, S. Associations Between Built Environment Factors and SARS-CoV-2 Infections at the Neighbourhood Level in a Metropolitan Area in Germany. *J. Urban Health* **2023**, *100*, 40–50. [CrossRef]
47. Wang, J.; Wu, X.; Wang, R.; He, D.; Li, D.; Yang, L.; Yang, Y.; Lu, Y. Review of Associations between Built Environment Characteristics and Severe Acute Respiratory Syndrome Coronavirus 2 Infection Risk. *Int. J. Environ. Res. Public. Health* **2021**, *18*, 7561. [CrossRef]
48. Li, C.; Li, B.; Lei, S.; Long, L.; Ke, Q.; Li, L. Research progress on prevention and treatment of novel coronavirus pneumonia infection. *China Pharm.* **2020**, *23*, 1168–1174.
49. Yang, J.; Xu, J.; Hu, T.; Cao, J. Satisfaction and Demands of Indoor Space in the High-Density Residential Areas in the COVID-19 Era. *Buildings* **2022**, *12*, 660. [CrossRef]
50. Wu, Y.; Zuo, P.; Peng, C.; Zhu, X. Resilience Assessment and Its Association Characteristics of Urban Community for Public Health Emergencies: A Case Study of the Central District of Wuhan City. *Int. Urban Plan.* **2023**, *38*, 11–20. [CrossRef]
51. Navaratnam, S.; Nguyen, K.; Selvaranjan, K.; Zhang, G.; Mendis, P.; Aye, L. Designing Post COVID-19 Buildings: Approaches for Achieving Healthy Buildings. *Buildings* **2022**, *12*, 74. [CrossRef]
52. Wang, L.; Jia, Y.; Li, X.; Yang, X.M. Urban Spatial Intervention Strategies for Infectious Disease Prevention and Control. *Urban Plan.* **2020**, *44*, 13–20+32.
53. Hishan, S.S.; Ramakrishnan, S.; Qureshi, M.I.; Khan, N.; Hasan, N.; Al-Kumaim, S. Pandemic thoughts, civil infrastructure and sustainable development: Five insights from COVID-19 across travel lenses. *J. Talent. Dev. Excell.* **2020**, *12*, 1690–1696.
54. Dietz, L.; Horve, P.F.; Coil, D.A.; Fretz, M.; Eisen, J.A.; Van Den Wymelenberg, K. 2019 Novel Coronavirus (COVID-19) Pandemic: Built Environment Considerations To Reduce Transmission. *mSystems* **2020**, *5*, e00245-20. [CrossRef] [PubMed]
55. Qin, T.; Ruan, X.; Duan, Z.; Cao, J.; Liang, J.; Yang, J.; Jiang, Y.; Shi, M.; Xu, J. Wildlife-Borne Microorganisms and Strategies to Prevent and Control Emerging Infectious Diseases. *J. Biosaf. Biosecurity* **2021**, *3*, 67–71. [CrossRef]

56. Shenzhen City 2022 National Economic and Social Development Statistical Bulletin. Available online: http://tjj.sz.gov.cn/zwgk/zfxxgkml/tjsj/tjgb/content/post_10577661.html (accessed on 14 September 2023).
57. Fang, Y.; Zhou, X.; Zhang, Q. The Construction of Densely-populated Communities in Shenzhen: Practice and Rethinking in the Post-pandemic Era. *Urban Insight* **2022**, *4*, 64–74. [CrossRef]
58. Shenzhen Municipal Government Open Data Platform. Available online: https://opendata.sz.gov.cn/data/dataSet/toDataDetails/29200_05800007 (accessed on 28 April 2023).
59. Futian District Government Online. Available online: http://www.szft.gov.cn/ (accessed on 28 April 2023).
60. Wang, L.; Hu, Z.; He, L. Research on the Influencing Factors of Community Resilience in Shanghai Based on Analysis of the Pandemic Situation. *Urban Dev. Stud.* **2023**, *30*, 60–66.
61. Wang, L.; Liao, S.; Zhao, X.; Qian, F. Exploration of Approaches and Factors of Healthy City Planning. *Int. Urban Plan.* **2016**, *31*, 4–9. [CrossRef]
62. Harden, S.R.; Schuurman, N.; Keller, P.; Lear, S.A. Neighborhood Characteristics Associated with Running in Metro Vancouver: A Preliminary Analysis. *Int. J. Environ. Res. Public. Health* **2022**, *19*, 14328. [CrossRef]
63. Sokadjo, Y.M.; Atchadé, M.N. The Influence of Passenger Air Traffic on the Spread of COVID-19 in the World. *Transp. Res. Interdiscip. Perspect.* **2020**, *8*, 100213. [CrossRef]
64. Zhao, S.; Shen, M.; Musa, S.S.; Guo, Z.; Ran, J.; Peng, Z.; Zhao, Y.; Chong, M.K.C.; He, D.; Wang, M.H. Inferencing Superspreading Potential Using Zero-Truncated Negative Binomial Model: Exemplification with COVID-19. *BMC Med. Res. Methodol.* **2021**, *21*, 30. [CrossRef] [PubMed]
65. Nakaya, T.; Fotheringham, A.S.; Brunsdon, C.; Charlton, M. Geographically Weighted Poisson Regression for Disease Association Mapping. *Stat. Med.* **2005**, *24*, 2695–2717. [CrossRef] [PubMed]
66. Waang, C. Selection of Optimal Significance Level α on Statistical Hypothesis Test. Master's Thesis, Lanzhou University of Finance and Economics, Lanzhou, China, 2013. Available online: https://kns.cnki.net/KCMS/detail/detail.aspx?dbcode=CMFD&dbname=CMFD201302&filename=1013212553.nh&v= (accessed on 20 April 2022).
67. O'Brien, R.M. A Caution Regarding Rules of Thumb for Variance Inflation Factors. *Qual. Quant.* **2007**, *41*, 673–690. [CrossRef]
68. Frank, L.D.; Wali, B. Treating Two Pandemics for the Price of One: Chronic and Infectious Disease Impacts of the Built and Natural Environment. *Sustain. Cities Soc.* **2021**, *73*, 103089. [CrossRef] [PubMed]
69. Li, T.; Wang, J.; Huang, J.; Gao, X. Exploring Temporal Heterogeneity in an Intercity Travel Network: A Comparative Study between Weekdays and Holidays in China. *J. Geogr. Sci.* **2020**, *30*, 1943–1962. [CrossRef]
70. Yin, C.; Shao, C.; Wang, X.; Xiong, Z.; Cui, W. Influence of land use mix on non-work travel behavior based on modified entropy index. *J. Beijing Jiaotong Univ.* **2018**, *42*, 92–97.
71. Li, B.; Zhang, L. The revelation of SARS outbreak to urban planning in China. *City Plan. Rev.* **2003**, *7*, 71–72.
72. Yang, J.; Shi, B.; Shi, Y.; Li, Y. Construction of A Multi-Scale Spatial Epidemic Prevention System in High-Density Cities. *City Plan. Rev.* **2020**, *44*, 17–24.
73. Zhou, Y.; Zhang, J.; Guo, S. Exploration and Reflection On the Research of Medical and Health Facilities Planning Standards In the Case of COVID-19 Epidemic. *City Plan. Rev.* **2020**, *44*, 55–60+84.

buildings

MDPI

Article

Exploring the Impact of Campus Landscape Visual Elements Combination on Short-Term Stress Relief among College Students: A Case from China

Hui He, Tong Zhang, Qinghao Zhang *, Sheng Rong, Yihe Jia and Fengqian Dong

School of Architecture and Urban Planning, Huazhong University of Science and Technology, Wuhan 430074, China; hehui66@163.com (H.H.); 18303693806@163.com (T.Z.)
* Correspondence: m202173945@hust.edu.cn

Abstract: Although the effect of campus landscape space on stress relief among college students has been confirmed, few existing studies have considered the impact on stress recovery from the perspective of factor combination, and the key visual elements and the most effective combination of visual elements to relieve stress are still unclear. This study attempts to conduct a natural experiment within Chinese campuses, measuring physiological indicators of stress such as heart rate (HR), frequency domain index of heart rate variability (LF/HF), skin conductance level (SCL), skin temperature (SKT), and respiratory rate (RESP) using physiological instruments. It explored the effects of visual elements and their combinations in campus landscape spaces on short-term stress relief among college students through semantic segmentation, multifactorial analysis of variance, and post hoc multiple comparison methods. Research results demonstrate that the presence of water elements in the field of vision can effectively improve the stress relief effects of landscape spaces. Reasonable combinations of natural landscape elements and artificial landscape elements in the design can also effectively promote stress relief among students. Building facade area and sky area, water area and sky area, and plant species and pavement area are three combinations of factors with the strongest interactive effects. "Natural water scenery" and "exquisite artificial" are two campus landscape design patterns most conducive to short-term stress relief.

Keywords: campus; landscape; visual elements; stress relief; element combinations

Citation: He, H.; Zhang, T.; Zhang, Q.; Rong, S.; Jia, Y.; Dong, F. Exploring the Impact of Campus Landscape Visual Elements Combination on Short-Term Stress Relief among College Students: A Case from China. *Buildings* **2024**, *14*, 1340. https://doi.org/10.3390/buildings14051340

Academic Editor: Nikos A. Salingaros

Received: 19 March 2024
Revised: 10 April 2024
Accepted: 24 April 2024
Published: 9 May 2024

1. Introduction

Studies have shown that the transition from late adolescence to early adulthood is a high-risk period for psychological issues [1,2], coinciding with the stage of university education [3]. Evidence from a study involving 15 countries during the COVID-19 pandemic indicates that 31.2% of college students exhibit symptoms of depression, with a stress morbidity rate of 26.0% [4]. A study in China shows that over 30% of Chinese college students suffer from depression [5,6]. Entering university entails facing multiple pressures, such as social role transitions, changes in the living environment, detachment from previous social relationships, and academic demands [7]. Numerous pieces of evidence indicate that accumulated stress during this period can lead to negative psychological states such as pessimism, anxiety, and depression among adolescents who are sensitive in their psychosocial development [8–11]. This, in turn, affects various aspects of academic success, including academic performance, career development, interpersonal relationships, work efficiency, and academic functioning. Without a timely intervention, it may also lead to the onset of mental illnesses. As college students represent the reserve force for driving future technological development and social progress, a healthy psychological state is the primary prerequisite for fully realizing their own value and potential. This suggests that the mental health status of college students will also, to some extent, influence the future development and health trends of society, necessitating focused attention and collaborative efforts from all sectors of society [12]. Currently, there are increasing numbers of psychological counseling and intervention methods aimed at helping college students alleviate mental stress. However, due to the hidden nature of psychological stress and feelings of shame associated with mental illness, this measure has not achieved significant results. The psychological pressure within the college student population continues to grow and become more acute, making it a focal point for the prevalence of mental health issues [1,13,14]. As the most direct and frequently used natural spaces for college students, people are beginning to pay attention to the numerous benefits that campus landscape spaces provide to students. Several studies have shown that campus landscape spaces have benefits, such as stress relief, attention restoration, physical activity and social interaction promotion, and academic performance improvement [15–19]. These spaces offer an environmental intervention measure for alleviating stress among college students. How to construct campus landscape environments that promote the mental health of college students and fully utilize the natural healing effects of landscapes has become a topic of interest for many scholars.

2. Literature Review

2.1. Theoretical Basis

The biophilia hypothesis [20], attention restoration theory [21–23], and stress restoration theory [24,25] suggest that the health benefits of stress recovery in most natural environments can also promote improvements in the emotional state of perceivers, restoration of directed attention, and beneficial physiological changes [26–30]. These theories provide a solid theoretical foundation for research. The stress restoration theory posits that stress is the process by which individuals respond to threatening or challenging external stimuli in psychological, physiological, and behavioral forms. When individuals' coping abilities fail to adapt to environmental demands, it can lead to the development of negative emotions and physiological responses and even evoke avoidance behaviors. Visual landscapes can influence individuals' emotional states; observing natural or urban landscapes may affect personal perceptions and have significant positive or negative effects on the body. This theory emphasizes the visual characteristics of the environment and the aesthetic and emotional responses they evoke. In his book *The Hidden Dimension*, Hall analyzed proportion data of the human senses [31], revealing that vision accounts for 78% of the sensory system. Due to the difficulty in capturing attributes such as olfaction, taste, and touch, along with the subjective limitations of the human body [32], vision is considered the most important sensory form for perceiving the objective world. It can simultaneously convey environmental stimuli with good temporal and spatial resolutions [33]. Therefore, investigating the impact of visual landscape elements on the stress recovery of college students is of significant importance.

2.2. Research Progress

Most existing research on landscape and stress recovery primarily focuses on quantitative studies or a combination of quantitative and qualitative research types, with fewer purely qualitative studies. Previous qualitative studies often used descriptive words to express people's perception of landscapes by constructing an evaluation index system. Although it has the advantage of easy operation, it is difficult to summarize accurate conclusions and effectively explain landscape attributes. Consequently, many scholars have attempted to use quantitative experimental methods to explore the impact of visual landscapes on stress recovery, primarily divided into two categories. One category primarily focuses on comparative studies of the differences in stress recovery effects among different landscape spaces, involving various landscape types (plazas, woodlands, flower fields, lake views, etc.) [34,35] and vegetation structures (open, semiopen, enclosed, etc.). For example, Ping Zhan and others selected three types of landscape spaces on university campuses as experimental environments, comparing the differences in physiological indicators, emotional states, and perceptual recovery among them. They found that all three types of spaces had some impact on relaxing participants' mental states [12]. Mohamed Elsadek and others chose landscapes in Vancouver, Canada, and Japanese and architectural gardens as experimental sites, discovering that the Japanese garden attracted more attention, and the duration of attention varied depending on the garden style [36]. However, these studies mainly explore which types of landscape spaces have better stress recovery effects from the overall perspective of environmental scenes; the duration of the gaze varies depending on the garden style. Another category involves attempting to establish measurable indicators describing the objective features of landscapes and using mathematical methods to explore the mathematical relationships between landscapes and specific health indicators. Some scholars believe that taking a detailed perspective on environmental elements makes it possible to identify which specific landscape elements or features positively impact stress recovery. This approach can better provide scientific evidence for practice. Additionally, advancements in technologies such as computer vision and image processing enable the precise measurement of landscape visual features, promoting the transition of research from traditional qualitative evaluation to technological, quantitative, and scientific development. A few studies have employed the construction of multivariate models to explore the relationships between landscape elements and psychological recovery, whether promoting or inhibiting. These studies involve visual elements such as plant diversity, tree coverage density, and water bodies [37–39]. For example, Huang Q measured the perceived stress differences among three types of courtyards: grass, trees, and concrete [40]. Jaeyoung Ha and others confirmed the relationship between the composition ratio of trees, grass, and water elements in green spaces in Chicago and the reported level of psychological distress in individuals [41].

2.3. Insufficient Research

Existing research has produced relatively rich research results but has focused mainly on the relationship between the effects of individual landscape elements and stress recovery. In fact, landscape spaces are not formed and presented in isolation by individual landscape elements alone. Instead, they are composed of combinations of multiple landscape elements, which evoke physiological and psychological responses in individuals through the overall picture presented by the objective environment. Solely focusing on the positive or negative effects of a single element often overlooks the interaction and combination relationships among elements, making it difficult to grasp the general principles of how landscape spaces function fully. This limitation restricts the application scenarios and effectiveness of research conclusions. Considering the comprehensive benefits of visual element combinations on stress recovery is more practically meaningful in the design and management processes. However, current research rarely considers the interaction of multiple landscape elements or examines the impact of stress recovery from the perspective of element combinations. Therefore, we need more examples to increase the detailed understanding of visual landscape elements for stress recovery.

Exploring the impact of campus visual landscape element combinations on short-term stress relief in college students is important in providing scientific evidence for creating psychologically healthy university campus environments. This paper takes the Huazhong University of Science and Technology campus as the research area. Physiological experiments are conducted by recruiting experimental subjects, measuring psychological indicators, and designing experimental procedures. Multiple factor analysis and other methods are employed to explore the key visual elements and combinations of elements that are most effective in relieving student stress, supplementing research on the impact of stress recovery from the perspective of visual element combinations. This paper aims to achieve three main objectives: First, explore the most beneficial visual elements in campus landscapes for promoting stress relief in college students. Second, introduce interaction terms to investigate the effects of element combinations. Finally, explore the most effective landscape design patterns for alleviating stress in college students. The research hopes to enrich theoretical studies related to restorative environments while providing landscape designers with scientific evidence and design guidance for updating campus landscape spaces, thereby providing more directions for efforts to maintain the physical and mental health development of college students.

3. Materials and Methods

3.1. Landscape Space Selection

The Huazhong University of Science and Technology has two campuses, old and new, and its diverse built environment enables it to cover most types of campus landscapes. The campus covers an area of about 470 hectares, with a green coverage rate of up to 72%, earning it the nickname "Forest University", and it boasts a considerable amount of natural space. Moreover, as a well-known high-level university in China, the Huazhong University of Science and Technology has more than 50,000 undergraduate and graduate students, who face heavy academic pressure and need more campus landscapes for relaxation. Whether from the campus environment or student needs, selecting this campus for research is highly representative.

In order to make the research results have authenticity and practical application value, this paper carefully combed through the existing scholars' methods of dividing typical campus landscape types in China [42]. Taking 20% of the total number of each landscape space type as standard, 45 landscape spaces with different scales, forms, and design features were selected as samples, which comprehensively covered various landscape types of the campus (Figure 1).

Figure 1. Distribution of landscape space samples.

3.2. Participant Recruitment

Participants were recruited voluntarily for landscape experience both online and offline in the university. A total of 54 participants from different majors and grades were recruited. Personal information surveys were conducted, which included gender, age, academic stage, major, and initial stress levels. The initial stress level was assessed using the College Student Stress Scale developed by Li Hong in 2002 [43]. Group consistency tests were performed when assigning participants to each experimental site to eliminate potential group differences that could impact the study results. This was performed to mitigate external factors and enhance the internal validity of the experiment. The number of participants at each experimental site ranged from 4 to 5 individuals, and the same student could participate in multiple experiments. In total, 228 independent experimental participants were involved. Detailed descriptive statistics of participant attributes for all experimental sessions are presented in Table 1.

Table 1. Recruitment of students' sociodemographic characteristics.

Indicator	Quantity
Gender	Male (1) = 78; Female (2) = 150
Age	Mean = 23; Median = 23; Minimum = 16; Maximum = 26; Standard Deviation = 2.53
Academic Stage	Undergraduate (1) = 75; Master's (2) = 143; Ph.D. (3) = 10
Major	Humanities (1) = 40; Science (2) = 10; Engineering (3) = 178
Initial Stress Level	Mean = 28; Median = 25; Minimum = 2; Maximum = 65; Standard Deviation = 14.72

Source: this study (own contribution).

3.3. Experimental Procedure Design

The experiment was scheduled from 27 October to 3 November 2022, during early autumn, with pleasant weather and rich vegetation displaying diverse color changes. The average temperature and wind speed during the experiment were 22 to 29 °C and 2 to 3 levels, respectively, maintaining stable external weather conditions.

We conducted 45 repeated experiments over 9 consecutive days, with five experimental locations each day. The duration at each location was controlled to be approximately 50 min, avoiding exposure to harsh midday sunlight and insufficient evening light. The basic experimental procedure is as follows (Figure 2): First, we arranged participants to sit indoors to calm their emotions and measure the baseline physiological stress level. Next, following a predetermined route, participants were led to various landscape spaces, where they engaged in short periods of sitting, while simultaneous measurements of their momentary physiological stress indicators were taken. To mitigate the cumulative effects of landscape exposure and emotional influences during the journey, participants were given sufficient time to adjust their emotions before entering the next testing point. Once the physiological indicators stabilized, data collection was conducted. During the experiment, a staff member was responsible for capturing five on-site photos from the participant's perspective (top, front, rear, left, and right). These photos would be used to calculate visual environment indicators (refer to Figure 3). This design facilitated a comprehensive understanding of participants' physiological responses in different landscape environments, ensuring the accuracy and reliability of experimental data.

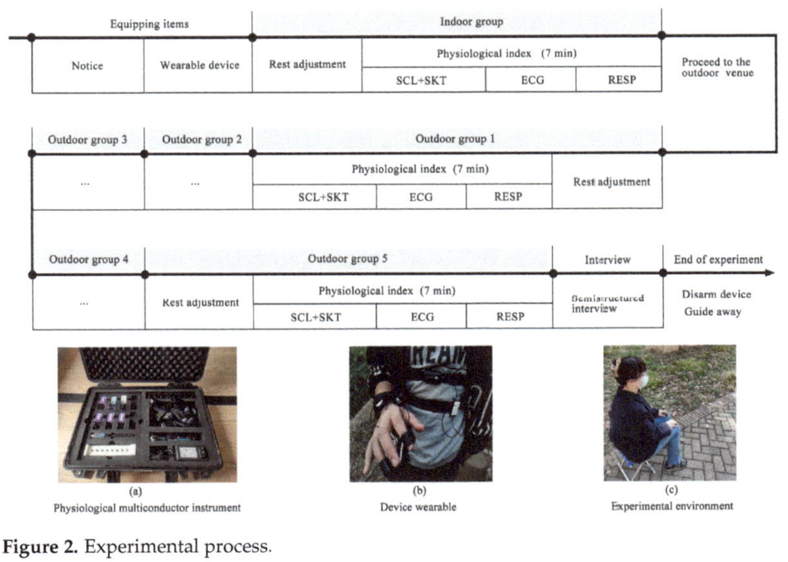

Figure 2. Experimental process.

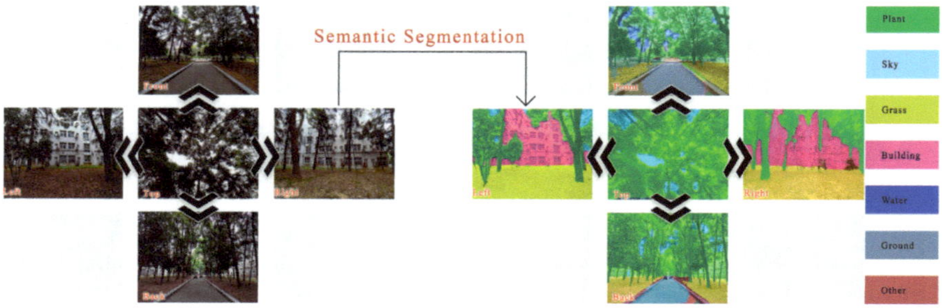

Figure 3. Semantic segmentation of on-site captured photos.

3.4. Measurement of Experimental Indicators

(1) Physiological Indicator Measurement

Subtle physiological changes induced by the environment can also be accurately and rapidly recorded and measured through physiological sensors, facilitating the objective recording of subconscious cognition and experiences that are challenging to measure directly. Electrocardiographic (ECG) signals represent the periodicity within a certain range of cardiac excitation and conduction, with the electrophysiological depolarization and repolarization of cardiac cells facilitating normal heart function. Stress induces changes in the autonomic nervous system, leading to an acceleration of the heart rate, an increase in blood pressure, and accelerated respiration. In a stressed state, the excitability of the heart is enhanced. Heart rate variability has been confirmed to reflect individual psychological stress changes in different situations and is a key indicator characterizing psychological stress [44,45]. The LF/HF frequency domain index of heart rate variability is commonly used to reflect the balance between the sympathetic and parasympathetic nervous systems [46]. Skin temperature is used to reflect changes in skin surface temperature in response to environmental stimuli. Some studies suggest that the sympathetic nervous system is activated in tense situations, leading to an increased metabolic rate and elevated body temperature. After reducing tension and anxiety, the body temperature naturally returns to normal. Respiration refers to the process of gas exchange between the human body and the external environment, and to some extent, it can reveal individual emotions, tension, and other psychological states. Deep, fast breathing and increased oxygen supply to the body to improve gas exchange in the lungs also occurs when there is emotional excitement or psychological stress. The integration of physiological parameters for psychological stress assessment is a developing trend based on evaluating stress through physiological parameters. Karthikevan et al. utilized various physiological parameters, including ECG, EEG, skin conductance, and blood pressure, measured using stimuli such as colored cards, public speaking, and games [47]. They demonstrated that the integration of these physiological parameters is more effective in assessing stress. Healey et al. demonstrated that, compared with electromyography (EMG) and respiration, ECG and skin conductance more effectively reflect the psychological stress status of drivers [48]. Costin et al. collected two physiological signals, EEG and ECG, and confirmed that the morphological variability of heart rate variability and EEG signals effectively detects psychological stress status [49]. The comprehensive measurement of various physiological indicators provides a more comprehensive understanding of participants' physiological responses in different environments, offering crucial objective data support for the field of environmental psychology.

The Ergolab wearable physiological multichannel recorder produced by Jinfa Technology was used to collect participants' physiological indicators. The selected modules include EDA, PPG, RESP, and SKT. Apart from the respiratory belt dedicated to testing breathing waves and the temperature probe for measuring body surface temperature, all other physiological indicators were measured using standard (Ag/AgCl) skin electrodes. The collection of physiological signals was analyzed using Ergolab 3.0 software. The selected physiological indicators included HR, LF/HF, SCL, SKT, and RESP. Due to individual variations in physiological responses, a relative percentage change was used for standardization to compare the observed physiological responses in the natural environment (baseline) and the original data. The formula is as follows:

$$((\text{Raw value} - \text{Baseline value})/\text{Baseline value}) \times 100.$$

(2) Measurement of Visual Element Indicators

The quantity and area ratio of various elements in the field of view are considered important factors affecting visual perception and are also the focus of landscape space design. Drawing from existing research on landscape visual elements and specific methods of landscape construction, we determined 10 key visual elements of interest in this study from the dimensions of "quantity" and "proportion" (Table 2). The measurement of quantity-based elements was conducted by on-site audits by staff. The measurement of proportion-based elements was conducted using semantic segmentation, which divides landscape photos taken from the perspective of the subjects into seven types: vegetation, grassland, pavement, water bodies, sky, buildings, and others. The proportions of different visual elements in the field of view were calculated. The specific process is illustrated in Figure 3. The semantic segmentation task was completed by an open program provided by the High-Performance Computing Laboratory (CUG.HPSCIL). The program is based on a deep learning fully convolutional network (FCN) and was trained on the ADE_20K dataset, demonstrating the ability to segment 150 object labels. During testing, the program achieved a pixelwise accuracy of 0.81 on the training dataset and an accuracy of 0.67 on the testing dataset, indicating that the program exhibits high accuracy when performing image segmentation tasks, while also demonstrating good applicability and generalization capabilities.

Table 2. Composition indicators of visual elements in campus landscape space.

Type	Visual Element	Measurement Method	Content	Mean	Variance
Quantity element	Number of facilities	Manual audit	Number of furniture items, such as tables, chairs, pavilions, fountains, pools, flower beds, sculptures, landscape walls, etc.	7.91	41.95
	Number of plant species		Number of visible plant species in the field of view	6.33	17.95
Proportion element	Percentage of vegetation area	Semantic segmentation of on-site captured images	$\frac{\sum_{i=1}^{4} \text{Vegetation pixels}_i}{\sum_{i=1}^{4} \text{Total pixels}_i}$	0.58	0.04
	Percentage of paving area		$\frac{\sum_{i=1}^{4} \text{Pavement pixels}_i}{\sum_{i=1}^{4} \text{Total pixels}_i}$	0.23	0.03
	Percentage of water body area		$\frac{\sum_{i=1}^{4} \text{Water pixels}_i}{\sum_{i=1}^{4} \text{Total pixels}_i}$	0.11	0.04
	Percentage of sky area		$\frac{\sum_{i=1}^{4} \text{Sky pixels}_i}{\sum_{i=1}^{4} \text{Total pixels}_i}$	0.16	0.02
	Percentage of building facade area		$\frac{\sum_{i=1}^{4} \text{Buliding pixels}_i}{\sum_{i=1}^{4} \text{Total pixels}_i}$	0.10	0.01

Source: this study (own contribution).

3.5. Experimental Data Analysis

(1) Descriptive Statistics

Descriptive statistics were derived on the changes in stress physiological indicators before and after the experiment for all participants (Table 3). It can be observed that, except for SCL, the changes in the other four stress physiological indicators are all negative, indicating a trend of stress relief during short periods of sitting in campus landscape spaces.

This reflects the potential positive impact of campus landscapes on students' psychological well-being. However, there may be individual differences in increasing SCL.

Table 3. Changes in stress physiological indicators (postexperiment–pre-experiment).

Physiological Indicators	Maximum	Minimum	Mean	Median	Metric
HR	10.482	−51.758	−10.677	−9.256	92.251
LF/HF	1.919	−6.488	−0.967	−0.728	1.209
RESP	2.536	−8.746	−1.642	−1.381	2.425
SCL	45.073	−11.252	1.799	1.616	24.89
SKT	8.932	−42.708	−8.411	−7.35	57.241

Source: this study (own contribution).

(2) Data Processing

First, the environmental indicators were categorized. The natural breakpoint method was employed to classify elements into three levels: high, medium, and low (Table 4). This approach enables a more accurate assessment of different levels of environmental indicators.

Table 4. Results of visual element reclassification.

Type	ID	Visual Element	Type 1	Type 2	Type 3
Quantity element	X1	Number of Facilities	1 (0~5)	2 (6~13)	3 (14~26)
	X2	Number of plant species	1 (1~5)	2 (6~13)	3 (14~17)
Proportion element	X3	Percentage of vegetation area	1 (0.051~0.440)	2 (0.441~0.656)	3 (0.656~0.868)
	X4	Percentage of paving area	1 (0~0.209)	2 (0.210~0.418)	3 (0.419~0.781)
	X5	Percentage of water body area	1 (0~0.077)	2 (0.078~0.364)	3 (0.365~0.766)
	X6	Percentage of sky area	1 (0.003~0.123)	2 (0.124~0.280)	3 (0.281~0.470)
	X7	Percentage of building facade area	1 (0~0.116)	2 (0.117~0.388)	3 (0.389~0.736)

Source: this study (own contribution).

Second, physiological indicators were selected. Using the "bidirectional selection/random consistency" method, the intraclass correlation coefficient (ICC) was calculated to test the consistency. The results indicated that the changes in various physiological indicators of the participants were not entirely consistent within the same scene. Specifically, the consistency between HR and RESP was good (ICC > 0.75), while the consistency between HR and SKT, RESP and SCL, and LF/HF and RESP was at a moderate level (ICC > 0.50). There was considerable inconsistency among other indicators. Based on the consistency of physiological indicators reflecting stress (such as the highly consistent changes between HR and RESP) and considering data collection quality issues (such as sensors measuring SCL being significantly affected by external temperature), one of these factors was chosen for analysis. Ultimately, HR and LF/HF, two indicators that are both representative and of good data quality, were selected for further analysis.

(3) Analysis Process

We employed a multifactorial analysis of variance to investigate the impact of visual elements on stress relief, following the steps below: first, a unifactorial analysis was conducted; second, interaction terms were introduced to explore the effects of element combinations; and third, we sought the optimal combination of visual elements for stress recovery. To avoid the complexity and interpretational challenges associated with too many interaction terms, this analysis considered interactions involving up to three elements (Figure 4).

In the process of exploring element combinations, we listed all possible combinations of two and three elements, constructing n analysis of variance models (where n is the total number of possible combinations). Each model included seven individual visual elements and an interaction term to be explored. Additionally, five personal attribute indicators were introduced as covariates to enhance the accuracy and interpretability of research effects by controlling for other variables that might influence the results. Significant ($p < 0.05$) element combinations were selected, and the constituent features of these significant combinations were analyzed. Finally, through post hoc multiple comparisons, we identified landscape visual element combinations that exhibited the most effective stress relief. The entire analysis process was conducted using the Python programming language and the Statsmodels statistical analysis module.

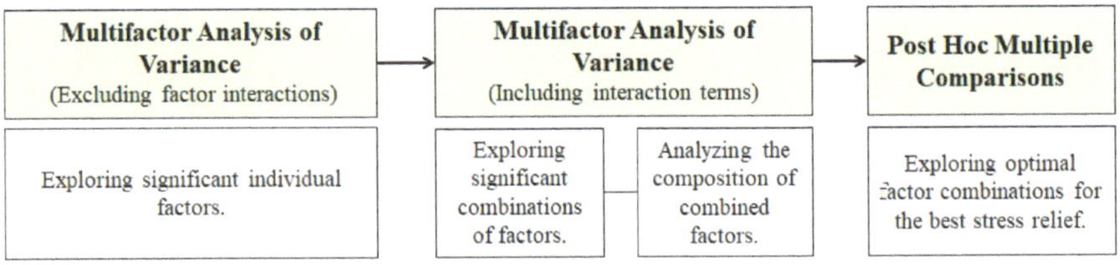

Figure 4. Data analysis process.

4. Results

4.1. Stress Relief at Different Experimental Locations

The locations exhibiting the most effective stress relief were primarily situated near water bodies (e.g., Figure 5(a_4,e_5)) and expansive campus green areas (e.g., Figure 5(e_1)). Additionally, some carefully designed landscape spaces, such as the campus entrance (e.g., Figure 5(b_5)) and the academic exchange center (e.g., Figure 5(d_5)), also demonstrated notable stress reduction effects. In comparison, the stress relief effects in less artificially intervened mountainous and forested areas (e.g., Figure 5(e_2,e_3)) were relatively limited. Small green spaces (e.g., Figure 5(h_2,i_1)) and courtyards between teaching buildings or dormitories (e.g., Figure 5(f_1,h_4)) displayed the least effective stress recovery outcomes (Figure 5).

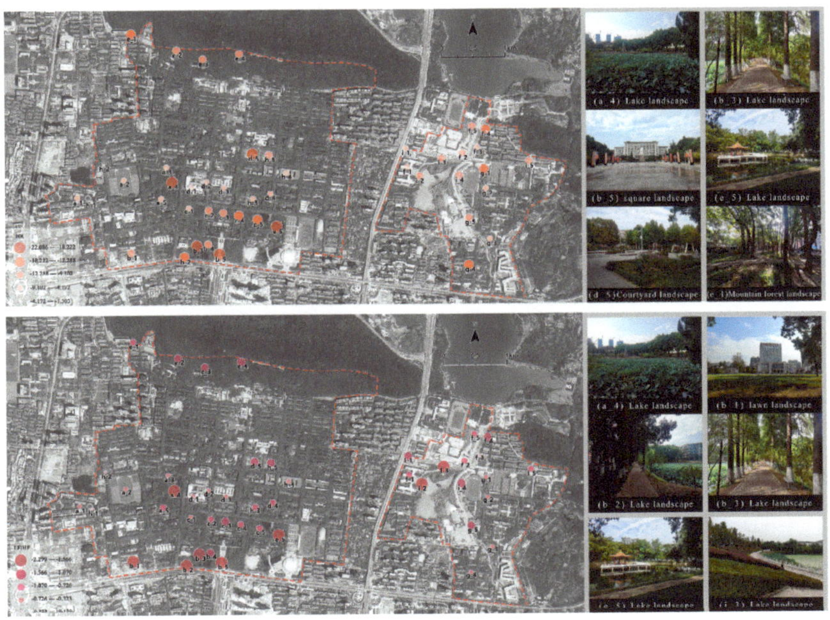

Figure 5. Multifactorial analysis results for HR and LF/HF indicators.

4.2. Influence of Single Factors on Stress Relief

According to the results in Table 5, it was found that only the water feature among all landscape visual elements had a significant impact on the changes in HR and LF/HF stress physiological indicators ($p < 0.05$). As depicted in Figure 6, it is evident that, with an increase in the area of water elements in the field of view, university students experience better stress relief effects in the landscape space.

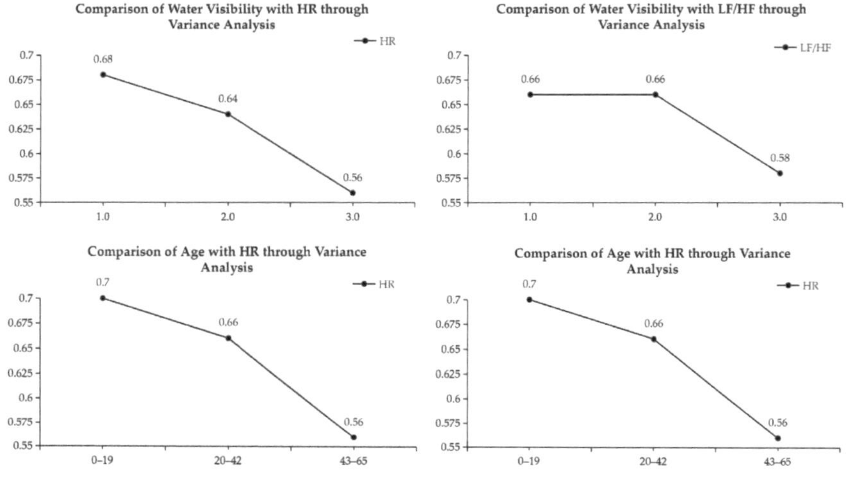

Figure 6. Single-factor analysis of significant indicators.

Regarding personal attributes, the initial stress level significantly influenced HR ($p < 0.001$), while age significantly affected LF/HF ($p < 0.05$). A higher initial stress level in individuals resulted in better stress relief effects. Compared with individuals aged 16–20, the landscape space had a better stress relief effect on those aged 21–26.

Table 5. Stress relief in landscape spaces.

Source	HR					LF/HF				
	Sum of Squares	Degrees of Freedom	Mean Square	F	*p*	Sum of Squares	Degrees of Freedom	Mean Square	F	*p*
Corrected model	1.077 a	19	0.057	2.723	0.000 ***	0.413 a	19	0.022	1.302	0.185
Intercept	1.413	1	1.413	67.916	0.000 ***	1.163	1	1.163	69.715	0.000 ***
Number of facilities	0.036	2	0.018	0.857	0.426	0.025	2	0.012	0.736	0.48
Number of plant species	0.004	2	0.002	0.093	0.911	0.001	2	0.000	0.021	0.98
Percentage of vegetation area	0.093	2	0.046	2.226	0.111	0.050	2	0.025	1.502	0.225
Percentage of paving area	0.113	2	0.056	2.709	0.069	0.063	2	0.031	1.879	0.155
Percentage of water area	0.135	2	0.068	3.244	0.041 *	0.114	2	0.057	3.415	0.035 *
Percentage of sky area	0.003	2	0.001	0.067	0.936	0.016	2	0.008	0.484	0.617
Percentage of building facade area	0.027	2	0.013	0.642	0.527	0.029	2	0.014	0.869	0.421
Gender	0.002	1	0.002	0.105	0.746	0.015	1	0.015	0.881	0.349
Age	0.006	1	0.006	0.274	0.601	0.070	1	0.070	4.185	0.042 *
Education level	0.007	1	0.007	0.351	0.554	0.026	1	0.026	1.530	0.218
Major	0.053	1	0.053	2.525	0.114	0.001	1	0.001	0.076	0.782
Level of stress	0.375	1	0.375	18.040	0.000 ***	0.006	1	0.006	0.384	0.536
Error	4.329	208	0.021			3.470	208	0.017		
Total	104.738	228				102.195	228			
Corrected total	5.406	227				3.882	227			
	a. R square = 0.199 (adjusted R square = 0.126)					a. R square = 0.106 (adjusted R square = 0.025)				

Source: this study (own contribution).

4.3. Influence of Element Combinations on Stress Relief

In the variance analysis model with 112 included interaction terms, 33 element combinations significantly affected the HR physiological indicator, and 11 element combinations significantly affected the LF/HF physiological indicator. This phenomenon demonstrates the significant importance of artificially designing and combining landscape visual elements, in addition to the natural element of water, for stress relief among university students.

Analyzing the frequency of occurrence of individual elements in the interaction terms, the results (Figure 7) indicate that all visual elements exhibit certain interaction effects. Among them, building facade (X7) and sky area (X6) are the two elements that participate in the most interactions. Additionally, plant species (X2), paving area (X4), and water area (X5) also show some interaction effects with other elements. From the frequency of occurrence of each element in the combinations, it can be observed that combinations involving building facade (X7) and sky area (X6), water area (X5) and sky area (X6), and plant species (X2) and paving area (X4) consistently occur together. This suggests that these elements are more inclined to collectively contribute to stress relief. Through post hoc multiple comparison analysis of the model, five landscape visual element combinations were selected from the models related to the HR index and LF/HF index, respectively, which effectively maintained the stress physiological indicators at lower levels. In theory, these combinations of elements are expected to have the best stress relief effects for university students.

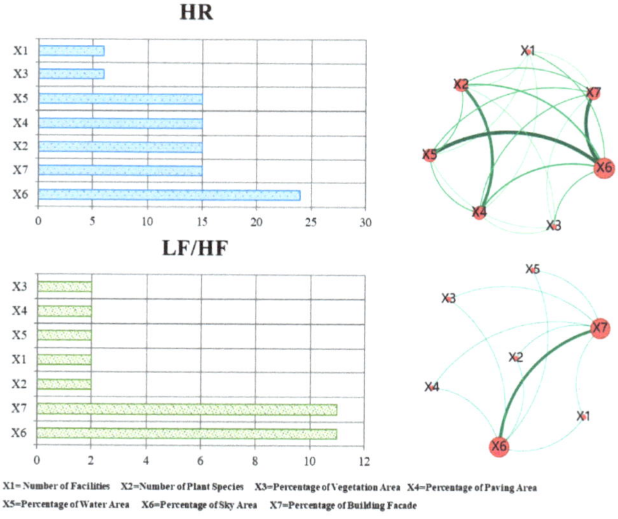

Figure 7. Significant interactive elements and their connections.

Table 6 indicates that, in the analysis models related to the HR index, the combination that achieves the best stress relief consists of "fewer building structures + moderate plant species + more paving area." In the analysis models related to the LF/HF index, several combinations are identified as achieving the best stress relief effects. These combinations include "fewer building structures + moderate sky area + moderate building area", "less water area + moderate sky area + moderate building area", "more paving area + moderate sky area + moderate building area", and "less vegetation area + moderate sky area + moderate building area." It can be observed that moderate sky area and building area are common elements in all of these combinations.

Table 6. Element combinations achieving the best stress relief effects.

Metric	Combination	Value			Mean
HR	X1 × X2 × X4	X1 = 1 (0~5)	X2 = 2 (6~13)	X4 = 3 (0.42~0.78)	0.48
	X4 × X5 × X6	X4 = 2 (0.21~0.42)	X5 = 2 (0.08~0.36)	X6 = 1 (0.01~0.12)	0.49
	X3 × X5 × X6	X3 = 2 (0.44~0.66)	X5 = 2 (0.08~0.36)	X6 = 1 (0.01~0.12)	0.49
	X5 × X6	X5 = 3 (0.37~0.77)	X6 = 1 (0.01~0.12)		0.50
	X2 × X5 × X6	X2 = 2 (6~13)	X5 = 3 (0.37~0.77)	X6 = 1 (0.01~0.12)	0.50
LF/HF	X1 × X6 × X7	X1 = 1 (0~5)	X6 = 2 (0.12~0.28)	X7 = 2 (0.12~0.39)	0.54
	X5 × X6 × X7	X5 = 1 (0~0.08)	X6 = 2 (0.12~0.28)	X7 = 2 (0.12~0.39)	0.54
	X4 × X6 × X7	X4 = 3 (0.42~0.78)	X6 = 2 (0.12~0.28)	X7 = 2 (0.12~0.39)	0.54
	X3 × X6 × X7	X3 = 1 (0.05~0.44)	X6 = 2 (0.12~0.28)	X7 = 2 (0.12~0.39)	0.54
	X2 × X6 × X7	X2 = 2 (14~26)	X6 = 2 (0.12~0.28)	X7 = 2 (0.12~0.39)	0.55

Source: this study (own contribution). X1 = number of facilities. X2 = number of plant species. X3 = percentage of vegetation area. X4 = percentage of paving area. X5 = percentage of water area. X6 = percentage of sky area. X7 = percentage of building façade.

5. Discussion

5.1. The Impact of Individual Factors on Stress Relief

Our experiment shows that water bodies significantly promote stress relief among university students during periods of sitting, and the stress relief effect is positively correlated with the visibility of water bodies in the field of view. Christopher Alexander made multiple uses of pattern languages associated with water bodies in designing "EISHIN GAKUEN" in Japan, thereby creating a campus environment conducive to stress relief for students [50]. Furthermore, numerous studies have affirmed the significance of water elements in alleviating stress. The importance of the "water" element for stress relief has been confirmed by several studies. For example, White et al. [51] in a restorative experiment comparing natural and built scenes, found that scenes with water, whether natural or built, were preferred, had a greater positive impact, and were perceived as more restorative. Jie Y [34] through on-site experiments with subjects, discovered that urban environments with water induce noticeable physiological and psychological relaxation effects. Yan C's [52] study on waterfront landscape characteristics and mental health also demonstrated that blue visibility is more likely to evoke positive emotions compared with green visibility or other elements. What sets this analysis apart from other studies is that water bodies are the only element found to have a significant impact. We speculate that this difference may be due to the experimental method commonly used in similar studies, which often involve comparing multiple scenes. This method makes it difficult to discern whether the element alone is causing an effect or if it is combined with other elements. Another important factor may be related to the short duration of time participants spend sitting and observing in the landscape space. When participants suddenly arrive in a landscape space with water, their attention is easily drawn to the water, temporarily overlooking the surrounding environment.

Furthermore, this experiment found significant associations between certain personal attributes and stress relief effects, indicating that individuals with high initial stress levels and older-age students are more susceptible to the influence of landscape elements. This suggests that, for some university students who are under high levels of stress for prolonged periods, taking a brief sit in campus landscape spaces can be an effective way to relax. Generally, older students tend to have better emotional regulation abilities, so they are more likely to use the landscape environment to quickly adjust their negative emotions.

5.2. The Impact of Element Combinations on Stress Relief

The experimental results reveal that, when incorporating interaction terms of landscape elements into the analysis of variance model, the number of elements significantly affecting stress relief increases noticeably. This indicates that when university students sit and observe campus landscapes, most visual elements work together in combination to alleviate stress. Many studies have demonstrated the beneficial effects of landscape visual elements and scenes on mental health. Our study complements these findings by suggesting that the therapeutic effect of landscape spaces may also result from the combination of several specific elements within the scene. A similar view has been proposed by other studies before; for example, Zhao et al. suggest that audiovisual interactions can have a combined effect on stress relief [53], while Lin et al. emphasize the importance of considering the proportion of landscape elements [54]. The difference between our study and these previous ones is our use of quantitative analysis to identify landscape visual elements that may interact. Among these, the combinations with the strongest interaction effects are building facades and sky area, water area and sky area, and plant species and paving area.

In summary, our study suggests that, in campus spaces, designing a combination of natural elements such as water bodies and plants with artificial elements such as buildings and sculptures can have a better stress-relieving effect on university students. This may be related to the specificity of such urban spatial environments. Universities, as spaces that accommodate a large concentration of higher education communities, have ubiquitous and meaningful artificial elements such as sculptures, patterns, symbols, and buildings, in addition to natural landscapes. These elements can also serve as cultural symbols influencing the emotions and thoughts of university students experiencing stress. Carefully designed landscape forms are more easily perceived and understood by observers with abundant knowledge and rich ideas. Similar findings were observed in studies conducted by Michael W. M. et al., which noted that individuals tend to pay more attention to architectural facades and some artificial architectural decorations when appreciating campus scenery [55]. Xu et al. found in an experiment comparing the recovery capabilities of Chinese cultural and natural landscapes that cultural landscapes have a similar impact on mental recovery as natural landscapes. Moreover, combining humanistic elements with natural environments to create a relaxed and tranquil cultural atmosphere is believed to enhance the environmental recovery capability [56].

5.3. The Best Combination of Elements to Promote Stress Relief

Through post hoc multiple comparisons, we identified the optimal combinations of elements that promote stress relief. Our study found that landscape element combinations related to "a smaller number of facilities + a moderate amount of plant species + more paved area" and "moderate sky area + moderate building facade area" had the greatest promoting effect on stress relief.

Combining the participants' experiential feedback allows for a reasonable interpretation of these results: When the number of facilities in the participants' field of view is controlled to within five, carefully designed sculptures, landscape walls, fountains, and seating can effectively serve as visual focal points. The appropriate variety of plant species further enhances spatial depth without causing the participants to feel overwhelmed by clutter. When the paved area in the field of view ranges from 42% to 78%, or the proportion of the building facade area is between 12% and 39%, and the sky area is between 12% and 28%, these indicators suggest that the location where the participants are seated is likely to be a small plaza surrounded by buildings. The artificial environment provides a sense of place and security for observers, ensuring that they can comfortably enjoy the diverse flora, fauna, and beautiful landscape features nearby. This type of landscape space likely allows university students to momentarily forget about their stress through attention diversion. This work complements the findings of Lin W et al. [53], who demonstrated through virtual experiments that the optimal stress relief effect is achieved when the proportion of sky and tree leaves seen in forests is around 33%. Similarly, we demonstrate that when the proportion of building facades to sky in university campuses is similar, a similar effect is observed.

Through further summarization, the landscape element combination patterns that achieve the best stress-relieving effects can be roughly categorized into two types: "natural water scenes" and "exquisite artificial" modes. The first type is the "natural water scene" mode: primarily focusing on rivers, ponds, and lakes within the campus. By enhancing the surrounding greenery and increasing the variety of plant species to maintain the site's natural attributes, along with incorporating more water-friendly facilities, students can have a more extensive contact with water bodies, thereby alleviating their stress. This model has been validated through its application in experimental areas, such as the Youth Park and Lake Creek River sites. The second type is the "exquisite artificial" mode: creating a pleasant outdoor space through well-organized architectural layouts. Within the enclosed area of the site, various plants are appropriately planted, and facilities such as sculptures, flower beds, and seats are installed, while architectural facades and pavements are carefully designed to enhance the cultural atmosphere of the location. Campus entrance squares and academic exchange centers, due to their unique characteristics, attract many teachers and students to visit frequently for exchanges, becoming ideal places for relaxation and leisure after class.

5.4. Application

The above research conclusions provide more meaningful assistance for the design of campus landscape spaces.

(1) To construct campus landscape spaces with stress-relieving effects, priority should be given to increasing water features. Encouraging the introduction of water features such as artificial lakes and fountains within or around the campus area is essential to provide the most effective stress-relief effects.

(2) It is important to emphasize the organic integration of natural landscape elements and cultural landscape elements and to strengthen the stress-relief effects of campus landscape spaces through combined designs. For example, cultural elements such as sketches, sculptures, and other artistic installations can be introduced into natural landscapes to enhance the environmental restoration potential. Throughout this process, particular attention should be paid to controlling the proportions of visible architectural facades, sky, and water surfaces and controlling the combination of plant species and paving forms.

(3) Landscape designers can refer to these two design models, "natural water scenery" and "exquisite artificial", to select certain areas within the campus to create professional healing gardens.

5.5. Limitations

This study has some shortcomings that need to be addressed in future research. First, the recruitment of participants faced practical challenges, such as time constraints, budget limitations, communication difficulties, and participant enthusiasm, leading to a higher proportion of females and graduate students in the sample. Although the study did not find significant effects of gender and education level on stress relief, the uneven distribution of the sample may still lead to some biases in the conclusions. Therefore, it is necessary to control the composition of participants further to enhance the rigor of the research process and ensure the relative fairness of experimental conclusions if conditions permit.

Second, the experiment observed inconsistent changes in the six stress physiological indicators of the human body when facing the same scene. This phenomenon requires the integration of more professional medical knowledge for accurate explanation. In order to make related research more realistic and reasonable, efforts should be made in the future to standardize the effectiveness of various physiological indicators and their ability to explain psychological stress in different usage scenarios. Furthermore, considering personal privacy and student willingness, some personal attribute indicators considered relevant, such as sleep duration, physical activity preferences, and BMI, were not included in the analysis process. In the future, it is necessary to supplement the collection of more prosperous and more scientific personal attribute characteristics to improve the model's effectiveness further.

Finally, due to limitations in the number of instruments, this study used an experiential route experimental approach. Despite taking necessary measures to avoid the potential effects of cumulative effects, the goodness of fit of the regression analysis model and the significance level of various factors have yet to reach the ideal state. In future research, it is necessary to increase the number of participants and instruments to improve the accuracy and reliability of the model.

6. Conclusions

This study explores the effects of campus landscape visual elements on short-term stress relief among college students using physiological instruments, semantic segmentation techniques, and statistical methods, identifying the most effective elements and combinations for promoting stress relief through experimental analysis. The findings of this study enrich the theory of restorative environments and can be applied to the design of university campus landscapes.

The study hopes to continue adding various experimental cases of university campus landscapes and student psychological stress in the future to verify the reliability of the conclusions. It also calls for continued attention to landscape perceptual elements in related research, especially the role of the combination ratio of elements in mental health, to provide more directions for the maintenance of the physical and mental health development of college students.

Author Contributions: Conceptualization, H.H., T.Z. and Q.Z.; methodology, H.H., T.Z. and Q.Z.; software, T.Z. and Q.Z.; validation, H.H., T.Z. and Q.Z.; formal analysis, T.Z. and Q.Z.; investigation, H.H., Q.Z., F.D., T.Z. and Y.J.; resources, H.H. and T.Z.; data curation, H.H., Q.Z., F.D., T.Z. and Y.J.; writing—original draft, Q.Z., T.Z. and F.D.; writing—review and editing, H.H., Q.Z. and T.Z.; visualization, Q.Z., T.Z., Y.J. and F.D.; supervision, H.H., S.R., Q.Z. and T.Z.; project administration, H.H., S.R., Q.Z. and T.Z.; funding acquisition, H.H. All authors have read and agreed to the published version of the manuscript.

Funding: This research was supported by the National Natural Science Foundation of China (Hui, H., Grant No. 51978300 and Grant No. 52178039).

Data Availability Statement: The data supporting the findings of the study are available from the first author upon request.

Acknowledgments: This study received equipment support from the School of Architecture and Urban Planning of the Huazhong University of Science and Technology for the project team, for which we would like to express our deepest appreciation.

Conflicts of Interest: The authors declare no conflicts of interest.

References

1. Eisenberg, D.; Gollust, S.E.; Golberstein, E.; Hefner, J.L. Prevalence and Correlates of Depression, Anxiety, and Suicidality among University Students. *Am. J. Orthopsychiat.* **2007**, *77*, 534–542. [CrossRef] [PubMed]
2. Kessler, R.C.; Chiu, W.T.; Demler, O.; Walters, E.E. Prevalence, Severity, and Comorbidity of 12-Month DSM-IV Disorders in the National Comorbidity Survey Replication. *Arch. Gen. Psychiatry* **2005**, *62*, 617. [CrossRef] [PubMed]
3. Arnett, J.J. Emerging Adulthood: A Theory of Development from the Late Teens through the Twenties. *Am. Psychol.* **2000**, *55*, 469–480. [CrossRef] [PubMed]
4. Batra, K.; Sharma, M.; Batra, R.; Singh, T.P.; Schvaneveldt, N. Assessing the Psychological Impact of COVID-19 among College Students: An Evidence of 15 Countries. *Healthcare* **2021**, *9*, 222. [CrossRef] [PubMed]
5. Dong, Y.; Li, H. The Relationship between Social Support and Depressive Symptoms among the College Students of Liaoning, China: A Moderated Mediated Analysis. *Psychol. Health Med.* **2020**, *25*, 368–378. [CrossRef] [PubMed]
6. Gao, W.; Ping, S.; Liu, X. Gender Differences in Depression, Anxiety, and Stress among College Students: A Longitudinal Study from China. *J. Affect. Disord.* **2020**, *263*, 292–300. [CrossRef]
7. Taneja, N.; Sachdeva, S.; Dwivedi, N. Assessment of Depression, Anxiety and Stress among Medical Students Enrolled in a Medical College of New Delhi. *Indian J. Soc. Psychiatry* **2018**, *34*, 157–162. [CrossRef]
8. Bruffaerts, R.; Mortier, P.; Kiekens, G.; Auerbach, R.P.; Cuijpers, P.; Demyttenaere, K.; Green, J.G.; Nock, M.K.; Kessler, R.C. Mental Health Problems in College Freshmen: Prevalence and Academic Functioning. *J. Affect. Disord.* **2018**, *225*, 97–103. [CrossRef]

9. Goldman-Mellor, S.J.; Caspi, A.; Harrington, H.; Hogan, S.; Nada-Raja, S.; Poulton, R.; Moffitt, T.E. Suicide Attempt in Young People: A Signal for Long-Term Health Care and Social Needs. *JAMA Psychiatry* **2014**, *71*, 119. [CrossRef]
10. Kerr, D.C.R.; Capaldi, D.M. Young Men's Intimate Partner Violence and Relationship Functioning: Long-Term Outcomes Associated with Suicide Attempt and Aggression in Adolescence. *Psychol. Med.* **2011**, *41*, 759–769. [CrossRef]
11. Webster, J.R.; Beehr, T.A.; Christiansen, N.D. Toward a Better Understanding of the Effects of Hindrance and Challenge Stressors on Work Behavior. *J. Vocat. Behav.* **2010**, *76*, 68–77. [CrossRef]
12. Zhang, P.; He, Q.; Chen, Z.; Li, X.; Ma, J. An Empirical Study on the Promotion of Students' Physiological and Psychological Recovery in Green Space on Campuses in the Post-Epidemic Era. *Int. J. Environ. Res. Public Health* **2022**, *20*, 151. [CrossRef]
13. Dyson, R.; Renk, K. Freshmen Adaptation to University Life: Depressive Symptoms, Stress, and Coping. *J. Clin. Psychol.* **2006**, *62*, 1231–1244. [CrossRef] [PubMed]
14. Ibrahim, A.K.; Kelly, S.J.; Adams, C.E.; Glazebrook, C. A Systematic Review of Studies of Depression Prevalence in University Students. *J. Psychiatr. Res.* **2013**, *47*, 391–400. [CrossRef] [PubMed]
15. Li, D.; Sullivan, W.C. Impact of Views to School Landscapes on Recovery from Stress and Mental Fatigue. *Landsc. Urban Plan.* **2016**, *148*, 149–158. [CrossRef]
16. Gulwadi, G.B.; Mishchenko, E.D.; Hallowell, G.; Alves, S.; Kennedy, M. The Restorative Potential of a University Campus: Objective Greenness and Student Perceptions in Turkey and the United States. *Landsc. Urban Plan.* **2019**, *187*, 36–46. [CrossRef]
17. Boyd, F. Between the Library and Lectures: How Can Nature Be Integrated into University Infrastructure to Improve Students' Mental Health. *Front. Psychol.* **2022**, *13*, 865422. [CrossRef]
18. Stepansky, K.; Delbert, T.; Bucey, J.C. Active Student Engagement within a University's Therapeutic Sensory Garden Green Space: Pilot Study of Utilization and Student Perceived Quality of Life. *Urban For. Urban Green.* **2022**, *67*, 127452. [CrossRef]
19. Hodson, C.B.; Sander, H.A. Green Urban Landscapes and School-Level Academic Performance. *Landsc. Urban Plan.* **2017**, *160*, 16–27. [CrossRef]
20. Byrne, J. Biophilia. In Proceedings of the SIGGRAPH06: Special Interest Group on Computer Graphics and Interactive Techniques Conference, Boston, MA, USA, 30 July–3 August 2006. [CrossRef]
21. Kaplan, R.; Kaplan, S.; Ryan, R. With People in Mind: Design and Management of Everyday Nature. *Landsc. J.* **1999**, *18*, 99–101.
22. Kaplan, R. The Nature of the View from Home: Psychological Benefits. *Environ. Behav.* **2001**, *33*, 507–542. [CrossRef]
23. Kaplan, S. The Restorative Benefits of Nature: Toward an Integrative Framework. *J. Environ. Psychol.* **1995**, *15*, 169–182. [CrossRef]
24. Ulrich, R.S.; Simons, R.F.; Losito, B.D.; Fiorito, E.; Miles, M.A.; Zelson, M. Stress Recovery during Exposure to Natural and Urban Environments. *J. Environ. Psychol.* **1991**, *11*, 201–230. [CrossRef]
25. Ulrich, R.S. Aesthetic and Affective Response to Natural Environment. In *Behavior and the Natural Environment*; Altman, I., Wohlwill, J.F., Eds.; Springer: Boston, MA, USA, 1983; pp. 85–125. [CrossRef]
26. Lee, A.C.K.; Maheswaran, R. The Health Benefits of Urban Green Spaces: A Review of the Evidence. *J. Public Health* **2011**, *33*, 212–222. [CrossRef] [PubMed]
27. Barton, J.; Pretty, J. What Is the Best Dose of Nature and Green Exercise for Improving Mental Health? A Multi-Study Analysis. *Environ. Sci. Technol.* **2010**, *44*, 3947–3955. [CrossRef]
28. Woo, J.; Tang, N.; Suen, E.; Leung, J.; Wong, M. Green Space, Psychological Restoration, and Telomere Length. *Lancet* **2009**, *373*, 299–300. [CrossRef]
29. Van den Berg, A.E.; Maas, J.; Verheij, R.A.; Groenewegen, P.P. Green Space as a Buffer between Stressful Life Events and Health. *Soc. Sci. Med.* **2010**, *70*, 1203–1210. [CrossRef]
30. White, M.; Alcock, I.; Wheeler, B.; Depledge, M. Would You Be Happier Living in a Greener Urban Area? A Fixed-Effects Analysis of Panel Data. *Psychol. Sci.* **2013**, *24*, 920–928. [CrossRef]
31. Hall, E.T. *The Hidden Dimension*; Anchor: New York, NY, USA, 1990.
32. Fang, W.-T.; Hassan, A.; LePage, B.A. Environmental Psychology. In *The Living Environmental Education: Sound Science toward a Cleaner, Safer, and Healthier Future*; Fang, W.-T., Hassan, A., LePage, B.A., Eds.; Springer Nature: Singapore, 2023; pp. 127–149. [CrossRef]
33. Qiu, M.; Zhang, J.; Zhang, H.; Zheng, C. Is Looking Always More Important than Listening in Tourist Experience? *J. Travel Tour. Mark.* **2018**, *35*, 869–881. [CrossRef]
34. Yin, J.; Ramanpong, J.; Chang, J.; Wu, C.-D.; Chao, P.-H.; Yu, C.-P. Effects of Blue Space Exposure in Urban and Natural Environments on Psychological and Physiological Responses: A within-Subject Experiment. *Urban For. Urban Green.* **2023**, *87*, 128066. [CrossRef]
35. Li, H.; Dong, W.; Wang, Z.; Chen, N.; Wu, J.; Wang, G.; Jiang, T. Effect of a Virtual Reality-Based Restorative Environment on the Emotional and Cognitive Recovery of Individuals with Mild-to-Moderate Anxiety and Depression. *Int. J. Environ. Res. Public Health* **2021**, *18*, 9053. [CrossRef]
36. Elsadek, M.; Sun, M.; Sugiyama, R.; Fujii, E. Cross-Cultural Comparison of Physiological and Psychological Responses to Different Garden Styles. *Urban For. Urban Green.* **2019**, *38*, 74–83. [CrossRef]
37. Jiang, B.; Chang, C.-Y.; Sullivan, W.C. A Dose of Nature: Tree Cover, Stress Reduction, and Gender Differences. *Landsc. Urban Plan.* **2014**, *132*, 26–36. [CrossRef]

38. Deng, L.; Li, X.; Luo, H.; Fu, E.-K.; Ma, J.; Sun, L.-X.; Huang, Z.; Cai, S.-Z.; Jia, Y. Empirical Study of Landscape Types, Landscape Elements and Landscape Components of the Urban Park Promoting Physiological and Psychological Restoration. *Urban For. Urban Green.* **2020**, *48*, 126488. [CrossRef]
39. Zhou, X.; Cen, Q.; Qiu, H. Effects of Urban Waterfront Park Landscape Elements on Visual Behavior and Public Preference: Evidence from Eye-Tracking Experiments. *Urban For. Urban Green.* **2023**, *82*, 127889. [CrossRef]
40. Huang, Q.; Yang, M.; Jane, H.; Li, S.; Bauer, N. Trees, Grass, or Concrete? The Effects of Different Types of Environments on Stress Reduction. *Landsc. Urban Plan.* **2020**, *193*, 103654. [CrossRef]
41. Ha, J.; Kim, H.J.; With, K.A. Urban Green Space Alone Is Not Enough: A Landscape Analysis Linking the Spatial Distribution of Urban Green Space to Mental Health in the City of Chicago. *Landsc. Urban Plan.* **2022**, *218*, 104309. [CrossRef]
42. Review of the Experience of Nature: A Psychological Perspective. *Contemp. Psychol. J. Rev.* **1990**, *35*, 1098. [CrossRef]
43. Hong, L. Development of Stress Scale for College Student. *Chin. J. Appl. Psychol.* **2002**, *8*, 27–32.
44. Thayer, J.F.; Åhs, F.; Fredrikson, M.; Sollers, J.J.; Wager, T.D. A Meta-Analysis of Heart Rate Variability and Neuroimaging Studies: Implications for Heart Rate Variability as a Marker of Stress and Health. *Neurosci. Biobehav. Rev.* **2012**, *36*, 747–756. [CrossRef]
45. Reyes Del Paso, G.A.; Langewitz, W.; Mulder, L.J.M.; Van Roon, A.; Duschek, S. The Utility of Low Frequency Heart Rate Variability as an Index of Sympathetic Cardiac Tone: A Review with Emphasis on a Reanalysis of Previous Studies. *Psychophysiology* **2013**, *50*, 477–487. [CrossRef] [PubMed]
46. Lovallo, W.R.; Thomas, T.L. Stress Hormones in Psychophysiological Research: Emotional, Behavioral, and Cognitive Implications. In *Handbook of Psychophysiology*, 2nd ed.; Cambridge University Press: New York, NY, USA, 2000; pp. 342–367.
47. Karthikeyan, P.; Murugappan, M.; Yaacob, S. A Review on Stress Inducement Stimuli for Assessing Human Stress Using Physiological Signals. In Proceedings of the 2011 IEEE 7th International Colloquium on Signal Processing and Its Applications, Penang, Malaysia, 4–6 March 2011; pp. 420–425. [CrossRef]
48. Healey, J.A.; Picard, R.W. Detecting Stress during Real-World Driving Tasks Using Physiological Sensors. *IEEE Trans. Intell. Transp. Syst.* **2005**, *6*, 156–166. [CrossRef]
49. Costin, R.; Rotariu, C.; Pasarica, A. Mental Stress Detection Using Heart Rate Variability and Morphologic Variability of EeG Signals. In Proceedings of the 2012 International Conference and Exposition on Electrical and Power Engineering, Iasi, Romania, 25–27 October 2012; pp. 591–596. [CrossRef]
50. Alexander, C.; Neis, H.; Alexander, M.M. *The Battle for the Life and Beauty of the Earth: A Struggle between Two World-Systems*, 1st ed.; Oxford University Press: New York, NY, USA, 2012.
51. White, M.; Smith, A.; Humphryes, K.; Pahl, S.; Snelling, D.; Depledge, M. Blue Space: The Importance of Water for Preference, Affect, and Restorativeness Ratings of Natural and Built Scenes. *J. Environ. Psychol.* **2010**, *30*, 482–493. [CrossRef]
52. Yan, C.; Cai, X.; Wu, Y.; Tang, X.; Zhou, Y.; Yang, Q.; Li, F.; Lan, S.; Lin, L. How Do Urban Waterfront Landscape Characteristics Influence People's Emotional Benefits? Mediating Effects of Water-Friendly Environmental Preferences. *Forests* **2024**, *15*, 25. [CrossRef]
53. Zhao, J.; Xu, W.; Ye, L. Effects of Auditory-Visual Combinations on Perceived Restorative Potential of Urban Green Space. *Appl. Acoust.* **2018**, *141*, 169–177. [CrossRef]
54. Lin, W.; Zeng, C.; Bao, Z.; Jin, H. The Therapeutic Look up: Stress Reduction and Attention Restoration Vary According to the Sky-Leaf-Trunk (SLT) Ratio in Canopy Landscapes. *Landsc. Urban Plan.* **2023**, *234*, 104730. [CrossRef]
55. Mehaffy, M.W.; Salingaros, N.A.; Lavdas, A.A. The "Modern" Campus: Case Study in (Un)Sustainable Urbanism. *Sustainability* **2023**, *15*, 16427. [CrossRef]
56. Xu, W.; Zhao, J.; Ye, L. Culture Is New Nature: Comparing the Restorative Capacity of Cultural and Natural Landscapes. *Int. J. Environ. Stud.* **2018**, *75*, 847–865. [CrossRef]

buildings

MDPI

Article

Applicability of Vegetation to Reduce Traffic-Borne PM2.5 Concentration in Roadside User Zones in Hot Arid Climates: The Case of Central Doha, Qatar

Soujanya Mogra * and Mohd Faris Khamidi

Architecture and Urban Planning Department, College of Engineering, Qatar University, Doha P.O. Box 2713, Qatar; mohd.khamidi@qu.edu.qa
* Correspondence: sm1513220@student.qu.edu.qa

Abstract: The 'Beautification of Roads and Parks in Qatar' is an urban development project that intends to provide space for exercising in roadside greenery in central Doha due to a lack of accessible open spaces. Considering the potential health risks associated with inhaling traffic-borne PM2.5, this study investigated the efficacy of four common road vegetation scenarios in reducing traffic-borne PM2.5 concentration in roadside user zones using ENVI-met. It examined Spearman's rank correlation between air temperature, relative humidity, traffic emission rate, and PM2.5 concentration in roadside user zones. Based on the results, (1) hedgerows lower PM2.5 concentrations in roadside user zones, while trees significantly increase the concentration. (2) There is a strong association between air temperature and relative humidity and the PM2.5 concentration. The PM2.5 concentration decreases as air temperature increases but it increases as relative humidity increases. (3) There is a moderately negative association between the traffic emission rate and the PM2.5 concentration; however, this association is not found to be statistically significant. The ENVI-met simulation showed a slight overestimation of PM2.5 concentration compared to the wind tunnel simulation. These findings provide insight into planning road vegetation to reduce traffic-borne PM2.5 in roadside user zones in the local hot arid climate.

Keywords: road vegetation; traffic-borne PM2.5; meteorological conditions; Spearman's rank correlation; ENVI-met

Citation: Mogra, S.; Khamidi, M.F. Applicability of Vegetation to Reduce Traffic-Borne PM2.5 Concentration in Roadside User Zones in Hot Arid Climates: The Case of Central Doha, Qatar. *Buildings* **2024**, *14*, 1388. https://doi.org/10.3390/buildings14051388

Academic Editors: Xia Zhang and Hui He

Received: 20 March 2024
Revised: 26 April 2024
Accepted: 2 May 2024
Published: 13 May 2024

1. Introduction

Air pollution is a significant environmental threat to human health. Today, 99 percent of the world's population inhales poor air [1]. Qatar is one of the countries with the highest ambient PM2.5 concentrations [2]. PM2.5 refers to fine particulate matter with a diameter of 2.5 microns or less and is a commonly used standard measurement of outdoor air quality [3]. The annual average PM2.5 in urban Qatar in 2023 is classified between 'moderate' to 'unhealthy for sensitive groups' that range between 12.1 and 55.4 $\mu g/m^3$, which is higher than the 'good' classification of the United States Environmental Protection Agency (EPA) guidelines [4,5]. Studies have revealed that road vehicular traffic is one of the dominant sources of particulate matter pollution in Doha [6–8]. The literature shows that the canyon-like configuration of buildings along urban roads, known as street canyons, impedes the dispersion of pollutants emitted by traffic and increases road users' exposure to the pollutants [9,10]. Hence, densely built urban areas are more vulnerable to traffic-borne air pollution.

Implementing vegetation along urban roads is a passive strategy to reduce roadside users' exposure to traffic-borne air pollutants [9,11]. Vegetation reduces traffic-borne PM2.5 exposure via absorption, dispersion, and deposition mechanisms [12]. It diverts the pollutants' stream from the source away from the roadside user zone via dispersion. Vegetation filters the pollutant particles through the leaves and branches before reaching the pedestrians via deposition. The leaves and bark absorb gaseous pollutants and fine particulate matter. Figure 1 is a graphical presentation of traffic-borne pollution dispersion, deposition, and absorption by road vegetation. However, road vegetation can increase traffic-borne air pollution exposure for roadside users if not carefully aligned with the dispersion and deposition mechanisms. For example, trees can block vertical air exchange and increase the pollutant concentration in street canyons [13,14]. Therefore, road vegetation planning is crucial to reducing roadside users' exposure to traffic-borne pollutants.

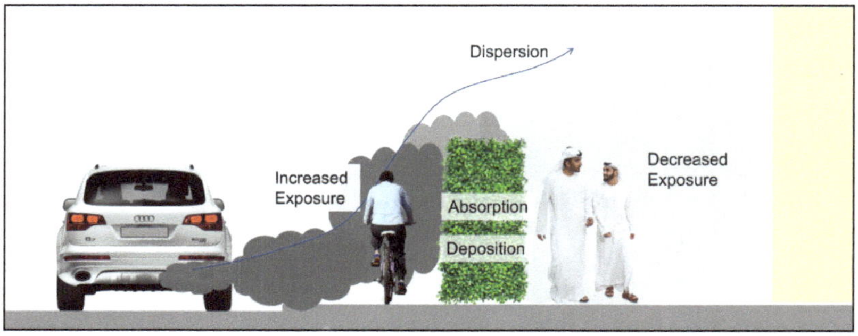

Figure 1. A graphical representation of street vegetation mechanisms in the reduction in traffic-borne pollutant exposure.

Various factors, such as wind conditions, air temperature, relative humidity, road geometry, traffic emission rate, and the physical characteristics of the vegetation, govern the dispersion of pollutants by road vegetation [9,11]. In particular, these wind directions are thought to have different flow fields. For example, Table 1 shows studies that looked at how plants affected pollutant levels in street canyons when the wind came from 90°, 45°, or 0° to the street's main axis.

Buccolieri et al. [13] investigated the influence of avenues like tree planting pollutant concentration in a street canyon with a height (H) to width (W) ratio of 1 (H/W = 1). The findings showed that the average concentration of pollution on the leeward side of the wall increased significantly for perpendicular approaching wind in comparison to obliquely approaching wind to the main axis of the street. Conversely, an investigation by Gromke and Ruck [14] on avenue-like tree planting showed the highest concentration along the leeward wall of a street canyon, with an aspect ratio (H/W) of 0.5 for oblique wind compared to perpendicular approaching wind. A study by Y.-D. Huang et al. [15] looked at seven wind directions with equal gaps between 0° and 90°. It found that the average wall and maximum wall concentrations were highest on the leeward side for 75° approaching wind in a long street canyon, which means that the length (L) to height (H) ratio was more than seven. Overall, the literature showed that street canyons with orientations of 45° and 75° to the direction of the prevailing wind significantly degrade the air quality in the roadside user zone.

Several studies, including Li et al. [16], Jia Wanga et al. [17], Zhu et al. [18], and Wania et al. [19], have investigated the influences of wind speed on vegetation-induced pollutant exposure reduction. Researchers discovered that wind speed has a significant influence on vegetation pollutant concentration, particularly on obliquely oriented urban roads facing the prevailing wind direction [20]. On the other hand, researchers found that wind speed had less of an influence on the vegetation effect on perpendicularly oriented urban roads [16]. Therefore, it is crucial to explore the impact of vegetation on pollutant concentration in roadside user zones of urban roads oriented at an oblique angle to the wind, as it significantly affects the exposure of roadside users to traffic-borne pollutants.

Trees and hedgerows are commonly considered for pollutant exposure reduction studies in roadside areas, as shown in Table 1. Zhang et al. [20] discovered that sideways shrub arrangements work well for urban roads oriented parallel and oblique to the wind. However, wind angles other than 45° to the main axis of the canyon-like urban roads are rarely considered in the literature. This can provide a deep understanding of how to apply suitable vegetation scenarios for pollutant concentration reduction in roadside user zones.

Table 1. Summary of the literature on the effect of vegetation on traffic-borne pollutants' concentration in street canyons for various wind conditions.

Researchers	Wind Directions	Wind Speed [m/s]	Aspect Ratio [H/W]	Vegetation Configurations
Gromke and Ruck [21]	90°	4.65	1	Trees
Buccolieri et al. [13]	90°	4.7	0.5	Trees
Li et al. [16]	90°	1, 2, 3, 20	0.39, 0.18, 0.78	Trees and vegetation barriers
Jia Wanga et al. [17]	90°	1, 3	0.5, 0.9, 1.2	Vegetation barriers
Huang et al. [22]	90°	4.7	0.5	Trees
Zhang et al. [23]	90°	4.7	0.5	Trees
Zhu et al. [18]	90°	2.45, 6.7, 9.35	0.6, 0.67, 1, 2	Trees
Shen et al. [24]	90°	4.7	0.5	Trees, hedgerows, and Tree-hedge
Buccolieri et al. [25]	90°, 45°	4.7	0.5, 1	Trees
Wania et al. [19]	90°, 45°	1, 3	1	Hedgerows and trees
Vos et al. [26]	90°, 45°	3	Asymmetrical	Trees, hedgerows, and green barriers
Morakinyo and Lam [27]	90°, 45°	1, 3	1, 2, 3, 4	Trees
Amorim et al. [28]	45°, 0°	4–7, 5–10	0.33, 0.75	Trees
Gromke et al. [29]	90°, 0°	4.65	0.5	Hedgerows
Hong et al. [30]	90°, 0°	3	0.5, 1, 2	Trees
Gromke and Ruck [14]	90°, 45°, 0°	4.65	0.5, 1	Trees
Zhang et al. [20]	90°, 45°, 0°	4.7, 2	0.5	Trees, hedgerows, tree-hedgerow combinations

Most of the studies are concentrated on the Koppen Climate classifications (Cfb), which indicate a temperate climate with warm summers but no dry season. It includes Buccolieri et al. [13], Buccolieri et al. [25], Chen et al. [31], Gromke et al. [29], Gromke and Ruck [14,21], and Vranckx et al. [32]. Amorim et al. [28] conducted some studies in regions with hot dry summers and moderate wet winters (CSa); Li et al. [16] conducted a few studies in a warm temperate climate region (Cfa); Hong et al. [30] and Wania et al. [19] conducted certain studies in a humid continental climate (Dwa); and Zhu et al. [18] conducted a few studies in a warm and temperate climate region.

Most of the previous studies have investigated the thermal effect of street canyon surfaces on pollutant dispersion [33–36]. Only Chen et al. [31] examined the influence of relative humidity on particulate matter pollutant reduction by open road vegetation in Wuhan, China, in a field experiment. The study revealed that relative humidity has the most influence on pollutant removal by road vegetation compared to wind speed and air temperature. Due to hygroscopic particles increasing in size when they absorb water, which affects their deposition, it is important to consider the influence of the relative humidity. There is a lack of literature on the effects of urban roadside vegetation on roadside users' exposure to traffic-borne pollutants in street canyons in hot arid climate (BWh) regions, where Qatar is located. This study aims to investigate the impact of BWh climatic conditions on the relationship between road vegetation and traffic-related PM2.5 concentrations in roadside user zones in Doha, Qatar.

In addition, it is crucial to determine the relationship where the presence of vegetation will impact the pollutant concentration due to the traffic emission rate. Based on this relationship, vegetation scenarios can be implemented according to the road hierarchy. Chen et al. [31] previously demonstrated, in the context of open roads, that major roads suit shrubs and small trees, sub-arterial roads suit shrubs and grass, and branch roads prefer parallel trees. However, there are no studies that have investigated the association between traffic emission rate and pollution concentration in the roadside user zone in the presence of vegetation on canyon-like urban roads.

A recent pilot survey conducted by Mogra et al. [37] in central Doha revealed that aesthetic enhancement of streets, micro-climate regulation, and air quality enhancement benefits from road vegetation are highly preferred by roadside users. Roadside users most prefer trees because they perceive them to provide shade. People perceive hedgerows as enhancing the safety of children and pets. Qatar Urban Design Compendium [38] also considers trees an integral element of the streetscape in local hot and arid climatic conditions. In this regard, four common road vegetation scenarios, i.e., center hedgerows (CH), sidewise hedgerows (SH), center trees (CT), and sidewise trees (ST), are considered.

Researchers extensively use Computational Fluid Dynamics (CFD) to evaluate the effects of vegetation on pedestrians' exposure to traffic-borne pollutants. A few studies by Wania et al. [19], Vos et al. [26], Morakinyo and Lam [27], and Jia Wanga et al. [17] adopted ENVI-met numerical simulation. ENVI-met, unlike general CFD tools, specifically designs itself for microscale modeling of urban and environmental meteorology. Analyzing air interaction within and between vegetation, surfaces, and buildings is a specific application of ENVI-met. Moreover, ENVI-met has a built-in traffic tool that automatically calculates traffic emission rates based on annual average daily traffic (AADT). A study by Sun et al. [39] showed that ENVI-met performs better compared to the most frequently used CFD software, ANSYS Fluent, in assessing the correlation between PM concentrations and intersectional and meteorological factors. Therefore, this study employs ENVI-met to examine the impact of four common urban road vegetation scenarios on the exposure of roadside users to traffic-borne PM2.5.

This study seeks answers to the following research questions:

1. What are the effects of the four vegetation scenarios, i.e., CH, SH, CT, and ST, on traffic-borne PM2.5 concentrations in the roadside user zone in comparison with the base case scenario?
2. How are variables like air temperature, relative humidity, and traffic emission rate correlated to traffic-borne PM2.5 concentrations in roadside user zones?

2. Materials and Methods

2.1. Conceptual Framework

This study is based on a numerical simulation. The research parameters include both environmental and built features. The main variables under investigation are air temperature, relative humidity (RH), traffic emission rate, and vegetation. Other parameters are the traffic emission rate, wind speed, wind direction, and road geometry. As shown in Figure 2, the conceptual framework includes collecting data, making models, validating them, and evaluating four vegetation scenarios. It also looks at how meteorological variables and traffic emission rate affect PM2.5 reduction at roadside user zones when vegetation is present, in order to learn more about how to plan vegetation for urban roads.

2.2. The Site

Four-ring roads with a concentric street pattern encircle central Doha. The arterial roads radiate from the center area. The center area is the densest area of Doha in terms of the resident population [40]. Bank Street and Ali Bin Abi Talib Street in central Doha were chosen as case study sites. The meteorological data were observed at the Old Doha Airport weather station, the nearest weather station to the two sites. The locations of the two sites relative to the weather station are shown in Figure 3A.

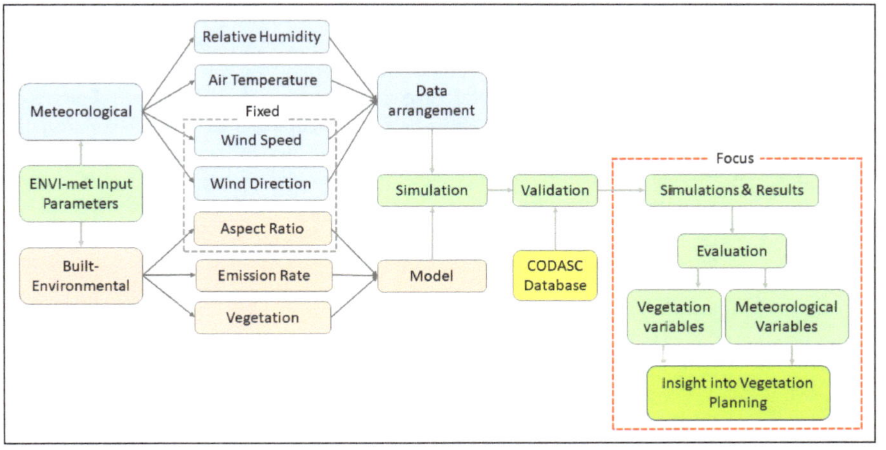

Figure 2. A conceptual framework for the evaluation of urban road vegetation scenarios to reduce roadside users' exposure to traffic-borne PM2.5 exposure.

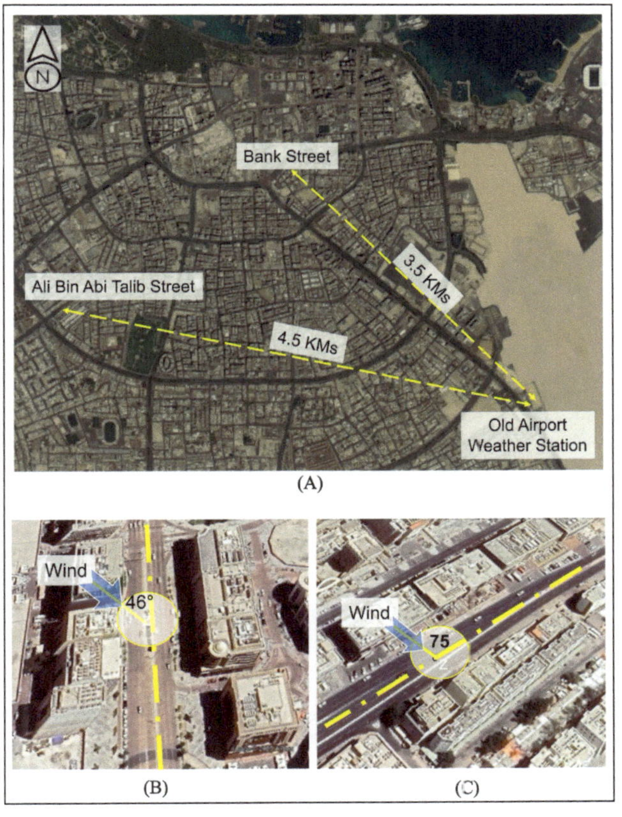

Figure 3. Locations of the sites and the weather station on the map of central Doha (**A**). Images of Bank Street (**B**) and Ali Bin Abi Talib Street (**C**). Source: Qatar GIS Portal and Google Earth. Accessed on 29 October 2023.

The sites were chosen according to aspect ratio and orientation to the prevailing wind direction. Both sites have an identical aspect ratio of approximately 0.43. The prevailing wind direction in central Doha is 310° north-facing. The first road selected for this study, as shown in Figure 3B, is Bank Street, consisting of buildings with heights between 28 and 30 m on both sides, each almost six meters apart. The road has a width of approximately 66 m. The orientation of the first road is 266° north-facing, which is 46° in the wind direction. Secondly, Ali Bin Abi Talib Street, as shown in Figure 3C, consists of building heights between 17 and 21 m on both sides, except for the leeward corner building, which has a height of 12 m. The road has a width of 41 m. For this case, road orientation is 235° north-facing, i.e., 75° to the prevailing wind direction. Both sites are six-lane traffic roads.

2.3. ENVI-Met Modeling

This study used ENVI-met software to examine the effects of urban road vegetation on traffic-borne PM2.5 concentrations in roadside user zones. The numerical description of ENVI-met can be found at Bruse [41]. The ENVI-met model was validated using the CODASC database for an idealized street canyon scenario with an aspect ratio of 0.5.

2.3.1. ENVI-Met Validation

ENVI-met model performance for an idealized street canyon was validated using the CODASC (Concentration DAta of Street Canyons) data. Several researchers, including Abhijith and Gokhale [42], Buccolieri et al. [43], Jeanjean et al. [44,45], and Morakinyo and Lam [27] have used CODASC to validate their study. CODASC data are developed in the Laboratory of Building and Environmental Aerodynamics at the Institute for Hydromechanics (IfH) at the University of Karlsruhe, Germany, through a series of wind tunnel experiments. It provides a normalized pollutant concentration distribution at street canyon walls. Each wall of the street canyon has 700 grid points of normalized concentration measurement at positions x+ = 0.04167 x/H. 4. Figure 4 shows the x, y, and z axes of the street canyon. More details are on the CODASC webpage [46].

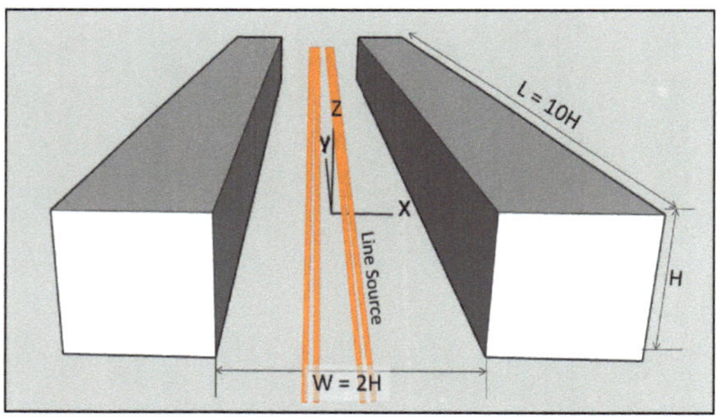

Figure 4. The axis x, y, and z of the street canyon. Source: CODASC database webpage (Re-drawn).

The ENVI-met model performance showed a slight overestimation of the PM2.5 concentration for the base-case scenario, with a FB value of 0.49. However, in the case of the vegetation scenario, the FB value was within the acceptable range ($< |0.3|$). This study applied a grid resolution of 3.6 m × 3.6 m × 2 m. Previously, a study by Morakinyo and Lam [27] considered a grid resolution of 0.5 m × 0.5 m × 0.5 m. The study revealed an overestimation of PM2.5 concentration for both the base-case scenario and the scenario with trees with FB values of 0.5 and 0.8, respectively.

ENVI-met student license version 5.5.1 is used in this study. A cartesian resolution of 3.6 m × 3.6 m × 2 m was used in ENVI-met modeling. ENVI-met 3D models of street canyons with an aspect ratio of 0.5 were generated, both with and without tree planting, like the models used in wind tunnel experiments to develop CODASC data. Figure 5 shows the sections of the wind tunnel street canyon models, both without and with tree planting.

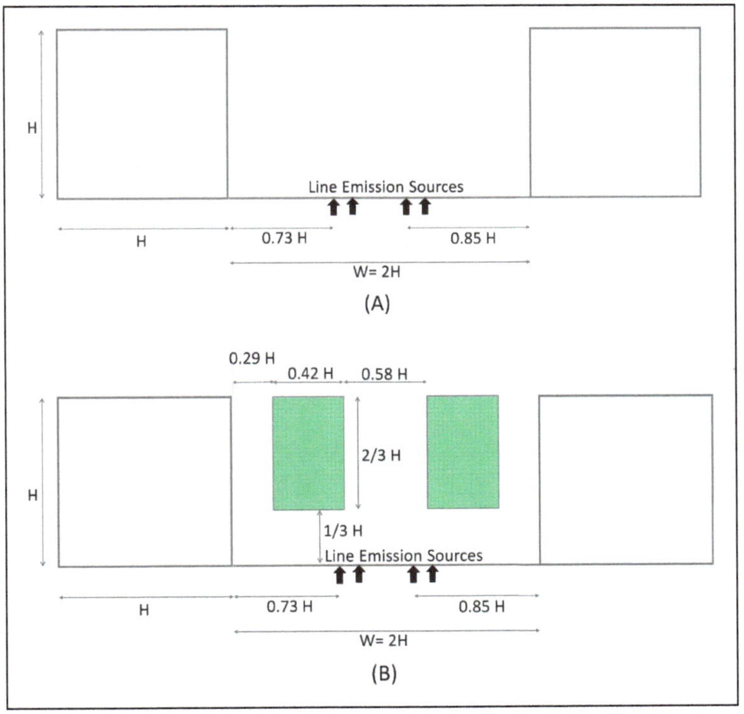

Figure 5. Sections of the wind tunnel experiment model the street canyons with an aspect ratio of 0.5 without trees (**A**) and with trees (**B**). Source: CODASC, KIT (Re-drawn).

The ENVI-met model replicated the wind tunnel model with a building height and width of 18 m. The distance between the buildings was set to 36 m to match the aspect ratio of 0.5. The line emission sources in the ENVI-met models were positioned to closely align with the locations of line emission sources in the wind tunnel experiment models. The porosity of the trees in the wind tunnel experiment model is 96.0%. The leaf area density (LAD) of 1.5 m^2 m^{-3} was used in the ENVI-met tree modeling to correspond with the porosity of the wind tunnel's tree model. Ten nested grids were implemented for the models. Plans of the ENVI-met model layouts are provided in Supplementary Figures S1 and S2.

The selected sites for this study have an oblique orientation to the prevailing wind direction. Hence, to investigate ENVI-met model performance, a CODASC dataset on wind direction 45° to the street axis was considered. The wind speed used in the wind tunnel experiment is 4.7 m/s at the building roof height, i.e., 18 m. The annual average wind speed in central Doha is 3.76 m/s, measured at a 10-m height, which is 4.7 m/s at a height of 18 m above the ground. Meteorological data input for the ENVI-met model was provided in Table 2 in Section 2.3.3 and Table S1.

Table 2. Normalized pollutant concentration at the street canyon walls when oriented at 45° to the wind direction.

BOOT Criteria	Ideal Range	Acceptable Range	Base Case Model	Model with Trees
NMSE	0	<1.5	0.27	0.03
R	1	>0.8	0.995	0.966
FA2	1	>0.5	1	1
$\lvert FB \rvert$	0	$< \lvert 0.3 \rvert$	0.49	−0.156

The PM2.5 concentration data of ENVI-met models were extrapolated from all 13 vertical and 50 horizontal grid points at both walls of the street canyons. The PM2.5 concentration data obtained from ENVI-met simulations were normalized using the following formula:

$$C^{+} = [C \times U_H \times H]/Q_1 \quad (1)$$

where H is the height of the building, Q_1 is the intensity of the PM2.5 emission line source, which is a uniform line emission of 15 μg/s, U_H is the wind speed at height H, C is the PM2.5 concentration at the grids, and C^{+} is the normalized concentration at the grids.

Chang and Hanna [47] developed a statistical analysis tool (known as BOOT) to evaluate pollutant dispersion modeling of the street canyons. The ASTM (American Society for Testing and Materials) procedure was used to evaluate the ENVI-met model's performance using BOOT software. The BOOT has four criteria for model acceptance: NMSE Normalized Mean Squared Error (NMSE), correlation coefficient (R), Fractional Absolute Error 2 (FA2), and Absolute Fractional Bias (FB). Table 2 refers to a street canyon with an aspect ratio (H/W) of 0.5, with and without avenue trees.

Table 2 shows that for the base case model, the NMSE, R, FA2, and FB values are 0.27, 0.995, 1, and 0.49, respectively. The model slightly overestimated the pollutant concentration. Previously, a performance evaluation of an ENVI-met model for a street canyon with an aspect ratio of 1 and a grid resolution of 0.5 m × 0.5 m × 0.5 m based on CODASC by Morakinyo and Lam [27] also showed an overestimation of the pollutant concentration. For the model with avenue trees, the NMSE, R, FA2, and FB values are 0.03, 0.966, 1, and −0.156, respectively, all within the acceptable range.

Overall, the ENVI-met models of the base case and avenue, like the tree planning case, were in reasonable agreement with the wind tunnel experiment of the CODASC database. Therefore, this study applied a grid resolution of 3.6 m × 3.6 m × 2 m for the site models.

2.3.2. Model Scenarios

The ENVI-met models of the two case study sites included five scenarios: base case, center hedgerow (CH), sidewise hedgerow (SH), center trees (CT), and sidewise trees (ST). Figure 6 presents 3D views of the site models. The base case is the scenario before vegetation implementation.

In the CT and ST scenarios, the height and width of the tree models are 10 and 7 m, respectively, with a spherical crown of leaf area density (LAD) of 1.5 m^2 m^{-3}. In the CH and SH scenarios, the height and width of the hedgerows are two and one meters, respectively, with a LAD of 3 m^2 m^{-3}. The vegetation models were created in Albero with a grid resolution of 1-m × 1-m × 1-m and then imported to the models with a grid resolution of 3.6-m × 3.6-m × 2-m.

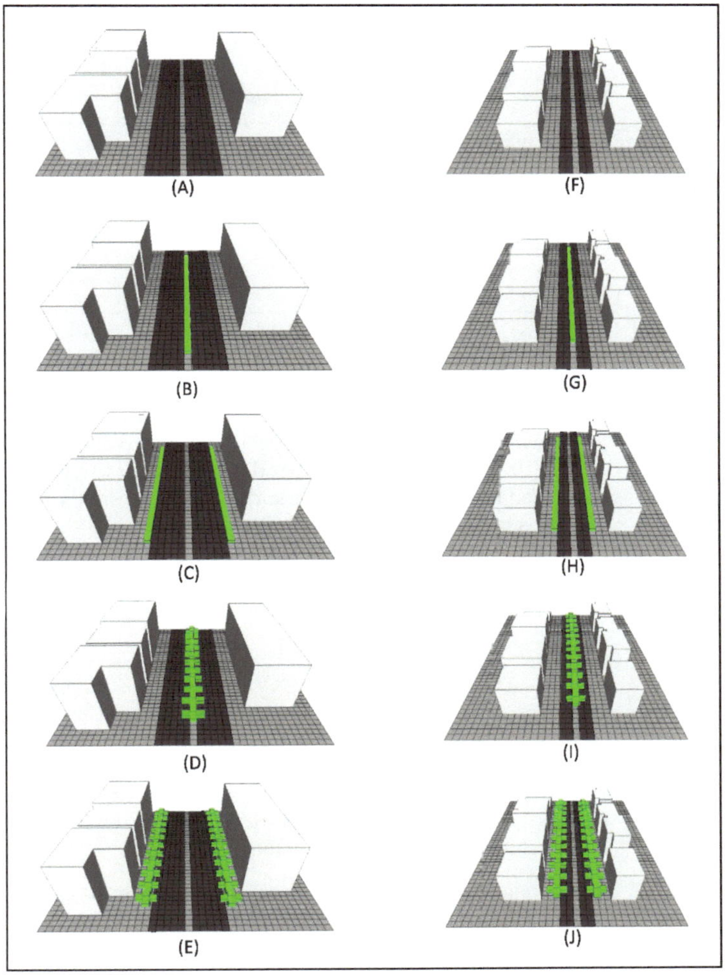

Figure 6. ENVI-met models of the five scenarios, i.e., (**A–E**) are respectively for Base Case, Centre Hedgerow (CH), Sidewise Hedgerows (SH), Centre Trees (CT), and Sidewise Trees (ST) scenarios of Bank Street (**left**) and (**F–J**) are, respectively, for Base Case, Centre Hedgerow (CH), Sidewise Hedgerows (SH), Centre Trees (CT), and Sidewise Trees (ST) scenarios of Ali Bin Abi Talib Street (**right**).

2.3.3. Meteorological Data Collection and Input

The meteorological dataset was extrapolated from Iowa Environmental Management (IEM) [48]. Observations recorded at Old Doha Airport from midnight on 1 January 2016 to midnight on 1 January 2022 on wind speed, air temperature, relative humidity, and specific humidity at 2500 m were considered. Averaged hourly data from 12 a.m. to 10 p.m. for relative humidity (RH), air temperature, and specific humidity at 2500 m were calculated. The prevailing wind direction of the six-year-old, which is 310° north-facing, and the yearly average wind speed of 3.76 m/s were used.

The performance simulation of the ENVI-met model commenced at zero hours and continued for six hours, concluding at 6 a.m. The PM2.5 concentration output values from the sixth hour were utilized. Typically, the initialization time and the initial six hours are required to establish atmospheric stability; therefore, they are disregarded. Table 3 shows the meteorological data input.

Table 3. Meteorological data input for the ENVI-met simulation file.

Input Parameters	Values	Attributes
Wind Speed at a height of 10-m	3.76 m/s	Fixed
Wind direction	310° North bearing	Fixed
Specific humidity in 2500 m	12.9 g/kg	Fixed
Relative humidity (RH) in percentage (%)	Hourly average	Variable (Supplementary Table S1)
Air temperature in Celsius°	Hourly average	Variable (Supplementary Table S1)

2.3.4. Traffic Emission Rate Calculation and Input

During the time of this study, annual average daily traffic (AADT) of the case study roads was not readily available. Therefore, the Transportation Research Board'sthe Highway Capacity Manual, 6th Edition was referred. The manual considers AADT between 29,100 and 55,300 vehicles per day as a rule of thumb for a six-lane urban road with signals [49]. As a result, we assumed that a six-lane traffic road with three lanes in each direction, left turn lanes at busy intersections, and coordinated signals would have an AADT of 55,300 vehicles per day.

Hourly traffic emission rates at a height of 0.5 m were estimated using the ENVI-met default configuration for the traffic distribution of the in-built traffic tool.

2.3.5. Data Analysis

The Leonardo tool in ENVI-met was used to extrapolate the PM2.5 concentration data. Given the focus of this study on roadside users' exposure to traffic-borne PM2.5 concentration, PM2.5 concentration data were extracted from vertical grid number three, which corresponds to the 1.4-m height of the 1:1 model scenario. The grid provides the closest match to pedestrians' breathing height. Horizontally, PM2.5 concentration was considered from both the leeward and windward sides of the urban roads. Figure 7 shows a schematic representation of the roadside user zone.

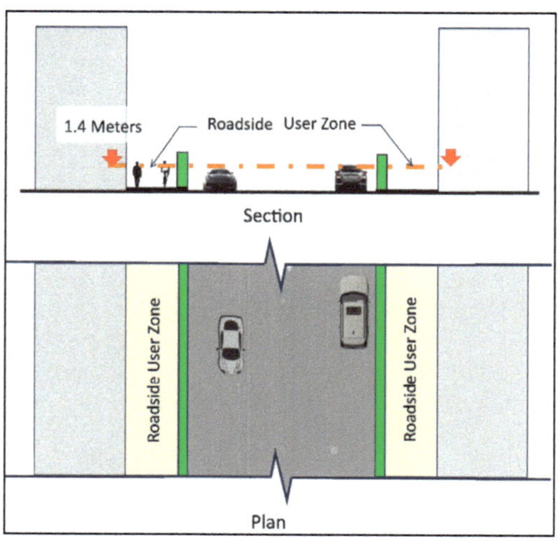

Figure 7. Schematic plan and section of the road model showing roadside user zone at the height of 1.4-m above the ground.

Data on the PM2.5 concentration were extrapolated for the base case as well as the four vegetation scenarios. All PM2.5 concentration data obtained from Leonardo were in micrograms per cubic meter (μg/m^3). PM2.5 concentrations on the roadsides were extrapolated from the 6th hour to the 22nd hour since ENVI-met requires the first few hours to achieve stability. The absolute percentage difference in PM2.5 between treeless urban road cases and scenarios with vegetation was calculated for the comparison.

The data obtained were analyzed using Spearman's rank correlation. The Spearman's rank correlation coefficient, commonly represented as r_s (rho), serves as a non-parametric statistical tool for evaluating the strength and direction of monotonic relationships between two variables [45]. Unlike Pearson correlation, which assumes a linear relationship, Spearman's rank correlation is based on the ranks of the data rather than their actual values. This makes it suitable for variables with non-linear associations or data that may not follow a normal distribution. Spearman's rank correlation assesses association, not causation [50].

The procedure entails assigning ranks to each variable's observations and converting the data into a set of ranks. The correlation coefficient is then calculated based on the differences between these ranks. A positive correlation coefficient indicates a monotonic increase in both variables, while a negative correlation coefficient suggests a monotonic decrease. The coefficient ranges from −1 to 1, where −1 represents a perfect negative correlation, 1 represents a perfect positive correlation, and 0 indicates no correlation.

3. Results

This section presents the results of simulations performed on Bank Street and Ali Bin Abi Talib Street. The impacts of the four vegetation scenarios, i.e., CH, SH, CT, and ST, on traffic-borne PM2.5 concentration in roadside user zones. Additionally, this section presents Spearman's rank correlation between PM2.5 concentration in the presence of the optimal vegetation scenarios in the roadside user zone and air temperature, humidity, and traffic-emitted PM2.5.

3.1. Bank Street

3.1.1. PM2.5 Concentration Reduction in Roadside User Zone

The line graph drawn (Figure 8) shows the temporal variations in PM2.5 concentration from 6 a.m. to 10 p.m. It reveals that SH shows lower PM2.5 concentrations than the base case from 8 a.m. to 4 p.m. However, the concentrations are higher during the early morning and evening hours compared to the base case. In the CH scenario, the temporal variations in PM2.5 concentrations do not differ much compared to the base-case scenario. The CT and ST scenarios showed higher PM2.5 concentrations throughout compared to the base case scenario and the ST scenario showed the highest PM2.5 concentrations.

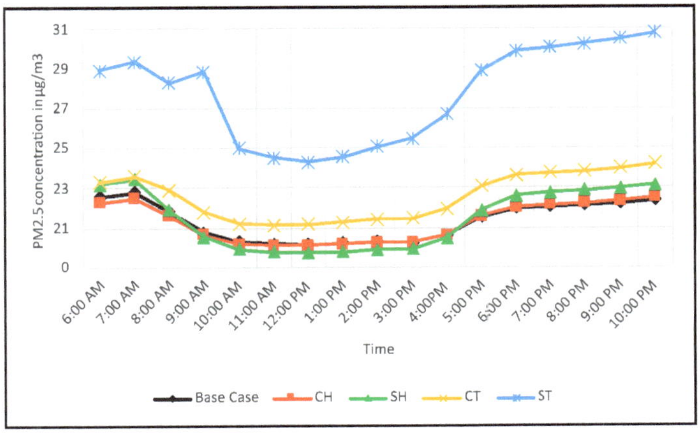

Figure 8. Temporal variations in PM2.5 concentrations in the five scenarios of the Bank Street models.

As shown in Figure 9, the boxplot illustrates the PM2.5 concentrations in the roadside user zone for the base case, CH, SH, CT, and ST scenarios. It shows that the median values of the base cases, CH, SH, CT, and ST, are 21.54, 21.57, 21.88, 22.90, and 28.83, respectively. The base-case scenario has the lowest median value compared to the vegetation scenarios. The mean values of base case, CH, SH, CT, and ST are 21.31, 21.26, 21.46, 22.57, and 27.71. For the SH scenario, the mean value is the lowest. It indicates that in most of the hours between 6 a.m. and 10 p.m., the base case scenario has the lowest PM2.5 concentration in roadside user zones; however, on average, the PM2.5 concentration is lower in the CH scenario compared to the base case scenario. The mean and median values for the ST scenario are the highest among all five scenarios. It shows that the ST scenario significantly increases the PM2.5 concentration in roadside user zones. Overall, CT and ST scenarios showed higher PM2.5 concentrations compared to CH and SH scenarios, with higher mean and median values.

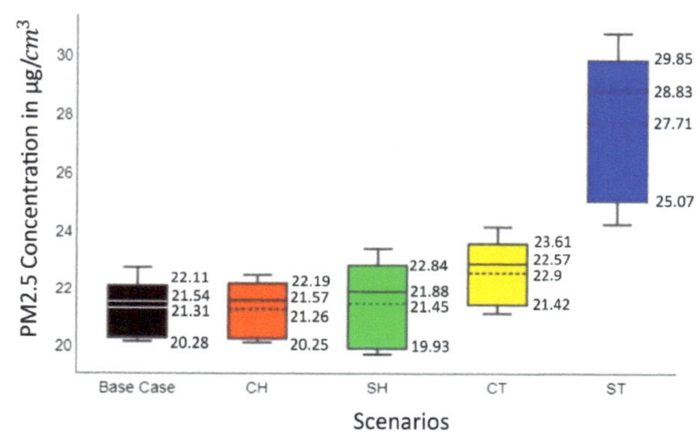

Scenarios / Values	Base Case	CH	SH	CT	ST
Sample size (n)	17	17	17	17	17
Q3	22.11	22.19	22.84	23.61	29.85
Mean (x̄)	21.31	21.26	21.88	22.57	27.71
Median	21.54	21.57	21.45	22.9	28.83
Q1	20.28	20.25	19.93	21.42	25.07

Figure 9. Boxplot shows the effects of the four vegetation scenarios on traffic-borne PM2.5 concentration in roadside user zones in Bank Street.

3.1.2. Spearman's Rank Correlation between Air Temperature, Relative Humidity, Traffic Emission Rate, and the PM2.5 Concentration

Spearman's rank correlation was calculated between air temperature, relative humidity, traffic emission rate, and PM2.5 concentration in roadside user zones for the base case scenario and the four vegetation scenarios. Table 4 presents the results.

Spearman's rank correlation coefficients between the PM2.5 concentrations in the roadside user zones and air temperature is strongly associated with correlation coefficients of −0.98, −0.99, −0.95, −0.86, and −0.88 for the base case, CH, SH, CT, and ST scenarios, respectively. The results are statistically significant, with p-values less than 0.05 at a 95% confidence interval. It means that when the air temperature is higher, the PM2.5 concentration in the roadside user zone is lower, and when the air temperature is lower, the PM2.5 concentration in the roadside user zone is higher in the presence of CH.

Conversely, Spearman's rank correlation coefficient between the PM2.5 concentration and roadside user relative humidity is strong and positive, with correlation coefficients of

0.98, 0.99, 0.96, 0.89, and 0.90, respectively, for the base case, CH, SH, CT, and ST scenarios. The results are statistically significant, with p-values less than 0.05 at a 95% confidence interval. It means that when the relative humidity is higher, the PM2.5 concentration in the roadside user zone is also higher, whereas when the relative humidity is lower, the PM2.5 concentration in the roadside user zone is also lower in the presence of CH.

Table 4. Spearman's rank correlation (r_s) and significance (p) between PM2.5 concentration in roadside user zones in the five scenarios, air temperature and relative humidity, and traffic emission rate.

Variables	r_s and p	PM2.5 Concentration in Roadside User Zone				
		Base Case Scenario	CH Scenario	SH Scenario	CT Scenario	ST Scenario
Air Temperature [°C]	r_s	−0.98	−0.99	−0.95	−0.86	−0.88
	p	<0.001	<0.001	<0.001	<0.001	<0.001
Relative Humidity [%]	r_s	0.98	0.99	0.96	0.89	0.90
	p	<0.001	<0.001	<0.001	<0.001	<0.001
Traffic Emission Rate [μg/s]	r_s	−0.47	−0.44	−0.46	−0.37	−0.40
	p	0.06	0.07	0.06	0.14	0.11

For the base case, CH, SH, CT, and ST scenarios, the PM2.5 concentration in the roadside user zone weakly and negatively correlates with the traffic emission rate, with correlation coefficients of −0.47, −0.44, −0.46, −0.37, and −0.40, respectively. The results are not statistically significant, with p-values greater than 0.05 at a 95% confidence interval. It means that when the traffic emission rate is higher, the PM2.5 concentration in the roadside user zone is slightly lower, whereas when the traffic emission rate is lower, the PM2.5 concentration is slightly higher.

For the CH scenario, scatterplots were drawn to visualize Spearman's correlation coefficient between air temperature, relative humidity, traffic emission rate, and PM2.5 concentration in roadside user zones. Figure 10A displays scatter plots that correlate PM2.5 concentration rankings with air temperature rankings. Figure 10B displays a scatterplot correlating PM2.5 concentration rankings with relative humidity ranks. Figure 10C displays a scatterplot correlating PM2.5 concentration ranks with traffic emission rate ranks.

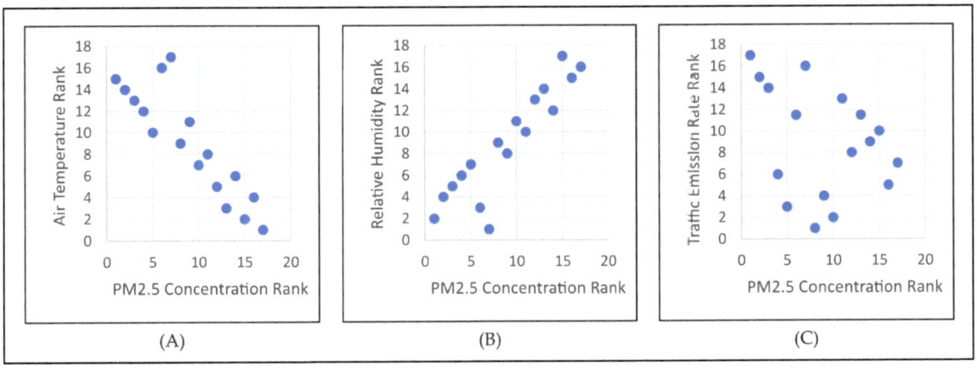

Figure 10. The scatterplot shows the rank distributions of PM2.5 concentration in the roadside user zone and air temperature (**A**), PM2.5 concentration in the roadside user zone and relative humidity (**B**), and PM2.5 concentration in the roadside user zone and traffic emission rate (**C**) in the Bank Street in CH scenario.

3.2. *Ali Bin Abi Talib Street*

3.2.1. PM2.5 Concentration Reduction in Roadside User Zone

The line graph drawn (Figure 11) shows the temporal variations in PM2.5 concentration from 6 a.m. to 10 p.m. It reveals that SH has lower PM2.5 concentrations than the base case during the early morning and evening hours. However, the concentrations are higher from 8 a.m. to 4 p.m. compared to the base case. Temporal variations in the PM2.5 concentrations in the CH scenario do not vary much compared to the base case scenario; however, the concentrations are slightly higher than the base case scenario throughout. The CT and ST scenarios showed higher PM2.5 concentrations throughout compared to the base case scenario and the ST scenario showed the highest PM2.5 concentrations.

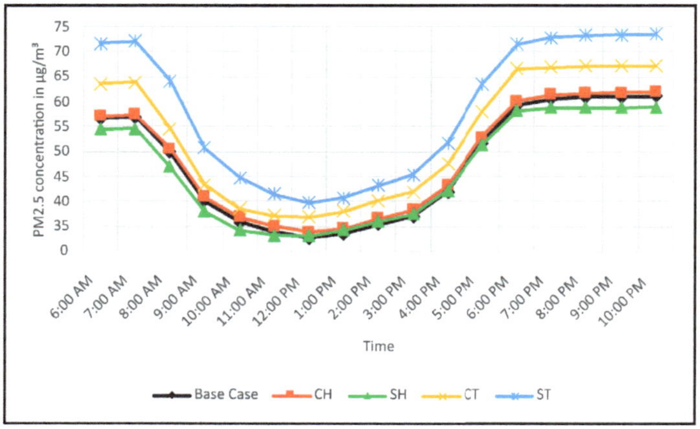

Figure 11. Temporal variations in PM2.5 concentrations in the five scenarios of Ali Bin Abi Talib Street models.

Boxplot analysis was conducted for the PM2.5 concentrations in roadside user zones in five scenarios, i.e., base case, CH, SH, CT, and ST, as shown in Figure 12.

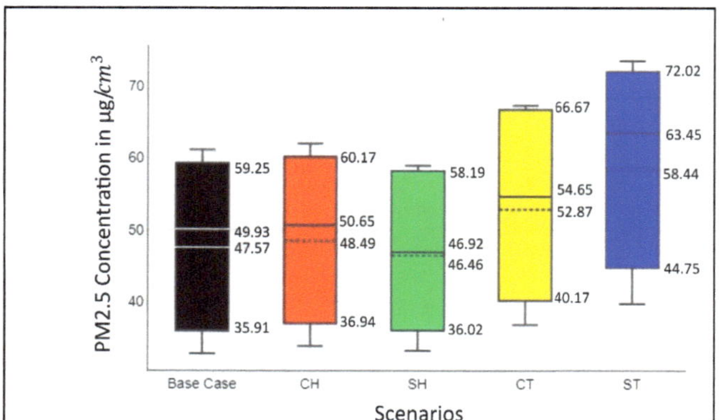

Scenarios / Values	Base Case	CH	SH	CT	ST
Sample size (n)	17	17	17	17	17
Q3	59.25	60.17	58.19	66.67	72.02
Mean (x̄)	47.57	48.49	46.46	52.87	58.44
Median	49.93	50.65	46.92	54.65	63.45
Q1	35.91	36.94	36.02	40.17	44.75

Figure 12. Boxplot shows the effects of the four vegetation scenarios on traffic-borne PM2.5 concentration in roadside user zones in Ali Bin Abi Talib Street.

The boxplot analysis shows that the median values of the base case, CH, SH, CT, and ST are 49.93, 50.65, 46.92, 54.65, and 63.45, respectively. The mean values of the base cases, CH, SH, CT, and ST are 47.57, 48.49, 46.46, 52.87, and 58.44. The median and mean values are the lowest for the SH scenario. It indicates that in most of the hours between 6 a.m. and 10 p.m., the SH scenario has the lowest PM2.5 concentration in roadside user zones and on average, PM2.5 concentration is lower in the SH scenario compared to the base case scenario. The mean and median values for the ST scenario are the highest among all five scenarios. It shows that the ST scenario significantly increases the PM2.5 concentration in roadside user zones. Overall, CT and ST scenarios showed higher PM2.5 concentrations compared to CH and SH scenarios, with higher mean and median values.

3.2.2. Spearman's Rank Correlation between Air Temperature, Relative Humidity, Traffic Emission Rate, and the PM2.5 Concentration

Spearman's rank correlation was calculated between PM2.5 concentration in the roadside user zone in the presence of SH, air temperature and relative humidity, and traffic emission rate. Table 5 presents the results.

Table 5. Spearman's rank correlation (r_s) and significance (p) between PM2.5 concentration in roadside user zones in the presence of SH, air temperature and relative humidity, and traffic emission rate.

Variables	r_s and p	PM2.5 Concentrations in Roadside User Zones				
		Base Case Scenario	CH Scenario	SH Scenario	CT Scenario	ST Scenario
Air Temperature [°C]	r_s	−0.89	−0.89	−0.88	−0.88	−0.93
	p	<0.001	<0.001	<0.001	<0.001	<0.001
Relative Humidity [%]	r_s	0.91	0.91	0.90	0.90	0.94
	p	<0.001	<0.001	<0.001	<0.001	<0.001
Traffic Emission Rate [μg/s]	r_s	−0.34	−0.34	−0.36	−0.36	−0.40
	p	0.18	0.18	0.154	0.15	0.11

Scatterplots were drawn to visualize the Spearman's correlation coefficients between the PM2.5 concentrations in the roadside user zones, and air temperature. It shows that the PM2.5 concentrations are strongly and negatively correlated with air temperature, with correlation coefficients of −0.89, −0.89, −0.88, −0.88, and −0.93 for the base case, CH, SH, CT, and ST scenarios, respectively. The results are statistically significant, with *p*-values less than 0.05 at a 95% confidence interval. It means that in the presence of SH, when the air temperature is higher, the PM2.5 concentration in the roadside user zone is lower and when the air temperature is lower, the PM2.5 concentration in the roadside user zone is higher.

Conversely, for the base case, CH, SH, CT, and ST scenarios, the Spearman's rank correlation coefficient between the PM2.5 concentration in the roadside user zone and relative humidity is strongly and positively correlated, with correlation coefficients of 0.91, 0.91, 0.9, 0.9, and 0.94, respectively. The results are statistically significant, with *p*-values less than 0.05 at a 95% confidence interval. It means that when the relative humidity is higher, the PM2.5 concentration in the roadside user zone is also higher, whereas when the relative humidity is lower, the PM2.5 concentration in the roadside user zone is also lower.

The traffic emission rate and PM2.5 concentration in the roadside user zone have a moderately negative correlation, with correlation coefficients of −0.34, −0.34, −0.36, and −0.40, respectively. However, the results are statistically not significant, with *p*-values greater than 0.05 at a 95% confidence interval. This means that when the traffic emission rate is higher, the PM2.5 concentration in the roadside user zone is slightly lower, whereas when the traffic emission rate is lower, the PM2.5 concentration is slightly higher.

For the SH scenario, scatterplots were drawn to visualize Spearman's correlation coefficient between air temperature, relative humidity, traffic emission rate, and PM2.5 concentration in roadside user zones. Figure 13A displays scatter plots correlating PM2.5 concentration rankings with air temperature ranks. Figure 13B displays a scatterplot correlating PM2.5 concentration rankings with relative humidity ranks. Figure 13C displays a scatterplot correlating PM2.5 concentration ranks with traffic emission rate ranks.

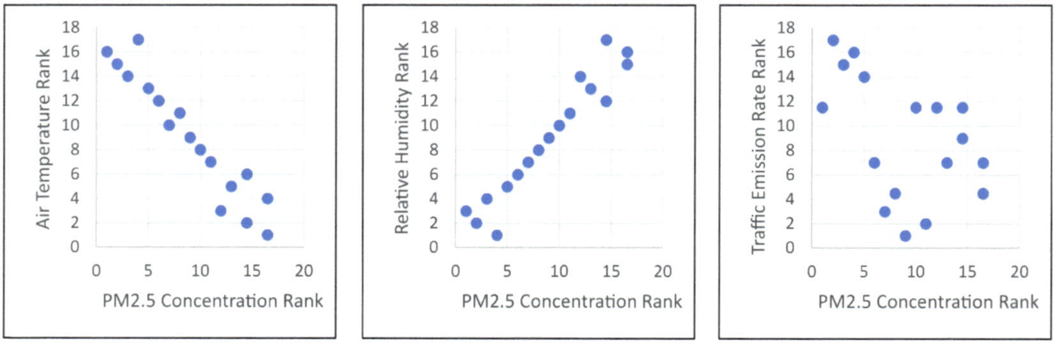

Figure 13. The scatter plot shows the rank distributions of PM2.5 concentration in the roadside user zone and air temperature (**A**), PM2.5 concentration in the roadside user zone and relative humidity (**B**), and PM2.5 concentration in the roadside user zone and traffic emission rate (**C**) in Ali Bin Abi Talib Street in SH scenario.

4. Discussion

The research looked at how four common types of road vegetation—CH, SH, CT, and ST—affected the amount of PM2.5 that was carried by traffic in two streets in central Doha, Qatar: Bank Street and Ali Bin Abi Talib Street. It also examined the influences of air temperature, relative humidity, and traffic emission rate on the PM2.5 concentration in roadside user zones in the presence and absence of vegetation scenarios.

According to the results, the PM2.5 concentration in Ali Bin Abi Talib Street is much higher than that in Bank Street. The primary distinctions between the two road models lie in the total volume the road occupies and the orientation of the road relative to the prevailing wind direction. According to the literature, the pollutant concentration in roadside areas is significantly higher on a road oriented at 75° than on a road oriented at 45° to the prevailing wind direction. It is also a well-known fact that as volume increases, pollutant concentration dilutes.

The CH scenario on Bank Street, oriented at 46° to the prevailing wind direction, is evident in the results. Zhang et al. [20] demonstrated, as previously mentioned, that sidewise vegetation arrangement is appropriate for urban roads oriented parallel to and perpendicular to the wind. The street aspect ratio is the main difference between this study and Zhang et al. [20]. Zhang et al. [20] applied a street aspect ratio of 1, while this study applied a ratio of 0.5. It indicates that the effects of vegetation on PM2.5 concentrations in roadside user zones also vary according to aspect ratios. In this study, both roads exhibit an oblique tilt toward the prevailing wind direction, albeit at varying angles. The study found that Ali Bin Abi Talib Street, oriented at 75° to the prevailing wind direction, is the most suitable for reducing PM2.5 concentration. It indicates that within the obliquely oriented roads to the prevailing wind direction, differences in the angles can have a substantial influence on vegetation effects on PM2.5 concentration in roadside user zones.

The temporal variation in the PM2.5 concentrations in the roadside user zone showed that the CH scenario effectively reduced the concentrations from 8 a.m. to 4 p.m. on Bank Street. In Ali Bin Abi Talib Street, the SH scenario reduced the concentration during the early morning and evening hours. It suggests that the effects of air temperature and relative humidity on vegetation and PM2.5 concentration may vary over time.

Results show that trees (CT and ST) consistently increase PM2.5 concentration in both streets and the ST scenario appears to show the highest increase. Other studies, such as those by Buccolieri et al. [13] and Gromke and Ruck [14], also showed a similar result for ST. Hedgerows (CH and SH) showed better performance than trees. This is primarily because hedgerows impede wind velocity in street canyons less than trees, thereby contributing more to pollution dispersion [19].

The results revealed that in all five scenarios, air temperature has a robust negative correlation with PM2.5 concentration in roadside user zones, while relative humidity has a robust positive correlation. It indicates that as the air temperature increases, the PM2.5 concentration decreases while increasing with an increase in relative humidity. Therefore, we can anticipate an increase in PM2.5 concentrations in roadside user zones during cooler months with higher relative humidity. On the other hand, we can anticipate a decrease in PM2.5 concentrations during the hotter months with lower relative humidity.

Previously, studies such as Sini et al. [35], Baik and Kim [34], and Xie et al. [33] showed that airflow regimes in street canyons of various aspect ratios change with an increase in surface heating and influence traffic-borne pollutant dispersion. Particularly, regarding the aspect ratio of 0.5, Xie et al. [33] revealed that the spatial behaviors of air pollutant movement in the street canyon were similar in wall and ground heating scenarios. The heating scenarios formed two vertices, comprising a primary vortex that extended beyond the roof level and a secondary vortex positioned at the windward corner at ground level. The pollutants were carried toward the leeward side shortly after being emitted from the source. Consequently, pollutant concentration increased at the leeward wall side, which was then uplifted along the leeward wall by the primary vortex, and the pollutant gradually dissipated or re-entered the street canyon at the roof level. In summary, the surface heating scenarios strengthened the primary vertex, shifted pollutants upward along the leeward wall, and gradually dissipated the pollutants.

Studies have shown that relative humidity plays an important role in the dispersion and dry deposition of particulate matter in relation to relative humidity [51,52]. Previously, Chen et al. [31] in Wuhan, China, reported that the sizes of hygroscopic particles grow with an increase in relative humidity, which impedes particle movement, thus increasing particulate matter concentration in the form of haze. On average, over 70% of RH and 28% of no-wind weather occur in Wuhan each year. A gray relational analysis of air temperature, relative humidity, and wind speed revealed that relative humidity significantly influences the efficiency of vegetation-induced particulate matter removal. The study showed that relative humidity has the greatest impact on vegetation-induced particulate matter reduction in roadside user zones. In general, as the wind velocity increases, the rate of particle diffusion accelerates. An increase in relative humidity facilitates the condensation of airborne particles, making a significant contribution to the formation of fog and haze. Due to the Brownian motion of particles, temperature has a less significant impact on particulate removal efficiency in regions with lower wind speeds than relative humidity or wind speed [53]. Compared to Wuhan, the wind speed in Doha is much higher, with a yearly average wind speed of 3.76 m/s. Therefore, wind speed might have played a role in having an almost equal influence of both air temperature and relative humidity on the PM2.5 dispersion in roadside user zones in this study.

Finally, this study showed a moderate and negative correlation between the PM2.5 concentration in the roadside user zone and traffic emission rate. However, we found no statistical significance in the correlation. It indicates that road vegetation's benefit in reducing the average concentration of PM2.5 in roadside user zones may be greater on roads with higher vehicular traffic flow. Since the correlation is not statistically significant, further investigations are necessary.

To summarize, this study provides the following insights into road vegetation design in central Doha

- Hedgerows have the potential to reduce roadside users' exposure to traffic-borne PM2.5;

- Trees significantly increase the PM2.5 concentrations in roadside user zones;
- Air temperature and relative humidity both have a significant influence on the PM2.5 concentration in the roadside user zone. In the roadside user zone, the air temperature is negative, while relative humidity is positively associated with the PM2.5 concentration;
- Traffic emissions may have a moderate influence on the PM2.5 concentration in the roadside user zone;
- The influence of air temperature and relative humidity on the PM2.5 concentration in roadside user zones does not differ substantially in the presence of vegetation;
- The influence of air temperature and relative humidity on the PM2.5 concentration may vary slightly depending on wind directions.

Overall, this study successfully demonstrated that the road's orientation to the prevailing wind direction plays a crucial role in the selection of vegetation to reduce traffic-borne PM2.5 exposure for roadside users. It also showed that air temperature and relative humidity both have a significant influence on vegetation effects on PM2.5 concentration in roadside user zones in the local climate. In general, in all five scenarios, i.e., base-case, CH, SH, CT, and ST, traffic-borne PM2.5 concentration is lowest when the air temperature is the highest and the relative humidity is the lowest.

However, this study has limitations in several areas. Firstly, the roads selected for the case study are the same in terms of aspect ratio but different in the volume occupied. Secondly, we assumed the AADT during the modeling process due to the lack of readily available data. Thirdly, the LAD of trees and hedgerows was assumed based on the literature. Finally, this study confines itself to the four prevalent vegetation scenarios.

5. Conclusions

This study revealed the applicability of the four common road vegetation scenarios in reducing roadside users' exposure to traffic-borne PM2.5 concentrations. It revealed that hedgerows are more effective than trees in reducing the PM2.5 concentration in roadside user zones. Trees have the potential to significantly increase the concentration in the roadside user zone. Both air temperature and relative humidity have a robust correlation with the PM2.5 concentration in roadside user zones. Vegetation might be beneficial in reducing roadside user exposure to traffic-borne PM2.5 concentrations on heavy-traffic roads.

Future research endeavors can assess the impact of vegetation on traffic-borne PM2.5 concentrations across varying wind directions. Furthermore, we can conduct investigations in various geographical locations, as microclimate and street geometry have the potential to influence the effects of vegetation on the PM2.5 concentrations in roadside user zones. Additionally, we can conduct investigations on various arrangements and sizes of urban road vegetation to identify the most effective configurations for pollutant reduction. By systematically investigating these aspects, future studies can contribute valuable knowledge to inform the design and implementation of vegetation-based strategies for improving air quality in roadside user zones.

Supplementary Materials: The following supporting information can be downloaded at https://www.mdpi.com/article/10.3390/buildings14051388/s1. Figure S1: Base Case Model with grid resolution 3.6 m × 3.6 m × 2 m modeled using ENVI-met, Figure S2: Model with Avenue Trees with grid resolution 3.6 m × 3.6 m × 2 m modeled using ENVI-met, Table S1: ENVI-met simulation file input values for relative humidity (RH) and air temperature.

Author Contributions: Conceptualization, S.M. and M.F.K.; methodology, S.M. and M.F.K.; software, S.M.; validation, S.M.; formal analysis, S.M. and M.F.K.; investigation, S.M. and M.F.K.; resources, S.M. and M.F.K.; data curation, S.M.; writing—S.M.; writing—review and editing, S.M. and M.F.K.; visualization, S.M. and M.F.K.; supervision, M.F.K.; project administration, S.M. and M.F.K. All authors have read and agreed to the published version of the manuscript.

Funding: This research received no specific grant from any funding agency in the public, commercial, or not-for-profit sector.

Data Availability Statement: The dataset of the study is available from the author upon reasonable request.

Conflicts of Interest: The authors declare that there are no conflicts of interest.

References

1. WHO. Ambient (Outdoor) Air Quality and Health. Fact Sheet. 2018. Available online: https://www.who.int/en/news-room/fact-sheets/detail/ambient-(outdoor)-air-quality-and-health (accessed on 4 April 2024).
2. World Health Organization. Concentration of Fine Particulate Matter (PM2.5). 2022. Available online: https://www.who.int/data/gho/data/indicators/indicator-details/GHO/concentrations-of-fine-particulate-matter-(pm2-5) (accessed on 4 April 2024).
3. United States Environmental Protection Agency. Particulate Matter (PM) Basics. 2024. Available online: https://www.epa.gov/pm-pollution/particulate-matter-pm-basics (accessed on 6 April 2024).
4. IQAir. Air Quality in Qatar: Air Quality Index and PM2.5 Air Pollution in Qatar. 2024. Available online: https://www.iqair.com/qa/qatar (accessed on 24 April 2024).
5. United States Environmental Protection Agency. Revised Air Quality Standards for Particle Pollution and Updates to the Air Quality Index (AQI). 2024. Available online: https://www.epa.gov/sites/default/files/2016-04/documents/2012_aqi_factsheet.pdf (accessed on 24 April 2024).
6. Javed, W.; Iakovides, M.; Garaga, R.; Stephanou, E.G.; Kota, S.H.; Ying, Q.; Wolfson, J.M.; Koutrakis, P.; Guo, B. Source apportionment of organic pollutants in fine and coarse atmospheric particles in Doha, Qatar. *J. Air Waste Manag. Assoc.* **2019**, *69*, 1277–1292. [CrossRef] [PubMed]
7. Javed, W.; Guo, B. Chemical characterization and source apportionment of fine and coarse atmospheric particulate matter in Doha, Qatar. *Atmos. Pollut. Res.* **2021**, *12*, 122–136. [CrossRef]
8. Al-Thani, H.; Koc, M.; Isaifan, R.J. Investigations on Deposited Dust Fallout in Urban Doha: Characterization, Source Apportionment and Mitigation. *Environ. Ecol. Res.* **2018**, *6*, 493–506. [CrossRef]
9. Abhijith, K.V.; Kumar, P.; Gallagher, J.; McNabola, A.; Baldauf, R.; Pilla, F.; Broderick, B.; Di Sabatino, S.; Pulvirenti, B. Air pollution abatement performances of green infrastructure in open road and built-up street canyon environments—A review. *Atmos. Environ.* **2017**, *162*, 71–86. [CrossRef]
10. Vardoulakis, S.; Fisher, B.E.A.; Pericleous, K.; Gonzalez-Flesca, N. Modelling air quality in street canyons: A review. *Atmos. Environ.* **2003**, *37*, 155–182. [CrossRef]
11. Janhäll, S. Review on urban vegetation and particle air pollution–Deposition and dispersion. *Atmos. Environ.* **2015**, *105*, 130–137. [CrossRef]
12. Garland, J.A. On the Size Dependence of Particle Deposition. *Water Air Soil Pollut. Focus* **2001**, *1*, 323–332. [CrossRef]
13. Buccolieri, R.; Gromke, C.; Di Sabatino, S.; Ruck, B. Aerodynamic effects of trees on pollutant concentration in street canyons. *Sci. Total Environ.* **2009**, *407*, 5247–5256. [CrossRef]
14. Gromke, C.; Ruck, B. Pollutant Concentrations in Street Canyons of Different Aspect Ratio with Avenues of Trees for Various Wind Directions. *Bound.-Layer Meteorol.* **2012**, *144*, 41–64. [CrossRef]
15. Huang, Y.-D.; Hou, R.-W.; Liu, Z.-Y.; Song, Y.; Cui, P.-Y.; Kim, C.-N. Effects of Wind Direction on the Airflow and Pollutant Dispersion inside a Long Street Canyon. *Aerosol Air Qual. Res.* **2019**, *19*, 1152–1171. [CrossRef]
16. Li, X.-B.; Lu, Q.-C.; Lu, S.-J.; He, H.-D.; Peng, Z.-R.; Gao, Y.; Wang, Z.-Y. The impacts of roadside vegetation barriers on the dispersion of gaseous traffic pollution in urban street canyons. *Urban For. Urban Green.* **2016**, *17*, 80–91. [CrossRef]
17. Wang, J.; Gao, H.; Lv, C. Analysing the influence of different street vegetation on particulate matter dispersion using microscale simulations. *Desalin. Water Treat.* **2018**, *110*, 319–327. [CrossRef]
18. Zhu, X.; Wang, X.; Lei, L.; Zhao, Y. The influence of roadside green belts and street canyon aspect ratios on air pollution dispersion and personal exposure. *Urban Clim.* **2022**, *44*, 101236. [CrossRef]
19. Wania, A.; Bruse, M.; Blond, N.; Weber, C. Analysing the influence of different street vegetation on traffic-induced particle dispersion using microscale simulations. *J. Environ. Manag.* **2012**, *94*, 91–101. [CrossRef] [PubMed]
20. Zhang, L.; Zhang, Z.; Feng, C.; Tian, M.; Gao, Y. Impact of various vegetation configurations on traffic fine particle pollutants in a street canyon for different wind regimes. *Sci. Total Environ.* **2021**, *789*, 147960. [CrossRef]
21. Gromke, C.; Ruck, B. Influence of trees on the dispersion of pollutants in an urban street canyon—Experimental investigation of the flow and concentration field. *Atmos. Environ.* **2007**, *41*, 3287–3302. [CrossRef]
22. Huang, Y.-D.; Li, M.-Z.; Ren, S.-Q.; Wang, M.-J.; Cui, P.-Y. Impacts of tree-planting pattern and trunk height on the airflow and pollutant dispersion inside a street canyon. *Build. Environ.* **2019**, *165*, 106385. [CrossRef]
23. Zhang, L.; Zhang, Z.; McNulty, S.; Wang, P. The mitigation strategy of automobile generated fine particle pollutants by applying vegetation configuration in a street-canyon. *J. Clean. Prod.* **2020**, *274*, 122941. [CrossRef]

24. Shen, J.; Cui, P.; Huang, Y.; Wu, Y.; Luo, Y.; Sin, C.H.; Guan, J. New insights on precise regulation of pollutant distribution inside a street canyon by different vegetation planting patterns. *Environ. Sci. Pollut. Res.* **2023**, *30*, 63148–63174. [CrossRef]
25. Buccolieri, R.; Salim, S.M.; Leo, L.S.; Di Sabatino, S.; Chan, A.; Ielpo, P.; de Gennaro, G.; Gromke, C. Analysis of local scale tree-atmosphere interaction on pollutant concentration in idealized street canyons and application to a real urban junction. *Atmos. Environ.* **2011**, *45*, 1702–1713. [CrossRef]
26. Vos, P.; Maiheu, B.; Vankerkom, J.; Janssen, S. Improving local air quality in cities: To tree or not to tree? *Environ. Pollut.* **2013**, *144*, 41–64. [CrossRef] [PubMed]
27. Morakinyo, T.E.; Lam, Y.F. Study of traffic-related pollutant removal from street canyon with trees: Dispersion and deposition perspective. *Environ. Sci. Pollut. Res.* **2016**, *23*, 21652–21668. [CrossRef] [PubMed]
28. Amorim, J.H.; Rodrigues, V.; Tavares, R.; Valente, J.; Borrego, C. CFD modelling of the aerodynamic effect of trees on urban air pollution dispersion. *Sci. Total Environ.* **2013**, *461–462*, 541–551. [CrossRef] [PubMed]
29. Gromke, C.; Jamarkattel, N.; Ruck, B. Influence of roadside hedgerows on air quality in urban street canyons. *Atmos. Environ.* **2016**, *139*, 75–86. [CrossRef]
30. Hong, B.; Lin, B.; Qin, H. Numerical Investigation on the Effect of Avenue Trees on PM2.5 Dispersion in Urban Street Canyons. *Atmosphere* **2017**, *8*, 129. [CrossRef]
31. Chen, X.; Pei, T.; Zhou, Z.; Teng, M.; He, L.; Luo, M.; Liu, X. Efficiency differences of roadside greenbelts with three configurations in removing coarse particles (PM_{10}): A street scale investigation in Wuhan, China. *Urban For. Urban Green.* **2015**, *14*, 354–360. [CrossRef]
32. Vranckx, S.; Vos, P.; Maiheu, B.; Janssen, S. Impact of trees on pollutant dispersion in street canyons: A numerical study of the annual average effects in Antwerp, Belgium. *Sci. Total Environ.* **2015**, *532*, 474–483. [CrossRef] [PubMed]
33. Xie, X.; Liu, C.-H.; Leung, D.Y.C. Impact of building facades and ground heating on wind flow and pollutant transport in street canyons. *Atmos. Environ.* **2007**, *41*, 9030–9049. [CrossRef]
34. Baik, J.-J.; Kim, J.-J. A Numerical Study of Flow and Pollutant Dispersion Characteristics in Urban Street Canyons. *J. Appl. Meteorol.* **1999**, *38*, 1576–1589. [CrossRef]
35. Sini, J.-F.; Anquetin, S.; Mestayer, P.G. Pollutant dispersion and thermal effects in urban street canyons. *Atmos. Environ.* **1996**, *30*, 2659–2677. [CrossRef]
36. Baik, J.-J.; Park, R.-S.; Chun, H.-Y.; Kim, J.-J. A Laboratory Model of Urban Street-Canyon Flows. *J. Appl. Meteorol.* **2000**, *39*, 1592–1600. [CrossRef]
37. Mogra, S.; Khamidi, M.F.; Fadli, F. Insight into vegetation inclusion along urban roads: A pilot study on the preferences of expatriate roadside users in downtown Doha, Qatar. *Landsc. Online* **2023**, *98*, 1108. [CrossRef]
38. QNMP. *Qatar Urban Design Compendium (QUDC)*; QNMP: Doha, Qatar, 2019.
39. Sun, D.; Wu, S.; Shen, S.; Xu, T. Simulation and assessment of traffic pollutant dispersion at an urban signalized intersection using multiple platforms. *Atmos. Pollut. Res.* **2021**, *12*, 101087. [CrossRef]
40. PSA. *Residence for Households by Zone 2020*; PSA: Doha, Qatar, 2020; Available online: https://psaqatar.maps.arcgis.com/apps/dashboards/4e2cf1451a8f43c2b258751609752cae (accessed on 24 April 2024).
41. Bruse, M. ENVI-Met Implementation of the Gas/Particle Dispersion and Deposition Model PDDM. 2007. Available online: http://envi-met.com/documents/sources (accessed on 24 April 2024).
42. Abhijith, K.V.; Gokhale, S. Passive control potentials of trees and on-street parked cars in reduction of air pollution exposure in urban street canyons. *Environ. Pollut.* **2015**, *204*, 99–108. [CrossRef]
43. Buccolieri, R.; Jeanjean, A.P.R.; Gatto, E.; Leigh, R.J. The impact of trees on street ventilation, NO_x and $PM_{2.5}$ concentrations across heights in Marylebone Rd street canyon, central London. *Sustain. Cities Soc.* **2018**, *41*, 227–241. [CrossRef]
44. Jeanjean, A.P.R.; Hinchliffe, G.; McMullan, W.A.; Monks, P.S.; Leigh, R.J. A CFD study on the effectiveness of trees to disperse road traffic emissions at a city scale. *Atmos. Environ.* **2015**, *120*, 1–14. [CrossRef]
45. Jeanjean, A.P.R.; Buccolieri, R.; Eddy, J.; Monks, P.S.; Leigh, R.J. Air quality affected by trees in real street canyons: The case of Marylebone neighbourhood in central London. *Urban For. Urban Green.* **2017**, *22*, 41–53. [CrossRef]
46. CODASC. CODASC Laboratory of Building-and Environmental Aerodynamics Karlsruhe Institute of Technology KIT. Available online: https://www.umweltaerodynamik.de/bilder-originale/CODA/CODASC.html (accessed on 16 November 2023).
47. Chang, J.C.; Hanna, S. *Technical Descriptions and User's Guide for the BOOT Statistical Model Evaluation Software Package, Version 2.0*; George Mason University and Harvard School of Public Health: Fairfax, VA, USA, 2005.
48. IEM. Computed Daily Summary of Observations. 2023. Available online: https://mesonet.agron.iastate.edu/request/daily.phtml?network=QA__ASOS (accessed on 24 September 2023).
49. National Research Council; Transportation Research Board. *Highway Capacity Manual: A Guide for Multimodal Mobility Analysis*, 6th ed.; Transportation Research Board: Washington, DC, USA, 2016.
50. Dodge, Y. Spearman Rank Correlation Coefficient. In *The Concise Encyclopedia of Statistics*; Springer: New York, NY, USA, 2008; pp. 502–505. [CrossRef]

51. Chen, L.; Peng, S.; Liu, J.; Hou, Q. Dry deposition velocity of total suspended particles and meteorological influence in four locations in Guangzhou, China. *J. Environ. Sci.* **2012**, *24*, 632–639. [CrossRef]
52. Zhang, M.; Chen, S.; Zhang, X.; Guo, S.; Wang, Y.; Zhao, F.; Chen, J.; Qi, P.; Lu, F.; Chen, M.; et al. Characters of Particulate Matter and Their Relationship with Meteorological Factors during Winter Nanyang 2021–2022. *Atmosphere* **2023**, *14*, 137. [CrossRef]
53. Dimbour, J.P.; Dandrieux, A.; Dusserre, G. Reduction of chlorine concentrations by using a greenbelt. *J. Loss Prev. Process Ind.* **2002**, *15*, 329–334. [CrossRef]

buildings

MDPI

Article

Enhancing Dementia Nursing Homes in South Korea: Lessons from German Building Standards

Soo In Jee

Department of Architecture and Architecture Engineering, Seoul National University, Seoul 08826, Republic of Korea; lion3190@snu.ac.kr

Abstract: South Korea is an aging society with a rapidly increasing number of people with dementia. In that context, this study aimed to compare dementia nursing home building standards in South Korea and Germany, based on the minimum physical and architectural requirements specified by both countries' relevant laws, to gain insights into improving South Korean building standards. I examined South Korea's Welfare of Senior Citizens Act and its Enforcement Regulations, and Germany's Federal Nursing Home Act (HeimG), Regulation on Minimum Standards for Nursing Homes (HeimMindBauV), and the corresponding state regulations. The analysis revealed differences regarding the basic requirements, facility sizes, composition of necessary rooms, and detailed regulations pertaining to these rooms. South Korea emphasized autonomy, including barrier-free and compensatory environments, as well as safety and security. Meanwhile, Germany enhanced similar aspects of autonomy to those in South Korea, including the barrier-free standard DIN 18040-2. Additionally, Germany incorporated features such as familiarity, sensory stimulation, legibility, and social interaction, reflecting aspects that offer orientational cues for autonomy. Improving South Korean building standards requires stronger regulations on the autonomy of individuals with dementia. Further, additional regulations on familiarity, sensory stimulation, legibility, and social interactions should be considered. The results provide foundational data for cross-national comparisons to establish building standards for dementia-friendly built spaces in dementia nursing homes in South Korea. Further surveys on spatial utilization, discussions, and the application of proposed improvements in Korea would contribute to enhancing the dementia-friendly nursing home design in the country.

Keywords: dementia nursing homes; dementia-friendly design; building standards; Welfare of Senior Citizens Act; The Enforcement Regulations of the Welfare of Senior Citizens Act; HeimG; HeimMindBauV; cross-national comparison

Citation: Jee, S.I. Enhancing Dementia Nursing Homes in South Korea: Lessons from German Building Standards. *Buildings* **2024**, *14*, 1427. https://doi.org/10.3390/buildings14051427

Academic Editor: Xia Zhang

Received: 13 February 2024
Revised: 23 April 2024
Accepted: 2 May 2024
Published: 15 May 2024

1. Introduction

1.1. Background and Objectives

Global social, ecological, political, and economic changes have led to demographic changes such as aging populations due to declining birth and death rates and increased life expectancy [1]. Improvements in economic conditions and advancements in medicine continue to extend the average lifespan of humans, leading to a global increase in the elderly population [2]. Consequently, an increase in the number of people with dementia is expected. According to a United Nations (UN) report, the global population aged 65 years and above, which was estimated at 730 million in 2019, is projected to double to 1.5 billion by 2050 [3]. The proportion of the population aged 65 years and above is expected to increase from 6% in 1990 and 9% in 2019 to 16% by 2050 [3], suggesting a serious issue of aging. This aging problem is directly linked to an explosive increase in the number of people with dementia. Alzheimer's Disease International (ADI) has forecasted a threefold increase in this number worldwide from 2015 to 2050 (refer to Table 1). According to the World Health Organization (WHO), this number worldwide is anticipated to rise from 47 million in 2015 to 75 million by 2030 and further increase to 132 million by 2050 [4].

Table 1. Trends in the number of people with dementia globally [5].

Year	2015	2020	2025	2030	2035	2040	2045	2050
Number of people with dementia (million)	49.92	58.66	69.20	82.05	97.45	114.83	133.28	152.24

As part of the international effort to address dementia, the WHO designated it as a public health priority in its *Global Action Plan on the Public Health Response to Dementia 2017–2025* published in 2017 [4]. In line with this, the ADI's 2020 report was titled *Design, Dignity, Dementia: Dementia-related Design and the Built Environment*, focusing on the protection of the dignity of people with dementia [6].

The UN has observed that given the speed of aging, South Korea is projected to have the highest increase in the proportion of the global population aged 65 years and above between 2015 and 2030 [3]. South Korea is expected to rank second in country-specific rankings of the proportion of the population aged 65 years and older by 2050 (Table 2). According to domestic research, South Korea, which is currently classified as an aged country, is expected to become a super-aged society, with the proportion of the population aged 65 years and above rising from 15.7% in 2020 to 19.9% by 2025 and 37% by 2050 (Table 3I). Concurrently, the prevalence of dementia among those aged 65 years and older in South Korea is predicted to increase from 9.36% in 2013 to 16.09% by 2050 (Table 3II). The growth rate of the number of people with dementia in South Korea is estimated to be approximately 4.5 times faster than that in other countries from 2015 to 2050 [2], with 41.2 cases of dementia per 1000 people by 2050, ranking fifth globally [7].

Table 2. Country-specific rankings of the proportion of people aged 65 and above [8].

Year	1980		2022		2050	
Ranking	**Country**	**People Aged 65 and Older (%)**	**Country**	**People Aged 65 and Older (%)**	**Country**	**People Aged 65 and Older (%)**
1	Sweden	16.3	Japan	29.8	Hongkong	40.6
2	Germany	15.7	Italy	23.7	South Korea	39.4
3	Austria	15.4	Finland	22.9	Japan	37.5
4	UK	14.9	Portugal	22.6	Italy	37.1
5	Norway	14.8	Greece	22.5	Spain	36.6
6	Belgium	14.4	Bulgaria	22.4	Taiwan	35.3
7	Denmark	14.4	Puerto Rico	22.4	Greece	34.8
8	France	14.0	Germany	22.2	Portugal	34.5

Table 3. Projected trends of population aging and dementia prevalence in South Korea *1 [9–12], *2 [13,14], *3 [15], *4 [2].

Year	I. People Aged 65 or Older		II. Dementia Prevalence Rate among Individuals Aged 65 and Above	
	People Aged 65 or Older (Persons)	**Population Ratio (%)**	**People Aged 65 or Older Living with Dementia (Persons)**	**Dementia Prevalence Rate among Individuals Aged 65 and Above (%)**
2050	-	37 *4 (c)	3,026,593 *4	16.09 *4
2025	-	19.9 *4 (b)	1,083,977 *4	10.32 *4
2020	8,125,000 *1	15.7 *1 (b)	832,795 *3	10.24 *3
2019	7,685,000 *1	14.9 *1 (b)	794,280 *3	10.33 *3
2018	7,381,000 *1	14.3 *1 (b)	750,488 *2	10.16 *2
2017	7,076,000 *1	13.8 *1 (a)	702,436 *2	9.94 *2

(a) Aging society, (b) Aged society, (c) Super-aged society.

Within sustainable development, addressing optimal living standards in rapidly urbanizing cities poses a paramount challenge, considering economic, ecological, social, and demographic factors [1,16]. Demographically sustainable buildings ensure safe and independent habitation for all, with a focus on accommodating the increasing elderly population, including those with dementia, while upholding human dignity [17]. Especially in South Korea, given the rapid aging and increase in dementia cases, the demand for dementia-friendly built spaces designed to accommodate this vulnerable group of individuals is urgent. Therefore, research aimed at identifying improvements in the current building standards for dementia nursing homes to promote their transformation into dementia-friendly built spaces is both multifaceted and essential.

To enhance nursing homes in South Korea, research on nursing home building standards from the perspective of dementia-friendly built spaces is needed, particularly studies comparing South Korea's standards with those of other countries. However, these studies were not sufficiently comprehensive, lacking detailed qualitative comparisons, and did not adequately explore the perspective of dementia-friendly built spaces. Previous research addressing building standards for dementia nursing homes in South Korea (median age as of 2020: 42.8 years [18]) has focused predominantly on comparisons with Japan (median age: 48 years [18]), the United Kingdom (median age: 39.5 years [18]), the United States (median age: 37.5 years [18]), and Australia (median age: 36.7 years [18]), [19–22]. These studies were limited in number and scope, providing only a general overview within a restricted framework. Studies conducted by the South Korea Health Industry Development Institute [23] and Jin Young-ran [24] contained more detailed analyses of building standards, focusing on Japan, the United States, and Australia, but did not sufficiently reflect the perspective of dementia-friendly built spaces. In a comparative previous study with Germany (median age: 44.9 years [18]), the approach was more general, with a focus only on the minimal net floor area for bedrooms [25].

Therefore, this study targets the legal building standards specified in the current laws of South Korea and Germany. South Korea entered the aged society phase around 2018 and is set to become a super-aged society by 2025. By contrast, Germany, as a representative super-aged society, transitioned to this demographic phase in 1972 and to a super-aged society only in 2007 [26]. Germany is among the G7 countries with a highly advanced social welfare system, and it is also an aging society [27,28]. As of 2019, it ranked seventh globally in terms of the number of dementia patients (1,691,221 individuals), and it is projected to rank ninth by 2050 (2,796,783 individuals), making it a high-risk country for dementia [29,30]. Also, the consistently high ranking in the proportion of elderly individuals places Germany at the forefront of aging societies in the EU. The percentages of individuals aged 65 years and above in Germany compared to the EU average over the last few decades are as follows: 2000: 17% (EU-27: 15.6%), 2010: 21% (EU-27: 17.6%), and 2020: 22% (EU-27: 20.7%) [31,32]. Along with that, Germany responded proactively to dementia policies in 2020 by developing National Dementia Plans (NDPs), which include action plans for creating a safe environment for people with dementia, improving residential settings, and introducing alternative housing models [33]. For these reasons, Germany is an apt candidate for comparison for the purposes of this study. Accordingly, this study is a foundational research that aims to compare the current building standards for dementia nursing homes in South Korea with those in Germany, revealing the differences and implications by analyzing the basic requirements, facility sizes, necessary rooms and detailed regulations to these rooms as indicated by the current standards in both countries.

1.2. Research Methods

To illustrate disparities between building standards in South Korea and Germany, and provide insights into South Korea's standards, the first chapter explores contextual background including dementia data, concepts of dementia-friendly spaces, and nursing homes in both countries. This is accomplished by utilizing preceding studies on dementia and dementia-friendly built spaces, as well as statistical data and preceding studies related to the current status of nursing homes in South Korea and Germany, along with regulatory information. Then, the following chapter analyzes the current laws related to dementia nursing homes in South Korea and Germany. A previous comparative study with Germany focused solely on the Federal Nursing Home Act in Germany, with only a brief mention of the minimal net floor area for bedrooms [25]. In contrast, this analysis encompasses the South Korean Welfare of Senior Citizens Act [34] and its subsidiary Enforcement Regulations [35], as well as the German Nursing Homes Act (HeimG) [36] and its subsidiary Regulation on Minimum Standards for Nursing Homes (HeimMindBauV) [37], along with 10 German state laws. The legal and building standard data were collected from various sources, including the internet, government websites, documents, research papers, and reports. Using these data, relevant building standards in both South Korea and Germany were meticulously analyzed. For South Korea, building standards from the 1990s up to the most recent revisions in 2022 were examined, while for Germany, the focus was on the current revisions building standards set by both the federal and state governments. Finally, in the context of dementia-friendly spaces, the last chapter compares and analyzes the key aspects of building standards in both countries, detailing basic requirements, facility sizes, the composition of necessary rooms, and specific regulations related to these rooms. In this chapter, the analysis specifically focuses on regulatory data derived from the Enforcement Regulations of the Welfare of Senior Citizens Act in South Korea, and the Regulation on Minimum Standards for Nursing Homes (HeimMindBauV), as well as ten state laws in Germany that adhere to substantive building standards. The preceding research data on the utilization status of nursing homes in South Korea are also utilized to reflect the actual environment of nursing homes.

2. Background Status

2.1. Dementia and the Concept of Dementia-Friendly Built Space

2.1.1. Symptoms of Dementia

The term "dementia" is derived from the Latin words "*de* (out of) + *mens* (mind) + *ia* (state of)", signifying a state of being out of the mind [33]. According to the WHO's International Classification of Diseases, 10th edition (ICD-10), dementia is defined "as a syndrome resulting from chronic or progressive diseases of the brain, characterized by a multitude of disorders in higher cortical functions, including memory, thinking, orientation, comprehension, calculation, learning ability, language, and judgment" [38]. Individuals with dementia exhibit a decline in overall cognitive and intellectual functions, and Alzheimer's disease, accounting for 50–60% of dementia cases, is associated with a broad range of cognitive, behavioral, and daily living ability changes (Table 4).

Table 4. Disorders in the elderly with dementia (adapted from [39]).

Disorder		Associated Symptoms
Cognitive disorder		Memory disorder, orientation disorder, language disturbance, space–time perceptual disturbance, agnosia, apraxia, prefrontal functional disorders causing judgment disorder
Behavior-psychological disorder	Behavior disorder	Aggression, wandering without meaning, inappropriate sexual behavior, screaming, cursing, insomnia, bulimia
	Psychological disorder	Anxiety, agitation, apathy, insensibility, depression, hallucination, delusion
Changes in activities of daily living	Changes in P-ADL [1]	Defecating, dressing, bathing, etc.
	Changes in I-ADL [2]	Shopping, money management, household, food preparation, etc.

[1] Physical activities of daily living, [2] Instrumental activities of daily living.

2.1.2. Dementia Nursing Homes as Dementia-Friendly Built Spaces

Research on dementia nursing homes has emphasized design concepts from a therapeutic environmental perspective since the 1980s. Concepts such as therapeutic physical environments [40], therapeutic goals [41,42], therapeutic buildings [43], environment as treatment [44], and therapeutic architecture [45] have been central to describing spaces for individuals with dementia. Since the 2000s, the design concept of therapeutic environments for individuals with dementia has evolved into terms such as dementia-friendly architecture [46,47] and dementia-sensitive built spaces [16,48]. The design criteria for dementia-friendly architecture are classified by Marquardt into five categories (autonomy, familiarity, sensory stimulation, legibility, and social interaction) and are implemented through three design principles each, as shown in Table 5.

Table 5. Criteria and design principles of dementia-friendly architecture (adapted from [47]).

Criteria	I. Autonomy	II. Familiarity	III. Sensory Stimulation	IV. Legibility	V. Social Interaction
Design principles	Barrier-free, compensatory environment	Biographical reference	Encouragement	Logical room syntax	Privacy
	Safety and security	Homogenous and small groups	Avoidance of overstimulation	Furnishing	Belonging
	Orientational cues	Noninstitutional character	Fixtures and fittings	Fixtures and fittings	Communication

2.2. Dementia Nursing Homes in South Korea

The first senior nursing home in South Korea, established in 1981 through the enactment of the Welfare of Senior Citizens Act, was a public nursing home aimed at providing free or partially subsidized meals, treatment, and other amenities for seniors in need of caregiving [34]. Private fee-based elderly nursing homes were legalized in 1989. In 1997, an amendment to the same law introduced specialized public and private nursing homes for seniors with dementia, stroke, or other severe conditions. At that time, public specialized nursing homes offered inmates daily necessities at either no cost or reduced fees [34]. Meanwhile, private specialized nursing homes collected all associated costs from residents. But, with subsequent legal revisions in 2007, the five types of facilities, that is, public and private nursing homes as well as specialized nursing homes, were ultimately integrated into a single type for seniors with conditions such as dementia and stroke. This integration was in line with the implementation of the long-term care insurance system that began in 2008 and has been effective to date. Furthermore, through a legal revision in 2016, it became mandatory to establish dementia units in nursing homes specializing in dementia care, aimed at providing tailored care for individuals with dementia. This involved either installing dementia units in some nursing homes or converting the nursing homes themselves into dementia-specific ones.

As of 2019, there were 772,430 recipients of long-term care benefits for older adults [49]. Among them, 28.3% (218,240) received facility-based care, whereas 71.7% (554,190) opted for outpatient care [49]. The number of elderly nursing homes at the time was 3604 (with a capacity of 174,634), and elderly care group homes numbered 1939 (with a capacity of 17,065) [49]. The operational structure of elderly nursing homes was dominated by private establishments (2303 facilities), followed by corporate-run facilities (1188), local government facilities (104), and others (9) [49]. According to the current Welfare of Senior Citizens Act, facilities accommodating residents with conditions such as dementia and stroke are categorized as elderly nursing homes. Residents of these nursing homes include patients with Alzheimer's dementia, numbering 56,000 individuals, which constitutes 63% of the total number of people with Alzheimer's dementia [50].

2.3. Dementia Nursing Homes in Germany

In Germany, elderly nursing homes originated from almshouses (Spitäler) built near churches and monasteries in the early modern period, providing a place to sleep and care for the seniors, sick, and impoverished [51]. Toward the end of the 19th century, independent elderly nursing homes emerged, replacing these earlier institutions [51]. In the aftermath of World War II, during the economic reconstruction periods of the 1950s and 1960s, a significant number of elderly nursing homes were established [51]. The Federal Nursing Homes Act (HeimG) was enacted in 1974 and regulated the rights of facilities for older people, adults in need of care, or disabled adults. After more than 20 years of active discussions recognizing the need for independent long-term care, the long-term care insurance system (Pflegeversicherung) was introduced in 1995 as the first of its kind in the world [52]. The Nursing Homes Act lost its significance due to a constitutional amendment in 2006 (Federalism Reform I) and the associated transfer of legislative authority from federal law to state laws [52]. Starting in 2008, all federal states gradually established regulations regarding the Nursing Homes Act. As of 2019, the number of long-term care users in Germany was 4.12 million, with 80% (3.3 million) utilizing outpatient care (ambulante Pflegedienst), four times the 20% (810,000) who opted for facility-based care [53]. The utilization rate of facility-based care is slightly lower compared to Korea. There were 15,380 operating facility institutions in 2019 [54], which were more than three times the quantity compared to Korea. Among residents aged 65 and older in elderly care facilities, the prevalence of dementia was reported as 50% [55] or 69% [56], with the dementia rate rising to 64.3% among those aged 75 and older [57].

3. Current Legislation on Dementia Nursing Homes in South Korea and Germany

In South Korea, the current legislation governing dementia nursing homes for elderly residents is rooted in the welfare-oriented Welfare of Senior Citizens Act, established in 1981, along with its subsidiary legislation, the Enforcement Regulations of the Welfare of Senior Citizens Act, which was enacted in 1982. In Germany, the legal framework is centered on the Federal Nursing Homes Act (HeimG, 1974) and its subsidiary legislation, the Regulation on Minimum Standards for Nursing Homes (HeimMindBauV, 1978). Germany operates under a continental legal system with a federal structure comprising 16 states forming the Federal Republic of Germany, which is divided into federal law and state laws. The Federal Nursing Homes Act serves as the foundation for state-specific Nursing Home Acts, and the Federal Regulation on Minimum Standards for Nursing Homes guides state-specific Regulations on Minimum Standards for Nursing Homes (Table 6). State laws in Germany vary in nomenclature across states according to each one's characteristics.

Table 6. Current legal frameworks for elderly nursing homes in South Korea and Germany.

	South Korea	Germany	
		Federal Law (Bundesgesetz)	**State Law (Landesgesetz)**
Superior law	Welfare of Senior Citizens Act	Nursing Homes Act (Heimgesetz: HeimG)	State law for Nursing Home Act * (Länderheimgesetze)
Inferior law	The Enforcement Regulations of the Welfare of Senior Citizens Act	Regulation on Minimum Standards for Nursing Homes (Heimmindestbauverordnung: HeimMindBauV)	State law for Regulations on Minimum Standards for Nursing Homes * (Heimmindestbauverordnung der Bundesländer)

* Different names in each state.

3.1. Relevant Laws in South Korea

3.1.1. Superior Law: Welfare of Senior Citizens Act

The classification system and definitions outlined divide elderly nursing homes into the upper category of elderly welfare facilities and middle category of elderly medical welfare facilities [34]. These facilities are defined as establishments that admit seniors with significant mental and physical disabilities resulting from conditions such as dementia, stroke, and other geriatric illnesses [34] (Table 7I). Dementia is defined as an acquired composite disorder that affects a person's daily life due to a degenerative brain disorder or cerebrovascular disease, resulting in the deterioration of faculties, such as memory, language, orientation, judgment, and performance [58]. Geriatric diseases, as classified by the South Korean Standard Classification of Diseases, include Alzheimer's disease, intracranial hemorrhage, cerebral infarction, other cerebrovascular diseases, Parkinson's disease, post-stroke sequelae, and other degenerative diseases of the basal ganglia [59]. Consequently, residents of South Korean elderly nursing homes are limited to seniors suffering from major geriatric diseases such as dementia and stroke. The nursing homes are further categorized into those accommodating more than 10 residents and those with between 5 and 9 residents, known as elderly care group homes [34].

Table 7. Comparison of laws related to nursing homes in South Korea.

<table>
<tr><th>Regulation</th><th colspan="2">Legal Provisions</th><th colspan="2">Main Contents</th></tr>
<tr><td rowspan="2">I. Welfare of Senior Citizens Act</td><td>Art. 31</td><td>Types of elderly welfare facilities</td><td colspan="2">Upper classification: Welfare facilities for the elderly
Middle classification: Medical welfare facilities for the elderly
Subclassification: Elderly nursing homes</td></tr>
<tr><td>Art. 34(1)</td><td>Elderly medical welfare facilities</td><td colspan="2">Definition of elderly nursing home: Facility that accommodates seniors with significant mental and physical disabilities arising from conditions such as dementia, stroke, and other geriatric illnesses, and provides meals, care, and other necessities of daily life</td></tr>
<tr><td rowspan="3">II. The Enforcement Regulations of the Welfare of Senior Citizens Act</td><td rowspan="3">Art. 22(1)</td><td rowspan="3">Building standards of medical welfare facilities for the elderly</td><td>Common matters</td><td>Facility size, structure, and equipment</td></tr>
<tr><td>Facility standards</td><td>Necessary rooms: Bedroom, kitchen, dining room, washroom, bathroom, program room, physical (occupational) therapy room, medical/nursing room, and other rooms</td></tr>
<tr><td>Facility equipment standards</td><td>Facility equipment for necessary rooms</td></tr>
</table>

3.1.2. Inferior Law: The Enforcement Regulations of the Welfare of Senior Citizens Act

The subordinate law specifies detailed provisions on common matters, facility standards, and equipment standards required for the accreditation of elderly nursing homes and elderly care group homes through building standards for elderly medical welfare facilities (Table 7II).

3.2. *Relevant Laws in Germany*

3.2.1. Superior Law: Nursing Homes Act and State Laws

In the German Federal Nursing Homes Act, the scope of the law encompasses paid nursing homes (Heime) that provide care services, such as residence, care, and meals, to the elderly and adults in need of care or with disabilities [36]. It comprehensively describes some necessary rooms [36] (Table 8I). Based on the Nursing Homes Act, individual states in Germany have legislated their own primary care laws, outlining the law's purpose and also extending its application to nursing homes and various other facility types of ambulatory shared care homes and assisted living facilities (Table 8II).

Table 8. Legislation on nursing homes in Germany (1): HeimG and state laws.

		Legal Provisions		Main Contents
I. Nursing Homes Act (HeimG)		Art. 1	Scope of application	Nursing homes (Heime) that provide a residential space, medical care, and meals to the elderly or adults who need care or are disabled
		Art. 3(2)	Performance	Necessary rooms: Residential space, treatment spaces, household management spaces, corridor, bathrooms and toilets/Sanitary and technical equipment
II. State laws for the Nursing Homes Act				
State		State law	Legal provisions	Scope of application
1	BW	WTPG [60]	Art. 2–6	For adults in need of care or with disabilities: Nursing home (1), ambulatory assisted living community (3A)
2	Bay	PfleWoqG [61]	Art. 2	For seniors or adults in need of care: Nursing home (1), ambulatory assisted living community (3A), special housing form (6C), assisted living group (5A)
3	Berl	WTG [62]	Art. 2–7	For adults in need of care or with disabilities: Nursing home (1), care facility (2A), assisted living community (5B), special housing form (6C)
4	Ham	HmbWBG [63]	Art. 2	For seniors or adults in need of care or with disabilities: Service housing facility (7A), living community (8), assisted living community (5C), residential facility (9), guest facility (10), ambulatory services (4E)
5	Hess	HGBP [64]	Art. 2	For seniors or adults in need of care or with disabilities: Ambulatory care and nursing services (4D)
6	MV	EQG [65]	Art. 2	For seniors or adults in need of care, as well as those with mental health issues or disabilities: Day and night care (11), ambulatory assisted living community (3A), training living groups for individuals with mental health challenges (12)
7	NW	WTGNW [66]	Art. 2–5	For seniors or adults in need of care or with disabilities: Facility with comprehensive service offerings (13B), residential community with care services (5D), serviced living (7B), ambulatory services (4E), guest facility (10)
8	RP	LWTG [67]	Art. 3–6	For seniors or adults in need of care or with disabilities: Facility with increased self-determination and participation (13A), facility with comprehensive service offerings (13B), self-determined living community (5F)
9	Sach	SächsBeWoG [68]	Art. 2	For seniors or adults in need of care, as well as those with mental health issues or disabilities: Nursing home (1), ambulatory assisted living community (3A), assisted living group (5A)
10	SH	SbStG [69]	Art. 6–10	For adults in need of care or with disabilities: Nursing home (1), special forms of housing, care, and support (14), assisted living (6A), ambulatory supported residential and house community (3B)
11	BB	BbgPBWoG [70]	Art. 1, 4, 5	For adults in need of care or with disabilities: Supported housing (6B), housing with limited self-responsibility (6D)
12	BR	BremWoBeG [71]	Art. 2, 5, 7–9	For seniors or adults in need of care or with disabilities: Guest facility (10), mobile support services (4F), serviced living (7B), living community with support services (5G), care and support facility (2B)
13	NS	NHeimG [72]	Art. 1	For seniors or adults in need of care or with disabilities: Non-self-determined living communities (5E), assisted living (6A)
14	Sar	LHeimGS [73]	Art. 1	For adults in need of care or with disabilities: Nursing home (1), ambulatory assisted living (4A), ambulatory care services (4C)

Table 8. *Cont.*

II. State laws for the Nursing Homes Act				
State		**State law**	**Legal provisions**	**Scope of application**
15	SA	WTG-LSA [74]	Art. 2–4	For seniors or adults in need of care or with disabilities: Nursing home (1), non-self-organized housing forms (6E)
16	Tuer	ThürWTG [75]	Art. 1–3	For seniors or adults in need of care or with disabilities: Nursing home (1), ambulatory care housing forms (4B)

BW (Baden-Württemberg); Bay (Bavaria); Berl (Berlin); Ham (Hamburg); Hess (Hesse); MV (Mecklenburg-Vorpommern); NW (Nord Rhine-Westphalia); RP (Rhineland-Palatinate); Sach (Saxony); SH (Schleswig–Holstein); BB (Brandenburg); BR (Bremen); NS (Lower Saxony); Sar (Saarland); SA (Saxony-Anhalt); Tuer (Thuringia). (1) Stationäre Einrichtungen; (2A) Pflegeeinrichtungen; (2B) Pflege- und Betreuungseinrichtungen; (3A) Ambulant betreute Wohngemeinschaften; (3B) Ambulant betreute Wohn- und Hausgemeinschaften; (4A) Ambulant betreuten Wohnens; (4B) Ambulant betreute Wohnformen; (4C) Ambulante Pflegedienste; (4D) Ambulante Betreuungs- und Pflegedienste; (4E) Ambulante Dienste; (4F) Mobile Unterstützungsdienste; (5A) Betreute Wohngruppe; (5B) Betreute Wohngemeinschaften; (5C) Wohnassistenzgemeinschaften; (5D) Wohngemeinschaften mit Betreuungsleistungen; (5E) Nicht selbstbestimmte Wohngemeinschaften; (5F) Selbstbestimmte Wohngemeinschaften; (5G) Wohngemeinschaften mit Unterstützungsleistungen; (6A) Betreutes Wohnen; (6B) Unterstützter Wohnform; (6C) Besondere Wohnform; (6D) Wohnformen mit eingeschränkter Selbstverantwortung; (6E) Nicht selbstorganisierte Wohnform; (7A) Servicewohnanlagen; (7B) Service Wohnen; (8) Wohngemeinschaften; (9) Wohneinrichtungen; (10) Gasteinrichtungen; (11) Tages- und Nachtpflege; (12) Trainingswohngruppen; (13A) Einrichtungen mit höherer Selbstbestimmung und Teilhabe; (13B) Einrichtungen mit umfassendem Leistungsangebot; (14) Besonderen Wohn- Pflege und Betreuungsform.

3.2.2. Inferior Law: Regulation on Minimum Standards for Nursing Homes and State Laws

The building standards for German nursing homes are detailed through the Federal Nursing Homes Act's sublaw, the Federal Regulation on Minimum Standards for Nursing Homes, according to which nursing homes are classified into three types: elderly nursing homes (Altenheim), elderly residential homes (Altenwohnheim), and adult nursing homes (Pflegeheim fuer Volljährige). Building standards are categorized into general regulations that are simultaneously applied to all three types, as well as specific regulations applied to each type (Table 9I). Among the 16 federal states, 10 autonomously legislate and enforce their own regulations based on the Regulation on Minimum Standards for Nursing Homes (Table 9II). Starting with Hesse in 2010, followed by Mecklenburg-Vorpommern (2010), Baden-Wuerttemberg (2011), Bavaria (2011), Schleswig–Holstein (2011), Hamburg (2012), Berlin (2013), Rhineland-Palatinate (2013), Nord Rhine-Westphalia (2014), and Saxony (2014), the states have implemented regulations for nursing homes. Meanwhile, the remaining six states continue to apply the federal regulation directly, as their own state laws have not yet been enacted.

Table 9. Legislation on nursing homes in Germany (2): HeimMindBauV and state laws.

Regulation		**Legal Provisions**		**Main Contents**	
I. Regulation on Minimum Standards for Nursing Homes (HeimMindBauV)	Art. 1–18	Minimum requirements	Common regulations	Scope of application	Art. 1
				Residential rooms, care rooms	Art. 2
				Corridors, stairs, elevators, floor, lighting	Art. 3–6
				Emergency alarm devices, phones, exits	Art. 7–9
				Sanitary equipment, household space	Art. 10–11
				Heating, building entrance	Art. 12–13
			Special regulations	Residential room, functional auxiliary room	Art. 14–15
				Common areas, treatment room	Art. 16–17
				Sanitary equipment	Art. 18

II. State laws for Regulation on Minimum Standards for Nursing Homes							
State	**State law**	**State**	**State law**	**State**	**State law**	**State**	**State law**
BW	LHeimBauVO [76]	Bay	APfleWoqG [77]	Berl	WTG-BauV [78]	Ham	WBBauVO [79]
Hess	HGBPAV [80]	MV	EMindBauVOMV [81]	NW	WTGDVO [82]	RP	LWTGDVO [83]
Sach	SächBeWoGDVO [84]	SH	SbStG-DVO [85]	BB	-	BR	-
NS	-	Sar	-	SA	-	Tuer	-

4. A Comparison of Building Standards for Dementia Nursing Homes in South Korea and Germany

As discussed in Section 3, the current laws in South Korea and Germany regarding the building standards for dementia nursing homes are primarily outlined in South Korea's Enforcement Regulations of the Welfare of Senior Citizens Act and Germany's Regulation on Minimum Standards for Nursing Homes (HeimMindBauV), a sublaw under the Nursing Homes Act, along with the Minimum Building Regulations of the individual federal states.

4.1. Regulations Regarding Basic Requirements for Dementia Nursing Homes

4.1.1. South Korea

The two countries' building standards specify 11 basic requirements for dementia nursing homes. Six of these requirements are specified in South Korea's building standards: a barrier-free environment, safety lock devices, light, thermal heating system, indoor ventilation, and community interactions (Tables 10I and 11I). They emphasize mobility, safety, and health and sanitation facilities, and can be traced back to the revised building standards in 1998, which emerged with the legal establishment of specialized nursing homes for dementia patients in 1997. In other words, most of these basic requirements of specialized nursing homes have largely remained unchanged since 1998 to the present day. Upon closer examination, this continuity follows the integration of five types of nursing homes into the current unified structure in 2007, coinciding with the implementation of Long-Term Care Insurance for the Elderly in 2008, prompted by South Korea's aging population. In 2016, regulations mandated entrance doors at dementia unit entrances to separate spaces and allow emergency access. This measure aimed to ensure the stability and safety of elderly individuals with dementia, meeting their specific needs.

Table 10. Comparison of the basic requirements for dementia nursing homes.

Basic Requirements	(A) Barrier-Free Environment	(B) Safety Lock Device	(C) Light		(D) Thermal Heating System	(E) Indoor Ventilation	(F) Spatial Orientation	(G) Home-like Environment	(H) Personal Belongings and Furniture	(I) Emergency Alarm Communication Device	(J) Protected Outdoor Space	(K) Community Interaction
			Daylight	Lighting								
I. South Korea	○	○	○	○	○	○	-	-	-	-	-	○
II. Germany												
HeimMindBauV (federal law)	○	○	-	○	○	-	-	-	-	○	-	-
BW	○	-	○	○	○	○	○	○	○	-	○	-
Bay	○	-	-	-	-	-	-	-	-	○	-	-
Berl	○	○	-	○	○	-	-	-	-	○	-	-
Ham	○	-	○	○	-	-	○	○	-	○	○	-
Hess	○	○	○	○	○	-	-	-	○	○	○	-
MV	○	-	-	-	-	-	-	-	-	○	-	-
NW	-	-	○	-	○	-	-	-	-	○	○	-
RP	-	-	-	-	○	-	○	-	○	○	○	-
Sach	○	○	○	○	○	○	○	○	○	○	○	-
SH	○	-	-	-	○	-	-	○	○	○	○	-

○: Regulations for basic requirement exist.

Table 11. Comparison of the details of the basic requirements for dementia nursing homes.

Basic Requirements		I. South Korea	II. Germany
A-1	Stairs, steps	■ Gradual stairs.	■ Entrance and hallway are level (1). ■ Sloped if height difference (1).
A-2	Corridor width		■ Ensuring the mobility of the bed (1).
A-3	Anti-slip floor	■ Soft and non-slippery flooring. ■ Non-slip floors in corridors, bedrooms, toilets, washrooms, and bathrooms.	■ Prevent slipping in living space and hallway (1).
A-4	Elevator	■ Passenger elevators or ramps.	■ Elevator needed: minimum one location (1).
A-5	Wheelchair usability	■ Allow wheelchair access: corridor, bedrooms, bathrooms, and toilets. ■ No thresholds between the corridor, bedrooms, and bathroom.	■ Allow wheelchair access: bathrooms and toilets (1).
A-6	Safety handle	■ Safety handles in the hallway, bedrooms, and bathrooms. ■ Move freely on and off the bed. ■ Safety equipment around the bed. ■ Auxiliary rods and vertical pillars for bathtub.	■ Safety handles on both sides of the hallway (1), and around bathtubs, showers, and toilets (1). ■ Bathtub entry safety (1).
A-7	Barrier-free standard specification		■ Statements of German barrier-free standards (2, 3, 4, 5, 6, 10, 11).
B-1	Outside entrance	■ Door-lock device. ■ Automatically opens during emergencies like fires.	
B-2	Stairs	■ A lock on the door of the stairs. ■ Automatically opens during emergencies like fires.	
B-3	Living room, bedroom, and bathroom		■ Emergency doors openable from outside: common areas, bedrooms, toilets (1, 4, 6, 10).
B-4	Dementia unit	■ Install emergency doors for fire safety.	
B-5	Kitchen	■ Install lockable doors in kitchens for fire safety.	
C-1	Daylight	■ Consider daylight for health and hygiene. ■ At least 1/7th of the bedroom floor area should be windows. ■ Daylight suitable in bedroom.	■ Use daylighting (2). ■ Window to allow a view from bed (7). ■ Sunlight control devices in bedrooms, residence, and treatment spaces (4). ■ Avoid facing north (7).
C-2	Lighting	■ Night lights in corridors and toilets. ■ Lighting suitable for bedrooms.	■ Lighting suitable for the residents' needs (2). ■ Even and bright lighting (2). ■ Adequate, uniform, glare-free lighting always (4). ■ Night lights on stairs and in corridors (1). ■ Residents cannot turn off corridor lighting (4). ■ Reading lights in bedrooms and common living rooms (1,4). ■ Adjust room lighting from bed (4). ■ Night lights for night care (4).
D-1	Heating system	■ Consider heating for health and hygiene. ■ Heating suitable for bedrooms.	■ Proper heating and temperature throughout all rooms, corridors, toilets, bathrooms, and stairs (1). ■ An indoor climate suitable for residents' needs (2).
E-1	Ventilation	■ Consider ventilation for health and hygiene. ■ Ventilation suitable for bedrooms.	■ Exhaust facilities suitable for residents' needs (2).
F-1	Space arrangement	-	■ Consider visual perception decline due to aging and dementia for better understanding of room functions (5). ■ Spatial orientation support for residents (2,10). ■ Central living room and kitchen layout visible at a glance (5). ■ Bedroom layout mimics typical household bathroom placement and design (5).

Table 11. *Cont.*

	Basic Requirements	I. South Korea	II. Germany
F-2	Orientation element	-	■ Safe navigation cues for finding personal spaces (5). ■ Auxiliary elements for spatial orientation in corridors, staircases, and elevators (9).
G-1	Size of the space		■ Common living room size reflects personal unit characteristics (5).
G-2	An atmosphere like home		■ Residential unit simulates a home-like environment (11, 2). ■ A private atmosphere (10). ■ Amplifying personal feelings through space arrangement, usage characteristics, and decorations suitable for age group of typical houses (5). ■ Organize the decoration of bedroom with residents (11). ■ Common living room decoration reflects personal unit characteristics (5).
H-1	Past personal belongings		■ Use familiar furniture, belongings, and decorations in the bedroom (2, 6, 9, 10).
I-1	Alarm device		■ An alarm on all beds (1, 3, 4, 8). ■ Alarm devices at an appropriate level in bedrooms, toilets, rest areas, treatment spaces, and public areas (3, 4, 5, 6, 9, 10, 11).
I-2	Communication device		■ At least one telephone line throughout the facility (1). ■ Radio, TV, phone, internet communication (3, 4, 6, 7, 9, 11).
J-1	Outdoor space Access		■ Always be accessible (5). ■ Connected to central common living room (5).
J-2	Outdoor space safety		■ Residents must remain safe at all times (5). ■ Exclude outdoor spaces accessible to outsiders from the surrounding environment (9).
J-3	Self-use of outdoor spaces		■ Residents have independent access to outdoor spaces (11). ■ Wheelchair users have independent use (8, 10, 2).
J-4	Adequate outdoor space		■ Sufficient size (2, 6, 9, 10).
J-5	Solar control device in outdoor spaces		■ Such as shades and parasols (5).
K-1	Additional cultural and sports facilities	■ Such as libraries, and sports and recreation facilities. ■ Open but not intrusive to residents.	
K-2	Day and night care facility	■ Such as day and night care services for seniors	

(1) HeimMindBauV (federal law); (2) BW; (3) Bay; (4) Berl; (5) Ham; (6) Hess; (7) MV; (8) NW; (9) RP; (10) Sach; (11) SH.

4.1.2. Germany

In Germany's building standards, similar to South Korea, five common requirements focusing on mobility, safety, health, and sanitation facilities are included. Additionally, Germany adds specifically the barrier-free standard DIN18040-2:2011-9 [86]. Beyond these, five more items are regulated, encompassing emergency alarm and communication devices, spatial orientation, creating homelike environments, managing personal belongings and furniture, as well as providing protected outdoor spaces (Tables 10II and 11II). Except for emergency alarm and communication devices, these items have been gradually integrated into the new state building standard, including the barrier-free standard. This transition was a result of the federalism reform in 2006, where federal government regulations shifted to state regulations [87], leading to the establishment of state building standards between 2010 and 2014. Consequently, since 2010, state-level measures have been implemented to address wandering issues and create a more home-like environment in facilities, with a focus on enhancing emotional stability for elderly individuals with dementia. State building standards have increasingly emphasized dementia-friendly design principles since 2010, surpassing previous federal standards.

4.1.3. Summary

The results indicate that South Korean building standards have emphasized specific principles supporting dementia patient autonomy and fostering social interaction within community contexts since 1998 (Table 12). Notably, certain standards present in German state building regulations since 2010, such as the barrier-free standard DIN18040-2:2011-9, are absent in South Korea. These standards, particularly within the autonomy criterion, promote principles of barrier-free and compensatory environmental design.

Table 12. Design criteria inherent in the basic requirements for dementia nursing homes.

Design Criteria	I. Autonomy			II. Familiarity			III. Sensory Stimulation			IV. Legibility			V. Social Interaction		
	I-1	I-2	I-3	II-1	II-2	II-3	III-1	III-2	III-3	IV-1	IV-2	IV-3	V-1	V-2	V-3
	Barrier-Free and Compensatory Environment	Safety and Security	Orientational Cues	Biographical Reference	Small Groups	Noninstitutional Character	Encouragement	Avoidance of Overstimulation	Access to the Outdoors	Logical Room Syntax	Furnishing	Fixtures and Fittings	Privacy	Belonging	Communication
(A)	○ [1,2]	○ [1,2]	-	-	-	-	-	-	-	-	-	-	-	-	-
(B)	-	○ [1,2]	-	-	-	-	-	-	-	-	-	-	-	-	-
(C)	○ [1,2]	-	-	-	-	-	-	-	-	-	-	-	-	-	-
(D)	○ [1,2]	-	-	-	-	-	-	-	-	-	-	-	-	-	-
(E)	○ [1,2]	-	-	-	-	-	-	-	-	-	-	-	-	-	-
(F)	-	-	○ [2]	-	-	-	-	-	-	○ [2]	-	-	-		-
(G)	-	-	-	-	○ [2]	○ [2]	-	-	-	-	-	-	○ [2]	○ [2]	-
(H)	-	-	-	○ [2]	-	○ [2]	-	-	-	-	○ [2]	-	○ [2]	-	-
(I)	○ [2]	○ [2]	-	-	-	-	-	-		-	-	-	-	-	-
(J)	○ [2]	○ [2]	-	-	-	-		-	○ [2]	○ [2]	-	-	-	-	
(K)	-	-	-	-	-	-	-	-	-	-	-	-	-	-	○ [1]

(A) Barrier-free-like environment; (B) Safety lock device; (C) Light; (D) Thermal heating system; (E) Indoor ventilation; (F) Spatial orientation; (G) Home-like environment; (H) Personal belongings and furniture; (I) Emergency alarm and communications device; (J) Protected outdoor space; (K) Community interaction. [1] South Korea, [2] Germany. ○: Regulations for basic requirement exist.

Furthermore, Germany enhances patient autonomy by providing detailed guidelines on emergency alarm and communication devices, as well as spatial orientation for better wayfinding. Apart from autonomy, Germany addresses aspects that enrich the familiarity criterion, including design principles involving autobiographical reference points, small groupings, and non-facility-specific characteristics. Additionally, it reinforces the sensory stimulation criterion through the accessibility of outdoor spaces.

Moreover, German regulations exhibit features supporting the legibility criterion, such as logical spatial syntax and furniture design, as well as the social interaction criterion, encompassing privacy protection and a sense of belonging. In contrast, South Korea currently lacks regulations addressing these aspects. A 2019 study examining environmental factors in nursing homes found that residents perceived the physical environment as significantly different from their own homes, lacking a sense of "home-like" feeling [88]. Furthermore, changes in regulations regarding access to outdoor spaces, particularly the rescinding of the mandate for sunlight rooms in 2008, have reduced legal reinforcement for individuals with dementia who rely on sensory stimulation.

4.2. Regulations Regarding the Facility Size of Dementia Nursing Homes

4.2.1. South Korea

The building standards in both South Korea and Germany comprehensively regulate various aspects (from A to I) concerning the sizes of dementia nursing homes (Table 13). South Korean standards specify admission capacity, floor area, and other requirements for both general and dementia units. These include bedroom capacity and net floor area, as well as net floor area of common living room space, with a particular focus on capacity of dementia units (Table 13I).

The maximum capacity of bedrooms in both general and dementia units has been reduced from six individuals, as specified in the revised standards in 1998, to four individuals currently since 2006. The current minimum net floor area for a bedroom in general units remains unchanged at 6.6 m^2 per person, as it was for specialized nursing homes in 1998. Regarding the bedroom floor area in dementia units, with the introduction of dementia units in 2016, the proposed minimum net floor area for a single bedroom has been 9.9 m^2, which has remained unchanged to the present day. However, despite concerns from caregivers, the capacity of dementia units increased from 12 individuals in 2016 to 16 in 2019, and it has remained at that level since.

Additionally, the 2019 building standard revision updated the minimum floor area for common living rooms in dementia units to 1.65 m^2 per person up to now. Before 2019, a mandate was in place requiring that over 25% of the total area be allocated for common living spaces in dementia units, introduced in 2016, while no such regulations existed for common living room areas in general units prior to 2016.

4.2.2. Germany

The German standards cover similar items, excluding those related to dementia units. German bedrooms typically accommodate one to two residents and have a higher minimum net floor area for single bedrooms compared to Korea. Federal law mandates 12 m^2 per person, while state laws stipulate 12–14 m^2 per person (Table 13II).

The living room area in Germany varies significantly depending on the state. Each person requires a minimum of 0.75–2.5 m^2, and the total area must be at least 18–20 m^2, or alternatively regulated at 3–5 m^2 per person, as stipulated by certain state laws to enhance a sense of belonging. Therefore, common living rooms in Germany are generally larger than those in Korea. Moreover, the capacity for common living rooms is approximately 12–15 individuals, slightly smaller than the regulation of 16 individuals specified for dementia unit living rooms in Korea.

Additionally, Germany introduces additional criteria such as the ratio of single bedrooms and minimum space width of single bedrooms. These criteria were newly regulated according to state standards established between 2010 and 2014 to enhance personal living spaces in nursing homes for individuals with dementia, providing significantly more privacy.

4.2.3. Summary

The results suggest a close relationship between facility sizes and design criteria such as autonomy, familiarity, sensory stimulation, legibility, and social interaction (Table 14). In both South Korea and Germany, regulations governing the capacity and net floor area of bedrooms, as well as common living rooms, influence design criteria related to autonomy, familiarity, and social interaction for individuals with dementia.

Table 13. Comparison of the size of facilities in dementia nursing homes.

Facility Sizes	The Entire Facility			Residential Unit									
				General Unit						Dementia Unit			
				Resident's Bedroom				Common Living Room		The Unit	Resident's Bedroom		Com. Living Room
	A–B		C	D	E	F	G	H	I	B	D	F	I
	Capacity (p.)		Min. Floor Area (m^2/p.)	Capacity (p.)	Single-Bedroom Ratio (%)	Min. Net Floor Area (m^2)	Min. Width (m)	Capacity (p.)	Min. Net Floor Area (m^2/p.)	Capacity (p.)	Capacity (p.)	Min. Net Floor Area (m^2)	Min. Net Floor Area (m^2/p.)
	Min.	Max.		Max.				Max.		Max.	Max.		
I. South Korea													
	10	-	23.6	4	-	6.6 per person	-	-	-	16	4	9.9 (1p.) 16.5 (2p.) 23.1 (3p.) 29.7 (4p.)	1.65
II. Germany													
HeimMindBauV	6	-	-	2 [6]	-	12 (1p.) 18 (2p.) 24 (3p.) 30 (4p.)	-	-	(20 + 1) [1]				
BW	-	100	-	1	100	14 (1p.) [2]	3.2	15	5				
Bay	6	-	-	1, 2	-	14 (1p.) [2] 20 (2p.) [2]	-	-	(20 + 1.5) [1,3]				
Berl	-	-	-	1, 2	60	14 (1p.) [2] 22 (2p.) [2]	-	-	5 [3]				
Ham	-	-	-	1	-	14 (1p.) [2]	3.2	12	(18 + 1) [1]				
Hess	6	-	-	1, 2	-	14 (1p.) [2] 24 (2p.) [2]	3.2	-	(20 + 2.5) [1]				
MV	-	-	-	1, 2	-	12 (1p.) 18 (2p.)	-	-	-				
NW	-	36	45	1, 2	80–100	14 (1p.) [2] 24 (2p.) [2]	-	-	5				
RP	-	-	-	1, 2	-	14 (1p.) 20 (2p.)	-	-	3				
Sach	9	-	-	1, 2	-	12 (1p.) [2] 18 (2p.) [2]	-	-	(20 + 0.75) [1,3]				
SH	-	-	-	1, 2	75	14 (1p.) 20 (2p.)	-	-	-				

(A) Min. capacity of the facility; (B) Max. capacity of the facility; (C) Min. floor area for the facility; (D) Bedroom capacity; (E) Single-bedroom ratio; (F) Min. net floor area for bedrooms; (G) Min. space width for a single bedroom; (H) Common living room capacity; (I) Min. net floor area for common living rooms. [1] Each person requires a minimum of 0.75–2.5 m^2, but the total area must be at least 18–20 m^2. [2] Non-calculation of private bathroom and its front room. [3] Non-calculation of outdoor spaces as greenhouses, etc. [4] A max. of four beds are allowed only with the permission of the competent authority.

Table 14. Design criteria inherent in the facility sizes of dementia nursing homes.

Design Criteria	I. Autonomy			II. Familiarity			III. Sensory Stimulation			IV. Legibility			V. Social Interaction		
	I-1	I-2	I-3	II-1	II-2	II-3	III-1	III-2	III-3	IV-1	IV-2	IV-3	V-1	V-2	V-3
	Barrier-Free and Compensatory Environment	Safety and Security	Orientational Cues	Biographical Reference	Small Groups	Noninstitutional Character	Encouragement	Avoidance of Overstimulation	Access to the Outdoors	Logical Room Syntax	Furnishing	Fixtures and Fittings	Privacy	Belonging	Communication
(A)	-	-	-	-	-	-	-	-	-	-	-	-	-	-	-
(B)	-	-	-	-	-	-	-	-	-	-	-	-	-	-	-
(C)	-	-	-	-	-	-	-	-	-	-	-	-	-	-	-
(D)	○ [1,2]	○ [1,2]	-	-	○ [1,2]	○ [1,2]	-	-	-	-	-	-	○ [1,2]	-	-
(E)	○ [2]	○ [2]	-	-	○ [2]	○ [2]	-	-	-	-	○ [2]	-	○ [2]	-	-
(F)	○ [1,2]	○ [1,2]	-	-	-	○ [2]	-	-	-	-	-	-	○ [2]	-	-
(G)	○ [2]	○ [2]	○ [2]	○ [2]	-	○ [2]	-	-	-	-	○ [2]	-	○ [2]	-	-
(H)	○ [2]	○ [2]	-	-	○ [2]	○ [2]	-	○ [2]	-	-	-	-	○ [2]	○ [2]	○ [2]
(I) *	○ [1,2]	○ [1,2]	-	-	-	-	-	-	○ [2]	-	-	-	-	○ [2]	○ [2]

[1] South Korea, [2] Germany; (A) Min. capacity of the facility; (B) Max. capacity of the facility; (C) Min. floor area for the facility; (D) Bedroom capacity; (E) Single-bedroom ratio; (F) Min. net floor area for bedrooms; (G) Min. space width for a single bedroom; (H) Common living room capacity; (I) Min. net floor area for common living rooms. * Non-calculation of outdoor spaces: greenhouses, balconies, loggias, terraces, and rooftop gardens. ○: Regulations for facility sizes exist.

Decreasing bedroom capacity and increasing the minimum net floor area of both bedrooms and common living rooms contribute to creating an accessible and compensatory environment, enhancing safety, security, and patient autonomy. This approach also fosters familiarity and supports individual privacy and a sense of belonging in social interactions.

A comparison between South Korea's maximum capacity of four residents and the minimum net floor area of bedrooms, in both general and dementia units, and Germany's standards indicates that German regulations are more conducive to autonomy, familiarity, and social interaction. Additionally, in some states (BW, Berlin, NW, RP), common living room areas are larger than those in Korea, which is advantageous for these characteristics. Notably, in Germany, bathroom and surrounding corridor areas are typically excluded from the bedroom area calculation, further contributing to this advantage.

Germany's specifications for the ratio of single bedrooms and the minimum space width of single bedrooms are particularly noteworthy, optimizing autonomy, familiarity, legibility, and individual privacy protection in social interactions. For instance, Germany's regulations include a single-bedroom ratio of 60–100% (BW, Berlin, NW, SH) and a minimum space width of 3.2 m for single bedrooms (BW, Ham, Hess), aimed at maximizing personal space and facilitating the arrangement and use of personal furniture by individual residents.

Furthermore, Germany's approach to calculating the net floor area of common living rooms, excluding the outdoor space area, enhances the potential for maximizing the net floor area of common living rooms while ensuring accessibility to outdoor spaces for sensory stimulation.

In summary, South Korea's regulations on facility size primarily support autonomy with less emphasis on familiarity, whereas Germany's regulations encompass autonomy, familiarity, sensory stimulation, legibility, and social interaction.

4.3. Necessary Rooms in Dementia Nursing Homes and Detailed Regulations for Necessary Rooms

4.3.1. Configuration of Necessary Rooms within the Residential Unit

1. South Korea

The system for configuring necessary rooms in the residential units of dementia nursing homes, as regulated by the standards of both countries, requires multiple rooms classified into seven spatial areas, as shown in Table 15. Since 2002, the configuration of the necessary rooms differs based on the scale of the facility, categorized either as having a capacity of 30 or more residents, or 10–30 residents.

Table 15. A system for the spatial areas and configuration of necessary rooms in dementia nursing homes.

A. Individual Spaces		B. Communal s.		C. Therapy s.		D. Hygiene s.		E. Functional Auxiliary s.		F. Household Management s.		G. Staff Duty s.	
Bedroom	○ [1,2]	Common living room/ pro-gram room	○ [1,2]	Physical occupational therapy room	○ [1,2]	Ensuite bathroom/ toilet	○ [2]	Waste disposal room	○ [2]	Storage room	○ [2]	Nursing station	○ [1,2]
Special bedroom	○ [1,2]	Kitchen/ cooking room	○ [1,2]	-	-	Common toilet	○ [1,2]	Morgue	○ [2]	Laundry room	○ [1,2]	Staff office	○ [1,2]
-	-	Dining room	○ [1,2]	-	-	Common washbasins/ bathrooms	○ [1,2]	-	-	Drying room	○ [1,2]	Staff room	○ [1,2]
		Outdoor space *[2]										Volunteer room	○ [1]

[1] South Korea, [2] Germany. * Garden, terrace, shared balcony, and loggias. ○: Necessary room is identified.

The basic and additional configurations are applied on different capacities accordingly. The basic configurations are commonly applicable for both types and include essential rooms for individual, communal, treatment, hygiene, and staff duty spaces (Table 16I). Related to additional configuration, facilities with a capacity of 30 or more must include additional rooms such as staff offices, caregiver rooms, and volunteer rooms (staff duty spaces) along with shared washbasins/bathrooms (hygiene spaces) and laundry/drying rooms (household management spaces). However, for the latter facilities, two or three of these rooms can be consolidated into one room.

Table 16. Comparison of necessary room configuration in dementia nursing homes.

Spatial Areas	A. Individual Spaces		B. Communal s.				C. Therapy s.	E. Functional Auxiliary s.		D. Hygiene s.			F. Household Management s.			G. Staff Duty s.		
Necessary Room	**B**	**SB**	**LR**	**PR**	**KD**	**OS**	**PT**	**WD**	**M**	**EB**	**CT**	**CB**	**LD**	**SR**	**SO**	**SR**	**VR**	**NS**
I. South Korea																		
>30	○ [1]	○ [1]	-	○ [1]	○ [1]	-	○ [1]	-	-	-	○ [1]	○ [2]	○ [2]	-	○ [2]	○ [2]	○ [2]	○ [1]
10–30	○ [1]	○ [1]	-	○ [1]	○ [1]	-	○ [1]	-	-	-	○ [1]	○ [2]		-		○ [2]		○ [1]
Dementia units	○		○	-	○	-	-	-	-	-	○	○		-		-		
II. Germany																		
HeimMindBauV	○	○	○	-	-	○ [5]	○	-	○	-	○	○ [4(A)]	- ○ [7]	○		-		
BW	○	-	○	-	○	○ [6]	-	- ○ [7]		○ [3]	-	○ [4(A)]	- ○ [7]	○ [7]		-		
Bay	○	○	○	-	-	-	○ [7]	-	○	○ [3]	-	○ [4(B)]	-			-		
Ber	○	○	○	-	-	-	○ [7]	○ ○ [7]	-	○ [3]	○	○ [4(C)]	○ [7]			○		
Ham	○	-	○	-	○	○ [6]	-	- ○ [7]		○ [3]	○	○ [4(B)]	- ○ [7]	○ [7]		○ [7]		

Table 16. *Cont.*

Spatial Areas	A. Individual Spaces		B. Communal s.				C. Therapy s.	E. Functional Auxiliary s.		D. Hygiene s.			F. Household Management s.			G. Staff Duty s.		
Necessary Room	**B**	**SB**	**LR**	**PR**	**KD**	**OS**	**PT**	**WD**	**M**	**EB**	**CT**	**CB**	**LD**	**SR**	**SO**	**SR**	**VR**	**NS**
Hess	○	○	○	-	-	-	○ [7]	○ ○ [7]	○	○ [3]	-	○ [4]	- ○ [7]	○		○ [7]		
MV	○	○	○	-	-	○ [6]	○	○ ○ [7]	-	○ [3]	○	○ [4(D)]	○ [7]			○ [7]		
NW	○	○	○	-	○	○ [6]	-	-		○ [3]	○	○ [4(E)]	-	○		-		
RP	○	○	○	-	-	○ [6]	○	○	-	○ [3]	○	○ [4]	- ○	○		-		
Sachs	○	○	○	-	-	-	-	○ [7]		○ [3]	-	-	-			○ [7]		
SH	○	○	○	-	○	○	○	○ ○ [7]	-	○ [3]	○	○ [4]	- ○ [7]	○ [7]		-		

B: Bedroom; SB: Special bedroom; LR: Common living room; PR: Program room; KD; Kitchen/dining room; OS: Outdoor space; PT: Physical occupational therapy room; WD: Waste disposal room; M: Morgue; EB: Ensuite bathroom/toilets; CT: Common toilet; CB: Common bathroom; LD: Laundry/drying room; SR: Storage room; SO: Staff office; SR: Staff room; VR: Volunteer room; NS: Nursing station. [1]: Basic configurations; [2]: Additional configurations; [3]: Inclusion of washbasins, toilets, and shower; [4]: Common bathroom ([A]: One unit per facility; [B]: At least one unit is required for a max. of 40 people; [C]: At least one unit is required for a max. of 30 people; [D]: At least one unit is required for a max. of 32 people; [E]: At least one unit is required for a max. of 20 people); [5]: Shared balconies, loggias, terraces; [6]: Garden, shared balconies, terraces; [7]: Sufficient number and size of necessary facilities. ○: Necessary room is identified.

The current building standards do not require ensuite bathrooms/toilets for either general or dementia units, while installing them is essential for meeting the needs of fully dependent seniors with dementia, ensuring convenient and frequent access.

Moreover, since 2016, dementia units have transitioned to common living rooms instead of program rooms found in general residential units. Notably, regarding outdoor spaces, sunrooms were introduced from 1998 to 2008, and outdoor spaces for dementia units were recommended in revised building standards from 2016 to 2018, with a mandate for dementia-oriented care group homes to have outdoor spaces of at least 15 m^2. However, both mandates and recommendations for outdoor spaces have since been removed, reducing the possibility for elderly individuals with dementia in nursing homes to access outdoor areas.

2. Germany

The overall scope of room configurations in Germany, which may vary based on federal and state laws, seems very similar to that of Korea, but there are also differences (Table 16II). All German state laws mandate the inclusion of ensuite bathroom/toilets to support the autonomy of elderly individuals with dementia.

Additionally, Germany includes outdoor spaces, waste disposal rooms, morgues, and storage rooms in the list of necessary rooms, which are not covered by South Korea's building standards. Outdoor spaces, in particular, are crucial areas for enhancing sensory stimulation for elderly individuals with dementia, whether in independent or non-independent states, through elements such as sunlight, fresh air, and nature [89].

Furthermore, Germany stipulates the installation of treatment, functional auxiliary, household management, and staff duty spaces in sufficient numbers and sizes.

4.3.2. Detailed Regulations for Necessary Rooms

1. Individual spaces

   a. Bedrooms and special bedrooms
   In South Korea, detailed regulations focus on various types of bedrooms, including single, multi-person, and couple-shared occupancy types. These regulations primarily emphasize the autonomy design criterion, focusing on aspects such as bed safety, window size, health and safety equipment, anti-slip flooring, wheelchair access, and daylight. There is only a brief mention of personal storage facilities to enhance the sense of familiarity and privacy (Table 17I). Within the same criterion of autonomy, Germany's building standards pertain to size, accessibility, door control, dimensions, external views, shading, lighting, etc., in bedrooms (Table 17II). Furthermore, they promote familiarity and personal privacy within the criterion of social interaction. This is achieved by arranging diverse furniture, providing adequate personal storage facilities, and ensuring space adaptability, such as allowing the use of a single room as a double room in the case of spousal admission. Both South Korean and German dementia nursing homes require a specific number of special bedrooms to temporarily separate residents in multi-person rooms.

Table 17. Comparison of detailed regulations for individual spaces.

Regulation		I. South Korea	II. Germany
(A-1) Bedroom			
Autonomy	Barrier-free and compensatory environment; Safety and security	Room type: Single, multi-person, and couple-shared bedrooms. Bed safety: Adjustable height. Window size: 1/7 or more of bedroom floor area. Health equipment: Adequate heating, ventilation, lighting, and moisture-proofing. Safety equipment. Safety handle: Move freely on and off the bed. Anti-slip floor. Wheelchair access: No thresholds. Daylight.	Size: Spacious for wheelchair movement (4). Accessibility: Easy access from corridor or communal spaces (1, 3, 6, 8). Room type: Avoid designs requiring crossing through bedroom (4, 6, 8, 10). Door control: Lockable from inside and unlockable from outside in an emergency (1, 3, 4, 6, 10). Basement bedroom: Prohibited (6). Outside view: Bed should offer outside view (8). Direction: Avoid facing north (8). Shading device: For indoor temperature control (4). Lighting: Bedside reading stand (1, 6). Night lights installed (6).
Familiarity	Biographical reference; Noninstitutional character	Personal storage: Separate storage by multi-person type.	Furniture: Bed, bedside table, closet, TV, chair, and table arranged for sleep and daily use (4). Spatial variability: Combining single rooms into doubles (5, 6, 2). Personal storage: Ensure at least one location on each floor (6, 9), tailored to resident count (9). Adequate space (1, 2, 10). At least 1 m^2 per person near or within private living space, like a closet or shelf, with lockable options (1,5).
Social interaction	Privacy		
(A-2) Special bedroom			
		Size: Less than 5% of total capacity.	Numbers: Single room for temporary use by multi-person room residents (1, 3, 6, 7, 8, 9, 10, 11). At least one single room designated as special for every 30 residents in a double bedroom (8).

(1) HeimMindBauV (federal law); (2) BW; (3) Bay; (4) Berl; (5) Ham; (6) Hess; (7) MV; (8) NW; (9) RP; (10) Sach; (11) SH.

2. Communal space
    a. Common living room
    In South Korea, regulations primarily focus on entry control for dementia units and the provision of space usage programs for program rooms in general units, aiming to support the autonomy of seniors with dementia (Table 18I), albeit with relatively limited requirements. On the other hand, German common living rooms have numerous and detailed requirements within the autonomy criterion, such as size, accessibility, space usage program, treatment integration, bed participation, shading, and reading lights (Table 18II). Furthermore, Germany has specifications on privacy and living space to enhance familiarity and a sense of belonging within the social interaction criterion. Additionally, it provides orientational cues and logical room syntax, such as location, thereby supporting autonomy and legibility criteria.
    b. Kitchens, dining rooms, and outdoor space
    In Korea, kitchen–dining rooms are considered necessary, but kitchens are closed off to patients with dementia, and specific regulations regarding dining rooms are lacking. Between 2016 and 2018, some revisions mandated the installation of a simple kitchenette and dining area within dementia units. However, as of 2019, this requirement is no longer enforced, leaving elderly individuals with dementia in both dementia and general units unable to physically or sensorily engage in kitchen activities.
    In contrast, in Germany, kitchen and dining rooms are designated as open spaces accessible to patients with dementia and with wheelchair, and there are regulations regarding their size, numbers, storage, and locations. Consequently, these rooms support residents' independence and enhance the readability of the dining space, aligning with the autonomy and legibility criteria. Moreover, outdoor spaces in Germany are regulated in terms of size, accessibility, location, protection, and shading to address the autonomy, legibility, and sensory stimulation needs of dementia patients.

Table 18. Comparison of detailed regulations for communal spaces.

Regulation		I. South Korea	II. Germany
(B-1) Common living room			
Autonomy	Barrier-free and compensatory environment	Dementia unit: Entrance door ensures space separation and emergency access. Program room: Cultural and entertainment facilities available for free use.	Size: Consider facility concept, resident count, wheelchair use (11). Accessibility: Good access from bedrooms (3, 4, 6, 10). Single-story layout: (4). Space usage program Functions as living, dining, group activity, socializing, and play areas (4, 7, 9). Treatment integration: Doubles as therapy space (1), ensuring resident privacy (3, 4, 6, 7). Participation from bed: All residents participate (3, 4, 6, 7, 11), even from bed (4, 10). Shading device: (4). Reading lights: (1).
	Safety and security		
	Orientational cues	-	Location: Centrally located within units (5).
Legibility	Logical room syntax		
Familiarity	Noninstitutional character	-	Privacy: Residence units emphasize individuality (5).
Social interaction	Belonging		Living space: Living area corridor and greenhouse (11).

Table 18. *Cont.*

Regulation		I. South Korea	II. Germany
(B-2) Kitchen/dining room			
Autonomy	Barrier-free and compensatory environment	Kitchen flooring: Water-resistant. Kitchen cleaning and drainage: Structured for easy cleaning and drainage. Kitchen door lock: Door locks installed for dementia patients.	Size: Adequate for meal prep and resident participation, considering furniture (5). Numbers: At least one kitchen, based on facility needs (1, 2, 8). The location, size, and furniture layout based on resident count (9). Storage: Sufficient for prep and food, with extra per facility needs (4, 8). Wheelchair users: Sink height adjustable for wheelchair access, promoting kitchen participation (5).
	Safety and security		
	Orientational cues	-	Location: Centrally located within units (5).
Legibility	Logical room syntax		
(B-3) Outdoor space			
Sensory stimulation	Access to the outdoors	-	Size: Adequate (2, 8, 9). Accessibility: Safe and unrestricted resident access (5), even for wheelchair users (2, 8, 9). Location: Proximity to central communal area (5). Protection: Design to ensure protection (2, 8, 11). Shading device: (5).

(1) HeimMindBauV (federal law); (2) BW; (3) Bay; (4) Berl; (5) Ham; (6) Hess; (7) MV; (8) NW; (9) RP; (10) Sach; (11) SH.

3. Therapy and hygiene spaces
    a. Physical occupational therapy room and ensuite bathroom
    In South Korea, there are no regulations related to physical therapy rooms and ensuite bedrooms in South Korea (Table 19I), with the latter lacking specific mandates for necessary rooms. German regulations address the autonomy criterion in the physical occupational therapy room by specifying numbers, sink usage, and shading. Furthermore, they provide various detailed regulations for ensuite bathroom/toilets, including specific guidelines related to autonomy, familiarity, and legibility criteria (Table 19II), such as numbers, accessibility, entrance space, entrance, door control, sanitary equipment, barrier-free, and home-like configuration.
    b. Common toilets and bathrooms
    South Korean regulations for common toilets and bathrooms include features such as anti-slip floors, wheelchair access, night lights, safety handles, bath tubs, and hot water, aiming to enhance the autonomy of patients with dementia. In contrast, German regulations include more detailed provisions such as numbers, door controls, visitor toilets, and mechanical baths, focusing on autonomy and personal privacy within the criterion of social interaction

Table 19. Comparison of the detailed regulations for therapy and hygiene spaces.

Regulation		I. South Korea	II. Germany
(C-1) Physical occupational therapy room			
Autonomy	Barrier-free and compensatory environment	-	Numbers: Minimum of one per facility (1, 7, 11), with sufficient size and location determined by resident count and specific needs (3, 4, 6, 7, 9). Sink: At least one sink (4, 6). Shading device: (4).
	Safety and security		

Table 19. *Cont.*

Regulation		I. South Korea	II. Germany
(D-1) Ensuite bathroom/toilets			
Autonomy	Barrier-free and compensatory environment Safety and security		Numbers: Required for 100% single-bedroom ratio (2).Accessibility: Direct connection from bedroom or adjacent room (3, 10, 7). Entrance space: Connecting two bedrooms and one ensuite between a common space and a private space (2). In front of the ensuite (2). Entrance: Avoid two entrances (5, 8). Exceptionally, one ensuite may serve two single rooms (11). Door control: Lockable door, emergency openable (1, 10, 4). Sanitary equipment: Double bedroom requires two sinks (1). At least one of each item: washbasin, shower, and toilet (2–11). Barrier-free: Safety handles in all sanitary facilities (3, 4, 6, 10). Anti-burn devices in washbasins, showers, and bathtubs (3, 6).
Familiarity	Noninstitutional character	-	Home-like configuration: Arrangement of bathroom, sanitary equipment, and openings to resemble those found at home (5).
Legibility	Logical room syntax		
(D-2) Common toilet			
Autonomy	Barrier-free and compensatory environment Safety and security	Anti-slip floor: Non-slip floors. Wheelchair access. Night lights.	Numbers: At least one per eight individuals on the same floor (1). Adequate number near common bathroom (4). Wheelchair user: Enough for wheelchair users (1). Safety handle: (4). Door control: Emergency door opening (1). Visitor toilet: Sufficient for visitors (4). At least one installation place (5), up to 40 people (8).
(D-3) Common bathroom			
Autonomy	Barrier-free and compensatory environment Safety and security	Anti-slip floor: Non-slip floors. Wheelchair access. Safety handle. Bathtub: Shallow with at least one grip rail and vertical handle lights Hot water: Temperature under 40 degrees Celsius with automatic thermostat use. Night lights.	Numbers: One facility per 20 residents (8), 30 or 32 people (4, 7), and 40 residents (3, 5, 7). One bathtub or shower per 20 people (1) and consider resident count (9). Wheelchair user: Available (6), sufficient quantities (1). Door control: Emergency door opening (1). Sanitary equipment: At least one tub, shower, and sink. Bathtub: Safe entry and exit (1, 4). Safety handle: In the bathtub, shower, and toilet (1, 4, 6). Mechanical bath: For medical use (6), tailored to individual needs (9).
Social interaction	Privacy		Privacy: Modesty protection (1,6) with curtains around tub and shower (4)

(1) HeimMindBauV (federal law); (2) BW; (3) Bay; (4) Berl; (5) Ham; (6) Hess; (7) MV; (8) NW; (9) RP; (10) Sach; (11) SH.

4. Functional auxiliary, household management, and staff duty spaces

Spaces that typically do not permit the entry and accessibility of patients with dementia, such as functional auxiliary, household management, and staff duty spaces, are crucial from the perspective of effective caregiving. These spaces generally do not reflect their correlation with design criteria for patients with dementia. In South Korea, detailed regulations exist only regarding the installation of medical supplies and equipment in medical and nursing staff rooms (Table 20I). Germany has slightly more detailed regulations for them (Table 20II), which describe the number, location, morgue, and storage.

Table 20. Comparison of detailed regulations for functional auxiliary, household management, and staff duty spaces.

Regulation	I. South Korea	II. Germany
(F) Functional auxiliary spaces		
Waste disposal room		Numbers: Installed on each floor (6).
Morgue		Morgue: Required if quick body transport is not ensured in double-bedroom facilities (1, 3, 6).
(G) Household management spaces		
Storage room		Numbers: At least one in-house management space (2). If unable to outsource external household services, provide adequate facilities internally (1, 4, 6, 7). Storage: Accommodate machinery, appliances, in-house items, wheelchairs, beds, and medical devices (9, 4). Include a warehouse for household goods, laundry, and medicines (4, 5).
Laundry room, drying room		
(H) Staff duty spaces		
Nursing station	Medical supplies and devices: Installation of necessary medical equipment and pharmaceutical hygiene materials.	Numbers: As many locations and sizes as necessary (2, 5, 6, 7, 10). Location: Essential for resident care (5, 7), positioned in a common area within the residential unit (5).
Staff offices		
Staff changing room, staff break room		Location: Place outside the residential unit (5).
Caregiver's room		
Volunteer room		

(1) HeimMindBauV (federal law); (2) BW; (3) Bay; (4) Berl; (5) Ham; (6) Hess; (7) MV; (9) RP; (10) Sach.

4.3.3. Summary

The analysis of the necessary rooms configurations reveals that in South Korea, the configuration is differentiated based on facility size, with basic and additional configurations applied depending on whether the facility accommodates 10–29 residents or 30 or more residents. The ensuite bathroom/toilets in bedrooms are not mandatory for both general and dementia units, which interferes with the autonomy and privacy of elderly individuals with dementia. There are no recommendations or mandatory regulations for outdoor spaces, which were maintained in the form of a sunroom in 1998 and then removed in 2008, weakening the sensory stimulation for residents. However, Germany has regulations for ensuite bathroom/toilets, outdoor spaces, waste disposal rooms, morgues, and storage rooms. In terms of the detailed regulations for each space, both South Korea and Germany emphasize specific design principles aimed at enhancing the autonomy of dementia patients. While there are some detailed regulations related to autonomy in individual, communal, and hygiene spaces in Korea, regulations for treatment and other areas of spaces are few or largely excluded. Germany primarily focuses on autonomy in the majority of its regulations and additionally includes provisions that enhance familiarity, sensory stimulation, legibility, and social interaction. The total characteristics observed in the detailed regulations for necessary rooms are summarized in Table 21.

Table 21. Design criteria inherent in the detailed regulations for necessary rooms of dementia nursing homes.

Design Criteria	I. Autonomy			II. Familiarity			III. Sensory Stimulation			IV. Legibility			V. Social Interaction		
	I-1	I-2	I-3	II-1	II-2	II-3	III-1	III-2	III-3	IV-1	IV-2	IV-3	V-1	V-2	V-3
	Barrier-Free and Compensatory Environment	Safety and Security	Orientational Cues	Biographical Reference	Small Groups	Noninstitutional Character	Encouragement	Avoidance of Overstimulation	Access to the Outdoors	Logical Room Syntax	Furnishing	Fixtures and Fittings	Privacy	Belonging	Communication
(A) Individual spaces															
Bedrooms	○ [1,2]	○ [1,2]	-	○ [2]	-	○ [2]	-	-	-	-	-	-	○ [2]	-	-
Special bedrooms	-	-	-	-	-	-	-	-	-	-	-	-	-	-	-
(B) Communal s.															
Common living rooms	○ [1,2]	○ [1,2]	○ [2]	○ [2]	-	○ [2]	-	-	-	○ [2]	-	-	-	○ [2]	-
Kitchens/dining rooms	○ [1,2]	○ [1,2]	○ [2]	-	-	-	-	-	-	○ [2]	-	-	-	-	-
Outdoor spaces	○ [2]	○ [2]	○ [2]	-	-	-	-	-	○ [2]	-	-	-	-	-	-
(C) Therapy s.															
Physical occupational therapy rooms	○ [2]	-	-	-	-	-	-	-	-	-	-	-	-	-	-
(D) Hygiene s.															
Ensuite bath-room/toilets	○ [1,2]	○ [1,2]	-	-	-	○ [2]	-	-	-	○ [2]	-	-	-	-	-
Common toilet	○ [1,2]	○ [1,2]	-	-	-	-	-	-	-	-	-	-	-	-	-
Common bathrooms	○ [1,2]	○ [1,2]	-	-	-	-	-	-	-	-	-	-	○ [2]	-	-

[1] South Korea, [2] Germany. ○: Detailed regulations for necessary rooms exist.

5. Conclusions

This study began by addressing the issue of rapid aging in South Korea that is turning the country into a super-aged society with a growing population of individuals with dementia. The focus was on the architectural and regulatory importance of dementia nursing homes as dementia-friendly built spaces by comparing the building standards for dementia nursing homes in South Korea and Germany. This study is a foundational research and aimed to analyze differences through a comparison of building standards for dementia nursing homes in South Korea and Germany, and proposing improvement points for South Korea by examining the current laws of both countries: the Welfare of Senior Citizens Act and its enforcement regulations in South Korea, as well as the Federal Nursing Home Act (HeimG), the Regulation on Minimum Standards for Nursing Homes (HeimMindBauV), and its state laws in Germany. The results of this study are summarized as follows.

First, Germany introduced the Federal Nursing Home Act in 1974 and its subsidiary legislation, the Regulation on Minimum Standards for Nursing Homes in 1978, along with the Long-Term Care Insurance System in 1995. Similarly, South Korea enacted the Welfare of Senior Citizens Act in 1981, along with its secondary legislation, the Enforcement Regulations of the Welfare of Senior Citizens Act in 1982, and additionally established the Long-Term Care Insurance System in 2008.

Second, the scope of application of the superior laws differed between South Korea and Germany. In South Korea, laws such as the Welfare of Senior Citizens Act vary in their application based on facility size, distinguishing between nursing homes (with 10 or more residents) and care group homes (with between 5 and 9 residents), both targeting elderly individuals and integrated into elderly medical welfare facilities. In Germany, the Federal Nursing Home Act and state laws apply not only to nursing homes but also to various forms of ambulatory shared care homes and assisted living facilities, catering not only to the elderly but also to adults in need of care and individuals with disabilities. Furthermore, inferior laws that specify building standards for dementia nursing homes differ in their structure. In South Korea, these standards are outlined in the Enforcement Regulations of the Welfare of Senior Citizens Act, a subordinate regulation of the Welfare of Senior Citizens Act. In Germany, these regulations are consolidated at the federal level in the Regulation on Minimum Standards for Nursing Homes, along with 10 corresponding regulations at the state level, following the federalism reform in 2006.

Third, the significance of planning dementia nursing homes arises from the need to incorporate design criteria related to autonomy, familiarity, sensory stimulation, legibility, and social interaction within dementia-friendly built spaces. Regarding the basic requirements for dementia nursing homes, current building standards in South Korea primarily focus on supporting autonomy (barrier-free and compensatory environments, safety, and security) and social interaction (communication with the local community), which have remained unchanged since 1998. However, items related to oriental cues, alarms, communication devices under the autonomy criterion, and design principles within other criteria aiding elderly individuals with dementia have not been discussed.

In contrast, Germany's state laws, since 2010, additionally emphasize their barrier-free standards and oriental cues, focusing on autonomy, as well as familiarity (biographical references, small groups, noninstitutional characteristics), sensory stimulation (outdoor accessibility), legibility (logical room syntax, furniture), and social interaction (privacy, sense of belonging) (Table 22I). To enhance South Korean standards, integrating the existing Barrier Free (BF) Living Environment Certificate into building standards is recommended. This voluntary participation system was introduced in 2007 and became mandatory for children, senior welfare, and disability welfare facilities in 2015. Furthermore, detailed regulations on autonomy criterion and new regulations addressing Korean residents' perception of a lack of "home-like" features in nursing homes were identified as areas for South Korea's further development for discussion, covering familiarity, sensory stimulation, legibility, and social interaction (Table 23I).

Table 22. Comprehensive comparative analysis of building standards for dementia nursing homes in South Korea and Germany.

Design Criteria	I. Autonomy			II. Familiarity			III. Sensory Stimulation			IV. Legibility			V. Social Interaction		
	Barrier-Free and Compensatory Environment	Safety and Security	Orientational Cues	Biographical Reference	Small Groups	Noninstitutional Character	Encouragement	Avoidance of Overstimulation	Access to the Outdoors	Logical Room Syntax	Furnishing	Fixtures and Fittings	Privacy	Belonging	Communication
(1) Basic requirements															
South Korea	○	○													○ [1]
Germany	●	●	○	○	○	○			○	○	○		○		
(2) Facility sizes															
South Korea	○	○			○	○							○		
Germany	●	●	○	○	●	●		○	○		○		○	○	○

Table 22. *Cont.*

Design Criteria	I. Autonomy			II. Familiarity			III. Sensory Stimulation			IV. Legibility			V. Social Interaction		
	Barrier-Free and Compensatory Environment	Safety and Security	Orientational Cues	Biographical Reference	Small Groups	Noninstitutional Character	Encouragement	Avoidance of Overstimulation	Access to the Outdoors	Logical Room Syntax	Furnishing	Fixtures and Fittings	Privacy	Belonging	Communication
(3) Necessary rooms															
South Korea													○		
Germany									○ [2]				○ [3]		
(4) Detailed regulations of necessary rooms															
South Korea	○	○													
Germany	●	●	○	○		○			○	○			○	○	

○ Regulations are in place, ● More stringent regulations; [1] Community communication, [2] Outdoor spaces, [3] Ensuite bathroom/toilets.

Table 23. Implications for building standards for dementia nursing homes in South Korea.

	Design Criteria					Implications
	I	II	III	IV	V	
(1) Basic requirements	○					- Specification of Korean Barrier Free (BF) Living Environment Certificate.
	○					- Description of natural lights (an external view from bed, sun control devices).
						 [90]  [91]
	○					- Description of artificial lighting (reading lights, night lights for nighttime care).
	○					- Description related to orientational cues.
						 [92]  [93]
	○					- Description of emergency alarm on beds and in various spaces, as well as communication devices such as radios, TVs, phones, internet access for residents.
	○				○	- Description of personal privacy, as well as aspects related to home environment and past personal furniture and items in bedrooms.

Table 23. *Cont.*

	Design Criteria					Implications
	I	II	III	IV	V	
(1) Basic requirements						 [94]  [95]  [96]  [94]
	○		○			- Description of secure outdoor spaces (safety, access, solar control devices).
						 [96]  [97]
			○			- Description of logical room syntax (spatial arrangement, orientation elements).
						 [98]  [99]
(2) Facility sizes	○	○		○	○	- Reducing the occupancy in bedrooms (currently a max. of four individuals).
	○	○			○	- Increasing the min. net floor area for bedrooms.
	○	○	○		○	- Introducing the occupancy for common living rooms in general units (currently 16 persons in dementia units)
	○	○	○		○	- Expanding the net area for common living rooms (currently min. 1.65 m^2 per person within dementia units).
	○	○			○	- Introducing the ratio of single bedrooms.
	○	○		○	○	- Introducing the min. room width for single bedrooms.

Table 23. *Cont.*

	Design Criteria					Implications	
	I	II	III	IV	V		
(3) Necessary rooms			○			- Organizing outdoor spaces for residents.	
						 [96]  [100]	
	○	○			○	- Installation of an ensuite bathroom.	
						 [91]  [94]	
						- Inclusion of functional auxiliary spaces such as a waste disposal room and a morgue.	
(4) Detailed regulations of necessary rooms	○	○			○	Bedrooms	- Reviews on size, accessibility, door control, outside view, shading device, furniture, spatial variability, etc.
	○	○		○	○	Common living room	- Reviews on size, accessibility, participation from bed, shading device, reading lights, location, privacy, living spaces, etc.
	○			○		Kitchen and dining room	- Reviews on size, wheelchairs users, location, etc.
	○		○	○		Outdoor spaces	- Reviews on size, accessibility, location, protection, shading device, etc.
	○					Physical occupational therapy room	- Reviews on numbers, sink, shading device, etc.
	○	○		○		Ensuite bathrooms	- Reviews on numbers, accessibility, entrance space, entrance, door control, sanitary equipment, home-like configuration, etc.
	○					Common toilets	- Reviews on numbers, wheelchair users, etc.
	○				○	Common bathrooms	- Reviews on numbers, wheelchair user, door control, mechanical bath, etc.

○: Improvement points are identified.

Fourth, parameters in both countries regarding facility size, such as capacity and minimum net floor area for bedrooms and common living rooms, highlight Germany's greater emphasis on design criteria related to autonomy, familiarity (biographical references, small groups), and social interaction (privacy, sense of belonging) (Table 22(2)). For instance, Germany's standards even include specific ratios and minimum widths for single bedrooms, reflecting this emphasis. Research in Korea has shown that elderly residents in nursing homes who use single rooms exhibit higher levels of self-respect, suggesting the importance of privacy and personal space [101]. However, meeting this demand remains challenging, especially with a maximum of four people per room currently. Previous studies in Korea have recommended adhering to single-room standards to achieve a more adequate bedroom area, which can support individual living space and prolonged bed rest or extended stays in the bedroom after admission [102–104]. Moreover, there has been a persistent demand in Korea to expand the area of common living rooms, as indicated by earlier research [105–108]. This demand aims to diversify living spaces outside of bedrooms, provide optional extensions of spaces, and integrate functions of open integrated living rooms combining living, dining, and program rooms for both personal and social interactions. Therefore, adjustments such as reducing the capacity per bedroom and increasing the minimum net floor area of bedrooms and common living rooms are necessary. South Korea could benefit from ongoing discussions on parameters such as capacity per bedroom and minimum net floor area, and consider introducing additional parameters like the ratio and minimum width of single bedrooms (Table 23(2)).

Fifth, in South Korea, ensuite bathrooms/toilets, common outdoor spaces, and functional auxiliary spaces such as waste disposal rooms, morgues, and storage rooms are not explicitly specified as required rooms (Table 22(3)). In contrast, Germany has regulations for these rooms to enhance the health and well-being of dementia patients and the efficiency of caregivers. Given the importance of ensuite bathrooms/toilets for autonomy and privacy [109], as well as outdoor spaces for sensory stimulation [89,110], South Korea should consider incorporating these features into dementia nursing homes (Table 23(3)). Discussions regarding these enhancements should address how ensuite bathrooms and outdoor spaces contribute to autonomy, individual privacy, social interactions, and sensory stimulation for elderly individuals with dementia. Thorough review of outdoor spaces, as well as kitchenette and dining areas in dementia units, which were temporarily introduced from 2016 to 2018, should be part of the discussion.

Sixth, both South Korea and Germany prioritize specific provisions related to the design criterion of autonomy for seniors with dementia in their detailed regulations for necessary rooms (Table 22(4)). In Korea, these regulations gradually decrease across personal, communal, and hygiene spaces, with few or no regulations for treatment, functional auxiliary, household management, and staff duty spaces. In contrast, Germany's building standards for these rooms go beyond autonomy to include provisions enhancing familiarity, sensory stimulation, legibility, and social interaction. Therefore, when revising South Korea's detailed regulations for necessary rooms, it is essential to conduct research and consider discussing and potentially incorporating these additional provisions in the future (Table 23(4)).

In summary, while South Korea's building standards for nursing homes catering to seniors with dementia primarily emphasize autonomy, particularly focusing on features such as barrier-free and compensatory environments, safety and security measures, Germany's building standards delve more specifically into characteristics related to alarms, communication devices, and support for orientational cues within the same criteria. Beyond autonomy, Germany's standards encompass numerous design criteria including familiarity, sensory stimulation, legibility, and social interaction. Consequently, for future enhancements to South Korea's building standards, conducting analyses that specifically focus on spatial usage within Korean nursing homes should be the next step in this foundational research.

6. Discussion

Given the limited research on spatial utilization in preceding studies, further exploration in this area is warranted. After conducting the analyses, it is crucial to engage in discussions and incorporate the findings to accurately address the current needs of residents in Korean dementia nursing homes while avoiding overestimating these requirements. Moreover, during the analysis of spatial usage and subsequent discussions and implementations, it is imperative to thoroughly review regulations pertaining to autonomy, familiarity, sensory stimulation, legibility, and social interaction.

In both Korea and Germany, autonomy emerged as the foremost design criterion in building standards, while regulations concerning familiarity, sensory stimulation, legibility, and social interaction were relatively scarce. Hence, both countries must undertake a comprehensive review and supplementation of detailed items within these design criteria for future development.

This analysis of the disparities in building standards between Korea and Germany provides valuable insights for enhancing South Korea's building standards for dementia nursing homes. By considering the potential impact of adopting German standards, this study lays the groundwork for establishing more dementia-friendly environments in Korean nursing homes. Furthermore, it serves as a valuable resource for future reviews of building standards in South Korea, aiming to provide pertinent data to relevant institutions and policymakers to enhance the built environments of dementia care facilities in the country.

While the study's strength lies in its focus on dementia-friendly built space design and cross-national comparison with Germany's building standards, there are some limitations. These include the insufficient consideration of spatial utilization characteristics in current Korean nursing homes, neglect of cultural–historical backgrounds, regional focus, and limited international perspectives due to data availability.

Therefore, conducting more comprehensive analyses on spatial usage within cultural and historical contexts, expanding the geographical scope, and increasing sample sizes in future studies would enhance the validity and generalizability of findings.

Funding: This work was supported by the Ministry of Education of the Republic of Korea and the National Research Foundation of Korea (NRF-2017S1A5A8020076).

Data Availability Statement: Data is contained within the article.

Conflicts of Interest: The authors declare no conflict of interest.

References

1. Ebert, T.; Eßig, N.; Hauser, G. *Green Building Certification Systems*; Institut für Internationale Architekturdokumentation: München, Germany, 2011.
2. National Institute of Dementia Center. *2016 Nationwide Survey on the Dementia Epidemiology of Korea (NIDR-1603-0015)*; Ministry of Health and Welfare: Seoul, Republic of Korea, 2016; p. 19. Available online: https://www.nid.or.kr/info/dataroom_view.aspx?bid=182 (accessed on 2 February 2024).
3. Department of Economic and Social Affairs, Population Division. *World Population Ageing 2017 (ST/ESA/SER.A/408)*; United Nations: New York, NY, USA, 2017. Available online: https://www.un.org/development/desa/pd/sites/www.un.org.development.desa.pd/files/files/documents/2020/May/un_2017_worldpopulationageing_report.pdf (accessed on 2 February 2024).
4. World Health Organization. *Global Action Plan on the Public Health Response to Dementia 2017–2025*; World Health Organization: Geneva, Switzerland, 2017. Available online: https://www.who.int/publications/i/item/global-action-plan-on-the-public-health-response-to-dementia-2017---2025 (accessed on 2 February 2024).
5. Alzheimer's Disease International. Numbers of People with Dementia Worldwide: An Update to the Estimates in the World Alzheimer Report 2015. 2020. Available online: https://www.alzint.org/resource/numbers-of-people-with-dementia-worldwide/ (accessed on 2 February 2024).
6. Zeisel, J.; Bennett, K.; Fleming, R. *World Alzheimer Report 2020: Design, Dignity, Dementia: Dementia-Related Design and the Built Environment*; Alzheimer's Disease International: London, UK, 2020; Available online: https://www.alzint.org/u/WorldAlzheimerReport2020Vol1.pdf (accessed on 2 February 2024).
7. OECD. *Health at a Glance 2021: OECD Indicators*; OECD Publishing: Paris, France, 2021. [CrossRef]
8. Department of Economic and Social Affairs. *World Social Report 2023: Leaving No One Behind in an Ageing World*; United Nations: New York, NY, USA, 2023. Available online: https://www.un.org/development/desa/dspd/wp-content/uploads/sites/22/2023/01/2023-wsr-tablecontents.pdf (accessed on 2 February 2024).
9. Statistics Korea. Statistics on the Aged. 2017. Available online: https://kostat.go.kr/board.es?mid=a10301060500&bid=10820&act=view&list_no=363362 (accessed on 2 February 2024).
10. Statistics Korea. Statistics on the Aged. 2018. Available online: https://kostat.go.kr/board.es?mid=a10301060500&bid=10820&act=view&list_no=370779 (accessed on 2 February 2024).
11. Statistics Korea. Statistics on the Aged. 2019. Available online: https://kostat.go.kr/board.es?mid=a10301060500&bid=10820&act=view&list_no=377701 (accessed on 2 February 2024).
12. Statistics Korea. Statistics on the Aged. 2020. Available online: https://kostat.go.kr/board.es?mid=a10301060500&bid=10820&act=view&list_no=385322 (accessed on 2 February 2024).

13. National Institute of Dementia Center. *Annual Report 2017*; Ministry of Health and Welfare: Seoul, Republic of Korea, 2018. Available online: https://www.nid.or.kr/notification/notice_view.aspx?board_seq=1337 (accessed on 2 February 2024).
14. National Institute of Dementia Center. *Annual Report 2018*; Ministry of Health and Welfare: Seoul, Republic of Korea, 2018. Available online: https://www.nid.or.kr/info/dataroom_view.aspx?bid=183 (accessed on 2 February 2024).
15. National Institute of Dementia Center. *Korea Dementia Observatory 2020 (NIDR-2002-0031)*; Ministry of Health and Welfare: Seoul, Republic of Korea, 2020. Available online: https://www.nid.or.kr/info/dataroom_view.aspx?bid=221 (accessed on 2 February 2024).
16. Kuliga, S.; Berwig, M.; Roes, M. Wayfinding in people with Alzheimer's disease: Perspective taking and architectural cognition—A vision paper on future dementia care research opportunities. *Sustainability* **2021**, *13*, 1084.
17. Marquardt, G. *MATI: Mensch-Architektur-Technik-Interaktion für Demografische Nachhaltigheit*; Fraunhofer IRB Verlag: Stuttgart, Germany, 2015.
18. United Nations. UN Population Division Data Portal. Available online: https://population.un.org/dataportal/home?df=9b8e3b44-ac1a-4677-9dc4-51ed06e03943https://population.un.org/dataportal/home?df=60c689d5-2de8-4029-8649-b41708b74958 (accessed on 2 February 2024).
19. Statistisches Bundesamt (Destatis). Bevölkerung in Deutschland. Available online: https://service.destatis.de/bevoelkerungspyramide/index.html#!y=1951&a=20,65&v=2&g (accessed on 2 February 2024).
20. Kwon, S.J. A proposal for the improvement of facility regulation in the Act for Welfare of the Elderly. *J. Korean Inst. Healthc. Archit.* **2000**, *6*, 130–136. Available online: https://www.riss.kr/link?id=A3011451 (accessed on 2 February 2024).
21. Lee, S.O.; Jeon, C.H. Residential environment of nursing homes: Busan suburbs (Busan, Gimhae, Yangsan)—A case study. *J. Reg. Assoc. Archit. Inst. Korea* **2000**, *2*, 27–34. Available online: https://www.earticle.net/Article/A33747 (accessed on 2 February 2024).
22. Kim, K.M.; Jung, T.W.; Kim, H.H. A study on the design methods of the dementia facilities. *J. Korean Inst. Healthc. Archit.* **2000**, *6*, 97–106. Available online: https://www.riss.kr/link?id=A3011435 (accessed on 2 February 2024).
23. Korea Health Industry Development Institute. Development of the Guidelines for the Design of the Long-Term Care Facilities in Korea. 2003. Available online: http://lib.nhis.or.kr/search/detail/CATNHZ000000009973 (accessed on 2 February 2024).
24. Korea Health Industry Development Institute. International Comparison of Operating Standards and operating Status Of Nursing Facilities for the Elderly. 2007. Available online: https://www.khidi.or.kr/board/view?pageNum=50&rowCnt=10&no1=288&linkId=100434&menuId=MENU00085&maxIndex=00001005649998&minIndex=00001004329998&schType=0&schText=&schStartDate=&schEndDate=&boardStyle=&categoryId=&continent=&country= (accessed on 2 February 2024).
25. Chin, Y.R.; Lee, H.Y. Comparing standards and guidelines of long-term care facilities based on physical environment and manpower in Korea, Japan, USA, and Australia. *Health Policy Manag.* **2012**, *22*, 403–426. [CrossRef]
26. Statista. Deutschland: Anteil der Bevölkerung ab 65 Jahre an der Gesamtbevölkerung von 1950 bis 2049. Available online: https://de.statista.com/statistik/daten/studie/14135/umfrage/deutschland-anteil-senioren-an-der-bevoelkerung/ (accessed on 2 February 2024).
27. National Geography Information Institute (NGII). The National Atlas of Korea III 2021, Proportion of Expenditure on Social Welfare to GDP by Countries 2019. Available online: http://nationalatlas.ngii.go.kr/us/index.php (accessed on 2 February 2024).
28. Statista. Age Distribution of Inhabitants of G7 Countries in 2022, by Country. Available online: https://www.statista.com/statistics/1372598/g7-country-age-distribution/ (accessed on 2 February 2024).
29. Statista. Länder mit den Meisten Demenzfällen Weltweit im Jahr 2019. Available online: https://de.statista.com/statistik/daten/studie/1330371/umfrage/laender-mit-den-meisten-faellen-von-demenz-weltweit/ (accessed on 2 February 2024).
30. Statista. Länder mit den Meisten Prognostizieren Demenzfällen Weltweit im Jahr 2050. Available online: https://de.statista.com/statistik/daten/studie/1330423/umfrage/laender-mit-den-meisten-prognostizierten-faellen-von-demenz-weltweit/ (accessed on 2 February 2024).
31. OECD. Elderly Population (Indicator). Available online: https://www.oecd-ilibrary.org/social-issues-migration-health/elderly-population/indicator/english_8d805ea1-en (accessed on 2 February 2024).
32. Choi, H.W. An international comparative study on the installation and operation standards of nursing homes for the elderly with dementia. *J. Korean Hous. Assoc.* **2017**, *29*, 175. Available online: http://www.dbpia.co.kr/journal/articleDetail?nodeId=NODE07159072. (accessed on 2 February 2024).
33. Seong, S.J.; Kim, B.N.; Kim, K.W. Comparative Analysis on the Policy Approaches in the National Dementia Plans of G7 Countries and Korea and Their Implementation. *J. Korean Med. Sci.* **2023**, *38*, 1–13. [CrossRef]
34. Ministry of Health and Welfare, South Korea. Welfare of Senior Citizens Act. Available online: https://elaw.klri.re.kr/eng_mobile/viewer.do?hseq=59915&type=sogan&key=10 (accessed on 2 February 2024).
35. Ministry of Health and Welfare, South Korea. The Enforcement Regulations of Welfare of Senior Citizens Act. Available online: https://www.lawnb.com/Info/ContentView?sid=L000006940 (accessed on 2 February 2024).
36. Bundesministerium der Justiz und für Verbraucherschutz. Heimgesetz. Available online: https://www.gesetze-im-internet.de/heimg/ (accessed on 2 February 2024).
37. Bundesministerium der Justiz und für Verbraucherschutz. HeimMindBauV: Verordnung Über Bauliche Mindestanforderungen Für Altenheime, Altenwohnungen und Pflegeheime für Volljährige. Available online: https://www.gesetze-im-internet.de/heimmindbauv/ (accessed on 2 February 2024).

38. World Health Organization. International Statistical Classification of Diseases and Related Health Problems (10th version). Available online: https://icd.who.int/browse10/2016/en (accessed on 2 February 2024).
39. Korean Dementia Association. *Dementia A Clinical Approach*; Academy: Anyang, Republic of Korea, 2011.
40. Calkins, M.P. *Design for Dementia: Planning Environments for the Elderly and the Confused*; National Health Publishing: Owings Mills, MD, USA, 1988.
41. Cohen, U.; Weisman, G.D. *Holding On to Home: Designing Environments for People with Dementia*; Johns Hopkins University Press: Baltimore, MD, USA, 1991.
42. Brawley, E.C. *Designing for Alzheimer's Disease: Strategies for Creating Better Care Environments*; Wiley: New York, NY, USA, 1997; Volume 1.
43. Judd, S.; Marshall, M.; Phippen, P. *Design for Dementia*; Hawker Publications: London, UK, 1998.
44. Brawley, E.C. *Design Innovations for Aging and Alzheimer's: Creating Caring Environments*; Wiley: New York, NY, USA, 2006.
45. Chalfont, G.E. Connection to Nature at the Building Edge: Towards a Therapeutic Architecture for Dementia Care Environments. Ph.D. Thesis, University of Sheffield, Sheffield, UK, 2006.
46. Marquardt, G. *Kriterienkatalog Demenzfreundliche Architektur*; Logos: Berlin, Germany, 2007.
47. Marquardt, G.; Schmieg, P. Dementia-friendly architecture: Environments that facilitate wayfinding in nursing homes. *Am. J. Alzheimers Dis. Other Demen.* **2009**, *24*, 333–340.
48. Dietz, B. Demenzsensible Architektur. *Heilberufe* **2023**, *75*, 19–21. [CrossRef]
49. National Health Insurance Service. *Long-Term Care Insurance Statistical Yearbook 2019*; National Health Insurance Service: Seoul, Republic of Korea, 2020. Available online: https://www.nhis.or.kr/nhis/together/wbhaec07200m01.do?mode=view&articleNo=138244&article.offset=0&articleLimit=10 (accessed on 2 February 2024).
50. Lee, J.S.; Park, J.; Park, S. Factors related to institutionalization in elderly with dementia: Analysis of nationwide population-based data. *J. Korean Geriatr. Psychiatry* **2022**, *26*, 70–75. [CrossRef]
51. Heinzelmann, M. *Das Altenheim-Immer Noch Eine "Totale Institution"? Eine Untersuchung des Binnenlebens Zweier Altenheime*; Cuvillier Verlag: Göttingen, Germany, 2004.
52. Na, Y.S. A study on the long-term care insurance system for the elderly with focus on Germany, Japan, and Korea. *Rev. of Eurasian Stud.* **2011**, *8*, 255–281. Available online: https://www.riss.kr/link?id=A103204068 (accessed on 2 February 2024).
53. Statistisches Bundesamt (Destatis). Pflegestatistik 2019: Pflege im Rahmen der Pflegeversicherung Deutschlandergebnisse. Available online: https://www.destatis.de/DE/Themen/Gesellschaft-Umwelt/Gesundheit/Pflege/Publikationen/Downloads-Pflege/pflege-deutschlandergebnisse-5224001199004.pdf?__blob=publicationFile (accessed on 2 February 2024).
54. Statistisches Bundesamt (Destatis). Pflegeheime und Ambulante Pflegedienste. 2019. Available online: https://www.destatis.de/DE/Themen/Gesellschaft-Umwelt/Gesundheit/Pflege/Tabellen/pflegeeinrichtungen-deutschland.html (accessed on 2 February 2024).
55. Statista. Anteil Demenzkranker in Heimen, die in Heimen Leben in Deutschland, Österreich und der Schweiz. 2008. Available online: https://de.statista.com/statistik/daten/studie/180629/umfrage/anteil-demenzkranker-in-heimen-2008 (accessed on 2 February 2024).
56. Schäufele, M.; Köhler, L.; Hendlmeier, I.; Höll, A.; Weyerer, S. Prävalenz von Demenzen und ärztliche Versorgung in deutschen Pflegeheimen: Eine bundesweite repräsentative Studie. *Psychiatr. Prax.* **2013**, *40*, 200–206. [CrossRef]
57. Jakob, A.; Busse, A.; Riedel-Heller, S.G.; Pavlicek, M.; Angermeyer, M.C. Prävalenz und Inzidenz von Demenzerkrankungen in Alten-und Altenpflegeheimen im Vergleich mit Privathaushalten. *Z. Gerontol. Geriatr.* **2002**, *35*, 474–481. [CrossRef]
58. Ministry of Health and Welfare, South Korea. Dementia Management Act. Available online: https://elaw.klri.re.kr/eng_mobile/viewer.do?hseq=55099&type=sogan&key=10 (accessed on 2 February 2024).
59. Ministry of Health and Welfare, South Korea. The Enforcement Decree Of The Long-term Care Insurance Act. Available online: https://elaw.klri.re.kr/eng_service/lawView.do?hseq=63236&lang=ENG (accessed on 2 February 2024).
60. Ministerium für Soziales, Gesundheit und Integration Baden-Württemberg WTPG: Wohn, Teilhabe- und Pflegegesetz. Available online: https://www.landesrecht-bw.de/bsbw/document/jlr-WohnteilhGBWrahmen (accessed on 2 February 2024).
61. Bayerisches Staatsministerium für Arbeit und Sozialordnung, Familie und Frauen PfleWoqG: Pflege-und Wohnqualitätsgesetz. Available online: https://www.gesetze-bayern.de/Content/Document/BayPfleWoqG (accessed on 2 February 2024).
62. Berliner Senatsverwaltung für Integration, Arbeit und Soziales WTG: Wohnteilhabegesetz. Available online: https://www.berlin.de/sen/soziales/service/berliner-sozialrecht/kategorie/rechtsvorschriften/wtg_pflege-573411.php (accessed on 2 February 2024).
63. Behörde für Soziales, Familie, Gesundheit und Verbraucherschutz Hamburg HmbWBG: Hamburgisches Wohn und Betreuungsqualitätsgesetz. Available online: https://www.biva.de/dokumente/gesetze/HH-Hamburgisches-Wohn-und-Betreuungsqualit%C3%A4tsgesetz-HmbWBG.pdf (accessed on 2 February 2024).
64. Hessisches Ministerium für Arbeit, Familie und Gesundheit HGBP: Hessisches Gesetz über Betreuungs und Pflegeleistungen. Available online: https://hlfgp.hessen.de/sites/hlfgp.hessen.de/files/2022-12/hgbp_bf_das_gesetz.pdf (accessed on 2 February 2024).
65. Ministerium für Soziales und Gesundheit Mecklenburg-Vorpommern EQG: Einrichtungenqualitätsgesetz. Available online: https://www.landesrecht-mv.de/bsmv/document/jlr-EinrQualGMVrahmen (accessed on 2 February 2024).

66. Ministerium für Arbeit, Gesundheit und Soziales des Landes Nordrhein-Westfalen WTG NW: Wohn und Teilhabegesetz. Available online: https://recht.nrw.de/lmi/owa/br_text_anzeigen?v_id=10000000000000000678 (accessed on 2 February 2024).
67. Ministerium für Arbeit, Soziales, Gesundheit, Familie und Frauen Rheinland-Pfalz LWTG: Landesgesetz über Wohnformen und Teilhabe. Available online: https://www.mainz.de/medien/internet/downloads/dezernate/Rheinland-Pfalz_LWTG_02.pdf (accessed on 2 February 2024).
68. Sächsisches Staatsministerium für Soziales und Verbraucherschutz SächsBeWoG: Sächsisches Betreuungs und Wohnqualitätsgesetz. Available online: https://www.revosax.sachsen.de/vorschrift/12537-Saechsisches-Betreuungs-und-Wohnqualitaetsgesetz (accessed on 2 February 2024).
69. Ministerium für Arbeit, Soziales und Gesundheit Schleswig-Hollstein SbStG: Selbstbestimmungsstärkungsgesetz. Available online: https://www.gesetze-rechtsprechung.sh.juris.de/bssh/document/jlr-SbStGSH2009V2IVZ (accessed on 2 February 2024).
70. Ministerium für Arbeit, Soziales, Frauen und Familie Brandenburg BbgPBWoG: Brandenburgisches Pflege und Betreuungswohngesetz. Available online: https://bravors.brandenburg.de/de/gesetze-212541 (accessed on 2 February 2024).
71. Die Senatorin für Arbeit, Frauen, Gesundheit, Jugend und Soziales Bremen BremWoBeG: Bremisches Wohn und Betreuungsgesetz. Available online: https://www.biva.de/dokumente/gesetze/HB-Bremisches-Wohn-und-Betreuungsgesetz-BremWoBeG.pdf (accessed on 2 February 2024).
72. Niedersächsisches Ministerium für Soziales, Frauen, Familie, Gesundheit und Integration NHeimG: Niedersächsisches Heimgesetz. Available online: https://www.biva.de/dokumente/gesetze/NI_Nieders%C3%A4chsisches-Gesetz-ueber-unterstuetzende-Wohnformen_NuWG.pdf (accessed on 2 February 2024).
73. Ministerium für Arbeit, Familie, Prävention, Soziales und Sport Saarland LHeimGS: Saarländisches Wohn, Betreuungs-und Pflegequalitätsgesetz. Available online: https://www.biva.de/dokumente/gesetze/SL-Landesheimgesetz-Saarland-LHeimGS.pdf (accessed on 2 February 2024).
74. Ministerium für Gesundheit und Soziales Sachsen-Anhalt WTG-LSA: Wohn und Teilhabegesetz. Available online: https://www.landesrecht.sachsen-anhalt.de/bsst/document/jlr-WohnteilhGSTrahmen (accessed on 2 February 2024).
75. Ministerium für Soziales, Familie und Gesundheit Thüringen ThürWTG: Thüringer Wohn und Teilhabegesetz. Available online: https://landesrecht.thueringen.de/bsth/document/jlr-WohnteilhGTHrahmen (accessed on 2 February 2024).
76. Ministerium für Arbeit und Sozialordnung, Familien und Senioren Baden-Württemberg LHeimBauVO:Verordnung des Sozialministeriums zur baulichen Gestaltung von Heimen und zur Verbesserung der Wohnqualität in den Heimen Baden-Württembergs. Available online: https://www.landesrecht-bw.de/bsbw/document/jlr-HeimBauVBW2011rahmen/part/R (accessed on 2 February 2024).
77. Bayerisches Staatsministerium für Arbeit und Sozialordnung, Familie und Frauen AVPfleWoqG: Verordnung zur Ausführung des Pflege und Wohnqualitätsgesetzes. Available online: https://www.gesetze-bayern.de/Content/Document/BayAVPfleWoqG (accessed on 2 February 2024).
78. Berliner Senatsverwaltung für Integration, Arbeit und Soziales WTG-BauV: Wohnteilhabe-Bauverordnung. Available online: https://www.berlin.de/sen/soziales/service/berliner-sozialrecht/kategorie/rechtsvorschriften/wtg_bauv-573419.php (accessed on 2 February 2024).
79. Behörde für Soziales, Familie, Gesundheit und Verbraucherschutz Hamburg WBBauVO: Wohn und Betreuungsbauverordnung. Available online: https://www.luewu.de/docs/gvbl/2012/08.pdf (accessed on 2 February 2024).
80. Hessisches Ministerium für Arbeit, Familie und Gesundheit HGBPAV: Ausführungsverordnung zum Hessischen Gesetz über Betruungs- und Pflegeleistungen. Available online: https://www.rv.hessenrecht.hessen.de/bshe/document/jlr-BetrPflGAVHEV1IVZ (accessed on 2 February 2024).
81. Ministerium für Soziales und Gesundheit Mecklenbug-Vorpommern EMindBauVO M-V: Einrichtungenmidestbauverordnung. Available online: https://www.landesrecht-mv.de/bsmv/document/jlr-EinrMindBauVMVpP10 (accessed on 2 February 2024).
82. Ministerium für Arbeit, Gesundheit und Soziales des Landes Nordrhein-Westfalen WTG DVO:Wohn und Teilhabegesetz-Durchführungsverordnung. Available online: https://recht.nrw.de/lmi/owa/br_text_anzeigen?v_id=10000000000000000512 (accessed on 2 February 2024).
83. Ministerium für Arbeit, Soziales, Gesundheit, Familie und Frauen Rheinland-Pfalz LWTGDVO: Verordnung zur Durchfüehrung des Landesgesetzes über Wohnformen und Teilhabe. Available online: https://msagd.rlp.de/fileadmin/msagd/Soziale_Teilhabe/Teilhabe_Dokumente/LWTGDVO.pdf (accessed on 2 February 2024).
84. Sächsisches Staatsministerium für Soziales und Verbraucherschutz SaechBeWoGDVO: Verordnung des Saechsischen Staatsministeriums für Soziales und Verbraucherschutz zur Durchfüehrung des Sächsischen Betreuungs und Wohnqualitätsgesetzes. Available online: https://www.revosax.sachsen.de/vorschrift/14128-SaechsBeWoGDVO (accessed on 2 February 2024).
85. Ministerium für Arbeit, Soziales und Gesundheit Schleswig-Hollstein SbStG-DVO: SbStG-Durchführungsverordnung. Available online: https://www.gesetze-rechtsprechung.sh.juris.de/bssh/document/jlr-SbStGDVSHrahmen (accessed on 2 February 2024).
86. Bayerisches Staatsministerium des Innern, für Bau und Verkehr, DIN 18040-1 und DIN 18040-2—Planungsgrundlagen des barrierefreien Bauens. Available online: https://www.stmb.bayern.de/assets/stmi/buw/baurechtundtechnik/planungsgrundlagen_barrierefreies_bauen.pdf (accessed on 2 February 2024).
87. Dickmann, F. Heimmindestbauverordnung. 2024. Available online: https://www.socialnet.de/lexikon/Heimmindestbauverordnung (accessed on 2 February 2024).

88. Lee, E.H. Environmental Factors for Making Residents Feel 'at home' in Long-Term Care Facilities in South Korea. *Soc. Sci. Res. Rev.* **2019**, *35*, 131–159. [CrossRef]
89. Jeon, S.M.; Kwon, S.J. A Basic Study on the Architectural Planning of Outdoor Space for the Elderly with Dementia. *J. Korea Inst. Healthc. Archit.* **2007**, *13*, 61–71. Available online: https://www.riss.kr/link?id=A102048806 (accessed on 2 February 2024).
90. Pflegeheim Hospital zum Heilligen Geist in Wangen. Available online: https://www.german-architects.com/de/projects/view/pflegeheim-hospital-zhl-geist (accessed on 2 February 2024).
91. Jakob-Sigle-Heim in Kornwestheim. 2017. Available online: https://www.baunetzwissen.de/daemmstoffe/objekte/wohnen/jakob-sigle-heim-in-kornwestheim-5338842/gallery-1/7 (accessed on 2 February 2024).
92. Fürstlich Fürstenbergischen Pflegeheim in Hüfingen. Available online: https://www.dbz.de/artikel/dbz_Mit_Licht_und_Farbe_planen_Barrierefreie_Farbgestaltung_fuer_Demenzkranke-1383039.html (accessed on 2 February 2024).
93. Senioren-zentrum in Schorndorf. Available online: https://ifgroup.org/de/projekt/736/pflegeheim-schorndorf/ (accessed on 2 February 2024).
94. Heeg, S.; Bäuerle, K. *Heimat für Menschen mit Demenz*; Mabuse Verlag: Frankfurt am Main, Germany, 2008; Gradman Haus in Stuttgart. 136.
95. Fedderson, E.; Lüdtke, I. *Raum Verloren: Architektur and Demenz*; Birkhäuser: Basel, Switzerland, 2014; Kompetenzzentrum für Menschen mit Demenz in Nürnberg. 130.
96. Kompetenzzentrum für Menschen mit Demenz in Nürnberg. Available online: https://www.dbz.de/artikel/bildersuche-kompetenzzentrum-fuer-menschen-mit-demenz-nuernberg-1383135.html (accessed on 2 February 2024).
97. Seniorenzentrum Heideweg in Magdeburg (2002). Available online: https://www.ak-lsa.de/objekt/seniorenzentrum-heideweg/ (accessed on 2 February 2024).
98. Seniorenpflegeheim Stavnagerstrasse in Berlin. Available online: https://www.feddersen-architekten.de/portfolio/seniorenpflegeheim-stavangerstrasse-berlin/ (accessed on 2 February 2024).
99. Betreutes Wohnen in Baiersbronn. Available online: https://www.db-bauzeitung.de/architektur/gesundheit/raeume-zum-atmen/ (accessed on 2 February 2024).
100. DRK Pflegeheim in Stuttgart. Available online: https://www.german-architects.com/en/projects/view/drk-pflegeheim#image-6 (accessed on 2 February 2024).
101. Chung, S.D. Predictors of Self-Esteem Among Older Adults in Nursing Home. *J. Korean Gerontol. Soc.* **2004**, *24*, 107–122. Available online: https://www.riss.kr/link?id=A100124297 (accessed on 2 February 2024).
102. Yang, K.S. Study on the Spatial Use Patterns of the Elderly in Skilled Nursing Facilities. *J. Korean Inst. Rural Archit.* **2004**, *6*, 21–31. Available online: https://www.riss.kr/link?id=S40000833 (accessed on 2 February 2024).
103. Yang, K.S. A Study on the Utilization of Living Space in Geriatric Hospital and Nursing Home. *J. Korean Inst. Rural Archit.* **2007**, *9*, 103–110. Available online: https://www.riss.kr/link?id=A104092085 (accessed on 2 February 2024).
104. Chu, Y.C.; Lee, D.S.; Youn, C.Y. A Study on the spatial configuration & Use Characteristics of the Elderly Nursing Institutions. *J. Korean Inst. Rural Archit.* **2010**, *12*, 57–66. Available online: https://www.riss.kr/link?id=A82403421 (accessed on 2 February 2024).
105. Lee, M.A. A Study of Activiy Room Planning in Skilled Nursing Facilities for the Elderly through the Analysis of Spatial Characteristics and Using Behavior. *Korean Inst. Inter. Design J.* **2005**, *14*, 148–156. Available online: https://www.riss.kr/link?id=A99632984 (accessed on 2 February 2024).
106. Lee, M.A. The Analysis of the Elements for the Lively Use of Activity Areas in the Elderly Nursing Facilities. *Korean J. Hum. Ecol.* **2009**, *18*, 971–984. Available online: https://www.riss.kr/link?id=A100482622 (accessed on 2 February 2024). [CrossRef]
107. An, B.Y.; Cho, J.Y.; Yang, N.W. A Study on the Residents' Behaviors in Living Area of Special Nursing Homes for the Elderly. *J. Korea Inst. Healthc. Archit.* **2008**, *14*, 31–38. Available online: https://www.riss.kr/link?id=A75071548 (accessed on 2 February 2024).
108. Yang, K.S. A Case Study on the Living Activities and the Use of Spaces of the Recuperating Elderly in Nursing homes at Small City. *J. Korean Inst. Rural Archit.* **2010**, *12*, 109–116. Available online: https://www.riss.kr/link?id=A82344173 (accessed on 2 February 2024).
109. Hwang, E.K.; Jang, H.W. An Analysis on the Convenient Facilities and Utilization in the Elderly Nursing Institutions. *J. Archit. Inst. Korea* **2009**, *25*, 39–46. Available online: https://www.riss.kr/link?id=A77024637 (accessed on 2 February 2024).
110. Lee, J.H.; Lee, J.S. A Study on the Outdoor Spaces Introduced in Paid Retirement Homes. *Soc. People Plants Environ.* **2004**, *7*, 18–27. Available online: https://www.riss.kr/link?id=A100388873 (accessed on 2 February 2024).

buildings

MDPI

Article

Research on the Design Strategies of Public Outdoor Spaces in Government Centers from a Health Perspective

Yilin Wang [1], Wenjun Peng [1], Yanyan Huang [1,*], Tiancheng Yang [2], Xinyu Du [3], Xinyan Wang [2], Zhengming Zhang [4] and Liang Zhu [5]

1 School of Civil Engineering Architecture and Environment, Hubei University of Technology, Wuhan 430068, China; wyl_678@126.com (Y.W.); wenjun_p@hbut.edu.cn (W.P.)
2 China Railway 18th Bureau Group Building and Installation Engineering Co., Ltd., Tianjin 300072, China; z1581454352@163.com (T.Y.); jagskjb@163.com (X.W.)
3 College of Landscape Architecture, Sichuan Agricultural University, Chengdu 611130, China; duxinyu920303@126.com
4 Ausspec Landscape Planning and Design Co., Ltd., Beijing 100089, China; zyj1099113033@126.com
5 CITIC General Institute of Architectural Design & Research Co., Ltd., Wuhan 430000, China; zhuliang@citic.com
* Correspondence: 19960014@hbut.edu.cn

Abstract: This study delves into an investigation of urban public outdoor spaces (POSs) from a health-oriented perspective, recognizing varied health needs encompassing physical, psychological, social, and environmental aspects. In this study, POSs of two typical government service centers (GSCs) were analyzed based on their structure, user demographics, and user satisfaction, revealing several problems with the current GSC POS designs. To address these problems, principles for GSC POS design were proposed for natural, playing, and social spaces and applied to redesign the Wuchang GSC. Firstly, through on-site surveys, questionnaire surveys, and data analysis, the existing problems in promoting residents' health in the GSC POSs were revealed, such as the insufficient greening of natural spaces, lack of interest in playing spaces, and unreasonable design of the scale of social spaces. Based on the above analysis, a health design optimization principle based on Maslow's theory is proposed. Firstly, improvement solutions were proposed and implemented for green spaces, such as using more diversified natural elements, zone differentiation, and landscape improvements to promote the health of users. Secondly, the leisure and sports needs of different age groups can be met simultaneously by diversifying the layout and functional settings of playing spaces. Finally, public spaces suitable for social interaction were redesigned to promote the psychological health of citizens in social activities by optimizing the scale of communication spaces. The proposed design optimization strategies for GSCs not only provide theoretical support for the healthy design of POSs but also provide useful references for the healthy development of urban public spaces.

Keywords: health-oriented design; government service center; open space design; design strategy

Citation: Wang, Y.; Peng, W.; Huang, Y.; Yang, T.; Du, X.; Wang, X.; Zhang, Z.; Zhu, L. Research on the Design Strategies of Public Outdoor Spaces in Government Centers from a Health Perspective. *Buildings* **2024**, *14*, 1505. https://doi.org/10.3390/buildings14061505

Academic Editor: Adrian Pitts

Received: 10 April 2024
Revised: 16 May 2024
Accepted: 21 May 2024
Published: 23 May 2024

1. Introduction

The concept of public open spaces (POSs), introduced in the sociological and philosophical literature in the 1950s and subsequently integrated into urban planning and design in the 1960s, has evolved to contain a variety of social attributes and has become increasingly focused on social function and improving the quality of life of its users [1].

Residential buildings stand at the core of contemporary urban advancements according to increasing standards and requirements for healthy cities [2–4]. The publication of the Athens Charter in 1939 sparked extensive global deliberations on housing and health, culminating in the formulation of a 15-point guideline for a healthy housing concept by the World Health Organization in the 1990s. The health-centric urban concept revolves around the implementation of healthy structures, embodies a people-centric philosophy, and strives to meet the health requirements (encompassing physiological, psychological, social, and environmental aspects) of inhabitants [5,6]. Following the declaration of the Healthy China Initiative in the report to the Nineteenth National Congress of the Communist Party of China in 2017, the health-oriented design of POSs has received increased attention.

In the current process of urbanization, the design quality of POSs, as important places for citizens to live, directly affects the physical and mental health of citizens. However, the current design of POSs gives little consideration to health factors, neglecting the potential use of public spaces as a major tool for promoting and improving the population's health [7].

In order to make up for these shortcomings, in recent years, numerous studies have been conducted on this subject, concentrating on the functional diversity and spatial planning that caters to the needs of individuals across different age groups, genders, and abilities [8–10]. These studies and practices aim to create a healthier and more humane spatial environment through specific design techniques.

Previous research has advocated for urban POS design to emphasize the creation of varied spaces for local residents, introducing the concept of urban POS resilience design to adapt to evolving social norms, psychological requirements, and travel patterns [11–14]. Before proposing an optimization design, determining the factors that affect the health status of residents through data analysis is necessary. Hu et al. [15] and Li et al. [16] collected data on the built environment and epidemic trends through a generalized linear regression analysis, aiming to understand the key built environment factors that affect user health. Specifically, conducting sufficient and specific questionnaire surveys can provide effective and authentic data. For example, in the past few years, many studies have analyzed the impact of buildings and outdoor environments on COVID-19 through questionnaires to residents [17–19].

After identifying the factors that need to be optimized, there have been some widely adopted design methods in urban public spaces: some studies [20,21] have shown that increasing the green vegetation, water features, etc. in public spaces can provide opportunities to interact with nature, promoting user physical and mental health. Public spaces should meet the needs of different groups of people and set up diverse functional areas [22,23]. Building upon the aforementioned design principles and methodologies, Tao et al. [24] proposed a wellbeing-centric life cycle assessment framework to support decision-making and improve energy efficiency in relation to human well-being and socioeconomic considerations. Additionally, Yan et al. [25] and Reinhart et al. [26] also presented models for evaluating the health metrics of public spaces.

While the current urban spatial design underscores the functionality and aesthetics of public spaces and combines strategies to optimize the fulfillment of the daily needs of local residents, the planners paid scant attention to health considerations, thereby overlooking the potential of public spaces as a key instrument for promoting and enhancing public health. In contemporary urban developments, the POSs of government service centers (GSCs) play a crucial role as hubs for interactions between local residents and the government, serving as "city saloons" that offer city dwellers venues for social gatherings, physical activities, and relaxation.

In this paper, based on a case study of two established GSCs, the problems of GSC POS design are analyzed from a health perspective, and health perspective-based strategies and principles for the design optimization of POSs are proposed. Finally, these strategies are applied to the redesign of the Wuchang GSC, aiming to introduce a novel health-centered framework to enhance the well-being of urban POS users.

2. Methodology

The POSs of two typical GSCs, the Wuhan Citizens Home (Wuhan GSC) and the Chengdu High-Technology District GSC (Chengdu GSC), were systematically investigated to understand the design problems of existing GSCs, determine user satisfaction, and provide benchmark data for the subsequent POS design optimization of the Wuchang GSC. The investigation was conducted by combining an internet search, a site survey, and mapping. The demographic structure and user satisfaction level were investigated using a questionnaire survey, with 100 questionnaires distributed for each GSC. The users were divided into three age groups according to the latest age grade system of the National Bureau of Statistics of China (2022): children (0–17), the young and middle-aged (18–65), and the elderly (above 65). The same method was used to evaluate the optimization effect of the Wuchang GSC.

In 1943, the American psychologist Abraham Maslow proposed the well-known behavioral science theory of Maslow's hierarchy of needs, which classifies human needs into five levels, namely from most to least required, physiological, safety, social, esteem, and self-actualization [27]. POS design is generally affected by multiple factors, such as their required functions, land use planning, financial budget, climate environment, social environment, and urban development planning [28]. In this paper, a POS design approach inspired by Maslow's hierarchy is proposed, emphasizing the health perspective and guided by the principles of promoting well-being, ecological sustainability, and long-term viability. The subsequent analysis focuses on the design principles for natural, recreational, and social spaces within POSs.

3. Investigation of Existing GSC POSs

3.1. User Demographics

Figure 1 illustrates a comparison of the user age distributions at the Wuhan and Chengdu GSCs. The predominant user demographic at the Wuhan GSC comprises young and middle-aged individuals, accounting for 68% of the total users, whereas at the Chengdu GSC, the largest user group was the elderly, constituting 46% of the user population.

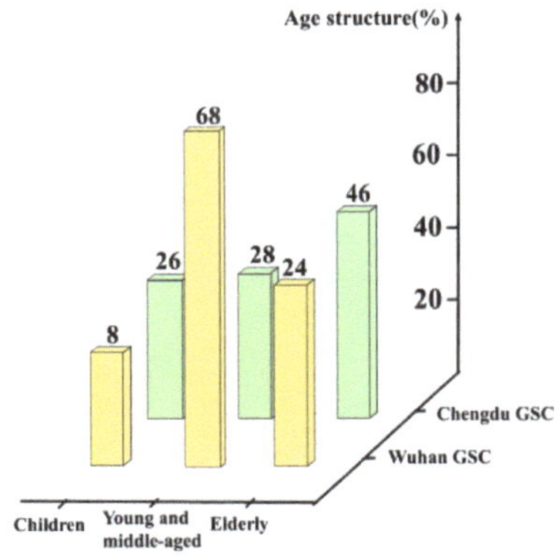

Figure 1. Age structures of users of Wuhan and Chengdu GSCs.

3.2. Spatial Structure Analysis

The Assessment Standard [29] for Healthy Building was utilized to elucidate the functionality of GSC POSs, encompassing elements such as comfort, fitness, social interaction, and service. Subsequently, the GSC POSs were categorized into four zones: nature, relaxation, social interaction, and service. These zones were further subdivided into four subzones in the questionnaire survey: hard pavement (pedestrian paths), green space (areas with vegetation and water features), utilities (parking lots, waste containers, bulletin boards, smoking booths, and other service facilities), and aesthetic structures (sculptures, pavilions, pagodas, and other artistic elements). A spatial analysis of the Wuhan and Chengdu GSCs is depicted in Figure 2, indicating that hard pavements constitute the largest proportion of the spatial layout, followed by green spaces, utilities, and aesthetic structures. The on-site investigation revealed similarities in the POS design between these two GSCs. For instance, both designs aim to portray a government accessible to the public through open spaces and to offer recreational spaces for local residents.

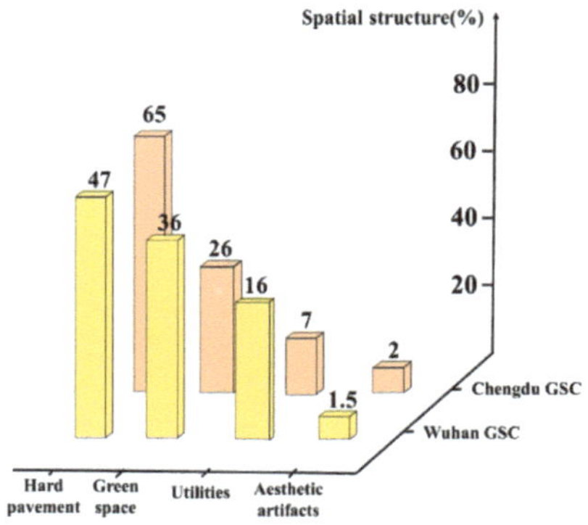

Figure 2. Spatial structures of Wuhan and Chengdu GSC POSs.

3.3. User Satisfaction Survey

User satisfaction of the two GSCs was examined from the six following aspects: government services, education and publicity, playing space, utilities, government–citizen interaction, and green space quality, as illustrated in Figure 3. The majority of the survey respondents expressed very high satisfaction levels with the government services provided at both GSCs. Education and publicity garnered the second-highest satisfaction ratings (72% and 68% for the Wuhan and Chengdu GSCs, respectively). Notably, there was a significant disparity in satisfaction ratings for playing spaces, with 85% and 62% satisfaction rates reported for the Wuhan and Chengdu GSCs, respectively. The lower user satisfaction rating for the green spaces at the Chengdu GSC may be attributed to a reduced green space proportion and distinct user demographics (children and the elderly constituted 32% and 72% of the Wuhan and Chengdu GSC POS users, respectively. Due to the significant disparity in personnel structure, the demand for fun in playing spaces had become more urgent, resulting in relatively negative evaluations). Social and green spaces at both GSCs received the lowest user satisfaction ratings, primarily due to a high prevalence of hard pavements and insufficient seating, walkways, and other structures essential for fostering social interaction, as well as limited green space allocation, which rendered the GSCs less appealing for relaxation, strolling, and cooling down.

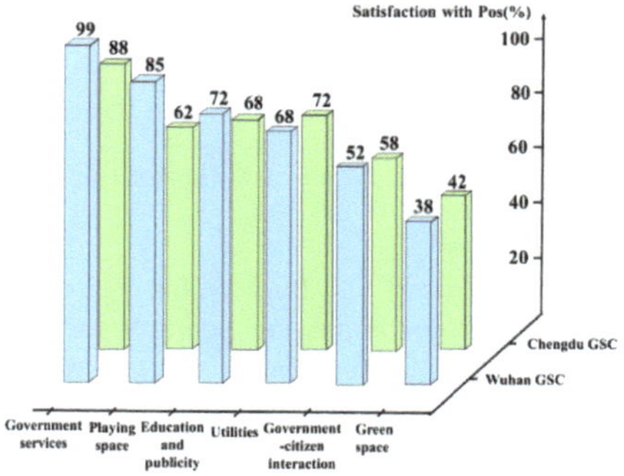

Figure 3. User satisfaction of Wuhan and Chengdu GSC POSs.

Based on the survey findings, the existing GSC POSs exhibited the following issues:

(1) Green spaces: The availability of green spaces is closely linked to user well-being. The survey indicated that the users preferred the GSC POSs with a higher proportion of green areas, which offer a closer connection to nature.

(2) Playing spaces: Approximately 50% of the GSC POS users resided nearby and utilized the POSs for leisurely activities and walking. These individuals expressed a desire for POSs with more engaging playing spaces.
(3) Social spaces: The users expressed low satisfaction with the spatial layout of social areas. Therefore, the design of social spaces should take into account an appropriate spatial structure to enhance user experience.

4. Healthy Design Principles for POSs

4.1. Design Principles for Natural Spaces

Natural spaces integrate vegetation, natural terrain features, and water elements, serving as integral components of urban ecology. By providing a visually appealing landscape and a comfortable environment, these spaces can address both physiological and psychological health needs [30,31]. Furthermore, the creation of a microclimate through diverse plants not only enhances occupant well-being but also contributes to the improvement of the urban ecological system [32,33]. The design of natural spaces within a GSC POS can adopt various basic forms, such as symmetrical, open, or semi-open layouts, integrating natural elements to establish spatial boundaries that define distinct zones and contribute to the aesthetic appeal of the space (Table 1).

Table 1. Basic forms of natural spaces.

Basic Form	Example Image	Health Characteristics
Symmetrical		Emphasizing symmetry, order, and balance, the design features a prominent central axis framed by a symmetrical arrangement [33]. Plants are meticulously organized in evenly spaced rows or recurring patterns.
Open		Offers an expansive, inclusive, and unobstructed vista to facilitate social interactions and various activities. Vegetation occupies a minimal portion of the area, primarily serving decorative and shading purposes, while seamlessly integrating with the natural surroundings or existing landscape.
Semi-open		Creates balance between openness and seclusion; akin to the open layout but with reduced openness; incorporates high plants to enclose one or more sides, creating a more private ambiance with limited visibility.

4.2. Design Principles for Playing Spaces

Playing spaces specifically refer to pedestrian paths for pedestrian traffic as well as designated sports areas, providing places and facilities for walking, exercising, and engaging in various activities, as depicted in Figure 4. Pedestrian pathways should accommodate high foot traffic, enhance access to the main structure, feature an aesthetically pleasing layout, and lead to enticing areas that inspire local residents to engage in physical activities [34,35]. By strategically situating sports fields and fitness apparatus, a comfortable and health-conscious environment for sports enthusiasts can be established, naturally encouraging exercise and bolstering user health through immunity enhancement and positive mental well-being.

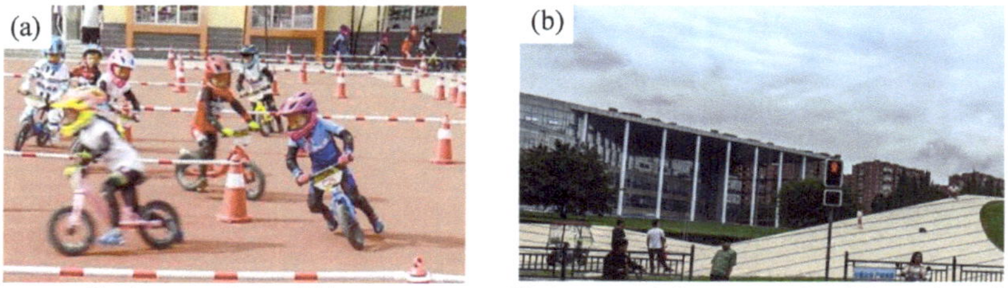

Playing grounds for children. Playing grounds for young and middle-aged people.

Figure 4. Different types of playing spaces.

For children, playing spaces should be designed with a spatial layout that matches the body size of children and with attractive colors. Child-friendly games that are both safe and captivating should be incorporated into the design of children's playing spaces. For young and middle-aged individuals, whose schedules are typically packed during work hours, playing spaces should be designed to provide a spatial experience that is relaxing and can be used for different types of sports activities. Conversely, for the elderly, play spaces should be crafted with their limited mobility in mind, providing venues for gentle exercises like square dancing and tai chi. In essence, playing spaces should be thoughtfully designed to address the distinct requirements of various age groups and societal backgrounds [36], fostering a secure and inclusive setting for a wide range of communal engagements.

4.3. Design Principles for Social Spaces

Social spaces are meticulously crafted environments dedicated to fostering interpersonal communication, interaction, and relaxation. These spaces are created by harmoniously blending elements such as steps, seating areas, walkways, pavilions, and other architectural features with the natural surroundings, providing communal hubs for public gatherings and events [37,38]. By establishing expansive open areas conducive to group activities of varying sizes, the potential for collective interaction is heightened, enriching the vibrancy of local communities [39]. Social spaces should be structured in a manner that encourages healthy interpersonal connections. When the height (H) of buildings matches the spacing (D) between them, users of these public open spaces experience a sense of comfort and security; when $D < H$, users may feel confined, whereas if $D > H$, users might perceive the area as overly spacious.

5. POS Design Optimization of Wuchang GSC

5.1. Project Description and Problems

The Wuchang GSC is situated in the Wuchang District of Wuhan, in close proximity to the Xinhejie station on metro line #7. This project is a part of the Wuchang Ecological and Cultural Corridor initiative and features a collection of strategically positioned public structures. The overall site area of the project, encompassing commercial facilities and green areas, spans approximately 24,017.57 m^2. The architectural design consists of a unified building complex, standing 16 stories above ground and 4 stories below ground level, with a total floor space of 141,314.97 m^2. Within this complex are facilities for the GSC, administrative offices, commercial spaces, a community health service center, and the Wuchang District Information Center. The building's POS features a ground-level plaza, a rooftop garden, and a sunken courtyard, as illustrated in Figure 5.

Figure 5. Wuchang GSC and the surrounding space.

Based on the above analysis, the existing POS has shortcomings in satisfying basic and reachable health needs. Specifically, the natural space has insufficient green space; for example, the roof garden has low aesthetic quality and insufficient shaded areas. These characteristics affect the aesthetic and ecological balance of the city and negatively impact the psychological and physiological health of users. Furthermore, the existing playing spaces lack diversity and fail to cater to various age groups and health conditions, thereby falling short of providing suitable venues for sports activities. As for social spaces, the current layout is congested, characterized by enclosed spatial configurations that constrain user movement and impede opportunities for social interaction.

5.2. Optimization of Natural Spaces

In order to tackle the issues related to inadequate green spaces and lack of variety, a set of enhancement strategies is proposed. Firstly, it is recommended that public green spaces be conceptualized as either partially enclosed or open areas, incorporating a layered layout featuring short trees and bushes to establish a welcoming and peaceful ambiance. The optimized design is dominated by large green trees, brightly colored small trees, and low shrubs. Furthermore, foliage and fruit-bearing trees with harmonious color schemes have been introduced to generate striking color contrasts. As an illustration, in the northern sector of the GSC, predominantly ginkgo and golden rain trees have been included, complemented by verdant tree borders predominantly consisting of young plants.

Tree planting techniques encompass clustered, sparse, and decorative forms. Clustered trees are predominantly positioned along building perimeters and corners, functioning to obstruct direct views into the structures, soften architectural lines, and establish secluded areas. Sparse tree arrangements are commonly found on lawns, offering expansive vistas and fostering an open ambiance. Decorative trees are typically situated within flower beds, tree basins, and containers. By predominantly incorporating these natural elements and a symbolic waterscape, the richness and variety of the landscape are enhanced, as illustrated in Figure 6.

Figure 6. Computer rendering of the entry plaza. (**a**) before and (**b**) after redesign.

The design of the green spaces incorporates a symmetrical layout to achieve a visually pleasing and well-balanced environment. Given the significance of the logo wall at the entrance of the POS for facilitating communication between the landscape facility and its users, and considering the natural surroundings of mountains and rivers in the Wuchang district, the logo wall was conceptualized with representations of these elements, utilizing curved forms. Constructed from granite, the wall is complemented by flower beds and artificial mountain features. Three layers of green borders adorn the logo wall, while symmetrical green hedges and designated parking areas for non-motorized vehicles flank its lateral sides. This new design enhances both the aesthetics and functionality of the spaces, providing users with a comprehensive understanding of the local natural landscape and traditional culture, as depicted in Figure 7.

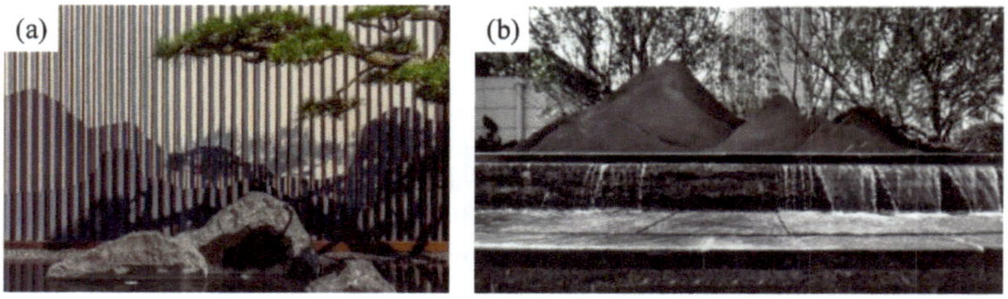

Figure 7. Computer rendering of the logo wall at the entry area. (**a**) considering the natural surroundings of mountains and (**b**) considering the natural surroundings of rivers.

As illustrated in Figure 8, the original roof garden lacks a contemplative area and visually appealing or shaded zones. The defining characteristic of roof gardens is that they are built on top of buildings. In contrast to conventional green spaces, inappropriately designed roof gardens can lead to serious safety hazards. Therefore, roof gardens should be planned following the safety-first principle and as natural, green, ecologically friendly, and sustainable spaces. Due to the limited soil depth available on rooftops, the design should primarily feature shrubs, flowers, and grass. This concept prioritizes structural soundness while aiming to establish an aesthetically pleasing and ecologically diverse green space. Moreover, pedestrian pathways are crafted using a blend of stepping stones and lawns. This holistic redesign notably enhances both the visual appeal and functionality of the roof garden, providing users with a healthier and more enjoyable outdoor environment. Considering that the roof garden has sufficient sunlight exposure, in order to improve the comfort of users resting or socializing in this space, shade or rain cover facilities should be added to the current optimization plan in the future.

Figure 8. Computer rendering of roof garden (**a**) before and (**b**) after redesign.

5.3. Optimization of Playing Spaces

As outlined in Section 3, nearly half of the visitors to the Wuhan and Chengdu GSC POSs did not go there for business. They were mainly passersby, residents in the surrounding communities, and tourists. Feedback from users indicated a low satisfaction level with the GSCs, particularly regarding the availability of areas for group activities.

The original design of the ground plaza at the Wuchang GSC shares similarities with the Wuhan and Chengdu GSCs, being spacious but lacking sufficient space for recreational activities. Therefore, it was optimized by focusing on providing space for healthy sports for different age groups.

For the elderly, a comfortable and safe environment has been created for recreational activities, such as tai chi, chatting over tea, and chess. For the young and middle-aged, a place is provided for them to participate in sports activities, such as power walking, jogging, and cycling, and multi-functional facilities for sports activities is also provided. For children, the playing space is designed to avoid overcrowding and facilitate the tracking of children, and bright and safe materials are used. Specifically, two open grounds of the Wuchang GSC were redesigned into two small playing grounds with grounds for sports, social, and relaxation activities and facilities for both high- and low-intensity exercises. Each playing ground faces a water-side landscape wall, which is fronted by a water pool and decorated with sculptures, thereby creating a more layered space. The attractiveness of the recreational, sports, and relaxation facilities has been improved by adding plants, thereby creating a space where users can feel close to nature. By integrating recreational facilities with a wavering green landscape, a more attractive playing space has been created. Furthermore, the pedestrian paths in the landscape are paved with dark and light grey natural stones, thereby increasing the color contrast, as shown in Figures 9 and 10.

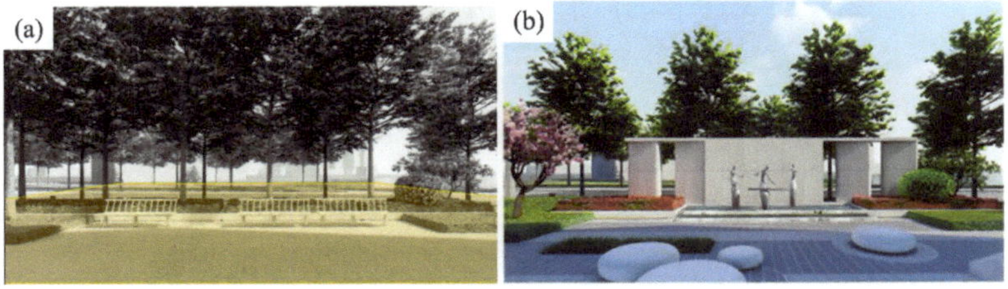

Figure 9. Comparison of playing space (leisure area #1) (**a**) before and (**b**) after redesign.

Figure 10. Comparison of playing space (leisure area #2) (**a**) before and (**b**) after redesign.

5.4. Optimization of Social Spaces

To rectify the issues of an unbalanced spatial layout, overcrowding, and enclosed spatial design observed in the original social spaces, a departure from traditional spatial morphology was undertaken. This transformation aims to establish a low-density open layered space, enhancing the visual appeal of the landscape and cultivating a more visually pleasing POS.

The spatial structure, encompassing both visual and functional dimensions, serves as a pivotal element in evaluating space quality and was fully considered during the redesign of the Wuchang GSC POS. The visual scale is defined based on visual perception. Generally, the vertical field of view for an average individual spans about 120°, with 50° above and 70° below the apparent horizon considered within the comfortable range, as depicted in Figure 11. A combined consideration of visual and functional scales in the design process culminated in an enhanced visual and functional experience within the Wuchang GSC POS.

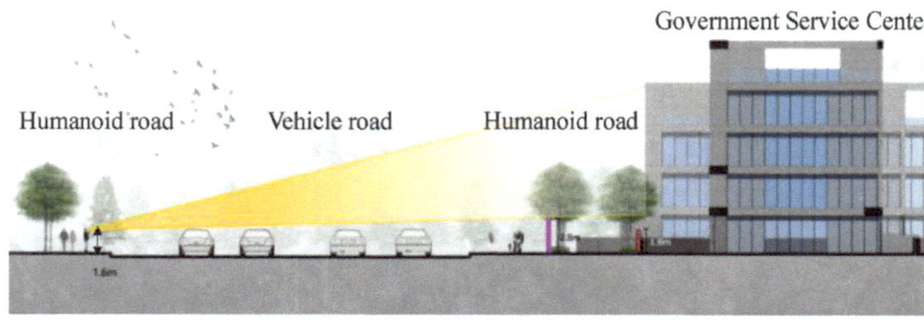

Figure 11. Illustration of comfortable spatial scale.

Figure 12 shows the array of public amenities and ornamental structures implemented within the Wuchang GSC. Essential public facilities, such as outdoor signage, communal seating, street lighting, pavilions, and other foundational elements, were strategically placed to foster social engagement. In crafting the design of directional signs, informational boards and site markers were strategically positioned at entry points and intersections along the pathways. The site markers at entry points are prominently displayed, succinct, and lucid, detailing the functions, services, departments, and pathways within the GSC.

Figure 12. Computer rendering of public facilities and decorative structures. (**a**) Round seats, (**b**) square seats, and (**c**) long seats.

Public seating, a pivotal feature of outdoor communal areas, was meticulously designed with ergonomic considerations in mind. Long, square, and circular seating options have been made available to cater to different preferences. The design of the elongated seats drew inspiration from the U-shaped concrete structure, with a unique configuration mirroring the building's form. Square seating arrangements have been integrated into the green spaces, constructed by assembling multiple granite slabs with greenery at the center. Circular seating designs, reminiscent of water droplets, have been strategically positioned in front of sculptures, offering users a vantage point to admire the surroundings and unwind in comfort. By setting up a reasonable resting stone bench, users are ensured to be at a comfortable social distance, and the form of the stone bench can be well integrated with the natural landscape, improving the interaction between users and the natural environment.

The illumination scheme of the POS encompasses a combination of road lamps and ornamental lighting fixtures. Road lamps have been strategically deployed not only as a safety measure but also to guide users along the pathways, positioned on the edges of roads, in expansive areas, and lining pedestrian walkways. On the other hand, decorative lights adorned the external façades of the building and the lawns. Given the governmental service character of the Wuchang GSC, modestly designed lamps with a height of approximately 2.5 m have been utilized, spaced at intervals of 10 m.

The decorative lights mounted on the building's exterior walls are discreetly nestled within wall recesses, predominantly featuring a harmonious blend of cool hues in a minimalist color palette. Within the lawns, visually appealing low-level decorative lights have been incorporated to create a striking contrast against the tranquil and inviting ambiance of the rooftop garden, effectively delineating the pathways at night and bridging the transition between soft and hard landscaping elements. The optimized lighting design of social spaces has been added to meet the user's needs for walking and resting in public spaces at night, as depicted in Figure 13.

Figure 13. Computer rendering of lighting. (**a**) Decorative lights on walls, (**b**) road lamps, and (**c**) decorative lights on lawns.

5.5. Optimization Effect and Discussion

Another questionnaire survey on the spatial design before and after optimization was conducted to evaluate the effectiveness of optimization. The results are provided as follows. In the initial design, the ground plaza of the Wuchang GSC featured extended hard pavement, offering space but failing to cater to users of all ages and lacking space for social activities. Furthermore, the existing public activity spaces lacked diversity and failed to provide suitable areas for sports activities, rendering them unsuitable for users of varying ages. As for the social spaces, the original layouts have relatively enclosed spatial zones, constraining user mobility and opportunities for social interaction. The optimized survey results are shown in Figure 14. Through the optimization design of each space, the user's satisfaction with the functionality of providing social and entertainment activities in the space is greatly increased, and the green space ratio on site is also significantly improved. As shown in Figure 14, the statistical data prove that the above design concept and implementation plan are effective and can provide references for relevant cases.

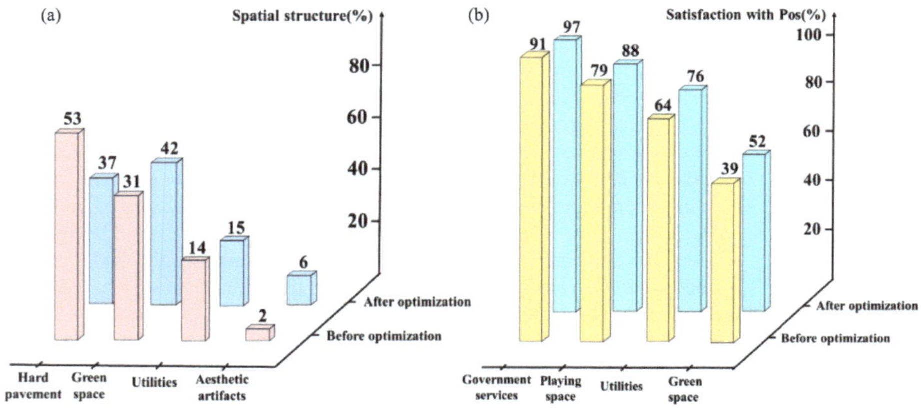

Figure 14. The optimization effects on the Wuchang GSC: (**a**) spatial structure, and (**b**) decorative lights on lawns.

6. Conclusions

In this study, a thorough investigation and analysis of two typical GSCs revealed several key problems with the existing GSC POS designs: deficiency in green spaces, inadequate appeal and functionality of playing spaces, and unsuitable spatial configurations for social interactions. In response to these design challenges, a comprehensive set of modifications is recommended to revamp the Wuchang GSC.

(1) In terms of natural space design, the problems of insufficient greenery and a monotonous landscape have been overcome. By increasing the green area and introducing diverse plant configurations, the new design not only enhances the natural atmosphere and aesthetics of the space but also provides citizens with more opportunities to get close to nature. Compared to the previous design, the new natural space places more emphasis on human experience and feelings, creating a beautiful and pleasant environment.
(2) The renovation of playing spaces places greater emphasis on age-friendly design concepts. Diverse activity facilities and recreational areas according to the needs of different age groups have been set up. Compared to the originally open space that lacks vitality, the new activity space is more diverse and can meet the sports and entertainment needs of various groups of people, greatly enhancing the attractiveness and practicality of the space.
(3) In the optimization of social spaces, the original closed layout has been broken and an open, low-density communication environment has been created. The new design is not only optimized for the visual and functional scales but also greatly enhances the comfort and social interaction of the space by adding humanized public facilities, such as signage and public seating.

In summary, by integrating Maslow's theory into the comprehensive optimization of the outdoor public space of the Wuchang GSC, a healthier, more functional, and humanized public space has been created, overcoming the shortcomings of the original design. These improvements not only enhance the quality of life of citizens but also make positive contributions to the sustainable development of the city. The optimized effect was validated through user satisfaction surveys. However, the evaluation of the effectiveness of the current design concept and scheme was only based on a relatively simple scoring system. In the future, more comprehensive data analysis methods will also be used to track and investigate the psychological and health status of users in order to evaluate the feasibility of optimizing design concepts and provide a basis for the widespread promotion of this design method.

Author Contributions: Conceptualization, W.P.; Formal analysis, W.P. and X.D.; Investigation, L.Z.; Project administration, Y.H.; Resources, Z.Z., Y.W. and X.D.; Software, X.W.; Supervision, T.Y.; Visualization, L.Z.; Writing—original draft, Y.W., Y.H. and T.Y.; Writing—review & editing, Y.W., X.W. and Z.Z. All authors have read and agreed to the published version of the manuscript.

Funding: Research on landscape optimization and low-carbon design of urban public space: G23-12-S.

Data Availability Statement: The original contributions presented in the study are included in the article, further inquiries can be directed to the corresponding authors.

Conflicts of Interest: Authors Tiancheng Yang and Xinyan Wang were employed by the company China Railway 18th Bureau Group Building and Installation Engineering Co., Ltd. Author Zhengming Zhang was employed by the company Ausspec Landscape Planning and Design Co., Ltd. Author Liang Zhu was employed by the company CITIC General Institute of Architectural Design & Research Co., Ltd. The remaining authors declare that the research was conducted in the absence of any commercial or financial relationships that could be construed as a potential conflict of interest.

References

1. Jacobs, J. *The Death and Life of Great American Cities*; Penguin Books: Harmondsworth, UK, 1964.
2. Giles-Corti, B.; Vernez-Moudon, A.; Reis, R.; Turrell, G.; Dannenberg, A.L.; Badland, H.; Foster, S.; Lowe, M.; Sallie, J.F.; Stevenson, M.; et al. City planning and population health: A global challenge. *Lancet* **2016**, *388*, 2912–2924. [CrossRef] [PubMed]
3. Guite, H.F.; Clark, C.; Ackrill, G. The impact of the physical and urban environment on mental well-being. *Publ. Health* **2006**, *120*, 1117–1126. [CrossRef] [PubMed]
4. Phelan, P.; Wang, N.; Hu, M.; Roberts, J.D. Sustainable, Healthy Buildings & Communities. *Build. Environ.* **2020**, *174*, 106806.
5. Zhang, J.; Yu, Z.; Cheng, Y. Evaluating the disparities in urban green space provision in communities with diverse built environments: The case of a rapidly urbanizing Chinese city. *Build. Environ.* **2020**, *183*, 107170. [CrossRef]
6. Anderson, J.; Ruggeri, K.; Steemers, K.; Huppert, F. Lively social space, well-being activity, and urban design: Findings from a low-cost community-led public space intervention. *Environ. Behav.* **2017**, *49*, 685–716. [CrossRef]
7. Allam, Z.; Jones, D.S. Pandemic Stricken Cities on Lockdown. Where Are Our Planning and Design Professionals [Now, Then and into the Future]? *Land Use Policy* **2020**, *97*, 104805. [CrossRef] [PubMed]
8. Carmona, M. Contemporary public space, Part Two: Classification. *J. Urban Des.* **2010**, *15*, 157–173. [CrossRef]
9. Giles-Corti, B.; Broomhall, M.H.; Knuiman, M. Increasing walking: How important is distance to, attractiveness, and size of public open space? *Am. J. Prev. Med.* **2005**, *28*, 169–176. [CrossRef]
10. Villanueva, K.; Badland, H.; Hooper, P. Developing indicators of public open space to promote health and wellbeing in communities. *Appl. Geogr.* **2015**, *57*, 112–119. [CrossRef]
11. Foster, S.; Hooper, P.; Duckworth, A.; Bolleter, J. An evaluation of the policy and practice of designing and implementing healthy apartment design standards in three Australian cities. *Build. Environ.* **2022**, *207*, 108493. [CrossRef]
12. Bluyssen, P.M. Towards new methods and ways to create healthy and comfortable buildings. *Build. Environ.* **2010**, *45*, 808–818. [CrossRef]
13. Francis, J.; Wood, L.J.; Knuiman, M.; Giles-Corti, B. Quality or quantity? Exploring the relationship between public open space attributes and mental health in Perth, Western Australia. *Soc. Sci. Med.* **2012**, *74*, 1570–1577. [CrossRef] [PubMed]
14. Turel, H.S.; Yigit, E.M.; Altug, I. Evaluation of elderly people's requirements in public open spaces: A case study in Bornova District (Izmir, Turkey). *Build. Environ.* **2007**, *42*, 2035–2045. [CrossRef]
15. Hu, Y.; Lin, Z.; Jiao, S.; Zhang, R. High-Density Communities and Infectious Disease Vulnerability: A Built Environment Perspective for Sustainable Health Development. *Buildings* **2023**, *14*, 103. [CrossRef]
16. Li, H.; Liu, Z. Temporal and Spatial Evolution of Urban Density in China and Analysis of Urban High Density Development: From 1981 to 2014. *Urban Dev. Studies.* **2019**, *26*, 46–54.
17. Zhao, S.; Shen, M.; Musa, S.S.; Guo, Z.; Ran, J.; Peng, Z.; Zhao, Y.; Chong, M.K.C.; He, D.; Wang, M.H. Inferencing Superspreading Potential Using Zero-Truncated Negative Binomial Model: Exemplification with COVID-19. *BMC Med. Res. Methodol.* **2021**, *21*, 30. [CrossRef] [PubMed]
18. Bin Kashem, S.; Baker, D.M.; González, S.R.; Lee, C.A. Exploring the Nexus between Social Vulnerability, Built Environment, and the Prevalence of COVID-19: A Case Study of Chicago. *Sustain. Cities Soc.* **2021**, *75*, 103261. [CrossRef] [PubMed]
19. Li, B.; Peng, Y.; He, H.; Wang, M.; Feng, T. Built Environment and Early Infection of COVID-19 in Urban Districts: A Case Study of Huangzhou. *Sustain. Cities Soc.* **2021**, *66*, 102685. [CrossRef]
20. Lau, S.S.Y.; Gou, Z.; Liu, Y. Healthy campus by open space design: Approaches and guidelines. *Front. Archit. Res.* **2014**, *3*, 452–467. [CrossRef]

21. Lau, S.S.Y.; Yang, F. Introducing healing gardens into a compact university campus: Design natural space to create healthy and sustainable campuses. *Landsc. Res.* **2009**, *34*, 55–81. [CrossRef]
22. Payne, S. Open space: People space. *J. Environ. Psychol.* **2009**, *29*, 532–533. [CrossRef]
23. Velarde, M.D.; Fry, G.; Tveit, M. Health effects of viewing landscapes—Landscape types in environmental psychology. *Urban For. Urban Green.* **2007**, *6*, 199–212. [CrossRef]
24. Tao, Y.X.; Zhu, Y.; Passe, U. Modeling and data infrastructure for human-centric design and operation of sustainable, healthy buildings through a case study. *Build. Environ.* **2020**, *170*, 106518. [CrossRef]
25. Yan, D.; O'Brien, W.; Hong, T.; Feng, X.; Gunay, H.B.; Tahmasebi, F.; Mahdavi, A. Occupant behavior modeling for building performance simulation: Current state and future challenges. *Energy Build.* **2015**, *107*, 264–278. [CrossRef]
26. Reinhart, C.; Dogan, T.; Jakubiec, J.A.; Rakha, T.; Sang, A. Umi-an urban simulation environment for building energy use, daylighting and walkability. In Proceedings of the 13th Conference of International Building Performance Simulation Association, Chambery, France, 25–28 August 2013.
27. Sugiyama, T.; Gunn, L.D.; Christian, H. Quality of public open spaces and recreational walking. *Am. J. Publ. Health* **2015**, *105*, 2490–2495. [CrossRef] [PubMed]
28. Gehl, J. *Life between Buildings: Using Public Space*; Island Press: Washington, DC, USA, 2011.
29. *T/ASC 02-2016*; Health Building Evaluation Standards. The Architectural Society of China: Beijing, China, 2017.
30. Evans, G.W. The built environment and mental health. *J. Urban Health* **2003**, *80*, 536–555. [CrossRef] [PubMed]
31. Lloyd, K.; Auld, C. Leisure, public space and quality of life in the urban environment. *Urban Pol. Res.* **2003**, *21*, 339–356. [CrossRef]
32. Wood, L.; Hooper, P.; Foster, S. Public green spaces and positive mental health–investigating the relationship between access, quantity and types of parks and mental wellbeing. *Health Place* **2017**, *48*, 63–67. [CrossRef]
33. Chen, L.; Wen, Y.; Zhang, L. Studies of thermal comfort and space use in an urban park square in cool and cold seasons in Shanghai. *Build. Environ.* **2015**, *94*, 644–653. [CrossRef]
34. Kumar, S.; Underwood, S.H.; Masters, J.L.; Manley, N.A.; Konstantzos, I.; Lau, J.; Haller, R.; Wang, L.M. Ten questions concerning smart and healthy built environments for older adults. *Build. Environ.* **2023**, *244*, 110720. [CrossRef]
35. Liu, Y.; Guo, Y.; Lu, S.; Chan, O.F. Understanding the long-term effects of public open space on older adults' functional ability and mental health. *Build. Environ.* **2023**, *234*, 110126. [CrossRef]
36. Sugiyama, T.; Thompson, C.W.; Alves, S. Associations between neighborhood open space attributes and quality of life for older people in Britain. *Environ. Behav.* **2009**, *41*, 3–21. [CrossRef]
37. Peng, Y.; Feng, T.; Timmermans, H. A path analysis of outdoor comfort in urban public spaces. *Build. Environ.* **2019**, *148*, 459–467. [CrossRef]
38. Zeng, X.; Luo, P.; Wang, T. Screening visual environment impact factors and the restorative effect of four visual environment components in large-space alternative care facilities. *Build. Environ.* **2023**, *235*, 110221. [CrossRef]
39. Xie, Y.; Wang, X.; Wen, J. Experimental study and theoretical discussion of dynamic outdoor thermal comfort in walking spaces: Effect of short-term thermal history. *Build. Environ.* **2022**, *216*, 109039. [CrossRef]

buildings

MDPI

Review

A Review of Research Progress on the Impact of Urban Street Environments on Physical Activity: A Comparison between China and Developed Countries

Yu Wen *,†, Bingbing Liu †, Yulan Li and Lin Zhao

School of Arts and Design, Yanshan University, 438, West Section of Hebei Avenue, Qinhuangdao 066004, China; lbbcxy0807@stumail.ysu.edu.cn (B.L.)

* Correspondence: wenyu@ysu.edu.cn; Tel.: +86-150-0335-3043

† These authors contributed equally to this work.

Abstract: Creating healthy street environments to encourage physical activity is an effective strategy against non-communicable diseases exacerbated by rapid urbanization globally. Developing countries face more significant health challenges than developed ones. However, existing research predominantly focuses on the perspective of developed countries. To address the health challenges in developing nations, studies should not only draw on the findings from developed countries but also clearly define unique research processes and pathways. Consequently, this study conducts a comparative analysis between China, representing developing countries, and developed nations, using databases like China National Knowledge Infrastructure (CNKI) and Web of Science (WOS) and tools such as Citespace, Bicomb, and Statistical Package for the Social Sciences (SPSS) to explore research hotspots, developmental trajectories, thematic categories, and trends. The findings reveal a shift in developed countries from macro-material to micro-environmental elements under multidisciplinary scrutiny, while future topics may include street space evaluations and psychological healing. In China, research has been dominated by different disciplines at various stages, starting with medical attention to chronic disease prevention, which then shifted to traffic engineering's focus on constructing green travel environments, and finally expanded to disciplines like landscape architecture examining the impact of street environment elements on pedestrian behavioural perceptions. Future themes will focus on promoting elderly health and urban health transport systems. Generally, research in developed countries exhibits a "bottom-up" approach, with practical issues at a "post-evaluation" stage, primarily based on the "socio-ecological model" and emphasizing multidisciplinary collaboration. Chinese research shows a "top-down" characteristic, driven by national policies and at a "pre-planning" stage, integrating theories such as Maslow's hierarchy of needs and attention restoration theory, with relatively loose disciplinary cooperation. Overall, research is shifting from macro to human-centric scales and is progressively utilizing multi-source and multi-scale big data analysis methods. Based on this, future research and development recommendations are proposed for developing countries, with China as a representative example.

Keywords: urban streets; physical activity; healthy cities; city development; China

Citation: Wen, Y.; Liu, B.; Li, Y.; Zhao, L. A Review of Research Progress on the Impact of Urban Street Environments on Physical Activity: A Comparison between China and Developed Countries. *Buildings* **2024**, *14*, 1779. https://doi.org/10.3390/buildings14061779

Academic Editor: Konstantina Vasilakopoulou

Received: 4 May 2024
Revised: 5 June 2024
Accepted: 8 June 2024
Published: 13 June 2024

1. Introduction

The coronavirus disease 2019 (COVID-19) pandemic has severely impacted global health, hindering progress toward the Sustainable Development Goals [1]. In 2023, the first Healthy City Partnership Summit in London brought together mayors and officials from over fifty major cities. The summit highlighted the renewed focus on non-communicable diseases and injuries three years after the COVID-19 pandemic outbreak, noting that cities have unique advantages in combating non-communicable diseases and reducing injuries by implementing policies that significantly reduce exposure to risk factors, thus helping the world return to the trajectory of the Sustainable Development Goals [2]. The global urbanization rate is expected to rise to 68% by 2050 [3], with an additional 2.5 billion urban inhabitants. Most of this urban population growth will occur in developing countries, especially in China, India, and Nigeria [4]. Issues such as poor sanitation, air pollution, road safety, and limited access to healthy food and spaces during urbanization elevate the risk of non-communicable diseases [5]. Non-communicable diseases cause 41 million deaths annually, accounting for 74% of all deaths globally, with 77% occurring in low- and middle-income countries [6]. These countries face more severe health challenges than developed nations due to several factors: a lack of quality healthcare services, inadequate social welfare systems, insufficient dissemination of disease prevention knowledge, and the rapid progression of urbanization and aging populations [7]. The World Health Organization (WHO) has identified urbanization as one of the leading public health challenges of the 21st century [8]. Evidence-based medical research has proven that a lack of physical activity is one of the top ten global death risk factors [9] and increasing physical activity and reducing sedentary behaviour could prevent at least 3.2 million deaths related to non-communicable diseases globally each year [10]. WHO has emphasized that urban planning should actively contribute to addressing health challenges, highlighting that "health must be a primary focus for urban planners" [11]. Streets, as a crucial component of urban spaces, can directly influence human health by providing residents with comfortable venues for physical activity [12,13]. They can also indirectly promote health by reducing factors that potentially impact traffic accidents, thereby ensuring the safety of residents' activities [14,15]. Health-oriented street design is an effective way to foster the creation of healthy cities [16].

As the world's largest developing country, over the past 40 years, China has seen its urbanization rate increase from 17.92% to 65.2% [17], which exceeds significantly that of other countries during the same period, making it one of the fastest urbanizing countries globally [18]. According to United Nations projections, China's urbanization rate will reach 74% by 2035, meaning that, in the next decade, over 100 million rural inhabitants will move to urban areas [19], resulting in a vast urban population. Like other developing countries, China faces health challenges due to shifts in disease patterns and an aging population. Non-communicable diseases have become the most significant health threat to Chinese citizens, accounting for approximately 88.5% of all deaths [20]. Additionally, due to the rapid development of the automotive industry, the physical environment of streets, including air quality and urban transportation infrastructure, complicates the risk of incidence and disease prevention efforts. The task of building healthy cities in China is unprecedented in its enormity and complexity. To address this challenge, the Chinese government has steadfastly implemented the policy of "integrating health into all policies" to promote the implementation of health governance measures. In 2016, the Chinese State Council issued the "Healthy China 2030" blueprint outline, placing health as a strategic priority in national development. This plan significantly measures China's active participation in global health governance and fulfilling its commitments to the United Nations' "2030 Agenda for Sustainable Development" [21]. In 2018, China established the "National Health City Evaluation Indicator System" and began health city evaluations based on this system [22]. In 2019, the "Opinions of the State Council on Implementing the Healthy China Action" were promulgated, which specified measures in the area of national fitness and health to prevent and control road traffic injuries. Other related documents have also proposed strengthening the construction of urban greenways, fitness trails, and other health-oriented living environments [23]. In 2021, the State Council issued the "National Fitness Plan (2021–2025)", which set forth requirements for enhancing the public service system for national fitness and making sports and fitness activities more accessible to the public [24]. This reflects the emphasis, on a national scale, on creating a healthy urban environment that promotes residents' physical activity.

Physical activity (PA) refers to any bodily movement produced by skeletal muscles that requires energy expenditure. The urban street environment includes the physical environment (natural and artificially modified built environment) and social environment (such as social norms and community cohesion). It is an essential factor that affects people's physical activity and is also an important entry point for cities to intervene in people's health proactively. Scholars from various fields, including public health and preventive medicine [25], ecology [26], urban planning [27], and sports science [28], have extensively and deeply researched the impacts of land mix use [29], street connectivity [30], walkability [27], and the built environment [31] on physical activity. However, current research is primarily concentrated in developed countries, such as the United States (which accounts for 44.1% of publications) and Australia (11.9% of publications). Among developing countries, China produces the most research in this field, with its output second only to the United States (16.9% of publications). As these research interests span multiple disciplines, the related literature is extensive and dispersed across various journals. Considering the differences in development stages and planning governance methods between countries, the practical directions, pathways, and depth of research on the impacts of street environments on physical activity vary, yet discussions on this topic have not been previously synthesized. Additionally, existing review articles mainly organize and evaluate research on the impact of urban architectural environments and green spaces on physical activity among different age groups. For instance, Zhang et al. used a narrative review method to summarize existing reviews on the association between architectural environments and physical activity among children, adults, and the elderly [32]. Prince et al. systematically described the current status, intensity, and quality of research on the association between the built environment and adult physical activity in high-income countries [33]. Ran et al. used Citespace and VOSviewer software for a bibliometric analysis of the association between urban green space features and the occurrence of leisure physical activity [34].

In conclusion, compared to developed countries, developing nations face more pronounced health challenges. Current research on the "impact of street environments on physical activity" is primarily conducted from the perspective of developed countries. However, due to differences in political systems, socioeconomic environments, urban infrastructure, and demographic structures between developing and developed countries, the research findings based on developed countries' contexts are not entirely applicable to developing nations. To address the health challenges faced by developing countries, it is necessary to both draw on the research outcomes from developed countries and to clearly define national research agendas and approaches, thereby more actively addressing the foreseeable health challenges in the process of urbanization. Therefore, the study selects China as representative of developing countries to conduct comparative research with developed countries. Using a systematic review methodology and applying software tools such as Citespace (version 6.2.R6), Bibliographic Items Co-occurrence Matrix Builder (Bicomb) 2.0, and Statistical Package for the Social Sciences (SPSS) 27.0, the study performs a systematic summary of research hotspots for studies in China and developed countries, organizes the development trajectories and thematic categories, and scientifically predicts research trends. This analysis deeply explores the differences between research in China and developed countries and the reasons for these differences (as shown in Figure 1). The findings provide a reliable basis and inspiration for future research by Chinese scholars, helping international scholars to deeply understand the current state of research on the impact of street environments on physical activity within the Chinese context.

Accordingly, the structure of this paper is as follows: Section 2 provides an overview of the research materials and methods; Section 3 presents the research hotspots, development contexts, thematic categories, and trends in China and developed countries; Section 4 delves into the similarities and differences between the research in China and developed countries, explores the reasons for these differences, and proposes future directions for research in China; Section 5 concludes the paper.

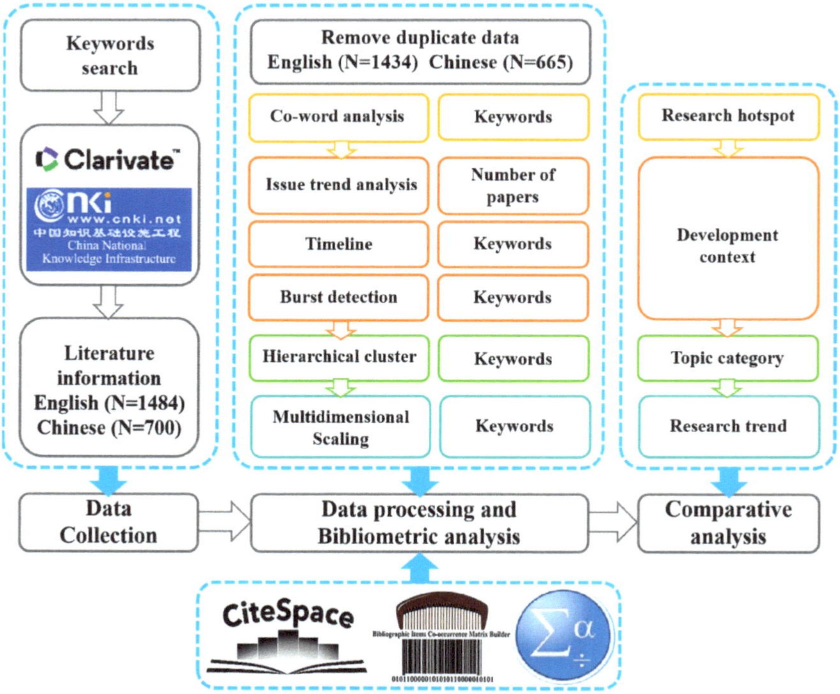

Figure 1. Research pathway.

2. Research Methods

2.1. Literature Selection

The study separately conducted literature searches on the impact of street environments on physical activity in developed countries and China. For English-language literature, the search was conducted using the Web of Science (WOS) core database with the search formula ((TS = (street)) AND TS = (physical activity)) AND PY = (1996–2023). The search included the Science Citation Index-Expanded (SCI-E), the Social Sciences Citation Index (SSCI), and the Arts & Humanities Citation Index (A&HCI), which were limited to English-language literature. Subsequently, by excluding literature irrelevant to the study of street environments' impact on physical activity based on titles and abstracts, 1484 entries were retrieved. For Chinese-language literature, the China National Knowledge Infrastructure (CNKI) database was utilized, which is the largest continuously updated database of Chinese academic literature [35], effectively supplementing the English database and providing a comprehensive view of research in China. To ensure the quality of the papers, journals indexed by Science Citation Index (SCI) or Engineering Index (EI) and core journals from Peking University, Chinese Social Sciences Citation Index (CSSCI) journals, and Chinese Science Citation Database (CSCD) journals were selected, covering the period from 1993 to 2023. The search terms used were "street" AND "physical activity", initially yielding 17 papers. Further searches with expanded thematic keywords, such as "urban" AND "physical activity", "street" AND "slow traffic", and "urban" AND "slow traffic", were conducted, and titles and abstracts were again screened to exclude irrelevant studies, ultimately resulting in 700 papers.

2.2. Research Steps

Firstly, data handling was carried out, wherein the literature data from WOS and CNKI were preprocessed using the Citespace (version 6.2.R6) built-in data converter. To ensure accuracy, after the format conversion, data from WOS and CNKI were filtered to remove duplicates, with 1434 English-language papers and 665 Chinese-language papers determined to be valid. In the second step, research hotspots were identified. The filtered data were imported into Citespace (version 6.2.R6) for bibliometric analysis, setting the time slice to one year and selecting "Keyword" as the node type, and keyword co-occurrence analysis was conducted separately for international and Chinese research fields related to the impact of street environments on physical activity. Furthermore, development trajectories were constructed. Both English- and Chinese-language literature annual publication volumes were visualized using Excel. Based on this, along with the keyword time zone distribution map and keyword prominence map generated by Citespace (version 6.2.R6), the development trajectories of research on the impact of street environments on physical activity in international and Chinese contexts were established. With regard to fourth step, a keyword clustering analysis was undertaken. Using Bicomb2.0, the data from both English- and Chinese-language literature were processed to identify high-frequency keywords and generate term–document matrices. After processing the term–document matrix data in Excel, they were imported into SPSS 27.0 software, where hierarchical clustering analysis was conducted to obtain dendrograms, followed by keyword clustering analysis. Finally, research trends were interpreted. By utilizing the multidimensional scaling feature in SPSS 27.0 software, multidimensional scaling analysis maps were constructed. These maps facilitate a scientific interpretation of research trends based on the distribution quadrants of keywords.

3. Research Progress and Trends on the Impact of Street Environment on Physical Activity in China and Developed Countries

Keywords play a crucial role in succinctly representing a paper's core content and important concepts [36]. The literature can be transformed into quantifiable analytical data by analysing information such as keyword co-occurrence, clustering, temporal distribution, and emergent trends. Particularly, the refinement of high-frequency keywords can, to some extent, reflect scholars' attention to specific topics, characterizing the hotspots and frontier trends in the research field.

3.1. Research Hotspots

The co-word analysis of keywords can reveal the hotspots and internal connections of specific research topics. The basic principle involves counting the occurrences of keywords appearing simultaneously in the same document, and through hierarchical clustering, it exposes the affinity between keywords and the main theme [37]. Co-occurrence frequency and centrality are two key indicators, the former reflecting the intensity of attention to keywords and the latter indicating the importance of keywords in the co-occurrence network.

After conducting a co-occurrence analysis of keywords in 1434 English articles, 530 keyword nodes were obtained. In the keyword co-occurrence network, the size of each node represents the frequency of occurrence of the corresponding keyword. The larger the node, the higher the frequency of occurrence. As shown in Figure 2, keywords such as "physical activity", "built environment", "walking", "health", and "obesity" have larger nodes, indicating greater attention and making them the main research topics in English-language literature regarding the impact of street environment on physical activity. Hotspot analysis not only considers the frequency of keyword occurrence but also comprehensively evaluates keyword centrality. Based on the centrality of keywords shown in Table 1, it is evident that the keyword "public health" is the most significant (0.07), followed by "health" (0.04) and "land use" (0.04). This indicates their central roles in the co-occurrence network and widespread attention among international scholars conducting related research.

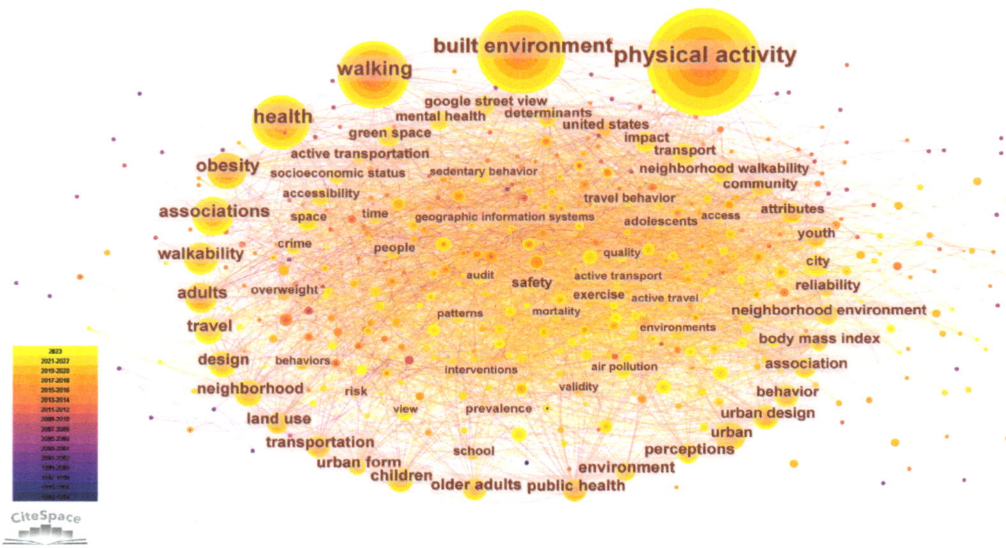

Figure 2. Co-occurrence network map of keywords in English-language literature.

Table 1. Top 20 hotspot words in English-language literature.

Serial Number	Keyword	Frequency of Occurrence	Centrality	Initial Appearance Year
1	physical activity	940	0.02	2002
2	built environment	604	0.02	2004
3	walking	389	0.03	2004
4	health	346	0.04	2002
5	obesity	204	0.03	2005
6	associations	179	0.01	2005
7	walkability	162	0	2009
8	adults	143	0.02	2005
9	travel	129	0.02	2005
10	design	120	0.02	2005
11	neighbourhood	116	0.02	2005
12	land use	108	0.04	2005
13	transportation	103	0.02	2005
14	urban form	103	0.02	2005
15	children	99	0.02	2005
16	older adults	98	0.03	2008
17	public health	94	0.07	2004
18	environment	93	0.03	2005
19	perceptions	91	0.03	2007
20	urban	89	0.02	2007

After conducting a co-occurrence analysis of 665 Chinese-language articles, a total of 372 keyword nodes were obtained. The co-occurrence frequency, ranked from high to low, includes "physical activity", "built environment", "landscape architecture", "non-motorized transport", "healthy city", "elderly", "public health", urban design", and so forth (as shown in Figure 3). Additionally, "physical activity" (0.37), "built environment" (0.16), "landscape architecture" (0.14), and "non-motorized transport" (0.15) are also significantly central keywords (as shown in Table 2), indicating that these keywords are hotspots in Chinese-language literature regarding the impact of the street environment on physical activity.

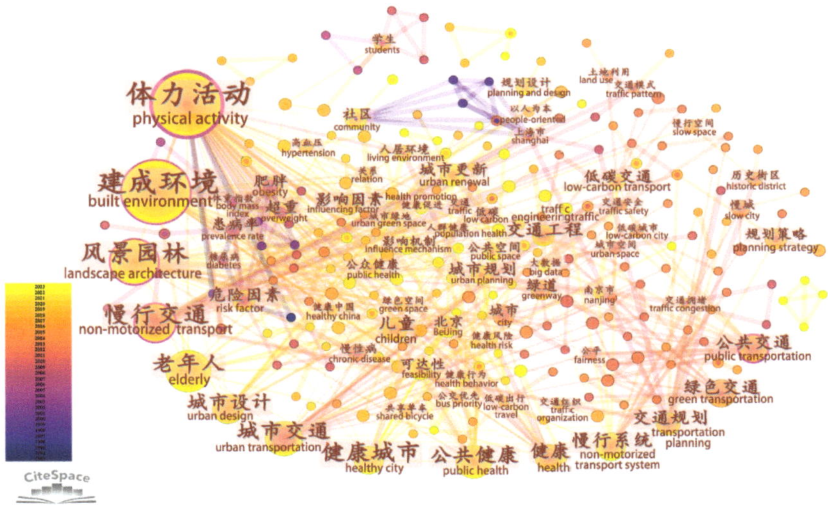

Figure 3. Co-occurrence network map of keywords in Chinese-language literature.

Table 2. Top 20 hotspot words in Chinese-language literature.

Serial Number	Keyword	Frequency of Occurrence	Centrality	Initial Appearance Year
1	physical activity	78	0.37	1994
2	built environment	67	0.16	2014
3	landscape architecture	35	0.14	2011
4	non-motorized transport	33	0.15	2008
5	healthy city	26	0.05	2013
6	elderly	24	0.03	2018
7	urban transportation	20	0.08	2010
8	public health	18	0.03	2015
9	urban design	16	0.07	2013
10	non-motorized transport system	14	0.06	2012
11	healthy	14	0.02	2010
12	children	12	0.02	2011
13	transportation planning	12	0.06	2008
14	public transportation	11	0.04	2005
15	traffic engineering	11	0.04	2010
16	green transportation	11	0.07	2010
17	risk factor	10	0.02	1997
18	urban renewal	9	0.06	2010
19	low-carbon transport	8	0.05	2011
20	influencing factor	8	0.01	2008

3.2. Development Context

The study integrates the annual publication volume of English- and Chinese-language literature, distribution maps of keyword temporal trends, and keyword emphasis maps to comprehensively analyse the development context of research on the impact of street environments on physical activity. Academic output is one of the important indicators for assessing the development status of a research field, and it is significant for predicting development trends and dynamics [38]. From the statistical graphs of publication volume, English-language literature (as shown in Figure 4) shows an overall increasing trend in annual publications, entering a growth pattern in 2004, with fluctuating increases observed after 2016, and with a significant surge in 2019. Chinese-language literature (as shown in Figure 5) exhibits a gradual-to-rapid growth trend in annual publications, notably increasing in 2008, experiencing a slight decline in 2013, and followed by rapid growth leading to a peak in 2019. After 2019, the annual publication volume gradually decreased, contrasting with the trend observed in English-language literature publications. Through literature retrieval and review, it was found that the outbreak of the COVID-19 pandemic at the end of 2019 had an impact, with the Chinese government actively utilizing communities as basic governance units in epidemic prevention and control processes [39,40]. In the post-pandemic era, Chinese scholars have placed greater emphasis on healthy city construction, starting from community environments. This has led to a decrease in research publications focusing on urban street environments.

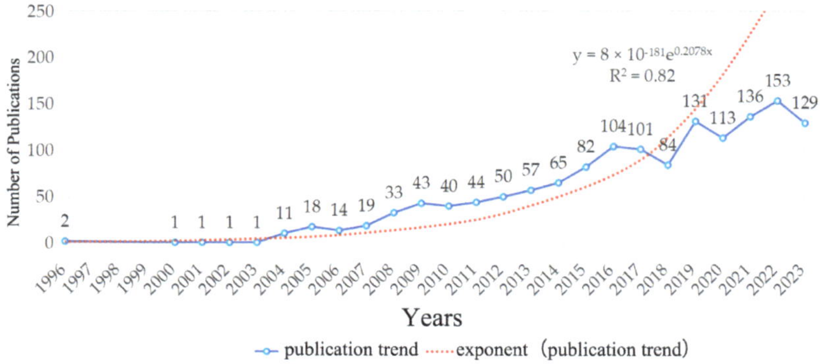

Figure 4. Annual publication volume of studies on the impact of street environments on physical activity in English-language literature.

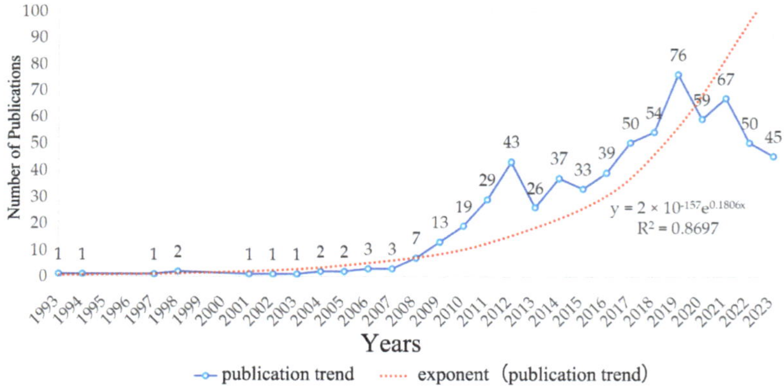

Figure 5. Annual publication volume of studies on the impact of street environments on physical activity in Chinese-language literature.

The timeline graph of keywords clearly demonstrates the evolution of keywords over different time spans and the interrelationships among them [41]. The burst detection graph of keywords with significantly increased frequencies over a short period reflects shifts and progress in research during different periods [42]. Therefore, by integrating the changes in the annual publication volume of English literature (as shown in Figure 4) and based on the timeline graph of keywords (as shown in Figure 6) and the burst detection graph of keywords (as shown in Figure 7), we summarize the evolutionary process of research on the impact of street environments on physical activity in developed countries. This process can be divided into two stages.

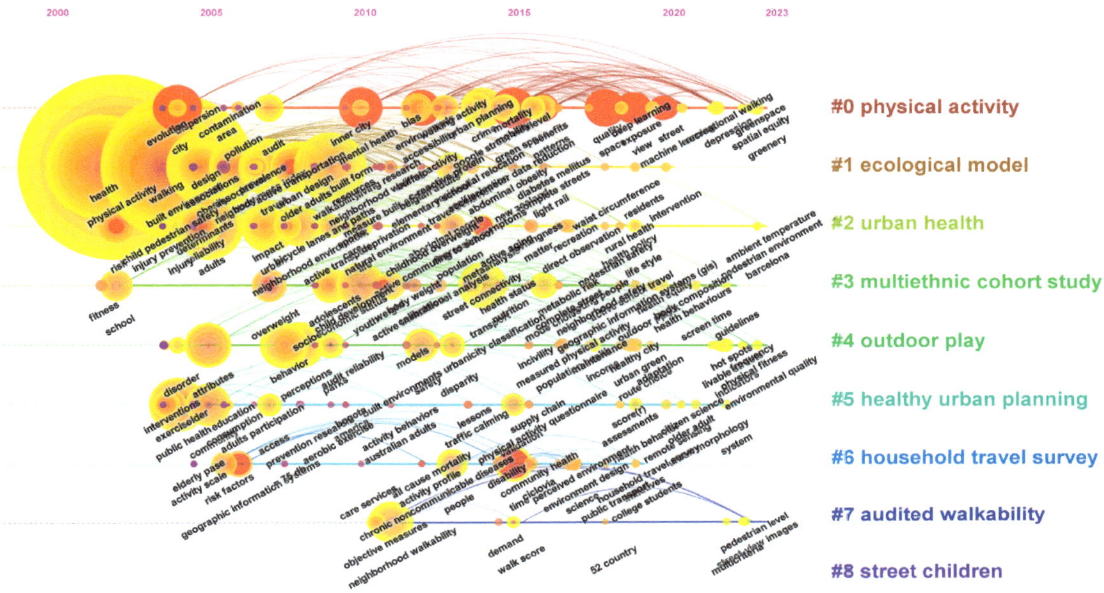

Figure 6. Timeline view of keywords in English-language literature.

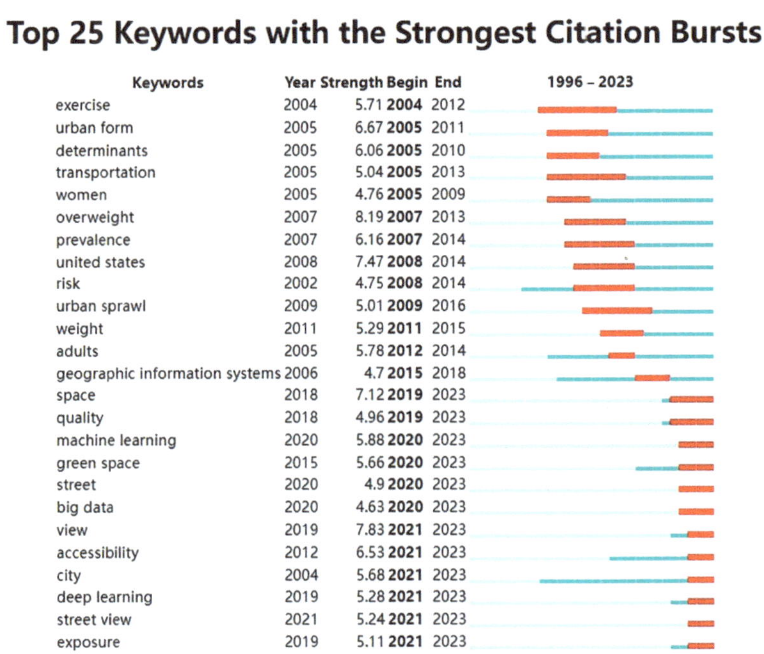

Figure 7. Keyword burst in English-language literature. (Note: the blue blocks denote time slices measured in years, in which the light blue interval indicates that the keyword has not appeared, the dark blue interval indicates the interval of the keyword, and the red blocks represent the start and end times of the emergence of a specific keyword).

Phase 1 (2004–2016): Since the publication, in 2004, of the globally pioneering and influential "Streetscape Guidance: A Guide to Better London Streets" [43], scholars from various countries worldwide have gradually initiated research on street environments. In 2007, street space governance in New York focused on enhancing user interaction and practical exploration [44], adhering to the governance concept of "streets for all" [45], effectively meeting the needs of street users, and publishing the "Street Design Manual" in 2009. Until 2016, research in this phase was predominantly represented by American research teams, with a publication count of 397 papers, accounting for 54.6% of the total publications. The research focused on exploring decisive factors affecting human health during urban morphological changes and city expansion processes from a macro perspective. The burst keywords "urban form" and "urban sprawl" highlighted in Figure 7, along with the high-frequency keywords "land use" and "urban form" used since 2005, as indicated in Table 1, illustrate this focus. For example, Brownson evaluated the impact of urban built environments on physical activity from four aspects: land use, transportation, aesthetics, and safety [46]. Ewing and Cervero incorporated accessibility and distance factors into a 3D model (density, land use, and design), ultimately forming a widely used 5D model (density, land use, design, accessibility, and distance) for the quantitative analysis of street-based built environments [47]. These two highly cited papers in the field have profoundly impacted subsequent research, becoming the foundation for many scholars' studies. Research methods often included interviews or questionnaires [48], systematic observation methods [46], and the use of high-resolution remote sensing imagery [49] for qualitative and quantitative analyses, dissecting the relationship between two-dimensional urban built environment characteristics and user activity behaviours.

Phase 2 (2016 to present): At the 9th Global Conference on Health Promotion in 2016, mayors from over 100 cities worldwide committed to prioritizing health in urban governance, taking measures to eliminate environmental pollution, improve urban infrastructure and public service systems, aiming to construct inclusive, safe, disaster-resistant, sustainable, and healthy cities. The Mayor's Transport Strategy released in London in 2018 [50] and the Global Street Design Guide published in the same year in the United States [51] both proposed the concept of "healthy streets" and ten key indicators, summarizing in detail the research process on urban streets. The core principle of "prioritizing pedestrians in street design" has profoundly impacted the subsequent development of street design guidelines in many cities. Subsequently, scholars' research perspectives shifted from a macroscopic view to focusing on the relationship between micro-level built environment elements of streets and physical activity from a human-centric perspective, expanding the research field centred on street space quality and accessibility. The burst keywords "quality" and "accessibility" from Figure 7 can attest to this shift. For instance, Yin objectively measured street-level urban design quality in Buffalo, New York, using 2D and 3D geographic information systems (GIS), and tested the correlation between micro-environmental characteristics of streets and observed pedestrian counts and walkability scores, aiding in a better understanding of the impact of street-level urban design features on walkability [52]. Larkin et al. employed open-source crowdsourced data, street view images, geographic information systems (GIS), and deep learning methods to identify architectural environmental factors associated with safety, vibrancy, and aesthetics perception in 56 cities, offering a new research method for measuring street space quality [53]. With advancements in measurement technologies (geographic information systems, global positioning systems, street view imagery, and machine learning), the proliferation of individual measurement instruments (eye-tracking devices, electroencephalograms, and virtual reality technologies), and the development of complex models, this phase of research began to emphasize the application of street-level big data, machine learning, and other new technologies to support quantitative indicator measurement, significantly improving research measurement efficiency and accuracy.

The keyword timeline graph (as shown in Figure 6) shows that research hotspot keywords were concentrated in the early part of the 21st century. The nodes of keywords such as "health", "physical activity", "walking", and "built environment" are more significant, transitioning in colour from purple to yellow, indicating sustained scholarly attention.

These keywords are clustered under the "ecological model", and the literature review reveals the profound impact of the "social-ecological theory model" on the study of street environments and physical activity relationships. The "social-ecological theory model" is a representative theory in studying human behaviour influenced by the environment, placing individuals at the centre of ecosystems and comprehensively considering influences on physical activity participation across individual, interpersonal, organizational, community, and policy levels, leading to a more comprehensive and detailed understanding of the factors influencing activity behaviours [54,55]. Since Corti [56] proposed the initial "social-ecological model" in 1998, scholars have subsequently conducted research and created enhancements based on this theoretical model: in 2003, Pikora [57] used the "social-ecological theory model" to establish a theoretical framework for the relationship between physical activity and the environment; in 2005, Zimring et al. [58] further refined the "social-ecological theory model" and studied its relationship with physical activity across four different spatial scales within cities, regions, neighbourhoods, and communities, deepening the study of environmental influences on physical activity and providing a more comprehensive theoretical framework. Subsequently, under the guidance of the "social-ecological theory model", scholars have conducted a substantial amount of empirical research and validation and modification of the model [59,60]. As research content diversified, scholars shifted focus from macro-level material spatial elements, such as road network design [61], land use [62], and their relationship with physical activity, to specific micro-environmental elements, such as street furniture [63], streetscape greening [64], specific populations such as children [65] and older adults [66], and various types of street environment assessments, including health impact assessments [67] and walkability index evaluations [68]. Research has gradually integrated theories from other disciplines, such as Maslow's hierarchy of needs theory [69], self-determination theory [70], attention restoration theory [71], and risk compensation theory [72], to further explore the mechanisms through which street environments influence physical activity.

The keyword time zone graph clearly displays the distribution of research hotspots during different time periods. Observing these hotspots can summarize the developmental trends of research content over time, which helps identify the distribution and evolution of research frontiers [73]. Combining the changes in the annual publication volume of Chinese-language literature mentioned above (as shown in Figure 5), along with the keyword time zone graph (as shown in Figure 8) and keyword burst detection graph (as shown in Figure 9), it is evident that research on the impact of street environments on physical activity in China can be broadly divided into three stages.

Phase 1 (1997–2009): China's focus on healthy cities originated from public health concerns and corresponded with the developmental stage of China's urban and rural socioeconomics. China's focus on physical activity began with medical research on diseases such as obesity and diabetes, primarily involving interviews and surveys of patients within specific streets or neighbourhoods, exploring factors in urban environments that lead to chronic diseases, and identifying physical activity as a significant determinant of human health [74,75]. In this research direction, keywords such as disease prevalence, risk factors, and diabetes received prolonged attention, stemming from Chinese doctors and scientists proposing in the "Daqing Diabetes Prevention Study" that simple lifestyle interventions (diet and exercise management) could reduce the incidence of diabetes by 51% in specific high-risk populations [76]. The conclusion that "diabetes can be prevented" was quickly recognized worldwide, sparking global scholarly interest in diabetes prevention from 1997 onwards. Subsequently, discussions in the public health field shifted from passive disease treatment to proactive health interventions, with unprecedented attention paid to urban material space and social space as critical determinants of health impact [77].

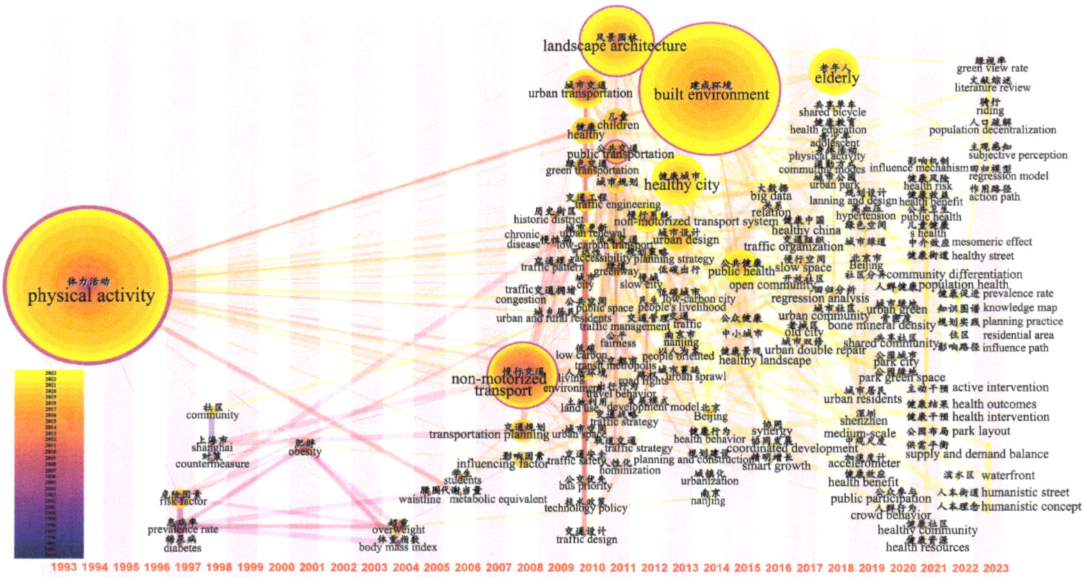

Figure 8. Time zone view of keywords in Chinese-language literature.

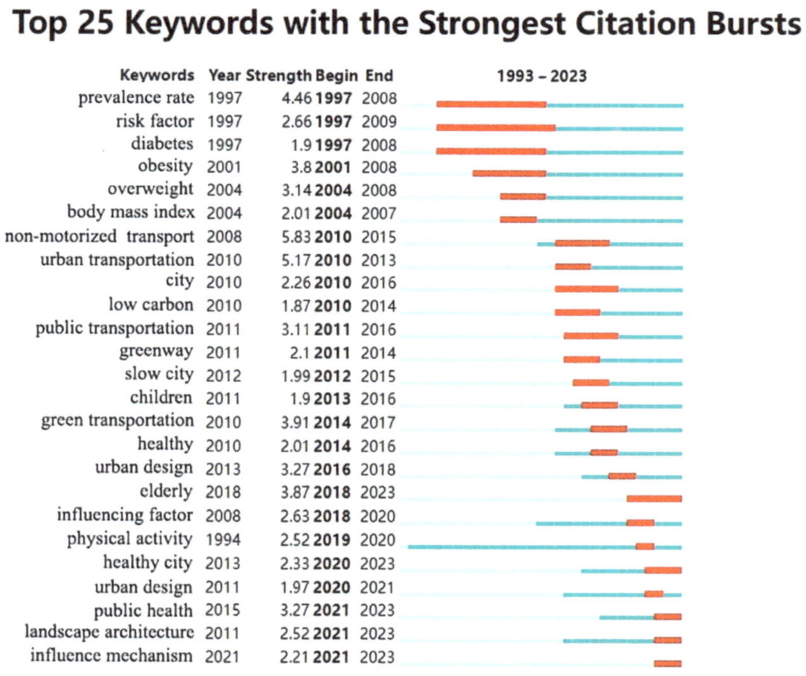

Figure 9. Keyword burst in Chinese-language literature. (Note: the blue blocks denote time slices measured in years, in which the light blue interval indicates that the keyword has not appeared, the dark blue interval indicates the interval of the keyword, and the red blocks represent the start and end times of the emergence of a specific keyword).

Phase 2 (2010–2016): With the rapid urbanization in China, factors such as air pollution, traffic congestion, and road traffic injuries have steadily increased, posing significant health risks related to transportation. Consequently, the impact of transportation on health became a focal point for both the government and scholars during this phase. Beijing introduced the "Beijing Urban Pedestrian and Non-Motorized Vehicle Traffic System Design Guidelines (2010)" [78], and in the same year, the National Development and Reform Commission initiated pilot projects for national low-carbon provinces and cities in Beijing, emphasizing the establishment of an industrial system characterized by low carbon emissions and advocating for low-carbon green lifestyles and consumption patterns [79]. This phase highlights the urgent need to meet urban expansion needs, urban traffic decongestion demands, and urban climate regulation requirements by developing green travel environments. Encouraging urban residents to use walking or cycling as low-carbon transportation methods helps mitigate environmental issues and promotes human health. This has driven extensive research led by transportation engineering professionals to enhance urban public transportation and pedestrian systems. For example, Xiangmin et al. analysed the design of slow transportation systems in Canada and proposed planning strategies for China's healthy city slow transportation systems [80]. Since then, the planning concepts of urban transportation practitioners and infrastructure development have entered a new era, shifting from a primary focus on main routes and long-distance, high-efficiency motorized travel to a concern for the local area's healthy environment and quality, as well as the travel experience of different groups within the city.

Phase 3 (2016 to present): During this period, China's urban development has transitioned from "expansion based on quantity" to a new era of "enhancement based on quality", whereby street space construction has shifted from prioritizing speed to pursuing quality. Enhancing urban space quality to provide residents with a healthy and comfortable living environment has become a significant focus for scholars. Scholars in human settlements are increasingly exploring research on urban health planning and design strategies, emphasizing urban ecological construction, green infrastructure, urban green space accessibility, physical activity, urban open spaces, and forest healing [81]. In 2016, Shanghai issued the "Shanghai street design guideline", the first guideline to systematically explore urban street design from the perspective of "complete streets" in China [82]. That same year, the State Council issued the "Healthy China 2030 Planning Outline", emphasizing "popularizing healthy lifestyles, optimizing health services, and building a healthy environment". Under the guidance of national policies, disciplines such as landscape architecture and urban planning have become the primary research forces. These focus on enhancing the quality of urban street spaces to attract people to engage in physical activities and promote public health development actively. They aim to explore the mechanisms by which street environment elements influence human behaviour, thus providing a theoretical basis for the practice of healthy street design. For instance, Chinese scholars Ying et al. [83] first proposed new approaches to analysing and planning the quality of street spaces in the context of new data environments; Leiqing et al. [84] and Yan et al. [85], respectively, introduced theories and methods for healing-oriented and healthy street design.

The development trajectory of research on the impact of China's street environment on physical activity exhibits characteristics of different stages led by different disciplines. It began with a focus on chronic disease prevention and completed a transformation from passive disease treatment to active health behaviour interventions led by the medical field. Subsequently, with the rapid expansion and development of Chinese cities, urban environmental issues gained attention at the national strategic level. The academic community began to focus on street environment construction for human service, achieving a shift in thinking from "vehicle-centric" to "people-centric", led by transportation engineering. Alongside increasing demands from residents for street space quality, the Chinese government actively promoted the construction of healthy urban environments. Currently, this effort is led by disciplines such as landscape architecture and urban planning in the field of human settlement environments, exploring how street environment elements influence human behaviour perception to provide scientific foundations for developing healthy city plans, improving urban environments, and promoting public health.

3.3. Theme Categories

Data extraction and analysis were conducted using the Bicomb2.0. Bicomb2.0 was utilized to extract 65 keywords with a frequency of 15 times or more from 1434 English-language articles. The keyword threshold was determined based on the Price's Law formula, which was initially used to identify core authors within a research field and is now applied to define high-frequency keywords [86,87]. The calculation formula is as follows:

$$m = 0.749 * \sqrt{n_{\max}} = 0.749 * \sqrt{367} = 14.348 \quad (1)$$

where m represents the threshold boundary and $n_{\max}$ represents the maximum frequency count.

Based on this, 65 keywords with a frequency of ≥15 times were identified as high-frequency keywords (as shown in Table 3). Compared with the high-frequency keywords (as shown in Table 1) analysed by Citespace software above, it was found that, although the frequency of keywords was different due to the different file formats of Citespace and Bicomb analysis, the keyword segments were consistent, which verified the reliability of the analysis results.

Table 3. High-frequency keywords (≥15 occurrences) in English-literature.

Serial Number	Keyword	Frequency of Occurrence
1	built environment	367
2	physical activity	365
3	walkability	174
4	walking	166
5	obesity	75
6	environment	62
7	children	61
8	neighbourhood	59
9	urban design	53
10	GIS	47
11	exercise	46
12	older adults	43
13	public health	40
14	active transportation	34
15	urban form	31
16	active travel	31
17	neighbourhood	29
18	urban planning	29
19	health	29
20	cycling	29
21	google street view	28
22	active transport	28
23	mental health	28
24	geographic information systems	27
25	deep learning	26
26	transportation	26
27	safety	25
28	accelerometer	24
29	land use	24
30	China	23
31	green space	23
32	accessibility	23
33	pedestrians	23
34	urban health	22

Table 3. *Cont.*

Serial Number	Keyword	Frequency of Occurrence
35	active living	22
36	machine learning	21
37	pedestrian	21
38	youth	21
39	mobility	21
40	space syntax	21
41	crime	20
42	COVID-19	20
43	street greenery	20
44	parks	20
45	adolescents	19
46	open streets	19
47	adolescent	19
48	reliability	18
49	neighbourhoods	18
50	aging	18
51	gender	17
52	bicycling	17
53	transport	17
54	air pollution	17
55	recreation	17
56	physical environment	17
57	elderly	16
58	policy	16
59	social environment	16
60	urban environment	16
61	street connectivity	16
62	public space	16
63	neighbourhood environment	16
64	epidemiology	15
65	measurement	15

Topic categories of the impact of the street environment on physical activity were analysed through cluster analysis using SPSS 27.0 software. Cluster analysis can aggregate topic groups representing certain research areas with solid interrelations among them [88], effectively compensating for the limitation of keyword co-occurrence networks in presenting connections between each pair of keywords.

Using Bicomb2.0, we constructed a dissimilarity matrix for 65 high-frequency keywords from English-language literature. This dissimilarity matrix was then imported into SPSS 27.0. Under the "Analysis" menu, we selected "Hierarchical Cluster Analysis" and chose "Within-groups linkage" as the clustering method, with measurement set to "Dice" under the "Binary" option. This process generated a dendrogram for the cluster analysis (as shown in Figure 10). Based on this analysis, the high-frequency keywords were grouped into nine clusters (as shown in Table 4). The thematic categories include the following:

(1) Research on the walkability index of the urban built environment. This topic encompasses 14 high-frequency keywords such as "built environment", "physical activity", and "walking". The focus of this theme revolves around the spatial quality of street-built environments, exploring the visual–spatial quality of built environments regarding urban vitality, public health, and environmental perception. Studies in this area investigate correlations such as walkability and urban vitality [89], walkability and public health [12], and the relationship between walkability and environmental perception [90]. Subjective variables in these studies often use social media, location-based services (LBS) data, or public ratings combined with machine learning methods for measurement. Objective variables frequently rely on OpenStreetMap (OSM) data, street view images, and semantic segmentation through deep learning to quantify

visual elements of the built environment, thus constructing relevant indicators for spatial quality assessment. Notably, studies focusing on establishing direct or indirect relationships between urban greenery features and walkability using survey research, statistical analysis, or spatial analysis methods [64,91] are among the highly prioritized topics within this theme.

(2) Exploring factors that promote active travel among the public. This topic consists of eight high-frequency keywords such as "neighbourhood", "active travel", and "urban planning". The primary focus of this theme is to explain the relationship between public engagement in active travel (dependent variable) and social demographic and geographical factors (independent variables) through the construction of binary regression models. Scholars are particularly interested in adolescents, and the research area of significant interest is the concept of the 15 min neighbourhood [92,93].

(3) Studying the impact of urban environmental hygiene on outdoor physical activity among the public. This topic includes seven high-frequency keywords such as "exercise", "public health", and "transportation". Urban environmental hygiene primarily refers to natural and social environmental factors including air, water, soil, urban and rural planning, and living conditions. Research in this area focuses on issues such as climate change [94], sky view factor (SVF), an effective indicator describing urban radiation and the thermal environment [95], air quality [96], and other aspects of urban physical environments to analyse their impact on outdoor physical activity among the public.

(4) Exploring street environment factors influencing physical activity among the elderly population. This topic primarily includes five high-frequency keywords such as "old adults", "health", and "safety". Engaging in physical activity is an essential factor for promoting healthy ageing, and the elderly population is particularly susceptible to environmental influences due to declining physical function. Street traffic safety and personal safety related to crime are major factors affecting activity frequency among this demographic, drawing significant attention from scholars [31,97]. Research in this area primarily employs interviews, surveys, and behavioural investigations to identify the perceptions of older adults towards street spaces and the street-level elements impacting these perceptions. With rapid advancements in computer vision and deep learning technologies within the field of artificial intelligence, studies are now utilizing street view images and machine learning methods to identify street facilities and building facade elements affecting the perceived safety of elderly individuals. Examples include traffic signs [98], road surface conditions [99], vehicle traffic and parking situations [100], and buildings' window-to-wall ratios [101], among other factors.

(5) Conducting environmental audits focusing on the micro-scale landscapes of streets. This topic primarily includes eight high-frequency keywords such as "reliability", "pedestrian", and "youth". After macro-scale built environment factors were shown to be correlated with physical activity, there has been continued contemplation among ordinary citizens and expert scholars on "what makes a good street landscape?" Many scholars have attempted to find scientific methods to assess pedestrian perspectives on street landscapes. Among these efforts, Microscale Audit of Pedestrian Streetscapes (MAPS) has emerged as an extensively validated observational field audit tool used in health research. MAPS is primarily employed to evaluate whether street environments support physical activity and to identify streetscape elements correlated with physical activity [102,103]. With the application of street view images, virtual audit tools like virtual assessment tools (Virtual-STEPS) [104] and Computer-Assisted Neighborhood Visual Assessment Systems (CANVAS) [105] are also being validated. Within this research theme, scholars focus significantly on adolescents, particularly assessing whether environmental elements along school-commuting routes attract adolescents to engage in physical activity voluntarily [106,107].

(6) Exploring the impact of "Active Living by Design (ALbD)" on promoting community physical activity. This topic is primarily composed of six high-frequency keywords

such as "open streets", "active living", and "bicycling". ALbD is a national program initiated by the Robert Wood Johnson Foundation (RWJF) in 2002, which serves as a community grant program dedicated to exploring and improving environments that support active living. It aims to assist communities in creating environments that facilitate active living using the "social-ecological model" [108]. Research within this theme primarily utilizes experimental methods and has validated in multiple American cities that the ALbD community action model provides effective strategies (the 5Ps: preparation, promotion, planning, policy, and physical projects) to promote community physical activity and support community change [109–112].

(7) Analysing landscape elements that influence pedestrian activity preferences using street view big data. This topic includes four high-frequency keywords: "Google Street View", "deep learning", "accessibility", and "pedestrians". The physical attributes and quality of urban environments can impact people's perceptions of urban spaces. Given that panoramic images captured from a human perspective can comprehensively capture details within the environment, they have become important data sources for perception assessments. Studies of this nature typically involve offline surveys or crowdsourcing via the internet to pose brief evaluation questions about street view images, examining public perceptions of urban environmental quality [113]. Subsequently, computer vision techniques are employed to extract street view image features annotated with emotional biases, followed by the use of machine learning algorithms to analyse perceived attributes or types of image scenes [114]. This process aims to identify urban environments that better align with pedestrian preferences and behaviours.

(8) Studying the impact of urban green spaces on people's mental health. The high-frequency keywords include "mental health", "green space", "mobility", and "China". In recent years, the influence of urban green spaces on the physiological, psychological, and social health of the public, as well as the health benefits of green spaces, has received attention in international research on urban environmental health, urban planning, and urban sustainability fields. Regarding mental health, studies have shown that urban green spaces can alleviate stress among nearby residents [115] and reduce the risk of mental illness [116]. Such research is often based on theories related to the health benefits of nature, such as attention restoration theory (ART) [117], stress reduction theory (SRT) [118], and the biophilia hypothesis [119]. For data acquisition, methods such as emotion response scales, sustained attention response tests, stress perception scales, and self-assessment are used to measure psychological health outcomes [120]. Additionally, with advancements in experimental methods, researchers have begun using techniques like biofeedback devices, functional magnetic resonance imaging (fMRI), and electroencephalography (EEG) to assess physiological indicators such as blood pressure, blood flow, skin conductance, and salivary cortisol levels among participants [121,122]. Furthermore, many studies on this topic have focused on China as the research area [123–125].

(9) Analysing the impact of urban built environment morphology on public behavioural activities. This topic includes nine high-frequency keywords such as "space syntax", "land use", and "crime". Urban built environment morphology indicators encompass land use accessibility and the accessibility of street networks based on space syntax models. This theme focuses on studying the effects of green space accessibility [126], transportation infrastructure accessibility [127], street network accessibility [128], and land use characteristics [129] on public behaviour. This analysis can lead to insights into urban crime issues and the examination of spatial development equity differences within the same city or between different cities, enriching the theoretical framework of environmental justice and socioeconomic impact mechanisms.

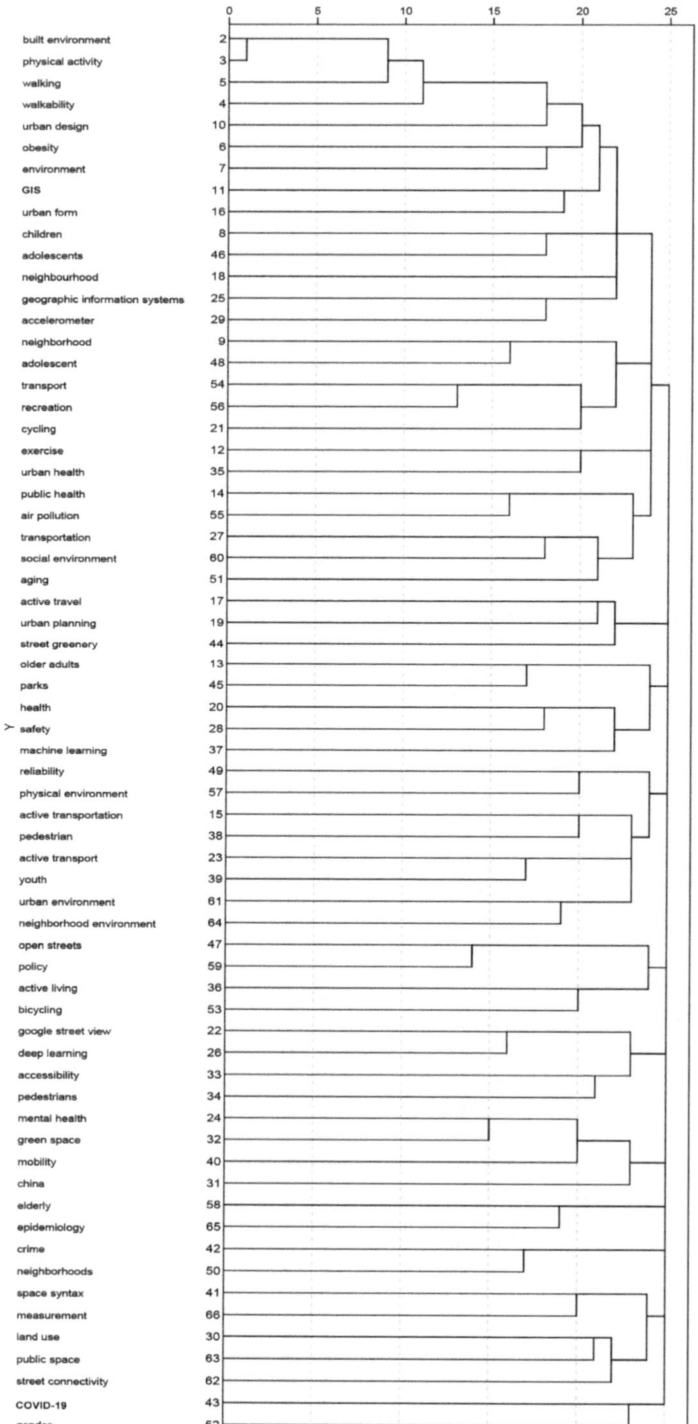

Figure 10. Cluster analysis diagram of high-frequency keywords in English-language literature.

Table 4. Cluster analysis of keywords in English-language literature.

Cluster	Number of Keywords	Cluster Analysis
1	2, 3, 5, 4, 10, 6, 7, 11, 16, 8, 46, 18, 25, 29	built environment, physical activity, walking, walkability, urban design, obesity, environment, GIS, urban form, children, adolescents, neighbourhood, geographic information systems, accelerometer
2	9, 48, 54, 56, 21, 17, 19, 44	neighbourhood, adolescent, transport, recreation, cycling, active travel, urban planning, street greenery
3	12, 35, 14, 55, 27, 60, 51	exercise, urban health, public health, air pollution, transportation, social environment, aging
4	13, 45, 20, 28, 37	older adults, parks, health, safety, machine learning
5	49, 57, 15, 38, 23, 39, 61, 64	reliability, physical environment, active transportation, pedestrian, active transport, youth, urban environment, neighbourhood environment
6	47, 59, 36, 53, 58, 65	open streets, policy, active living, bicycling, elderly, epidemiology
7	22, 26, 33, 34	google street view, deep learning, accessibility, pedestrians
8	24, 32, 40, 31	mental health, green space, mobility, China
9	41, 66, 30, 63, 62, 42, 50, 43, 52	space syntax, measurement, land use, public space, street connectivity, crime, neighbourhoods, COVID-19, gender

Using Bicomb2.0, 42 keywords with a frequency of seven or higher were extracted from 665 Chinese-language articles. The threshold for these keywords was determined based on Price's Law formula, which is calculated as follows:

$$m = 0.749 * \sqrt{n_{\max}} = 0.749 * \sqrt{81} = 6.741 \quad (2)$$

where m represents the threshold boundary and $n_{\max}$ represents the maximum frequency count.

Based on this, 42 words with a frequency of seven or more were determined to be high-frequency keywords in the Chinese-language literature, as shown in Table 5.

Table 5. High-frequency keywords (frequency ≥ 7) in Chinese-language literature.

Serial Number	Keyword	Frequency of Occurrence
1	physical activities	81
2	built environment	68
3	landscape architecture	38
4	landscape architecture	38
5	elderly	27
6	healthy city	26
7	urban transportation	25
8	non-motorized transport system	21
9	urban design	20
10	public health	18
11	green transportation	17
12	health	15
13	urban renewal	14
14	planning strategy	13
15	urban plan	13
16	transportation planning	13
17	obesity	12
18	overweight	12
19	children	12
20	public transportation	12
21	sustainable development	11

Table 5. *Cont.*

Serial Number	Keyword	Frequency of Occurrence
22	traffic engineering	11
23	risk factor	11
24	outdoor physical activity	10
25	greenway	10
26	influencing factor	10
27	planning and design	9
28	accessibility	9
29	recreational physical activity	8
30	new urbanization	8
31	low carbon transportation	8
32	prevalence rate	8
33	community	8
34	city	8
35	rail	8
36	transit-oriented development (TOD)	7
37	big data	7
38	Beijing	7
39	slow travel space	7
40	green travel	7
41	chronic	7
42	ecology	7

Using Bicomb 2.0, a dissimilarity matrix was constructed for 42 high-frequency keywords extracted from Chinese-language literature. This dissimilarity matrix was then imported into SPSS 27.0 to generate the dendrogram for cluster analysis, as shown in Figure 11. Based on the dendrogram, the high-frequency keywords were clustered into five categories (as shown in Table 6). The thematic categories include the following:

(1) Studies on the promotional effect of physical activity on chronic disease prevention. This topic is primarily composed of six high-frequency keywords, including "obesity", "overweight", and "risk factors". The development of urbanization has led to changes in people's lifestyle habits, resulting in an increasing prevalence of chronic diseases, particularly drawing attention from scholars to issues such as obesity and diabetes. Researchers mainly obtain population health indices and relevant influencing factors through questionnaire surveys and physical examinations, exploring urban natural environments (water, light, wind, etc.), built environments (aggregation of man-made constructions and modifications), and social demographic factors that trigger chronic diseases, thereby formulating intervention measures for chronic disease prevention and control [130]. The literature included in the analysis primarily focuses on selecting five healthy lifestyle factors: moderate alcohol consumption, non-smoking, maintaining a healthy weight (not overweight or obese), healthy diet, and regular physical activity [74,131]. Additionally, factors such as sleep [132], prolonged sitting [133], social interactions [134], and mental health [135] are also within the scope of scholars' research.

(2) Exploring community environments conducive to physical activity for children. This topic primarily composed of four high-frequency keywords: "children", "outdoor physical activity", "Beijing", and "community". Insufficient physical activity among children leads to physical deterioration, decreased independence, obesity, myopia, and other issues, drawing significant attention from scholars to children's growth-related problems. Researchers such as Sisi et al. [136] have provided detailed reviews of the current research on urban children's outdoor activity space allocation, exploring the impact of outdoor environments in different urban communities on children's outdoor activities [137–140]. Children's physical activity is closely linked to environmental spaces; actively enriching and natural environments can attract children to engage in a large amount of active play [141]. Existing studies often utilize methods such as

tracking surveys [142], Global Positioning System (GPS), and accelerometers [143] to obtain children's commuting routes and the surrounding sports venues and community facilities, using quantitative research methods to reveal the relationship between the two.

(3) Studies on the impact of urban built environment on the health of elderly populations. This topic primarily composed of nine high-frequency keywords such as "public health", "urban planning", "elderly individuals", and "leisure physical activities". Against the backdrop of an ageing population, concerns arise about the health vulnerability of elderly groups. Various complex factors influence the health of elderly individuals. In addition to the threshold of activity capacity due to individual differences, there is also a strong correlation between urban-built environments and active physical activities among the elderly. In Chinese cities, particularly in Hong Kong, comprehensive research has been conducted on the influence of street environments on physical activities among elderly individuals [144,145]. Existing studies have found that the accessibility of infrastructure, street connectivity, and land use diversity are critical factors in stimulating elderly participation in walking and recreational activities [146,147], demonstrating that physical activities can serve as influential mediating variables in the impact of built environments on health. Linchuan et al. [148] demonstrated a non-linear relationship between street greening and walking behaviour among the elderly in Hong Kong. Furthermore, research has shown that objective-built environments affect the subjective perceptions of elderly individuals, thereby influencing their health behaviours [149], providing a clear insight into the mechanism through which objective-built environment elements impact the health of the elderly.

(4) Using big data methods to analyse the impact of street environments on physical activity. This topic is primarily composed of five high-frequency keywords, including "urban design", "big data", and "non-motorized transport system". The transformation and planning of urban streets towards a more humane and refined design pose higher demands for research and analysis on the human scale. The combination of street view images and machine learning enables the accurate deep processing of street view images, effectively identifying various street landscape elements such as sky, sidewalks, lanes, buildings, and greenery [150], improving the previous challenges of acquiring primary street data and efficiently utilizing street view images. Additionally, movement trajectories and check-in data from mobile applications provide substantial data support for scholars to understand public activity conditions [151], commonly used for studying residents' exercise behaviour characteristics [152] and health level assessments [153]. Furthermore, the practice of utilizing big data to assist in planning and design is also beginning to emerge [154].

(5) Exploring urban transportation systems that promote human health development. This topic primarily encompasses 18 high-frequency keywords such as "planning strategies", "transportation planning", and "public transportation". Streets are essential components of urban transportation systems. Comprehensive urban transportation in China is currently undergoing a comprehensive transformation from focusing on speed to emphasizing development quality, from efficiency to resilience, from macro-level transportation and spatial structures to the integration of transit-oriented development (TOD) and station cities, and from a construction emphasis to comprehensive transformation through inventory renewal and retrofitting [155]. More and more scholars in transportation engineering and urban planning are paying attention to the impact of street environments on public health. Among the factors affecting public health in transportation systems, the expansion of road networks and the sharp increase in private vehicle ownership have led to significant issues of traffic noise and air pollution, which are important causes of chronic diseases [156]. The safety, connectivity, and walkability of streets not only promote physical activity but also have positive impacts on air quality and noise levels [157].

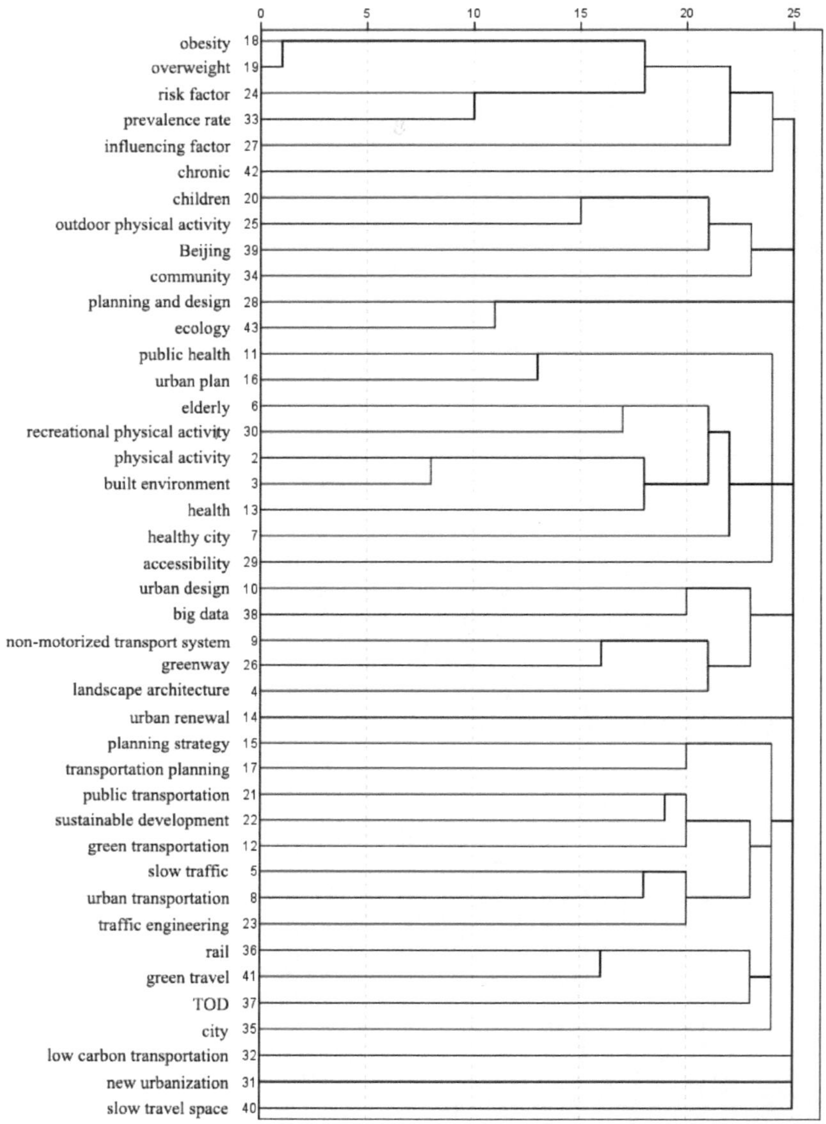

Figure 11. Cluster analysis diagram of high-frequency keywords in Chinese-language literature.

Table 6. Cluster analysis of keywords in Chinese-language literature.

Cluster	Number of Keywords	Cluster Analysis
1	18, 19, 24, 33, 27, 42	obesity, obesity, risk factor, prevalence rate, influencing factor, chronic
2	20, 25, 39, 34	children, outdoor physical activity, Beijing, community
3	11, 16, 6, 30, 2, 3, 13, 7, 29	public health, urban plan, elderly, recreational physical activity, physical activities, built environment, health, healthy city, accessibility

Table 6. *Cont.*

Cluster	Number of Keywords	Cluster Analysis
4	10, 38, 9, 26, 4	urban design, big data. non-motorized transport system, greenway, landscape architecture
5	15, 17, 21, 22, 12, 5, 8, 23, 36, 41, 37, 35, 28, 43, 14, 32, 31, 40	planning strategy, transportation planning, public transportation, sustainable development, green transportation, slow traffic, urban transportation, traffic engineering, rail, green trave, TOD, city, planning and design, ecology, urban renewal, low carbon transportation, new urbanization, slow travel space

3.4. Research Trends

Multidimensional scaling (MDS) analysis was conducted using SPSS 27.0 software. Under the "Analysis" menu, the "Scaling" option was selected to access "Multidimensional Scaling". The scaling model chose "Euclidean distance". This produced quadrant diagrams representing the research hotspots on the impact of street environments on physical activity in international and Chinese contexts, as shown in Figures 12 and 13.

The multidimensional scaling analysis presents a more intuitive graphical representation, displaying the relationships and similarities among keywords in a two-dimensional space. The closer the distance between keywords on the plane, the stronger the relationship between them [158]. The multidimensional scaling chart is divided into four quadrants counterclockwise. The origin represents the mean value; the horizontal axis (Dimension (1)) represents centrality and the external cohesion index, indicating the interaction strength of keywords with other keywords in different clusters; and the vertical axis (Dimension (2)) represents focus density and the internal cohesion index, indicating the internal association strength within the keywords composing the cluster, representing the degree of cluster development [159]. Keywords located in the first quadrant exhibit strong centrality and high density, occupying a central position in the research field. Keywords in the second quadrant, while not central topics, have high density, indicating relatively mature research. Keywords in the third quadrant have weaker density and insufficient centrality, typically indicating emerging or declining topics that are still immature in research. Keywords in the fourth quadrant exhibit strong centrality but lack internal maturity, requiring further in-depth research [160].

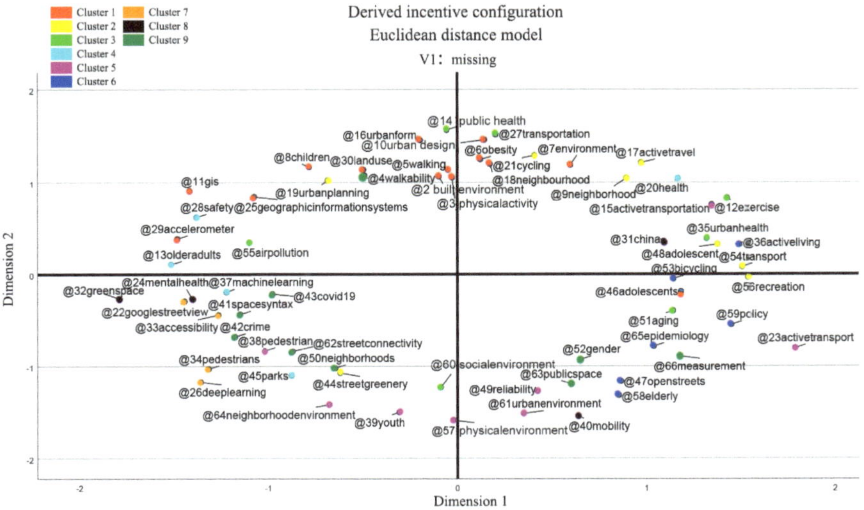

Figure 12. Multidimensional scaling plot of high-frequency keywords from English-language literature.

In the multidimensional scaling plot, clusters are represented by different colours and distributed across different quadrants based on their internal and external cohesion (referring to density and centrality, respectively). As shown in Figure 12, Cluster 1 in English-language literature, "Research on the walkability index of the urban built environment", primarily occupies Quadrants I and II, indicating close internal connections within this research theme and extensive links with other themes, forming a particular research scale. Cluster 1 is closer to Clusters 2 and 3, suggesting relatively tight connections. Keywords like "neighbourhood", "walking", "environment", and "obesity" are located in Quadrant I with high centrality, representing central concepts in this research theme. Cluster 2, "Exploring factors that promote active travel among the public", predominantly occupies Quadrant I, signifying a highly central and dense topic indicating mature research with extensive connections to other themes. This cluster represents a core focus on the impact of street environments on physical activity research. Cluster 3, "Studying the impact of urban environmental hygiene on outdoor physical activity among the public", shows dispersed keyword distribution across all four quadrants, indicating a loosely structured theme lacking systematic research. Keywords like "exercise", "urban health", "transportation", and "elderly" are distributed in the first and fourth quadrants, indicating relatively high centrality. These keywords serve as central terms for this theme. Cluster 4, "Exploring street environment factors influencing physical activity among the elderly population", mainly occupies Quadrants II and III with lower centrality. Research concerning elderly mobility safety is relatively mature, but studies using machine learning to identify street obstacles are less developed and require further exploration. Cluster 5, "Conducting environmental audits focusing on the micro-scale landscapes of streets", primarily occupies Quadrants III and IV, indicating less depth in research. However, with the development and application of big data technologies, there is considerable potential for virtual auditing of street environments, warranting further investigation. Cluster 6, "Exploring the impact of "Active Living by Design (ALbD)" on promoting community physical activity", mainly occupies Quadrants I and IV with higher centrality, suggesting ALbD has profound implications for subsequent research and public environment planning practices. Cluster 7, "Analysing landscape elements that influence pedestrian activity preferences using street view big data", is positioned in Quadrant III with weaker centrality and density. Based on the initial appearance time of keywords, as shown in the timeline of keywords from the English-language literature in the knowledge graph (as shown in Figure 6), it is inferred that the research on this theme is in its initial phase, requiring scholars to conduct more systematic and in-depth studies. Cluster 8, "Studying the impact of urban green spaces on people's mental health", with its four keywords dispersed across Quadrants I, III, and IV, reflects limited systematic research on this topic. However, there is considerable interest among researchers, suggesting that there is room for further development. Cluster 9, "Analysing the impact of urban built environment morphology on public behavioural activities", primarily occupies Quadrants III and IV with a less cohesive internal structure, indicating immature research. Based on the time of appearance of keywords (as shown in Figure 6), research in this category is relatively mature regarding urban crime issues. However, further deepening is needed, particularly concerning issues related to the fairness of urban resources.

Overall, conducting more in-depth and systematic research on urban street safety, environmental assessment, behavioural preferences, psychological healing, and issues of resource equity using extensive data analysis methods is the future trend of research in this field.

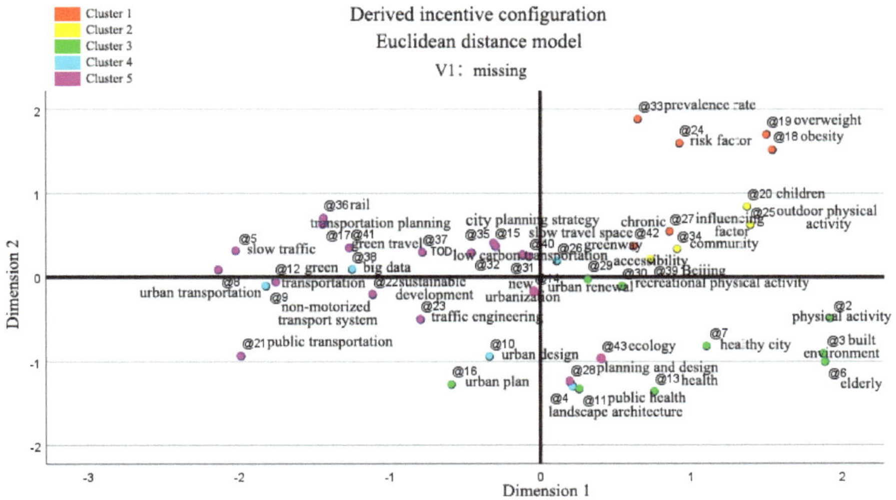

Figure 13. Multidimensional scaling plot of high-frequency keywords from Chinese-language literature.

As shown in Figure 13, in Chinese-language literature, Cluster 1, "Studies on the promotional effect of physical activity on chronic disease prevention", and Cluster 2, "Exploring community environments conducive to physical activity for children", are located in the first quadrant. This indicates that these two topics have strong centrality and high density, suggesting mature research trends with extensive connections to other themes, becoming the core topics in the Chinese-language literature regarding the impact of street environments on physical activity. The keywords "chronic" and "influencing factor" in Cluster 1 and "community" and "Beijing" in Cluster 2 are closely linked. Cluster 3, "Studies on the impact of urban built environment on the health of elderly populations", has its keywords distributed in the fourth quadrant, except for "urban planning", indicating researchers' interest but lacking internal maturity, requiring further study. The proximity of "public health" to "landscape architecture" in Cluster 4 and to "planning and design" in Cluster 5 indicates a close connection among them. Cluster 4, "Using big data methods to analyse the impact of street environments on physical activity", has keywords scattered across all quadrants, indicating a lack of close connections and immature research. The keywords show that the theme revolves around research methods, specifically big data, with relatively close connections to Cluster 3 and Cluster 5. According to the time zone view of keywords (as shown in Figure 8), this cluster is inferred to be a newly emerging research topic that requires further deepening by scholars. Cluster 5, "Exploring urban transportation systems that promote human health development", distributes keywords across the second, third, and fourth quadrants. Most keywords are located in the second quadrant, indicating close internal relevance of the theme and forming a particular scale of research, but with less connection to other themes. "Ecology" and "planning and design" are located in the fourth quadrant, serving as central terms for this theme. "Public transportation", "sustainable development", "urban renewal", "green transportation", and "traffic engineering" are in the third quadrant, representing future research trends for the theme.

Overall, utilizing extensive data methods to explore the impact of the urban built environment on the health of the elderly and employing extensive data methods to assist in constructing urban transportation systems to promote public health development are promising research directions for the future.

4. Comparative Analysis and Discussion

4.1. Development Context

Overall, the global research into and practice of healthy streets have transformed passive responses to survival challenges into proactive interventions for health benefits. A comparative analysis of research hotspots and development contexts in developed countries and China reveals differences in research perspectives, stages, theoretical foundations, and disciplinary backgrounds, while similarities are observed in research scales and methods (as shown in Table 7).

Research perspectives: A comparison of high-frequency critical terms in research from developed countries (Table 1) and China (Table 2) reveals that, in addition to directly relevant keywords such as "physical activity", "built environment", and "public health", high-frequency and high-centrality keywords in Chinese-language research also include "landscape architecture", and "slow traffic". Upon further examination of the relevant literature, it is found that the research perspective in China exhibits a "top-down" characteristic, with scholars predominantly guided by national policies. They analyse existing urban street design cases from a practical perspective or explore factors influencing physical activity through empirical research. Based on this, they provide optimization directions for design decisions, street design guidelines, transportation planning, and other policies, thereby guiding design implementation. Nevertheless, high-frequency keywords in international research also include "walking", "obesity", and "associations". Combined with relevant literature, it is observed that research perspectives in developed countries demonstrate a "bottom-up" feature, with scholars focusing on developing targeted research questions from social life experiences. They primarily conduct causal mechanism studies on street space users, attempt to audit and evaluate already built street environments, and provide feedback on scientific issues, thereby influencing the formulation of urban planning and design policies.

Research stages: The differences in research perspectives indirectly reflect the different stages of research between China and developed countries. Research in China is mainly at the "pre-planning" stage, with the primary purpose being to provide the scientific basis for the planning guidelines and design decisions of healthy streets. Research in developed countries is mainly at the "post-assessment" stage, focusing on assessing and reviewing the healthy streets environment better to understand the mechanisms between street environments and physical activity.

Theoretical foundation: When reviewing the developmental context of research in developed countries and China, it is found that the theoretical foundation for environmental impacts on physical activity mainly originates from the Western academic realm's "social-ecological theory model". Internationally, this theoretical model has been utilized extensively for empirical research [59,60], demonstrating that multiple factors, including individual characteristics, material environmental factors, and social environmental factors, influence physical activity. However, Chinese scholars have applied this theoretical model less frequently, primarily remaining at the stage of literature review research [161–163] and lacking relevant empirical studies. Chinese scholars primarily rely on psychological theories such as Maslow's hierarchy of needs [164], attention restoration theory [165], and Alfonso's five-level system proposed in environmental psychology for walking experience—feasibility, accessibility, safety, comfort, and pleasure [166]—to conduct research on healthy streets environments. Due to the multitude of elements involved in the "social-ecological theory model" and the constantly changing factors affecting specific types of physical activity or under specific background conditions, there is often diversity in the influencing factors. Chinese scholars need to rely on the "social-ecological theory model" to explore the interactive effects and degrees of influence of objective material environments, social environments, and psychological and perceptual environments on behavioural activities within the Chinese context to clarify the direction of the theoretical model further.

Table 7. Comparative analysis of research between developed countries and China.

Research Comparison	Developed Countries	China
Research perspectives	Bottom-up	Top-down
Research stages	Post-assessment	Pre-planning
Theoretical foundation	Sociology: social–ecological theory model; psychology: Maslow's hierarchy of needs, attention restoration theory, self-determination theory, risk compensation theory, and others.	Psychology: Maslow's hierarchy of needs, attention restoration theory; environmental psychology theory.
Disciplinary background	human settlement environment, medicine, transportation engineering, social sciences, psychology, criminology, sports science, and others.	medicine, transportation engineering, and human settlement environment, and other.
Research scope	Macro-scale → human-centred scale	Macro-scale → human-centred scale
Research methods	Small data methods → big data methods	Small data methods → big data methods

Disciplinary background: Research in developed countries emphasizes interdisciplinary collaboration starting from disciplinary issues. For example, the most representative initiative is the "Active Living by Design" (ALbD) national program launched by the United States to address public health issues. Sponsored sites establish interdisciplinary community partnerships, including public health, urban planning, traffic engineering, architecture, pedagogy, and other disciplines, to collectively assess existing policies and environmental conditions, develop strategic plans, identify favourable resources, and collaborate with governments, space users, and others to achieve the sharing of academic research and social practice network resources [167]. In contrast, research in China is influenced by national strategic development, exhibiting a phased transition dominated by different disciplines with loose interdisciplinary cooperation. This is manifested by applying research methods from other disciplines while the research content still belongs to the respective disciplines, representing a borrowing relationship between disciplines [168]. For instance, sports science employs GIS research methods from urban planning to explore the relationship between the density, distance, accessibility, and physical activity of built environment facilities [169].

Research scope: By comparing the keyword timeline graphs of research in developed countries (Figure 6) and China (Figure 8), it is found that there is a trend towards transitioning from a macro-scale to a human-centred scale in the research scope. Specifically, there is a shift from focusing on the influence of aspects such as urban expansion, infrastructure, building density, land use mix, etc., on population physical activity [61,62,170] to paying attention to the impact of factors such as street design elements, street space scale, distance to public facilities, etc., on population physical activity [63,64,142].

Research methods: Both developed countries and China actively utilize multi-source and multi-scale big data, such as government open data, satellite remote sensing images, street view images, etc., combined with machine learning methods for large-scale data quantification analysis. However, research primarily focuses on the perspective of human visual perception, with fewer studies exploring the influence of olfactory and auditory landscape elements in street environments on physical activity. Moreover, existing empirical research mainly consists of static data analysis. It primarily explores the independent effects of single or multiple elements of the built environment on physical activity, neglecting the interactions between elements and temporal changes.

The reasons for the similarities and differences above are mainly reflected in the following three aspects: (1) Different urbanization development processes. Developed countries entered the late stage of urbanization by the end of the last century [171] and had earlier encountered urban public health challenges, leading to corresponding research and practices. On the other hand, China has been undergoing rapid urban expansion and development over the past 40 years, with a relatively late focus on constructing a healthy urban environment, resulting in a lag in research compared to developed countries. Furthermore, as China is in the transitional period of urban construction, many planning guidelines are needed, which require scientific theories and empirical support for their formulation. Additionally, the government attaches great importance to urban health planning and construction, proposing policies and initiatives to promote urban health development, including strengthening urban pollution control, increasing urban greenery levels, promoting nationwide fitness activities, and improving pedestrian transportation systems. These directly or indirectly contribute to the development of research on the impact of urban street environments on physical activity in China, thus presenting a "top-down" research characteristic. (2) Changes in urban planning perspectives. The implementation and management of urban design in various countries are shifting from the "two-dimensional plane" to the "three-dimensional space", focusing on the human scale, which refers to the urban scale closely related to the human body that people can see, touch, and feel. This is a deepening and necessary supplement to the current scales such as grids, blocks, and plots [172]. (3) Development and application of new technologies. Big data methods such as street view images and machine learning, as well as the application of accelerometers and wearable devices, are less time-consuming and cheaper than traditional methods (questionnaires, behavioural observation), allowing for larger sample sizes and providing technical support for human-scale research.

4.2. Theme Categories

Research on the impact of street environments on physical activity in developed countries is categorized into nine themes: "Studies on the promotional effect of physical activity on chronic disease prevention", "Exploring factors that promote active travel among the public", "Studying the impact of urban environmental hygiene on outdoor physical activity among the public", "Exploring street environment factors influencing physical activity among the elderly population", "Conducting environmental audits focusing on the micro-scale landscapes of streets", "Exploring the impact of "Active Living by Design (ALbD)" on promoting community physical activity", "Analysing landscape elements that influence pedestrian activity preferences using street view big data", "Studying the impact of urban green spaces on people's mental health", and "Analysing the impact of urban built environment morphology on public behavioural activities". Research in China is categorized into five themes: "Studies on the promotional effect of physical activity on chronic disease prevention", "Exploring community environments conducive to physical activity for children", "Studies on the impact of urban built environment on the health of elderly populations", "Using big data methods to analyse the impact of street environments on physical activity", and "Exploring urban transportation systems that promote human health development".

Research on the impact of urban street environments on physical activity fundamentally explores the causal relationship between the environment and individuals. Through the comparison of thematic categories, the differences in content between research in developed countries and China are analysed from two perspectives: "individual" (research subjects, activity categories) and "environment" (research areas, environmental factors) (as shown in Table 8). Regarding research subjects, studies in developed countries have a broader focus on the main population groups. In addition to the elderly [97], children [173], adolescents [174], and women [175], attention is also paid to special groups such as low-income groups and minorities [176,177]. Research in China primarily focuses more on the elderly [147], followed by children [142], and women have become the focus in recent studies [178], but there is less attention paid to adolescents, especially college students. However, statistics from the National Health Commission of China show that the prevalence of chronic diseases is becoming increasingly common among younger populations, with hypertension affecting 25% of residents aged 18 and above and abnormal blood lipids affecting 40% [179]. The health issues of adolescents are becoming increasingly prominent, requiring scholars to engage in targeted discussions about this population group. Regarding activity categories, research in developed countries categorizes physical activity in more detail. Currently, international physical activity studies are mainly categorized according to research purposes, with walking being one of the most common activities studied. It comprises the following four types of walking activities: work, transportation, leisure, household, and moderate-intensity physical activity [180]. In addition to walking, activities such as jogging [181], other sports [182], and cycling [183] are also studied, second only to walking. In Chinese-language research, physical activity is not subdivided in detail. However, there is a greater focus on walking activities, with research primarily focusing on the walkability [184], vitality [185], and spatial quality [172] of streets. Regarding research areas, developed countries typically divide study areas into urban areas [186], suburbs [187], and exurbs [188], considering factors such as the level of development, population density, housing type, and income. On the other hand, research in China mainly focuses on the urban core area as the research area [151], with few comparative studies combining different urban development regions. Regarding environmental factors, research in developed countries mainly focuses on the following four aspects: natural ecological environment (blue–green spaces, temperature, sunshine, etc.), the built environment (land use, sidewalk design, facility accessibility, etc.), the sociocultural environment (population density, crime density, etc.), and environmental psychological perception (sense of security, aesthetics, etc.). In contrast, Chinese-language research focuses more on the ecological environment, built environment, and environmental psychological perception elements, lacking investigation into sociocultural and environmental factors.

Table 8. Comparative analysis of research content between studies in developed countries and China.

Research Comparison		Developed Countries	China
Individual	Research subjects	Elderly, children, adolescents, women, low-income groups, and minorities	The elderly, children, and women
	Activity categories	Walking (for work, transportation, leisure, and household purposes), cycling, jogging, and other sports	Walking, cycling, and jogging
Environment	Research areas	Areas, suburbs, and exurbs, considering factors such as the level of development, population density, housing type, and income	Urban core area
	Environmental factors	The natural ecological environment (blue–green spaces, temperature, sunshine, etc.), the built environment (land use, sidewalk design, facility accessibility, etc.), the sociocultural environment (population density, crime density, etc.), and environmental psychological perception (sense of security, aesthetics, etc.)	The natural ecological environment, the built environment, and environmental psychological perception

4.3. Research Trends

Based on the multidimensional scaling chart of research from developed countries and China (Figures 12 and 13), as well as the comparative analysis of research hotspots, development context, and theme categories, the following three considerations and suggestions are proposed for the future research trends of China and other developing countries undergoing urbanization transition: (1) Macro level: National governments should further introduce planning guidelines with strong practical applicability and encourage interdisciplinary collaboration among urban planning, landscape architecture, medicine, transportation engineering, psychology, sociology, and other fields to discuss existing policies and environmental conditions. In addition, the government should encourage researchers, practitioners, and various stakeholders to jointly explore the best research schemes and pathways for healthy streets, effectively realizing the interaction between theory and practice. (2) Mesoscopic level: Research should focus on integrating local characteristics and shifting towards "localization" while fully drawing on research results and practical experiences from developed countries. Urban types and regions should be classified based on population size, geographical location, cultural functions, etc., and targeted assessments of the healthy street environment, should be conducted identifying the pain points and difficulties in street environment construction, and promoting the comprehensive implementation of public policies. (3) Micro level: Scholars should further study the causal mechanism between street environment elements and physical activity. Future research should, at the "individual" level, differentiate population attributes (children, elderly, disabled, special groups, etc.) and expand types of physical activity (running, cycling, leisure activities, etc.). At the "environmental" level, environmental factors from different sensory perspectives (olfactory comfort, auditory comfort, safety perception, etc.) and environmental characteristics of different street types (residential, transportation-oriented, commercial, etc.) should be examined for their impact on physical activity. Additionally, social environment factors such as social norms and community cohesion should be investigated. Integrated use of small data (surveys, interviews, participatory observations, etc.) and big data (street view images, urban points of interest, remote sensing images, etc.) methods should be emphasized to improve survey accuracy and effectiveness, promote the integration of temporal and spatial data with individual behaviour, and mainly focus on the long-term tracking and evaluation of influencing factors, examining the relationship between changes in built environment and physical activity for the same population.

5. Conclusions

The study selects China as representative of developing countries to conduct comparative research with developed countries. It uses relevant literature from the CNKI and WOS databases on the impact of street environments on physical activity as analysis samples. The research tools Citespace, Bicomb, and SPSS are utilized for quantitative and comparative analysis of research hotspots, development contexts, thematic categories, and research trends. The results indicate that, from a research perspective, China exhibits a "top-down" characteristic dependent on policy guidance, whereas developed countries display a "bottom-up" characteristic driven by scientific issues. In the research phase, studies in China are mainly in the "pre-planning" stage, while those in developed countries are primarily in the "post-assessment" stage. Regarding theoretical foundation, the theoretical basis for the impact of environment on physical activity mainly comes from the "social-ecological theoretical model" in Western academic circles, while Chinese scholars often introduce Maslow's hierarchy of needs, attention restoration theory, etc. Regarding disciplinary background, research in developed countries focuses on interdisciplinary cooperation, starting from disciplinary issues, while research in China, influenced by national strategic development, shows a phased transition dominated by different disciplines with relatively loose disciplinary cooperation. On the research scale, both developed countries and Chinese-language studies show a trend of shifting from a macro-scale to a human-centred scale. In terms of research methods, scholars actively apply multi-source and multi-scale big data. Compared to Chinese-language studies, international research examines research subjects, activity types, regions, and environmental impact factors more meticulously and comprehensively in thematic categories. Based on this, the study summarizes three development directions for future research on the impact of street environments on physical activity in China: (1) Macro level: the government can encourage academic exchanges and practical activities on health street environment creation issues between disciplines and stakeholders to enhance the implementation of street planning guidelines; (2) Meso level: research and practice should integrate local characteristics and conduct targeted evaluations for different city types and regions to implement public policies fully; (3) Micro level: refine the influencing factors at the "human" and "environment" levels, conduct long-term tracking and evaluation, and further sort out the causal mechanisms between street environments and physical activity.

The study aids scholars in comprehensively understanding the current state and developmental directions of research on the impact of street environments on physical activity. From a comparative research perspective, it contrasts the similarities and differences between China and developed countries. From the perspective of street environments, it offers pathways and insights for addressing health issues in developing countries with backgrounds similar to China's. It also assists international scholars in accurately understanding the status of research on the impact of street environments on physical activity within the Chinese context, thereby promoting global health welfare and equity. However, it is important to note that this article analyses the hotspots, contexts, themes, and trends of research on the impact of street environments on physical activity in China and developed countries solely from the perspective of "keywords", lacking a comprehensive review of the global research network from the perspectives of "research institutions", "authors", and "publications". Future research should integrate methods such as institutional network analysis, author collaboration network analysis, and co-citation analysis to present the core knowledge framework of the field of healthy street environments to readers.

Author Contributions: Conceptualization, Y.W. and B.L.; methodology, Y.W. and B.L.; software, B.L.; validation, Y.W., B.L., Y.L. and L.Z.; formal analysis, B.L. and Y.L.; data curation, Y.L.; writing—original draft preparation, Y.W. and B.L.; writing—review and editing, Y.W., B.L., Y.L. and L.Z.; visualization, B.L.; supervision, Y.W. and L.Z.; project administration, L.Z.; funding acquisition, Y.W. All authors have read and agreed to the published version of the manuscript.

Funding: This research was funded by the Hebei Provincial Department of Education Scientific Research Project Humanities and Social Sciences General Project entitled "Translation and Expression of Traditional Health Culture in Contemporary Health Landscape Design", grant number SY2022003.

Data Availability Statement: All data supporting the reported results are included in this paper.

Conflicts of Interest: The authors declare no conflicts of interest.

References

1. Ghebreyesus, T.A. As WHO Turns 75, Let's Prescribe Science, Solutions and Solidarity for Our Future Health. World Health Organization. 2023. Available online: https://www.who.int/zh/news-room/commentaries/detail/as-who-turns-75--let-s-prescribe-science--solutions-and-solidarity-for-our-future-health (accessed on 28 April 2024).
2. Lewin, V.; Guerra, J.; Nelson, C.; Davis, A. Five Cities Recognized for Public Health Achievements at Partnership for Healthy Cities Summit. World Health Organization. 2023. Available online: https://www.who.int/zh/news/item/15-03-2023-five-cities-recognized-for-public-health-achievements-at-partnership-for-healthy-cities-summit (accessed on 28 April 2024).
3. World Health Organization. Urban Health. 2021. Available online: https://www.who.int/zh/news-room/fact-sheets/detail/urban-health (accessed on 28 April 2024).
4. United Nations. China's Urban Population to Increase by 255 Million by 2050. 2024. Available online: https://www.un.org/zh/desa/2018-world-urbanization-prospects (accessed on 31 May 2024).
5. Cyril, S.; Oldroyd, J.C.; Renzaho, A. Urbanisation, urbanicity, and health: A systematic review of the reliability and validity of urbanicity scales. *BMC Public Health* **2013**, *13*, 513. [CrossRef] [PubMed]
6. Rosenberg, P.; Kano, M.; Ludford, I.; Prasad, A.; Thomson, H. *Global Report on Urban Health: Equitable, Healthier Cities for Sustainable Development*; World Health Organization: Geneva, Switzerland, 2016.
7. Miranda, J.J.; Kinra, S.; Casas, J.P.; Davey Smith, G.; Ebrahim, S. Non-communicable diseases in low-and middle-income countries: Context, determinants and health policy. *Chin. J. Health Policy* **2013**, *10*, 12–20. [CrossRef] [PubMed]
8. World Health Organization. Noncommunicable Diseases. 2023. Available online: https://www.who.int/zh/news-room/fact-sheets/detail/noncommunicable-diseases (accessed on 28 April 2024).
9. Bush, C.L.; Pittman, S.; McKay, S.; Ortiz, T.; Wong, W.W.; Klish, W.J. Park-based obesity intervention program for inner-city minority children. *J. Pediatr.* **2007**, *151*, 513–517.e1. [CrossRef] [PubMed]
10. Seventh Plenary Meeting. WHO Global Action Plan on Physical Activity 2018–2030. Seventy-First World Health Assembly. 2018. Available online: https://apps.who.int/gb/ebwha/pdf_files/WHA71/A71_R6-ch.pdf (accessed on 2 December 2023).
11. Zaglio, A. The Jakarta Declaration on Leading Health Promotion into the 21st Century. *Ann. Ig. Med. Prev. Comunita* **1998**, *10*, 3–7.
12. Wang, R.; Lu, Y.; Zhang, J.; Liu, P.; Yao, Y.; Liu, Y. The relationship between visual enclosure for neighbourhood street walkability and elders' mental health in China: Using street view images. *J. Transp. Health* **2019**, *13*, 90–102. [CrossRef]
13. Wang, R.; Liu, Y.; Lu, Y.; Zhang, J.; Liu, P.; Yao, Y.; Grekousis, G. Perceptions of built environment and health outcomes for older Chinese in Beijing: A big data approach with street view images and deep learning technique. *Comput. Environ. Urban Syst.* **2019**, *78*, 101386. [CrossRef]

14. Macioszek, E.; Granna, A.; Krawiec, S. Identification of factors increasing the risk of pedestrian death in road accidents involving a pedestrian with a motor vehicle. *Arch. Transp.* **2023**, *65*, 7–25. [CrossRef]
15. Macioszek, E.; Granna, A. The Analysis of the Factors Influencing the Severity of Bicyclist Injury in Bicyclist-Vehicle Crashes. *Sustainability* **2022**, *14*, 215. [CrossRef]
16. Yu, Y.; Jiang, Y.Q.; Li, L. Health Pathways of Urban Public Spaces: Connotations, Elements, and Framework of Health Streets. *Chin. Landsc. Archit.* **2021**, *39*, 20–25.
17. National Bureau of Statistics, Urban Division. Significant Improvement in Urbanization Level, Urban Appearance Rejuvenated—Achievements in Economic and Social Development over 40 Years of Reform and Opening Up Series Report 11. National Bureau of Statistics. Available online: https://www.stats.gov.cn/zt_18555/ztfx/ggkf40n/202302/t20230209_1902591.html (accessed on 28 April 2024).
18. Guan, X.; Wei, H.; Lu, S.; Dai, Q.; Su, H. Assessment on the urbanization strategy in China: Achievements, challenges and reflections. *Habitat Int.* **2018**, *71*, 97–109. [CrossRef]
19. Li, G.P.; Sun, Y. Analysis of Regional Economic Review: Toward Urbanization in China by 2030 and Its Regional Disparities. *Reg. Econ. Rev.* **2020**, 72–81. [CrossRef]
20. Su, B.; Guo, S.; Zheng, X. Transitions in Chronic Disease Mortality in China: Evidence and Implications. *China CDC Wkly.* **2023**, *5*, 1131–1134. [CrossRef] [PubMed]
21. Song, B.; Ren, Y.Y.; Li, W.; Feng, M. Knowledge Graph Analysis of Interdisciplinary Research on Urban Planning and Public Health in English Literature. *J. Xi'an Univ. Archit. Technol. (Nat. Sci. Ed.)* **2021**, 568–576. [CrossRef]
22. Zhang, X.Y. Urban Public Health Risks from the Perspective of Urban Metabolism. *World Geogr. Res.* **2021**, *40*, 319–330.
23. Li, G.; Zhang, L. The Role and Strategy of Sports Industry in Promoting the Construction of Healthy China. *Sports Cult. Guide* **2020**, 67–72. [CrossRef]
24. State Council. The State Council Issues the National Fitness Plan (2021–2025). General Administration of Sport of China. Available online: https://www.sport.gov.cn/gdnps/files/c25531540/25531552.pdf (accessed on 3 December 2023).
25. Jefferis, B.J.; Sartini, C.; Lee, I.M.; Choi, M.; Amuzu, A.; Gutierrez, C.; Casas, J.P.; Ash, S.; Lennon, L.T.; Wannamethee, S.G.; et al. Adherence to physical activity guidelines in older adults, using objectively measured physical activity in a population-based study. *BMC Public Health* **2014**, *14*, 382. [CrossRef] [PubMed]
26. Tainio, M.; Jovanovic Andersen, Z.; Nieuwenhuijsen, M.J.; Hu, L.; de Nazelle, A.; An, R.; Garcia, L.M.T.; Goenka, S.; Zapata-Diomedi, B.; Bull, F.; et al. Air pollution, physical activity and health: A mapping review of the evidence. *Environ. Int.* **2021**, *147*, 105954. [CrossRef] [PubMed]
27. Yin, L.; Wang, Z. Measuring visual enclosure for street walkability: Using machine learning algorithms and Google Street View imagery. *Appl. Geogr.* **2016**, *76*, 147–153. [CrossRef]
28. Blodgett, J.M.; Ahmadi, M.N.; Atkin, A.J.; Chastin, S.; Chan, H.W.; Suorsa, K.; Bakker, E.A.; Hettiarcachchi, P.; Johansson, P.J.; Sherar, L.B.; et al. Device-measured physical activity and cardiometabolic health: The Prospective Physical Activity, Sitting, and Sleep (ProPASS) consortium. *Eur. Heart J.* **2024**, *45*, 458–471. [CrossRef]
29. Xia, G.O.A. Analyzing spatial relationships between urban land use intensity and urban vitality at street block level: A case study of five Chinese megacities. *Landsc. Urban Plan.* **2020**, *193*, 103669. [CrossRef]
30. Klein, S.; Brondeel, R.; Chaix, B.; Klein, O.; Thierry, B.; Kestens, Y.; Gerber, P.; Perchoux, C. What triggers selective daily mobility among older adults? A study comparing trip and environmental characteristics between observed path and shortest path. *Health Place* **2023**, *79*, 102730. [CrossRef] [PubMed]
31. Barnett, D.W.; Barnett, A.; Nathan, A.; Van Cauwenberg, J.; Cerin, E.; Council on Environment and Physical Activity (CEPA)—Older Adults working group. Built environmental correlates of older adults' total physical activity and walking: A systematic review and meta-analysis. *Int. J. Behav. Nutr. Phys. Act.* **2017**, *14*, 103. [CrossRef] [PubMed]
32. Zhang, Y.; Koene, M.; Chen, C.; Wagenaar, C.; Reijneveld, S.A. Associations between the built environment and physical activity in children, adults and older people: A narrative review of reviews. *Prev. Med.* **2024**, *180*, 107856. [CrossRef] [PubMed]
33. Prince, S.A.; Lancione, S.; Lang, J.J.; Amankwah, N.; de Groh, M.; Jaramillo Garcia, A.; Merucci, K.; Geneau, R. Examining the state, quality and strength of the evidence in the research on built environments and physical activity among adults: An overview of reviews from high income countries. *Health Place* **2022**, *77*, 102874. [CrossRef] [PubMed]
34. Zhang, R.; Shu, P. A Review of Urban Green Spaces Based on Recreational Physical Activity. *Landsc. Archit.* **2020**, *28*, 106–113.
35. Chen, Y. Comparative Study of Furniture Design in China and Abroad from the Perspective of Bibliometrics and Knowledge Graph. *Packag. Eng.* **2023**, 45–51. [CrossRef]
36. Cui, Y.Q.; Lin, X.Y. Quantitative Analysis and Evaluation of Research on National Education Policies in China. *J. Southwest Univ. (Soc. Sci. Ed.)* **2020**, 89–97, 195. [CrossRef]
37. Li, J.; Chen, C.M. *CiteSpace: Text Mining and Visualization in Scientific Literature*; Capital University of Economics and Business Press: Beijing, China, 2016.
38. Qiu, J.P.; Yang, S.L.; Song, Y.H. Visual Analysis of the Current Research Status of Knowledge Exchange. *J. Libr. Sci. China* **2012**, *38*, 78–89.
39. Duan, J.; Yang, B.J.; Zhou, L.; Zhang, J.X.; Ye, B.; Luo, H.M.; Yang, Y.Z. Planning to Enhance Urban Immunity: A Symposium on Responding to the Outbreak of Novel Coronavirus Pneumonia. *Urban Plan.* **2020**, *44*, 115–136.
40. Yan, D.R. Constructing a Resilient Community Emergency Governance System. *Adm. Forum* **2020**, *38*, 89–96.

41. Chen, Y. *Principles and Applications of Citation Spatial Analysis: A Practical Guide to CiteSpace*; Science Press: Beijing, China, 2014.
42. Zhou, F.; Luo, L.Y.; Chen, L.; Liu, Z.Y.; Zhang, J.; Dang, Q.Q.; Dong, C.M. Bibliometric and Visual Analysis of Sepsis and Microbiota Therapy. *Microbiol. Bull.* **2024**, 1–23. [CrossRef]
43. Ma, Q.; Wei, X.; Ren, G.N. Street design guidelines and the optimization and improvement of urban road system: The transformation from traffic capacity to space quality. *Urban Transp. China* **2021**, *19*, 1–16.
44. Chen, X.; Liu, Y.H.; Ding, Z.H. Research on Governance Pathways of Street Public Spaces Based on Action Planning: A Case Study of the Street Space Governance Process in New York City from 2007 to 2020. *Int. Urban Plan.* **2021**, *32*, 109–117.
45. Amos, D. Prytherch: Law, Engineering, and the American Right-of-Way: Imagining a More Just Street. *J. Am. Plan. Assoc.* **2020**, *86*, 384–385. [CrossRef]
46. Brownson, R.C.; Hoehner, C.M.; Day, K.; Forsyth, A.; Sallis, J.F. Measuring the built environment for physical activity: State of the science. *Am. J. Prev. Med.* **2009**, *36*, S99–S123.e12. [CrossRef] [PubMed]
47. Ewing, R.; Cervero, R. Travel and the Built Environment: A Meta-Analysis. *J. Am. Plan. Assoc.* **2010**, *76*, 265–294. [CrossRef]
48. Cerin, E.; Saelens, B.E.; Sallis, J.F.; Frank, L.D. Neighborhood Environment Walkability Scale: Validity and development of a short form. *Med. Sci. Sports Exerc.* **2006**, *38*, 1682–1691. [CrossRef]
49. Edwards, N.; Hooper, P.; Trapp, G.S.; Bull, F.C.; Boruff, B.J.; Giles-Corti, B. Development of a public open space desktop auditing tool (POSDAT): A remote sensing approach. *Appl. Geogr.* **2013**, *38*, 22–30. [CrossRef]
50. Caimotto, M.C. London Mayor's Transport Strategy. In *Discourses of Cycling, Road Users and Sustainability. Postdisciplinary Studies in Discourse*; Palgrave Macmillan: Cham, Switzerland, 2020.
51. Global Designing Cities Initiative; National Association of City Transportation Officials. *Global Street Design Guide*; Island Press: Washington, DC, USA, 2016.
52. Yin, L. Street level urban design qualities for walkability: Combining 2D and 3D GIS measures. *Comput. Environ. Urban Syst.* **2017**, *64*, 288–296. [CrossRef]
53. Larkin, A.; Gu, X.; Chen, L.; Hystad, P. Predicting Perceptions of the Built Environment using GIS, Satellite and Street View Image Approaches. *Landsc. Urban Plan.* **2021**, *216*, 104257. [CrossRef]
54. Owen, N.; Sugiyama, T.; Eakin, E.E.; Gardiner, P.A.; Tremblay, M.S.; Sallis, J.F. Adults' sedentary behavior: Determinants and interventions. *Am. J. Prev. Med.* **2011**, *41*, 189–196. [CrossRef]
55. Sallis, J.F.; Owen, N.; Fisher, E.B. Ecological models of health behavior. In *Health Behavior and Health Education: Theory, Research, and Practice*, 4th ed.; Glanz, K., Rimer, B.K., Viswanath, K., Eds.; Jossey-Bass: San Francisco, CA, USA, 2008; pp. 465–485.
56. Giles-Corti, W. The Relative Influence of, and Interaction between, Environmental and Individual Determinants of Recreational Physical Activity in Sedentary Workers and Homemakers. Ph.D. Thesis, The University of Western Australia, Perth, Australia, 1998.
57. Pikora, T. Developing a framework for assessment of the environmental determinants of walking and cycling. *Soc. Sci. Med.* **2003**, *56*, 1693–1703. [CrossRef] [PubMed]
58. Zimring, C.; Joseph, A.; Nicholl, G.L. Influences of building design and site design on physical activity: Research and intervention opportunities. *Am. J. Prev. Med.* **2005**, *28*, 186–193. [CrossRef] [PubMed]
59. Duncan, M.; Mummery, K. Psychosocial and environmental factors associated with physical activity among city dwellers in regional Queensland. *Prev. Med.* **2005**, *40*, 363–372. [CrossRef] [PubMed]
60. Cain, K.L.; Millstein, R.A.; Sallis, J.F.; Conway, T.L.; Gavand, K.A.; Frank, L.D.; Saelens, B.E.; Geremia, C.M.; Chapman, J.; Adams, M.A.; et al. Contribution of streetscape audits to explanation of physical activity in four age groups based on the Microscale Audit of Pedestrian Streetscapes (MAPS). *Soc. Sci. Med.* **2014**, *116*, 82–92. [CrossRef] [PubMed]
61. Ewing, R.; Cervero, R. Travel and the Built Environment: A Synthesis. *Transp. Res. Rec.* **2001**, *1780*, 87–114. [CrossRef]
62. Frank, L.D.; Schmid, T.L.; Sallis, J.F.; Chapman, J.; Saelens, B.E. Linking objectively measured physical activity with objectively measured urban form: Findings from SMARTRAQ. *Am. J. Prev. Med.* **2005**, *28* (Suppl. S2), 117–125. [CrossRef] [PubMed]
63. Ewing, R.; Hajrasouliha, A.; Neckerman, K.M.; Purciel-Hill, M.; Greene, W. Streetscape Features Related to Pedestrian Activity. *J. Plan. Educ. Res.* **2016**, *36*, 5–15. [CrossRef]
64. Lu, Y.; Sarkar, C.; Xiao, Y. The effect of street-level greenery on walking behavior: Evidence from Hong Kong. *Soc. Sci. Med.* **2018**, *208*, 41–49. [CrossRef]
65. Giles-Corti, B.; Wood, G.; Pikora, T.; Learnihan, V.; Bulsara, M.; Van Niel, K.; Timperio, A.; McCormack, G.; Villanueva, K. School site and the potential to walk to school: The impact of street connectivity and traffic exposure in school neighborhoods. *Health Place* **2011**, *17*, 545–550. [CrossRef] [PubMed]
66. Finlay, J.; Franke, T.; McKay, H.; Sims-Gould, J. Therapeutic landscapes and wellbeing in later life: Impacts of blue and green spaces for older adults. *Health Place* **2015**, *34*, 97–106. [CrossRef] [PubMed]
67. Sallis, J.F.; Kerr, J.; Carlson, J.A.; Norman, G.J.; Saelens, B.E.; Durant, N.; Ainsworth, B.E. Evaluating a Brief Self-Report Measure of Neighborhood Environments for Physical Activity Research and Surveillance: Physical Activity Neighborhood Environment Scale (PANES). *J. Phys. Act. Health* **2010**, *7*, 533–540. [CrossRef] [PubMed]
68. Duncan, D.T.; Aldstadt, J.; Whalen, J.; Melly, S.J.; Gortmaker, S.L. Validation of Walk Score® for Estimating Neighborhood Walkability: An Analysis of Four US Metropolitan Areas. *Int. J. Environ. Res. Public Health* **2011**, *8*, 4160–4179. [CrossRef] [PubMed]
69. Alfonzo, M.A. To walk or not to walk? The hierarchy of walking needs. *Environ. Behav.* **2005**, *37*, 808–836. [CrossRef]

70. Hang, T.; Solmon, M. Integrating self-determination theory with the social ecological model to understand students' physical activity behaviors. *Int. Rev. Sport Exerc. Psychol.* **2013**, *6*, 54–76.
71. Marselle, M.R.; Irvine, K.N.; Lorenzo-Arribas, A.; Warber, S.L. Moving beyond Green: Exploring the Relationship of Environment Type and Indicators of Perceived Environmental Quality on Emotional Well-Being following Group Walks. *Int. J. Environ. Res. Public Health* **2015**, *12*, 106–130. [CrossRef] [PubMed]
72. Schölmerich, V.L.N.; Kawachi, I. Translating the socio-ecological perspective into multilevel interventions: Gaps between theory and practice. *Health Educ. Behav.* **2016**, *43*, 17–20. [CrossRef] [PubMed]
73. Tang, Z.; Wei, C. Comparative study of street furniture at home and abroad based on analysis of scientific knowledge graphs. *Packag. Eng.* **2022**, *2022*, 276–285, 289.
74. Ma, W.; Xu, Y.; Li, J.; Xu, H.; Nie, S.; Chen, Z.; Deng, H.; Li, H. Epidemiological characteristics analysis of overweight and obesity among adults in Guangdong Province in 2002. *Chin. J. Epidemiol.* **2004**, *12*, 33–36.
75. Pan, X.; Li, G.; Hu, Y.; Wang, J.; Yang, W.; An, Z.; Liu, J.; Cao, H.; Hu, Z.; Pang, C.; et al. Influence of dietary and exercise interventions on the incidence of diabetes: A prospective observation of 530 individuals with impaired glucose tolerance. *Chin. J. Intern. Med.* **1995**, *2*, 108–112.
76. Pan, X.R.; Li, G.W.; Hu, Y.H.; Wang, J.-X.; Yang, W.-Y.; An, Z.-X.; Hu, Z.-X.; Lin, J.; Xiao, J.-Z.; Cao, H.-B.; et al. Effects of diet and exercise in preventing NIDDM in people with impaired glucose tolerance: The Da Qing IGT and Diabetes Study. *Diabetes Care* **1997**, *20*, 537–544. [CrossRef]
77. Yu, Y.F. Health city planning: From development concepts to planning practices. *Urban Plan. Forum* **2022**, *2022*, 44–49.
78. Beijing Municipal Institute of City Planning and Design. *Design Guidelines for Pedestrian and Non-Motorized Traffic Systems in Beijing Urban Areas;* China Plan Press: Beijing, China, 2010.
79. Xinhua News Agency. Development and Reform Commission Initiates Pilot Projects for National Low-Carbon Provinces and Cities. Central People's Government of the People's Republic of China. 18 August 2010. Available online: https://www.gov.cn/jrzg/2010-08/18/content_1683261.htm (accessed on 28 April 2024).
80. Guo, X.M.; Wang, D.X. Interpretation of Canadian environmental construction from the perspective of healthy cities. *Int. Urban Plan.* **2013**, *28*, 53–57.
81. Zhu, R.R.; Zhao, Y.; Zhang, A.; Gao, W.J. Literature review and analysis of research frontiers in the field of health in landscape architecture. *Chin. Landsc. Archit.* **2021**, *37*, 26–31.
82. Wu, W.Z.; Lu, J.S.; Zhao, B. Comparative study of street design guidelines at home and abroad. *Planner* **2022**, *38*, 58–65.
83. Tang, J.X.; Long, Y.; Zhai, W.; Ma, Y. Measuring quality of street space, its temporal variation and impact factors:an analysis based on massive street view pictures. *New Archit.* **2016**, *5*, 110–115.
84. Xu, L.Q.; Meng, R.X.; Huang, S.Q.; Chen, Z. Healing-oriented street design: Exploration based on VR experiments. *Urban Plan. Int.* **2019**, *34*, 38–45. [CrossRef]
85. Ge, Y.; Shen, X.; Cai, C.T. Theory, method and practice of healthy street design. *Shanghai Urban Plan.* **2020**, *2*, 49–56.
86. Ren, W.; Huang, J. Mapping the structure of interpreting studies in China (1996–2019) through co-word analysis. *Perspectives* **2022**, *30*, 224–241. [CrossRef]
87. Lu, W.; Liu, Z.; Huang, Y.; Bu, Y.; Li, X.; Cheng, Q. How do authors select keywords? A preliminary study of author keyword selection behavior. *J. Informetr.* **2020**, *14*, 101066. [CrossRef]
88. Xie, H.; Zhang, Y.; Wu, Z.; Lv, T. A Bibliometric Analysis on Land Degradation: Current Status, Development, and Future Directions. *Land* **2020**, *9*, 28. [CrossRef]
89. Wang, M.; He, Y.; Meng, H.; Zhang, Y.; Zhu, B.; Mango, J.; Li, X. Assessing street space quality using street view imagery and function-driven method: The case of Xiamen, China. *ISPRS Int. J. Geo-Inf.* **2022**, *11*, 282. [CrossRef]
90. Wang, R.; Liu, Y.; Lu, Y.; Yuan, Y.; Zhang, J.; Liu, P.; Yao, Y. The linkage between the perception of neighborhood and physical activity in Guangzhou, China: Using street view imagery with deep learning techniques. *Int. J. Health Geogr.* **2019**, *18*, 18. [CrossRef] [PubMed]
91. Sarkar, C.; Webster, C.; Pryor, M.; Tang, D.; Melbourne, S.; Zhang, X.; Jianzheng, L. Exploring associations between urban green, street design and walking: Results from the Greater London boroughs. *Landsc. Urban Plan.* **2015**, *143*, 112–125. [CrossRef]
92. Leung, K.Y.K.; Loo, B.P.Y. Determinants of children's active travel to school: A case study in Hong Kong. *Travel Behav. Soc.* **2020**, *21*, 79–89. [CrossRef]
93. Oliver, M.; McPhee, J.; Carroll, P.; Ikeda, E.; Mavoa, S.; Mackay, L.; Kearns, R.A.; Kyttä, M.; Asiasiga, L.; Garrett, N.; et al. Neighbourhoods for Active Kids: Study protocol for a cross-sectional examination of neighbourhood features and children's physical activity, active travel, independent mobility and body size. *BMJ Open* **2016**, *6*, e013377. [CrossRef] [PubMed]
94. Turrisi, T.B.; Bittel, K.M.; West, A.B.; Hojjatinia, S.; Hojjatinia, S.; Mama, S.K.; Lagoa, C.M.; Conroy, D.E. Seasons, weather, and device-measured movement behaviors: A scoping review from 2006 to 2020. *Int. J. Behav. Nutr. Phys. Act.* **2021**, *18*, 24. [CrossRef] [PubMed]
95. Carrasco-Hernandez, R.; Smedley, A.R.D.; Webb, A.R. Using urban canyon geometries obtained from Google Street View for atmospheric studies: Potential applications in the calculation of street level total shortwave irradiances. *Energy Build.* **2015**, *86*, 340–348. [CrossRef]
96. Wu, D.; Gong, J.; Liang, J.; Sun, J.; Zhang, G. Analyzing the influence of urban street greening and street buildings on summertime air pollution based on street view image data. *ISPRS Int. J. Geo-Inf.* **2020**, *9*, 500. [CrossRef]

97. Bracy, N.L.; Millstein, R.A.; Carlson, J.A.; Conway, T.L.; Sallis, J.F.; Saelens, B.E.; Kerr, J.; Cain, K.L.; Frank, L.D.; King, A.C. Is the relationship between the built environment and physical activity moderated by perceptions of crime and safety? *Int. J. Behav. Nutr. Phys. Act.* **2014**, *11*, 24. [CrossRef] [PubMed]
98. Campbell, A.; Both, A.; Sun, Q.C. Detecting and mapping traffic signs from Google Street View images using deep learning and GIS. *Comput. Environ. Urban Syst.* **2019**, *77*, 101350. [CrossRef]
99. Ren, M.; Zhang, X.; Chen, X.; Zhou, B.; Feng, Z. YOLOv5s-m: A deep learning network model for road pavement damage detection from urban street-view imagery. *Int. J. Appl. Earth Obs. Geoinf.* **2023**, *120*, 103335. [CrossRef]
100. Rita, L.; Peliteiro, M.; Bostan, T.-C.; Tamagusko, T.; Ferreira, A. Using deep learning and Google Street View imagery to assess and improve cyclist safety in London. *Sustainability* **2023**, *15*, 10270. [CrossRef]
101. Harvey, C.; Aultman-Hall, L.; Hurley, S.E.; Troy, A. Effects of skeletal streetscape design on perceived safety. *Landsc. Urban Plan.* **2015**, *142*, 18–28. [CrossRef]
102. Cain, K.L.; Gavand, K.A.; Conway, T.L.; Geremia, C.M.; Millstein, R.A.; Frank, L.D.; Sallis, J.F. Developing and validating an abbreviated version of the Microscale Audit for Pedestrian Streetscapes (MAPS-Abbreviated). *Transp. Res. Part F Traffic Psychol. Behav.* **2017**, *5*, 84–96. [CrossRef] [PubMed]
103. Kurka, J.M.; Adams, M.A.; Geremia, C.; Zhu, W.; Cain, K.L.; Conway, T.L.; Sallis, J.F. Comparison of field and online observations for measuring land uses using the Microscale Audit of Pedestrian Streetscapes (MAPS). *J. Transp. Health* **2016**, *3*, 278–286. [CrossRef]
104. Steinmetz-Wood, M.; Velauthapillai, K.; O'Brien, G.; Ross, N.A. Assessing the micro-scale environment using Google Street View: The Virtual Systematic Tool for Evaluating Pedestrian Streetscapes (Virtual-STEPS). *BMC Public Health* **2019**, *19*, 1246. [CrossRef] [PubMed]
105. Mooney, S.J.; Wheeler-Martin, K.; Fiedler, L.M.; LaBelle, C.M.; Lampe, T.; Ratanatharathorn, A.; DiMaggio, C.J. Development and validation of a Google Street View pedestrian safety audit tool. *Epidemiology* **2020**, *31*, 301–309. [CrossRef] [PubMed]
106. Vanwolleghem, G.; Ghekiere, A.; Cardon, G.; De Bourdeaudhuij, I.; D'haese, S.; Geremia, C.M.; Lenoir, M.; Sallis, J.F.; Verhoeven, H.; Van Dyck, D. Using an audit tool (MAPS Global) to assess the characteristics of the physical environment related to walking for transport in youth: Reliability of Belgian data. *Int. J. Health Geogr.* **2016**, *15*, 41. [CrossRef] [PubMed]
107. Rosenberg, D.; Ding, D.; Sallis, J.F.; Kerr, J.; Norman, G.J.; Durant, N.; Harris, S.K.; Saelens, B.E. Neighborhood Environment Walkability Scale for Youth (NEWS-Y): Reliability and relationship with physical activity. *Prev. Med.* **2009**, *49*, 213–218. [CrossRef]
108. Bors, P.; Dessauer, M.; Bell, R.; Wilkerson, R.; Lee, J.; Strunk, S.L. The Active Living by Design National Program: Community initiatives and lessons learned. *Am. J. Prev. Med.* **2009**, *37* (Suppl. S2), S313–S321. [CrossRef]
109. Sayers, S.P.; LeMaster, J.W.; Thomas, I.M.; Petroski, G.F.; Ge, B. Bike, walk, and wheel: A way of life in Columbia, Missouri, revisited. *Am. J. Prev. Med.* **2012**, *43* (Suppl. S4), S379–S383. [CrossRef] [PubMed]
110. Deehr, R.C.; Shumann, A. Active Seattle: Achieving walkability in diverse neighborhoods. *Am. J. Prev. Med.* **2009**, *37* (Suppl. S2), S403–S411. [CrossRef] [PubMed]
111. Gomez-Feliciano, L.; McCreary, L.L.; Sadowsky, R.; Peterson, S.; Hernandez, A.; McElmurry, B.J.; Park, C.G. Active Living Logan Square: Joining together to create opportunities for physical activity. *Am. J. Prev. Med.* **2009**, *37* (Suppl. S2), S361–S367. [CrossRef] [PubMed]
112. McCreedy, M.; Leslie, J.G. Get Active Orlando: Changing the built environment to increase physical activity. *Am. J. Prev. Med.* **2009**, *37* (Suppl. S2), S395–S402. [CrossRef] [PubMed]
113. Salleses, P.; Schechtner, K.; Hidalgo, C.A. The collaborative image of the city: Mapping the inequality of urban perception. *PLoS ONE* **2013**, *8*, e68400. [CrossRef] [PubMed]
114. Naik, N.; Philipoom, J.; Raskar, R.; Hidalgo, C.A. Streetscore–Predicting the Perceived Safety of One Million Streetscapes. In Proceedings of the 2014 IEEE Conference on Computer Vision and Pattern Recognition Workshops, Columbus, OH, USA, 23–28 June 2014; pp. 793–799.
115. Aspinall, P.; Mavros, P.; Coyne, R.; Roe, J. The urban brain: Analysing outdoor physical activity with mobile EEG. *Br. J. Sports Med.* **2015**, *49*, 272–276. [CrossRef] [PubMed]
116. Bratman, G.N.; Hamilton, J.P.; Hahn, K.S.; Daily, G.C.; Gross, J.J. Nature experience reduces rumination and subgenual prefrontal cortex activation. *Proc. Natl. Acad. Sci. USA* **2015**, *112*, 8567–8572. [CrossRef] [PubMed]
117. Kaplan, R. Some psychological benefits of an outdoor challenge programme. *Environ. Behav.* **1974**, *6*, 101–116.
118. Ulrich, R.S. Aesthetic and affective response to natural environments. In *Behavior and the Natural Environments*; Altman, I., Wohlwill, J.F., Eds.; Plenum Press: New York, NY, USA, 1983.
119. Wilson, E.O. *Biophilia*; Harvard University Press: Cambridge, UK, 1984.
120. Koordzij, J.M.; Beenackers, M.A.; Oude Groeniger, J.; Van Lenthe, F.J. Effect of changes in green spaces on mental health in older adults: A fixed effects analysis. *Epidemiol. Community Health* **2020**, *74*, 48–56. [CrossRef] [PubMed]
121. Beyer, K.M.M.; Kaltenbach, A.; Szabo, A.; Bogar, S.; Nieto, F.J.; Malecki, K.M. Exposure to neighborhood green space and mental health: Evidence from the Survey of the Health of Wisconsin. *Int. J. Environ. Res. Public Health* **2014**, *11*, 3453–3472. [CrossRef]
122. Chen, X.; Wang, B.; Zhang, B. Far from the madding crowd: The positive effects of nature, theories and applications. *Adv. Psychol. Sci.* **2016**, *24*, 270–281. [CrossRef]
123. Helbich, M.; Yao, Y.; Liu, Y.; Zhang, J.; Liu, P.; Wang, R. Using deep learning to examine street view green and blue spaces and their associations with geriatric depression in Beijing, China. *Environ. Int.* **2019**, *126*, 107–117. [CrossRef] [PubMed]

124. Wang, R.; Helbich, M.; Yao, Y.; Zhang, J.; Liu, P.; Yuan, Y.; Liu, Y. Urban greenery and mental wellbeing in adults: Cross-sectional mediation analyses on multiple pathways across different greenery measures. *Environ. Res.* **2019**, *176*, 108535. [CrossRef] [PubMed]
125. Wang, R.; Feng, Z.; Pearce, J.; Liu, Y.; Dong, G. Are greenspace quantity and quality associated with mental health through different mechanisms in Guangzhou, China: A comparison study using street view data. *Environ. Pollut.* **2021**, *290*, 117976. [CrossRef] [PubMed]
126. Fan, P.Y.; Chun, K.P.; Mijic, A.; Tan, M.L.; Liu, M.S.; Yetemen, O. A framework to evaluate the accessibility, visibility, and intelligibility of green-blue spaces (GBSs) related to pedestrian movement. Urban For. *Urban Green* **2022**, *69*, 127494. [CrossRef]
127. Mohamed, A.A.; Kronenberg, J.; Łaszkiewicz, E. Transport infrastructure modifications and accessibility to public parks in Greater Cairo. *Urban For. Urban Green* **2022**, *73*, 127599. [CrossRef]
128. Sarkar, C.; Gallacher, J.; Webster, C. Built environment configuration and change in body mass index: The Caerphilly Prospective Study (CaPS). *Health Place* **2013**, *19*, 33–44. [CrossRef] [PubMed]
129. Ozbil, A.; Gurleyen, T.; Yesiltepe, D.; Zunbuloglu, E. Comparative Associations of Street Network Design, Streetscape Attributes and Land-Use Characteristics on Pedestrian Flows in Peripheral Neighbourhoods. *Int. J. Environ. Res. Public Health* **2019**, *16*, 1846. [CrossRef] [PubMed]
130. Xue, B.; Pang, Z.; Bao, G.; Gao, W.; Nan, H.; Wang, S.; Ren, J.; Zhang, L.; Qiao, Q. Analysis of overweight and obesity status and influencing factors among residents in Qingdao City. *Chin. J. Public Health* **2008**, *5*, 585–586.
131. Deng, Y.; Gao, Y.L.; Wu, X.P.; He, J.; Ji, K.; Zhang, N.M.; Yuan, J.G. Epidemiological analysis of overweight and obesity among urban and rural adults in Sichuan Province. *Mod. Prev. Med.* **2007**, *18*, 3487–3489.
132. Feng, S.; Yang, Y.; Xia, N. Associations between healthy lifestyles and depressive symptoms among adults in China: A longitudinal study. *Lancet* **2018**, *392*, S59. [CrossRef]
133. Wu, M.Q.; Xiao, J.; Li, F.F.; Hu, R.R.; Liu, C.X. 24-hour activity behavior: A new direction in studying factors influencing the health of elderly people. *Chin. Gen. Pract.* **2024**, *27*, 1911–1916.
134. Shen, T.; Wang, Y.; Jin, W.; Lin, Z.H.; Yan, L. Analysis of risk factors for cognitive decline in elderly patients with chronic diseases and construction of a risk model. *J. Jilin Univ. (Med. Ed.)* **2023**, *49*, 1304–1309.
135. Pan, F.; Zhang, X.Y.; Gan, Y.D.; Gong, L.T.; Zhang, L.C.; Chang, C. Study on the current status and influencing factors of depression and anxiety symptoms among patients with different chronic diseases in Daxing District, Beijing. *Chin. Health Educ.* **2023**, *39*, 948–954.
136. Chen, S.S.; Yang, G.; Liu, L. Reflections on the study of outdoor activity space allocation for urban children in China. *J. Southw. Univ. (Nat. Sci. Ed.)* **2019**, *44*, 103–108.
137. Xiao, X.N.; Han, X.L. Characteristics of outdoor physical activity spaces for children in urban villages and environmental influencing factors: A case study of Pingshan Village, Shenzhen City. *Mod. Urban Res.* **2019**, *34*, 8–14.
138. Huang, L.; Yin, X.M. Study on the correlation between children's activity behavior and space in mountainous urban communities: A case study of Shangdatianwan Community in Yuzhong District, Chongqing. *Urban Plan.* **2022**, *46*, 87–98, 120.
139. Leng, H.; Gao, Z.Q.; Yuan, Q. Characteristics of outdoor activities in different seasons in cold urban areas for children and implications for spatial planning. *Chin. Landsc. Archit.* **2020**, *9*, 53–58.
140. Liu, X.H.; Yu, Y.F. Influence of community built environment on the health of the elderly under conditions of high-density living environment and intervention paths. *Urban Dev. Res.* **2023**, *8*, 35–42.
141. Wang, X.T.; Han, X.L. Influence of environmental factors diversity on children's physical activity in activity venues. *J. West. Hum. Settl.* **2018**, *6*, 100–105.
142. He, X.L.; Zhuang, J.; Zhu, Z.; Wang, C.; Chen, P.J. Built environment factors affecting high-intensity physical activity among children and adolescents: A study based on GIS objective measurement. *Sports Sci.* **2017**, *38*, 101–110+51.
143. Quan, M.H.; He, X.L.; Su, Y.Y.; Chen, P.J.; Zhuang, J. Tracking study on spatial characteristics of physical activity of children and adolescents based on GPS and accelerometer. *Sports Sci.* **2017**, *1*, 111–120.
144. Yang, Y.; He, D.; Gou, Z.; Wang, R.; Liu, Y.; Lu, Y. Association between street greenery and walking behavior in older adults in Hong Kong. *Sustain. Cities Soc.* **2019**, *51*, 101747. [CrossRef]
145. Zang, P.; Lu, Y.; Ma, J.; Xie, B.; Wang, R.; Liu, Y. Disentangling residential self-selection from impacts of built environment characteristics on travel behaviors for older adults. *Soc. Sci. Med.* **2019**, *238*, 112515. [CrossRef] [PubMed]
146. Yu, Y.; Lu, X. Behavioral variables and their mediating effects in the study of the health impact of built environment: A case study of the health behavior of the elderly in Shanghai community. *Sci. Technol. Rev.* **2021**, *39*, 8.
147. Wu, Z.J.; Wang, Z.Y.; Zhang, F.; Song, Y.L.Q.; Wang, H.L.; Gao, L. The impact of urban built environment on the health of the elderly: Model verification mediated by physical activity. *China Sport Sci. Technol.* **2019**, *55*, 41–49.
148. Yang, L.; Ao, Y.; Ke, J.; Lu, Y.; Liang, Y. To walk or not to walk? Examining non-linear effects of streetscape greenery on walking propensity of older adults. *J. Transp. Geogr.* **2021**, *94*, 103099. [CrossRef]
149. Feng, J.; Huang, X.; Tang, S. Study on the influence mechanism of objective and subjective built environment on different physical activities of the elderly: A case study of Nanjing. *Shanghai Urban Plan. Rev.* **2017**, *3*, 17–23.
150. Shen, Q.; Zeng, W.; Ye, Y.; Arisona, S.M.; Schubiger, S.; Burkhard, R.; Qu, H. StreetVizor: Visual exploration of human-scale urban forms based on street views. *IEEE Trans. Vis. Comput. Graph.* **2018**, *24*, 1004–1013. [CrossRef]

151. Zhu, C.; Zheng, S.; Zhen, R.; Rong, Q. Research on the optimization of healthy street environment based on motion trajectory data: A case study of the core area of Beijing. *Planner* **2023**, *39*, 72–79.
152. Cui, Z.; He, L.; Wu, L.; Zhang, X. Voting with feet: Performance evaluation and diagnosis of running spaces in Beijing based on individual trajectories. *Planner* **2023**, *39*, 68–75.
153. Wu, J.; Lin, Y.; Zhou, Z. Research on the development of "Internet+" sports health products. *Packag. Eng.* **2017**, *38*, 16–19.
154. Huang, H.; Gan, X.; Chen, Y. Optimization research on slow travel system in Lugu Lake scenic area based on big data of sports trajectory. *Planner* **2020**, *22*, 19–24.
155. Kong, L. Review and reflection on the development of urban transportation in the past 20 years: Witness of the 20th anniversary of the founding of Urban Transportation. *Urban Transp.* **2023**, *21*, 16–25.
156. Wang, Y. Influence of health-supportive environment on regular physical activity of urban and rural residents: From the perspective of social ecology. *Mod. Urban Res.* **2021**, *10*, 111–117.
157. Song, Y.; Wang, Z.; Wu, Z. Fuzzy evaluation of urban community built environment for leisure physical activity of elderly people. *J. Xi'an Inst. Phys. Educ.* **2018**, *35*, 309–317.
158. Ding, Y.; Xue, H. Retrospective analysis of hot topics in educational management research in China: Based on bibliometric and co-word analysis of "Modern Educational Management" (2009–2016). *Mod. Educ. Manag.* **2017**, *12*, 1–7.
159. Li, H.Y.; Cui, L.; Cui, M. Hot topics in Chinese herbal drugs research documented in PubMed/MEDLINE by authors inside China and outside of China in the past 10 years: Based on co-word cluster analysis. *Altern. Complement. Med.* **2009**, *15*, 779–785. [CrossRef] [PubMed]
160. Zhao, F.; Shi, B.; Liu, R.; Zhou, W.; Shi, D.; Zhang, J. Theme trends and knowledge structure on choroidal neovascularization: A quantitative and co-word analysis. *BMC Ophthalmol.* **2018**, *18*, 86. [CrossRef]
161. Huang, Q.; Jin, H. Research progress and implications of the impact mechanism of green spaces on physical activity based on the social-ecological model. *Chin. Landsc. Archit.* **2023**, *39*, 93–98.
162. Wang, Y.; Sun, Y.; Dai, D. A review of research on physical activity of urban residents abroad based on the social-ecological model. *Mod. Urban Res.* **2020**, *4*, 27–35.
163. Wang, X.; Jiao, J. A review of research on built environment based on commuting trips. *Int. Urban Plan.* **2018**, *33*, 57–62, 109. [CrossRef]
164. Huang, Z.; Li, Z.; Lang, W. Measurement of street space quality and its impact on street vitality based on multi-source big data: A case study of historical urban areas in Guangzhou. *Shanghai Urban Plan.* **2023**, *6*, 122–130.
165. Zhang, J.; Yu, Z.; Zhao, B. The pathways through which urban green spaces promote public health: Theoretical framework and practical implications. *Landsc. Archit.* **2020**, *8*, 104–113.
166. Zhao, M.; Zhen, F.; Jiang, Y. Comprehensive evaluation and optimization strategies of urban community walking environment: A case study of the main urban area of Nanjing. *Mod. Urban Res.* **2021**, *2*, 41–48.
167. Wu, X.; Wang, H.; Zhang, Y.; Qian, C. Landscape architecture and public health: Consensus, boundaries, and integration—An examination of disciplinary relationships under the perspective of two public health revolutions. *Chin. Landsc. Archit.* **2021**, *37*, 6–13.
168. Wang, Z. *Research Methods in Landscape Design*; China Architecture & Building Press: Beijing, China, 2022.
169. Wu, Z.; Wang, Z.; Zhu, J.; Wang, H.; Zhang, F.; Hu, F. The influence of subjective and objective built environments on outdoor physical activity among older adults. *J. Cap. Inst. Phys. Educ.* **2023**, *35*, 317–325.
170. Liang, H. Research on the relationship between urban morphology and physical activity. *J. Chengdu Sport Univ.* **2016**, *42*, 32–36.
171. Ma, P.; Li, W.; Fang, W. Urbanization, agglomeration effect, and the development of the tertiary industry. *J. Financ. Econ.* **2010**, *8*, 101–108.
172. Ye, Y.; Zhang, Z.; Zhang, X.; Zeng, W. Measurement of street space quality at the human scale: A large-scale, high-precision evaluation framework combining street view data and new analysis techniques. *Int. Urban Plan.* **2019**, *34*, 18–27. [CrossRef]
173. Rahman, T.; Cushing, R.A.; Jackson, R.J. Contributions of Built Environment to Childhood Obesity. *Mt. Sinai J. Med.* **2011**, *78*, 49–57. [CrossRef] [PubMed]
174. Dzhambov, A.; Hartig, T.; Markevych, I.; Tilov, B.; Dimitrova, D. Urban Residential Greenspace and Mental Health in Youth: Different Approaches to Testing Multiple Pathways Yield Different Conclusions. *Environ. Res.* **2018**, *160*, 47–59. [CrossRef] [PubMed]
175. Shaoming, Z.; Yuan, Y.; Linting, W. Impacts of Urban Environment on Women's Emotional Health and Planning Improving Strategies: An Empirical Study of Guangzhou Based on Neuroscience Experiments. *China City Plan. Rev.* **2023**, *32*, 17–27.
176. Wolch, J.R.; Byrne, J.; Newell, J.P. Urban Green Space, Public Health, and Environmental Justice: The Challenge of Making Cities 'Just Green Enough'. Landsc. *Urban Plan.* **2014**, *125*, 234–244. [CrossRef]
177. James, P.; Hart, J.E.; Banay, R.F.; Laden, F.; Signorello, L.B. Built Environment and Depression in Low-Income African Americans and Whites. *Am. J. Prev. Med.* **2017**, *52*, 74–84. [CrossRef] [PubMed]
178. Chen, C.; Haili, T.; Yong, C. Study on Built Environment Influencing Factors and Planning Response of Obesity in Elderly Women. *Hum. Geogr.* **2018**, *33*, 76–81.
179. Pan, H.; Zuo, Q.; Yang, X.; Chen, S.; Zhao, L. The proactive health model integrating physical education, health, and hygiene is an important means of preventing and controlling chronic diseases in adolescents. *Chin. J. Med.* **2024**, *15*, 211–216.

180. Sugiyama, T.; Neuhaus, M.; Cole, R.; Giles-Corti, B.; Owen, N. Destination and route attributes associated with adults' walking: A review. *Med. Sci. Sports Exerc.* **2012**, *44*, 1275–1286. [CrossRef] [PubMed]
181. Lin, D.; Jiang, H.; Li, W.; Qiu, B.; Wang, H.; Qiu, W. Assessing impacts of objective features and subjective perceptions of street environment on running amount: A case study of Boston. *Landsc. Urban Plan.* **2023**, *235*, 104756.
182. Randers, M.B.; Petersen, J.; Andersen, L.J.; Krustrup, B.R.; Hornstrup, T.; Nielsen, J.J.; Nordentoft, M.; Krustrup, P. Short-term street soccer improves fitness and cardiovascular health status of homeless men. *Eur. J. Appl. Physiol.* **2012**, *112*, 2097–2106. [CrossRef]
183. Kellstedt, D.K.; Spengler, J.O.; Foster, M.; Lee, C.; Maddock, J.E. A scoping review of bikeability assessment methods. *J. Community Health* **2021**, *46*, 211–224. [CrossRef] [PubMed]
184. Long, Y.; Li, L.; Li, S.; Chen, L.; Pan, Z.; Yao, Y.; Chen, M.; Wang, Y.; Quan, J.; Zhang, L.; et al. Measurement of street walking environment index in China's urban vitality centers. *South. Archit.* **2021**, *1*, 114–120. [CrossRef]
185. Hao, X.; Long, Y.; Shi, M.; Wang, P. Street vitality in Beijing: Measurement, influencing factors, and planning design implications. *Shanghai Urban Plan.* **2016**, *3*, 37–45.
186. Sallis, J.F.; Cerin, E.; Conway, T.L.; Adams, M.A.; Frank, L.D.; Pratt, M.; Salvo, D.; Schipperijn, J.; Smith, G.; Cain, K.L.; et al. Physical activity in relation to urban environments in 14 cities worldwide: A cross-sectional study. *Lancet* **2016**, *387*, 2207–2217. [CrossRef] [PubMed]
187. Janeczko, E.; Bielinis, E.; Wójcik, R.; Woźnicka, M.; Kędziora, W.; Łukowski, A.; Elsadek, M.; Szyc, K.; Janeczko, K. When urban environment is restorative: The effect of walking in suburbs and forests on psychological and physiological relaxation of young Polish adults. *Forests* **2020**, *11*, 591. [CrossRef]
188. Mygind, L.; Bentsen, P.; Badland, H.; Edwards, N.; Hooper, P.; Villanueva, K. Public open space desktop auditing tool—Establishing appropriateness for use in Australian regional and urban settings. *Urban For. Urban Green.* **2016**, *20*, 65–70. [CrossRef]

buildings

Article

Experimental Study on the Universal Design of Signage Size and Brightness Contrast for Low Vision Individuals

Erke Zhang [1], Wei Zhao [1,*], Zihan Mei [2], Zhexi Yang [3], Fei Chen [1], Yuanyuan Xia [4] and Yihan Wang [1]

1 Tianjin Key Laboratory of Healthy Habitat and Smart Technology, School of Architecture, Tianjin University, Tianjin 300072, China; zhangerke@tju.edu.cn (E.Z.); fchen@tju.edu.cn (F.C.); iran_0908@tju.edu.cn (Y.W.)
2 Darwin College, University of Cambridge, Cambridge CB3 9EU, UK; zihanmeicn@gmail.com
3 Department of Architecture and Civil Engineering, City University of Hong Kong, Hong Kong 999077, China; zhexiyang2-c@my.cityu.edu.hk
4 School of Environment and Architectural Art, Tianjin Academy of Fine Arts, Tianjin 300402, China; xiayuanyuan@tjarts.edu.cn
* Correspondence: tjuzhaowei@tju.edu.cn; Tel.: +86-18649182139

Abstract: Signage is an important medium for individuals to obtain spatial environmental information conveniently and accurately. However, most previous studies focus on signage design for individuals with normal vision, neglecting the specific requirements of those with low vision. Therefore, these signage designs lacking universality restrict the activities of individuals with low vision in public spaces and increase their risks. This study aims to conduct quantitative research on signage size and brightness contrast for low vision individuals. A virtual simulation experimental platform investigated how individuals interpret and react to signage. Two sets of experiments were carried out in Tianjin Key Laboratory of Healthy Habitat and Smart Technology to evaluate the effects of signage size and brightness contrast on response time and accuracy among a total of 139 participants. The results showed that the signage size should be at least 7% of the reading distance to meet the wayfinding needs of low vision individuals. The impact of contrast on wayfinding was strongly dependent on signage size. This research provides valuable insights into the design of signage for low vision individuals.

Keywords: brightness contrast; low vision individuals; signage size; universal design

Citation: Zhang, E.; Zhao, W.; Mei, Z.; Yang, Z.; Chen, F.; Xia, Y.; Wang, Y. Experimental Study on the Universal Design of Signage Size and Brightness Contrast for Low Vision Individuals. *Buildings* **2024**, *14*, 2063. https://doi.org/10.3390/buildings14072063

Academic Editor: Derek Clements-Croome

Received: 4 June 2024
Revised: 21 June 2024
Accepted: 23 June 2024
Published: 5 July 2024

Correction Statement: This article has been republished with a minor change. The change does not affect the scientific content of the article and further details are available within the backmatter of the website version of this article.

1. Introduction

Signage is an abstract symbol delivering spatial orientation and environmental information through text and graphics. The functional and safety-related information about space and other aspects obtained from signage facilitates wayfinding activities in everyday life [1]. The proportion of people with low vision significantly increases with age [2], and those individuals are at a disadvantage in spatial recognition. "Low vision" is defined as one or several impairments to the normal function of the eye or visual nervous system. The concept of low vision was introduced by Eleanor Faye and Gerald Fonda in 1950 [3]. Scientists who focused on visual impairments during the same period emphasized the importance of residual vision and the need to address limitations in visual function [4,5].

Thus, a clear and effective signage system should be introduced to help low vision individuals to integrate into society freely and independently, thereby improving the accessibility and utilization of the spatial environment [6,7]. In the 1990s, the concept of "universal design" was introduced to the field of architectural design, and visual accessibility was defined as the efficient and safe use of vision to navigate through the environment, perceive key features of the spatial layout, and track position and orientation in the environment [8]. In 1989, Rubin conducted a series of experiments and demonstrated a strong correlation between the contrast of visual objects and low vision reading [9]. In

1992, Arthur proposed that signage is an information vector that clarifies spatial orientation through sequence, numbering, and area marking, enabling the understanding of functional and safety information about space [1]. The guideline 'Lighting and the Visual Environment for Senior Living (ANSI/IES RP-28-16)' [10] is one of the earliest guidelines focusing on how visual impairment affects people's perception of the environment. Then, in 2010, Schambureck et al. used typology of environmental scenarios to address low vision-related issues [11].

In recent years, new technologies for information acquisition for individuals with visual impairments have been developed [12]. In 2006, Walker and Lindsay proposed the use of virtual auditory displays for navigation and localization [13]. Wilson et al. designed a wearable audio navigation system [14]. Legge et al. suggested that "visually impaired individuals use digital signage systems for indoor navigation" [15]. Zhang [16] proposed an intelligent icon design platform; however, this platform only considers the color schemes without considering the size of the icons. Thus, the provided intelligent color schemes still do not meet environmental design needs. Although scholars have made great efforts to develop new technologies to provide equivalent information for individuals with visual impairments, almost all of them focus on utilizing non-visual media to compensate the impaired vision for spatial activities, with little research exploring how to exploit the residual vision [17]. In the foreseeable future, signage remains necessary for the general public and low vision population to capture spatial information for participating in independent activities. Poorly designed spatial signage can reduce users' recognition of the space, affect their wayfinding activities within the space, and, in severe cases, lead to injuries.

Thus, signage size and brightness contrast are taken as the main variables in this research to study universal signage design for low vision individuals. The brightness contrast is crucial for recognition of signage by low vision individuals. It has been widely recognized that a higher level of brightness contrast between an object and its surrounding environment enables low vision individuals to utilize their residual vision more effectively [6,17–20], thereby improving their ability to recognize signage [9,21,22]. In some countries, using black-white or white-black combinations for signage has even been recommended, emphasizing the importance of maximizing contrast for conveying information.

A common contrast ratio recommendation is 70% for color combinations between foreground and background [22–26]. However, quantitative research on brightness contrast is limited, and there is a lack of both a reliable theoretical framework and empirical evidence [10]. The calculation methods for contrast differ among countries, with the UK and US using Weber's formula:

$$C = \frac{B_1 - B_2}{B_1 \times 100} \tag{1}$$

where C is the brightness contrast, B_1 is the light reflectance value (LRV) of the brighter area, and B_2 is the LRV of the darker area. Germany and Spain use Michelson's formula:

$$C(\%) = \frac{L_{max} - L_{min}}{L_{max} + L_{min}} \tag{2}$$

where L_{max} is the brightness of the brightest color, and L_{min} is the brightness of the darkest color. The standard Australian signage formula is expressed as follows:

$$C = \frac{125(Y_{max} - Y_{min})}{25 + Y_{max} + Y_{min}} \tag{3}$$

where Y is the luminous reflectivity.

In subjective evaluation experiments, although brightness contrast or luminous reflectance is often used as the main indicator, there is still controversy over the results. Various studies have shown different minimum requirements of brightness contrast for effective recognition of signage by individuals with low vision. For example, the Japanese researcher Hayanouchi claimed that the brightness contrast should be greater than 60% for designing the color schemes of tactile paths [27]. A study by the Georgia Institute of Technology in 1985 found that the minimum brightness contrast for effective recognition of signage by low vision individuals should be 70% [26], which was later included in the ADA (Americans with Disabilities Act) Accessibility Guidelines. Yoshida [28] showed that elderly adults required a minimum contrast ratio of 33–50%. Lomperski [29] found through empirical research that the recommended minimum brightness contrast of 70% was significantly higher than the appropriate value. The study of Bright et al. [30] on architectural interfaces in the UK proposed a minimum brightness contrast of 30%.

Although many scholars have recognized the importance of brightness contrast in signage design, there is no consensus on the optimal level, and existing research on brightness contrast in signage design varies widely in terms of both design standards and calculation methods [31]. Table 1 summarizes the lightness contrast requirements for signage in different countries. Moreover, the diverse types of visual impairment necessitate different design requirements for different groups of low vision individuals [32]. In this study, we focus on adjusting brightness contrast to better ensure the independent activity of individuals with low vision.

Table 1. Requirements for brightness contrast of signages in different countries [21,33–36].

Standard	The Requirements for Brightness Contrast
Germany	Appropriate combinations are characterized by one or more of the following: high brightness contrast, achromatic components, and combinations of complementary colors. Among them, <0.16 is low contrast, <0.64 is medium, and ≥0.64 is high contrast; for the visually impaired, the contrast should be more than 0.83.
Spain	A minimum of 60% brightness contrast between shapes and backgrounds for the low vision individual.
The Netherlands	For normal groups, the difference in reflectance of signage and background of signage and background must be ≥0.30, and the reflectance values of white and black are 0.85 and 0.04, respectively. However, the rationale for the values is not provided.
Sweden	Due to weathering and discoloration, the brightness contrast diminishes overtime. Therefore, for normal groups, some institutions believe that a brightness contrast of 60% is not recommended and should be set it to 70%.
UK	For low vision individuals, the brightness contrast between the wall and the signboard should be 70%. However, no quantitative experimental data are provided.
USA	The brightness contrast should be more than 70%.
Canada	A recommendation of 70% brightness contrast between the foreground and background (NCS color system) is provided without any further information.
China	The background color of the pedestrian directional signages and the brightness contrast between brightness and dark of the layout elements should not be less than 30%.
ISO 21542 [36]	The proposed brightness contrast for signage is 60%.

The signage size affects people's recognition of signage, and appropriate signage size can help individuals better understand signage information. Research on signage size mainly focuses on two aspects: the size of the characters and the size of the signage icons. In the literature collected from 10 countries or regions in this study (Table 2), the relationship between character size and distance ranges from 1.8% to 9%, with a mean of 4.28%. The United States is the birthplace of "universal design," and research and regulations on accessible signage are more advanced. Peters and Smith propose formulas for user-friendly signage font design [37,38]. Loomis provides indicators for the near-sight font size for low vision individuals [39]. Holmes-Siedle suggested several installation heights for accessible signage [40]. British standards stipulate a minimum height of 2300 mm for suspended signage, and font heights should be 5.7% of the viewing distance [41]. Japan's "Accessibility

Transportation Law" simply specifies the installation height of accessible signage as the arithmetic mean of adult eye height and wheelchair eye height, which is 1.35 m [42].

Table 2. The calculation method for the size of signage (and its characters or letters) [21,32,43].

Standards	Recommended Letter Scale	Recommended Character Scale	Size and Reading Distance	Remark
Belgium	—	1.8–3.5%	2–4%	Depending on the importance of information displayed Text information should be supplemented by symbols for understanding
The Netherlands	—	5% of critical reading distance	5%	Contrasting letters must have a minimum ratio of 1:3 to the background
Luxemburg	—	—	3.75%	
Ireland	—	At least 15 × 15 cm	6%	Absolute minimum of 1.5–2.5 cm Emergency signs should be sized larger, as they may need to be followed in smoky conditions and/or without good lighting
United Kingdom	Minimum Size (1.5–2.5 cm) Medium (5–10 cm) Long distance (>15 cm)	The overall height of the symbol should be at least 10 cm, if space permits	Average of 5%	Symbols are generally better than words
Spain	—	—	2.75% (minimum 1.4%)	—
Sweden	The character size cannot be increased indefinitely because the text is no longer readable when low vision individuals get too close to the signage	—	2% (minimum 1.5%)	—

Compared to characters, only a few countries have regulated symbol sizes, and the regulations on the sizes that meet the needs of low-vision people are also relatively messy. Moreover, there are significant structural differences between Latin alphabet characters and Chinese characters. Regarding the requirements for Chinese characters, their use for individuals with visual impairments are likely to be significantly different, and further research is needed. Therefore, this study also seeks to investigate signage size in the context of Chinese characters.

2. Experimental Design

2.1. Enrollment and Determination of Visual Status

The experimental design focuses on the visual evaluation characteristics of individuals with clinically diagnosed low vision [44]. Upon enrollment, participants were grouped based on a combination of subjective assessments and objective experimental methods [45]. Visual acuity tests were conducted on the subjects at the experimental site, and they were divided into four groups according to their visual acuity status, including normal vision, low vision, visual field defect, and color vision anomalopia, as illustrated in Figure 1 [22].

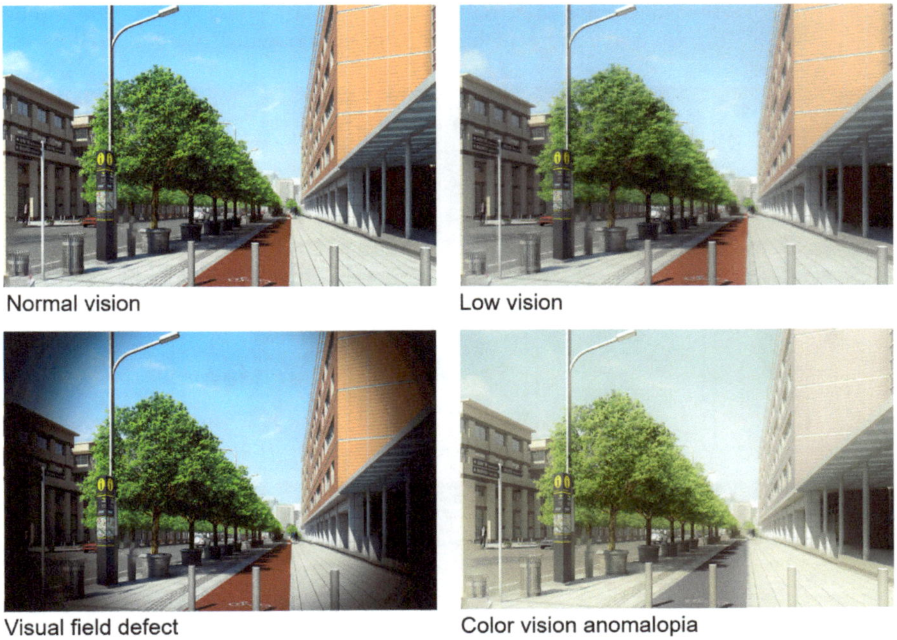

Figure 1. The same scene observed through different vision impairments (created by authors).

2.2. Experiment Setup

Figure 2 presents the experiment setup. The experiment variables included shape, size, and brightness contrast, and spatial environmental variables such as uniform lighting are controlled.

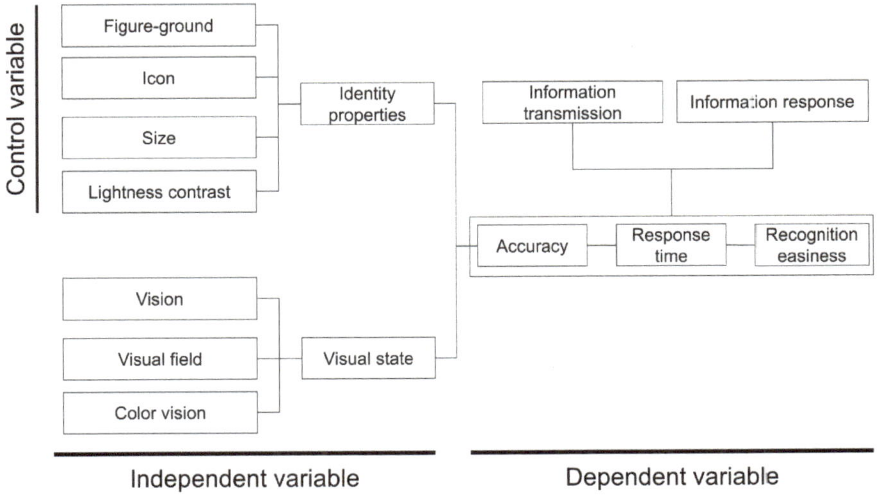

Figure 2. Configuration of experimental variables.

(1) Signage Selection: Ten common signs were selected from the Chinese National Standard Library [35], including five white-on-black signs and five black-on-white signs shown in the bottom two rows in Figure 3. The texts were in sans-serif font and were divided into five white-on-black and five black-on-white signs. In addition, the top two rows in Figure 3 present the ten image signs included, resulting in a total of twenty signs.

Figure 3. Signage for experiments.

(2) Signage Size: Five different sizes, which were determined by the height of signage and word, were chosen that equaled sizes of 4.5, 13.5, 22.5, 31.5, and 40.5 cm, which corresponded to 1%, 3%, 5%, 7%, and 9% of the reading distance (4.5 m) from the screen respectively. (3) Brightness Contrast: Five levels were chosen as illustrated in Figure 4. The contrast intensity used in Experiments 1 and 2 were predetermined according to Table 1, and the values were 14%, 21%, 33%, 60%, and 76%. The selection of signage size and brightness contrast was based on the guidelines provided in Reference [22].

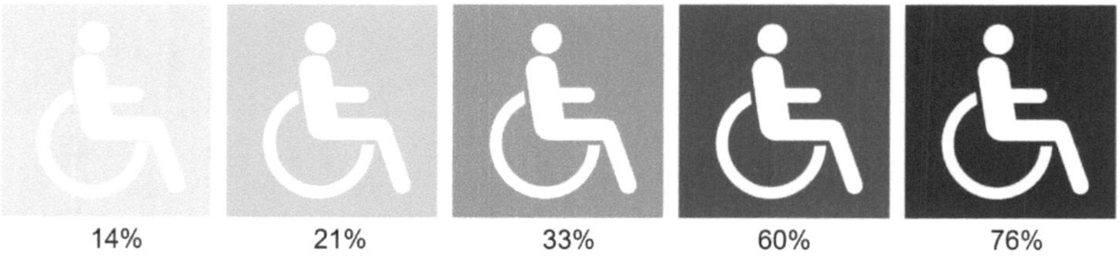

Figure 4. Examples of the different brightness contrast signage.

2.3. Procedure

After the participants entered the experiment room, the research objectives and experimental procedures were introduced. Clinical visual function tests (visual acuity measurement, visual field testing, and vision color measurement) and medical history inquiries were conducted first, followed by the execution of the experiments. The entire protocol is described in Figure 5. The experiment was divided into two sets of tasks. One was called information transmission, and the other was called information response. Two primary aspects of visual accessibility were studied; the first was the ability to locate the position of the signs in space, and the second was the accuracy of identifying or reading the signage's information [46].

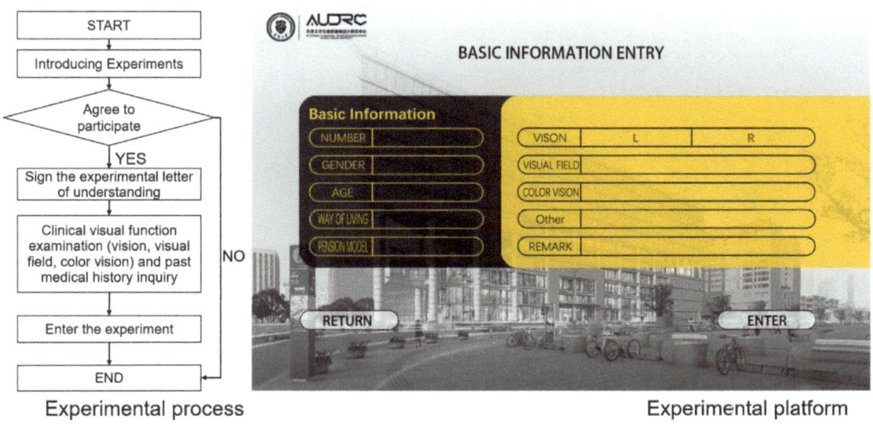

Figure 5. Experimental process.

Experiment 1: Information transmission experiment

In the first experiment (Figure 6), there are 20 signages with 5 sizes and 5 brightness contrasts, resulting in 500 different signages. The participants conducted the experiment at a distance of 4 m from the screen.

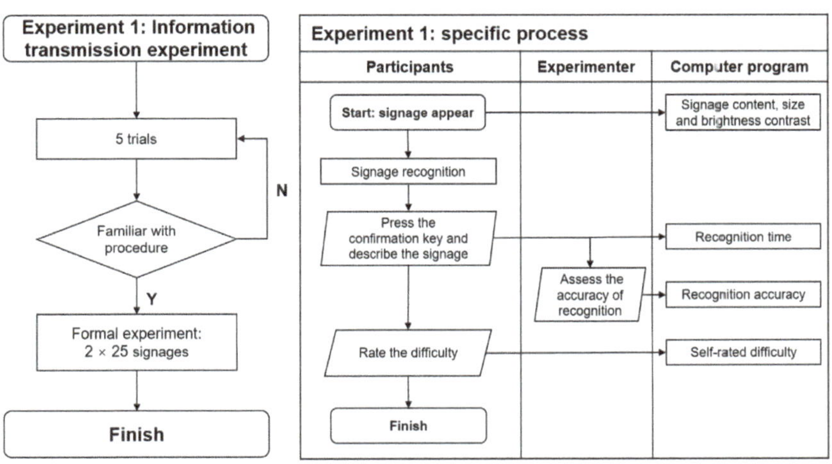

Figure 6. Flowchart of Experiment 1.

Before starting the experiment, participants performed five trials to learn the procedure. When a signage appeared, if the participants recognized it, they pressed the confirmation key and verbally described the meaning of the signage. They were also required to rate the difficulty of their information recognition process on a scale of 1 (very difficult) to 5 (very easy). Additionally, the experimenter assessed the accuracy of the participants' recognition and recorded the time taken for recognition of each signage.

Each participant completed two sets of experiments, with 25 instances of signage recognition in each set. There was a 3-min break between the two sets to mitigate the potential impact of visual fatigue.

Experiment 2: Information response experiment

In real-world scenarios, visual elements, sounds, and other perceptual factors can affect the recognition of signage information. Therefore, Experiment 2 (Figure 7) embedded the signages used in Experiment 1 into images of real-life scenes to simulate the real world.

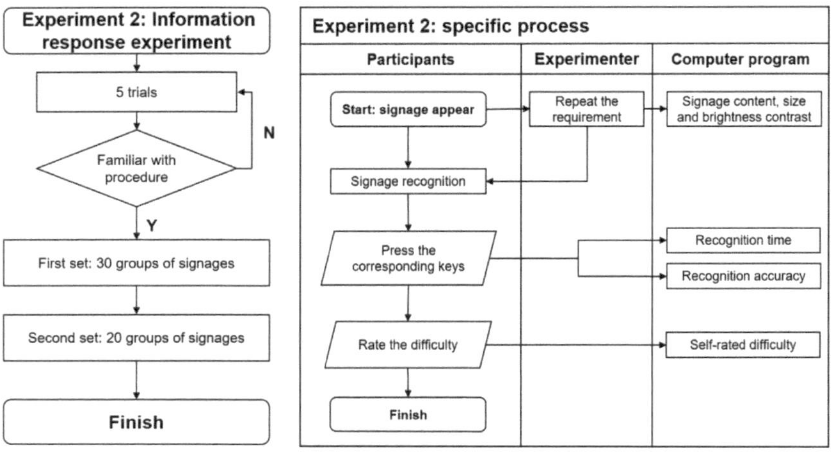

Figure 7. Flowchart of Experiment 2.

Similarly at the beginning of the experiment, participants performed five trials to familiarize themselves with the procedure. Nine random signages with the same size and brightness contrast but different contents were displayed, and participants were asked to identify a particular signage. They had to press a corresponding key on a 9-key keyboard representing the position of that target signage and rate the difficulty level as described for Experiment 1.

Each participant completed two sets of experiments. The first set consisted of 30 groups of random signages (with sizes of 1%, 3%, and 5% and brightness contrasts of 14%, 21%, 33%, 60%, and 76%), and the second set consisted of 20 groups of random signages (with sizes of 7% and 9% and brightness contrasts of 14%, 21%, 33%, 60%, and 76%).

3. Experimental Results

3.1. Descriptive Statistics

3.1.1. Experiment 1

A total of 153 participants participated in the experiment, and 139 valid sets of experimental data were collected. The control group consisted of 30 individuals with normal vision, while the low vision group included 75 individuals with low vision, 16 individuals with visual field defects, and 17 individuals with color vision anomalopia. These participants were recruited through posters and social media platforms, and the experiment took place in Tianjin, China. Raw data were preprocessed using the percentile method in Python to remove the outliers. Further analysis, including data correlation and path analysis, was conducted using Statistical Product and Service Solutions (SPSS) software.

1. Accuracy

The size of the signage was positively related to its legibility. Figure 6 shows that as signage size increased from 1% to 3%, the accuracy increased rapidly, while the effect of increasing brightness contrast on accuracy was not profound. The low vision, visual field defects, and color vision anomalopia groups all showed significant improvements in response accuracy with larger signage sizes, but the growth rate of response accuracy gradually slowed down at 5%, 7%, and 9%.

Response accuracy for the four groups remained relatively constant at different contrast levels. Under a fixed brightness contrast condition, the response accuracy of the groups ranked from high to low was as follows: normal group, color vision anomalopia, visual field defect, and low vision group.

2. Response time

Only the response time of correct identification was used in the analysis. The larger the signage size was, the shorter the time needed for signage recognition. For the low vision and visual field defect groups, the response time was influenced by different signage sizes, with time decreasing as size increased. However, color vision anomalopia individuals did not exhibit a clear trend in changes. When the signage size reached 5%, the decreasing trend in response time slowed down. Among the four groups, the groups with low vision and visual field defects were more sensitive to changes in signage size.

Brightness contrast had a slight effect on response time for all four groups, but the effect was not significant and showed no trend. Brightness contrast had a more significant impact on low vision and color vision anomalopia groups. For the low vision group, brightness contrast values of 21% and 76% had a significant effect on response time, while other values were not significant (Figure 8).

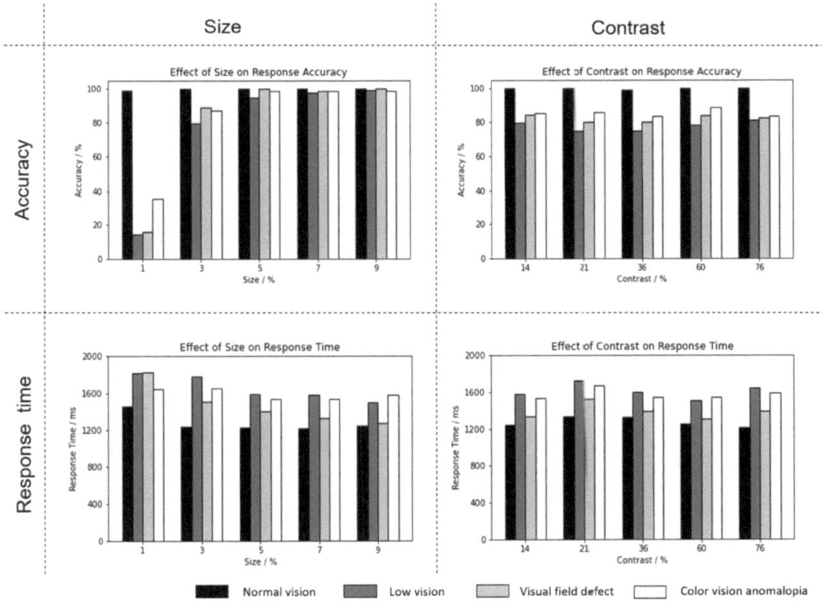

Figure 8. The accuracy and response time results for Experiment 1.

3. Difficulty level

All groups exhibited a decrease in response difficulty as signage size increased. On the one hand, effective growth was observed from 1% to 5% signage size, and the growth trend slowed down after 5%. On the other hand, brightness contrast had a relatively small impact on response difficulty. There were differences in average response difficulty under different brightness contrasts, but the impact was not profound and did not show a trend. Under a fixed signage contrast, the order of response difficulty was normal vision group > color vision anomalopia > visual field defect > low vision group (Figure 9).

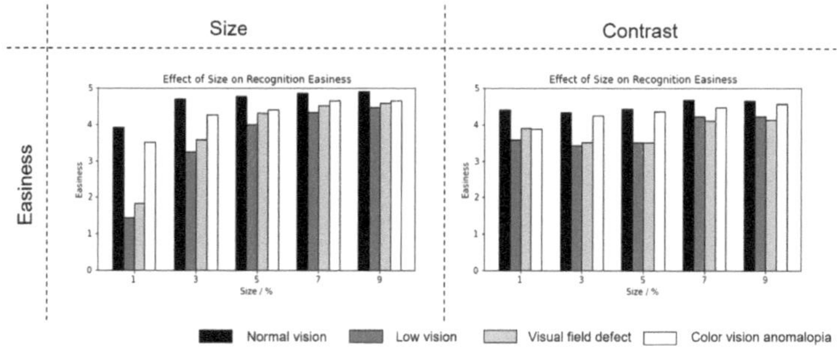

Figure 9. The difficulty level results for Experiment 1.

3.1.2. Experiment 2

(1) Accuracy

All groups exhibited a significant improvement in response accuracy with increasing signage size (Figure 10). The low vision and visual field defect groups had much higher accuracy than the normal group at 7% and 9% signage sizes. Based on the follow-up interview, this might be due to the fact that the low vision group had a strong desire to get correct results, thereby spending more time. In the response experiment, at smaller sizes, the response accuracies of the low vision, visual field defect, and color vision anomalopia groups were much higher than those obtained in Experiment 1 and only slightly lower than that of the normal group. There was no significant difference in response accuracy among the four groups at different contrast levels in the response experiment. The response accuracy of the normal vision, visual field defect, and color vision anomalopia groups ranged from 80% to 86%, while the response accuracy of the low vision group was slightly lower, ranging from 76% to 79%, representing the lowest values.

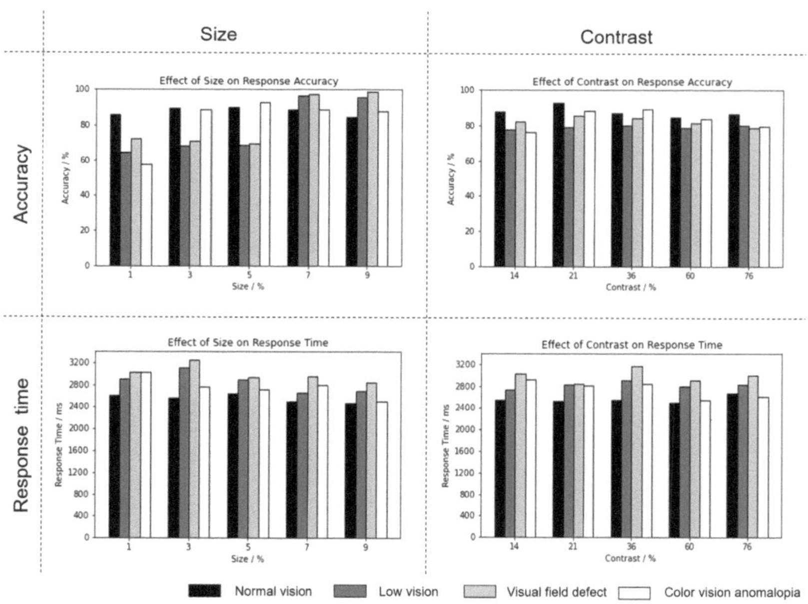

Figure 10. The accuracy and response time results for Experiment 2.

(2) Response time

As in Experiment 1, only the response times of correct identification were used in the analysis. Overall, as the size of the signage increased from 3% to 9%, the time spent searching and identifying the signage decreased. Among the low vision, visual field defect, and color vision anomalopia groups, the response time decreased as the signage size increased. However, for signage sizes of 1% and 3%, the response time of the low vision and visual field defect groups increased with the increase in signage size. The brightness contrast had little effect on identification time. The response time for the four groups were relatively stable and did not differ significantly at different brightness contrasts. Under a fixed signage size and brightness contrast, the order of response time for the four groups from the lowest to highest was as follows: the normal group, color vision anomalopia group, visual field defect group, and low vision group (Figure 10).

(3) Difficulty level

As signage size increased, the response difficulty decreased. The impact of brightness on difficulty was smaller than that of size. Under different brightness contrasts, individuals with anomalopia felt that the tasks were easier to perform than those with low vision and visual field defects. However, the difference between the groups with low vision and visual field defects was not significant (Figure 11).

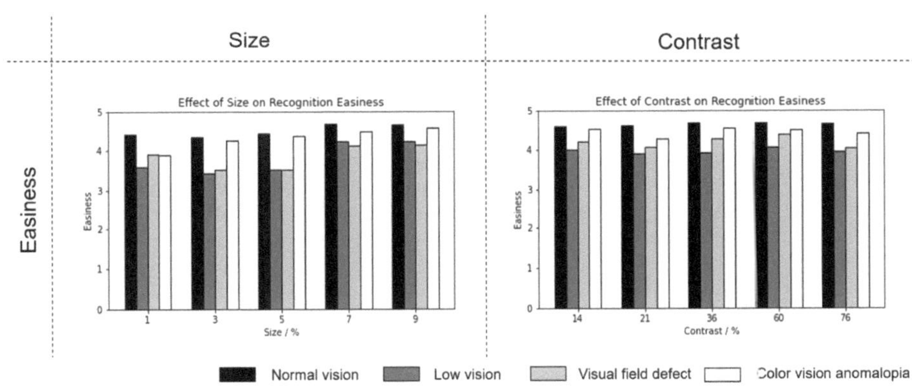

Figure 11. The difficulty level results for Experiment 2.

3.1.3. Correlation between Experiments 1 and 2

As the low vision group accounted for most of the participants and their response results were mostly at a lower level, this experiment further analyzed the interactive relationship between signage size and brightness contrast, specifically for the low vision group.

(1) Accuracy

In Experiment 1, signage size significantly impacted response accuracy, with larger signage sizes leading to higher response accuracy. When the signage size was below 5%, the brightness contrast was an important factor in improving recognition accuracy. At 1% and 3% contrast levels, brightness contrast still had a significant effect on improving the rate of correct identification. At the 1% contrast level, as compared to the three curves with the largest sizes (5%, 7%, and 9%), the response accuracy significantly increased as brightness contrast increased. In the three cases where the sizes were 5%, 7%, and 9%, the influence of brightness contrast on response accuracy was relatively weak.

In Experiment 2, for low vision individuals, similarly, larger signage sizes led to higher response accuracy. When the signage size reached 7%, the response accuracy increased to over 90%. In the cases when size was 1%, 3%, and 5%, the response accuracy was clustered between 60% and 75%.

Compared to Experiment 1, Experiment 2 required a higher signage size to improve response accuracy, as shown in Figure 12. The response accuracy for the 7% and 9% sizes reached over 90%. In the 5% case, the accuracy of Experiment 2 was significantly lower than that of Experiment 1, with a difference of 26.37%. In the 3% case, the accuracy of Experiment 2 also slightly decreased, with around 60% and 70% response accuracy. In the 1% case, however, the accuracy of Experiment 2 reached 50.29%.

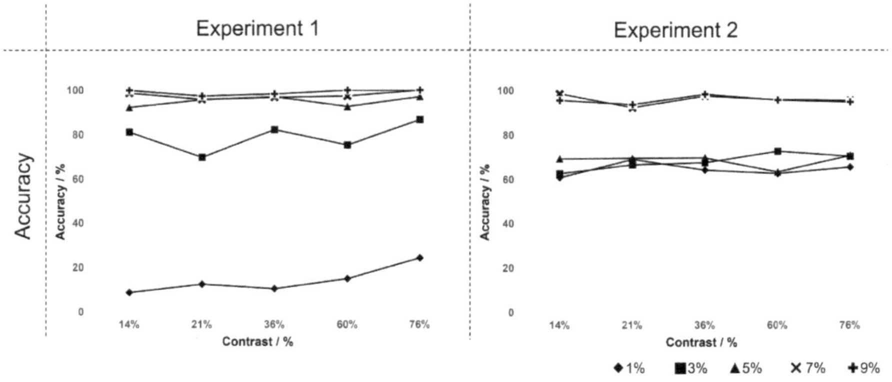

Figure 12. The comparison of accuracy results between Experiments 1 and 2.

(2) Response time

Data from Experiment 1 showed that the response time of the low vision group was clustered in a range between 1500 ms to 2300 ms, and as signage size increased, the response time decreased. For different signage sizes, the difference in response time was significant when the contrast increased from 14% to 21%. However, when the contrast increased to 36%, the response time decreased rapidly and reached a plateau. When the contrast increased to 60% and 76%, only little variation in response time was observed in terms of size variations.

Compared to Experiment 1, Experiment 2 showed a longer response time that ranged from around 2500 ms to 3200 ms, and as signage size increased, the response time decreased. For the same combination of brightness contrast and size, all response time data in Experiment 2 were greater than those in Experiment 1. It could be inferred that low vision individuals spent more time searching for signage during the recognition process than they did when the signage was already located (Figure 13).

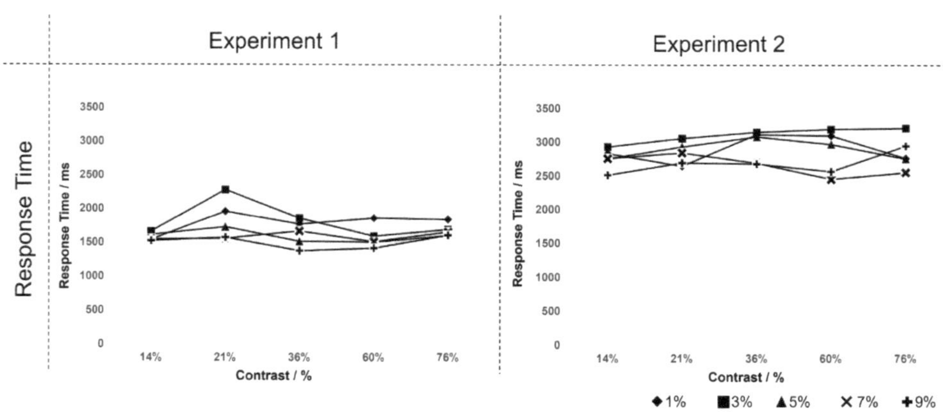

Figure 13. The comparison of response time results between Experiments 1 and 2.

3.2. Path Analysis

A statistical model was established to explore the different factors that affect accuracy and response time using the method of path analysis.

3.2.1. Experiment 1

The independent variables of Experiment 1 consisted of signage attributes and visual state. Signage attributes included both signage size and brightness contrast, while visual state included vision, visual field, and color vision (Table 3). The following conclusions could be drawn:

(1) When the dependent variable is accuracy, visual acuity (r = 0.272, z = 18.210, p = 0.000 < 0.01) and signage size (r = 0.517, z = 35.424, p = 0.000 < 0.01) are significantly positively correlated with accuracy.
(2) After excluding experimental data with erroneous response results in Experiment 1, analysis with correct response time as the dependent variable found that vision (z = −12.438, p = 0.000 < 0.01) and signage size (z = −6.885, p = 0.000 < 0.01) are significantly negatively correlated with response time. Color vision (z = −2.432, p = 0.015 < 0.05) is negatively correlated with response time. Moreover, the visual field (z = 5.575, p = 0.000 < 0.01) is significantly positively correlated with response time.
(3) When the dependent variable is self-rated difficulty, vision (z = 26.468, p = 0.000 < 0.01) and signage size (z = −3.369, p = 0.001 < 0.01) are significantly positively correlated with self-rated difficulty. Color vision (z = 44.096, p = 0.000 < 0.01) is negatively correlated with difficulty.

Table 3. Model regression coefficients (Experiment 1).

X	→	Y	z (CR Value)	p
Vision	→	Accuracy	18.210	0.000 **
Visual field	→	Accuracy	−1.567	0.117
Color vision	→	Accuracy	0.651	0.515
Signage size	→	Accuracy	35.424	0.000 **
Brightness contrast	→	Accuracy	1.143	0.253
Vision	→	Response time	−12.438	0.000 **
Visual field	→	Response time	5.575	0.000 **
Color vision	→	Response time	−2.432	0.015 *
Signage size	→	Response time	−6.885	0.000 **
Brightness contrast	→	Response time	−0.932	0.351
Vision	→	Difficulty	26.468	0.000 **
Visual field	→	Difficulty	−1.397	0.162
Color vision	→	Difficulty	−3.369	0.001 **
Signage size	→	Difficulty	44.096	0.000 **
Brightness contrast	→	Difficulty	1.271	0.204

Remarks: → means path influence relationship, * means p value < 0.05, ** means p value < 0.01, CR means critical ratio.

3.2.2. Experiment 2

The data model of Experiment 2 (Table 4) is identical to that of Experiment 1, yielding the following conclusions:

(1) When the dependent variable is accuracy, vision (z = 6.787, p = 0.000 < 0.01) and signage size (z = 12.904, p = 0.000 < 0.01) are significantly positively correlated with accuracy.
(2) After excluding experimental data from Experiment 1 with incorrect response results, analysis of correct response time as the dependent variable revealed that vision (z = −5.513, p = 0.000 < 0.01), visual field (z = −4.038, p = 0.000 < 0.01), and signage size (z = 12.904, p = 0.000 < 0.01) are significantly positively correlated with response time.
(3) When the dependent variable is self-rated difficulty, vision (z = 18.150, p = 0.000 < 0.01) and signage size (z = 19.523, p = 0.000 < 0.01) are significantly positively correlated

with self-rated difficulty. Color vision ($z = -3.092$, $p = 0.002 < 0.01$) is significantly negatively correlated with self-rated difficulty.

Table 4. Model regression coefficients (Experiment 2).

X	→	Y	z (CR Value)	p
Vision	→	Accuracy	6.787	0.000 **
Visual field	→	Accuracy	−0.752	0.452
Color vision	→	Accuracy	0.419	0.675
Signage size	→	Accuracy	12.904	0.000 **
Brightness contrast	→	Accuracy	0.006	0.995
Vision	→	Response time	−5.513	0.000 **
Visual field	→	Response time	−4.038	0.000 **
Color vision	→	Response time	0.149	0.881
Signage size	→	Response time	−5.646	0.000 **
Brightness contrast	→	Response time	−0.119	0.906
Vision	→	Difficulty	18.150	0.000 **
Visual field	→	Difficulty	0.599	0.549
Color vision	→	Difficulty	−3.092	0.002 **
Signage size	→	Difficulty	19.523	0.000 **
Brightness contrast	→	Difficulty	1.342	0.180

Remarks: → means path influence relationship, ** means p value < 0.01.

4. Discussion

The experiment has demonstrated that an increase in signage size improves information recognition accuracy and reduces the time required for information recognition. The response accuracy reached a stable high-accuracy state when the signage size was 7% of the reading distance. Therefore, to ensure greater universality, it is recommended that 7% of the reading distance be the minimum size for the signage design. However, the results of the response time analysis differed slightly from those of the response accuracy. In the case of signage recognition, low-vision individuals can quickly identify the correct signage when the signage size is 7% or greater than the reading distance. At 9%, the signage recognition level was mostly acceptable. Increasing the symbol size made identification more comfortable for low-vision individuals. Therefore, the experimental recommendation for signage size should not be lower than 7% of the reading distance. If it is lower than 7%, low vision individuals would face more problems in wayfinding. The relationship between the observation distance and the signage size is summarized in the following formula:

$$L = 0.07D \tag{4}$$

where L is the short side length of the signage (m), and D is the maximum observation distance (m).

This minimum size recommendation is larger than the recommendation for accessible signage in the Japanese "Barrier-Free Building Design Guide" [39] possibly because Chinese characters are more complex in structure than Japanese characters as an increase in the number of strokes may decrease the legibility of signage fonts. The increase in the number of strokes makes the Chinese characters more crowded, reduces the space between strokes, obscures the details and features of the characters, lowers their recognizability, and ultimately leads to an increase in cognitive processing time. The 7% size limit meets the standards proposed by Den Brinker et al. [47] and the United Kingdom Sign Design Guide and is greater than the average guidelines of European Union countries. The recommended signage value from this experiment is also greater than the ADA Standard for Accessible Design Act, Section 703.5.5 [43]. However, it is smaller than the recommended minimum scale of graphic symbols for signage boards in Article 41 of China's "Building

Construction-Accessibility and Usability of the Built Environment" (ISO 21542) [42], which suggests the following:

$$L = 0.09D \tag{5}$$

where L is the short side length of the signage (m), and D is the maximum observation distance (m).

Compared with Experiment 1, the recognition effect of Experiment 2 differed significantly. Based on interviews conducted after the experiment, it was evident that accurate communication of information and the search for responsive icons did not present equivalent levels of visual recognition difficulty. Moreover, factors such as the familiarity of symbols among the subjects could affect the accuracy of their response. Notably, a considerable number of participants found it challenging to identify signs that they were unfamiliar with or those that had high similarities in their morphology This underscores the critical need for a unified and clear design of signage graphics.

The impact of brightness contrast on response accuracy was strongly dependent on changes in signage size. When the size of the signage was less than 5%, the brightness contrast of the signage was a dominant factor affecting the correct recognition rate. In cases where the signage size was 1%, increasing the brightness contrast significantly improved the accuracy of the response. However, this did not reveal a clear quantitative relationship, which indicated that a signage size of approximately 1% reading distance was not sufficient to meet the wayfinding needs of individuals with low vision. Even though increasing brightness contrast could improve recognition efficiency, it still failed to satisfy the desired results for universal design. Furthermore, the experimental results showed that when the signage size reached 5% reading distance, the impact of brightness contrast on recognition accuracy rapidly decreased. However, it still helped to improve the recognition efficiency. These conditions are based on fixed-point signage recognition discussions. In actual spatial relationships, the specific relationship between individuals and signage is a variable that is not fixed. Therefore, it is recommended that the brightness contrast should be as high as possible to meet the wayfinding needs of different groups of people in different scenarios.

The conclusions drawn above apply to the general population with low vision ($0.3 > v \geq 0.05$). Among the diverse individuals with visual impairments, some need more time to identify signages than others. For example, in groups with visual field impairment, there was a linear decrease in response time with an increase in signage size. In Experiment 2, if the signage size exceeded the subjects' visual range, individuals with low vision must adjust their gaze through eye or head movements, resulting in a longer response time for signage recognition. Therefore, multiple factors must be considered in determining signage size. Additionally, it was found in the experiment that many commonly used symbols were composed of multiple parts, and the results of the experiment as well as post-interview feedback indicated that these details were difficult for individuals with low vision to recognize. Thus, a unified, concise, and accurate signage image is helpful for low vision individuals. In addition, the participants mostly believed that a combination of figures and text ensured the successful transmission of information.

This study provides reference data for the design of signage recognition size and brightness contrast for individuals with low vision. This research specifically focused on three subgroups: individuals with low vision, individuals with visual field defects, and individuals with color vision anomalopia. However, the variety of underlying causes and visual function types in these groups indicates a wide range of visual impairment characteristics. Consequently, the findings are still inefficient to be considered as guidelines applicable to low vision people at the individual level. The element interaction discussed in this study suggests that signage guidelines based on single-element experiments are not complete. Moreover, the placement of signage in a space is also very important. The consensus is that signage should be uniformly positioned in clearly visible locations throughout the entire architectural space, and it must not interfere with the activities of the users. Brightness contrast and size are only two factors that may affect visual information acquisition; therefore, before developing standard design guidelines, more experiments should be conducted in the future to further refine specialized research on lighting, positioning, signage design, and other relevant elements as well as their interactions. Moreover, due to inherent limitations in laboratory research, subsequent studies in real-world scenarios are essential to validate and assess the practicality of these experimental conclusions. The application of these findings will require verification in varied real-world settings, encompassing different lighting conditions, diverse environmental contexts, and a broader spectrum of user participation. This approach will facilitate the development of inclusive design guidelines that address both universal and specific needs.

5. Conclusions

Ensuring the legibility of signage recognition for individuals with low vision is crucial. However, developing reasonable, clear, actionable, appropriately scoped, and fair accessibility guidelines for all stakeholders poses significant challenges. This paper shows that signage size and brightness contrast have significant influence on signage legibility. The afore-provided signage design recommendations not only improve the effectiveness of conveying information but also enhance spatial accessibility for low vision individuals. Furthermore, these recommendations offer more viable design choices for graphic designers and signage manufacturers, benefiting the practices of universal design and architectural signage groups.

In addition, an intelligent experimental platform was developed to quantify the impact of brightness contrast and size on signage design in this study. By analyzing response accuracy and response time, it provides a mathematical and precise approach to signage design. The platform's applicability extends to universal signage design and other related studies, facilitating the quantitative evaluation of multiple signage design elements such as font, line spacing, color, size, icon, and brightness contrast. Its versatility and expandability make it a valuable tool in the field of signage design and beyond.

Author Contributions: Conceptualization, E.Z. and W.Z.; methodology, E.Z. and Z.Y.; software, E.Z. and Z.M.; validation, E.Z. and Z.Y.; formal analysis, E.Z. and Z.M.; investigation, Z.Y. and Y.X.; resources, W.Z. and Y.X.; data curation, E.Z. and F.C.; writing—original draft preparation, E.Z. and Z.Y.; writing—review and editing, Z.Y., W.Z., F.C. and Y.W.; visualisation, E.Z. and Z.M.; supervision, W.Z.; project administration, W.Z.; funding acquisition, W.Z. All authors have read and agreed to the published version of the manuscript.

Funding: This research was funded by National Natural Science Foundation of China (Research on color design theory and method of barrier-free environment in elderly care facilities, grant number 51808382) and the National Natural Science Foundation of Tianjin (Research on the color design method of barrier-free universal signage of elderly care facilities, grant number 20JCQNJC01930).

Institutional Review Board Statement: The study was conducted in accordance with the Declaration of Helsinki, and approved by the Ethics committee of School of Architecture, Tianjin University for studies involving humans. The approval No. is TJUE-2024-369.

Informed Consent Statement: Informed consent was obtained from all subjects involved in the study. Written informed consent has been obtained from the patients to publish this paper.

Data Availability Statement: The original contributions presented in the study are included in the article, further inquiries can be directed to the corresponding author.

Conflicts of Interest: The authors declare no conflicts of interest.

References

1. Arthur, P.; Passini, R. *Wayfinding: People, Signs, and Architecture*; McGraw-Hill Book Company: New York, NY, USA, 1992.
2. Calabrese, A.; Cheong, A.M.Y.; Cheung, S.H.; He, Y.; Kwon, M.; Mansfield, J.S.; Subramanian, A.; Yu, D.; Legge, G.E. Baseline MNREAD measures for normally sighted subjects from childhood to old age. *Investig. Ophthalmol. Vis. Sci.* **2016**, *57*, 3836–3843. [CrossRef] [PubMed]
3. Calder, L.; Woo, G.C. Clinical low vision. *Optom. Vis. Sci.* **1985**, *62*, 228–229.
4. Feinbloom, W. Report on the science information program. *Optom. Vis. Sci.* **1966**, *43*, 112–115. [CrossRef]
5. Hellinger, G. Vision rehabilitation for aged blind persons. *J. Vis. Impair. Blind.* **1969**, *63*, 175–177. [CrossRef]

6. Adeyemo, O.; Jeter, P.E.; Rozanski, C.; Arnold, E.; Dalvin, L.A.; Swenor, B.; Dagnelie, G. Living with ultra-low vision: An inventory of self-reported visually guided activities by individuals with profound visual impairment. *Transl. Vis. Sci. Technol.* **2017**, *6*, 10. [CrossRef] [PubMed]
7. Davies, T.; Beasley, K. *Accessible Design for Hospitality: Ada Guidelines for Planning Accessible Hotels, Motels and Other Recreational Facilities*; McGraw-Hill: New York, NY, USA, 1994.
8. Weisman, J. Evaluating architectural legibility: Way-finding in the built environment. *Environ. Behav.* **1981**, *13*, 189–204. [CrossRef]
9. Rubin, G.S.; Legge, G.E. Psychophysics of reading. VI—The role of contrast in low vision. *Vis. Res.* **1989**, *29*, 79–91. [CrossRef] [PubMed]
10. American National Standards Institute. *Accessible and Usable Buildings and Facilities*; Council of American Building Officials: Falls Church, VA, USA, 1992.
11. Schambureck, E.M.; Parkinson, S.F. Design for Sight: A Typology System for Low–Vision Design Factors. *J. Int. Des.* **2018**, *43*, 33–54. [CrossRef]
12. Giudice, N.A.; Legge, G.E. Blind navigation and the role of technology. In *The Engineering Handbook of Smart Technology for Aging, Disability, and Independence*; Helal, A., Mokhtari, M., Abdulrazak, B., Eds.; John Wiley & Sons: New York, NY, USA, 2008; pp. 479–500.
13. Walker, B.N.; Lindsay, J. Development and evaluation of a system for wearable audio navigation. *Proc. Hum. Factors Ergon. Soc. Annu. Meet.* **2005**, *49*, 1607–1610. [CrossRef]
14. Wilson, J.; Walker, B.N.; Lindsay, J.; Cambias, C.; Dellaert, F. SWAN: System for Wearable Audio Navigation. In Proceedings of the 2007 11th IEEE International Symposium on Wearable Computers, Boston, MA, USA, 11–13 October 2007; pp. 91–98.
15. Legge, G.E.; Beckmann, P.J.; Tjan, B.S.; Havey, G.; Kramer, K.; Rolkosky, D.; Gage, R.; Chen, M.; Puchakayala, S.; Rangarajan, A. Indoor navigation by people with visual impairment using a digital sign system. *PLoS ONE* **2013**, *8*, e76783. [CrossRef]
16. Zhang, E.; Yang, Z.; Zhao, W.; Mei, Z.; Xia, Y.; Chen, F. Unlocking Brilliance: A Smart Approach to Icon Colour Design Inspired by Universal Design Principles. *Buildings* **2024**, *14*, 1522. [CrossRef]
17. Arditi, A. Improving the design of the letter contrast sensitivity test. *Investig. Ophthalmol. Vis. Sci.* **2005**, *46*, 2225–2229. [CrossRef] [PubMed]
18. Regan, D.; Neima, D. Low-contrast letter charts as a test of visual function. *Ophthalmology* **1983**, *90*, 1192–1200. [CrossRef] [PubMed]
19. Scheiman, M.; Scheiman, M.; Whittaker, S. *Low Vision Rehabilitation: A Practical Guide for Occupational Therapists*; Slack Incorporated: West Deptford, NJ, USA, 2007.
20. Zhao, W.; Zhang, E.; Zhang, M. Interior design considerations for individuals with impaired vision. *Inter. Des. Constr.* **2021**, *12*, 126–128.
21. Arditi, A. Rethinking ADA signage standards for low-vision accessibility. *J. Vis.* **2017**, *17*, 8. [CrossRef] [PubMed]
22. ANEC. *New Standard for the Visual Accessibility of Signs and Signage for People with Low Vision*; Acute Necrotizing Encephalopathy of Childhood: Brussels, Belgium, 2010.
23. Lerner, N.; Singer, J.; Robinson, E.; Huey, R.; Jenness, J. *Driver Use of Enroute Real-Time Travel Time Information*; Federal Highway Administration: Washington, DC, USA, 2009.
24. NIBS. *Design Guidelines for the Visual Environment: Version 6*; National Institute of Building Sciences: Washington, DC, USA, 2015.
25. RTIG Inform. Bus Passenger Information Technology. Available online: https://theaccessofficer.n-somerset.gov.uk/2016/02/12/bus-passenger-information-technology/ (accessed on 21 May 2024).
26. Zimring, C.M.; Bostrom, J.; Wineman, J. *Multidisciplinary Assessment of the State of the Art of Signage for Blind and Low Vision Persons: Final Report*; Architectural and Transportation Barriers Compliance Board: Washington, DC, USA, 1985.
27. Hayanouchi, S. *Investigation of Color and Discriminability of Tactile Pavements for the Visually Impaired*; Japan Road Technical Association: Tokyo, Japan, 1974.
28. Yoshida, M.; Clara, A.; Shoko, S. The use of films to simulate age-related declines in yellow vision. *J. Occup. Rehabil.* **1996**, *6*, 119–134. [CrossRef]
29. Lomperski, T.J. Enhancing interior building sign readability for older adults: Lighting color and sign color contrast. *J. Int. Des.* **1997**, *23*, 17–27. [CrossRef]
30. Bright, K.; Cook, G.; Harris, J. *Colour, Contrast & Perception: A Design Guidance for Internal Built Environments*; University of Reading: London, UK, 2004.
31. Bullough, J. Factors affecting sign visibility, conspicuity, and legibility: Review and annotated bibliography. *Interdiscip. J. Signage Wayfinding* **2017**, *1*, 2–25. [CrossRef]
32. Yang, P.; Zhao, K. *Ophthalmology*; People's Medical Publishing House: Beijing, China, 2000.
33. Arthur, P. *Broadening Our Comprehension of the Principles of Environmental Communications in Public Buildings*; National Research Council Canada: Ottawa, ON, Canada, 1984.
34. *GBT 15566.1-2007*; Guidance System for Public Information—Setting Principles and Requirements—Part 1: General Principles. China Planning Publishing House: Beijing, China, 2007.
35. *GBT 10001.1-2012*; Public Information Graphic Symbols—Part 1: General Symbols. China Planning Publishing House: Beijing, China, 2012.

36. *ISO 21542:2011*; Building Construction—Accessibility and Usability of the Built Environment. International Organization for Standardization: Geneva, Switzerland, 2011.
37. Peters, G.; Adams, B. Three criteria for readable panel markings. *Prod. Eng.* **1959**, *30*, 123–131.
38. Smith, S.L. Letter size and legibility. *Hum. Factors* **1979**, *21*, 661–670. [CrossRef]
39. Loomis, J.M. A model of character recognition and legibility. *J. Exp. Psychol. Hum. Percept. Perform.* **1990**, *16*, 106–120. [CrossRef] [PubMed]
40. Holmes-Siedle, J.; Goldsmith, S. *Accessible Design*; Dalian University of Technology Press: Dalian, China, 2002.
41. BSI. *Design of Buildings and Their Approaches to Meet the Needs of Disabled People: Code of Practice*; British Standards Institution: London, UK, 2018; p. 58.
42. Architectural Institute Of Japan. *A Concise Collection of Accessible Building Design Information*; China Architecture & Building Press: Beijing, China, 2006.
43. Civil Rights Division of U.S. Department of Justice. *ADA Standards for Accessible Design*; U.S. Department of Justice: Washington, DC, USA, 2010.
44. Sun, B.; Hu, A. *Clinical Low Vision*; People's Medical Publishing House: Beijing, China, 2013.
45. Ding, X.; Zhu, R.; Li, J. A criterion of image quality assessment based on property of HVS. *J. Image Graph.* **2004**, *9*, 190–194.
46. Cook, G.; Webber, G.; Gillham, S.; Le Scouiller, S.; Moseley, E. The legibility and conspicuity of emergency escape route signage for people with visual impairments. *Int. Congr. Ser.* **2005**, *1282*, 1016–1020. [CrossRef]
47. Den Brinker, B.; Wijk, M.; Smeets, J. In public environments character height of textual information should be at least 5 percent of critical reading distance. In Proceedings of the 9th International Conference on Low Vision, Montreal, QC, Canada, 7–11 July 2008.

buildings

MDPI

Article

Climate Adaptability Research of Vernacular Dwellings in Jiangxi Based on Numerical Simulation—An Example from Nanfeng County

Zhiyi Zhou, Yuxuan Xu, Cheng Ouyang, Mengyao Gui, Wanping Jiang, Chunlei Zhou, Kai Ma, Jiaxin Zhang and Jingyong Huang *

Architecture and Design College, Nanchang University, No. 999 Xuefu Avenue, Honggutan District, Nanchang 330031, China; zhouzhiyi@ncu.edu.cn (Z.Z.); 408800220009@email.ncu.edu.cn (Y.X.); 408800220006@email.ncu.edu.cn (C.O.); gmy@email.ncu.edu.cn (M.G.); jiangwanping@ncu.edu.cn (W.J.); zcl@email.ncu.edu.cn (C.Z.); makai@ncu.edu.cn (K.M.); jiaxin.arch@ncu.edu.cn (J.Z.)

* Correspondence: huangjingyong@ncu.deu.cn; Tel.: +86-138-0708-0432

Abstract: Energy conservation and carbon reduction in buildings have become important concerns and, at the same time, the value of low-tech approaches employed in indigenous architecture is increasingly acknowledged as a pertinent reference for contemporary design practices. The research on vernacular dwellings in Jiangxi has many perspectives and fruitful results, but not enough attention has been paid to the research on climate adaptation. This article verifies the vernacular dwellings' climate adaptation and summarizes the low-tech methods embedded in vernacular dwellings, aiming to provide guidelines for future exploration of energy-saving and carbon-reducing practices in architecture. By selecting different types of vernacular dwellings in Nanfeng County, this article verifies three aspects of the ecological characteristics of vernacular dwellings: the light environment, wind environment, and energy consumption, by comparing them with those of local modern residential buildings. It is concluded that the average daylight factor of the hall area of vernacular dwellings is better than that of the modern residential buildings in rural areas, and the vernacular dwellings regulate the indoor wind environment and maintain indoor comfort through natural ventilation in winter and summer seasons. Also, the annual energy consumption of the vernacular dwellings per unit area per year can be reduced by up to about 32% in comparison with modern residential buildings. Subsequently, the article concludes that patio space has a positive impact on the indoor physical environment through comparative experiments. Vernacular dwellings are well adapted to the local climate in terms of form, structure, and materials, and these low-tech methods should be applied to the design of rural dwellings in the future.

Keywords: vernacular dwellings; digital simulation; indoor physical environment; Nanfeng county; climate adaptability; low-tech

Citation: Zhou, Z.; Xu, Y.; Ouyang, C.; Gui, M.; Jiang, W.; Zhou, C.; Ma, K.; Zhang, J.; Huang, J. Climate Adaptability Research of Vernacular Dwellings in Jiangxi Based on Numerical Simulation—An Example from Nanfeng County. *Buildings* **2024**, *14*, 2211. https://doi.org/10.3390/buildings14072211

Academic Editors: Derek Clements-Croome, Antonio Caggiano and Apple L.S. Chan

Received: 16 April 2024
Revised: 12 July 2024
Accepted: 16 July 2024
Published: 18 July 2024

1. Introductory

Global warming and the energy crisis are among the most prominent issues facing the world today. The construction sector accounts for 30% of the world's total greenhouse gas emissions [1]. In developed countries, the construction sector accounts for 40% of total greenhouse gas emissions [2]. In China, the carbon emissions from the whole life cycle of dwellings account for 51.2 percent of the country's total carbon emissions [3] and the construction industry accounts for 40 percent of the world's total energy consumption [4]. In developed countries, the building sector accounts for 40% of all energy consumption [2], of which 30–60% is used to regulate the indoor environment of dwellings [5]. The building sector accounts for 46.5 percent of total national energy consumption in China [6], with residential dwellings accounting for 61 percent of the total energy consumption of dwellings [7]. China has committed to peak carbon emissions by 2030 and become a carbon neutral society by 2060, while aiming to reduce energy consumption in dwellings to less than 1% to ensure comfort and health [8]. The use of low-tech methods in vernacular dwellings is an indispensable part of reducing energy consumption and achieving sustainable development. This is an important direction that cannot be ignored. In the process of this profound change, Chinese vernacular dwellings show the energy-saving potential and low-tech, low-cost passive designs, which are worth in-depth excavation, practical transformation, and application.

1.1. Relevant Research on Climate Adaptation

Vernacular architecture refers to both traditional and contemporary residential dwellings constructed spontaneously in rural areas, under a system of non-professional architects [9]. It is worth learning from the experience of vernacular architecture, as the main architectural system in rural areas, in reducing energy consumption and waste, using resources rationally and adapting to local climatic conditions [10,11].

Many scholars have studied climate-appropriate vernacular architectural design practices around the world. In Europe, the vernacular architecture coped with the high summer temperatures typical of the Mediterranean climate thanks to the materials of their construction elements, the thickness of their envelopes, and the orientation of their façades [12]. In Cyprus, vernacular architecture in its three different climatic zones employed a variety of design strategies to achieve optimal thermal comfort [13]. Simulations of the vernacular architecture of Pompeii show that the Romans felt comfortable in the summer but required heating in the winter [14]. In Asia, the semi-outdoor spaces of traditional Japanese houses acted as thermal buffers for promoting cross-ventilation as well as pre-cooling to provide "warm but breezy" conditions to the connected indoor spaces that can effectively cope with the hot and humid summer climate [15]. In India, Assam-type houses and stilt houses are highly responsive to the local climate, utilize locally available materials, and reflect the living style, customs, and socio-economic conditions [16]. Vernacular architecture in the Himachal Pradesh region of India adheres to the foundations of traditional Indian architecture while adopting new measures to adapt to nature [17]. In Malaysia, the courtyard forms that characterized their thermal environments were classified into five types and the different types of courtyards performed different functions with respect to improving the indoor thermal comfort in the CSHs [18].

Chinese ancestors constructed a diverse array of dwellings according to their geographical location and climatic conditions [19]. Due to the diversity of building materials, local people developed a wide range of house types, construction techniques, and climatic appliances to adapt to the local climate [20]. In China, there are a wealth of research findings on the climate adaptability of vernacular dwellings in different regions, and different emphasis is placed on the research level and the geographical and climatic areas.

(1) Research on vernacular dwellings in the Huizhou region.

Huang et al. [21] measured the physical environment of traditional vernacular buildings in Huizhou and concluded that slope roof and roof structure as well as the hollow brick wall construction in vernacular dwellings are of significance in heat shielding in summer. Zhu et al. (2020) [22] employed the climate adaptability technology of Huizhou residential dwellings to measure both the indoor and outdoor temperature and humidity of a residential property and took 24 solar terms as the cycle to change the indoor and outdoor temperature and humidity. The variation law of indoor and outdoor temperature and humidity with solar terms in hot summer and cold winter areas were studied. Gao et al. (2023) [23] combed the light aspects of vernacular dwellings in Huizhou to improve indoor requirements. Through experimental comparison, it was concluded that the height of the window side can improve the lighting effect of the window side, and the width of the window and patio can improve the overall lighting of the room.

(2) Research on vernacular dwellings in the Lingnan region.

Gao et al. (2015) [24] took the vernacular dwellings in the Lingnan area as a research object and conducted a digital simulation analysis of their natural ventilation. They concluded that the vernacular buildings rely on the combination of "patio—cold alley—courtyard" to provide good ventilation in the interior. Yao et al. (2018) [25] simulated the indoor lighting and wind environment of vernacular dwellings in Guangdong Province using Ecotect, Phoneics, and found that the living space of vernacular dwellings has a good lighting environment. Due to the reasonable layout of vernacular dwellings, ventilation can be effectively organized through the courtyard in summer. Xiao et al. (2018) [26] analyzed the characteristics of climate adaptation of villages in the Lingnan area, pointing out that cold lanes and courtyards form the spatial system as the main spatial elements. Then, they summarized the reasonable scale range and organization pattern.

(3) Research on vernacular dwellings of ethnic minorities.

Jin et al. (2021) [27] monitored the physics of Dong-nationality vernacular dwellings in real time. The results demonstrated that the Dong nationality's vernacular dwellings have higher thermal comfort adaptability than urban residential dwellings under natural ventilation. Zhang et al. (2022) [28] summed up the advantages and disadvantages of residential physics to climate characteristics by using the indoor physics of Dong vernacular dwellings, measured and simulated the vernacular dwellings in the Dong-nationality area, and concluded that they are well adapted to the local climate in form, material, and structure. Dong et al. (2021) [29] measured the indoor and outdoor heat of typical Tujia dwellings in Chongqing during the extremely hot period. The results demonstrated that the ventilation roof structure of Tujia dwellings is unique and effectively utilizes the valley wind, which not only improves the heat insulation ability during the day but also is beneficial to the ventilation and refrigeration of the house at night.

1.2. Overview of Vernacular Dwellings in Jiangxi

The vernacular dwellings of Jiangxi represent a pivotal area of research within the field of Chinese vernacular dwellings, exerting a profound impact on the surrounding region. From the perspective of migration, Jiangxi has historically played a key role as the birthplace and transit point of many large-scale migrations. The three wars in the Central Plains, namely the Yongjia Rebellion, the Anshi Rebellion, and the Jingkang Rebellion, triggered large-scale population migration, and Jiangxi became one of the important destinations. The spatial distribution of Jiangxi dwellings as well as the innovation of dwelling types were greatly influenced by this. In the movement of "filling the lake with Jiangxi", the migrants from Jiangxi not only brought the customs, language, and culture of Jiangxi to new settlements, but also spread the architectural style and construction techniques of dwellings [30].

Jiangxi is a relatively closed and complete geographic unit with the east, south, and west of Jiangxi surrounded by mountains and with the Yangtze River in the north. Since the Song and Yuan Dynasties, Jiangxi's political changes have been relatively stable, ensuring the stability of the regional characteristics of dwellings [31]. In addition, some researchers [32] have sought to demonstrate the evolution of the dwelling form through a process of deduction. This approach offers more convincing proof of their assertions.

1.3. Present Situation of Research on Vernacular Dwellings in Jiangxi Province

Lu et al. [33] reviewed and summarized the research on Chinese dwellings from 1957 to 2007, and divided the research on dwellings into three periods; namely, the establishment period of dwellings in the 1950s, the surveying and mapping period in the 1960s, and the planned and organized research period from the 1980s to the beginning of the 21st century. Because of the inconvenient transportation and relatively backward economic conditions in the early stage, the research on the vernacular dwellings of Jiangxi started relatively late. In 1989, with the support of the policy, many scholars conducted in-depth research on vernacular dwellings in Jiangxi. Their research findings are mainly manifested in the form, characteristics, technology, culture, geography, and climate adaptability of vernacular dwellings, and so forth.

(1) Research on the form and characteristics of vernacular dwellings

Mr. Hao Huang's research on Jiangxi dwellings is the earliest and most mature. In his book [34], he conducted profound research and exploration on the composition laws and architectural art characteristics of Jiangxi patio dwellings, summarizing the three types of dwellings in Jiangxi. And, he traced back the degradation and disappearance of patio dwellings from the plan form and elements. This is a useful exploration of the inheritance and development of dwellings. Pan et al. [35] pointed out the reasons for the formation of ancient Jiangxi dwellings in the Ming and Qing Dynasties and thought that ancient Jiangxi dwellings were formed in adaptation to the natural ecology and social circumstances of the Jiangxi region. Yao et al. (2015) [36] systematically discussed the history, type, distribution, and characteristics of vernacular dwellings in Jiangxi. They provided a microscopic discussion of architectural techniques, decorative arts, and material use.

(2) Research on residential technology

Li et al. mainly focused on the characteristics and decorative differences between vernacular dwellings in Jiangxi [37]. Lan et al. (2017) [38] took the brick masonry structure in Jinxi County as an example to reveal the internal relationships and structural language of this mixed masonry. Yi et al. (2018) [39] studied the evolutionary patterns of hall structures in Ming Dynasty residences in the Fu River, especially the plan, section, and node construction features, and analyzed the similarities and differences between the structures. By comparing cases in other regions, they revealed the technological inheritance and fusion, and the comparison of hall-type structures in the Qing Dynasty showed the development of a more open spatial layout and simplified support structure. Li et al. (2018) [40] systematically analyzed the characteristics of the construction techniques of vernacular dwellings in the Lichuan area in terms of plan layout, structure, roof form, dimensions, and construction techniques. They tried to define the absolute scale of the local vernacular rulers and explain the diversity and complexity of the vernacular dwellings' ruler-making methods.

(3) Research on cultural geography of vernacular dwellings

Guo et al. studied the social background and architectural culture of ethnic areas in Hunan and Jiangxi based on the study of ethnic groups [41]. Pan et al. systematically extracted three types of elements: patio, courtyard, housing hall, and two types of basic units: row house unit, triple patio unit, and quadruple patio unit. From the perspective of typology and systematics, the level of vernacular dwellings has been divided into eight complex levels according to the multiple units involved in their construction and combination [42]. Kang et al. (2020) [31] systematically combed the formation of vernacular dwellings in Jiangxi since the Ming and Qing Dynasties by investigating the existing dwellings in Jiangxi and historical documents. The study revealed a close correlation between the distribution of Jiangxi dwellings and the water system, with dwellings exhibiting an asymmetrical distribution along the water system and watershed. Furthermore, this study highlighted a notable association between spatial distribution and population distribution.

(4) Research on climate adaptability of vernacular dwellings

Wang et al. clarified the livable characteristics of patio dwellings according to the form characteristics and spatial form composition of patio dwellings in different regions [43]. Chen et al. (2020) [44] studied the climate adaptability of vernacular dwellings along the Hunan–Jiangxi border from the material and structural level of architectural form and the architecture–climate interface. The study summarized the advantages and disadvantages of local climate adaptation from the thermal point of view by combining qualitative and quantitative analysis. Wu et al. (2023) [45] took the vernacular dwellings in Futian Town as the research object and summarized the climate adaptability strategy of local vernacular dwellings by actual measurement and simulation from the thermal perspective.

1.4. Research Review

In conclusion, the majority of domestic scholars engaged in research on Jiangxi dwellings have focused their attention on exploring the architectural aspects of the settlements, including their cultural characteristics, architectural forms, and so on. The research on the ecological characteristics of dwellings remains relatively limited. Moreover, the ecological performance of dwellings with different characteristics in this area is lacking. In terms of research methodology, the research on climate adaptability was initially limited to preliminary qualitative analyses of patio dwellings. In the last few years, although computer simulation has been introduced to study climate adaptability, some scholars focused on a single physical factor and ignored the overall comprehensive consideration. Not only that, the research lacks a comparative analysis between vernacular dwellings and modern residential buildings, making it challenging to combine the advantages of vernacular dwellings and modern residential buildings.

Therefore, this study takes vernacular dwellings in Jiangxi as the research object and employs digital simulation technology to verify the climate adaptability characteristics of vernacular dwellings in three aspects: light, wind, and energy consumption. This is accomplished by comparing the vernacular dwellings with local modern residential buildings. The purpose of this study is to refine ecological wisdom and provide practical guidance for the improvement of local vernacular dwellings and the construction of new residential dwellings in local villages. In addition, this study aims to reveal the scientific nature of vernacular dwellings and provide theoretical support for the scientific nature of construction ideas, while scientifically analyzing climate adaptability. Figure 1 illustrates the framework of this study.

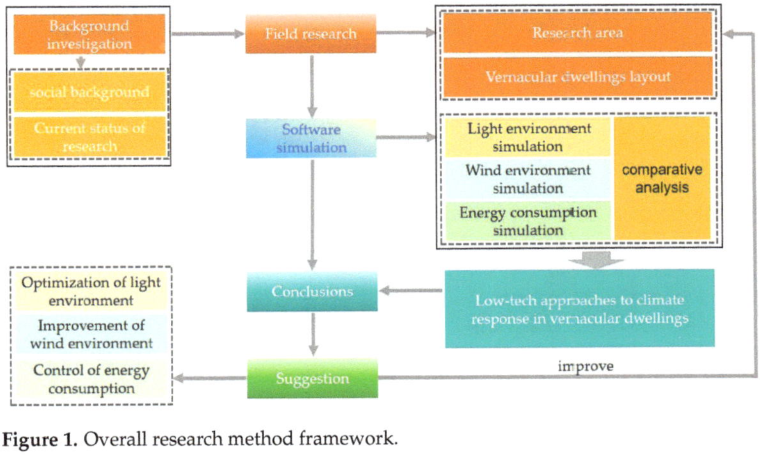

Figure 1. Overall research method framework.

2. Overview of The Study Population

Although there are many different types of vernacular dwellings in Jiangxi, such as patio-style dwellings, patio courtyard-style dwellings, halls, etc., the patio dwellings are more widely distributed with a substantial number in existence (Figure 2). The object of this paper is to study the patio-style dwellings.

Nanfeng County is known as one of the "cultural counties" in Jiangxi Province (Figure 3). It currently has 12 well-preserved, distinctive traditional villages, 11 national traditional villages, and 10 provincial traditional villages, accounting for 3.23% of the number of traditional villages in Jiangxi Province in China. The traditional villages in Fuzhou ranked fourth in terms of the number of traditional villages in Jiangxi. Nanfeng County ranks fourth in Fuzhou and eighth in Jiangxi in terms of the number of traditional villages, and Nanfeng Ancient City has been rated as a provincial famous historical and cultural city. Therefore, the traditional villages in Nanfeng County are typical representatives of the traditional villages in the whole of Jiangxi Province, both in terms of quantity and culture. This paper includes field research and consults classic works [33,34,36]. The patio-style dwellings are divided into three typical types: three openings with one entry and one patio, three openings with one entry plus half a patio, and three openings with multiple entries. (As Table 1).

Figure 2. Distribution map of different vernacular dwellings in Jiangxi.

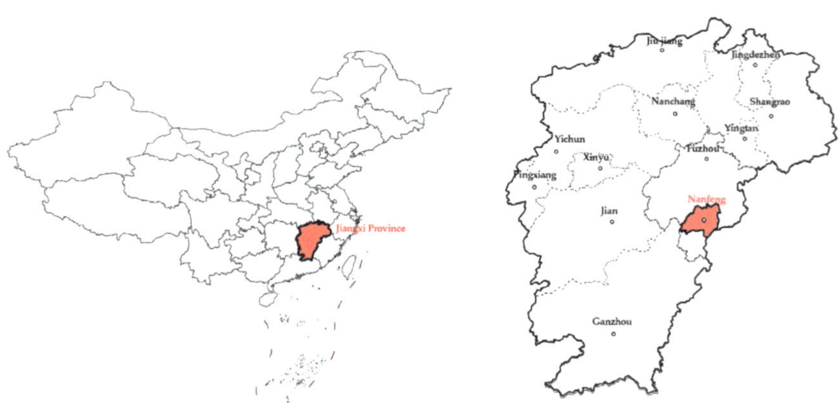

Figure 3. Geographic location of Nanfeng County, Jiangxi, China.

2.1. Geography and Climate Conditions of Nanfeng County

Nanfeng County is mainly characterized by hilly landscapes, accounting for about 65% of the total area of the county. Nanfeng County has a subtropical monsoon climate, with an average annual temperature of 19.8 °C, the coldest month being January with an average temperature of 6.2 °C, and the hottest month being July with an average temperature of 29.3 °C. The annual precipitation is 1791.8 mm, and the main precipitation in the region is concentrated in April–June. Therefore, local vernacular dwellings are usually required to adapt to the needs of sun shading, ventilation, and cooling in summer, while also taking into account cold protection and warmth in winter.

Table 1. Summary of different types of vernacular building plans and modern residential building plans.

Typology	Plan Map Prototype	Typical Instance	
Three-room, single-entry, single-patio type		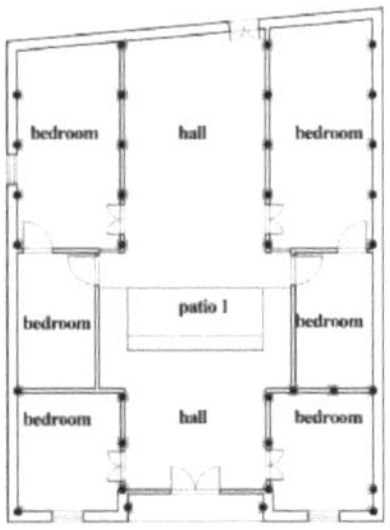 Kangdu Village Guard Camp Dwelling	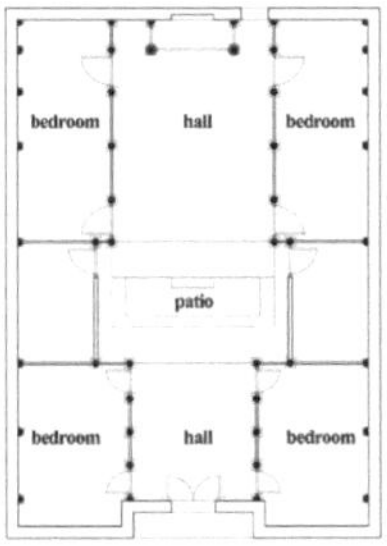 Lumin No. 130 in Shanggan Dwelling
Three-room, single-entry plus half-well type		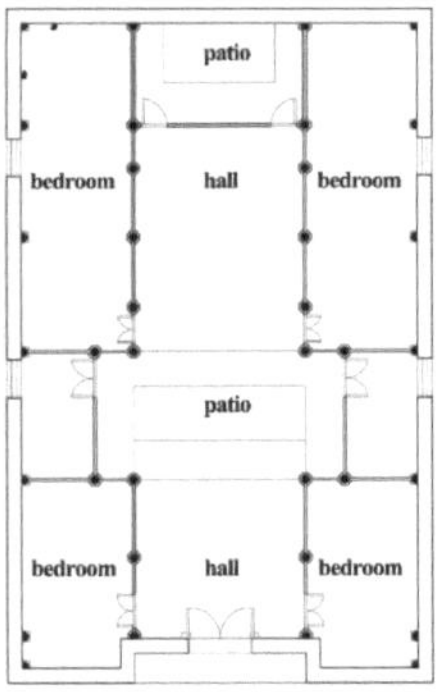 Meikengs' Dwelling	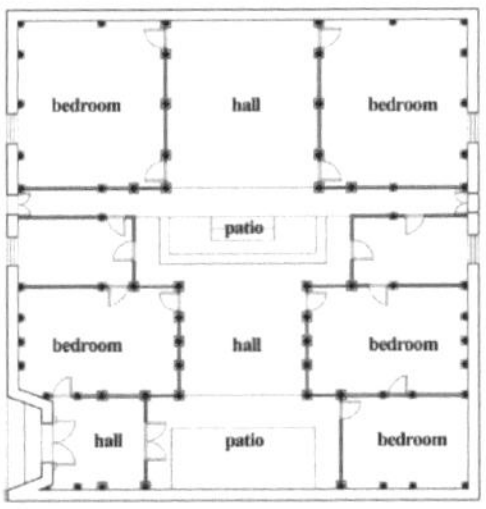 Shiyou villagers' Dwelling
Three-room, multi-entry type		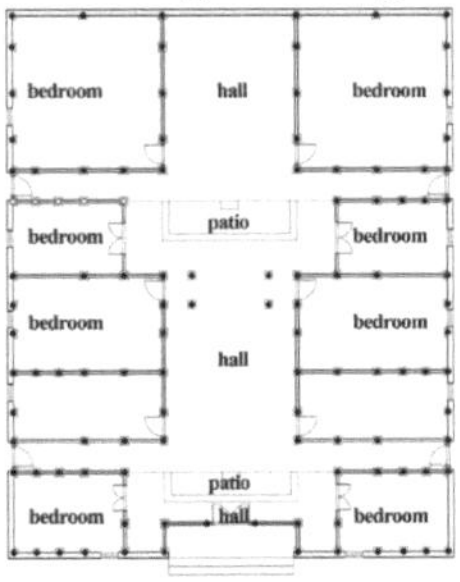 Former site of the Red Army in KangDu	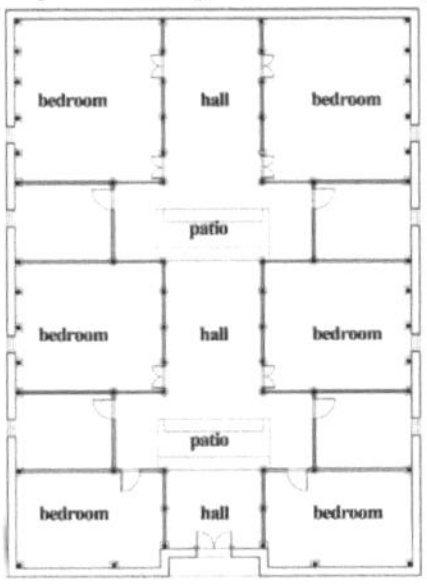 Yaopu Village Zengjia Dwelling
Modern residential building type		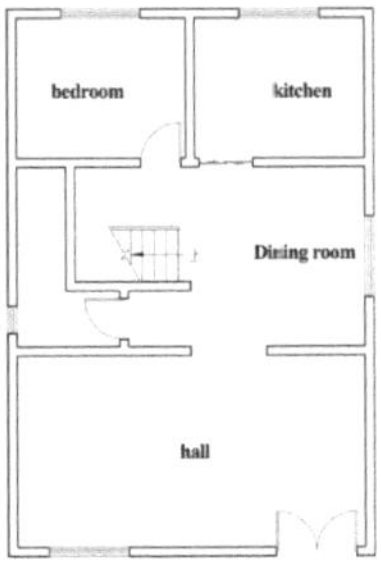	

2.2. *Vernacular Dwellings Layout of Nanfeng County*

2.2.1. Plan Form

The combination plan form of vernacular dwellings in Nanfeng County is centered on the patio and symmetrically distributed on the central axis. The upper and lower halls are sequentially placed on the central axis, and the upper and lower rooms are on both sides of the halls. The upper room has the highest rank and is usually inhabited by the elders, while the compartments on both sides of the patio have a lower rank and are usually inhabited by the children. This is in line with the social relationship of the feudal family and the feudal hierarchy. The patio carries the functions of drainage, ventilation, sunlight, and lighting inside the building. According to the system of vernacular dwellings in Nanfeng County, there are one-entry, two-entry, and multi-entry types of layouts. Among them, the three rooms with one entry and one patio type is the smallest type in Nanfeng County, which is characterized by simple composition and neat layout, and is widely distributed in Nanfeng County.

2.2.2. Structures

The structure of vernacular dwellings in Nanfeng County is mainly of the pierced-double type (Figure 4), followed by the inserted-beam type, which is generally used in higher-grade dwellings. And, the outer walls of vernacular dwellings are mostly cavity walls (Figure 5), their masonry is "one sleep one bucket", a brick outer skin layer and a space between the bricks to form an air layer. The inner walls are mostly wooden partition walls.

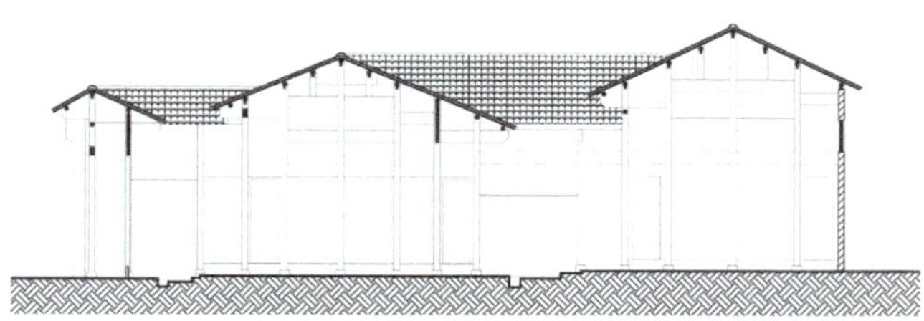

Figure 4. The structure of vernacular dwellings in Nanfeng County.

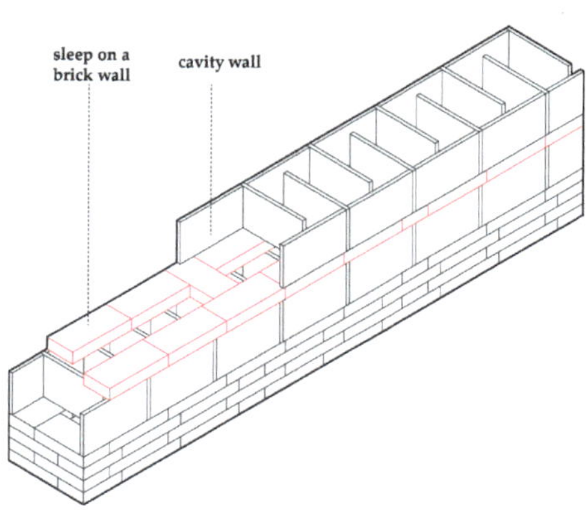

Figure 5. Schematic diagram of empty-cavity wall structure.

2.2.3. Materials

The vernacular dwellings in Nanfeng County mainly use easily accessible materials such as wood, brick, and stone. Nanfeng County is located in a red hilly area, and red stone is an important material for vernacular dwellings in Nanfeng County, being widely used for door and window frames, column bases, wainscotting, forming a unique architectural style, and overall harmonization.

2.3. Low Technology

Low tech is a mature and traditional technology compared to high-tech. Low technology in the field of architecture is a kind of concept, strategy, and idea in the macro perspective, and a specific technical form in the micro perspective. Low technology is characterized by reverence for nature, ease of implementation, and economic use. In terms of spatial function, low technology is manifested in the use of building pattern design to create energy-saving effects, in order to achieve the purpose of maintaining human comfort and reducing energy consumption [46]. From the aspect of economy, low-tech represents the use of cheap local materials and use of certain technical means to achieve good construction results. As a result of Nanfeng County's long history, the form of vernacular dwellings in Nanfeng County is constantly being perfected, containing low-tech methods such as plan layout, building structure, construction techniques, wall structure, and building materials.

3. Methods

This study constructs research ideas from climate adaptation, conducts research and mapping on six typical vernacular dwellings in Nanfeng County, and makes use of Ecotect (2011), Phoenics (2019) [47], and Energy plus (v8.7.0) software [48]. These programs are used to carry out mathematical simulation and analysis of the light environment, wind environment, and energy consumption in comparison with a selected modern residential building. By exploring the indoor physical environment of vernacular dwellings, the passive strategies involved in coping with the climate can be more accurately distilled.

3.1. Theoretical Basis

Climate adaptability originally refers to the interrelationship between living organisms and the climate environment in climatology and ecology, and belongs to the intersection of the two research fields. Later, the field of architecture quoted this research method, and formed the bioclimatological method of architecture [49].

Foreign exploration of the relationship between architecture and climate can be traced back to the Greek architect Vitruvius who compiled "Ten Books of Architecture", which is the earliest documented description of the principles of architectural design interacting with climate. Later, as a pioneer in the climate adaptation of traditional dwellings abroad, Rudovsky explored the relationship between architecture and climate in different regions of the world, and carried out corresponding research on traditional architecture from the perspective of the climatic environment. "Climatic adaptability" in architecture refers to the ability of architecture to resist and adapt to the natural climate environment. Climate refers to the weather conditions in general or over a long period of time, which are characterized by stability and regularity, and it is a different concept to that of weather, which refers to the changes in the hot and cold state of the near-surface atmosphere in a particular region within a short period of time.

3.2. Physical Modeling

The data of three kinds of vernacular dwellings were compiled through field research and mapping. The specific data are shown in Table 2.

Through field research, a modern residential building southwest of the guard battalion dwelling in Kandu Village, Taihe Town, Nanfeng County, was selected as the reference building, which is more similar in form to the residence of the Kangdu Village Guard Camp Dwelling with a sloping roof and the same orientation (Figure 6). The modern residential building has a two-story sloping roof and no internal patio, with a total width of 8.28 m and a total depth of 12.2 m. It adopts a brick–concrete structure, with steel security doors, single-glazed windows, and aluminum alloy frames, which is a relatively common form of modern residential building in the area. In order to simplify the model and accelerate the convergence speed of the calculation, this paper ignores all the minor bumps and various components in the simulation process, and mainly retains the main components of the building such as doors, windows, roofs, floor slabs, and walls [50]. Because the modern residential dwellings are located more towards the periphery of the village, the surrounding building density is low, while the vernacular dwellings are mostly concentrated in the middle of the village, with a high site coverage density. In order to control for the influence of the surrounding environmental conditions, the simulation experiments are carried out in an ideal environment with open and unobstructed surroundings for the calculation and analysis.

Table 2. Statistics of physical modelling indicators for vernacular and modern residential dwellings.

Serial Number	M1	M2	M3	M4
Typical instance	Kangdu village guard camp residence	Lumin No. 130 in Shanggan village	Meikengs' villagers	Shiyou villagers' residence
Average building height (m)	4.5	5.57	5.54	5.49
width (m) × depth (m)	7.5 × 10.2	10.5 × 15	9.8 × 15.5	14.2 × 15
Patio1 Length-width ratio	2.2:1	3.2:1	3.2:1	3.3:1
Patio1 Width–height ratio	1:3	1:2.9	1:3.3	1:3.2
Hall Width (m) × depth (m)	2.8 × 4.5	4.3 × 6.3	3.9 × 5.2	4.4 × 5.5
Bedroom Width (m) × depth (m)	2.1 × 3.8	3.0 × 6.3	2.9 × 7.9	4.4 × 5.1
Serial Number	**M5**	**M6**	**M7**	
Typical instance	The old site of the Red Army of Kangdu Village	Yaopu Village Zengjia grand house	Modern residential buildings in rural areas	
Average building height (m)	5.73	5.82	6.3	
Width (m) × depth (m)	14.5 × 19.8	15.7 × 19.2	8.2 × 12.2	
Patio1 Length–width ratio	3.4:1	3.1:1	—	
Patio1 Width–height ratio	1:3.1	1:3.5	—	
Hall Width (m) × depth (m)	4.6 × 5.5	4.1 × 5.7	8 × 4.5	
Bedroom Width (m) × depth (m)	5.2 × 5.4	5.2 × 5.7	4 × 3.3	

Figure 6. Location of guard camp houses and modern residential buildings in Kangdu Village.

3.2.1. Light Environment Modeling

The light environment simulation is analyzed by Ecotect software, and the three-dimensional spatial model is established according to the actual mapping drawings. As the vernacular dwellings are affected by a series of factors such as climate, historical deposits, living customs, construction technology, etc., wall features such as textures and being out of true, grille ash accumulation, uneven reflections, and other irregularities may appear. However, to simplify, these difficult-to-control, small variations can be ignored [50]. In the simulation, the bedrooms and halls of seven dwellings are selected to simulate the value of the daylight factor. The grid is set at 750 mm from the indoor floor height, and each room is divided by a 0.1 m × 0.1 m grid cell specification. Grid independence is verified by spacings of 1.2 m × 1.2 m, 0.1 m × 0.1 m, 0.08 m × 0.08 m, and 0.06 m × 0.06 m. The bedroom of M1 is selected as the experimental object in this experiment, and compared to the average lighting coefficient and the computing time of different grid numbers. It was found that through the comparison (Table 3), the grid cell specification of 0.08 m × 0.08 m is encrypted. The error of the average daylight factor is 2%, but the computation time is doubled. When the grid is encrypted again, the average daylight factor does not change, but the time is doubled. The grid cell size of 0.1 m × 0.1 m shows the same result, but the calculation time is shortened compared to the grid cell size of 0.08m*0.08m. Therefore, the simulation results using a 0.1 m × 0.1 m grid cell size can better reflect the light environment of the building and save computing costs.

Table 3. Grid Independence Verification Simulation.

Grid Cell Specifications	Daylight Factor Value (%)	Time (s)
1.20 m × 1.20 m	0.92	141
0.10 m × 0.10 m	0.92	206
0.08 m × 0.08 m	0.92	435
0.06 m × 0.06 m	0.94	920
0.04 m × 0.04 m	0.94	1929

3.2.2. Wind Environment Modeling

Since wind environment simulation uses Phoenics software, the study needs to set the corresponding calculation domain, based on the "DB22/T 5055-2021 green building design standards" [51], to ensure that the building coverage area is less than 3% of the area of the calculation domain. So, with the building as the center of the circle, the radius of the horizontal calculation area is not less than 5H (building height), and the calculation area of the upper part of the building is not less than 3H. After determining the calculation area, the built-in PARSOL function of Phoenics software creates a structured mesh division and, in order to improve the calculation accuracy, the mesh is encrypted in the space where the main body of the building is located and in the near-ground area. In order to verify the independence of the grid, which is divided into 100,000; 500,000; or 1 million cells, experiments show that a 500,000-cell grid can guarantee the accuracy of the calculation. We examined and compared the statistical analysis of the relative wind speed of the five measurement points in the three different experiments (Figure 7). Through comparison, it was found that when the unit number of the grid was increased for the first time (from 100,000 to 500,000), the relative wind speed of some measurement points changed remarkably. However, when further increasing the unit number (from 500,000 to 1 million), the error rate of the relative wind speed at each measurement point was less than 5%, which proved that a grid with 500,000 units could ensure computational accuracy. Therefore, this article employed a grid with 500,000 units for numerical simulation to improve the operational efficiency.

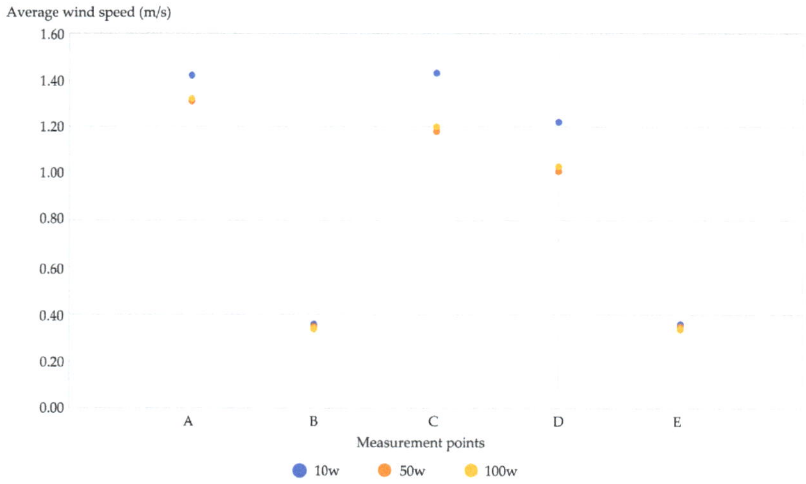

Figure 7. Grid independence verification.

3.2.3. Energy Consumption Modeling

The simulation of building energy consumption is analyzed by EnergyPlus software, and the three-dimensional spatial model of vernacular dwellings and modern residential buildings is established according to the actual mapping drawings, and sets different enclosure structures such as doors, windows, exterior walls, roofs, interior walls, floor slabs, and so forth. Then, the weather file (EnergyPlus Weather File, EPW) of Fuzhou City Meteorological Station in Nanfeng County is then downloaded from the official EnergyPlus website, and set as the year-round meteorological parameters for the vernacular dwellings and modern residential buildings. After completing the 3D modelling in Sketchup, the idf file was exported, and the EnergyPlus software's idf editor was used to define the delamination, thickness, material thermal conductivity, and light transmission properties of the building envelope, as well as to set the parameters of the lighting, personnel, electrical equipment, and HVAC system. The simulation of the building's energy consumption was then carried out.

3.3. Model Setup and Parameters

3.3.1. Light Environment Parameters and Mathematical Modeling

1. Parameter Setting

The light environment was modeled using Ecotect Analysis. According to the Chinese light climate zoning standard, Nanfeng County belongs to the class IV light climate zone. The lighting critical illuminance of its areas are all 4500 Lux, and the design illuminance value is 13,500 Lux. The most unfavorable lighting conditions were calculated by using the full-coverage model of the International Commission on Illumination. Vernacular dwellings with wood roofs have an albedo ratio of 0.2, brick wall interior surfaces have an albedo ratio of 0.6, and cinderblock floors have an albedo ratio of 0.23. Modern residential buildings have concrete roofs with an albedo ratio of 0.7, brick wall interior surfaces have an albedo ratio of 0.6, and concrete floors have an albedo ratio of 0.23.

2. Mathematical Modeling

Ecotect uses the Building Research Establishment (*BRE*) shunt method, a widely accepted and very useful technique for calculating daylight factors. The method assumes that direct sunlight is ignored and that three separate components of natural light arrive at any point inside the building:

Daylighting Factor (*DF*)—represents the daylighting factor at a given point;

Sky Component (*SC*)—represents the portion of the room that receives light directly from the sky through the window;

External Reflectance Component (*ERC*)—represents the portion that reflects back from the ground, trees, or other dwellings;

Internal Reflection Component (*IRC*)—represents the internal reflection of the first two components on the interior surface.

The Daylight Factor calculation uses the following expression:

$$DF = SC + ERC + IRC \tag{1}$$

3. Range of the light environment simulation

The simulation range of the light environment of the seven dwellings is mainly based on the regularity of the range of human daily activities, which can be divided into the bedroom and the hall (Figure 8). In this paper, the simulation range of the bedroom in the vernacular dwellings is in the upper main room area on the left side of the hall, and the simulation range of the hall in the vernacular dwellings is in the area of the hall, because the hall in the vernacular dwellings is the main space for people's daily activities.

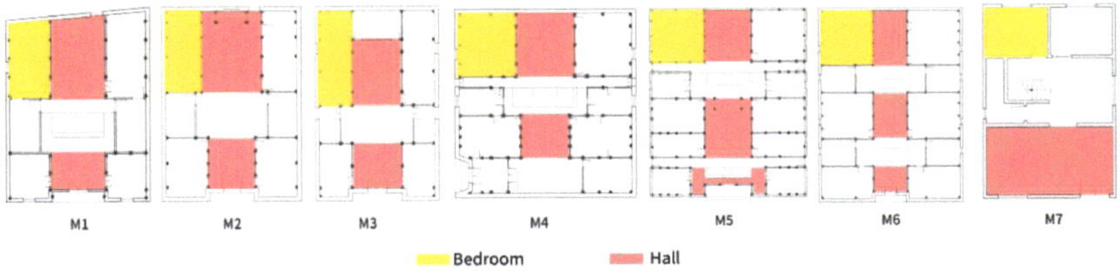

Figure 8. Scope of light environment simulation for different dwellings.

3.3.2. Wind Environment Parameters and Mathematical Modeling

1. Meteorological Parameters

The main evaluation of the wind environment conditions in summer and winter for the seven dwellings involves drawing up the main technical parameters by referring to the "Air Conditioning Design Manual (Third Edition)" and consulting the Nanfeng County Meteorological Bureau. The dominant wind direction in summer is south-easterly with a wind speed of 2.7 m/s and an outdoor design temperature of 28.3 °C whilst the dominant wind direction in winter is northerly with a wind speed of 2.1 m/s and an outdoor design temperature of 7.2 °C.

2. Mathematical Modeling

This article adopted the standard k-ε model. It has the characteristics of low cost, small fluctuations, and a high numerical calculation accuracy. When the iteration number of the standard k—ε model is set to 200, the residuals of the continuity and momentum equations are less than 10^{-4}, and the residuals of the k and ε equations are less than 10^{-7}, hence the model's judgment is credible. In this article, assumptions are made based on air flow: (1) room temperature, low velocity, incompressible fluid flow, (2) steady-state flow is studied, and (3) the fluid is considered isotropic. The numerical solution process is shown in Figure 9. All the control differential equations include the continuity, momentum, k and ε equations [52,53], and the simplified equations after considering the steady state incompressibility of the fluid are:

Mass conservation equations:

$$\frac{\partial u_i}{\partial x_i} = 0 \tag{2}$$

momentum equation (math.)

$$\frac{\partial(\rho u_i u_j)}{\partial x_i} = \frac{\partial}{\partial x_i}\left(\mu_{\text{eff}}\frac{\partial u_j}{\partial x_i}\right) - \frac{\partial p}{\partial x_j} + \frac{\partial}{\partial x_i}\left(\mu_{\text{eff}}\frac{\partial u_i}{\partial x_j}\right) \tag{3}$$

the conservation of energy equation

$$\frac{\partial(\rho u_j T)}{\partial x_i} = \frac{\partial}{\partial x_i}\left(\Gamma_{T,eff}\frac{\partial T}{\partial x_i}\right) + S_T \tag{4}$$

T represents the temperature; $\Gamma_{T,eff}$ represents the effective diffusion coefficient; S_T represents the heat generation term; ρ represents the fluid density; u_j represents the velocity vector; and x_i represents the coordinate position.

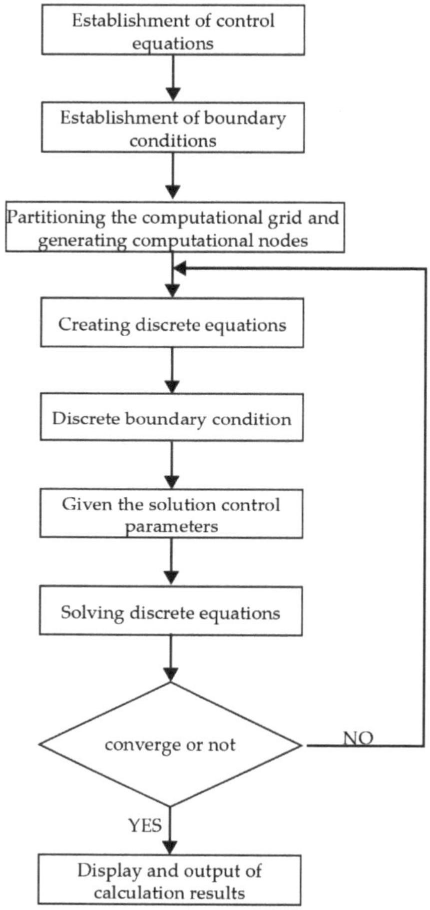

Figure 9. Basic ideas of numerical solution.

3. Calculation Area

The size of the calculation area was set according to the Green Building Design Standard and affects the speed and accuracy of the software calculation. The area covered by the building should be less than 3% of the whole calculation area, the size of the horizontal calculation area should be a radius of 5H (with the building as the center of the circle), and the calculation area above the building should be more than 3H (H is the height of the main body of the building).

4. Wind Environment Boundary Conditions
(1) Inlet boundary condition

In practice, the wind speed near the surface decreases with height due to friction at the surface. The pattern of change in wind speed at the incoming surface is expressed as an exponential rate:

$$\sqcap|_{inflow} = U_z \tag{5}$$

$$\frac{U_z}{U_o} = \left(\frac{Z}{Z_o}\right)^{\alpha} \tag{6}$$

U_z is the wind speed in the horizontal direction at height Z; U_o is the wind speed at Z_o of the reference height; and α is a power index determined by the roughness of the terrain. According to the relevant Chinese standards and norms, traditional villages in Jiangxi belong to the field, countryside, jungle, and hilly areas, and the value of α is taken as 0.16.

By referring to "Air Conditioning Design Manual (Third Edition)", consulting Nanfeng County Meteorological Bureau, and by other means, it is known that in the typical climate of Nanfeng County in summer the dominant wind direction is south-easterly, with a wind speed of 2.7 m/s; the dominant wind direction in winter is northerly, with a wind speed of 2.1 m/s. For example, for a northerly wind, the north side of the calculation area is set as the "velocity inlet", and the east, west, and south sides are set as the "pressure outlets".

(2) Outlet boundary condition

The exit boundary condition uses a chi-square Neumann boundary condition, implying that at the exit boundary the normal gradient of all variables is assumed to be zero, i.e., the flow is sufficiently developed to be no longer affected by the flow near the exit. The equation can be expressed as:

$$\frac{\partial u}{\partial \iota} = 0 \tag{7}$$

u represents the fluid velocity; $\frac{\partial}{\partial \iota}$ denotes the partial derivative of the flow direction along the outlet boundary.

By assuming that the flow on the outlet surface is sufficient and has returned to the normal conditions, free from obstruction by buildings, the outlet pressure is set as the standard atmospheric pressure. The equation can be expressed as:

$$P_{amb} = 101,325 \text{ Pa} \tag{8}$$

(3) Wall boundary conditions

The wall and underlay of the building are set up as no-slip movement conditions, conditions that correspond to actual physical phenomena. Due to the viscous interaction between the fluid and the wall, the velocity of the fluid immediately above the wall can be assumed to be zero. The equation can be expressed as:

$$u|_{wall} = 0 \tag{9}$$

5. Thermal Environment Boundary Conditions

In this simulation, the research object is placed in an ideal environment, and there are no other sources of heat and cold in the vicinity, so in the temperature field simulation, the first type of boundary conditions are used in this simulation. The equation can be expressed as:

$$T = T_{boundary} \tag{10}$$

$T_{boundary}$ is the constant temperature at the boundary.

By means of reference to the *Air Conditioning Design Manual* (3rd Edition) and consultation with the Nanfeng County Meteorological Bureau, it can be learned that the average outdoor temperature for a typical summer climate is 28.3 °C, and the average outdoor temperature for a typical winter climate is 7.2 °C.

6. Wind speed detection point setting

According to the flow field area of the wind environment of the building and the daily activities of people, the simulated wind speed detection points in different types of dwellings are determined (Figure 10). Point A is located at the main door of the building to detect the wind environment on the windward side; Point E is located at the doorway of the bedroom to detect the wind environment at the doorway of the bedroom; Point F is located in the center of the bedroom to detect the wind environment in the bedroom space; Points B, D, and G in the vernacular dwellings are located in the center of the hall to detect the wind environment in the main public activity space, and Point B in the modern residential building is to detect the main wind environment in the main public activity space.

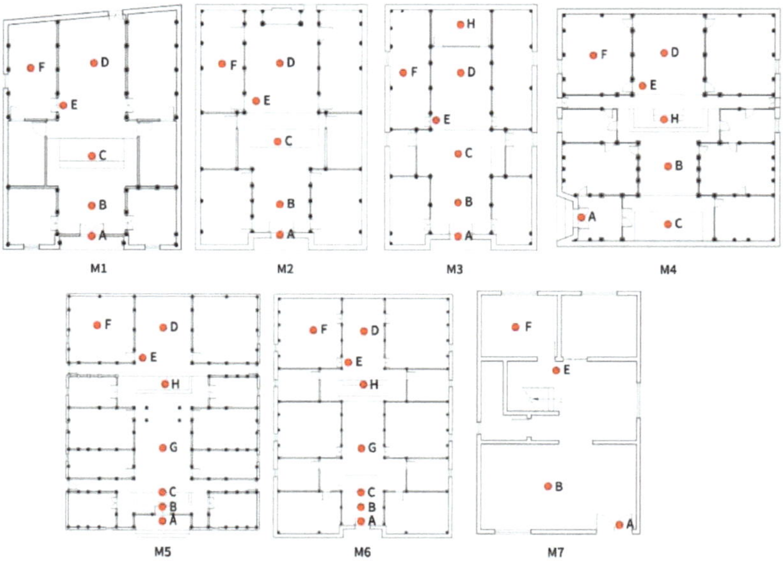

Figure 10. Location of simulated wind speed detection points.

3.3.3. Energy Consumption Parameters and Mathematical Modeling

1. Meteorological Parameters

In the simulation, the number of people using lighting and appliances and the electricity load are proportional to the building area. The population number is set to eight people, and according to the Energy Saving Design Standard for Residential Buildings in Jiangxi Province, it is known that the lighting and equipment loads are 5.0 W/m^2 and 3.8 W/m^2, respectively. The daily opening time is set to 6 h/day, the indoor summer air conditioning cooling temperature is set to 26 °C, and the intended winter indoor heating temperature is set to 18 °C. Combined with the local climatic conditions, the cooling time is set from 1 June to 1 October each year, with daily cooling times of 12:00–14:00 and 20:00–24:00, and the heating time is set from 15 December each year to 1 March of the following year. According to the "Energy-saving Design Standards for Residential Buildings in Jiangxi Province" and the results of the field research of the simulation object, the parameters of the enclosure structure of the vernacular building and the modern residential building are set separately.

Table 4 shows the building's established three-dimensional model. The specific values are shown in Table 5.

Table 4. Digital Physical Models of the Study Buildings in Energyplus Software.

Energyplus Model

M1

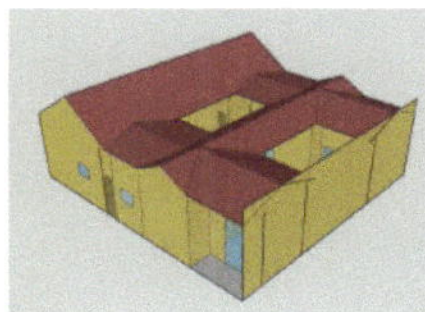

M3

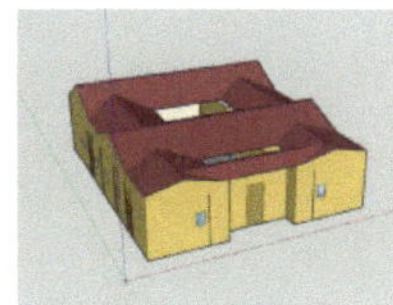

M5

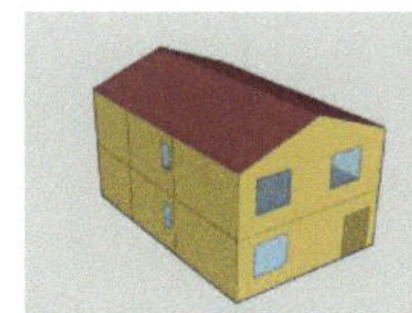

M7

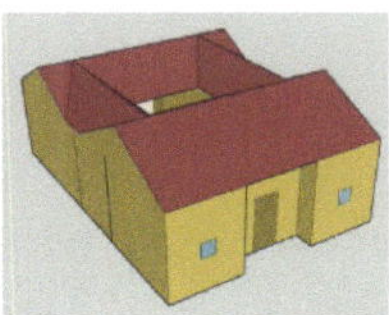

M2

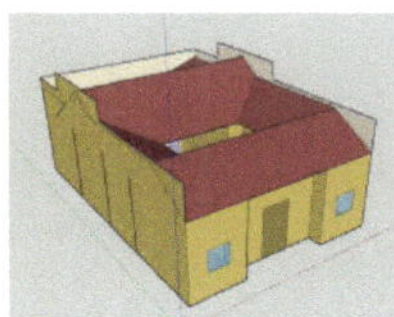

M4

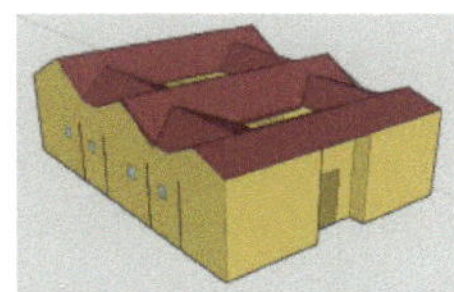

M6

Table 5. Parameter Settings for the Envelope Structure of the study buildings.

Building Envelope	Construction Method	Material Thermal Index			
		Thickness (mm)	Density (kg/m^3)	Specific Heat kJ/(kg·K)	Heat Conductivity W/(m·K)
cavity wall	Atmosphere		1.3	1005	0.0023
	Blue brick	240	1900	1050	0.265
Brick wall	1. Lime cement	15	1800	1050	0.93
	2. Clay brick	240	1800	1050	0.81
	3. Lime cement coating	15	1800	1050	0.93
Wooden Wall	Wooden board	30	500	2510	0.17
Door	Wooden door	40	550	2301	0.343
Floor	1. C20 fine aggregate concrete	60	2300	920	1.51
	2. Soil pile	150	1600	1010	0.81
	3. 3:7 Lime soil	300	1795.6	884	0.72
Window	Single-glass wooden board K = 4.7 (W/m^2 K)	6	2500		
Roof	1. 15 mm gray tile roof	15	2044	1050	0.96
	2. Rafters	80	2300	656.9	0.753
	3. Purlin	100	2300	656.9	0.753

2. Mathematical Modeling

In this study, Energy Plus is used to simulate the energy consumption of a residence to calculate the total energy consumption of the building in one year, and the energy consumption per unit area of the building is compared because of the different sizes of the building areas of different buildings. The total energy consumption of the building in Energy Plus is composed of Heating, Cooling, Lighting, Equipment, and Fans, respectively, Pumps and other energy consumption system, due to this paper using the ideal air conditioning system in the simulation, does not take into account the energy consumption of air conditioning systems or ventilation equipment, so the total energy consumption of the building is simplified into air conditioning load, personnel equipment, and indoor lighting.

The equation for energy consumption per unit area of a building is as follows:

$$Q_{Total} = \frac{Q_{HVAC}+Q_{Equ}+Q_{Light}}{A_{Total}} \quad (11)$$

Energy Plus is composed of a building module, an air module, and an HVAC system module, with real-time feedback between the three using an integrated synchronization method. The indoor air energy conservation equation and the building wall energy conservation equation are the two basic equations used in the calculations performed by the energy simulation software.

$$\sum_{i=1}^{N} q_{i,c}A_i + Q_{other} + G_aC_p(T_{a-out} - T_m) - Q_{heat-extra} = \rho VC_p\frac{dT_{in}}{dt} \tag{12}$$

The building wall energy conservation equation is as follows:

$$q_{i+}q_{ir} = \sum_{k=1}^{N} q_{ik} + q_{i,c} \tag{13}$$

i is the number of individual surfaces of the building envelope; $q_{i,c}$ is the convective heat transfer through the ith surface; N is the number of surfaces of the envelope; A_i is the actual heat transfer area of the ith surface of the envelope; Q_{other} is the latent heat caused by the convective part and evaporation of water in the heat balance equation of the surface, sunlight, equipment, lighting, and heat gain of the personnel; G_a is the sum of air volume of the fresh air and infiltration air; C_p is the constant pressure-specific heat capacity of the air; $\mathrm{T_m}$ is the temperature of the zone; $Q_{eat-extra}$ is the amount of heat lost from the inside air to the outside; C_p is the constant pressure-specific heat capacity of the air; $\rho VC_P\frac{dT_{in}}{dt}$ is the heat capacity of the room air; $T_{\mathrm{a.out}}$ is the outdoor air temperature; q_i is the heat transfer from the wall to the wall i; q_{ir} is the radiant heat from the internal heat source and the sun.

Energy Plus uses the heat balance method to calculate the air conditioning system load. Based on the first law of thermodynamics and the law of the conservation of energy, the air conditioning system supplies hot or cold air to each indoor area to satisfy the indoor heat or cold loads under non-linear outdoor conditions by assuming that the room air temperatures and the surface temperatures of the enclosure are balanced and consistent. The instantaneous load of the room is calculated by solving each surface and room air temperature through the heat balance equations for the external surface of the building, the internal surface of the building, the heat balance of the building body, and the outdoor heat balance.

The formula for calculating the instantaneous load of a building by the heat balance method is as follows:

$$q_i(t) = h_i(T_R(t) - T_i(t)) + \sum_{k=1}^{n} hr_{i,k}(T_K(t) - T_i(t)) + R_i \tag{14}$$

$$Q(t) = \sum_{i=1}^{n} s_ih_i(T_i(t) - T_R(t)) + m(t)C_P(T_o(t) - T_R(t)) + Q_S(t) \tag{15}$$

The energy consumption formula for building HVAC systems is as follows:

$$P_{HVAC,\ total} = \frac{Q(t)t_{HVAC}}{COP_{ref}} \tag{16}$$

The energy consumption formula for indoor lighting in buildings is as follows:

$$P_{Light} = P(t)t_{Light} \tag{17}$$

The energy consumption formula for building indoor equipment is as follows:

$$P_{Equ} = \sum_{e=1}^{n} n_eP\tau \tag{18}$$

$q_i(t)$ is the heat gain of the ith surface at moment t, (kw/m^2); h_i is the heat transfer coefficient of the ith surface; $T_R(t)$ is the calculated indoor temperature, (K); $T_i(t)$ is the temperature of the ith internal surface; R_i is the radiant absorption of the ith surface (kW/m^2); $Q(t)$ is the cooling load at moment t, (kW); s_i is the area of the ith surface, (m^2); $m(t)$ is the fresh air volume at time t, (kW/s); C_P is the constant pressure-specific heat capacity, (kJ/kg); $T_o(t)$ is the calculated temperature of the outdoor air, (K); $Q_S(t)$ is the heat generation from the indoor heat source at time t, (kW). t_{HVAC} is the HVAC turn-on time. COP_{ref} is the rated cooling factor of the building's HVAC system. t_{Light} is the indoor lighting runtime. P is the average power of a device in operation. e is the total number of device types. n is the quantity of a particular type of equipment. τ is the equipment runtime.

4. Results

4.1. Light Environment Simulation

According to the Chinese building lighting design standard (GB55016-2021 [54]), the lighting coefficient of bedrooms and halls in residential buildings should not be lower than 2.0%. According to the specification, the minimum value of daylight factor C_{min} = standard value × light climate coefficient K (K = 1.1) for residential lighting in Nanfeng County, the minimum value of daylight factor for bedrooms and halls shall not be less than 2.20%.

Comparing the light environment of the hall of seven dwellings (Tables 6 and 7, and Figures 11 and 12), M7's hall has a lower percentage of areas with a daylight factor value of not less than 2.20% of M1. The average daylight factor value (1.07%) is lower than M1's, in which the difference in the average value of daylight factor is 1.07%, so M1's hall light environment is better than that of M7. Due to M1 and M2 being of the same type with little difference in their daylight factor, it can be concluded that M2's hall light environment is better than M7. By comparing the light-environment-related indicators of different types of halls, it is clear that the light environment of M7 in the hall area is lower than that of the vernacular dwellings.

Table 6. Simulation of daylighting coefficients for different dwellings.

Typology	Three-Room, Single-Entry, Single-Patio Type	Three-Room, Single-Entry Plus Half-Well Type	Three-Room, Multi-Entry Type	Modern Residential Building Type
Light environment simulation diagram.	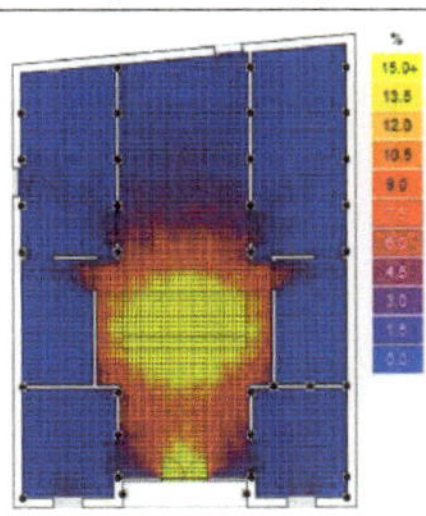 M1	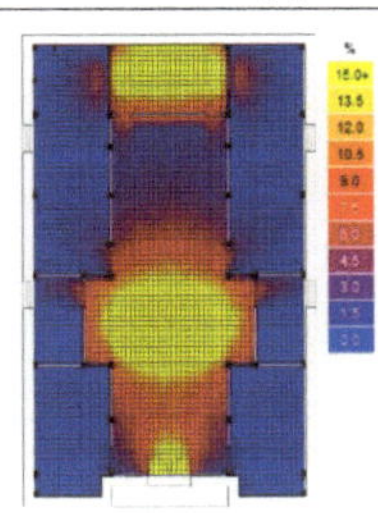 M3	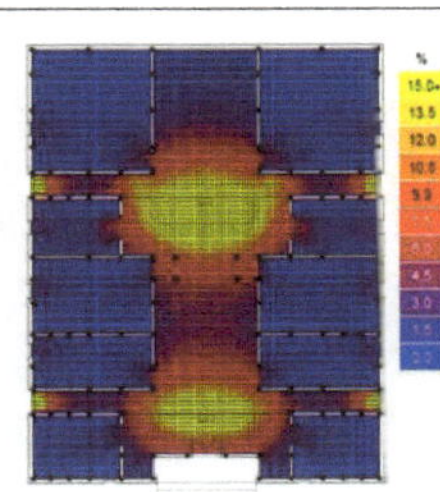 M5	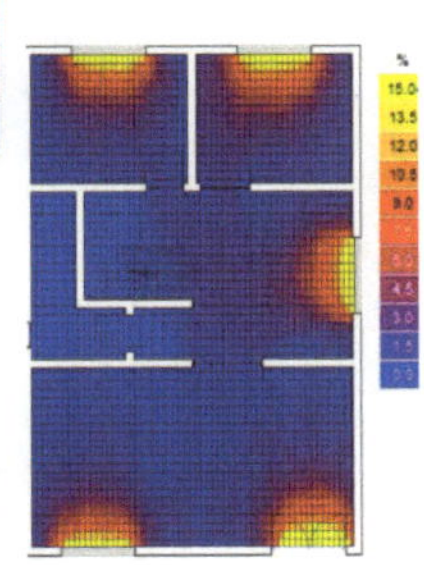 M7
	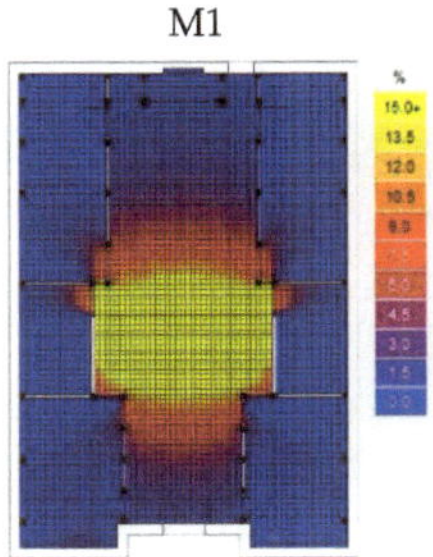 M2	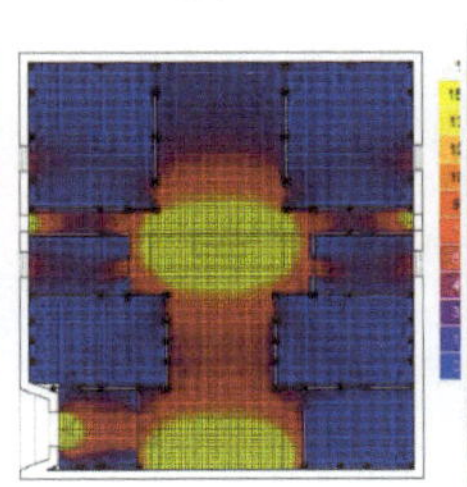 M4	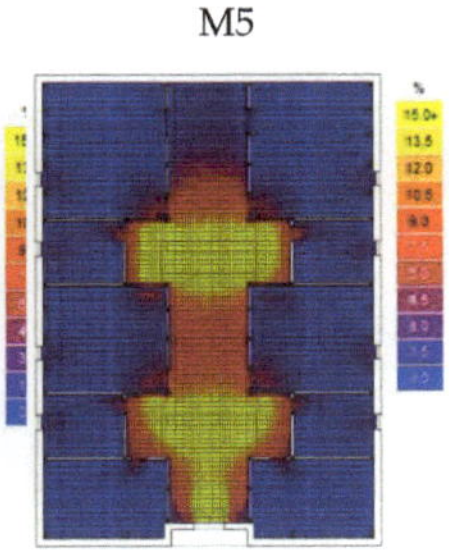 M6	

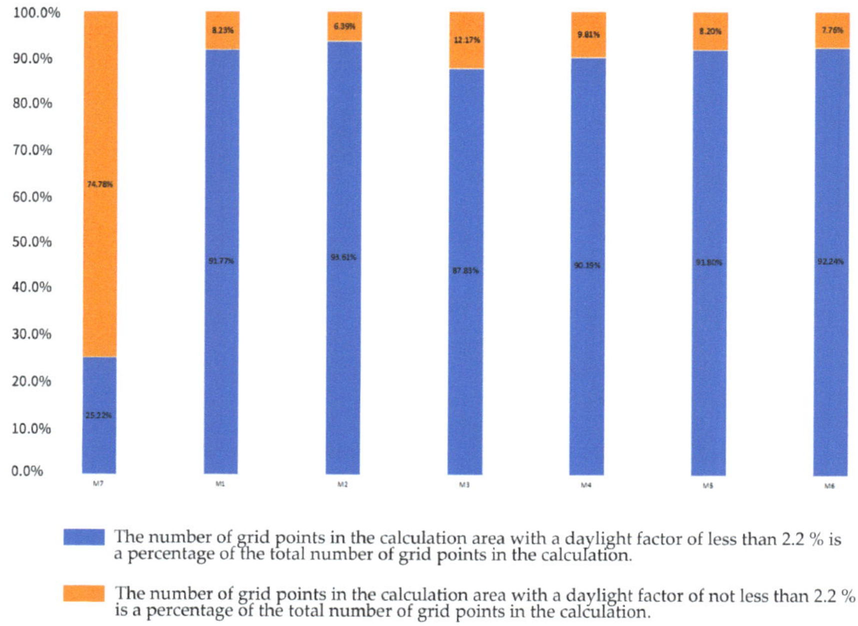

Figure 11. Comparison of simulation of daylight factor in hall areas of different dwellings.

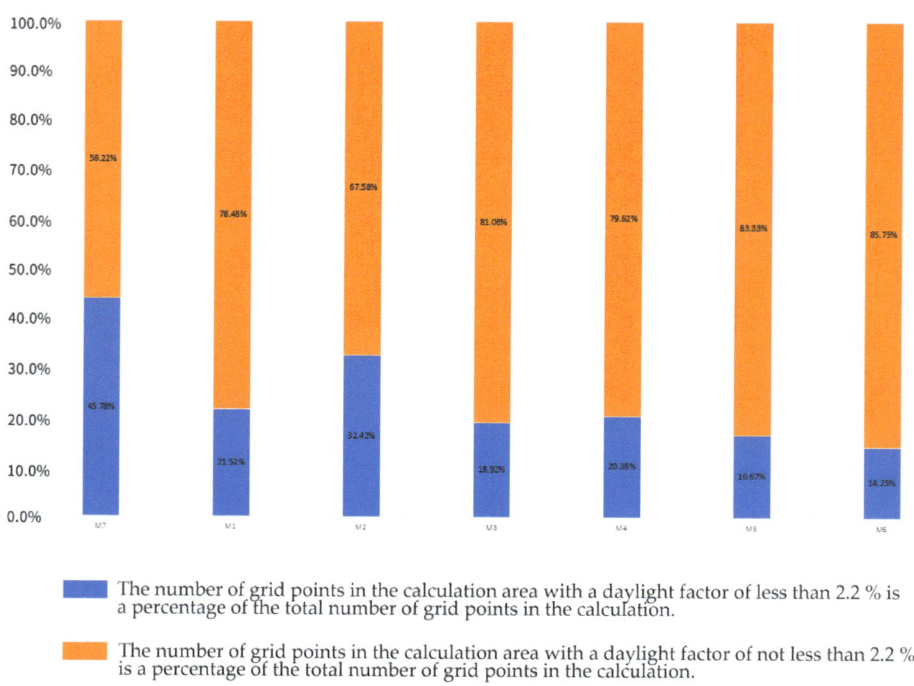

Figure 12. Comparison of simulated daylight factor in bedroom areas of different dwellings.

Table 7. Table of light environment indicators for bedroom and hall areas in different buildings.

	Area	M1	M2	M3	M4	M5	M6	M7
Hall area	Average light factor (%)	5.45	4.7	6.48	5.45	5.77	7.51	4.38
Bedroom area	Average light factor (%)	0.83	0.75	1.21	1.06	1.02	0.87	5.12

Comparing the light environment of the bedrooms of the seven dwellings, M7's bedrooms have a much higher percentage of areas with a daylight factor value not less than 2.20% than M1, and the average daylight factor value (4.29%,) is much higher than those of M1, so the difference in the average value of the daylight factor is 4.29%, so the light environment of the bedrooms of M7 is better than that of M1. In the same vein, it appears that the light environment of M7 in the bedroom area is better than that of the vernacular dwellings.

4.2. Wind Environment Simulation

Comparing the wind speeds at different points in the six vernacular dwellings under ideal conditions in summer (Tables 8 and 9), it was found that the indoor wind environment improves with the increase in the number of patios. M1 and M2 are of the three-room, single-entry and single-patio type, except at point A where the wind speed gap is obvious, but the wind speed difference in other areas is not significant. M3 and M4 are of the three-room, single-entry plus half-patio type; however, due to the main door of M4 being on the west side, the range of the wind speed of point A of M4 is much lower than that of M3, and also the wind speed difference between the two at point C is larger. Points B, F, and D have smaller differences in wind speed ranges. Comparing the wind speed situation of M1 and M2, the wind speed range of M3 and M4 at points E and F tends to be the same, the wind speed range of points A and C has a small difference, and the wind speed at point B is slightly higher. By comparing, it can be concluded that the wind environment of M3 and M4 in the hall area is better than that of M1 and M2. M5 and M6 are three-room multi-entry types and, by comparing the range of wind speeds at different points, the difference in the internal wind environment is not large, and the difference in wind speed is larger only at point D. This is because there are corridors on both sides of the upper hall of M5, which create a wind pressure that reduces the wind speed in the upper hall area. Comparing the wind speed situations of M1 and M2, the wind speed range of M5 and M6 at points A, B, and C is slightly higher, and the wind speed range of points E and F tends to be the same, while the wind speed at point D is lower than that of M1 and M2. This is due to the fact that there is an open door on the north side of M1 and M2 to form a through-traveling wind, and the wind speed at point H is higher than the wind speed of M1 and M2 at point D. By comparing these, it can be concluded that the wind environments of M5 and M6 in the hall area are better than those of M1 and M2.

Table 8. Table of wind speed values at each test point simulated in different dwellings in summer.

Serial Number	Point A (m/s)	Point B (m/s)	Point C (m/s)	Point D (m/s)	Point E (m/s)	Point F (m/s)	Point G (m/s)	Point H (m/s)
M1	1.2–1.4	0.6–0.8	0.6–0.8	0.4–0.6	0.4–0.6	0–0.2	/	/
M2	1.4–1.6	0.6–0.8	0.6–0.8	0.4–0.6	0.4–0.6	0–0.2	/	/
M3	1.4–1.6	0.8–1.0	0.8–1.0	0.6–0.8	0.2–0.4	0–0.2	0.4–0.6	/
M4	0.2–0.4	0.6–0.8	0.2–0.4	0.4–0.6	0.2–0.4	0–0.2	0.4–0.6	/
M5	1.4–1.6	0.6–0.8	1.2–1.4	0.2–0.4	0.2–0.4	0–0.2	0.2–0.4	0.8–1.0
M6	1.4–1.6	0.8–1.0	1.2–1.4	0.2–0.4	0.4–0.6	0–0.2	0.4–0.6	0.8–1.0
M7	1.4–1.6	0.2–0.4	/	/	0.6–0.8	0.2–0.4	/	/

Table 9. Simulated wind speed clouds for different dwellings indoor wind environments in summer.

Typology	Summer	
Three rooms with one entrance and one patio	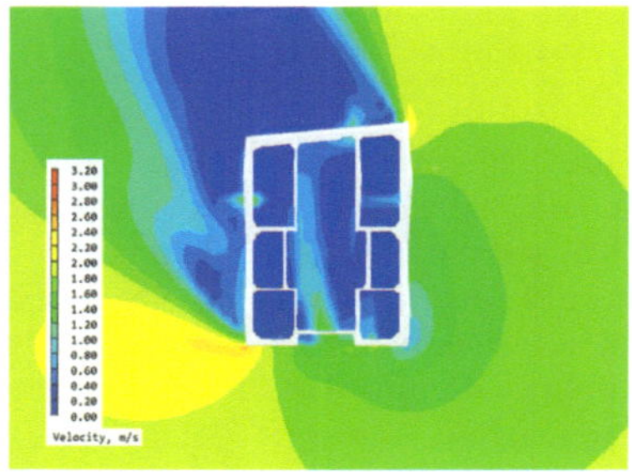 M1	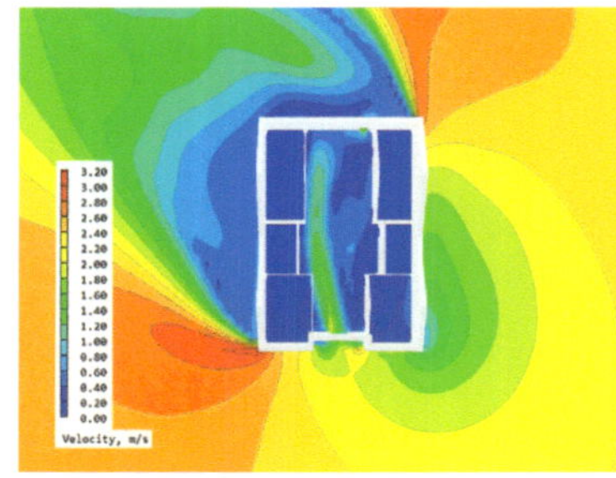 M2
Three-room, single-entry plus half-well type	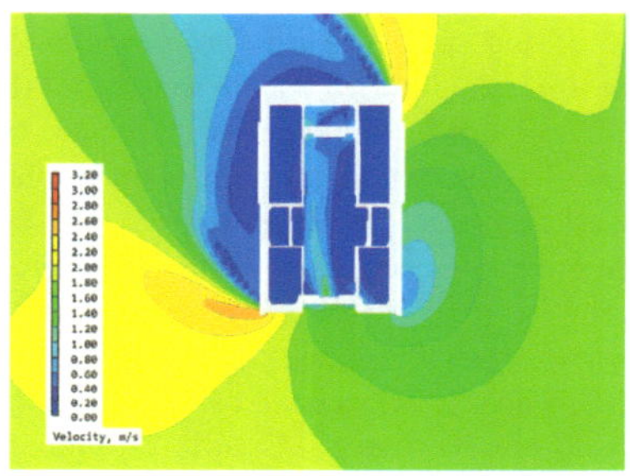 M3	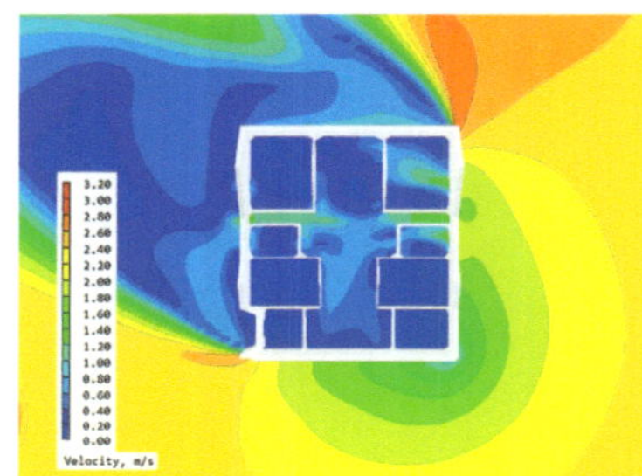 M4
Three-room multi-entry type	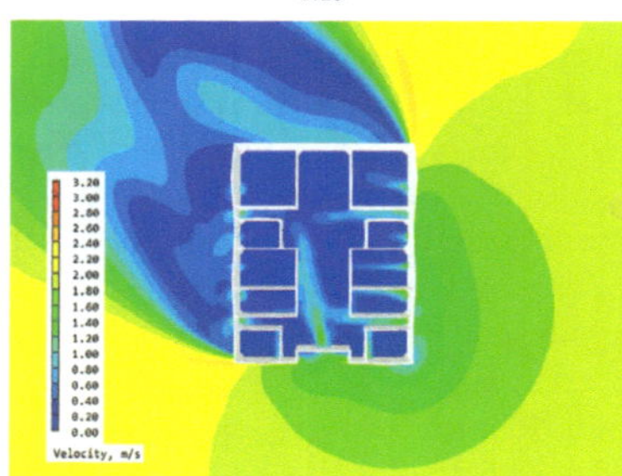 M5	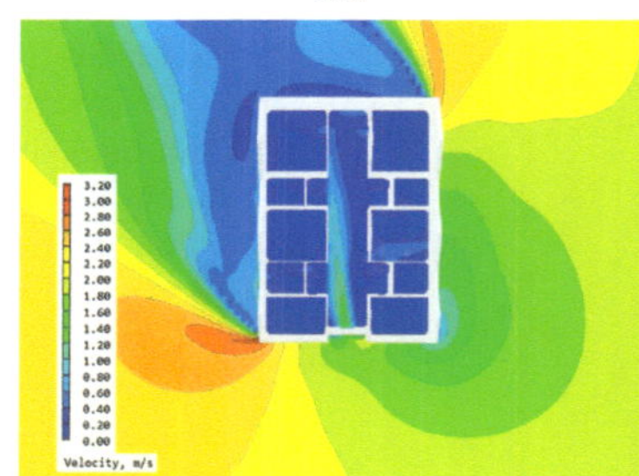 M6
Modern residential buildings type	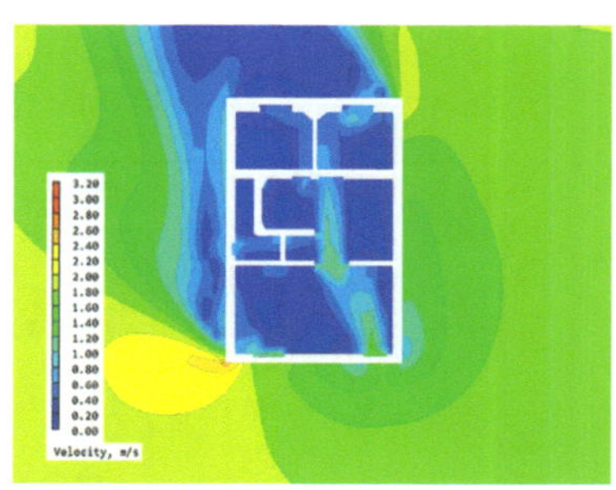 M7	

Comparing the vernacular dwellings with the modern residential buildings, the scale ratio of M1 and M7 is not much different. Comparing the wind speed range at point A, M7's wind speed range (1.4–1.6 m/s) is higher than the M1's wind speed range (1.2–1.4 m/s).

Comparing the wind speed at point B, M7's wind speed range (0.2–0.4 m/s) is slightly lower than M1's wind speed range (0.4–0.6 m/s). Comparing the wind speed range of M7 at points E and F, the wind speed range of M7 is higher than that of M1. In summary, although the wind speed is higher in M7's bedroom area than in M1's bedroom, its hall area wind speed is lower than that of M1's hall area.

Under the ideal conditions in winter, comparing the wind speeds at different points of the six vernacular dwellings (Tables 10 and 11), it was found that their internal wind environments tend to be the same. M1 and M2 are three-room, single-entry and single-well types so, except for the obvious difference in wind speed at point A, the difference between the points in other areas is not significant. M3 and M4 represent the three-room single-entry plus half-well type, and M5 and M6 represent the three-room multi-entry type. Although they are different types, by comparing the wind speeds at different points, it was found that the internal wind environment conditions tend to be similar. Compared with the three-room, single-entry, single-well type represented by M1 and M2, the greatest difference is that the wind environment at point B is slightly higher than that of M1 and M2.

Table 10. Table of wind speed values at each test point simulated in different dwellings during in winter.

Serial Number	Point A (m/s)	Point B (m/s)	Point C (m/s)	Point D (m/s)	Point E (m/s)	Point F (m/s)	Point G (m/s)	Point H (m/s)
M1	0.55–0.69	0.28–0.41	0.28–0.41	0.14–0.28	0.14–0.28	0–0.14	/	/
M2	0.55–0.69	0.28–0.41	0.28–0.41	0.14–0.28	0–0.14	0–0.14	/	/
M3	0.55–0.69	0.28–0.41	0.14–0.28	0.14–0.28	0.14–0.28	0–0.14	0.14–0.28	/
M4	0.14–0.28	0.41–0.55	0.55–0.69	0–0.14	0–0.14	0–0.14	0.14–0.28	/
M5	0.55–0.69	0.41–0.55	0.55–0.69	0–0.14	0.14–0.28	0–0.14	0.41–0.55	0.41–0.55
M6	0.55–0.69	0.55–0.69	0.55–0.69	0.14–0.28	0–0.14	0–0.14	0.14–0.28	0.41–0.55
M7	0.68–0.83	0.41–0.55	/	/	0.83–0.96	0.2–0.4	/	/

M1 and M7 were also compared. By comparing the wind speed at point A, the wind speed range of M7 is higher at 0.69–0.83 m/s than that of M1, which is 0.55–0.69 m/s. Comparing the wind speed at point B, the wind speed range of M7 is about twice as high as that of M1 (0.41–0.55 m/s). In particular, at point E, the wind speed range of M7 reaches 0.96–1.1 m/s, which is much higher than that of M1 at 0.14–0.28 m/s due to the formation of a penetrating wind. Comparing point F, the wind speed range of M7 (0.28–0.41 m/s) is higher than the wind speed range of M1. In summary, the overall indoor wind speed of M7 is higher than that of M1, especially the wind speed in the hall area and bedroom area of M7 is higher than in M1.

Table 11. Simulated wind speed clouds for different dwellings indoor wind environments in winter.

Typology	Winter	
Three rooms with one entrance and one patio	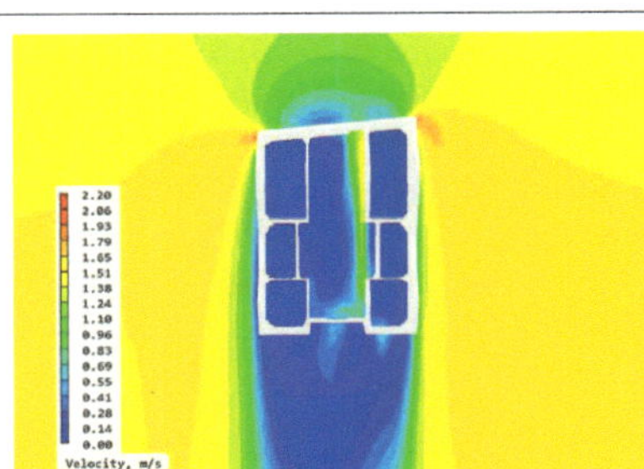 M1	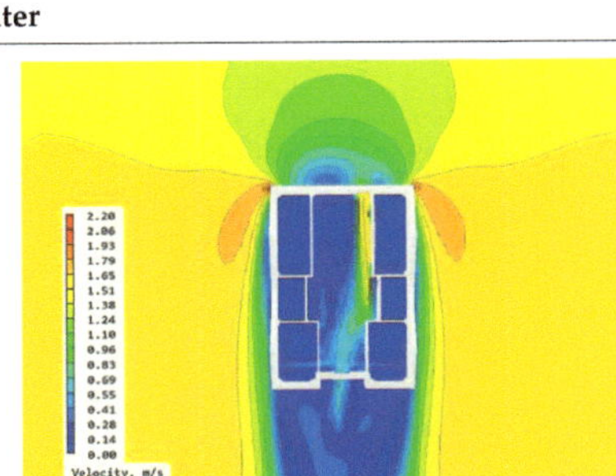 M2

Table 11. *Cont.*

Typology	Winter	
Three-room, single-entry plus half-well type	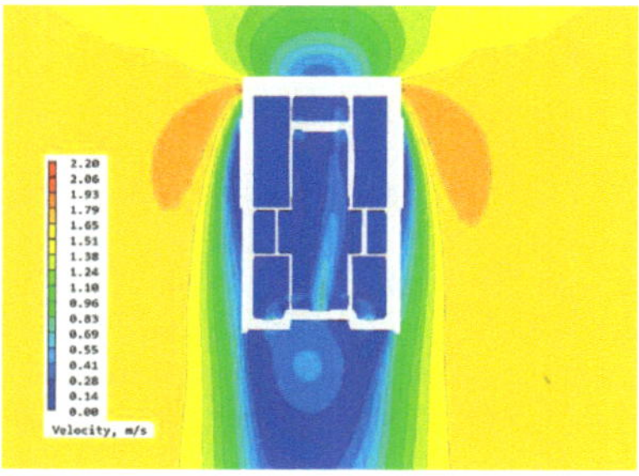 M3	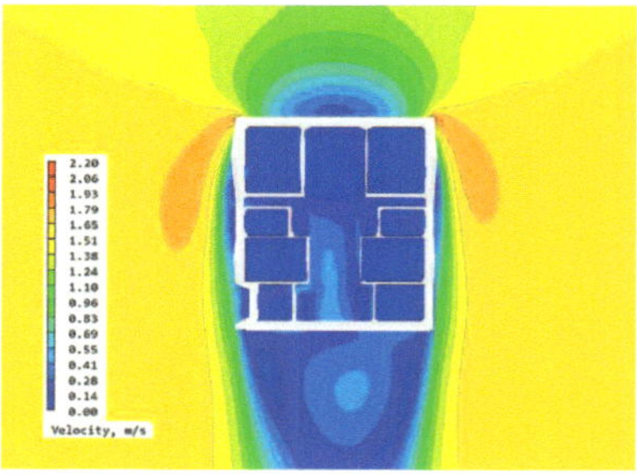 M4
Three-room multi-entry type	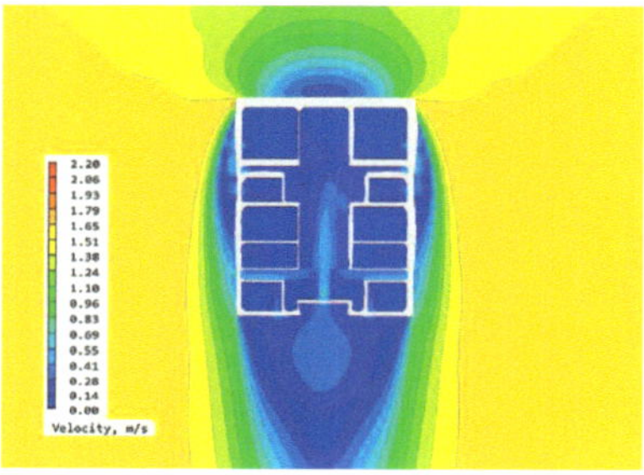 M5	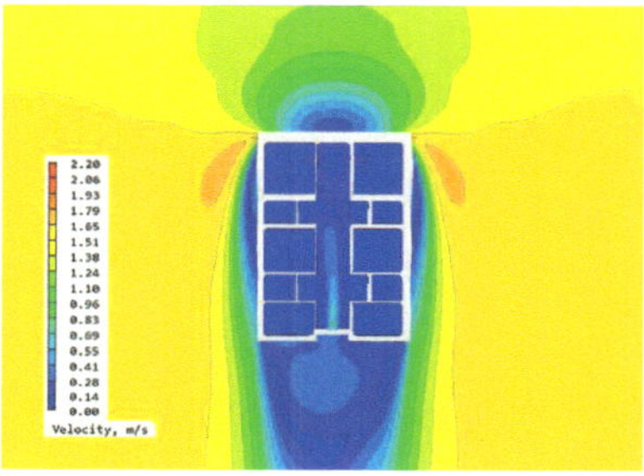 M6
Modern residential buildings type	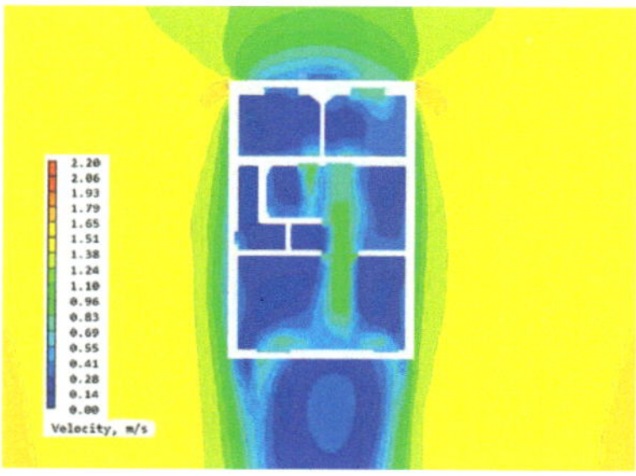 M7	

4.3. Energy Consumption Simulation

The comparison the six vernacular dwellings (Table 12) reveals that their annual energy consumption per unit area is relatively close. The highest annual energy consumption per unit area was in M2, with a difference between M2 and M1 of 2.16 kWh/m^2. The residential energy consumption for both is the highest for the unit area energy consumption of the HVAC system, and the lowest energy consumption is the unit area energy consumption of the lighting system. The energy consumption of M2 is 9.1% higher than that of M1, qwithin which the energy consumption per unit area of the lighting system is 12.2% higher. The energy consumption per unit area of the equipment system is 8.3% higher, and the energy consumption per unit area of the HVAC system is 8.5% higher.

Table 12. Different dwellings of energy consumption simulation comparison.

Serial Number	Unit Area Energy Consumption Annual Energy Consumption (kWh/m^2)	Lighting System Unit Area Energy Consumption (kWh/m^2)	Energy Consumption of Equipment System Unit (kWh/m^2)	HVAC Unit Area Energy Consumption (kWh/m^2)
M1	23.73	3.66	5.88	14.20
M2	25.89	4.11	6.37	15.41
M3	24.82	3.84	6.29	14.69
M4	23.94	3.61	5.77	14.56
M5	25.76	4.13	6.07	15.56
M6	25.47	4.31	6.74	14.42
M7	32.83	5.13	7.80	19.9

Comparing the annual energy consumption per unit area of the vernacular dwellings with the modern residential building, M1 and M7 are selected for comparison. The energy consumption of M1 is better than that of M7, with the main difference in energy consumption coming from the HVAC system. The total annual energy consumption of M7 is 32.83 kWh/m^2. The annual energy consumption per unit area of M7 is 1.32 times higher than that of M1; the energy consumption per unit area of the lighting system of M7 is 1.22 times higher than that of M1; the energy consumption per unit area of the equipment system of M7 is 1.24 times higher than that of M1; and the energy consumption per unit area of the HVAC system of M7 is 1.35 times higher than that of M1. In summary, the annual energy consumption per unit area of the six selected vernacular dwellings is relatively close, and the annual energy consumption per unit area of M1 is lower than that of M7. Thus, it can be concluded that the annual energy consumption per unit area of vernacular dwellings is lower than that of modern residential dwellings.

5. Discussion

Through the simulation of three aspects of light environment, wind environment, and energy consumption of vernacular dwellings and modern residential buildings, the following results are obtained:

(1) In comparison to modern residential buildings, vernacular dwellings exhibit a superior lighting effect in the living hall area, although the natural lighting effect in the bedroom area is suboptimal.
(2) In contrast to modern residential buildings, vernacular dwellings have limited natural ventilation in bedrooms. However, the main activity areas can regulate indoor wind conditions through natural ventilation in summer, adapting to climate change in hot and cold regions. This maintains indoor comfort.
(3) Vernacular dwellings are more energy-efficient than modern residential buildings.

5.1. Analyzed Aspects

5.1.1. Light Environment

The light environment simulation reveals that the vernacular dwellings use patios to provide good lighting for the hall area while avoiding direct sunlight. As the number of patios increases, the lighting effect of the hall area improves. However, due to the emphasis on security and privacy, the bedrooms of vernacular dwellings only have small windows to receive direct light from the outside, and indirectly receive weak light from the patio through the inner windows. Consequently, the lighting in the bedroom fails to meet the minimum requirements of the code. In contrast, modern residential buildings are more flexible in setting up windows and doors, and the windows and doors in each area are set up more reasonably, but the lighting effect in the hall area is far from being as good as that of vernacular dwellings. The hall area is the primary location for human activities, and thus requires specific lighting conditions. Good lighting uniformity not only provides a comfortable visual experience but also enhances the sense of openness and warmth in the space. The top lighting method is more straightforward than the side lighting method for achieving better lighting uniformity within the interior. The light environment in the bedroom area of modern residential buildings is superior to that of vernacular dwellings. While the light environment in the bedroom areas of modern dwellings is superior to that of vernacular dwellings, vernacular dwellings exhibit a superior light environment in the hall area due to their patios.

5.1.2. Wind Environment

The wind simulation reveals that the clever use of the patio in the vernacular dwellings significantly enhances the efficiency of natural ventilation in the hall area, effectively reducing the heat and stuffiness in the room. In contrast, although the bedrooms have internal windows facing the patio, the cooling effect is far less obvious than that of the halls due to the location and airflow channels. The principle behind the patio design lies in the fact that external airflow is first brought in through the main door, and then smoothly discharged through the patio under the dual driving effect of wind pressure and heat pressure to achieve heat exchange. Particularly in the daytime, when there is plenty of sunshine, the room temperature rises, prompting the hot air to rise and be released through the patio. If the external wind is strong at this time, the indoor air is driven to form a negative pressure in the area of the patio, and the air is discharged under the effect of wind pressure. On the other hand, modern residential buildings have more windows and doors and, if they are fully opened, it is easy to trigger the phenomenon of a through- wind, although it can improve the overall air circulation speed, but it may also bring the problem of uncontrolled indoor wind speed and living discomfort.

5.1.3. Energy Consumption

The energy simulation study reveals that the vernacular dwellings are superior in energy saving performance, and this advantage comes from the comprehensive effect of their unique construction features: the first factor is that the vernacular dwellings use low-thermal conductivity building materials, such as wood, bricks, and rammed earth, which have smaller thermal conductivity compared to the building materials used in modern buildings, effectively blocking the transfer of heat to reduce the loss of energy, and reduce the energy consumption of the HVAC system. Secondly, the wall masonry method also has a certain impact on the energy consumption of the dwelling. The walls of Nanfeng County are mostly hollow bucket walls, with a masonry method of one outside wall and one cavity wall. Secondly, the masonry of the building wall also has a certain impact on the energy consumption of the building, the walls of vernacular dwellings in Nanfeng County are mostly empty cavity walls with masonry for a sleep a bucket, an outer skin brick layer and a space between the bricks to leave an air layer between the bricks; that is, to maintain better structural stability, but also to enhance the structure of the adiabatic properties. Furthermore, the two-story building around the patio of vernacular dwellings creates a narrow and high space that facilitates heat and pressure ventilation. It means that the indoor temperature of the building will fall, and the HVAC energy consumption of the building will be reduced. Compared with modern residential buildings, the annual energy consumption per unit area can be reduced by up to 32%.

In particular, a study of Dong dwellings shows that the full use of wood-based materials can achieve energy savings of 39% compared to brick-based materials. However, the results obtained in this paper are lower than the above data, and the reason for the difference can be attributed to two points: firstly, the difference in building material selection strategy, i.e., the aforementioned study used brick for the external walls and wood for the internal walls, whereas in this paper, the same building material was used for both the internal and external walls; secondly, there are differences in the floor plan and wall masonry, which have an impact on the final results of the energy consumption simulation.

5.2. Comparison Experiment

By analyzing the results of the experiment, this paper finds that the patio has an impact on the indoor physical environment of vernacular dwellings. In order to investigate how it has an impact, this paper will conduct a comparative experiment. The experiment selects one of the examples from different types of vernacular dwellings. The indoor patio space is closed by a roof, and the light environment, wind environment, and energy consumption are analyzed and compared to explore how the patio space affects the indoor physical environment of the vernacular dwelling. Considering the light environment analysis, the material of the roof used to enclose the patio space is made of glass, which is a single-layer glass aluminum frame with a transmittance of 0.753. Table 13 illustrates the light environment results. Table 14 illustrates the wind environment results. Table 15 illustrates the energy consumption results.

Table 13. Comparison table before and after the light environment of the experimental subjects.

Light Environment Simulation		Pre-Laboratory	Post-Experimental
Diagram		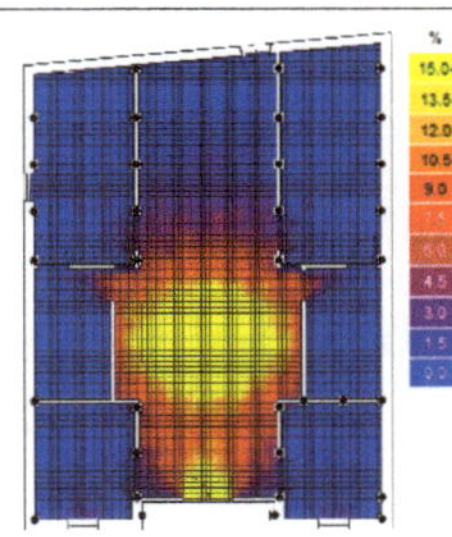	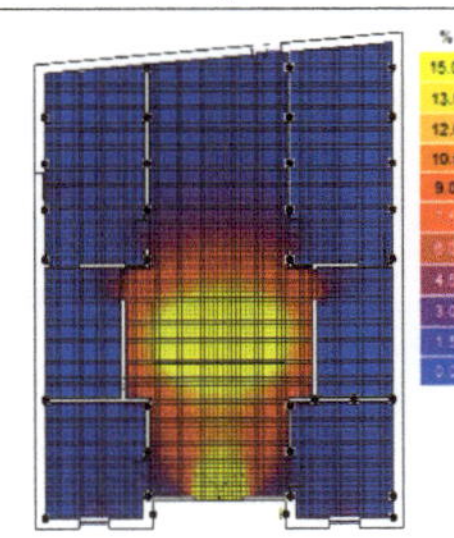
Hall area	Average light factor	5.45	4.54
Bedroom area	Average light factor	0.83	0.83

Table 14. Comparison table before and after wind environment of experimental subjects.

Wind Environment Simulation	Pre-Laboratory	Post-Experimental
Wind environment in summer	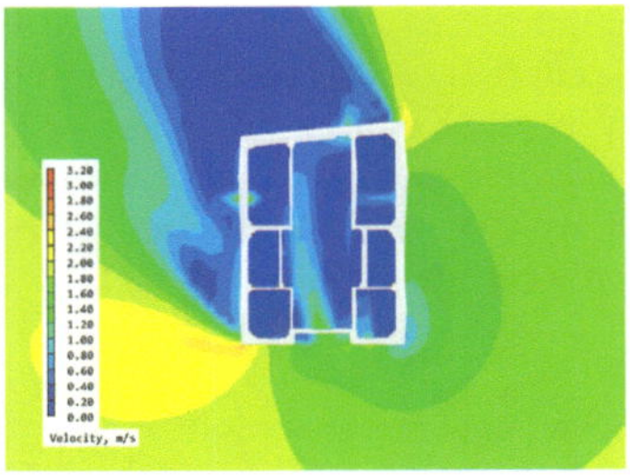	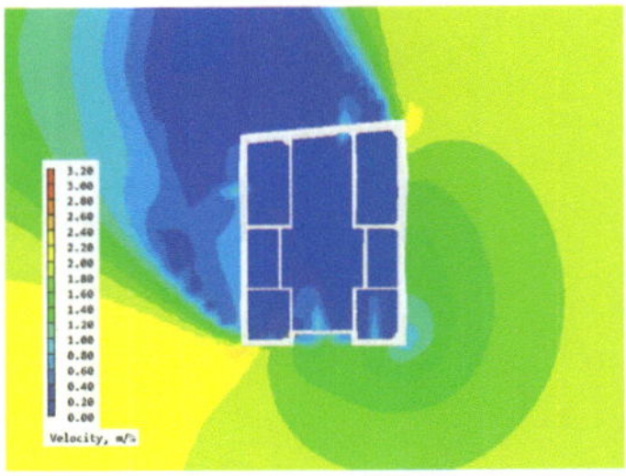
Wind Environment Simulation	**Pre-Laboratory**	**Post-Experimental**
Wind environment in winter	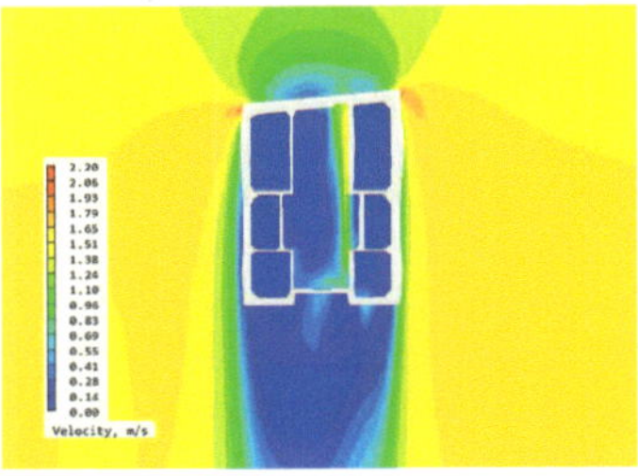	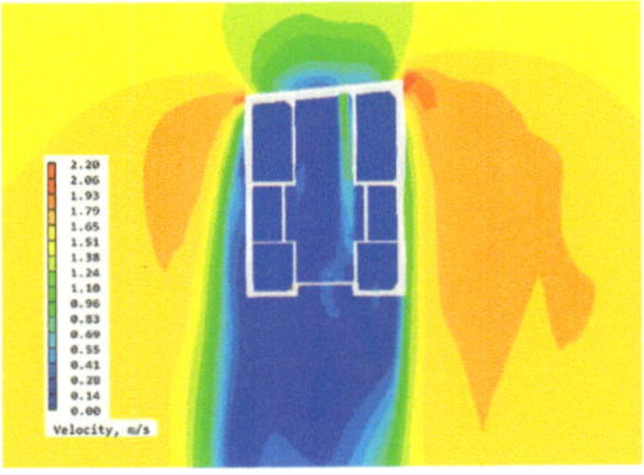

Table 15. Comparison table before and after energy consumption of experimental subjects.

Energy Consumption Simulation	Pre-Laboratory	Post-Experimental
Diagram	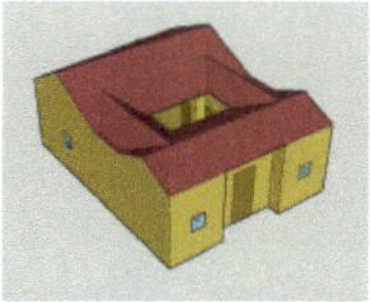	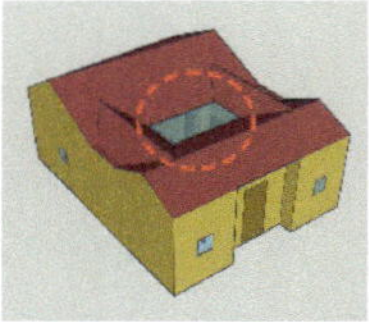
Unit area energy consumption annual energy consumption (kWh/m^2)	23.73	24.47
Lighting system unit area energy consumption (kWh/m^2)	3.66	3.54
Energy consumption of equipment system unit (kWh/m^2)	5.88	5.92
HVAC unit area energy consumption (kWh/m^2)	14.20	15.01

In a comparative experiment, this paper observes that the average light factor in the room decreases by 16% when the patio is enclosed with glass. This change stems from the reflectivity and light transmittance of the glass. On the one hand, some of the light is reflected from the glass surface to the outdoors; on the other hand, a part of the light passes through despite the high transmittance of the transparent glass used, which permits the penetration of most of the visible light.

In addition, the indoor wind environment of the building is more affected in summer and winter, especially in summer when the indoor wind speed decreases more. This phenomenon is attributed to two reasons. First, the natural ventilation path is restricted. In other words, the closure of the patio blocks the natural path for the free influx of outside air, thus reducing the indoor air flow rate. Secondly, the effect of thermal pressure ventilation is weakened. In other words, the hot air cannot be discharged through the patio, resulting in a reduction in the difference between indoor and outdoor heat pressure, and the power of natural ventilation is weakened, thus reducing the indoor air speed.

The energy consumption simulation experiment showed that the energy consumption increased by 3.1% after the experiment compared to before the experiment. The main reason behind this is the obstruction of ventilation. The patio is originally used as an important channel for natural ventilation, which could promote air circulation, a fall the indoor temperature of the building, and reduce the reliance on air conditioning. When enclosed, the effect of natural ventilation is reduced, which may increase the need to use mechanical ventilation or air conditioning systems, thus increasing energy consumption.

In summary, enclosing the patio spaces has a negative impact on the indoor physical environment, which not only reduces indoor living comfort but also leads to an increase in building energy consumption.

5.3. Innovation

The innovation of this paper is, first of all, at the research methodology level. In recent years, although the research on the climate adaptation of vernacular dwellings in Jiangxi has been introduced to computer simulation, the research focuses on a single physical factor, ignoring the consideration of the overall environment. This paper, on the other hand, analyses the light environment, wind environment, and energy consumption at three levels to better analyze the reasons behind them. Not only that, this paper uses comparative experiments to summarize the advantages and disadvantages of vernacular dwellings and modern residential buildings in Jiangxi. Vernacular dwellings in Jiangxi have a pivotal position, and the study of the climate adaptability of vernacular dwellings in Jiangxi helps to excavate the principal mechanism behind it.

Secondly, this paper concludes that the patio has a positive impact on the indoor physical environment of vernacular dwellings through a comparative experiment with and without an enclosed patio space. This low-tech approach not only effectively reduces energy consumption and costs, but also embodies the concept of sustainable development, demonstrating the unique value of the low-tech approach in contemporary Chinese rural construction. A low-tech approach achieves a more appropriate and sustainable construction model compared to high-tech paths with low cost, low environmental burden, and minimal interference with the local ecology.

5.4. Recommendations

The findings of the software simulation analysis indicate that vernacular dwellings offer several advantages in terms of energy efficiency and comfort. These advantages come from the layout, patio, traditional materials, and the design of doors and windows. These designs are worth learning and borrowing from modern residential buildings. However, vernacular dwellings also have some disadvantages, including inadequate sealing and an unfavorable humidity environment. It is necessary to enhance vernacular dwellings with modern architectural design methods to facilitate their effective accommodation of the ecological environment.

(1) In terms of optimizing the light environment, vernacular dwellings can take measures to improve indoor light conditions by increasing the brightness of roof tiles, adjusting the size of windows and patios, and enhancing the reflectivity of interior finish materials. Moderately lowering the elevation of windowsills can strengthen the effectiveness of natural lighting in the area adjacent to the windows. The enlargement of windows and patios not only increases the total amount of light in the room but also ensures an even distribution of light. The application of highly reflective materials to the interior walls serves to enhance the lighting effect within the interior space. When implementing these improvements, care needs to be taken to maintain a reasonable window-to-ground area ratio and to ensure consistency between window design and decorative style, maintaining the windy condition of the original facade.

(2) In terms of wind environment improvement, vernacular dwellings make skillful use of patio design to promote heat and pressure-induced natural ventilation, ensuring a stable airflow environment in the main living space. Modern residential buildings can learn from this experience by adding ventilated roofs or skylights in order to enhance the effect of non-powered ventilation. In addition, for poorly ventilated spaces in vernacular dwellings, the performance of natural ventilation can be enhanced by combining it with light optimization, such as increasing the number of windows and opening skylights on the roof to expand the ventilation area and optimize airflow channels.

(3) In terms of energy consumption control, vernacular dwellings are characterized by their thick peripheral walls, small and few windows, and lower energy consumption brought about by the structure of the cavity bucket wall, with the latter's internal air layer acting as a certain degree of thermal insulation. Nevertheless, there is still a considerable gap between the thermal conductivity of the cavity bucket wall and the current standard (thermal conductivity < 1.0 W/(m^2K)). In order to enhance the thermal insulation performance of external walls, several strategies may be employed. These include the incorporation of an additional thermal insulation layer within both the internal and external walls, or the filling of the interior of the cavity bucket wall with high-efficiency thermal insulation materials. The general lack of airtightness observed in vernacular dwellings, particularly in the gaps between windows and doors and in structural joints, can be effectively addressed through the adoption of sealing materials. Such methods have the potential to prevent the infiltration of cold air and reduce the dissipation of energy through air exchanges, thus contributing to the goal of energy saving and emission reduction.

This paper examines the climate adaptation of vernacular dwellings in Nanfeng County with the objective of providing a reference and database for the design of new and improved vernacular dwellings in rural areas of Nanfeng County. It should be noted that this study is not without shortcomings and limitations. Firstly, the research methods employed in this study are primarily field measurement and digital simulation. During the field measurement process, errors may occur due to human factors, which could result in deviations in the digital simulation. Secondly, the optimization strategy proposed in this paper has not been verified in real projects. In addition, the factors influencing the outcome of this study are limited to light environment, wind environment, and energy consumption. These factors warrant further investigation, particularly in relation to temperature, humidity, PET, and other variables. Future research should adopt advanced measurement tools and multi-faceted analysis of the physical environment to provide a comprehensive summary of the low-tech approach to vernacular dwellings.

6. Conclusions

The vernacular dwellings of Jiangxi represent a set of practices that have been developed and refined over centuries by Chinese ancestors. These practices aim to achieve a harmonious relationship between humans and nature. This paper employs a variety of simulation software to analyze the indoor physical environment and energy consumption of different types of vernacular dwellings. The aim is to compare these dwellings with local modern residential buildings in order to verify the climate adaptability of vernacular dwellings. The results of the simulation are analyzed in this paper, with particular attention paid to the reasons behind them. In addition, the article explores the low-tech methods embedded in vernacular dwellings.

1. The simulation results demonstrate that the vernacular dwellings in Jiangxi have a better climate adaptation capability. The vernacular dwellings represented by the guard camp dwellings in Kandu Village exhibit an average natural lighting illuminance in the hall area that is 21% higher than that in the living hall area of modern residential buildings in terms of the light environment. In terms of wind environment, the main activity area can adjust the indoor wind environment through natural ventilation in summer, thus adapting to climate change toward hot summer and cold winter areas. In terms of energy consumption, the vernacular dwellings consume 32% less energy than modern residential buildings.
2. The utilization of patio spaces, the application of wall masonry techniques, and the incorporation of local materials permit vernacular dwellings to be modified to accommodate local climatic conditions. Such low-tech methods of vernacular dwellings enable them to operate with low energy consumption while maintaining a satisfactory physical environment indoors.

Author Contributions: Conceptualization, Z.Z.; methodology, J.Z. and J.H.; software, Y.X. and C.O.; validation, M.G.; formal analysis, W.J. and C.Z.; investigation, Y.X., C.O. and K.M.; writing—original draft, Y.X.; writing—review and editing, Z.Z.; supervision, W.J. and K.M. All authors have read and agreed to the published version of the manuscript.

Funding: This research was funded by Jiangxi Province Culture and Art Science Planning Annual Project (grant number YG2023002) and Jiangxi Province 2023 Special Funds for Postgraduate Innovation Project (grant number YC2023-S038).

Data Availability Statement: The data presented in this study are available on request from the corresponding author. The data are no publicly available due to privacy. Please contact corresponding author before use.

Conflicts of Interest: The authors declare no conflicts of interest.

Abbreviations

Noun	Representation
T	the temperature
$\Gamma_{T,eff}$	the effective diffusion coefficient
S_T	the heat generation term
ρ	the fluid density
u_j	the velocity vector
x_i	the coordinate position
U_z	the wind speed in the horizontal direction at height Z
U_o	the wind speed at Z_o of the reference height
α	a power index determined by the roughness of the terrain
$q_{i,c}$	the convective heat transfer through the *i* surface
N	the number of surfaces of the enclosure
A_i	is the actual heat transfer area of the *i* surface of the enclosure
Q_{other}	the latent heat caused by the convective part and moisture evaporation of sunlight, equipment, lights, and occupants' heat gain in the heat balance equation of the surface
G_a	the sum of the air volume of fresh air and infiltration air
C_p	the specific heat capacity of the air at constant pressure
$T_{a\text{-}out}$	the temperature of outdoor air
$q_i(t)$	the heat gain of the *i*th surface at moment *t*
h_i	the heat transfer coefficient of the *i*th surface
$T_R(t)$	the calculated indoor temperature
$T_i(t)$	the temperature of the ith internal surface
R_i	the radiant absorption of the *i*th surface
$Q(t)$	the cooling load at moment *t*
s_i	the area of the *i*th surface
$m(t)$	the fresh air volume at time *t*
C_P	the constant-pressure specific heat capacity
$T_o(t)$	the calculated temperature of the outdoor air
$Q_S(t)$	the heat generation from the indoor heat source at time *t*
$q_{i,c}$	the convective heat transfer through the *i*th surface
N	the number of surfaces of the envelope
A_i	the actual heat transfer area of the *i*th surface of the envelope
Q_{other}	the latent heat caused by the convective part and evaporation of water in the heat balance equation of the surface, sunlight, equipment, lighting and heat gain of the personnel
G_a	the sum of air volume of the fresh air and infiltration air
C_p	the constant-pressure specific heat capacity of the air
T_m	the temperature of the zone
$Q_{eat-extra}$	the amount of heat lost from the inside air to the outside
C_p	the constant-pressure specific heat capacity of air
$\rho V C_P \frac{dT_{in}}{dt}$	the heat capacity of the room air
$T_{a.out}$	the outdoor air temperature
q_i	the heat transfer from the wall to the wall *i*
q_{ir}	the radiant heat from the internal heat source and the sun
t_{HVAC}	the HVAC turn-on time
COP_{ref}	the rated cooling factor of the building's HVAC system
t_{Light}	the indoor lighting runtime
P	the average power of a device in operation
e	the total number of device types
n	the quantity of a particular type of equipment
τ	the equipment runtime

References

1. Mostafavi, F.; Tahsildoost, M.; Zomorodian, Z.S. Energy efficiency and carbon emission in high-rise buildings: A review (2005–2020). *Build. Environ.* **2021**, *206*, 108329. [CrossRef]

2. Berardi, U. A cross-country comparison of the building energy consumptions and their trends. *Resour. Conserv. Recycl.* **2017**, *123*, 230–241. [CrossRef]
3. Building Energy Efficiency. *China Building Energy Consumption Annual Report 2020*; Building Energy Efficiency: Beijing, China, 2020.
4. United Nations. 2018 Global Status Report: Towards a zero-emission, efficient and resilient buildings and construction sector. In *International Energy Agency and the United Nations Environment Programme*; United Nations: Nairobi, Kenya, 2018. Available online: https://www.unep.org/resources/report/global-status-report-2018 (accessed on 15 April 2024).
5. Wang, J.; Huang, Y.; Teng, Y.; Yu, B.; Wang, J.; Zhang, H.; Duan, H. Can buildings sector achieve the carbon mitigation ambitious goal: Case study for a low-carbon demonstration city in China? *Environ. Impact Assess* **2021**, *90*, 106633. [CrossRef]
6. Wu, Y.; Shen, L.; Shuai, C.; Jiao, L.; Liao, S.; Guo, Z. Key driving forces on the development of low carbon city (LCC) in China. *Ecol. Indic.* **2021**, *124*, 107379. [CrossRef]
7. Huo, T.; Ren, H.; Zhang, X. China's energy consumption in the building sector: A Statistical Yearbook-Energy Balance Sheet based splitting method. *J. Clean. Prod.* **2018**, *185*, 665–679. [CrossRef]
8. Qi, Y.; Stern, N.; He, J.K.; Lu, J.Q.; Liu, T.L.; King, D.; Wu, T. The policy-driven peak and reduction of China's carbon emissions. *Adv. Clim. Change Res.* **2020**, *11*, 65–71. [CrossRef]
9. Bourdier, J.-P. *Dwellings Settlements and Tradition: Cross-Cultural Perspectives*; University Press of America: New York, NY, USA, 1989.
10. Zhai, Z.Q.; Previtali, J.M. A taxonomy of vernacular architecture an addendum to "Ancient vernacular architecture: Characteristics categorization and energy performance evaluation". *Energy Build.* **2015**, *110*, 71–78.
11. Nguyen, A.T.; Truong, N.S.H.; Rockwood, D. Studies on sustainable features of vernacular architecture in different regions across the world: A comprehensive synthesis and evaluation. *Front. Archit. Res.* **2019**, *8*, 535–548. [CrossRef]
12. Saez-Perez, M.P.; Ruiz, L.M.G.; Tajani, F. Assessment of the Thermal Properties of Buildings in Eastern Almeria (Spain) during the Summer in a Mediterranean Climate. *Sustainability* **2024**, *16*, 746. [CrossRef]
13. Philokyprou, M.; Michael, A.; Malaktou, E. Environmentally responsive design in Eastern Mediterranean. the case of vernacular architecture in the coastal, lowland and mountainous regions of Cyprus. *Build. Environ.* **2017**, *111*, 91–109. [CrossRef]
14. Rajapaksha, I.; Fiorito, F.; Lazer, E. Exploring thermal comfort in the context of historical conservation. a study of the vernacular architecture of Pompeii. *Archit. Sci. Rev.* **2018**, *61*, 4–14. [CrossRef]
15. Ayaz, F.H.; Tetsu, K. Effects of building microclimate on the thermal environment of traditional Japanese Houses during Hot-Humid Summer. *Buildings* **2019**, *9*, 22. [CrossRef]
16. Choudhury, A.S.; Chettry, V. Investigation of bioclimatic design features in vernacular architecture of Northeast India: Case studies of Assam-type houses and stilt houses (Chang Ghar) in Assam, India. *Indian J. Tradit. Knowl.* **2023**, *22*, 664–673.
17. Chandel, S.S.; Sharma, V.; Marwah, B.M. Review of energy efficient features in vernacular architecture for improving indoor thermal comfort conditions. *Renew. Sustain. Energy Rev.* **2016**, *65*, 459–477. [CrossRef]
18. Kubota, T.; Zakaria, M.A.; Abe, S.; Toe, D.H.C. Thermal functions of internal courtyards in traditional Chinese shophouses in the hot-humid climate of Malaysia. *Build. Environ.* **2017**, *112*, 115–131. [CrossRef]
19. Tang, S.Y.; Lee, Y.S. Comparisons and Analysis on Architectural Features of Seven Typical Traditional Dwellings in China. *KIEAE J.* **2014**, *14*, 5–13. [CrossRef]
20. Sun, F.F. Chinese Climate and Vernacular Dwellings. *Buildings* **2013**, *3*, 143–172. [CrossRef]
21. Huang, Z.J.; Zhang, H.; Gong, C.; Zhang, Y. Analysis on Field Measurement of Thermal Environment of Huizhou Traditional Vernacular Dwellings in Summer. *Build. Sci.* **2017**, *33*, 26–31.
22. Zhu, L.P.; Chen, X.L.; Huang, Z.J.; Zhang, Y. Thermal Environment of 24 Solar Terms in Huizhou Traditional Residences. *Build. Sci.* **2020**, *36*, 67–78.
23. Gao, R.; Liu, J.P.; Shi, Z.H.; Zhang, G.J. Patio Design Optimization for Huizhou Traditional Dwellings Aimed at Daylighting Performance Improvements. *Buildings* **2023**, *13*, 583. [CrossRef]
24. Gao, N.; Hu, W.B.; Wu, C.C. Numerical Simulation of Natural Ventilation of Traditional Residences in Lingnan Area. *Build. Energy Environ.* **2015**, *34*, 76–79.
25. Yao, Z.J.; Li, H. A Green Study on the Spatial Pattern of Traditional Houses with Three Rooms and Two Corridors in Lingnan Area. *Constr. Mater. Decor.* **2018**, *03*, 97–98.
26. Xiao, Y.Q.; Lin, H.K.; Hui, X.Y. Research on climate adaptive spatial system of traditional villages in Guangfu Area. *South Archit.* **2018**, *05*, 62–69.
27. Jin, Y.; Zhang, N. Comprehensive Assessment of Thermal Comfort and Indoor Environment of Traditional Historic Stilt House, a Case of Dong Minority Dwelling, China. *Sustainability* **2021**, *13*, 9966. [CrossRef]
28. Zhang, F.; Shi, L.; Liu, S.; Shi, J.; Ma, Q.; Zhang, J. Climate Adaptability Based on Indoor Physical Environment of Traditional Dwelling in North Dong Areas, China. *Sustainability* **2022**, *14*, 850. [CrossRef]
29. Dong, X.; Yang, Z.J.; Xu, Y.A.; Tang, M.Y. Research on Thermal Performance of Traditional Tujia Ventilation Roof in Southeast Chongqing. *Build. Sci.* **2021**, *37*, 207–214.
30. Sun, Y.F. Comparative Research on Folk Dwelling of Hubei & Hunan under the Influence of Jiangxi to Huguang Immigration during Ming to Qing Dynasty —In the case of Southeast Hubei & Northeast Hunan Area. Master's Thesis, Huazhong University of Science and Technology, Wuhan, China, 2008.

31. Kang, Y.W. Research on the Geography of Jiangxi Traditional Dwellings. Ph.D. Thesis, Hunan Normal University, Changsha, China, 2020.
32. Li, Y.H. *The Chinese Artisan: An Analysis of the Principles of Classical Chinese Architectural Design*, 2nd ed.; Tianjin University Press: Tianjin, China, 2014; pp. 47–51.
33. Lu, Y.D.; Yang, G.S. *Chinese Residential Dwelling*; South China University of Technology Press: Guangzhou, China, 2004; pp. 11–63.
34. Huang, H. *Jiangxi Dwelling*, 3rd ed.; China Architecture & Building Press: Beijing, China, 2008; pp. 147–151.
35. Pan, Y.; Shi, Y. A brief analysis of the reasons for the formation of traditional Dwelling in Jiangxi during the Ming and Qing Dynasties. *Agric. Archaeol.* **2006**, *3*, 179–181.
36. Yao, T.; Cai, Q. *Ancient Architecture of Jiangxi*, 1st ed.; China Architecture & Building Press: Beijing, China, 2015; pp. 152–154.
37. Li, G.X. Study on Vernacular Dwellings in Jiangxi Province and Their Districts. Master's Thesis, Southeast University, Nanjing, China, 2001.
38. Lan, C.J.; Yao, T. Study on the Form and Structure of a Traditional Mixed Masonry of Brick and Stone--The Case of Traditional dwellings in Jinxi County, Jiangxi Province. *Huazhong Archit.* **2017**, *35*, 26–31.
39. Yi, Y. Study on the Structural Characteristics of the Hall of the Ming Dynasty in the Fu River. Ph.D. Thesis, Wuhan University, Wuhan, China, 2018.
40. Li, Z.; Ding, X.M. Analysis of the characteristics of vernacular architectural construction techniques in the Lichuan region in the context of traditional construction techniques. *Herit. Archit.* **2018**, *04*, 29–37.
41. Guo, Q. *Research on the Architecture and Culture of Xiang-Gan Traditional Dwelling*, 2nd ed.; China Construction Industry Press: Beijing, China, 2005; pp. 57–60.
42. Pan, Y. Interpretation of the planar pattern of traditional Dwelling in Jiangxi. *Agric. Archaeol.* **2009**, *3*, 197–199.
43. Wang, X.Y.; Xu, X.W. Study of patios in traditional residential architecture. *J. Nanchang Univ.* **2010**, *S1*, 54–57.
44. Chen, S.Y. Research on the Climate Adaptability Strategy of Vernacular Architecture in Xiang-Gan Border Area—A Case Study of Jinggangshan. Master's Thesis, Southeast University, Nanjing, China, 2020.
45. Wu, X.D. Study on Climate Adaptability of Traditional Dwellings in Luling Area. Master's Thesis, East China Jiaotong University, Nanchang, China, 2023.
46. Li, Z.; Wang, Y.F. The Concept, Connotation and System of "Low Technology" in Vernacular Construction: Inspiration and Reflection on the Conservation Practice of Jinkeng Ancient Village in Shaowu, Fujian. *Architect* **2019**, *4*, 95–100.
47. Available online: http://www.cham.co.uk/phoenics.php (accessed on 20 June 2024).
48. Available online: https://energyplus.net/ (accessed on 20 June 2024).
49. Olgyay, V. *Design with Climate: Bioclimatic Approach to Architectural Regionalism*, 1st ed.; Princeton University Press: Princeton, NJ, USA, 1963.
50. Chen, Q.Q.; Qian, H.; Zuo, Y.H. Simulation of natural daylighting in patio type residential dwellings. *Light Light.* **2016**, *03*, 26–29.
51. *DB22/T 5055-2021*; Green Building Design Standards. Housing and Urban-Rural Development of Jilin Province: Changchun, China, 2021.
52. Zhou, Z.Y.; Deng, J.; Wang, P.F.; Zhou, C.L.; Xu, Y.X.; Jiang, W.P. Physical Environment Study of Traditional Village Patterns in Jinxi County, Jiangxi Province Based on CFD Simulation. *Processes* **2022**, *11*, 2453. [CrossRef]
53. Available online: http://www.cham.co.uk/casestudies.php (accessed on 20 June 2024).
54. *GB 55016-2021*; General Code for Building Environment. China Architecture & Building Press: Beijing, China, 2021.

buildings

Article

Research on Urban Community Street Environment Evaluation and Optimization Strategy under the Concept of a Healthy City: A Case Study of the Dingwangtai Area of Changsha City

Yichi Zhang [1], Hui Tang [1,2,3,*], Kecheng Huo [1] and Jiangfan Tang [1,3]

1 College of Architecture and Urban Planning, Hunan City University, Yiyang 413000, China; zhangyichi011022@163.com (Y.Z.); huokecheng0613@163.com (K.H.); ydqingyan@126.com (J.T.)
2 Key Laboratory of Key Technologies of Digital Urban-Rural Spatial Planning of Hunan Province, Yiyang 413000, China
3 Key Laboratory of Digital Technology in Hunan General Colleges and Universities, Yiyang 413000, China
* Correspondence: tanghui202086@163.com; Tel.: +86-151-7371-1708

Abstract: The World Health Organization (WHO) articulated a vision in 1986, hoping that countries around the world would actively promote the construction and development of healthy cities. In the context of China, a paradigm shift in urban development has been triggered by the deep implementation of the Healthy China Strategy and the gradual completion of the comprehensive national spatial planning. This shift emphasizes a micro-level focus, advocating a human-centered approach to urban space exploration. In this paper, the Dingwangtai Area of Changsha City is selected as a case study. A street space health evaluation index system called "5D+" is constructed from six dimensions, namely, human perception, degree of mixing, density, distance to transit, destination accessibility, and devise. This research adopts the community as the fundamental unit of analysis, employing the Entropy-weighting TOPSIS method for the computation of indicators. The results of this study show that the Dingwangtai Area exhibits relative deficiencies in the dimensions of density, devise, and destination accessibility. Specifically, at the community level, the Ma Wang Street Community and the Fanhou Street Community have scored particularly low. In response, this paper proposes targeted measures and detailed recommendations aimed at optimizing the design of the street, enhancing the human experience, enriching functional attributes, and refining the street network. The ultimate goal is to propose a Healthy Streets Evaluation Index System based on the concept of a healthy city and to explore its relationship with healthy streets so as to provide valuable insights for the development of healthy city streets.

Keywords: healthy city; street space; entropy-weighting TOPSIS method

Citation: Zhang, Y.; Tang, H.; Huo, K.; Tang, J. Research on Urban Community Street Environment Evaluation and Optimization Strategy under the Concept of a Healthy City: A Case Study of the Dingwangtai Area of Changsha City. *Buildings* **2024**, *14*, 2449. https://doi.org/10.3390/buildings14082449

Academic Editor: Konstantina Vasilakopoulou

Received: 11 July 2024
Revised: 2 August 2024
Accepted: 6 August 2024
Published: 8 August 2024

1. Introduction

Health stands as a universal aspiration for humanity in the 21st century, and the healthy city is an important carrier for realizing the physical and mental health of human beings. In 1986, the first global conference on health promotion provided a definition of a healthy city, which aims to continuously create and improve the environment and enrich the resources of a community so that the city can support people's lives in a way that further satisfies their need for self-fulfillment. However, with the globalization wave progressing gradually and urbanization developing rapidly, the urban space is expanding rapidly and disorderedly [1], neglecting the design of the inner-space environment of the city. The living environment of the residents has not been significantly improved [2], and the cities of each country are gradually paying attention to the conversion to a healthy city.

Globally, the World Health Organization (WHO) integrates quantifiable indicators, such as air quality, water quality, waste, etc., to construct a healthy city indicator system, which is continuously improved and periodically revised [3]. However, there is no unified standard for the construction of healthy cities, and China put forward an outline for the "Healthy China 2030" initiative in 2016, aligning with the nation's unique conditions and formally integrating the concept of healthy cities into the national strategic framework. In 2018, the National Healthy City Evaluation Indicator System (2018 Edition) was formulated, proposing a new index system containing five major categories, twenty medium categories and forty-two subcategories. These include healthy environment, healthy society, healthy service, healthy culture, and healthy people. The spatial scale of a healthy city consists of five levels: general urban space, public space, street space, living space, and architectural space. These spatial scales are gradually subdivided, and the human-centered concept is increasingly important as the Healthy China Strategy has been deepened and the preparation of the overall national spatial plan has been gradually completed. The construction of healthy cities has shifted from the macro to the micro level, requiring the planning and construction of urban space to be deeply explored using the human-centered concept. As an important part of the microspace of a healthy city, street space plays an important role in the healthy lives of residents [4,5] and the sustainable development of the city [6]. As a result, it has received extensive attention from scholars in various countries [7–9].

In international research, there is a lack of research literature on street space from the perspective of healthy cities. However, there is a rich scientific literature on designing street space directly or evaluating street environments by categorizing them as built environments. The London Street Design Guidelines, published in the United Kingdom in 2004, set out the classic structure of "policy vision + design guidance + management implementation" for subsequent street guidelines [10,11]. Subsequently, America, Canada, the Netherlands, and other western countries have introduced street design guidelines at the city or neighborhood level one after another. The idea has gradually evolved from "health and livability" to "humanity, vitality, and interaction", emphasizing the importance of human-centered street design [12–15]. Matthew Carmona et al. [16] point out that street space planning and design is undergoing a gradual evolution, from transportation to its place in the world. Evgeniya Prelovskaya [17] studied street design from a sustainability perspective, transforming existing urban streets in Russia and suggesting that it should increase green, shared, and convenient urban streets. Peter Jones [18] combined the functions of streets with the needs of users to categorize streets in the built environment, and he provided a human-centered concept of thought for subsequent street design. Chester Harvey et al. [19] used multi-source data to quantify street space and found that livability indicators of street space can play a decision-making role in the built environment of a city. As the depth of quantitative research advances, some scholars have begun to focus on the influence of urban street attributes on specific populations. Kaczynski, A. T. and Angela Curl [20,21] concluded that street space attributes have health-promoting effects on adolescents and the elderly.

In Chinese literature, studies relevant to street space focused on development following the introduction of the new urban construction concept in the early 21st century, and most scholars mainly focused on some of the street elements in the city and their own research methods. Firstly, in terms of street elements, Hao Xinhua [22] conducted a large number of measurements for Beijing's streets using street functional density, functional mixing, and transportation accessibility as indicators, and he explored some of the street- influencing factors. Ye Yu [23] carried out a refined multidimensional study of street elements, suggesting that the interaction between residents and urban spaces should be strengthened. Wang Lan [24] was the first to propose that urban planning's creation of street environments affects residents' health from the perspective of healthy cities. Secondly, in terms of research methodology, Long Ying and Shen Yao [25] took the lead in proposing Street Urbanism, and they carried out a quantitative comparison study of street quality in many cities in China based on streetscape images on a large scale and refined scale [26,27]. In the subsequent research on quantitative measurement of spatial quality in urban streets, it has been found that the streetscape images of high obtainability are the main data source in current research for street spatial measurement [28]. Some scholars [29–31] have explored the existing evaluation indexes of healthy city planning and proposed constructing a "multi-factor, multi-dimension and multi-scale" evaluation index system of healthy city planning. However, existing research objects are mostly Beijing, Shanghai, Nanjing, Shenzhen, and other relatively coastal cities, especially developed-area cities [32–36], and fewer studies are on cities in less developed areas.

In conclusion, although there is a large amount of existing literature on street space, research on street space from a healthy city perspective is still inadequate. Firstly, the current healthy city evaluation index system has a macro-spatial scale and lacks evaluation indexes for middle and micro scales such as street area space, especially the human-centered scale space. Secondly, previous explorations of the indexing system for healthy street space have included more quantitative research on objective attributes but have included fewer quantitative studies on subjective attributes.

The existing problems of the street, such as poor human perception, poor street traffic organization, lack of design, and low density of facilities, directly affect the healthy development of the street. In this study, we examined two research questions: (1) Characterize the spatial distribution of healthy streets under multidimensional evaluation. (2) Evaluate the relationship between evaluation indicators and street health development in different spaces. Therefore, in order to promote the construction of healthy streets, taking Dingwangtai as an example, this study constructed a street space Healthy Streets Evaluation Index System by screening the indicators using the human-centered concept. The Entropy-weighting TOPSIS method was used for comprehensive evaluation and ranking, and corresponding improvement strategies were provided to the community. This study aims to (1) explore the characteristics of the multidimensional distribution of healthy streets at the microspatial scale and (2) reveal the mechanisms by which each of the studied indicators affects healthy streets. The purpose of this study is to provide better guidance on planning practices for healthy streets and to promote healthy and sustainable people-centered urban development. The innovations and contributions of this paper are as follows:

(1) This paper takes Changsha's Dingwangtai Area as an example, combines domestic and international research progress, and constructs a street space health evaluation index system to make up for the vacancy in evaluation indexes for medium, micro, and human scale spaces, such as street area spaces.

(2) This paper tries to add quantitative indicators of human perception to the evaluation indexes of healthy street space for the purpose of quantitative assessment of the Dingwangtai Area from the perspective of a healthy city. It also conducts a comprehensive evaluation of the Dingwangtai Area's internal community environment from a comprehensive and refined perspective in order to propose corresponding quality enhancement countermeasures for the construction of healthy streets.

2. Materials and Methods

This paper selects Dingwangtai Street, a typical area in Changsha City, as the research object and obtains the research data through field research and network crawling, then processes and analyzes them. It synthesizes the contents of international references, constructs a Healthy Streets Evaluation Index System, and utilizes the Entropy-weighting TOPSIS method for comprehensive evaluation.

2.1. Study Area

The Dingwangtai Area is located in Changsha's commercial center zone (Figure 1). Within it, there are Changsha IFS, Huangxing Road Pedestrian Street and other commercial service points, Wuyi Square, Furong Square, and other square service points. The total area of this area is 1.48 square kilometers with a resident population of about 40,000 people. The street environment was complex, and crowd activities were frequent, which is a strong typicality. Therefore, this paper selects the Dingwangtai Area as the research object and uses the community as the evaluation unit to evaluate its street space health.

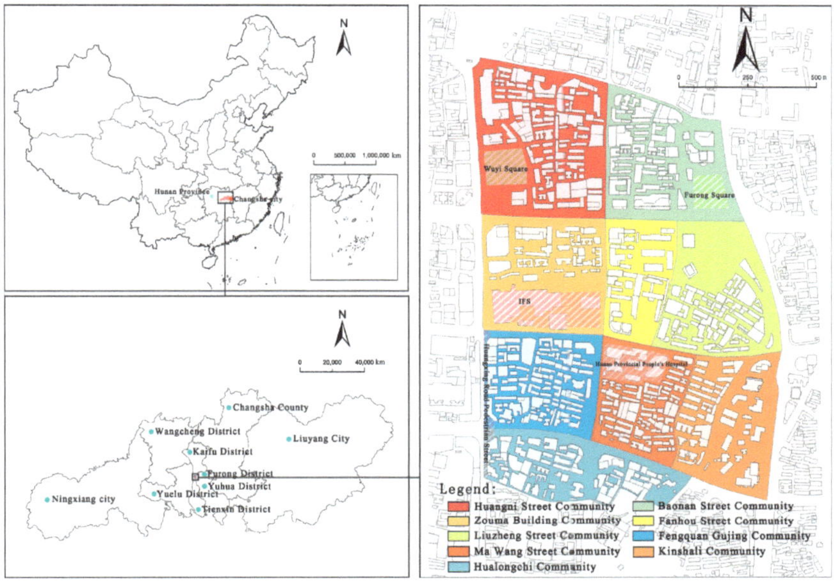

Figure 1. Study area.

2.2. Data Source and Processing

2.2.1. Data Source

Data acquisition methods included mainly Network Acquisition and field surveys. (1) The Network Acquisition data mainly included basic geographic data and web open-source data, which were obtained through official website downloads and web crawlers. (2) The field survey data were primarily objective scoring data. During the fieldwork, the site road network was measured by dividing sampling points at 50 m intervals, and the data obtained from the network were modified as appropriate. Details are shown in Table 1.

Table 1. Study data.

Data Type	Data Name	Data Source
Basic geographic data	China map vector data	National basic geographic information center
Web open-source data	Road network data	Open Street Map official website
	Building data	Web crawler
	POI data	Gaode map crawler
	Street view image data	Baidu map crawler
	Heat data	Baidu map crawler
Objective scoring data	Nighttime street brightness	Lumeno meter measurement
	Noise level	Decibel noise tester measurement
	Street furniture (seating, trash cans, signs)	Objective marking on the map
	Landscape level	Objective scoring of landscape vertical design
	Road speed limit	Objective marking on the map
	Sidewalk width	Field measurements
	Presence of barrier-free passage	Objective marking on the map
	Presence of non-motorized lane	Objective marking on the map

2.2.2. Data Processing

The raw data were normalized and then systematically analyzed using a GIS (10.6) platform and semantic segmentation. Finally, the comprehensive evaluation and ranking were carried out by the Entropy-weighting TOPSIS method. In this paper, the following representative thermal and image data are selected for analysis:

(1) The heat data were selected to be divided into four time segments for congestion studies in segment a (1:00–6:00), segment b (7:00–12:00), segment c (13:00–18:00), and segment d (19:00–24:00). We processed data on a GIS platform and classified the population congestion into 7 grades: very low, lower, low, average, high, higher, and very high, which are 7 levels (Figure 2).

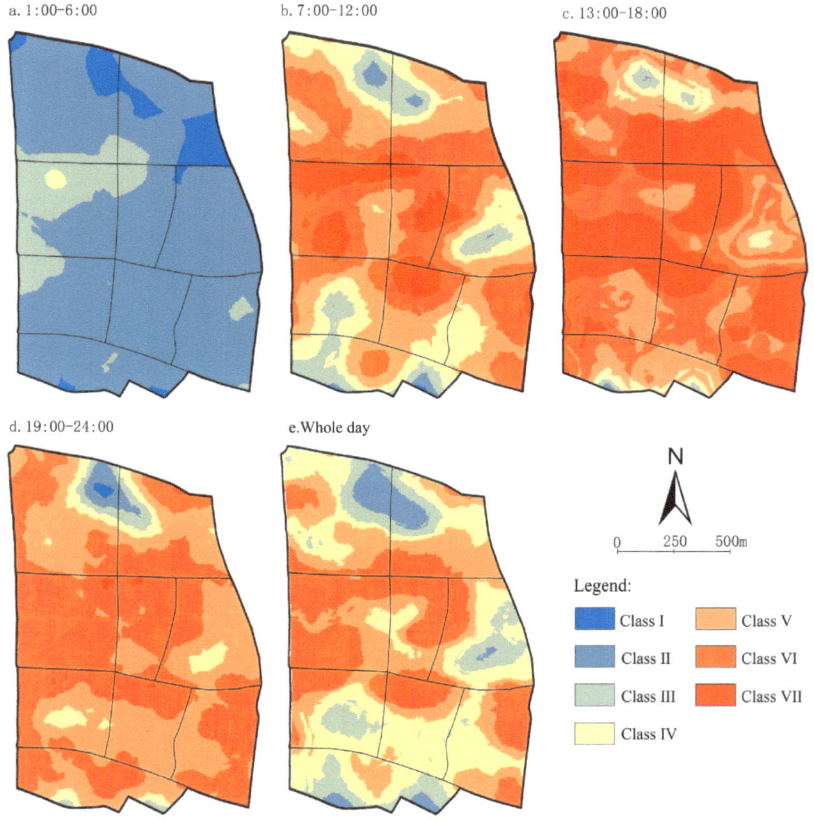

Figure 2. Time interval heat map.

(2) In this paper, the streetscape images were derived from web-scraped September 2022 data, which were sampled at 50 m intervals in four directions, 0°, 90°, 180°, and 270°, for a total of 1896 images. The image was converted from RGB to HSV, and the color pixels in the HSV channel from 60–180° were extracted and counted, and then systematically analyzed by formulae and MATLAB (2021a) software to calculate the percentage of green (Figure 3).

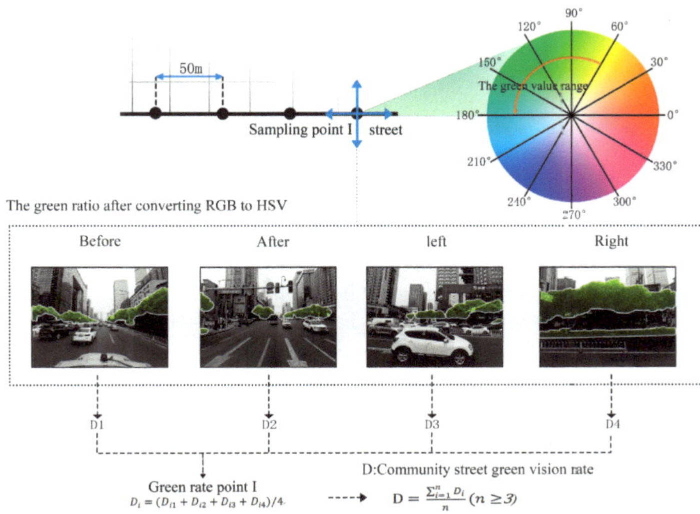

Figure 3. Green vision rate acquisition and analysis method.

2.3. Research Methods

2.3.1. Selection of Indicators

Domestic and foreign research on the street space environment evaluation index has gradually improved, and many scholars have put forward their own understanding of street space under the index system. The focus of the index system is more dispersed, and its content is more focused on a direction based on the research. Ewing et al. [37] provided another way of thinking for this paper's indicator system dimension construction, which consists of a "5D" index system for the urban built-space environment. The dimensions of the index system are more comprehensive, including degree of mixing, density, devise, destination accessibility, and distance to transit. However, modern health includes not only the traditional view of the absence of disease and infirmity but also an intact state of physical, mental, and social adaptation [38]. Increasing the ways in which the street environment interacts with urban residents and utilizes street components to support health is particularly important. Therefore, this paper adds the dimension of human perception to the "5D" evaluation dimensions to form a six-dimensional system according to the acquired data and based on the principles of availability, comparability, measurability and legality of indicators. The indicator layers under the dimensions of degree of mixing, destination accessibility, and distance to transit were appropriately deleted, and the indicator layers under the dimensions of human perception, density, and devise were organized and sorted.

In the human perception dimension, Dong Yu and other scholars [39] studied urban stressors and found that factors such as noise, crowding, and light affect residents' psychological emotions, while comparatively positive street spaces help alleviate residents' stress and loneliness [40]. Some scholars [41–44] believe that the Green View Ratio is more representative of residents' actual feelings in the street space than the green ratio and the green landscape ratio. The scale of the street and the building affects the residents' overall perception of the space. For the scale of its livability, Gao Kao and others [45] conducted in-depth research on the street width ratio and proposed a corresponding calculation method.

In the density dimension, Yang Junyan [46] believes that the street furniture is arranged so that the concept of sharing is implemented while improving the comfort of residents traveling. Therefore, this paper quantifies the density of the distribution of street furniture (seats, trash cans, signs). Some scholars [47,48] proposed a compact road network layout, which is similar to the idea of Shen Jie [49]. The high-density road network and high-density intersections will allow road accessibility to be improved, increase the probability of residents traveling, and enhance their physical function. Based on the qualitative research method, Tang Xiaoting [50] analyzed the relationship between space and physical activity through model construction. It was learned that reasonable building density can enhance the attractiveness of the area, and healthy and reasonable density of service facilities is the key to residents' activities [51,52].

In the devise dimension, Xie Yuwei [53] conducted an evaluation study on community jogging environments and found that within the same region, residents tended to exercise on paths with good shading conditions and rich surrounding landscapes. Li Hengxin [54] proposed the concept of a road connectivity index for urban road problems and provided a way of evaluating road patterns for sustainable livability. Chen Yu [55] believes that the speed limit design of the car line is related to the safety of the residents; the lower speed of the car caused less impact on the residents, thus improving the safety of the residents. The installation of dead-end roads in the city seriously affects traffic coherence and reduces the interest in travel. For streets and their surrounding buildings, some scholars have suggested that their design requires a coherent and unified street interface [56,57] and appropriate walking widths [58,59], and other scholars [60,61] have argued the need for increased consideration of user safety in design elements. The safety needs of bicyclists and special populations are fully considered in the design of its non-motorized path and accessible passageways.

By summarizing the research progress at home and abroad and combining the contemporary humanistic planning concept and the actual problems of this research object, this paper attempts to construct the "5D+" street space health evaluation index system (Table 2) by selecting thirty-one secondary indicators and six dimensions. The weighting data are calculated by objective evaluation with the Entropy-weighting method.

Table 2. "5D+" street space health evaluation index system.

Dimension Layer	Indicator Layer	Nature of Indicator	Weights (%)
Human perception	Green View Ratio	Positive indicators	1.979
	Crowding degree	Negative indicators	1.901
	Nighttime street brightness	Positive indicators	8.286
	Noise level	Negative indicators	2.431
	Street width ratio	Interval indicators	1.803
Degree of mixing	Functional mixing degree	Positive indicators	2.078
Density	Street furniture (seating, trash cans, signs)	Positive indicators	1.984
	Road network density	Positive indicators	2.200
	Building density	Negative indicators	2.565
	Scenic spots and historical sites density	Positive indicators	4.627
	Density of healthcare facilities	Positive indicators	5.283
	Density of sports and recreational facilities	Positive indicators	3.694
	Density of science, education, and cultural facilities	Positive indicators	4.895
	Density of transportation service facilities	Positive indicators	3.581
	Density of commercial service facilities	Positive indicators	2.438
	Density of intersections	Positive indicators	2.865
Distance to transit	Accessibility of commercial service facilities	Positive indicators	3.098
	Accessibility of healthcare facilities	Positive indicators	3.072
	Accessibility of scientific, educational, and cultural facilities	Positive indicators	3.446
	Accessibility of scenic and historical sites	Positive indicators	3.248
	Accessibility of transportation service facilities	Positive indicators	3.643
	Accessibility of sports and recreational facilities	Positive indicators	2.654
Destination accessibility	Transportation accessibility	Positive indicators	4.526

Table 2. *Cont.*

Dimension Layer	Indicator Layer	Nature of Indicator	Weights (%)
Devise	Landscape level	Positive indicators	2.454
	Road connectivity index	Positive indicators	3.424
	Road speed limit	Negative indicators	2.938
	Dead-end road	Negative indicators	3.091
	Line rate	Positive indicators	2.838
	Sidewalk width	Positive indicators	2.378
	Presence of barrier-free passage	Positive indicators	2.469
	Presence of non-motorized lane	Positive indicators	4.111

2.3.2. Entropy-Weighting TOPSIS Method

This paper uses the interquartile range normalization of polar deviation to pre-process the data and transform the data into dimensionless indexes. In order to make the data processing meaningful, the dimensionless indexes can be increased by 0.01 all the time in order to meet the requirements of the TOPSIS method of arithmetic.

(1) The entropy value (e) of the ith index is calculated as follows:

$$e_j = -\frac{1}{In^n}\sum\nolimits_{i=1}^{n} p_{ij}ln^{(p_{ij})},\ 0 \le e_j \le 1 \tag{1}$$

(2) The difference coefficient (g) is calculated as follows:

$$g_j = 1 - e_j \tag{2}$$

(3) The weight (W) is calculated as follows:

$$W_j = \frac{g_j}{\sum_{i=1}^{m} g_j}, j = 1, 2, 3 \ldots\ldots m \tag{3}$$

(4) The weighting of the indicators in the normalized matrix to form a weighting matrix is calculated as follows:

$$c_{ij} = b_{ij} * w_j \tag{4}$$

(5) To determine the positive ideal solution c^+ and the negative ideal solution c^-, we use the following:

$$c^+ = \left[c_1^+,\ c_2^+,\ \ldots,\ c_n^+\right]; c^- = \left[c_1^-,\ c_2^-,\ \ldots,\ c_n^-\right] \tag{5}$$

The positive ideal solution is as follows:

$$c_j^+ = \begin{cases} max\ c_{ij}, & j \text{ is a very large property} \\ minc_{ij}, & j \text{ is a very small property} \end{cases},\ \mathrm{j} = 1, 2,\ \ldots,\ \mathrm{n} \tag{6}$$

The negative ideal solution is as follows:

$$c_j^- = \begin{cases} min\ c_{ij}, & j \text{ is a very large property} \\ maxc_{ij}, & j \text{ is a very small property} \end{cases},\ \mathrm{j} = 1, 2,\ \ldots,\ \mathrm{n} \tag{7}$$

(6) To calculate the distance from each object to be evaluated to the positive and negative ideal solutions, we use the following:

To evaluate the distance from object a_i to the positive ideal solution, the equation is as follows:

$$d_i^* = \sqrt{\sum\nolimits_{j=1}^{n} (c_{ij} - c_j^+)^2},\ i = 1, 2, \ldots, m \tag{8}$$

To evaluate the distance from object a_i to the negative ideal solution, the equation is as follows:

$$d_i^0 = \sqrt{\sum_{j=1}^{n}(c_{ij} - c_j^-)^2}, \ i = 1, 2, \ldots, m \quad (9)$$

(7) To calculate the relative closeness of each object to be evaluated, the equation is as follows:

$$f_i = \frac{d_i^0}{d_i^0 + d_i^+}, \ i = 1, 2, \ldots, m \quad (10)$$

Then, f_i is arranged from large to small to discern the priority of each evaluation object.

3. Results

3.1. TOPSIS Method Results

The processed data were put through the TOPSIS algorithm to derive the health ranking of each community (Table 3). The Huangni Street Community and Baonan Street Community, which have square facilities, did not achieve the expected health rankings, but instead ranked fifth and seventh, respectively. The Zouma Building Community with the IFS commercial facility achieved the desired health rankings. But the Hualapai community ranked first in health, far exceeding the expected ranking. It can be seen that community health is a result of a combination of dimensions.

Table 3. TOPSIS method results.

Community Name	Positive Ideal Solution Distance (D+)	Negative Ideal Solution Distance (D−)	Composite Score Index	Arrange in Order
Hualongchi Community	3.131833	3.669372	0.539518	1
Kinshali Community	3.315379	3.030137	0.477524	2
Zouma Building Community	3.629652	3.307988	0.476818	3
Fengquan Gujing Community	3.639317	3.02641	0.454026	4
Huangni Street Community	3.535476	2.845015	0.445893	5
Liuzheng Street Community	3.660868	2.854895	0.438152	6
Baonan Street Community	3.678601	2.839215	0.435608	7
Fanhou Street Community	3.846028	2.810013	0.422175	8
Ma Wang Street Community	4.08249	2.94093	0.418732	9

3.2. The Dingwangtai Area Spatial Health Evaluation Results

The spatial health evaluation of the Dingwangtai Area is divided into two aspects: sub-dimensional evaluation and overall evaluation. The sub-dimensional evaluation includes six dimensions of evaluation: human perception, degree of mixing, density, distance to transit, devise, and destination accessibility. The results of each dimension are divided into five levels according to the natural breaks (Figure 4).

Considering the richness of indicators under these dimensions, the three dimensions of human perception, density, and devise were selected for specific analysis, and the results are as follows:

The general interval under the human perception dimension is (0.047–0.065], and scores higher than the general interval are good intervals, which can be categorized into two specific cases. One category is the high-value range, such as the Ma Wang Street Community. It has a high internal Green View Ratio, low noise levels, and well-designed street scales, which have a greater enhancement effect on residents' feelings. However, the installation of light facilities is missing, and residents lack a sense of security at night, which has a negative effect on residents traveling at night. The other category is the very high-value range of the Fengquan Gujing Community. The community is located in the vibrant location of the Wuyi business district, with pleasant street scale design and adequate safety for residents traveling at night, making it highly attractive to residents. Scores below the general interval are low intervals, categorized as lower- and very low-value ranges. Among them, the Fanhou Street Community is in the very low-value range. Although the internal noise level is well-controlled, the crowding degree of the interior and the dimness at night makes the residents feel unsafe to travel, which has a negative effect on their psychological situation. Overall, the dimension of human perception within the study area showed a distribution phenomenon of high scores in the southwest direction and low scores in the northeast direction.

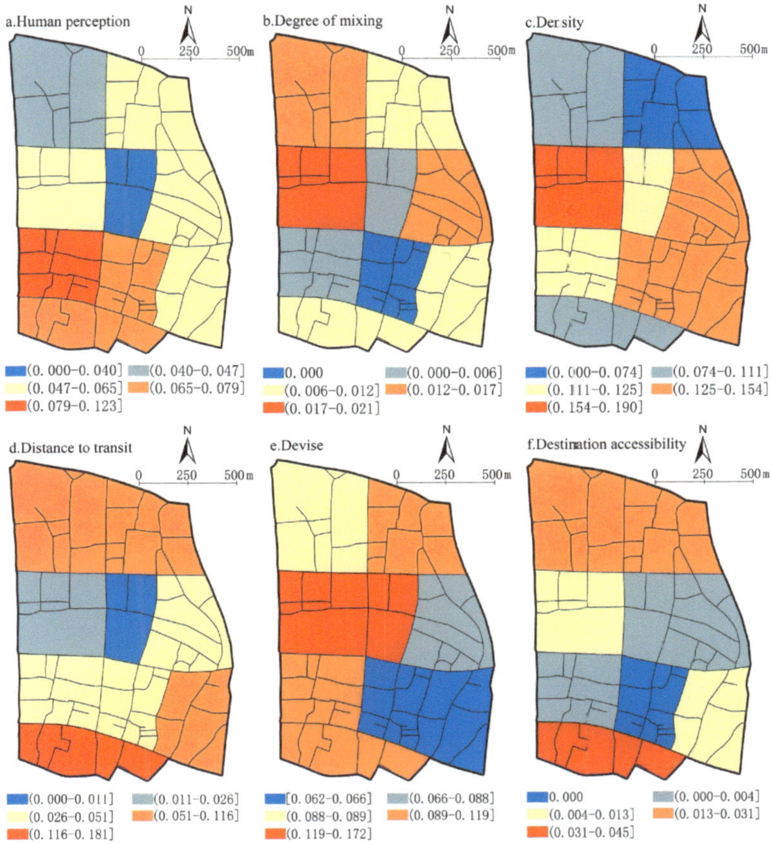

Figure 4. Results of evaluation through six dimensions.

The general interval under the density dimension is (0.111–0.125], and ratings above the general interval are good intervals, categorized into two categories. One category is the high-value range, such as the Kinshali Community, where road intersections are densely distributed. Residents have more options for traveling, and travel is more convenient. There is a reasonable building density, but the community lacks corresponding infrastructure on the street, reducing the quality of the travel experience for residents. The other category, the very high-value range, is the Zouma Building Community, which has a high density of internal street furniture amenities in terms of amenity density, which is a good guide for residents to travel. Among the amenities, the density of sports and leisure facilities, science facilities, education and cultural facilities, transportation facilities, and commercial facilities is high, providing residents with a richer and more convenient life and enhancing the attractiveness of the community. Ratings below the general interval are in the lower intervals, with the Baonan Street Community in the very low-value range, where the lack of street furniture as well as healthcare facilities reduces the quality of life for residents and affects their travel. Overall, within the study area, the density dimension showed an overall distribution of high scores in the central direction and low scores in the north and south directions.

The general interval under the devise dimension is (0.088–0.089], and scores above the general interval are good intervals, which are divided into two categories. One category is high-value ranges, such as the Hualongchi Community, whose higher internal road connectivity index and reasonable slow-moving system and street network design ensures that residents have more choices for traveling, enhancing residents' willingness to travel and promoting the healthy development of their physical functions. However, the interior of the community has a single landscape level and lacks vertical design. The other category is the very high-value range, such as the Zouma Building Community, which occupies a good spatial location with the internal establishment of the IFC Business Center. This community has a well-designed road network and wide pedestrian streets to significantly enhance the image of the community as well as its attractiveness. Scores below the general interval are lower intervals, with the Ma Wang Street Community in the very low-value interval. Although the continuity of the street interface within the community is good, the low landscape richness and the low provision of non-motorized lanes and narrow sidewalk widths have a negative effect on residents' travel. Overall, within the scope of this study, the devise dimension as a whole showed a distribution phenomenon of high scores in the south direction, low scores in the north direction, high scores in the west direction, and low scores in the east direction.

In order to visualize the evaluation results, the community scores for each dimension were ranked separately; communities with a difference of less than 0.002 in single-dimension scores were classified as a rank. Rankings 1–9 were given a new 9-1 score, which was visualized by radar charts (Figure 5).

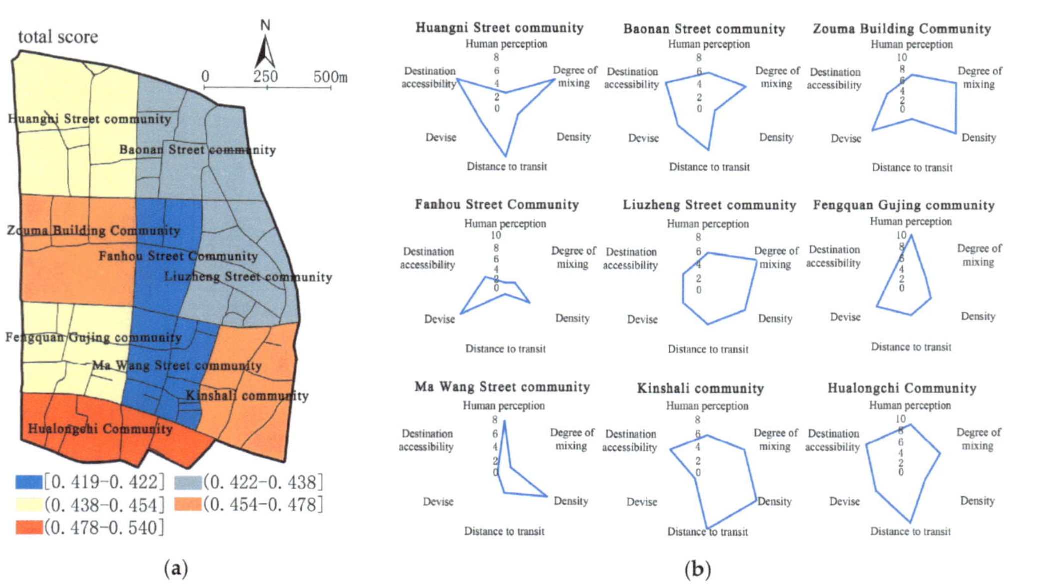

Figure 5. (**a**) Comprehensive results of street space health evaluation; (**b**) Radar chart of the scores of each community by dimension.

For the overall evaluation, the results of the above sub-dimensional evaluations were summarized and counted to produce the results of the street space health evaluation for each community; the very low-value range is (0.419–0.422], including the Fanhou Street Community and the Ma Wang Street Community. Its common problems are manifested in the two dimensions of destination accessibility and degree of mixing. While the Fanhou

Street Community places more emphasis on the devise dimension, it neglects the construction of the human perception dimension, whereas the Ma Wang Street Community is the opposite of the Fanhou Street Community. The low-value range is (0.422–0.438], including the Baonan Street Community and Liuzheng Street Community. Of these, the Liuzheng Street Community has issues with destination accessibility and the devise dimensions, while the Baonan Street Community is primarily characterized by the density dimension. The general-value range is (0.438–0.454], including the Huangni Street Community and the Fengquan Gujing Community. One of the main problems in the Huangni Street Community is reflected in the human perception and the density dimensions. The Fengquan Gujing Community places relative emphasis on the human perception dimension, but it lacks the construction of a degree of mixing dimension and is relatively deficient in the density and destination accessibility dimensions. The high-value range is (0.454–0.478], which includes the Zouma Building Community and the Kinshali Community. They have outstanding issues with distance to transit and the devise dimensions, respectively. The very high-value range is (0.478–0.540] for the Hualongchi Community, which is relatively lacking in the density dimension but scores relatively high on the remaining five dimensions. A spatial visualization of the results of this study results in a distribution of higher spatial health indexes for peripheral communities, lower spatial health indexes for internal communities, high scores in the western direction, and low scores in the eastern direction in the study area.

4. Strategy and Suggestion

4.1. Healthy Streets Space Design Strategy

According to the research results and the practical problems of the research scope, the ideal model of a healthy street is proposed, which includes the goals of good human feelings, complex functional types, a bright street environment, and a better walking and cycling experience (Figure 6). Three design optimization strategies are proposed for the healthy street space design of the Dingwangtai Area Community.

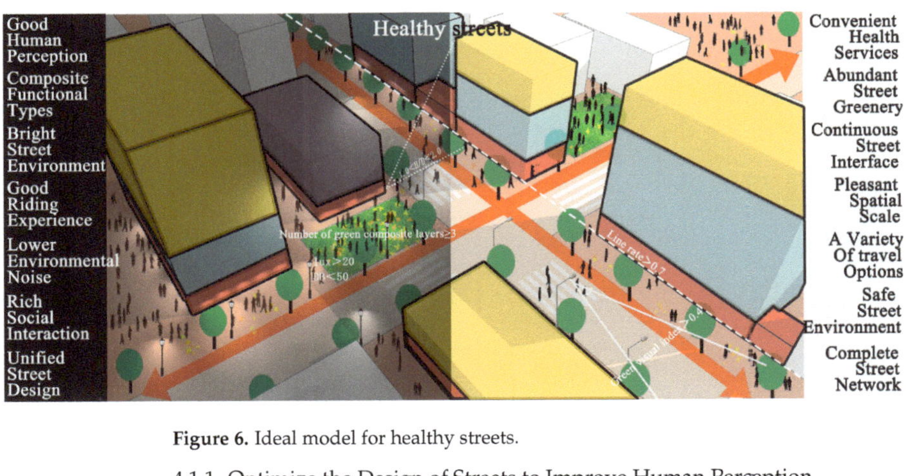

Figure 6. Ideal model for healthy streets.

4.1.1. Optimize the Design of Streets to Improve Human Perception

By planning the street layout properly, the human comfort of the neighborhood can be improved and the quality of the built environment of the city can be enhanced at the same time. First, increasing the density of vegetation and landscape levels around streets and alleys, as well as using crowd behavior analysis to reduce the degree of street crowding, can effectively improve the residents' sense of travel safety and experience using the street, and these changes can improve the overall impression of the street. As an example, the Zouma Building Community has significant deficiencies in these three areas. Pocket parks can be created by relying on wide street space, which not only enhances the richness of street greenery, but also effectively reduces the discomfort caused by overcrowding and the problem of emptiness due to overly open squares. The continuity of the street interface and its width are directly linked to the street image. Secondly, rational streets should ensure that the scale is rationalized in order to satisfy aesthetics and ensure the neatness and continuity of the street interface. Wide sidewalks directly affect the ability of a street to accommodate pedestrian traffic and the residents' sense of space. There are obvious problems in these two areas in the Kinshali Community: The first is the discontinuity of the street facade, which leads to visual confusion. The second is that the the sidewalks are not wide enough, which creates a depressing feeling for the residents. Due to structural limitations of the building, it is not possible to adjust the location of the existing building to widen the pedestrian walkway. Therefore, it is recommended to introduce architectural features on the building or adjust the appearance of the building. This can enhance the attractiveness of neighborhoods as a way to create distinctive living spaces, alleviating the current feelings of fragmentation and congestion in the street space and improving the overall functionality and aesthetics of the community.

4.1.2. Adjusting the Functional Structure to Enhance the Vitality of the Population

Optimizing the functional structure within the community can ensure that residents have greater convenience while achieving their daily travel purposes. Thus, residents are encouraged to increase the frequency of their trips, and the vitality of the community space is enhanced. In view of the low degree of mixing of functions in the Mawang Street Community, there may be practical challenges in changing the functional layout on a large scale. By creating podiums or replacing functions on the ground floor, the community's functional diversity can be enriched and its comprehensive service capacity can be enhanced. Reasonable functional organization can better meet the needs of residents and enhance the quality of life of community residents. In the Fan Hou Street Community, the scenic viewing value of the area can be enhanced through such measures as the construction of a high-rise podium for viewing. In response to the lack of healthcare facilities in the Baonan Street Community, underutilized space in the community can be used to build new healthcare facilities, or existing building functions can be remodeled, enhancing the residential environment and the health of the residents. In the Ma Wang Street Community, there is a relative lack of sports and recreational, scientific, educational, and cultural facilities. By providing residents with suitable recreational and leisure venues and emphasizing the provision of educational resources, we are able to meet the community's internal educational and cultural needs and promote the all-around development of the community's residents. The Liuzheng Street Community has serious problems with transportation services facilities, which could be improved by adding more parking lots and improving the parking system. The establishment of cab harbor parking stations could improve the travel experience of residents. The Hualongchi Community has a low density of commercial facilities and a high level of vitality within the parcel, which allows for the creation of a distinctive commercial-themed street to increase the density of commercial facilities and revitalize the community's economy.

4.1.3. Optimize Traffic Organization and Improve Travel Experience

Adjusting the road traffic within the community and constructing a suitable road system network could increase the opportunities for residents to make travel choices and enhance the attractiveness of travel. Enriching the street type increases the probability that residents will travel by bike or on foot. Roads within the Ma Wang Street Community are generally narrow, and traffic congestion is relatively high. It is difficult to set up non-motorized lanes to alleviate the opposite traffic flow congestion phenomenon, depending on the actual situation to increase the time-limited one-way roads. There are significant deficiencies in the design of the barrier-free passage to the pedestrian walkway in the Fengquan Gujing Community, and the barrier-free passage is frequently occupied. In order to optimize the situation, the degree of attention to barrier-free passages can be raised, and the walking experience of residents can be improved by increasing the design of barrier-free passages and strengthening the management strategy to ensure the safety and convenience of the use of the trail. The Huangni Street Community has a high number of dead-end roads within the community that limit the continuity of traffic flow and the efficiency of its use; by extending existing roadways and connecting them to neighboring roadways, residents are provided with more options for convenient travel. A high road connectivity index means that residents have more travel options. While the road network in the Baonan Street Community is dense, there is a lack of road connectivity, indicating that there are fewer road intersections within the community, which limits travel for residents. New connecting roads can be constructed to open up parallel road traffic, improve traffic flow and accessibility within the community, and improve the travel experience of residents.

5. Discussion

5.1. In Terms of Research Evaluation

This study proposes a "5D+" Healthy Streets Evaluation Index System, which is more comprehensive and refined than the traditional Healthy Streets Evaluation Index System. It aims to provide a quantitative tool that can be applied to the evaluation and updating of streets and help the designer understand the object of this study accurately and improve the quality of the streets. However, the evaluation of healthy streets is a multidimensional and complex subject, so this study refers to Ewing's "5D" evaluation index system and adds the human perception dimension to evaluate the healthy streets. From the results of this study, the problems found in the study area are similar to those found by the research team of Zhao Mengni [44] in the main urban area of Nanjing. If we can obtain the research data of the main urban area of Nanjing, we can use the "5D+" Healthy Streets Evaluation Index System proposed in this paper to conduct a study on it, so as to verify the generalizability of the indexing system in this paper. From the perspective of geographic location, the Healthy Streets Evaluation Index System should not be a fixed indicator; if data from other areas can be obtained, horizontal comparisons can be made for the Dingwangtai Area. Depending on the unique focus and needs of its different regions [62], the weights of the indicators are not the same. In this way, it is possible to know which indicators are geographical and changeable, thus obtaining a more detailed system of healthy city evaluation indicators at the medium scales and micro scales, such as the space of street wards.

5.2. In Terms of Optimization Strategies

Many authors have proposed improvement strategies for healthy streets from the traditional perspectives of functional structure [34], traffic organization [37], and design elements [59]. Their problem-solving strategies are mostly macroscopic coordination of the study area and its surroundings and sorting out the urban pulse, while the consideration of the human perception on the micro scale is comparatively weak. Therefore, this paper proposes three major corresponding improvement strategies for the problems of community health construction in the Dingwangtai Area, combined with the concept of a human-centered city, in order to achieve the ideal goal proposed in this paper for healthy streets. Since the study area is located in the center of Changsha City, with high pedestrian flow, high building density, and complex land ownership, large-scale demolition and renovation is not applicable. The strategy of this paper mostly focuses on the renovation of the existing environment and the functional supplementation of small spaces.

5.3. Insufficiency of Research

The basis for realizing the healthy street is based on the theoretical study of the street's microspace. This paper evaluates the health of the Dingwangtai Area, which still has relative deficiencies. Firstly, the evaluation of streets in this paper relies only on objective data and is missing a secondary subjective evaluation in the weighting of indicators. Secondly, there are problems in the indicator evaluation system, such as uneven division of indicator layers and the difficulty of obtaining or slow updating of some indicator data in small and medium-sized cities. Examples include questionnaire surveys as well as community population heat data and streetscape image data. Finally, there is a link between the community and its surroundings, and there are dynamic changes that make it difficult to evaluate them using fixed indicators.

5.4. Future Prospects

The city is a complex entity composed of multiple elements, and the interactions between its internal elements cannot be fully revealed by simple academic discussions. Healthy streets, as one of the bases for building a healthy city, are also intricately related to other elements in the city. It is hoped that the relevant teams will utilize new technologies and perspectives, combine multidisciplinary dimensions, fully integrate objective indicators with subjective evaluation, and promote the improvement and development of the Healthy Streets Evaluation Index System in their subsequent studies. This system is a way to guide the construction of healthy streets and promote the construction of healthy cities, carry out deeper research and exploration of healthy cities, and put forward more scientific and practical strategic recommendations for the construction of healthy cities.

6. Conclusions

In this study, we constructed a "5D+" street space health evaluation index system to evaluate the health of the Dingwangtai Area in Changsha City, with the community as the evaluation unit. The Entropy-weighting TOPSIS method was used to compute the indicator data to obtain the health ranking of the corresponding community. The conclusions of this study are as follows:

(1) At the area level, the Dingwangtai Area is comparatively weak under the three dimensions of density, devise, and destination accessibility. Four communities scored poorly in the corresponding dimensions, respectively.

(2) At the community level, the Ma Wang Street Community and the Fanhou Street Community scored lower. The Ma Wang Street Community is at a poor level in four dimensions: degree of mixing, distance to transit, devise, and destination accessibility. The Fanhou Street Community is at a poor level in four dimensions: human perception, degree of mixing, distance to transit, and destination accessibility.

(3) At the indicator level, comparing the weights of the indicator layers of each dimension, it was concluded that the indicators of nighttime street brightness, functional mixing degree, density of healthcare facilities, accessibility of transportation service facilities, presence of a non-motorized lane, and transportation accessibility have a significant impact on the development of healthy streets in the Dingwangtai Area.

(4) At the strategy level, this paper proposes three major transformation strategies for community problems based on the concept of humanism: optimize street design to improve human perception, adjust the functional structure to enhance the vitality of the population, and optimize traffic organization and improve travel experience.

Author Contributions: Conceptualization, Y.Z. and H.T.; methodology, Y.Z. and K.H.; software, Y.Z. and H.T.; validation, Y.Z., H.T. and J.T.; formal analysis, Y.Z.; investigation, Y.Z.; resources, Y.Z. and K.H.; data curation, Y.Z.; writing—original draft preparation, Y.Z.; writing—review and editing, H.T.; visualization, Y.Z.; supervision, J.T.; project administration, H.T.; funding acquisition, Y.Z. and H.T. All authors have read and agreed to the published version of the manuscript.

Funding: This research was funded by the National Student Innovation and Entrepreneurship Training Project (S202311527003) and the Project of Philosophy and Social Science Foundation of Hunan Province (20YBQ024).

Data Availability Statement: Data is contained within the article.

Acknowledgments: We would like to thank the reviewers for providing constructive comments on this manuscript.

Conflicts of Interest: The authors declare no conflicts of interest.

References

1. Li, X.; Zhao, M.W. Formation, Declination and Rebuilding of City Features. *Adv. Mater. Res.* **2011**, *1279*, 1527–1531. [CrossRef]
2. Zheng, B.H.; Guo, R.; Komi, B.B.; Xiang, Y.F. Quantitative Evaluation of Urban Style at Street Level: A Case Study of Hengyang County, China. *Land* **2022**, *11*, 453. [CrossRef]
3. Yang, Y.J.; Lei, H.C. Latest Development of Healthy Cities in Foreign Countries and Its Implications for China. *Med. Soc.* **2016**, *29*, 33–36. [CrossRef]
4. Hirsch, J.A.; Roux, A.V.D.; Moore, K.A.; Evenson, K.R.; Rodriguez, D.A. Change in walking and body mass index following residential relocation: The multi-ethnic study of atherosclerosis. *Am. J. Public Health* **2014**, *104*, 49–56. [CrossRef] [PubMed]
5. Saelens, B.E.; Handy, S.L. Built environment correlates of walking: A review. *Med. Sci. Sports Exerc.* **2008**, *40*, S550–S566. [CrossRef]
6. Girling, L.C. The Sustainable Street: The Environmental, Human and Economic Aspects of Street Design and Management. *Landsc. Urban Plan.* **2003**, *62*, 117–118. [CrossRef]
7. Ge, Y.; Shen, X.; Cai, C.T. Theory, Method and Practice of Healthy Street Design. *Shanghai Urban Plan. Rev.* **2020**, *2*, 49–56. [CrossRef]
8. Bai, M.; Liu, Y.; Zhang, J.J.; Zhang, X.L. Research on Healthy City and Street Space Renewal. *Urban. Archit.* **2020**, *17*, 11–13. [CrossRef]
9. Liang, H.N. Homo-urbanicus: Basic logic and operational procedure. *City Plan. Rev.* **2020**, *44*, 68–76.
10. Gallagher, M.R.; Wang, Z.Y. To Pursue Delicate Street Design-Discussion on Streetscape Guidance: A Guide to Better London Streets. *Urban Transp. China* **2015**, *13*, 56–64. [CrossRef]
11. He, H.; Fang, Y.X.; Zhang, T.; Zhang, Q.H.; Li, M.K.; Su, C. The Present and Future of Street Space Quality Research: Quantitative Analysis Based on the Visualization Literature at Home and Abroad in the Past Ten Years. *Shanghai Urban Plan. Rev.* **2022**, *6*, 73–81.
12. Badland, H.; Mavoa, S.; Boulangé, C.; Eagleson, S.; Gunn, L.; Stewart, J.; David, S.; Giles-Corti, B. Identifying, creating, and testing urban planning measures for transport walking: Findings from the Australian national live ability study. *J. Transp. Health* **2017**, *5*, 151–162. [CrossRef]
13. Deng, Y.L.; Guo, X.C.; Ye, M.; Guo, L.C.; Gong, X.L. The Experience and Inspiration from the Seattle Pedestrian Master Plan. *Mod. Urban Res.* **2012**, *9*, 17–22. [CrossRef]
14. Dai, J. Overview of International Urban Street Design Guidelines. *Urban Rural Dev.* **2020**, *2*, 76–79.
15. Jiang, Y.; Wang, Y.; Xie, J.H.; Liu, Y.; Zhao, J. Returning to Human-Centered Streets: Recent Developments of Street Design Guidelines in World Cities and Implications for Chinese Cities. *Urban Plan. Int.* **2012**, *27*, 65–72.
16. Carmona, M.; Gabrieli, T.; Hickman, R.; Laopoulou, T.; Livingstone, N. Street appeal: The value of street improvements. *Prog. Plan.* **2017**, *126*, 1–51. [CrossRef]
17. Prelovskaya, E.; Levashev, A. Modern Approach of Street Space Design. *Transp. Res. Procedia* **2017**, *20*, 523–528. [CrossRef]
18. Jones, P.; Marshall, S.; Boujenko, N. Creating more people-friendly urban streets through 'link and place' street planning and design. *IATSS Res.* **2008**, *32*, 14–25. [CrossRef]
19. Harvey, C.; Aultman-Hall, L. Measuring Urban Streetscapes for Livability: A Review of Approaches. *Prof. Geogr.* **2016**, *68*, 149–158. [CrossRef]
20. Kaczynski, A.T.; Besenyi, G.M.; Child, S.; Hughey, S.M.; Colabianchi, N.; McIver, K.L.; Dowda, M.; Pate, R.R.; Healthy Communities Study Team. Relationship of objective street quality attributes with youth physical activity: Findings from the Healthy Communities Study. *Pediatr. Obes.* **2018**, *13*, 7–13. [CrossRef]
21. Curl, A.; Thompson, W.C.; Aspinall, P. The effectiveness of 'shared space' residential street interventions on self-reported activity levels and quality of life for older people. *Landsc. Urban Plan.* **2015**, *139*, 117–125. [CrossRef]
22. Hao, X.H.; Long, Y.; Shi, M.; Wang, P. Beijing's Street Vitality: Measurement, Influencing Factors, and Implications for Planning and Design. *Shanghai Urban Plan. Rev.* **2016**, *3*, 37–45. [CrossRef]
23. Ye, Y.; Zhang, Z.X.; Zhang, X.H.; Zeng, W. Measuring Spatial Quality of Streets at the Human Scale—A Large-Scale, High-Precision Evaluation Framework Combining Streetscape Data and New Analytical Techniques. *Urban Plan. Int.* **2019**, *34*, 18–27. [CrossRef]
24. Wang, L.; Jia, Y.H.; Sun, W.Y.; Jiang, F.F. Quantitative Health Impact Assessment Study for Urban Planning Programs. *Planners* **2021**, *37*, 72–77. [CrossRef]
25. Long, Y.; Shen, Y. Street Urbanism and Data Enhanced Design. 2015. Available online: https://sc802d58b85de2333.jimcontent.com/download/version/1608969944/module/10814319899/name/%E9%BE%99%E7%80%9BDAD.pdf (accessed on 4 January 2024).
26. Long, Y.; Li, P.; Hou, J.X. Three-dimensional Urban Form at the Street Block Level for Major Cities in China. *Shanghai Urban Plan. Rev.* **2019**, *3*, 10–15.
27. Long, Y. Street Urbanism a New Perspective for Urban Studies and City Planning in the New Data Environment. *Time Archit.* **2016**, *2*, 128–132. [CrossRef]
28. Long, Y.; Tang, J.X. Large-scale quantitative measurement of the quality of urban street space: The research progress. *City Plan. Rev.* **2019**, *43*, 107–114.
29. Gao, Y.B. *A Study of Supportive Environments for Walking with Proactive Interventions for Resident Health*; Chongqing University: Chongqing, China, 2022. [CrossRef]

30. Wang, L.; Li, X.T.; Yang, X.M. Health in 15-minute Life Sphere: Community Response to Public Health Emergency. *Planners* **2020**, *36*, 102–106+120.
31. Bao, H.J.; Li, L.L. Evaluation indicator system for Healthy City Planning from the Perspective of Territorial Spatial Planning. *Acta Ecol. Sin.* **2024**, *10*, 1–11. [CrossRef]
32. Dong, X.F.; Yin, H.G.; Chen, L.; Zhang, Y.J. Healthy Cities Seek Solutions from the Community—Taking a Typical Community in Shang Di Street, Beijing as an Example. *Urban Dev. Stud.* **2023**, *30*, 97–103. [CrossRef]
33. Tang, J.X.; Long, Y. Metropolitan Street Space Quality Evaluation: Second and Third Ring of Beijing, Inner Ring of Shanghai. *Planners* **2017**, *33*, 68–73. [CrossRef]
34. Wu, W.S.; Niu, X.Y. Impact of Built Environmental Functional Diversity on Street Vitality: A Case Study of West Nanjing Road in Shanghai. *South Archit.* **2019**, *2*, 81–86. [CrossRef]
35. Fang, R.; Liu, B.Y. A Study on the Morphological Patterns of Living Streets and Their Motivation—Taking Nanjing as an Example. *Urban Dev. Stud.* **2022**, *29*, 129–136.
36. Si, R.; Lin, Y.Y.; Xiao, Z.P.; Ye, Y. Spatio-temporal Analysis of Built Environment and Street Vitality Relationship Based on Street-level Imagery: A Case Study of Futian District, Shenzhen. *Sci. Geogr. Sin.* **2021**, *41*, 1536–1545. [CrossRef]
37. Ewing, R.; Tian, G.; Goates, J. Varying influences of the built environment on household travel in 15 diverse regions of the United States. *Urban Stud.* **2015**, *52*, 2330–2348. [CrossRef]
38. Yang, T.T.; Peng, H.Y.; Tan, S.H.; Zeng, M.Y. From Urban Stress to Neuro-Urbanization: Review and Framing of Healthy Street Research from an Interdisciplinary Perspective. *Urban Plan. Int.* **2023**, *38*, 29–38. [CrossRef]
39. Dong, Y.; Li, Z.; Dong, W. The Influence of Perceived Living Street Environmental Characteristics on Mental Health of Residents: The Empirical Evidence from Old District in Harbin. *Chin. Landsc. Archit.* **2021**, *37*, 45–50. [CrossRef]
40. Wu, W.S.; Ma, Z.Y.; Guo, J.H.; Niu, X.Y.; Zhao, K. Evaluating the Effects of Built Environment on Street Vitality at the City Level: An Empirical Research Based on Spatial Panel Durbin Model. *Int. J. Environ. Res. Public Health* **2022**, *19*, 1664. [CrossRef]
41. Fumagalli, N.; Maccarini, M.; Rovelli, R.; Berto, R.; Senes, G. An Exploratory Study of Users' Preference for Different Planting Combinations along Rural Greenways. *Sustainability* **2020**, *12*, 2120. [CrossRef]
42. Xu, L.Q.; Meng, R.X.; Huang, S.Q.; Chen, Z. Healing Oriented Street Design: Experimental Explorations via Virtual Reality. *Urban Plan. Int.* **2019**, *34*, 38–45. [CrossRef]
43. Hao, X.H.; Long, Y. Street Greenery: A New Indicator for Evaluating Walkability. *Shanghai Urban Plan. Rev.* **2017**, *1*, 32–36+49. [CrossRef]
44. Zhao, M.N.; Zhen, F.; Jiang, Y.P. Study on Comprehensive Evaluation and Optimization Strategy of Urban Community Pedestrian Environment: A Case Study of Main Urban Area of Nanjing. *Mod. Urban Res.* **2021**, *2*, 41–48. [CrossRef]
45. Gao, K.; Chen, M.Y.; Liu, Y.; Wu, Z.H. Typical Case Study on the Street Which Takes The Height-Width Ratio and the Near-Line Rate As The Evaluation Indexes—Case Study: Qing Dao. *Intell. Build. Smart City* **2020**, 130–131+134. [CrossRef]
46. Yang, J.Y.; Wu, H.; Zheng, Y. Research on Characteristics and Interactive Mechanism of Street Walkability Through Multisource Big Data: Nanjing Central District as a Case Study. *Urban Plan. Int.* **2019**, *34*, 33–42. [CrossRef]
47. Yang, C.; Tan, S.H.; Li, M.M.; Dong, M.J. Research on active planning intervention strategies for healthy cities. *City Plan. Rev.* **2022**, *46*, 61–76.
48. Guo, L.L.; Tang, X.M. Evaluation of Walking Comfort of Community Roads under the Background of City Betterment and Ecological Restoration—A Case of Shanghai Cao Yang New Village. *Chin. Landsc. Archit.* **2020**, *36*, 70–75. [CrossRef]
49. Shen, J.; Chun, T.; Niu, Q.; Wei, W.; Peng, Y. Evaluation and Difference Analysis of Built Environment Walking Demands in Urban Residential Areas: A Case Study of Five Types of Residential Areas in Wuhan. *Planners* **2020**, *36*, 38–44.
50. Tang, X.T. *Research on Health Service Function of Urban Public Space Based on Linear Physical Activity VGI*; Harbin Institute of Technology: Harbin, China, 2020. [CrossRef]
51. Cao, Z.; Ren, S.B. Urban Street Space Walkability Measurement and Design Application by Conducting Street Design Guidelines: Taking Shanghai as an Example. *Urban Rural Plan.* **2022**, *1*, 1–12.
52. Jiang, G.Y.; Xia, Z.W.; Sun, X.D.; Wang, X.X.; Luo, S.X. Social Infrastructure and Street Networks as Critical Infrastructure for Aging Friendly Community Design: Mediating the Effect of Physical Activity. *Int. J. Environ. Res. Public Health* **2022**, *19*, 11842. [CrossRef]
53. Xie, Y.W.; Chen, S.Y.; Yan, Y.; Chen, N.P.; Zeng, Z.; Li, X.H. Classification of Urban Community Jogging Environment based on Multi-source Data: A Case Study based on Jinshan Street in Fuzhou City. *South Archit.* **2024**, *3*, 15.
54. Li, H.X. *Research on Neighborhood Scale and Road Patterns Based on Walking Principles of Compact City Concepts*; Nanjing University: Nanjing, China, 2016.
55. Chen, Y.; Huo, K.C.; Tang, H.; Tang, J.F. Study on health assessment system of old city street spaces in Ziyang District, China. *Proc. Inst. Civ. Eng. Smart Infrastruct. Constr.* **2023**, *40*, 1–19. [CrossRef]
56. Zhang, K.X.; Tan, Y. Research on the Spatial Form and Design Strategies of Street Corners Based on the Behavior Characteristics of Pedestrians. *Urban. Archit.* **2023**, *20*, 72–75. [CrossRef]
57. Yu, Y.F.; Hu, Y.T. Progress of International Research on Health Impact of the Built Environment of Communities Literature Review and Ref lections from a Perspective of Physical Activities. *Archit. J.* **2017**, *2*, 33–38. [CrossRef]
58. U.S. Department of Transportation. Pedestrian Road Safety Audit Guidelines and Prompt Lists. 2012. Available online: http://www.doc88.com/p-843810330614.html (accessed on 14 August 2015).

59. Shi, C.; Yuan, Q.; Pan, H.X.; Ye, Y. Measuring Walkability of Street Space and Its Design Control in the Context of New Analytical Techniques: A Case Study of Shanghai Jingan Temple Area. *Shanghai Urban Plan. Rev.* **2020**, *5*, 71–79.
60. Yu, F.; Sun, L.M.; Wang, S.F. Research on Optimization Strategy of Qingdao Coastal Pedestrian Path Based on Post Occupancy Evaluation. *Urban. Archit.* **2023**, *20*, 69–72. [CrossRef]
61. Deelen, I.; Janssen, M.; Vos, S.; Kamphuis, C.; Ettema, D. Attractive running environments for all? A cross-sectional study on physical environmental characteristics and runners' motives and attitudes, in relation to the experience of the running environment. *BMC Public Health* **2019**, *19*, 366. [CrossRef]
62. Li, M.Y.; Pan, J.H. Assessment of Influence Mechanisms of Built Environment on Street Vitality Using Multisource Spatial Data: A Case Study in Qingdao, China. *Sustainability* **2023**, *15*, 1518. [CrossRef]

buildings

MDPI

Article

Health-Oriented Evaluation and Optimization of Urban Square Space Elderly Suitability: A Case Study of Yiyang City Center

Yuhua Li [1] and Hui Tang [1,2,3,*]

1 College of Architecture and Urban Planning, Hunan City University, Yiyang 413000, China; liyuhua2025@163.com
2 Key Laboratory of Key Technologies of Digital Urban-Rural Spatial Planning of Hunan Province, Yiyang 413000, China
3 Key Laboratory of Digital Technology in Hunan General Colleges and Universities, Yiyang 413000, China
* Correspondence: tanghui202086@163.com; Tel.: +86-151-7371-1708

Abstract: Urban squares are the main place for older adults' daily activities. The spatial layout of the urban square is not only related to the construction of a healthy city and the equitable distribution of public resources, but also of great significance in enhancing the sense of gain, security, and happiness of older adults. This essay takes "health needs of older adults" as the perspective, and constructs a multi-dimensional evaluation index system of urban square space for elderly suitability in terms of physiological, psychological, and social health. It also takes six urban squares in Yiyang city center as an example and analyzes and evaluates them using GIS network analysis and the entropy weight method. The results of the study show that (1) Yiyang city center squares show the distribution characteristics of "dense in the middle and scattered around", and the spatial distribution is not balanced, which leads to a small service area. (2) The overall intervention value of urban square space is mainly distributed between 0.291 and 0.710, with an average value of 0.481, and the overall intervention effect of the square is good. The reasons for differences in health interventions for older adults in urban square spaces centered on six aspects: the quality of the square space, the variety and convenience of travel lines, the tidiness of the environment, physical activities, safety, and social interaction. Based on this, this essay proposes three optimization strategies for the renewal of urban squares, the construction of characteristic squares, and the reconstruction of the urban square space system. The aim is to provide a reference for the future design and renewal of the elderly suitability of urban squares, so as to promote the construction of a healthy city.

Keywords: healthy city; elderly suitability urban squares; health interventions for older adults; optimization strategies; Yiyang city center

Citation: Li, Y.; Tang, H. Health-Oriented Evaluation and Optimization of Urban Square Space Elderly Suitability: A Case Study of Yiyang City Center. *Buildings* **2024**, *14*, 2482. https://doi.org/10.3390/buildings14082482

Academic Editor: Tiago Miguel Ferreira

Received: 11 July 2024
Revised: 5 August 2024
Accepted: 9 August 2024
Published: 11 August 2024

1. Introduction

Urban square space is an area that provides ecological services for the city and leisure and open space for residents, and has a close connection with public health. In the background of global aging, the creation of elderly suitability square space can enhance urban health to a great extent. According to the United Nations Data Centre, the ageing of the world population is increasing the world population age is increasing year by year and it is predicted that by 2045, the world's population of people over 60 years of age will exceed 400 million, and by 2050 the world's older adults will outnumber young people. In recent years, the aging process in our country has developed faster and gained more momentum than in other aging countries. In 2023, the size of China's older adult population aged 60 and above reached 280 million, accounting for 19.8% of the total population [1], which means that China has entered a stage of deep aging. To cope with the problem of population aging, in 2023 the State Council issued the "14th Five-Year Plan for the Development of the National Aging Career and Pension Service System". A series of development goals have been proposed, such as making the social environment more friendly and livable for older adults, developing a multilevel elderly care service system and social participation. Higher demands for elderly suitability and upgrading of urban infrastructure. At the same time, the health of older adults in China is not optimistic. Due to the increase in age, consciousness, action, and sensory decline; mental illness; and other health problems are increasingly prominent [2], the demand for urban squares has increased, and the construction of an environment inclusive of older adults has become a focus of attention. As society's care for older adults has expanded from physical health to mental and social health, it has contributed to the overall development of "healthy aging". However, the current urban squares have problems such as insufficient elderly suitability functional facilities, a lack of public activity space, inadequate maintenance and management, conflict with other people's activities, and the irrational configuration of the number of fitness equipment, pavilions, corridors, and seats [3], which makes it difficult to meet older adults' activity needs. Therefore, the research on the elderly suitability of urban squares has an important impact on the development of the aging society.

In terms of elderly suitability research, most foreign scholars have explored the impact of external environmental factors on older adults from a public health perspective [4–7]. Some scholars use GIS spatial analysis and satisfaction research, IPA hierarchical analysis, and other methods to study the relationship between the behavioral characteristics of older adults and urban construction and the environment [8–10]. Most of the domestic scholars take urban public spaces, parks and green spaces, and neighborhoods as research objects [11–13]. Some scholars study spatial accessibility, safety, and comfort from the physiological and mental health needs or behavioral characteristics of older adults [14–17]. Generally speaking, from the perspective of large-scale urban public space, domestic and foreign scholars' research on elderly suitability mainly focuses on architecture, outdoor public space, and facilities for the elderly, but there are few studies on the elderly suitability of urban square space. At the same time, most of the cities in China are suffering from insufficient supply of basic public service facilities, including urban squares, and the rapid growth of the elderly population. In view of this, the Outline of the Fourteenth Five-Year Plan for the National Economic and Social Development of the People's Republic of China and the Vision 2035 proposed such development requirements as "implementing a national strategy for actively responding to the ageing of the population" and "improving the multi-level social security system". Incorporating the diversified health needs of older adults for urban square space, exploring the comprehensive evaluation index system of urban square space for elderly suitability has practical value for the design, renewal, and transformation of urban squares, and at the same time has positive significance for the construction of an age-friendly society.

To summarize, this essay will explore the relationship between urban square space and the city and older adults through preliminary theoretical and practical research, and construct a multi-dimensional evaluation index system for the elderly suitability of urban square space. Secondly, a comprehensive evaluation of the case of urban squares is conducted to derive the results of the health intervention, and finally, appropriate optimization strategies are proposed. The applied innovations and contributions of this essay are as follows: (1) It further enriches the research on urban square space's elderly suitability and provides a Chinese case study, which is helpful to provide a reference for the construction of squares in countries at the stage of aging. (2) This essay takes the health needs of older adults as an entry point, and uses GIS network analysis from a macro perspective to study the impact of differences in the spatial distribution and accessibility of urban squares on older adults. A multidimensional health evaluation index system is constructed from a micro perspective, and the indicators of the social health dimension are improved and supplemented. A comprehensive evaluation of urban square space for healthy city interventions for older adults is performed with corresponding mechanisms of action and optimization strategies to provide a scientific basis for the construction and renewal of urban squares for elderly suitability and to promote a healthy city for an aging society.

2. Materials and Methods

2.1. Theoretical Analysis

2.1.1. Relationship between Urban Square Space and Healthy City Construction

As an important urban public space, urban squares have an important role in promoting the health of residents [18]. The study of public space in the West started earlier, and the theoretical study of urban squares returned from geometric esthetics and visual esthetics to the exploration of human places. This can be summarized in three stages: (1) Initial stage (late 18th century–1945), due to the disorderly development of Western industry that led to overcrowding in the city and the deterioration of the physical environment, which caused the public to pay attention to the public space. This stage of the urban squares mainly considered the spatial form, public attributes [19,20], etc. Howard [21], Perry [22], and others hoped to fulfill the public's pursuit of a healthy life through the construction of urban public spaces that include urban squares. (2) The development phase (1945–2000), in which the city faced renewal and reconstruction after World War II, created a people-oriented spatial environment and strengthened the protection and construction of urban spaces such as city block texture and urban squares [23]. Toward the end of the 20th century, Monfortin [24], in *Streets and Squares*, analyzed in detail the elements of streets, squares, and buildings in cities, and studied the effects of the interactions among the elements on urban planning, focusing on the relationship between urban square space and crowd interaction activities, and exploring the characteristics and meaning of urban squares [25,26]. At that stage, in the face of the health challenges posed by urbanization, the concept of a "healthy city" was first introduced at a conference in Toronto, Canada, in 1984. In 1989, the World Health Organization (WHO) emphasized that health should include physical, mental, social, and moral health [27]. (3) The maturity stage (2000 to present), in which public space construction faced new challenges with the rise of the New Urbanism movement, was a stage in which public space construction was more closely linked to healthy cities. Research on urban squares has paid more attention to the social function and sustainability of public space [28,29]. More and more cities have begun to focus on the construction of green ecology, fitness for all, and health education, forming a healthy city development model with characteristics.

China's research on healthy cities was relatively late, beginning in 1989, and the creation of national health city activities laid the foundation for building healthy cities. Before 1993, the construction of a healthy city in China was in the exploratory and pilot stage, and the spatial planning of urban squares did not receive enough attention. Many urban squares were used as transportation hubs or parking lots, which lacked functionality and esthetics [30]. The concept of a healthy city was introduced in the late 1990s, and the construction of healthy cities in China entered a stage of comprehensive development. The increasing demand for urban space has led to a rise in the study of urban square space, with attention being paid to the design, function, and social significance of urban squares. Wang [31] included urban squares as typical urban spaces in Urban Design. After 2000, China's urban square space was researched gradually in-depth, mostly focusing on the spatial design and cultural connotation of urban squares [32]. After 2012, the city began to stock planning and renewal design, and the urban squares began the study of the current problem from the crowd use characteristics, exploring its strategy to meet the dual needs of older adults' physical and mental health space [14,33,34]. The 2016 Outline of the "Healthy China 2030" Plan made optimizing health services, improving health protection, and building a healthy environment a strategic goal. In the National Healthy City Evaluation Indicator System (2018 Edition), there were fewer evaluation indicators for small-scale spaces such as urban square spaces. Therefore, strengthening the assessment and construction of these small-scale public spaces and developing corresponding health indicators can help create a more vibrant and healthy city spatial environment.

2.1.2. Impact of Urban Square Space on Older Adults' Health

Relevant studies have shown that urban square space has an obvious intervention effect on the health of older adults, and the rational organization and creation of urban square space can effectively improve the daily activities of older adults and thus enhance their overall health [35,36]. Health interventions for older adults in urban square spaces can be categorized into three areas: physical health interventions, mental health interventions, and social health interventions (Figure 1).

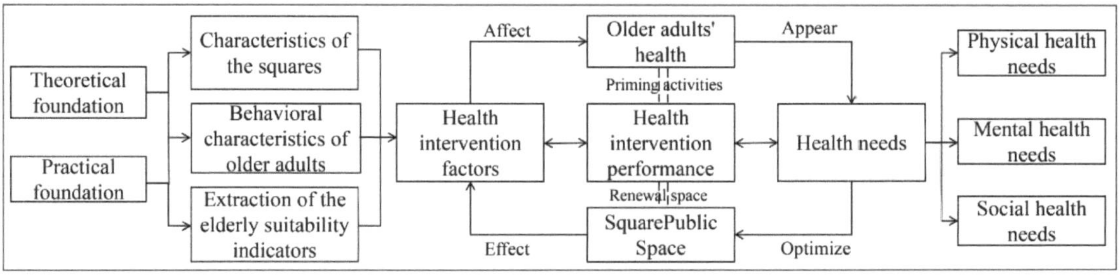

Figure 1. Map of relationship between urban square space and health interventions for older adults.

In terms of physical health interventions, the urban square space is a diverse space that meets the basic physical needs of older adults and has a positive effect of preventing disease and promoting their physical health. Improving the urban square space system, encouraging older adults to choose green travel, and increasing the time of daily walking activities are of great significance in improving the physical functions of older adults [32]. Relevant scholars have conducted research on the relationship between urban square space elements and physiological health, and have found that older adults begin to deteriorate in terms of physiological functions as they grow older. It is mainly characterized by decreased motor function, neurological decline, and decreased immunity [18,37], and has special needs for the simplicity of zoning, resting facilities, and the safety of materials in urban square space. Safety and convenience are important factors affecting the spatial elderly suitability of urban squares, and specific indicators include transportation accessibility, the degree of improvement of facilities, the diversity of travel lines, and spatial arrangement [14,38]. These mainly measure whether the spatial layout and distribution of urban squares are reasonable, and whether it is convenient for older adults to travel to reach the urban squares.

In terms of mental health interventions, the urban square space has a psychological healing effect, which can relax people through the visual, auditory, olfactory, and tactile senses; reduce stress; relieve fatigue; promote interaction; and enhance the sense of security. It has been found that urban square space can alleviate psychological problems such as anxiety, loneliness, and fatigue in older adults, and play a role in regulating and restoring their mental health [39]. Some scholars believe that the aroma of different plants has an aromatherapeutic effect to a certain extent, which produces significant health benefits through natural ecology and sensory experience, among other pathways [40]. The natural landscape has an important impact on improving the psychological stress of older adults, and blue–green space and a high green view index can effectively relieve the stress of older adults [41,42]. Sensory experience, which includes factors such as the positive elements of distraction and the tidiness of the environment, was used to measure the potential impact of urban square space on mental health. Analyzing the behavioral and psychological patterns of older adults, it was found that by introducing spatial elements, it is possible to promote a variety of experiential activities, effectively guide healthy behaviors, and improve the psychological activities of the population.

In terms of social health interventions, the urban square space is one of the key social benefits provided to older adults. Social health is defined as the population's ability to interact with society, to have good social relationships, and to realize their self-worth, i.e., social participation and social sustainability [43,44]. Participation in urban square space is closely related to publicness, while the diversity and continuity of urban square culture can effectively enhance the social connection of the urban squares and reduce the sense of distance [45]. Due to the complex attributes of the urban squares themselves, the sustainability of urban square space also has implications for social health. The construction of urban squares should be based on the theory of sustainable development, reflecting the intrinsic value of the urban squares in terms of environmental and landscape benefits and social benefits, and promoting the social integration of older adults' groups and the formation of good social relations [46]. Satisfying the healthy needs of the community is not limited to a specific space, but is a multifaceted integration of urban square spaces throughout the city. Enhancing the public nature and perpetuating the culture of the urban square, realizing the all-round inclusion of the older adult group, and reducing their sense of isolation contribute to the improvement of the social network of older adults.

Overall, the context of urban development in different periods and regions influences people's perceptions and needs for public space. The development of urban square space has gradually evolved from the early pursuit of morphological esthetics to the pursuit of a composite spatial system with multiple needs. Healthy city, on the other hand, focuses on the relationship between urban space and health, with urban public space as an important object of study. Although the existing studies have more systematically and comprehensively explained the connection between urban public space and health, less attention has been paid to the evaluation of older adults' multidimensional health needs and urban square spaces. Therefore, based on the theory related to healthy cities and elderly suitability, this essay utilizes GIS network analysis and the entropy weight method on the basis of constructing the evaluation index system, exploring the intervention performance of urban square spaces for older adults, and trying to propose corresponding optimization strategies.

2.2. Study Area

Located on the south shore of Dong Ting Lake and downstream of Zi Jiang River, Yiyang city is one of the important members of the city cluster in the middle reaches of the Yangtze River, as well as an important member of the Chang Zhu Tan 3 + 5 city cluster. According to the data released by the Yiyang Bureau of Statistics, at the end of 2023, older adults in Yiyang accounted for 22.6% of the total population, higher than the national average. The country no longer promotes the construction of municipal urban squares, and cultural and commemorative urban squares are no longer monofunctional. Urban squares are increasingly built with citizens' daily leisure, recreation, and entertainment in the center, and the urban squares are gradually diversified to the function of the development [47]. Therefore, studying and assessing the elderly suitability of Yiyang city center square space is not only crucial for enhancing the urban image, but also for promoting social interaction and economic development. It also meets the diverse needs of older adults, promotes sustainable development, and ensures the scientific and effective nature of urban planning. Based on the spatial distribution characteristics and the use of the urban squares in the Yiyang city center, this essay adopts a judgmental sampling method to select six leisure urban squares in the Yiyang city center that are adjacent to residential areas, have a large flow of people, and cover a large area as the main research objects (Figure 2).

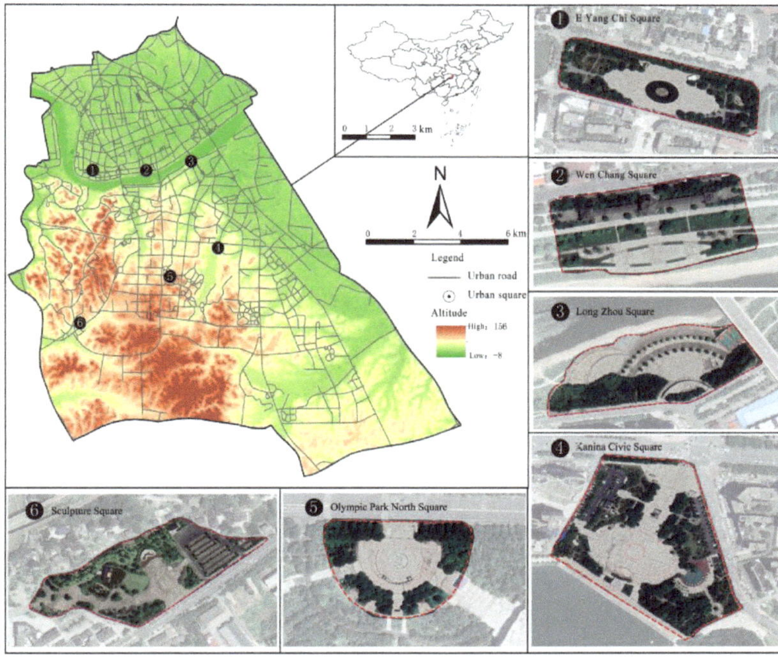

Figure 2. Study area.

2.3. Construction of Evaluation Indicator System

Elderly suitability evaluation indicators that were closely related to the health of older adults were screened based on relevant studies. Safety [16,48–50] was the most important aspect that scholars emphasized in elderly suitability design, listing it as one of the principles of elderly suitability design, followed by comfort [17,48–50], accessibility [15,51–53], socialization [15,51–53], ecology [16,53], humanism [11,54], and convenience [55], which also received high attention. Different scholars have presented their understanding of elderly suitability design from different fields, the focus of the research is more dispersed, and the content is more concentrated in one direction under the research. Therefore, in this essay, indicators related to convenience, safety, ecology, comfort, and socialization were selected from the three dimensions of physical, psychological, and social health. The concept of sustainable development in the construction of urban squares not only encourages the participation of older adults in the design and management of the squares and ensures that the space meets the practical needs of older adults, but also enhances the older adults' sense of belonging and participation and promotes social cohesion. Thus, this essay incorporates it into the social health dimension to further improve the completeness and systematization of the evaluation indicators. Six secondary indicators and 18 tertiary indicators were finally established to form the urban square space elderly suitability evaluation index system (Table 1).

Convenience refers to the ability to have smooth and unobstructed access to, entry to, and passage through, allowing one to reach as many urban square spaces as possible. The openness of space should ensure that their service radius can cover the mobility range of older adults around them [9], and that they can reach the urban squares more easily. Guide signs refer to the types of signage, such as safety and guidance, that can help the elderly to identify the characteristics of their environment [17]. Due to the differences in the mobility of older adults, the diversity of lines [56] facilitates the choice of suitable lines to reach the urban squares. The completeness of facilities refers to the availability of services such as public restrooms, sports equipment, and stores in urban squares [52].

Table 1. Evaluation index system of urban square space for elderly suitability.

Primary Indicators	Secondary Indicators	Tertiary Indicators	Nature of Indicators	Description of Indicators	Weights
Physical health	Convenience	Openness of space	Positive	The satisfaction score of older adults with the openness of urban square space	0.036
		Number of guide signs	Positive	The number of signs installed within the urban square	0.088
		Diversity of lines	Positive	The total number of lines reaching the urban square within the 800 m radius	0.039
		Completeness of facilities	Positive	The number of barrier-free facilities, sitting-out facilities, toilets, street lights	0.064
	Safety	Hygiene status	Positive	Trash cans, the number of cleaners, and the tidiness of roads	0.071
		Post-management maintenance	Positive	The number of management and maintenance staff and number of monitors	0.069
		Accessibility	Positive	The number of accessible ramps and handrails in urban squares	0.068
Mental health	Ecology	Air quality	Negative	Air quality AQI magnitude	0.054
		Noise levels	Negative	Noise level in decibels	0.046
		Green view index	Positive	Obtaining street pictures of the study area in four directions, front and back, left and right, for calculations, and selecting four study points for each case urban square and calculating the mean value of the urban square	0.055
	Comfort	Tidiness of the environment	Positive	The urban square environment is free of debris, regularly cleaned, and garbage-sorted	0.048
		Affinity of scale	Positive	Is the sense of scale pleasant and well integrated with the surrounding environment?	0.041
		Connectivity of lines	Positive	PRD = Practical length/Linear distance	0.060
Social health	Sustainability	Educational benefits	Positive	Public education sites and landscape environmental education	0.051
		Landscape benefits	Positive	The visual value of the landscape and perceived value of the city	0.062
	Socialization	Square cultural expression	Positive	The cultural characteristics of the place and cultural performance of the urban square	0.038
		Space accessibility	Positive	Event venues for people interaction	0.059
		Relaxing atmosphere	Positive	The richness of the urban square's interactions and the degree of crowd concentration	0.052

Safety is the first principle of elderly suitability design [53]. Good hygiene status minimizes the negative impact of urban squares on the health of older adults, who have weakened immunity and suffer from a variety of chronic diseases. Post-management maintenance ensures the safety of older adults using urban square activity facilities and reduces the occurrence of dangerous situations. Older adults with slow limbs and deteriorating mobility need to be assured of safe and smooth behavioral activities, and accessibility aids travel by providing mobility-impaired older adults with opportunities to get close to nature and participate in social activities [57].

Ecology means that the distribution of greening in the urban squares is appropriate, the elements are staggered and have both ornamental and practical value, which is conducive to improving the environment for older adults' activities. Air quality and noise levels reflect the suitability of urban squares for older adults, and good air quality and noise levels are beneficial to the physical and mental health of older adults [58]. The urban square plants are skillfully matched, with changing landscape design in all the seasons, and a high green view index can give older adults a fresh feeling and enhance their interest [41].

Comfort means that the urban square has a healthy city environment with convenient facilities that bring comfort to older adults. The tidiness of the urban square environment is conducive to a good activity experience for older adults [55]. Older adults and other age groups have different body scales and have their own unique activity characteristics therefore the extent to which the spatial site scale meets the activity needs of older adults in public activity spaces [15]. Next, PRD [56] is used to indicate the connectivity of lines within the urban square space, with a larger PRD indicating a smoother pathway for a good walking experience for older adults.

Sustainability means emphasizing the sustainability of the ecological environment, having low energy consumption, conforming to the requirements of the laws of nature, and adhering to the principle of sustainable development. Educational benefits motivate older adults to exchange cultural activities in urban squares. Landscape benefits [46] enrich the plant arrangement of the urban squares with good visual value of the landscape, and both together enhance the sustainability of the urban squares.

Socialization means that public spaces are designed from the outset to be participatory and to promote interactive interactions among groups of people. Enhancing the square cultural expression in urban square space can help to strengthen the connection and identity between the elderly and the environment, and form a more harmonious elderly suitability social space [54]. Space accessibility refers to the availability of various types of rich social and recreational facilities to meet the social needs of older adults and facilitate communication and fitness among them [50]. A relaxing atmosphere refers to the richness of the urban square space in the operation and management of the organization of all kinds of cultural and sports activities, to create a space that facilitates interaction to meet the pursuit of a rich life for older adults [48].

2.4. Data Sources

The research data in this essay are divided into two categories, one is qualitative data—the questionnaire was constructed through the Likert scale [56], and interviews and the distribution of questionnaires were conducted with older adults in the squares. To ensure that the survey was broad and representative, 50 copies were distributed in each urban square, and the questionnaires were filled out after screening their age of 60 years or older, and a total of 300 questionnaires were completed. The final result was 288 valid questionnaires (excluding incomplete questionnaires), with a validity rate of 96%. Data were collected on the scores of older adults on the evaluation indicators of the elderly suitability of each urban square, including the satisfaction scores on the openness of the space, square cultural expression, educational benefits, relaxing atmosphere, etc. The other type of data are quantitative data, which are required for road connectivity, core density, and accessibility, and are obtained from the Open Street Map website on 12 January 2024 for expressways, arterials, secondary roads, and side streets, combined with the Baidu map database, planning cloud-crawling urban square POI data. At the same time, field research was conducted on the squares at different times of the day (morning, midday, and evening), and data on the number of guide signs, air quality, noise levels, hygiene status, and accessibility in the six urban squares were collected.

2.5. Research Methodology

The entropy weight method (EWM) is an objective assignment method based on information entropy. It calculates the entropy value of each evaluation index to determine their degree of dissociation, and then determines the weight value of each index; the larger the entropy value, the smaller its weight. Intervention in older adults' health in the urban square space of the city encompasses three aspects of physical, psychological, and social health, and it is assumed that the total contribution of each evaluation indicator to the impact on the health of older adults is 100 percent. The steps for calculating the above three indicators are as follows: Firstly, the raw indicator data are standardized. Secondly, in order to eliminate subjective preferences in the assignment of indicators, the entropy weight method is applied to determine the weight of each indicator. Finally, the standardized values for each indicator are weighted and summed to obtain an index value, and the overall health intervention score is the sum of the scores for all primary indicators. Specific calculation steps are as follows:

(1) Constructing an indicator matrix with raw data from case urban squares.
(2) Normalization of raw indicator data by means of divergence normalization.
(3) Calculating the entropy value of each indicator:

$$e_j = -k\sum_{i=1}^{m}\left(r'_{ij} \times \ln r'_{ij}\right) \tag{1}$$

where $k = \frac{1}{\ln m}$, $k \geq 0, 0 \leq e_j \leq 1, e_j$ is the entropy value of indicator j, j and r'_{ij} is the weight of the value of indicator j for level i urban squares.

(4) Calculating the weight of each indicator:

$$d_j = 1 - e_j \tag{2}$$

$$W_j = \frac{d_j}{\sum_{j=1}^{n} d_j} \tag{3}$$

where d_j is the entropy redundancy of indicator j and W_j is the weight of indicator j.

(5) Calculating the tertiary indicator score for each case urban square:

$$S_{ij} = W_j \times r'_{ij} \tag{4}$$

where S_{ij} is the score for indicator j for level i urban squares.

(6) Calculating total health intervention scores for secondary and primary indicators for each case urban square:

$$S_j = \sum_{j=1}^{n} S_{ij}d_j \tag{5}$$

where S_j is the health intervention index for level j urban squares.

3. Results

3.1. Space Distribution of Urban Squares

Based on the POI data of Yiyang city squares, kernel density [59] and accessibility analysis [60] were performed using GIS network analysis (Figure 3). The results show that (1) Yiyang city square is in the overall spatial distribution of "One center with multiple dispersal points". That is, the gathering center consists of Wen Chang Square, Yiyang Celebrity Square, Da Yu Square, etc. on both sides of the Zi Jiang Scenic Belt, as well as Kanina Civic Square, Sculpture Square, and other squares that present a scattered distribution. (2) Calculating the range of service areas for different travel times based on the walking mode of older adults, it is found that when the walking time is less than 15 min, the service area coverage of the urban squares accounts for a total of 21.99%, which can only satisfy the use of older adults in a certain area. The percentage of urban square service area coverage corresponding to the three time periods of 0–5 min, 5–10 min, and 10–15 min were 3.06%, 7.97%, and 10.96%, respectively. The scope of the service area of the urban square is mainly concentrated in the north-central part of the city, located on both sides of the Zi Jiang wind belt in the urban center of the city. The northern part of Ziyang District and the southern part of Heshan District have few urban squares arranged, and the service area range is small in size. From the perspective of the needs of older adults, the uneven spatial arrangement of Yiyang city squares has led to differences in the enjoyment of urban square services by older adults.

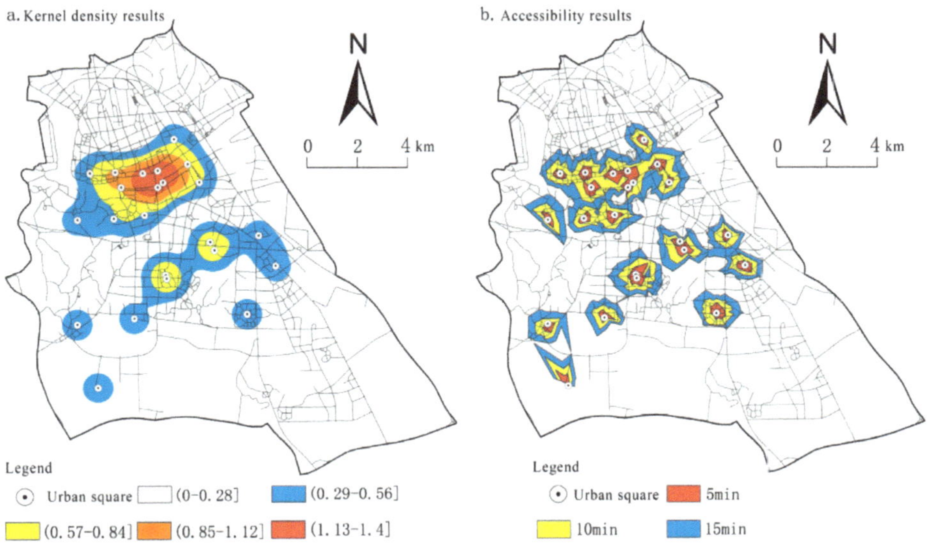

Figure 3. Analysis of urban square space (**a**) Kernel density results, (**b**) Accessibility results.

3.2. Urban Square Space Intervention Performance

Calculating the level of health interventions in the urban squares space by means of the indicators of the physical health, mental health, and social health needs of older adults and calculating the spatial health score of the urban squares, it is concluded that the overall intervention value of the urban square space is mainly distributed between 0.291 and 0.710, with an average value of 0.481. And, the overall intervention of the urban squares is better, but there is still room for improvement in some of the urban squares, such as Wenchang Square, the North Square of the Olympic Park, etc. (Table 2).

Table 2. Summary of scores for interventions in older adults' health by urban square space.

Name of the Square	Physical Health Interventions	Mental Health Interventions	Social Health Interventions	Holistic Interventions	Arrange in Order
Kanina Civic Square	0.369	0.188	0.153	0.710	1
E Yang Chi Square	0.334	0.131	0.162	0.628	2
Long Zhou Square	0.238	0.156	0.100	0.494	3
Sculpture Square	0.127	0.136	0.162	0.425	4
Wen Chang Square	0.127	0.145	0.019	0.291	6
Olympic Park North Square	0.068	0.088	0.179	0.335	5

3.2.1. Physical Health Intervention Impact

The mean physical health intervention score for the study case urban squares was 0.211. Urban squares rated above the mean are high intervention squares, including Kanina Civic Square, E Yang Chi Square, and Long Zhou Square. The relatively high density of the road network and the overall level of services and facilities in these urban squares affect the daily activities of older adults. This can be divided into two categories, a category such as the high diversity of lines within and around the Kanina Civic Square, with a high number of guide signs, making travel easier and more varied for older adults and promoting the healthy development of their physical functioning. Another category, such as E Yang Chi Square, is close to the main urban amenity road, with a suitable degree of openness, abundant activity space for older adults inside, and a high degree of completeness of facilities, attracting older adults for short restful walks and increasing their activity enrichment. Scores below the average for low-intervention squares, which have a single function, weak post-management and maintenance, poor accessibility to public facilities, and fewer gatherings of older adults resulting in unused space. Such as Olympic Park North Square, close to the Olympic Park, the surrounding industry is rich, but the square opening distance is far away, the square spatial expression forms monotony, and accessibility, seats, and other infrastructure are insufficient, resulting in lower length of stay and utilization by older adults and more trips to the interior of the park (Table 3).

Table 3. Comparison of physical health interventions in urban square space.

Typology	Urban Square Space Plan Form	Urban Square Space Characteristics	Impact of Space Interventions	Live Environment	Delegates' Square
High intervention square		(1) Higher density of road network (2) Strong space orientation (3) Abundant public space	High space openness Diversity of lines		Kanina Civic Square
Low intervention square	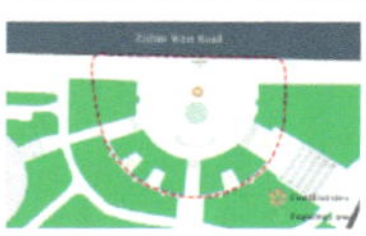	(1) Weak post-management (2) Single function of the urban square (3) Inadequate infrastructure	Waste of unused space Low security of the urban square		Olympic Park North Square

3.2.2. Mental Health Intervention Impact

The mean mental health intervention score for the study case urban squares was 0.141. The high-intervention squares include Kanina Civic Square, Long Zhou Square, and Wen Chang Square. These urban squares are better in terms of ecology and comfort, with pleasant spatial scales, and the natural landscape environment inside the squares has a positive effect on the mental health of older adults. A category such as Kanina Civic Square internal plant species, landscape green view index is the high, pleasant mood of older adults. At the same time, landscape vignettes alternating staggered arrangement, the urban square pavement material reasonable transition and change, so that the urban square steps, for older adults, have a soothing, relaxing role in the heart. Another category, such as Long Zhou Square, has good air quality and noise levels, and the comfortable venue prompts a range of recreational activities for older adults. Such as chess tournaments and winter swimming enhance the vitality of the urban square to create a comfortable and active atmosphere for older adults. Mental health low-intervention squares have poor environmental tidiness, low green view index and disjointed scale of the urban square's interior spaces, and a low sense of experience for older adults. For example, in E Yang Chi Square, due to the early construction time, the internal garbage of the urban square is piled up and thrown away, and there is a hygiene and monitoring dead corner. At the same time, the pool fountain in the center is too large in scale and single in shape, and the overall quality of the environment is reduced affecting the experience and mood of older adults (Table 4).

Table 4. Comparison of mental health interventions in urban square space.

Typology	Urban Square Space Plan Form	Urban Square Space Characteristics	Impact of Space Interventions	Live Environment	Delegates' Square
High intervention square		(1) Rich green landscape (2) Strong line connectivity (3) Pleasant space scale	Mental and emotional healing Physical activity enhancement		Long Zhou Square
Low intervention square	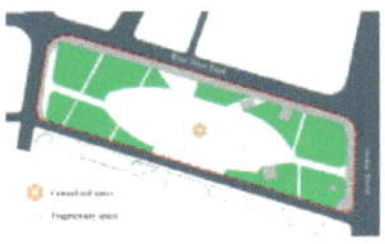	(1) Low green view index of the landscape (2) Low environmental cleanliness (3) Monotonous public space	Poor experience comfort Reduced environmental quality		E Yang Chi Square

3.2.3. Social Health Intervention Impact

The mean social health intervention score for the study case urban squares was 0.129. High intervention squares include Kanina Civic Square, Olympic Park North Square, E Yang Chi Square, Sculpture Square, etc. These types of squares have a relatively high degree of publicness and accessibility of the urban square space due to factors such as internal management and urban culture. One category, such as the Sculpture Square near Zhou Libo's former residence, set up a cultural promenade and sculpture in the urban square, to carry out a series of cultural activities to strengthen the construction of spiritual civilization in the urban square, and to enhance cultural exchanges among older adults. Another category of urban square, such as E Yang Chi Square, divides the internal space into several parts, such as social activities, cultural display, leisure, and exercise, prompting the gathering of a certain group of people. The urban square is more inclusive and effective in enhancing the intrinsic value of the urban square. Low intervention squares have poorer cultural performance and low space participation making older adults gather and stay for less time. Such as Wen Chang Square near the Wen Chang Pavilion has a deep cultural heritage, but the lack of corresponding development and the management of internal service facilities are weak, education and landscape benefits are poor, and the crowd mostly leisure walks through. At the same time, the sidewalks of the urban square are occupied by private cars, preventing older adults from walking in a sustainable development (Table 5).

Table 5. Comparison of social health interventions in urban square space.

Typology	Urban Square Space Plan Form	Urban Square Space Characteristics	Impact of Space Interventions	Live Environment	Delegates' Square
High intervention square	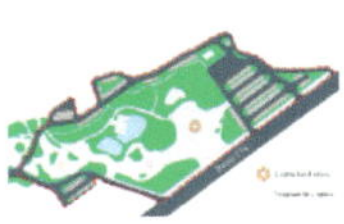	(1) High social benefits (2) Highly inclusive urban square (3) Variety of visual landscapes	Enhance public communication Creating a lively atmosphere		Sculpture Square
Low intervention square	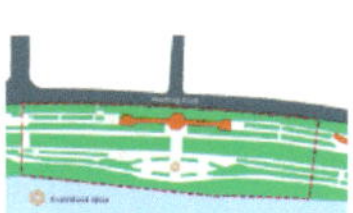	(1) Low space participation (2) Poor cultural performance (3) Low crowd gathering	Lack of leisure atmosphere Reduced sustainability		Wen Chang Square

In summary, the reasons why urban square space influences differences in the performance of health interventions for older adults centered on the following six areas: (1) The quality of the physical spatial quality of the urban square, including the established public space within the square, service facilities, and the degree of mixing of natural conditions and land use, affects the quality of older adults' travel and health. (2) The diversity and convenience of travel routes determine the richness of the daily activities of older adults. (3) The improvement of the environmental quality of the urban square space can increase the frequency and duration of activities for older adults. (4) The promotion of multiple categories of physical activity and ecological landscapes is beneficial to the physical and mental health of older adults. (5) The enhancement of the security of the urban square, including the creation of a safe and stable environment inside the square and post-management and maintenance. (6) The use of public space in squares to enhance the social interaction of older adults, including the construction of square culture campaigns and the organization of related activities, to create an atmosphere of leisure and bustle.

The renewal and renovation of health-oriented elderly suitability urban square space should be based on the connotation of physiological, psychological, and social health of older adults through the optimization of spatial quality, completion of infrastructure, sorting out of travel routes, creation of a cozy atmosphere, improving environmental quality, stimulating physical activity, enhancing safety and security, and promoting social interaction through the eight optimization pathways. Designing and evaluating the spatial elements of urban squares improves the theoretical foundation of healthy cities and elderly suitability urban squares, and provides lessons for updating the design of healthy urban square space in the future. The mechanism of action of elderly suitability urban square space reflects the impact of their spaces on the health of older adults, ensuring that the indicators are in line with the health connotations of elderly suitability urban square space (Figure 4). The space of healthy urban squares in line with elderly suitability should be oriented to the health of older adults, and be endowed with healthy functions in terms of spatial quality, landscape environment, safety and security, and social interaction in the urban squares, so as to create a relaxing and enjoyable atmosphere of the urban squares with regional characteristics.

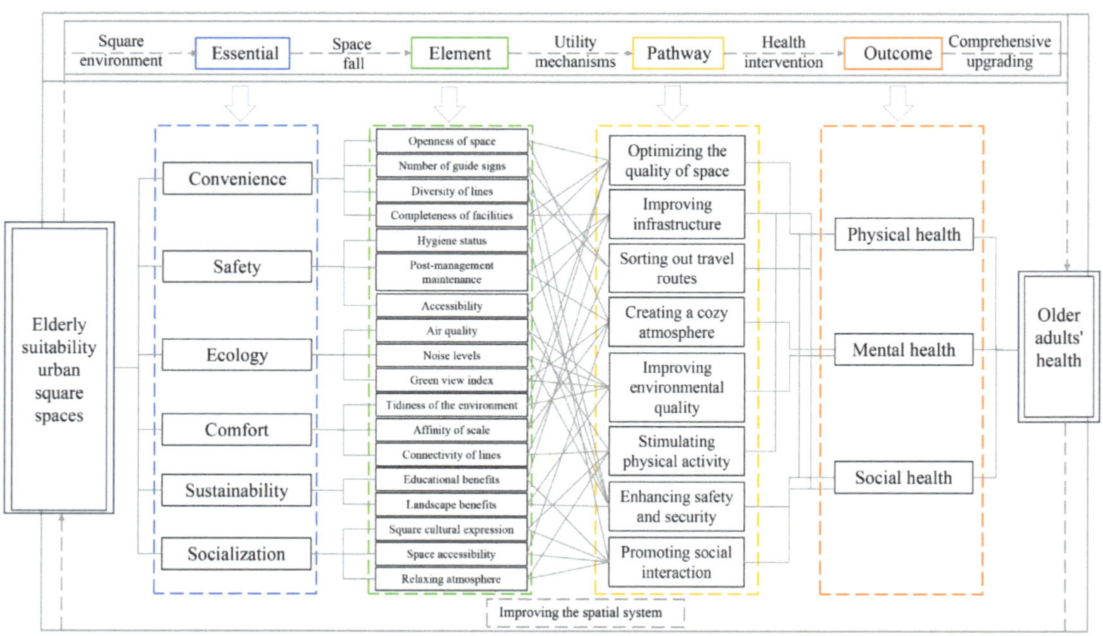

Figure 4. Mechanisms of spatial elderly suitability in health-oriented urban squares.

4. Discussion

4.1. In Terms of Research Findings

Compared with the existing studies, this essay finds that the spatial distribution of urban squares can visualize its density distribution in a certain area through kernel density analysis and accessibility analysis, revealing the rationality of its spatial distribution and its impact on the activities of older adults. These two analysis methods complement each other and together provide a scientific analysis basis for the spatial distribution of urban squares. Nevertheless, due to the limited availability of data, this essay does not incorporate the distribution characteristics of regional older adults to study the actual radiation range of urban squares, which has some limitations. Attention should be paid to overall accessibility, combining transportation planning for large urban spaces and small public spaces so that older adults can reach public spaces more easily. Due to the differences in the travel modes of older adults such as walking, biking, and public transportation, the various forms of transportation perform differently in terms of urban square space accessibility, which affects social equity and justice. The planning and layout of urban squares should be people-oriented, fully safeguarding the public interest and addressing the inequality in the enjoyment of square space by older adults and other disadvantaged groups.

The spatial intervention performance of urban squares categorizes the squares into two types of urban squares, high and low intervention, with differences mainly in terms of spatial diversity, connectivity of lines, completeness of infrastructure, and quality of the landscape environment. Older adults usually travel on foot and by public transportation, strengthening the transport connection between bus stops and open spaces in urban squares, and creating greenways that are safe and comfortable, with beautiful natural environments and smooth and convenient access.

4.2. In Terms of Optimization Strategies

As many urban squares have not been systematically planned, they have not been effectively integrated into the urban landscape network. Especially large and medium-sized urban squares, which often have difficulty attracting crowds due to the lack of organic links with residential areas. Meanwhile, many scholars basically propose improvement strategies for urban squares from traditional perspectives such as functional zoning [18], traffic organization [61], and landscape greening [62]. Its problematic strategy is mainly the macro-coordination of the study area and its surroundings, with less attention paid to the spatial system of the urban squares. And in the micro-scale is mostly focused on the completion of internal infrastructure, landscape, and green transformation, the consideration of urban square space design is relatively weak. Therefore, this essay focuses on the reconfiguration of the urban square spatial system, and internal spatial transformation to put forward the corresponding optimization strategy. At the same time, the optimization and renovation of urban squares need to be systematic and humanized, with full consideration of the needs of older adults. First of all harmonizing with the surrounding environment. Secondly, it optimizes the space for older adults in three aspects: upgrading the spatial character, stimulating the vitality of the, optimizing the landscape environment, creating a characteristic urban square, and constructing a networked healthy and synergistic urban square space system. Finally, the participation of many parties, so that older people are involved in the transformation process, the formation of a more perfect urban square space suitable for older adults to create a mechanism, so as to promote the construction of a harmonious and healthy urban square.

4.3. Limitations and Prospects

The research on the elderly suitability of urban squares still has some shortcomings: Firstly, due to the limited data and technology, the evaluation index system is constructed only from the physical, psychological, and social dimensions of older adults. In the future, further improvements can be made in terms of behavioral characteristics of older adults, natural conditions, etc. Secondly, the study subjects are typical recreational squares and do not cover other types of squares, which may have other elderly suitability problems. In future studies, expanding the scope of the study could be used to gain a more comprehensive understanding of the elderly suitability of different types of urban squares. Combining traditional research with big data and other methods, we are able to make a more scientific observation and record of the crowd behavior and distribution characteristics of urban square space as well as the comfort perception of users. Gradually improve research on the elderly suitability of urban square spaces to promote healthy city life and social interaction among older adults.

5. Conclusions and Suggestions

5.1. Conclusions

In the context of healthy city development, the urban square is not only the daily activity place of older adults but also one of the important media to promote their healthy life. GIS network analysis method and entropy weight method were used to analyze the distribution characteristics, spatial accessibility, and the influence of space on older adults. The results of the study show that:

(1) In terms of evaluation indexes, most of the existing studies have constructed an evaluation index system for elderly suitability squares from the dimensions of physiological, psychological, and behavioral activities of older adults. This essay further improves the evaluation index system of urban square space for elderly suitability by constructing it from the multi-dimensional aspects of physiological, psychological, and social health.

(2) In terms of spatial layout, the lack of a number of urban squares and their unbalanced layout have resulted in a smaller range of accessible service areas for urban squares. Lower accessibility does not meet the need for older adults to walk to the urban square for use.

(3) In terms of spatial intervention performance, the reasons for differences in health interventions for older adults in urban square spaces centered on the quality of the square space, the variety and convenience of travel lines, tidiness of the environment, the promotion of physical activities, safety, and social interaction.

5.2. Suggestions

Based on the performance of health interventions for older adults and the existing problems of the urban square, this essay proposes the following optimization suggestions from macro and micro perspectives, oriented to the health needs of older adults:

5.2.1. Improving the Quality of Space and Energizing the Urban Square

The urban square is one of the most important places for leisure and recreation for older adults, with enough space and venues to support the holding of various activities for older adults, such as chess and cards, tai chi, square dancing, and walking, which promote their physical health. In order to improve and energize the urban squares, the spatial quality of the urban square should be upgraded by transforming the design, updating the lost spaces, and utilizing the implementable space of the urban square and landscape node renovation [63] (Figure 5). First of all, integrate the existing space of the urban square and promote the orderly renewal of the urban square; the urban square space includes activity space, road space, building accessory space, etc. These spaces should cooperate with each other in order to give full play to their respective efficacy, such as activity space should be evenly distributed and diversified construction. Road space can be connected to various landscape nodes according to the road. And, the building accessory space can be set up with sports equipment for the construction of an interesting and interactive space. Secondly, constructing a large area of health trails, utilizing roads to connect various types of spaces, improving the linear flow of urban square space, and increasing the richness and convenience of walking activities for older adults [64]. Finally, with the support of the relevant government departments, the landscape nodes will be renovated, such as planting aquatic plants on the gentle slopes of the waterfront space, constructing water-friendly viewing platforms, etc., and the old facilities will be renewed. Through the renewal and renovation of negative and unused spaces, a large number of people are attracted to the area, enhancing the gathering of people and the vitality of the urban squares.

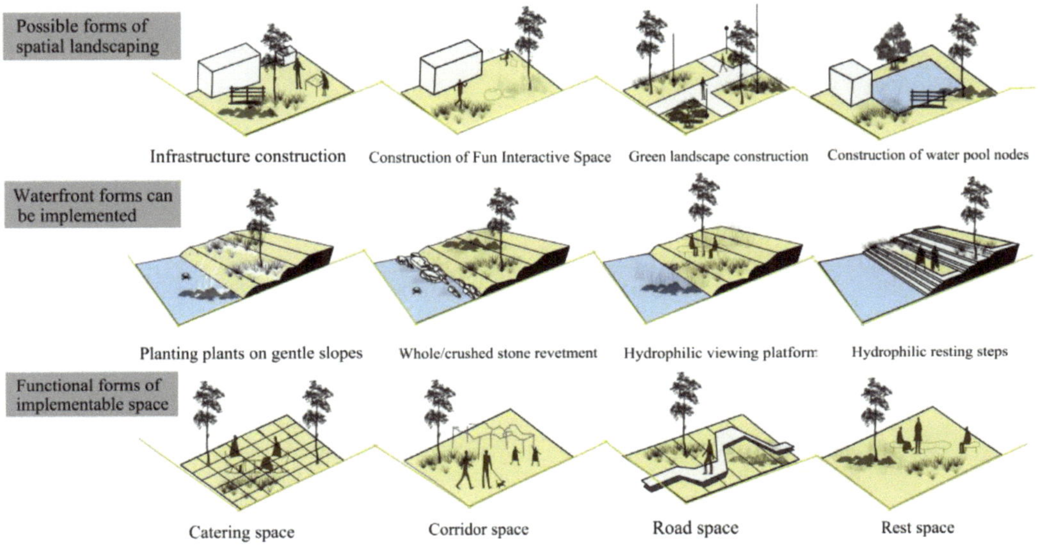

Figure 5. Renovation and renewal of urban square space.

5.2.2. Optimizing the Landscape Environment and Creating a Characteristic Urban Square

The elderly suitability urban squares not only focus on humanized design to meet the comfort of the older adults' sense and green travel needs, but should also emphasize local characteristics to urban squares with warmth and a sense of psychological belonging (Figure 6). The green environment not only affects the behavior of the older adults in terms of recreation and activities, but also isolates the street noise and other adverse effects from the outside world, and separates the space. Firstly, the creation of multi-layered landscapes enhances psychological healing. In the urban square open space planting lush plants, green planting in layers, using flowers, shrubs, and trees multi-level, multi-seasonal vegetation mixed planting. This approach optimizes the design of the spatial form and the sensory experience of older adults in order to enhance the space utilization and greening rate of the urban square [42]. Secondly, the urban square should show the environmental atmosphere of leisure and entertainment, and the various environmental elements of the urban square, such as green plants, structures, water features, and spatial colors should be coordinated with each other to create a lively and relaxing atmosphere. And creating a strong urban square culture is conducive to the shaping of local characteristics of the urban square, attracting older adults for leisure and recreational activities [65]. Finally, renewable landscaping is utilized to reduce waste of resources by paving the urban square with permeable gravel and lawns. At the same time, the use of participatory design methods for older adults and the addition of agricultural landscapes to the public space increases the interest of the older adults and enhances the attractiveness of the space while meeting ecological sustainability.

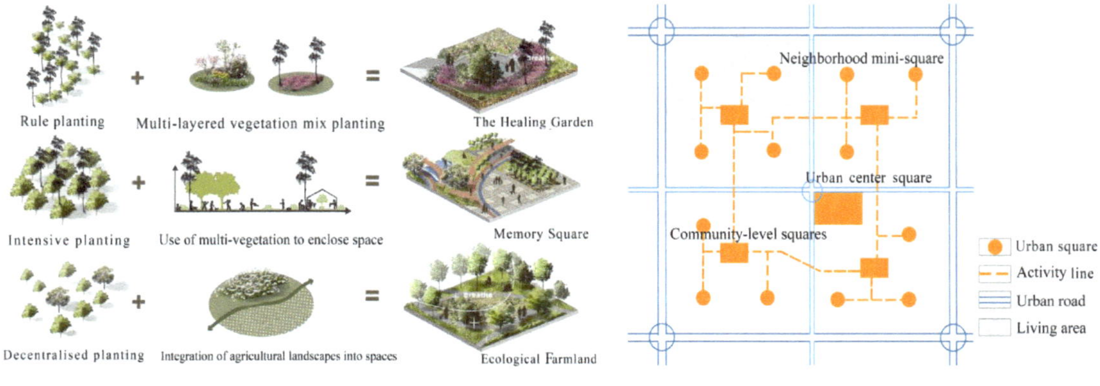

Figure 6. (**a**) Featured urban square creation; (**b**) construction of urban square space system.

5.2.3. Building a Networked Healthy and Synergistic Urban Square Spatial System

Building a healthy and synergistic network of urban square spaces is an important way for the health of urban squares to develop and further contribute to the health of society [56]. From the local to the whole, as the service radius changes, mutual assistance and sharing between different levels of urban squares facilitates the social interaction of older adults (Figure 6). The first level is the city center squares with the largest service radius, with a typical central urban square with regional characteristics as the core. The second level is the establishment of a system of community-level squares covering the entire region, with the residential community as the basic unit. The third level is a system of neighborhood mini-squares serving neighborhoods or blocks in the construction of the urban square space systems, the use of "point–line–surface" combination mode, and converting the urban square space system into a unified and complete system to better play its ecological landscape and social benefits. At the same time, a mutual assistance mechanism has been established to improve the monitoring system and management methods. The use of urban squares at different levels as firefighting grounds and alternate evacuation sites has been further promoted. Finally, it should also deeply excavate the city's unique history and culture, living customs, local customs, etc., integrating them into the urban green space system, so as to make it become a square space system integrating ecological environment, cultural characteristics, leisure and entertainment.

Author Contributions: Conceptualization, H.T. and Y.L.; methodology, H.T. and Y.L.; software, H.T. and Y.L.; validation, H.T. and Y.L.; formal analysis, Y.L.; investigation, Y.L.; resources, H.T.; data curation, H.T. and Y.L.; writing—original draft preparation, Y.L.; writing—review and editing, H.T. and Y.L.; visualization, H.T. and Y.L.; supervision, H.T.; project administration, H.T.; funding acquisition, H.T. All authors have read and agreed to the published version of the manuscript.

Funding: This research was funded by the National Student Innovation and Entrepreneurship Training Program Project (in China), grant number S202311527003; Hunan Provincial Philosophy and Social Science Fund Project (in China), grant number 20YBQ024.

Data Availability Statement: Data is contained within the article.

Acknowledgments: The authors would like to thank for teachers and classmates on the team, as well as parents supporting this research.

Conflicts of Interest: The authors declare no conflicts of interest.

References

1. National Bureau of Statistics of the People's Republic of China Issues the Statistical Bulletin of the People's Republic of China on National Economic and Social Development. 2022. Available online: https://www.stats.gov.cn/sj/zxfb/202302/t20230228_1919011.html (accessed on 2 November 2023).
2. Commission on Health, Ministry of Education, Ministry of Science and Technology, Ministry of Industry and Information Technology, Ministry of Finance, Ministry of Human Resources and Social Security, Ministry of Housing and Urban-Rural Development, Ministry of Veterans' Affairs, General Administration of Market Supervision, General Administration of Radio, Film and Television, General Administration of Physical Education and Sports, Medical Insurance Bureau of the Banking and Insurance Regulatory Commission, Traditional Chinese Medicine Bureau, China Disabled Persons' Federation on the issuance of the "14th Five-Year Plan" for Healthy Aging. Available online: https://www.gov.cn/gongbao/content/2022/content_5692863.htm (accessed on 2 November 2023).
3. Wang, J. Research on the Development Status of City Squares. *Farmers Consult.* **2020**, *3*, 210.
4. Harada, K.; Lee, S.; Lee, S.; Bae, S.; Harada, K.; Shimada, H. Environmental predictors of objectively measured out-of-home time among older adults with cognitive decline. *Arch. Gerontol. Geriatr.* **2019**, *82*, 259–265. [CrossRef] [PubMed]
5. Hosseini, M.M.; Beheshti, S.; Heydari, J.; Zangiabadi, M. Elderly care facility location in the face of the climate crisis: A case study in Canada. *Int. J. Disaster Risk Reduct.* **2024**, *108*, 104516. [CrossRef]
6. Yıldız, B.; Çağdaş, G. Fuzzy logic in agent-based modeling of user movement in urban space: Definition and application to a case study of a square. *Build. Environ.* **2020**, *169*, 106597. [CrossRef]
7. Najafi, P.; Mohammadi, M. Redefining Age-Friendly Neighbourhoods: Translating the Promises of Blue Zones for Contemporary Urban Environments. *Int. J. Environ. Res. Public Health* **2024**, *21*, 365. [CrossRef] [PubMed]
8. Todd, M.; Adams, M.A.; Kurka, J.; Conway, T.L.; Cain, K.L.; Buman, M.P.; Frank, L.D.; Sallis, J.F.; King, A.C. GIS-measured walkability, transit, and recreation environments in relation to older Adults' physical activity: A latent profile analysis. *Prev. Med.* **2016**, *93*, 57–63. [CrossRef] [PubMed]
9. Yung, E.H.K.; Wang, S.; Chau, C. Thermal perceptions of the elderly, use patterns and satisfaction with open space. *Landsc. Urban Plan.* **2019**, *185*, 44–60. [CrossRef]
10. Kim, J.W.; Kim, W.P. A Study on the Barrier-free Space through IPA Method for the Elderly in Multi-family Housing. *J. Korea Converg. Soc.* **2020**, *11*, 187–194. [CrossRef]
11. Ding, J.J. Research on Landscape Design Strategies for Elderly Suitability within Urban Public Space. *Beauty Times* **2022**, *02*, 6264.
12. Zhang, R.; Huang, L.; Wang, H. Accessibility Improvement and Renewal of Urban Park Green Space for the Elderly and the Disabled. *Forests* **2023**, *14*, 1801. [CrossRef]
13. Zhou, Y.M.; Wang, C.Y. Research on Exterior Environmental Design with Good Social Atmosphere of Elderly-friendly Communities: A Case Study of a Community in Beijing. *Shanghai Urban Plan. Rev.* **2020**, *06*, 15–21.
14. Zhao, Y.Q. The Discussion on the Humanization Design of Space and Activities of Urban Square Based on the Physical and Psychological Characteristics of the Older Adults. *Fujian Archit. Constr.* **2011**, *12*, 25–27.
15. Qin, W.K.; Wang, H. Evaluation Index System for Adaptation of City Greening for Elderly People Based on SEM. *J. Civ. Eng. Manag.* **2020**, *37*, 8. [CrossRef]
16. Huang, T.; Wei, M.; Zhi, Y.T.; Xi, J.C. Reconstruction of Aging Adaptability of Walking Space Network in Urban Leisure Green Space Based on Behavior Simulation: A Case Study of Hu Qiu Wetland Park in Suzhou. *Geogr. Geo-Inf. Sci.* **2022**, *38*, 79–88.
17. Zhao, Y.; Zhang, J.L. Design and Investigation for Age-friendliness of Outdoor Environment in Elderly-service Homes of Chongqing. *For. Inventory Plan.* **2018**, *43*, 205–212.
18. Wang, D.; LI, H. Research on the Design of Urban Square Reconstruction Based on Health Promotion Theory-Take the 1895 Square in Tang Zha Ancient Town as an Example. *Ind. Des.* **2023**, *09*, 43–47.
19. Liu, C.C.; Li, Y.H. Development of Public Spaces in Western Cities Based on Healthy Lifestyles. *South Archit.* **2021**, *05000*, 22–29.
20. Kahraman, M.D.; Türkoğlu, H. Evolution of Urban Squares and Transformation of Publicness. *A Z ITU J. Fac. Archit.* **2022**, *19*, 69–87. [CrossRef]
21. Ebenezer, H. *Tomorrow: The Path to Peace with Real Reform*; China Construction Industry Press: Beijing, China, 2020.
22. Wu, C.; Li, D. *Principles of Urban Planning*, 4th ed.; China Construction Industry Press: Beijing, China, 2010.
23. Szczepańska, B. Restoration of the Market Square in Opole as the Creation of a New Vision of the City's History. *Architects* **2021**, *67*, 35–44. [CrossRef]
24. Cliff, M. *Streets and Squares*, 2nd ed.; China Construction Industry Press: Beijing, China, 2004.
25. Jan, G. *Interaction and Space*; China Construction Industry Press: Beijing, China, 2002.
26. Lévy, B. Urban Square as the Place of History, Memory, Identity. *Mem. City Kult. Belgrade* **2012**, *156*, 173. Available online: https://www.academia.edu/1489197 (accessed on 23 January 2024).
27. World Health Organization. Constitution of WHO: WHO Remains Firmly Committed to the Principles Set Out in the Preamble to the Constitution. Available online: https://www.who.int/about/governance/constitution (accessed on 2 November 2023).
28. Soflaei, F.; Sanzighi, S.M. Public Squares, Social Interactions, and Urban Sustainability: Lessons Learned from Middle Easten Maidans. In Proceedings of the EAAE-ARCC International Conference & 2nd VIBR Arch: The Architect and the City, Universitat Politècnica de València, Valencia, Spain, 28 February 2021.

29. Dorigo, A.T.; Ferreira, L. Contributions of Gors' Environmental Perception About Squares and Parks in Brazil (2009–2013): Literature Review. *Rev. Gestão Ambient. E Sustent.* **2015**, *4*, 31–45. [CrossRef]
30. Zhang, C.L. Research on the Planning and Design of Urban Squares. *Jiangxi Build. Mater.* **2014**, *5*, 26.
31. Wang, J.G. *Urban Design*; China Construction Industry Press: Beijing, China, 2008.
32. Wan, Z. Creating a New Image of Urban Space—Diversified Design of Modern City Square. *New Archit.* **2000**, *6*, 20–23.
33. Zhang, L.; Wen, T.; Luo, Y.Y. A Study on the Evaluation of the Utilization Status of Urban Squares (POE)—The Case of Yiyang City Jia Ning Na Square. *Gansu Agric.* **2014**, *20*, 89–92. [CrossRef]
34. Gui, C.Z.; Song, S.; Bao, H.; Peng, Y.N.; Zhou, Z.Y.; Xiao, M. Using Space Syntax to Understand Spatial Vitality of City Square: A Case Study of Wuyi Square in Changsha. *Chin. Overseas Archit.* **2023**, *7*, 59–64. [CrossRef]
35. Jiang, Y.P.; Zhen, F.; Sun, H.H.; Wang, W.W. Research on the Influence of Urban Built Environment on Daily Walking of Older Adults from a Perspective of Health. *Geogr. Res.* **2020**, *39*, 570–584.
36. Massaro, A.; Birardi, G.; Manca, F.; Marin, C.; Birardi, V.; Giannone, D.; Galiano, A.M. Innovative DSS for intelligent monitoring and urban square design approaches: A case of study. *Sustain. Cities Soc.* **2021**, *65*, 102653. [CrossRef]
37. Han, Y.D.; Chen, L.; Liu, Y.X. Leisure and Successful Ageing of the Urban Elderly: Findings from X Community in Beijing and Implications for Geriatric Social Work in China. *Soc. Work Manag.* **2017**, *17*, 12–20.
38. Wei, Y.; Yang, D.F. Evaluation of Elderly Suitability of Built Environment in Neighborhood Destinations Based on Streetscape pictures—A Case Study of Dalian city. *Archit. J.* **2022**, *S1*, 24–30.
39. Li, Y.Q. *Evaluation of Elderly Suitability and Optimization of Renovation of Urban Leisure Squares in Shenyang*; Shenyang Architecture University: Shenyang, China, 2019. [CrossRef]
40. Acar, H.; Yavuz, A.; Eroglu, E.; Acar, C. Analysis of Activity, Space and User Relations in Urban Squares. *Indoor Built Environ.* **2020**, *30*, 1466–1485. [CrossRef]
41. Leng, H.; Yan, T.J.; Yuan, Q. Research Progress on Mental Health Effect of Blue-Green Space and Its Enlightenments. *Urban Plan. Int.* **2022**, *37*, 34–43+52. [CrossRef]
42. Yue, Y.F.; Yang, D.F.; Xu, D. Exploration and Examination of the Influence Mechanism of Urban Blue-Green Space on the Elderly's Mental Health. *Landsc. Archit.* **2022**, *29*, 71–77. [CrossRef]
43. Lee, H.J.; Lee, D.K. Do Sociodemographic Factors and Urban Green Space Affect Mental Health Outcomes Among the Urban Elderly Population? *Int. J. Environ. Res. Public Health* **2019**, *16*, 789. [CrossRef] [PubMed]
44. Wu, J.; Li, X.; Gao, B.; Su, X. The Effects of Serious Leisure Involvement on Subjective Well-being and Successful Aging: A Case Study of Young-old Participants in Chinese Seniors' Universities. *Sage Open* **2020**, *10*, 2158244020914559. [CrossRef]
45. Fu, X.Y.; Wang, Z.Z. Reflections on Cultural Theme Positioning and Landscape Expression in Urban Square Renovation—Taking Baoding People's Square Upgrading Design as an Example. *Intell. City* **2019**, *5*, 112–113. [CrossRef]
46. Zhou, L.; Zhang, S.F.; Liang, W.F.; Wu, W.T. Study on Social Benefits Evaluation and Optimization of Urban Square Landscape based on IPA Model: A Case Study of Hangzhou. *Archit. Cult.* **2021**, *5*, 59–62. [CrossRef]
47. Yu, S.H.; Su, Z.D. The Problems of City Squares in Planning and Construction. *Theor. Stud. Urban Constr. Electron. Version* **2013**, *00024*, 1–4.
48. Feng, Y.; Xu, T.; Gan, W.; Bai, G. Optimization Strategy and Aging Evaluation of Outdoor Sports Space in Old Community: Taking Seven Communities in Wuhan as an Example. *Huazhong Archit.* **2022**, *40*, 13–20. [CrossRef]
49. Gao, C.H.; Zhang, J.L.; Yang, H.F. Research on Elderly Suitability Design of Yue Ya Lake Park. *Contemp. Hortic.* **2021**, *44*, 163–165. [CrossRef]
50. Zhu, H.H.; Li, J. Evaluation of Rural Environmental Suitability Based on SVR Model—Taking Tian mu Village in Lin a District as an Example. *J. Southwest China Norm. Univ. (Nat. Sci. Ed.)* **2022**, *47*, 72–81. [CrossRef]
51. Yu, W.Y.; Hu, H. Research on the Evaluation Index System of Urban Leisure Greenland Adaptability for the Elderly. *For. Resour. Manag.* **2018**, *4*, 69–75. [CrossRef]
52. Luo, L.J.; Ding, C.Y.; Wang, J.Y.; Lin, F.J.; Yan, Y.Q. Evaluation of Adaptation of Urban Park Green Space for the Elderly on Structural Equation Model. *Science Technol. Eng.* **2022**, *22*, 10230–10237. [CrossRef]
53. Wang, F.Z.; Liu, H.; Luo, J.Y. Research on Activity Space Design for the Elderly in Residential Landscape. *Art Des.* **2017**, *2*, 69–71. [CrossRef]
54. Huang, Z.X.; Cao, Z.Y.; Xiong, F.; Zhang, L.S. Study on the Assessment of Urban Existing Community Public Space Aging—Taking the Yang jia Chang Community of Nanchang as an Example. *Jiangxi Sci.* **2022**, *40*, 1136–1141+1179. [CrossRef]
55. Li, G.H.; Wang, P.; Chen, T. Research on the Evaluation and Strategy of the Environmental Suitability for Aging in Public Space of Open Settlements—The Case of Guangzhou Swan Flower Garden. *Agric. Technol.* **2024**, *44*, 115–121. [CrossRef]
56. Kong, J.T.; Yang, N.; Xu, F.; Xing, Y.X. Research on Evaluation and Optimization of Community Public Space under the Multidimensional Health Orientation: A Case Study based on Jialing Road, Nankai District, Tianjin. *South Archit.* **2023**, *3*, 12–21.
57. Jiang, Z.P. "Barriers" to Accessibility. *East. Sword* **2018**, *10*, 1.
58. Jiang, Y.H. *A Study on the Impact and Trend of Leisure Sports on the Fitness of Older Adults in Changchun City*; Jilin Sports Institute: Changchun, China, 2024.
59. Yang, X.; Xing, P.F. Study on the Spatial Distribution of Park Green Space Use in the Main Urban Area of Chengdu City—An Empirical and Visualization Analysis Based on Big Data. *Urban. Archit.* **2024**, *21*, 196–199. [CrossRef]

60. Wang, Z.B.; Liu, D.; Li, H.Q.; Song, J.R. Analysis on the Accessibility and Fairness of Public Service Facilities from the Perspective of the Elderly. *J. Hunan City Univ. Nat. Sci.* **2023**, *32*, 27–32.
61. Pan, C.L. Optimization of Public Transport Organization of Station Square: Lianyungang Station As an Example. *China Transp. Rev.* **2023**, *45*, 180–187.
62. Bessmolnaia, M.I.; Imescenova, E.G.; Tatarnikova, V.L.; Kisova, S.V. Evaluation of green area in the squares of Ulan-Ude city. *Bull. NSAU (Novosib. State Agrar. Univ.)* **2018**, *04*, 7–17. [CrossRef]
63. Zhong, Y.Q.; Luo, Z.W. Optimization Strategy of Public Space in Old Community Based on Health Promotion of the Elderly. *Chin. Landsc. Archit.* **2021**, *37*, 56–61. [CrossRef]
64. Cassarino, M.; Bantry-White, E.; Setti, A. Cognitive and sensory dimensions of older people's preferences of outdoor spaces for walking: A survey study in Ireland. *Int. J. Environ. Res. Public Health* **2019**, *16*, 1340. [CrossRef] [PubMed]
65. Lin, L. Environmental Art Design of Feng Quan Square City Square. *Archit. J.* **2021**, *6*, 117.

buildings

MDPI

Article

Exploring the Influence of the Perceived Neighborhood Built Environment on the Fall Risks among Older Adults in China

Shichao Du [1], Chunyu Wang [2,3,*] and Longjian Mao [2]

[1] Department of sociology, Fudan University, Shanghai 200433, China; sdu@fudan.edu.cn
[2] School of Architecture and Urban Planning, Chongqing University, Chongqing 400044, China
[3] Key Laboratory of New Technology for Construction of Cities in Mountain Area, Chongqing University, Chongqing 400044, China
* Correspondence: chunyu@cqu.edu.cn

Abstract: Falls are the leading cause of accidental injury-related deaths among older adults, with approximately 50% of them occurring in the neighborhood built environment. This longitudinal study investigated the influence of the perceived built environment (PBE) on fall risks among Chinese older adults residing in neighborhoods. We utilized data from the 2018 and 2020 waves of the China Longitudinal Aging Social Survey (CLASS), comprising a sample size of 8686 respondents. A complementary log–log (cloglog) regression was used to effectively model falls because of their infrequent occurrence. The results revealed a significant U-shaped non-linear relationship between PBE and falls. As the PBE score increased from relatively low levels, there was a decrease in the probability of falls, indicating that enhancing PBE can effectively protect against fall risks. However, once an average PBE score threshold was reached (around a turning point score of 22), this association may slightly reverse. Living alone and living in urban areas are two major factors that increase the vulnerability of older adults to PBE, resulting in higher fall risks within their neighborhoods. The study enhances the understanding of how PBE affects fall risks among older adults.

Keywords: neighborhood; perceived built environment; fall; fall prevention; older adults

Citation: Du, S.; Wang, C.; Mao, L. Exploring the Influence of the Perceived Neighborhood Built Environment on the Fall Risks among Older Adults in China. *Buildings* **2024**, *14*, 2631. https://doi.org/10.3390/buildings14092631

Academic Editor: Bo Xia

Received: 24 July 2024
Revised: 20 August 2024
Accepted: 22 August 2024
Published: 25 August 2024

1. Introduction

Falls are the leading cause of accidental injury-related deaths among older adults, with approximately 30% of individuals aged 65 and above experiencing falls annually worldwide [1]. This has become a significant global public health challenge. It is worth noting that China bears the greatest burden of fall-related injuries in the world [2–4]. Falls can lead to a series of physiological and psychological problems, which can decrease the quality of life in later years and increase personal, familial, and societal burdens [5]. Consequently, it is crucial to implement corresponding strategies for preventing falls among older adults.

The occurrence of falls among older adults is attributed to multiple factors, primarily classified as individual-level and environmental-level factors [6]. Extensive research has focused on the individual-level risk factors associated with falls among older adults, such as socioeconomic status, health status, and lifestyle. However, there is an increasing emphasis on the contribution of environmental factors to fall risks, particularly within the built environment [7]. Scholars worldwide have conducted relevant studies demonstrating the impact of home environments on falls among older adults [8–10]. Specifically, it has been found that neighborhood-dwelling older adults are predominantly exposed to both their home environment (indoor) and the neighborhood built environment (outdoor), with approximately a 50% likelihood of experiencing falls in the latter setting [11]. Nevertheless, compared with studies conducted within home environments, less attention has been given to investigating the role of the neighborhood built environment in relation to fall risks.

Meanwhile, the majority of current studies on the factors of fall risk in the neighborhood built environment primarily have relied on measuring the objective built environment (OBE). The OBE primarily refers to the physical attributes of the built environment. To evaluate it, objective methodologies such as using geographic information system (GIS) technology or conducting on-site observations by auditors are required [12]. However, some existing literature suggests that these objective measurements may not align with older adults' subjective perceptions of their surrounding built environment [13,14]. This indicates that a model solely based on researchers' objective audits may be insufficient. The concept of the perceived built environment (PBE) offers a new perspective. PBE mainly refers to an individual's overall attitudes, behaviors, and satisfaction regarding their access to and utilization of the built environment [15]. We need to reveal whether and to what extent the quality of PBE can influence the fall risks of older adults. This would facilitate a closer alignment with older adults' subjective experiences.

Most residential areas in Chinese urban areas are gated neighborhoods, which provide an interesting context for study due to their high density, fenced boundaries, and unified property management. This distinguishes them from previous studies conducted in low-density residential areas in Europe and America. Meanwhile, the residential areas in rural China, despite being relatively dispersed in planning, still adhere to the principle of centralized community management by organizations such as village committees. The top-down environmental modification approach used in both urban and rural neighborhoods in China is more feasible and efficient, contributing to prevention of falls among older adults. Furthermore, it is essential to consider various factors, such as living arrangements and gender, when examining this issue. For instance, the existing literature indicates that older adults living alone are particularly vulnerable to the risks associated with the neighborhood built environment [16,17].

To fill the knowledge gaps mentioned above, the study focused on Chinese older adults residing in neighborhoods and revealed whether and to what extent PBE influences the fall risk among them. Furthermore, we explore the heterogeneity of this influence by examining the factors that may reinforce it. We utilized data from the 2018 and 2020 waves of China Longitudinal Aging Social Survey (CLASS). This study contributes to a deeper understanding of the role played by PBE in the fall risks among older adults and provides recommendations for environmental intervention measures targeting the prevention of falls.

2. Literature Review and Research Framework

2.1. Neighborhood Built Environment and Fall Risks

There has been an increasing number of studies addressing the impact of the neighborhood built environment on falls. Studies have identified various factors within the neighborhood built environment that contribute to increased fall risks, including the quality of pavements, the design of facilities, the transportation system, and the combined effects of natural environmental factors (Table 1). Among these factors, the pavement's quality stands out, and extensive evidence has suggested that the characteristics of the pavement, such as unevenness, slipperiness, the presence of obstacles, and the material's composition directly influence the walking patterns of older adults. This consensus has been supported by multiple studies [5,8,18]. Inadequate facilities can also contribute to falls among older adults, frequently occurring in seating areas, stairs, utility poles, and other locations. For instance, stairs with uneven heights tend to disrupt the stability of older adults' gait, causing them to lose balance and subsequently fall [19,20]. Furthermore, inadequate transportation systems contribute to falls due to factors such as high traffic speeds, the volume of vehicles, pedestrian density, etc. For example, congested pedestrian traffic not only increases the risk of older adults being knocked down but also compromises the stability of their gait, leading to an increased likelihood of falling [21].

Table 1. Key findings regarding the impact of factors of the neighborhood built environment on falls.

Major Factors	Key Findings	References
Quality of pavements	Unevenness, slipperiness, the presence of obstacles, and the material's composition directly influence the walking patterns of older adults	[5,8,18]
Design of facilities	Inadequate facilities such as seating areas, stairs, utility poles, and other locations contribute to falls	[19,20]
Transportation system	High traffic speeds, the volume of vehicles, pedestrian density, etc. contribute to falls	[21]
Natural environment	A superposition effect on risks related to the neighborhood built environment	[22–25]

In addition, the natural environment exhibits a superposition effect on the risks related to the neighborhood built environment. For instance, weather conditions that increase the probability of falls among older adults in rainy and snowy settings have been documented [22–24]. However, only a limited number of studies have quantitatively analyzed this phenomenon [25].

Other studies have explored potential variations in the fall risks among older adults based on gender, age, physical condition, and educational level within the context of the same neighborhood built environment. Moreover, examining how different living arrangements modulate the impact of the neighborhood built environment on fall risks among older adults represents an emerging research area. The literature suggests that older adults who live alone exhibit greater vulnerability to such hazards [26].

2.2. Perceived Built Environment

Lawton et al. (1982) proposed the person–environment (P-E) fit framework, which states that there needs to be a fit between persons (P) and their environment (E) (Figure 1) [27,28]. This implies that the environment encompasses not only OBE but also PBE, with the impact of objective environmental factors on older adults often being mediated by their subjective perceptions [29]. PBE is an essential yet frequently overlooked measure in studies investigating environmental behaviors.

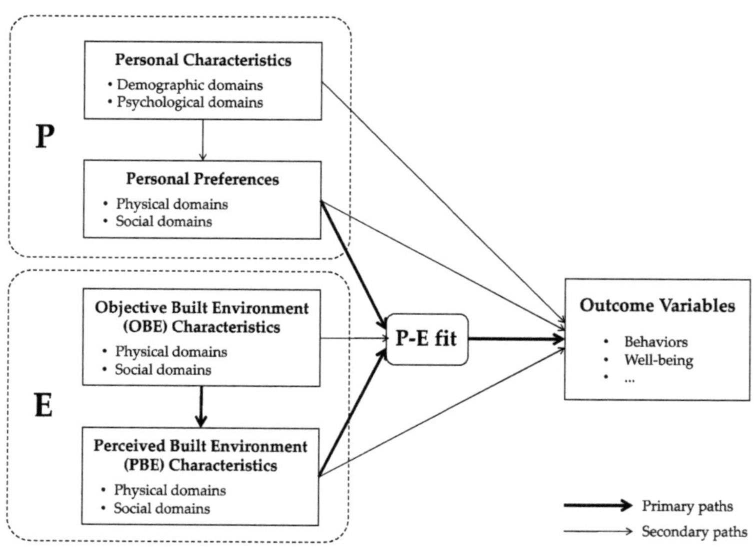

Figure 1. The objective built environment (OBE) and perceived built environment (PBE) in the person–environment fit model, adapted from Kahana et al. (2003) and Veeroja et al. (2024) [28,30].

The significance of PBE is further reflected by the potential inconsistency between OBE and individuals' subjective perceptions [13], leading to a disparity in measuring the built environment's actual impact. This discrepancy is particularly pronounced among older adults. For instance, when researchers conduct an audit of the brightness of lighting in neighborhoods at night, what they think to be sufficient lighting may still be insufficient for older adults with visual impairments. In contrast, older adults' ratings of the neighborhood lighting's quality obtained through questionnaires and interviews offer a more precise evaluation of the PBE's quality and align closely with the genuine impact of built environment on older adults. De Souza Moreira et al. (2020) discovered that the subjective perception of dimmer night lighting among older adults was associated with an increased frequency of falls. However, this study's limitations included its reliance on cross-sectional data and an analysis solely focused on falls in older adults from the previous year, which might have hindered an accurate assessment of the long-term cumulative impact of the built environment on fall risks in this population [31].

Additionally, the PBE is more closely associated with psychological effects on older adults compared with the OBE [32,33], which often manifest negatively [34,35]. For instance, a lower quality of the PBE leads to a fear of falling (FOF) among older adults, who express concerns about potential falls due to environmental factors [18,19]. Some studies have further explored the impact of the PBE on incidents of falls in older adults from a FOF perspective. For example, Lee et al. (2018) demonstrated that specific aspects of the PBE, such as adequate night lighting, enhance older adults' perceptions of their surroundings and reduce the likelihood of experiencing FOF [36].

2.3. Research Framework

We proposed a research framework to investigate the association between the neighborhood PBE and fall risks, as illustrated in Figure 2. Prior to data analysis, we explored how to evaluate PBE at the neighborhood scale. In the existing literature, the factors of PBE primarily encompass accessibility to destinations, the absence of sidewalks, high-speed traffic, heavy traffic, neighborhood safety, land use mix, infrastructure, aesthetics, etc. [30,37–39]. These factors constitute the fundamental framework of the PBE. Upon these, researchers have developed assessment tools such as the Neighborhood Environment Walkability Scale (NEWS) and the Senior Walking Environmental Assessment Tool (SWEAT). These tools use questionnaires, interviews, and other methods to evaluate subjective perceptions of the built environment within the neighborhood [14,19,29].

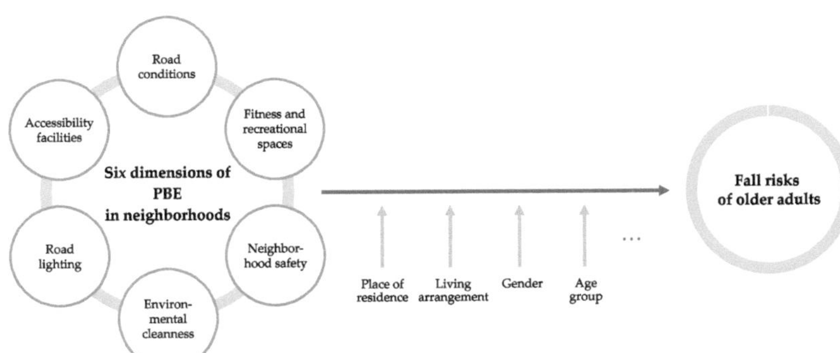

Figure 2. Research framework: the influence of the neighborhood PBE on fall risk.

Considering that some of the factors of the PBE at the macro- and meso-scales appear to have limited relevance to events of falling among older adults, we conducted further refinement and screening of the PBE framework from a micro-perspective. In terms of infrastructural factors, the existing literature indicated that infrastructure such as streetlights and activity spaces significantly contribute to older adults' walkability [3,40]. Therefore, we incorporated this measure into our framework. Meanwhile, the NEWS does not specifically evaluate accessibility facilities. China is a developing country with limited awareness of the construction of accessibility facility, inadequate supervision over accessibility facilities, and plenty of neighborhoods neglecting their accessible design. These problems pose risks to the walking activities of older adults with functional impairments [41]. Consequently, in addition to infrastructural factors, we also took into account the availability of accessibility facilities.

Additionally, we redefined the aesthetic indicators of the PBE as environmental cleanness, which more closely refers to pathology. Environments with poor cleanness are more likely to contribute to physical diseases and can also lead to depression [42,43], both of which may increase the fall risk. Moreover, plenty of old neighborhoods in China have problems with cleanness [44]. Therefore, it was imperative to assess the factor of environmental cleanness.

Finally, the evaluation of the quality of the PBE lay across six dimensions: road conditions, fitness and recreational spaces, neighborhood safety, environmental cleanness, road lighting, and accessibility facilities. The questionnaire settings of the CLASS comprehensively covered all six variables identified. Detailed information regarding the questionnaire's setup is provided in Section 3.

Data analysis was conducted on the basis of the framework of the PBE. Primarily, we hypothesized that there might be an association between the PBE and fall risks. Furthermore, we hypothesized that heterogeneity in terms of place of residence, living arrangement, gender, age group, etc., could strengthen this influence. The detailed procedures are described in Section 3.

3. Methods

3.1. Data

We utilized two waves of data (2018 and 2020) from the China Longitudinal Aging Social Survey (CLASS), a comprehensive nationwide social survey project that collects socioeconomic information on the aging population in China and their respective neighborhoods. The CLASS targets Chinese citizens aged over 60 years and has been officially implemented since 2014. The baseline wave encompassed 11,511 older adult respondents residing in 462 neighborhoods across 28 provinces or municipalities in China. Starting from 2018, the CLASS began gathering more detailed information on the neighborhood built environments, including the subjective perceptions of older adult respondents. Given our aims to examine the causal relationship between the perceived built environment and fall risks among older adults, we used panel information from CLASS, restricting our analysis to individuals who participated in both the 2018 and 2020 waves. We also excluded those who relocated residences between these observations. As the outcome variable for this study was the fall risk, individuals without walking abilities were excluded from the analysis. These data restrictions resulted in a final sample size of 8686 older adult respondents.

3.2. Measurement of Variables

3.2.1. Falling History

The respondents were asked to report any histories of falls within the past 12 months during both the 2018 and 2020 surveys. The response options included "none", "fell once", and "fell at least twice". To predict fall risks in 2020 while accounting for potential reversed causality, we dichotomized this measure into two categories: individuals with history of falling and those without. We utilized data from both waves to incorporate the lagged dependent variable approach, ensuring control over potential confounding factors. Further details regarding our analytical strategy are provided below.

3.2.2. PBE Score

The predicting variable, PBE, was measured by older adults' self-reported satisfaction with six dimensions of their neighborhood built environment. These six dimensions include road conditions, fitness and recreational spaces, neighborhood safety, environmental cleanness, road lighting, and accessibility facilities available in the neighborhood. The respondents were asked to rate their level of satisfaction with each item on a 5-point Likert scale ranging from "1" (very dissatisfied) to "5" (very satisfied). The alpha coefficient of the six dimensions was 0.83, showing a high consistency. Consequently, we added them up as a continuous variable representing the PBE. Additional details regarding the items are provided in Table 2. Given our hypothesis of a non-linear association between PBE and falls among older adults, we also included the squared term of PBE in our model's estimations.

Table 2. PBE scores assessed by respondents across six dimensions of the neighborhood built environment.

Dimensions	Items	Responses	Operationalization
Road conditions	What is the level of satisfaction regarding road conditions within your neighborhood?	Each item used a 5-point Likert-like scale: 1 = very dissatisfied 2 = relatively dissatisfied 3 = fair 4 = relatively satisfied 5 = very satisfied	Adding up the responses from the six dimensions to derive a composite score that reflected respondents' PBE of their neighborhood. The range of the PBE score was from 6 to 30.
Fitness and recreational spaces	What is the level of satisfaction regarding fitness and recreational spaces within your neighborhood?		
Neighborhood safety	What is the level of satisfaction regarding neighborhood safety in the environment within your neighborhood?		
Environmental cleanness	What is the level of satisfaction regarding environmental cleanness within your neighborhood?		
Road lighting	What is the level of satisfaction regarding road lighting within your neighborhood?		
Accessibility facilities	What is the level of satisfaction regarding accessibility facilities within your neighborhood?		

3.2.3. Covariates

The models were adjusted for individual-level covariates, including age, gender, marital status, retirement status, education, living arrangements (1 = living alone), and health status. Additionally, the respondents' indoor home environment was controlled by considering factors including brightness levels and the flatness of the flooring in their homes. Furthermore, we accounted for the necessity of using stairs; if the participants resided on the second floor or higher without access to an elevator, it was assumed that they would need to use stairs. These indoor environmental features have been identified as potential triggers for falls among older adults [45,46].

3.3. Statistical Analysis

Given that falls are rare events even among the aging population (in our sample, only around 7% of the respondents had experienced a fall in the past 12 months), we used complementary log–log (cloglog) regression to examine the relationship between the PBE and falls among older adults. The cloglog regression is more appropriate for rare events than regular logistic regression [47,48]. Specifically, we assumed that the hazard of falls takes the following form

$$Pr(y_i|x_i) = 1 - \exp\{-\exp(\beta x_i)\} \quad (1)$$

where y_i is the occurrence of a fall, x_i is a vector of the variables for individual i, and β is a vector of the parameters. The cloglog transformation of Equation (1) yielded

$$\log(-\log(Pr(y_i|x_i))) = \alpha_t + \beta x_i \quad (2)$$

where α_t is the cloglog transformation of the baseline hazard or the intercept. Since the relationship between PBE and falls can be reciprocal, we also controlled for the falling history in the previous survey year to fix the problem of reverse causality. This strategy is also known as lagged dependent variable cloglog regression.

4. Results

4.1. Descriptive Statistics

The descriptive statistics are presented in Table 3. The average PBE score was 22.21 on a scale of 6–30. The incidence of falls (occurring once or more than once in the past year) was 7.19% in the 2018 wave and increased to 7.71% in the 2020 wave. On average, respondents were aged 70.84 years (in 2018), with urban residents accounting for 50.07%. The low percentage of respondents living alone (11.73%) can be attributed to the survey's methodology, which involved visiting both spouses in older adult couples' households, resulting in a higher proportion of non-single older individuals being included. The respondents' health status was indicated by an average self-reported health score of 3.37 out of 5. Most respondents lived in well-equipped residential settings characterized by elevators or ground-floor locations, sufficient lighting, and flat flooring within their homes.

Table 3. Descriptive statistics.

	Mean/%	S.D.	Range
PBE score	22.21	3.61	[6, 30]
Incidence of falls (≥1) (wave 2018) %	7.19		
Incidence of falls (≥1) (wave 2020) %	7.71		
Age	70.84	6.25	[60, 98]
Gender (1 = male) %	50.34		
Place of residence (1 = urban) %	50.07		
Marital status (1 = married) %	72.86		
Retirement status (1 = retired) %	35.31		
Education			
Illiteracy %	25.13		
Primary school %	40.89		
Junior high school %	24.22		
Senior high school and above %	9.75		
Living arrangement (1 = living alone) %	11.73		
Health status (1 = poor, to 5 = excellent)	3.37	0.86	[1, 5]
Necessity of using stairs (1 = not necessary to use stairways in the home) %	70.93		
Brightness at home (1 = bright) %	87.75		
Flatness of the flooring at home (1 = flat) %	92.47		
N	8686		

Data source: CLASS2018 and 2020.

4.2. The Relationship between the PBE and Fall Risk among Older Adults

The results of the model's estimation are presented in Table 4. In Model 1, which included only the PBE score, there appeared to be no significant linear relationship between PBE and falls among older adults. However, this null association very likely obscured the true non-linear relationship between the two variables. In Model 2, where squared PBE scores were included, a significantly non-linear relationship between PBE and falls was revealed. To better illustrate the curvilinear relationship, Figure 3 plotted fall risks among older adults across different PBE scores. Fall risks decreased as PBE increased when it was relatively low. After reaching a turning point around 22, there seemed to be a slightly positive relationship between PBE and fall risks. However, the magnitude of this positive relationship after the turning point was smaller than that of the negative relationship before it.

Table 4. Results of the complementary log–log model: the relationship between PBE and fall risks among older adults.

	Model 1	**Model 2**
PBE score	−0.003	−0.212 *
	(0.012)	(0.085)
PBE score squared		0.005 *
		(0.002)
Falls in the previous survey	3.134 ***	3.137 ***
	(0.087)	(0.087)
Age	0.040 ***	0.041 ***
	(0.007)	(0.007)
Gender (1 = male)	0.007	0.005
	(0.087)	(0.087)
Place of residence (1 = urban)	0.213 *	0.215 *
	(0.106)	(0.107)
Marital status (1 = married)	0.027	0.024
	(0.105)	(0.105)
Retirement status (1 = retired)	0.285 *	0.301 **
	(0.113)	(0.114)
Education (rf. illiteracy)		
Primary school	−0.289 **	−0.281 **
	(0.103)	(0.104)
Junior high school	−0.025	−0.025
	(0.131)	(0.132)
Senior high school and above	−0.235	−0.233
	(0.173)	(0.172)
Living arrangement (1 = living alone)	−0.185	−0.196
	(0.139)	(0.139)
Health status (1 = poor, to 5 = excellent)	−0.189 ***	−0.189 ***
	(0.048)	(0.048)
Necessity of using stairs (1 = not necessary to use stairways in the home)	−0.051	−0.053
	(0.102)	(0.102)
Brightness in the home (1 = bright)	0.067	0.068
	(0.117)	(0.117)
Flatness of the flooring in the home (1 = flat)	−0.409 **	−0.405 **
	(0.140)	(0.140)
Intercept	−5.295 ***	−3.081 **
	(0.609)	(1.064)
Log-likelihood	−1535.11 ***	−1532.40 ***
N	8686	8686

Data source: CLASS 2018 and 2020. Note: standard errors are shown in parentheses; * $p < 0.05$, ** $p < 0.01$, *** $p < 0.001$.

Three batteries of robustness checks were conducted to verify the findings (Table 5). Firstly, we performed regular logistic regression modeling, and the results were consistent with those in our main analysis. Secondly, we also used Firth logit regression modeling, another popular estimation method for rare events, using a penalized maximum likelihood estimator to overcome potential bias. Compared with the cloglog approach, Firth logit regression is more suitable for small samples with rare events. The results obtained using this approach were also consistent with those in our main analysis. Thirdly, we used threshold regression to detect whether a significant turning point existed. On the basis of our hypothesis and findings, we set the number of thresholds as 1, and the results indicated that a significant threshold existed at a PBE score of 18.74. This score was very close to the estimated turning point of 22 in our main analysis. These three batteries of robustness tests collectively confirmed the U-shaped relationship between PBE and falls among older adults in China.

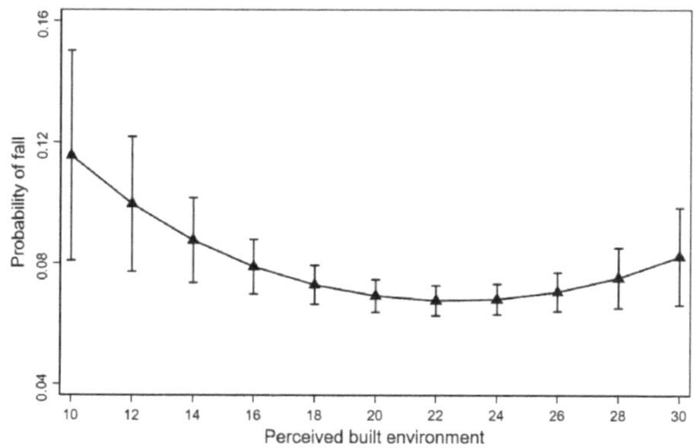

Figure 3. The predicted relationship between PBE and falls among older adults.

Table 5. Results of the robustness tests.

	Logit Regression	**Firth Logit Regression**	**Threshold Regression**
PBE score	−0.227 *	−0.236 *	
	(0.102)	(0.101)	
PBE score squared	0.005 *	0.005 *	
	(0.002)	(0.002)	
Threshold point			18.74
Controls	YES	YES	YES
N	8686	8686	8686

Data source: CLASS 2018 and 2020. Note: standard errors are shown in parentheses; * $p < 0.05$. The results of the threshold regression model are based on bootstrapping of 6950 regressions.

4.3. Heterogeneity

Four types of heterogeneity were tested in this study, as illustrated in Figure 4. Given the urban–rural dualism in China, we first examined the heterogeneity on the basis of the older adults' place of residence. As shown in the figure, there was a more pronounced curvilinear relationship between PBE and falls in urban areas compared with rural areas. For urban older adults, there existed a U-shaped relationship between their PBE and falls; whereas for rural older adults, their PBE was negatively associated with falls, indicating that increasing their PBE can significantly protect them against fall risks.

The relationship was also heterogeneous when assessed by the older adults' living arrangements. The impact of PBE on fall risks was more pronounced among older adults who live alone compared with their counterparts living with other family members. Increasing the PBE could significantly reduce the fall risks of older adults living alone, especially when their PBE was relatively low. However, for older adults not living alone, there was no significant association between their PBE and falls.

We did not detect any heterogeneity by gender or age. The curvilinear relationship between PBE and falls existed among both male and female older adults, although it was slightly more pronounced among the former. Similarly, this curvilinear relationship also existed among older adults aged 60–75 and those above 75, although the overall risk was higher among the latter group.

In summary, the findings indicated that older adults who live alone in urban neighborhoods are more vulnerable to PBE, particularly with regard to their fall risks.

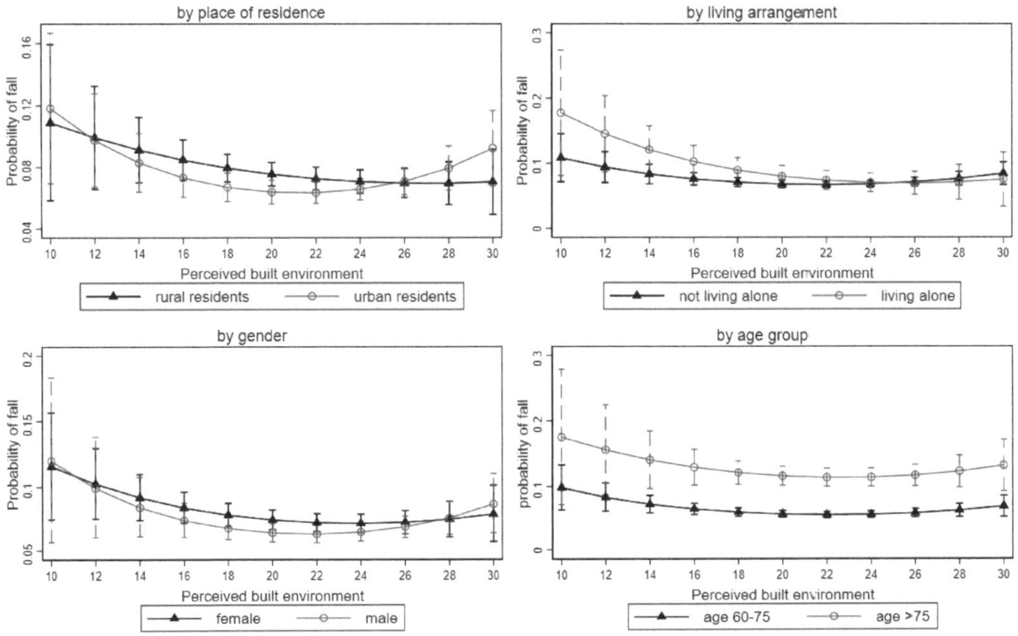

Figure 4. Heterogeneous effects of the perceived built environment on fall risk among older adults.

5. Discussion

This study represents a pioneering effort to investigate the impact and extent of the PBE on fall risks among older adults residing in Chinese neighborhoods. The findings demonstrated a significant non-linear relationship between the probability of falling and the PBE score; specifically, fall risks decreased as PBE improved when the PBE is relatively low. However, after reaching an average PBE score threshold, there appeared to be a slight positive association between PBE and fall risks. These results suggest that improving older adult residents' overall PBE can effectively reduce their fall risks when general PBE levels are not high. However, when older adult residents have excessively high levels of PBE, they are more likely to underestimate the potential risks of walking and exercising in their neighborhoods, thus increasing their probability of falls [10,18].

The results demonstrate a significant association between the neighborhood built environment and falls, which is consistent with previous studies [49,50]. However, by incorporating perception scales to assess respondents' neighborhood built environment, we have discovered that the relationship between PBE and fall risks is not simply linear but exhibits a distinct non-linear U-shaped pattern. This novel finding contributes significantly to the existing literature. The advantage of using PBE scales lies in their ability to uncover implicit information that combines both the physical and social factors of the built environment, which are challenging to obtain through systematic observations or geographic information systems [31]. Moreover, defining a neighborhood's boundaries objectively poses challenges due to variations in neighborhood activities among different older adults. Therefore, utilizing PBE scales provides a closer approximation of the true impact of the factors of the neighborhood built environmental on respondents [31,51].

In the heterogeneity analysis, there were two novel findings. Firstly, the impact of PBE on falls was more pronounced in urban areas compared with rural areas. This could be attributed to the high density in most urban neighborhoods in China, which leads older adults to walk more frequently rather than relying on transportation. Consequently, the built environment has a greater influence on the probability of falling. Conversely, rural areas in China have lower neighborhood density, resulting in a relatively smaller impact of the built environment on falls. The results from rural areas in China are similar to studies conducted in low-density cities in Western countries [18,19]. Secondly, improving the PBE can significantly reduce fall risks among older adults living alone, especially in neighborhoods with a low PBE. In contrast, there was no significant association between PBE and falls for older adults who do not live alone. This may be because family members serve as effective protection against falls for older adults. These conclusions are consistent with previous research utilizing OBE measurements [16,17,26].

The findings make several significant contributions to research, public policy, and clinical practice. Firstly, while current studies have primarily focused on the indoor home environment as a contributing external factor to fall risks among older adults, this study specifically highlights the impact of the neighborhood built environments. By utilizing nationwide longitudinal survey data, we provide evidence on the cumulative effects of factors of the built environmental on fall risks among older adults. These findings are crucial for informing future efforts to prevent falls targeting older adults, particularly through improvements in the neighborhood built environment. Secondly, this study introduced the innovative concept of perceived scales and revealed the non-linear impact of the PBE on fall risks among older adults. In line with the P-E fit theory, enhancing the OBE was identified as a prerequisite for improving the PBE. This established a theoretical foundation for future investigations on enhancing specific aspects of the OBE through design and renovation, thereby promoting the PBE for older adults and preventing falls. Efforts should be made to improve poor environments to reach an average level, while in excellent environments, educating older adults about fall prevention is important to prevent carelessness during physical activities. Thirdly, the heterogeneity analysis identified the higher fall risks among urban adults living alone. Greater policy attention, such as the provision of community-based care services and wearable fall detection devices, is imperative to mitigate falls among this vulnerable group.

Additionally, this study specifically focused on Chinese older adults living in neighborhoods with clearly defined boundaries. Among them, urban older adults predominantly reside in gated communities with community environments managed by developers or property management companies; whereas rural older adults live in villages under centralized management by village committees or other organizations. This centralized community governance model facilitates unified implementation of renewal of the built environment, thus rendering the policy implications of this study highly feasible.

Despite the contributions to the existing literature, the limitations of this study should be acknowledged. While the impact of the PBE on older adults was revealed, we were unable to measure the OBE in their neighborhoods due to confidentiality constraints regarding specific geographic information of the survey. Consequently, we were unable to further compare the differential effects of the fall risk between the OBE and PBE among older adults. This limited the discovery of how the OBE and PBE can predict each other. Future studies could address this issue by using alternative methodologies and datasets. For instance, conducting surveys within the same neighborhood to assess older adults' PBE, and integrating it with the researchers' observations to audit their OBE would be a valuable research direction.

6. Conclusions

This study innovatively introduced the concept of perceived scales and revealed the influence of the PBE on fall risks among Chinese older adults residing in neighborhoods. The results obtained from a cloglog regression model indicated a significant non-linear U-shaped relationship between the PBE and falls. Specifically, when the PBE is relatively low, improving it can effectively decrease the fall risk. However, once an average PBE score threshold is reached, this association may slightly reverse. Furthermore, the heterogeneity analysis identified that living alone and living in urban areas are two major factors that render older adults more vulnerable to the PBE, resulting in higher fall risks within their neighborhoods. These findings provide valuable evidence and insights for developing environmental intervention measures and policies aimed at preventing falls in China and in other countries with similar contexts. Future efforts should be made to improve poor environments to reduce falls, particularly through neighborhood renewal schemes. Meanwhile, educating older adults on fall prevention is important to prevent carelessness during physical activities. Policy attention should also be given to the higher fall risks among urban adults living alone, and should provide them with adequate community-based care services.

Author Contributions: Conceptualization, S.D. and C.W.; methodology, S.D. and C.W.; software, S.D.; validation, S.D. and C.W.; formal analysis, S.D. and C.W.; investigation, S.D. and C.W.; resources, S.D. and C.W.; data curation, S.D. and C.W.; writing—original draft preparation, S.D., L.M. and C.W.; writing—review and editing, S.D. and C.W.; visualization, S.D., L.M. and C.W.; supervision, C.W.; project administration, C.W.; funding acquisition, C.W. All authors have read and agreed to the published version of the manuscript.

Funding: This research study was supported by the National Natural Science Foundation of China (52308010), the Chongqing Municipal Construction Science and Technology Program (2023—No. 6-7), and the Fundamental Research Funds for the Central Universities (2024CDJXY014). The funding bodies did not influence this study in any way prior to circulation.

Data Availability Statement: The original data presented in the study are openly available in the China Longitudinal Aging Social Survey (CLASS) repository at http://class.ruc.edu.cn/ (accessed on 21 August 2024).

Acknowledgments: The authors acknowledge the support from the Institute of Gerontology and the National Survey Research Center at Renmin University of China for providing the CLASS data.

Conflicts of Interest: The authors declare no conflicts of interest.

References

1. Montero-Odasso, M.; Van Der Velde, N.; Martin, F.C.; Petrovic, M.; Tan, M.P.; Ryg, J.; Aguilar-Navarro, S.; Alexander, N.B.; Becker, C.; Blain, H.; et al. World Guidelines for Falls Prevention and Management for Older Adults: A Global Initiative. *Age Ageing* **2022**, *51*, afac205. [CrossRef]
2. Peng, K.; Tian, M.; Andersen, M.; Zhang, J.; Liu, Y.; Wang, Q.; Lindley, R.; Ivers, R. Incidence, Risk Factors and Economic Burden of Fall-Related Injuries in Older Chinese People: A Systematic Review. *Inj. Prev.* **2019**, *25*, 4–12. [CrossRef]
3. Wang, K.; Chen, M.; Zhang, X.; Zhang, L.; Chang, C.; Tian, Y.; Wang, X.; Li, Z.; Ji, Y. The Incidence of Falls and Related Factors among Chinese Elderly Community Residents in Six Provinces. *Int. J. Environ. Res. Public Health* **2022**, *19*, 14843. [CrossRef] [PubMed]
4. Zhou, M.; Wang, H.; Zeng, X.; Yin, P.; Zhu, J.; Chen, W.; Li, X.; Wang, L.; Wang, L.; Liu, Y.; et al. Mortality, Morbidity, and Risk Factors in China and Its Provinces, 1990–2017: A Systematic Analysis for the Global Burden of Disease Study 2017. *Lancet* **2019**, *394*, 1145–1158. [CrossRef] [PubMed]
5. Curl, A.; Thompson, C.W.; Aspinall, P.; Ormerod, M. Developing an Audit Checklist to Assess Outdoor Falls Risk. *Proc. Inst. Civ. Eng.-Urban Des. Plan.* **2016**, *169*, 138–153. [CrossRef]
6. Ganz, D.A.; Latham, N.K. Prevention of Falls in Community-Dwelling Older Adults. *N. Engl. J. Med.* **2020**, *382*, 734–743. [CrossRef] [PubMed]
7. Sousa, L.M.M.; Marques-Vieira, C.M.A.; de Caldevilla, M.N.G.N.; Henriques, C.M.A.D.; Severino, S.S.P.; Caldeira, S.M.A. Risk for Falls among Community-Dwelling Older People: Systematic Literature Review. *Rev. Gaúcha Enferm.* **2016**, *37*, e55030.
8. Li, W.; Keegan, T.H.; Sternfeld, B.; Sidney, S.; Quesenberry, C.P., Jr.; Kelsey, J.L. Outdoor Falls among Middle-Aged and Older Adults: A Neglected Public Health Problem. *Am. J. Public Health* **2006**, *96*, 1192–1200. [CrossRef]
9. Lyu, Y.; Forsyth, A.; Worthington, S. Built Environment and Self-Rated Health: Comparing Young, Middle-Aged, and Older People in Chengdu, China. *HERD Health Environ. Res. Des. J.* **2021**, *14*, 229–246. [CrossRef]
10. Curl, A.; Fitt, H.; Tomintz, M. Experiences of the Built Environment, Falls and Fear of Falling Outdoors among Older Adults: An Exploratory Study and Future Directions. *Int. J. Environ. Res. Public Health* **2020**, *17*, 1224. [CrossRef]
11. Kelsey, J.L.; Procter-Gray, E.; Hannan, M.T.; Li, W. Heterogeneity of Falls among Older Adults: Implications for Public Health Prevention. *Am. J. Public Health* **2012**, *102*, 2149–2156. [CrossRef] [PubMed]
12. Nordbø, E.C.A.; Nordh, H.; Raanaas, R.K.; Aamodt, G. Gis-Derived Measures of the Built Environment Determinants of Mental Health and Activity Participation in Childhood and Adolescence: A Systematic Review. *Landsc. Urban Plan.* **2018**, *177*, 19–37. [CrossRef]

13. Guo, Y.; Liu, Y.; Lu, S.; Chan, O.F.; Chui, C.H.K.; Lum, T.Y.S. Objective and Perceived Built Environment, Sense of Community, and Mental Wellbeing in Older Adults in Hong Kong: A Multilevel Structural Equation Study. *Landsc. Urban Plan.* **2021**, *209*, 104058. [CrossRef]
14. Zandieh, R.; Martinez, J.; Flacke, J.; Jones, P.; Van Maarseveen, M. Older Adults' Outdoor Walking: Inequalities in Neighbourhood Safety, Pedestrian Infrastructure and Aesthetics. *Int. J. Environ. Res. Public Health* **2016**, *13*, 1179. [CrossRef]
15. Parra, D.C.; Gomez, L.F.; Sarmiento, O.L.; Buchner, D.; Brownson, R.; Schimd, T.; Gomez, V.; Lobelo, F. Perceived and Objective Neighborhood Environment Attributes and Health Related Quality of Life among the Elderly in Bogotá, Colombia. *Soc. Sci. Med.* **2010**, *70*, 1070–1076. [CrossRef]
16. Stahl, S.T.; Beach, S.R.; Musa, D.; Schulz, R. Living Alone and Depression: The Modifying Role of the Perceived Neighborhood Environment. *Aging Ment. Health* **2016**, *21*, 1065–1071. [CrossRef] [PubMed]
17. Cornwell, E.Y. Social Resources and Disordered Living Conditions: Evidence from a National Sample of Community-Residing Older Adults. *Res. Aging* **2013**, *36*, 399–430. [CrossRef]
18. Nyman, S.R.; Ballinger, C.; Phillips, J.E.; Newton, R. Characteristics of Outdoor Falls among Older People: A Qualitative Study. *BMC Geriatr.* **2013**, *13*, 125. [CrossRef]
19. Chippendale, T.; Boltz, M. The Neighborhood Environment: Perceived Fall Risk, Resources, and Strategies for Fall Prevention. *Gerontologist* **2015**, *55*, 575–583. [CrossRef]
20. Schepers, P.; Brinker, B.D.; Methorst, R.; Helbich, M. Pedestrian Falls: A Review of the Literature and Future Research Directions. *J. Saf. Res.* **2017**, *62*, 227–234. [CrossRef]
21. Liu, J.Y.W. Fear of Falling in Robust Community-Dwelling Older People: Results of a Cross-Sectional Study. *J. Clin. Nurs.* **2015**, *24*, 393–405. [CrossRef] [PubMed]
22. Gazibara, T.; Kurtagic, I.; Kisic-Tepavcevic, D.; Nurkovic, S.; Kovacevic, N.; Gazibara, T.; Pekmezovic, T. Falls, Risk Factors and Fear of Falling among Persons Older Than 65 Years of Age. *Psychogeriatrics* **2017**, *17*, 215–223. [CrossRef] [PubMed]
23. Gyllencreutz, L.; Björnstig, J.; Rolfsman, E.; Saveman, B. Outdoor Pedestrian Fall-Related Injuries among Swedish Senior Citizens-Injuries and Preventive Strategies. *Scand. J. Caring Sci.* **2015**, *29*, 225–233. [CrossRef]
24. Morency, P.; Voyer, C.; Burrows, S.; Goudreau, S. Outdoor Falls in an Urban Context: Winter Weather Impacts and Geographical Variations. *Can. J. Public Health-Rev. Can. Sante Publique* **2012**, *103*, 218–222. [CrossRef]
25. Huynh, D.; Tracy, C.; Thompson, W.; Bang, F.; McFaull, S.R.; Curran, J.; Villeneuve, P.J. Associations between Meteorological Factors and Number of Emergency Department Visits Due to Unintentional Falls During Ontario Winters. *Health Promot. Chronic Dis. Prev. Can.-Res. Policy Pract.* **2021**, *41*, 441–453.
26. Lee, H.; Lim, J.H. Living Alone, Environmental Hazards, and Falls among U.S. Older Adults. *Innov. Aging* **2023**, *7*, igad055. [CrossRef]
27. Powell, L.M. Competence, Environmental Press, and the Adaptation of Older People. *Aging Environ. Theor. Approaches* **1982**, *7*, 33–59.
28. Kahana, E.; Lovegreen, L.; Kahana, B.; Kahana, M. Person, Environment, and Person-Environment Fit as Influences on Residential Satisfaction of Elders. *Environ. Behav.* **2003**, *35*, 434–453. [CrossRef]
29. Troped, P.J.; Tamura, K.; McDonough, M.H.; Starnes, H.A.; James, P.; Ben-Joseph, E.; Cromley, E.; Puett, R.; Melly, S.J.; Laden, F. Direct and Indirect Associations between the Built Environment and Leisure and Utilitarian Walking in Older Women. *Ann. Behav. Med.* **2017**, *51*, 282–291. [CrossRef]
30. Veeroja, P.; Foliente, G.; McCrea, R.; Badland, H.; Pettit, C. The Role of Neighbourhood Social and Built Environments-Including Third Places-in Older Adults' Social Interactions. *Urban Policy Res.* **2024**, *42*, 184–203. [CrossRef]
31. Moreira, B.d.S.; Andrade, A.C.d.S.; Xavier, C.C.; Proietti, F.A.; Braga, L.d.S.; Friche, A.A.d.L.; Caiaffa, W.T. Perceived Neighborhood and Fall History among Community-Dwelling Older Adults Living in a Large Brazilian Urban Area: A Multilevel Approach. *Int. J. Environ. Health Res.* **2020**, *32*, 522–534. [CrossRef]
32. Choi, Y.J.; Matz-Costa, C. Perceived Neighborhood Safety, Social Cohesion, and Psychological Health of Older Adults. *Gerontol.* **2018**, *58*, 196–206. [CrossRef] [PubMed]
33. Martin, K.R.; Shreffler, J.; Schoster, B.; Callahan, L.F. Associations of Perceived Neighborhood Environment on Health Status Outcomes in Persons with Arthritis. *Arthritis Care Res.* **2010**, *62*, 1602–1611. [CrossRef]
34. Delbaere, K.; Close, J.C.; Heim, J.; Sachdev, P.S.; Brodaty, H.; Slavin, M.J.; Kochan, N.A.; Lord, S.R. A Multifactorial Approach to Understanding Fall Risk in Older People. *J. Am. Geriatr. Soc.* **2010**, *58*, 1679–1685. [CrossRef]
35. Kao, S.; Wang, Y.-C.; Tzeng, Y.-M.; Liang, C.-K.; Lin, F.-G. Interactive Effect between Depression and Chronic Medical Conditions on Fall Risk in Community-Dwelling Elders. *Int. Psychogeriatr.* **2012**, *24*, 1409–1418. [CrossRef] [PubMed]
36. Lee, S.; Lee, C.; Ory, M.G.; Won, J.; Towne, S.D., Jr.; Wang, S.; Forjuoh, S.N. Fear of Outdoor Falling among Community-Dwelling Middle-Aged and Older Adults: The Role of Neighborhood Environments. *Gerontologist* **2018**, *58*, 1065–1074. [CrossRef]
37. Hou, Y.; Yap, W.; Chua, R.; Song, S.; Yuen, B. The Associations between Older Adults' Daily Travel Pattern and Objective and Perceived Built Environment: A Study of Three Neighbourhoods in Singapore. *Transp. Policy* **2020**, *99*, 314–328. [CrossRef]
38. McGinn, A.P.; Evenson, K.R.; Herring, A.H.; Huston, S.L.; Rodriguez, D.A. Exploring Associations between Physical Activity and Perceived and Objective Measures of the Built Environment. *J. Urban Health-Bull. N. Y. Acad. Med.* **2007**, *84*, 162–184. [CrossRef] [PubMed]

39. Yun, H.Y. Neighborhood Built Environments, Walking, and Self-Rated Health among Low-Income Older Adults in St. Paul, Minnesota. *Sustainability* **2021**, *13*, 3501. [CrossRef]
40. Wang, Y.; Shaw, D.; Yuan, K. Gated Neighborhoods, Privatized Amenities and Fragmented Society: Evidence from Residential Experience and Implications for Urban Planning. *Sustainability* **2018**, *10*, 4301. [CrossRef]
41. Ma, H.; Peng, Z. Study on Law of Barrier-Free Environmental Construction in China. In Proceedings of the International Conference on Education, Culture and Social Development (ICECSD), Wuhan, China, 20–21 May 2017.
42. Prüss-Ustün, A.; Bartram, J.; Clasen, T.; Colford Jr, J.M.; Cumming, O.; Curtis, V.; Bonjour, S.; Dangour, A.D.; De France, J.; Fewtrell, L.; et al. Burden of Disease from Inadequate Water, Sanitation and Hygiene in Low- and Middle-Income Settings: A Retrospective Analysis of Data from 145 Countries. *Trop. Med. Int. Health* **2014**, *19*, 894–905. [CrossRef]
43. Lu, Y.; Li, Z.; Qin, K.; Chen, J.; Zeng, N.; Yan, B.; Liu, D. Association between Perceived Neighborhood Environment and Depression among Residents Living in Mega-Communities in Guiyang, China: A Cross-Sectional Study. *BMC Public Health* **2024**, *24*, 343. [CrossRef]
44. Lu, T.; Zhang, F.; Wu, F. The Variegated Role of the State in Different Gated Neighbourhoods in China. *Urban Stud.* **2020**, *57*, 1642–1659. [CrossRef]
45. Decullier, E.; Couris, C.M.; Beauchet, O.; Zamora, A.; Annweiler, C.; Dargent-Molina, P.; Schott, A.M. Falls' and Fallers' Profiles. *J. Nutr. Health Aging* **2010**, *14*, 602–608. [CrossRef]
46. Chippendale, T. Development and Validity of the Outdoor Falls Questionnaire. *Int. J. Rehabil. Res.* **2015**, *38*, 263–269. [CrossRef]
47. Mwanga, D.M.; Kadengye, D.T.; Otieno, P.O.; Wekesah, F.M.; Kipchirchir, I.C.; Muhua, G.O.; Kinuthia, J.W.; Kwasa, T.; Machuka, A.; Mongare, Q.; et al. Prevalence of All Epilepsies in Urban Informal Settlements in Nairobi, Kenya: A Two-Stage Population-Based Study. *Lancet Glob. Health* **2024**, *12*, e1323–e1330. [CrossRef]
48. Penman, A.D.; Johnson, W.D. Complementary Log–Log Regression for the Estimation of Covariate-Adjusted Prevalence Ratios in the Analysis of Data from Cross-Sectional Studies. *Biom. J.* **2009**, *51*, 433–442. [CrossRef]
49. Nicklett, E.J.; Lohman, M.C.; Smith, M.L. Neighborhood Environment and Falls among Community-Dwelling Older Adults. *Int. J. Environ. Res. Public Health* **2017**, *14*, 175. [CrossRef] [PubMed]
50. Lee, S.; Lee, C.; Ory, M.G. Association between Recent Falls and Changes in Outdoor Environments near Community-Dwelling Older Adults' Homes over Time: Findings from the Nhats Study. *Int. J. Environ. Res. Public Health* **2019**, *16*, 3230. [CrossRef]
51. Célio, F.d.A.; Friche, A.A.d.L.; Jennings, M.Z.; Andrade, A.C.d.S.; Xavier, C.C.; Proietti, F.; Coulton, C.J.; Caiaffa, W.T. Contextual Characteristics Associated with the Perceived Neighbourhood Scale in a Cross-Sectional Study in a Large Urban Centre in Brazil. *BMJ Open* **2018**, *8*, e021445. [CrossRef] [PubMed]

buildings

MDPI

Review

Healing Spaces Improve the Well-Being of Older Adults: A Systematic Analysis

Xinyue Yan and Tao Geng *

College of Art and Design, Nanjing Forestry University, No. 159, Longpan Road, Nanjing 210037, China; yanxinyue@njfu.edu.cn
* Correspondence: nfugt@njfu.edu.cn

Abstract: As the global population ages, improving health statuses and quality of life has become crucial. Healing spaces enhance older adults' (OA) physical and mental health and well-being. However, the current research status and development trends must be systematically reviewed to guide future research directions. This study systematically examines the role of healing spaces in improving the health and well-being of OA through a bibliometric analysis. Results reveal three developmental stages: slow start, steady development, and in-depth refinement. Highly cited articles emphasize physical and environmental factors. The main research foundations encompass physical, chemical, biological, and ecological aspects of healing spaces and OA mental health. The research frontier is shifting towards a more comprehensive understanding of OA needs. Keyword analysis indicates a transition from a single physical environment focus to a holistic approach considering psychological, social, and spiritual needs across various disciplines. The field is progressing toward interdisciplinary integration, personalized demand, and systematic development. Future research should prioritize OA spiritual needs, develop scientific evaluation systems for spiritual healing, optimize personalized treatment environments using innovative technologies, and address diverse health and well-being needs. Strengthening international cooperation and integrating global research efforts are also essential for advancing the field.

Keywords: healing space; indoor air quality; well-being; environmental design; global research

Citation: Yan, X.; Geng, T. Healing Spaces Improve the Well-Being of Older Adults: A Systematic Analysis. *Buildings* **2024**, *14*, 2701. https://doi.org/10.3390/buildings14092701

Academic Editor: Paulo Amado-Mendes

Received: 26 July 2024
Revised: 24 August 2024
Accepted: 27 August 2024
Published: 29 August 2024

1. Introduction

The World Health Organization reports that the global population is aging rapidly, with 1 billion people expected to be 60 years or older by 2020, and 1.4 billion, or one-sixth of the world's population, expected to be 60 years or older by 2030 [1]. As the elderly population ages, elderly people experience reduced body function and activity ability and an increased likelihood of illness [2]. Consequently, population aging has become a central focus of architectural function transformation, highlighting the growing importance of healing spaces for OA. Population aging is a defining demographic reality of our time, associated with an increased social burden of providing care to older individuals with chronic diseases or frailty [3]. For older patients, hospitalization often leads to new or worsening disability, which can be attributed to the cumulative effects of aging, frailty, comorbidity, and the diseases that necessitate hospitalization, as well as issues in health-care management and the healing environment [4]. Healing is a holistic, transformative, physical, and spiritual process of repair and recovery that produces positive change, finds meaning, and moves toward self-fulfilling wholeness, regardless of disease [5]. Creating an environment with healing characteristics and utilizing the healing space's shape, level, and structure can stimulate the user's resilience and improve their healing ability from psychological and physiological perspectives [6]. In modern times, the term healing spaces acknowledges the impact of architectural spaces on patient health [7], and this connection has garnered the attention of researchers [8]. The role of individual elements in architectural spaces on health and well-being has also been explored [9,10], with studies investigating the environmental qualities of healing spaces from the perspectives of light [11], sound [12], air [13], and temperature [14]. Correlations have been found between specific features of the built environment and the well-being of building users [15]. Lemprecht [16] argued that well-designed spaces can promote rehabilitation and shorten hospital stays. Norberg–Schulz and Loci [17] describe architecture as the spirit of a place, a combination of multiple elements such as mental and environmental states. Effective therapeutic space design principles should include a homelike atmosphere, proximity to nature, sun exposure, noise control, and accessibility considerations [18].

Healing spaces provide an emerging research strategy for studying health and well-being in older populations [19]. Functional augmentation can be achieved by diversifying the synthesis of healing spaces [20,21] and updating spatial materials [22], enabling a more effective delivery of services to older populations. Healing space research is a well-established area of health research that can be applied to physical health outcomes [23,24]. In countries with severe population aging, the development of healing spaces for OA has driven many changes, including economic [25], cultural [26], and social concerns [27]. From a micro perspective, the healing space is a treatment dimension [10]. The healing environment benefits various health indicators, such as anxiety, blood pressure, postoperative recovery, analgesic use, and length of hospital stay [28]. A well-designed healing space may even promote recovery from certain psychological disorders [29]. While the healing space cannot directly lead to healing, it can facilitate the engagement of behaviors and emotions that support the healing process. The ideal environment fosters positive emotions such as happiness, contentment, and relaxation. This physical and mental health healing environment contributes to improved emotional regulation and is a prerequisite for physical and psychological healing [30]. Furthermore, in a well-designed healing space, the pain experienced by elderly individuals may be reduced [31]. Moreover, healing spaces significantly impact elderly populations with specific medical conditions [32]. For the terminally ill elderly, healing spaces provide home-like end-of-life care that mitigates the distress associated with death [33].

Current research on the impact of healing spaces on OA is gaining attention worldwide, particularly in countries with rapidly aging populations [34]. However, most attempts to review the scope of this research area have focused solely on the physiological healing response that healing spaces elicit in OA [10,35]. Previous literature has demonstrated a strong association between healing spaces and various aspects of health and well-being in older populations. Epidemiology, psychology, and architecture researchers are increasingly investigating the influence of healing spaces on a range of health outcomes among OA [36]. Prior studies have examined the role of healing spaces in the physical health of older populations. Several studies have explored the composition of the theory of perception and established the concept of the optimal healing environment (OHE), defining it as the social, psychological, physical, spiritual, and behavioral components of healthcare environments that support and stimulate the body's innate capacity for self-healing [37]. While these studies contribute to the understanding of healing spaces in the health of OA, they focus on only one of the many applications of healing spaces in research on elderly populations. Several rapidly emerging articles review the characteristics of healing spaces for OA [36] and explore the feasibility of their use [38]. Nevertheless, these approaches concentrate on physiological features such as pain reduction and enhanced healing. Such studies are primarily application-oriented and lack a systematic theoretical framework. Furthermore, the application of healing spaces in research on OA lags behind its development. Currently, there is no systematic review and evaluation of existing applications of healing spaces. This review aims to address this gap and identify areas for future research to capitalize on this new and expanding significant data source.

Due to the growing importance of healing spaces targeted at enhancing the well-being of older adults in the 21st century, a comprehensive and systematic review is necessary. This paper focuses on a multidimensional analysis of the different disciplines reflected in the literature using bibliometric software. The paper uses co-citation analysis and co-occurrence analysis to systematically review the trajectory of research in the therapeutic space for older adults. The main objective of this paper is to provide a thorough understanding and systematic overview of healing space research in OA. Quantitative and qualitative analyses were conducted using bibliometric methods and visualization tools to gain insight into the current status and future directions of healing space research in OA. First, the changes in the number of publications in different years were summarized, and the main subject areas were identified; second, the distribution of influential countries, authors, and disciplines in the field was found, and the main research directions and representative literature were identified. Meanwhile, keyword clustering analysis was conducted to reveal emerging research hotspots for promoting the transformation of well-being in old age through the therapeutic space. Finally, the future development trends in the field were outlined. This study is structured as follows:

- Section 1 is the introduction.
- Section 2 describes the main research methods and data sources, which include the research process.
- Section 3 explores the general characteristics of OA healing spaces, as well as the needs and main application areas of OA healing spaces in the knowledge phase, based on the conclusions of the data visualization.
- Section 4 provides an in-depth analysis of the significant impact of healing spaces in improving the well-being of older adults in a number of domains in response to the data in Section 3.
- Sections 5 and 6 summarize the discussion of current trends and discuss priorities for future evaluation research.

2. Research Methods and Data Sources

2.1. Data Sources

The bibliometric analysis began by selecting the Web of Science (WOS) platform as the primary database. Launched in 1997 by integrating SCI, SSCI, and AHCI, WOS is a multidisciplinary online literature database with over 20,000 high-impact academic journals and conference proceedings across diverse fields [39]. Its comprehensive coverage, authority, and scientific rigor render it an objective and reliable data source. Additionally, WOS is among the most authoritative and widely used scientific databases [40], justifying its choice for this study. Scopus was also utilized to provide citation data and integrated patent information from significant patent offices worldwide, offering a convenient and impactful interface [41]. While Google Scholar enables free access across document types, languages, and disciplines, its fragmented information, complex data, and lack of transparency for some content precluded its inclusion. A comprehensive perspective on international research landscapes and frontiers was obtained by complementing WOS and Scopus. A tailored search string further enhanced the scientific rigor [42].

Keyword selection was the second step, focusing on healing spaces and their role in enhancing elderly health and well-being, following the 2001 United Nations Second World Assembly on Ageing, which prioritized improving aging populations' health, supportive environments, and development [43]. Literature from January 2001 to December 2023 was collected. This 22-year timeframe captures research developments aligned with the UN's recommendations on healing spaces and increased societal attention to OA. Meanwhile, healing spaces relate to children, adults, and OA [44,45]. Considering the incompleteness of data in 2024, this study selected the 22-year period from 2001 to 2023 as the research timeframe. This time span can comprehensively reflect the recent development trends in this field, ensuring the timeliness and representativeness of the research results. By analyzing and reviewing the relevant literature within this period, it is conducive to clarifying the evolutionary trajectory of the research topic, grasping the cutting-edge dynamics of disciplinary development, and providing forward-looking suggestions for future research directions. This study concentrated on aging populations as healing spaces are more targeted in their health and well-being [46]. This article selects "Topic" as the retrieval type to retrieve titles, abstracts, authors, and keywords from the WOS core and Scopus collection. This paper uses the following search form to search (TS (theme simulation) = ("Healing" OR "Healing space" OR "Healing built" OR "Healing architecture" OR "Healing Environments" OR "Healthcare Design" OR "Holistic Healing environment" OR "Mental Healing") AND TS = ("The elder" OR "The old" OR "The aged" OR "older people" OR "elderly") AND DT = ("ARTICLE" OR "REVIEW") AND LA = ("ENGLISH")). The same keywords were used in Scopus to narrow or expand the search by finding synonyms, acronyms, and other alternative terms. This inclusiveness strengthens the critical search and ensures no related terms are ignored. Further searches were conducted for lead authors, essential sources, and important papers, and references to included articles were added after excluding rejected articles. Finally, these records were exported in a plain-text format and full-citation format. Then, the CiteSpace 6.2R7 software was imported for analysis (see Figure 1 for details).

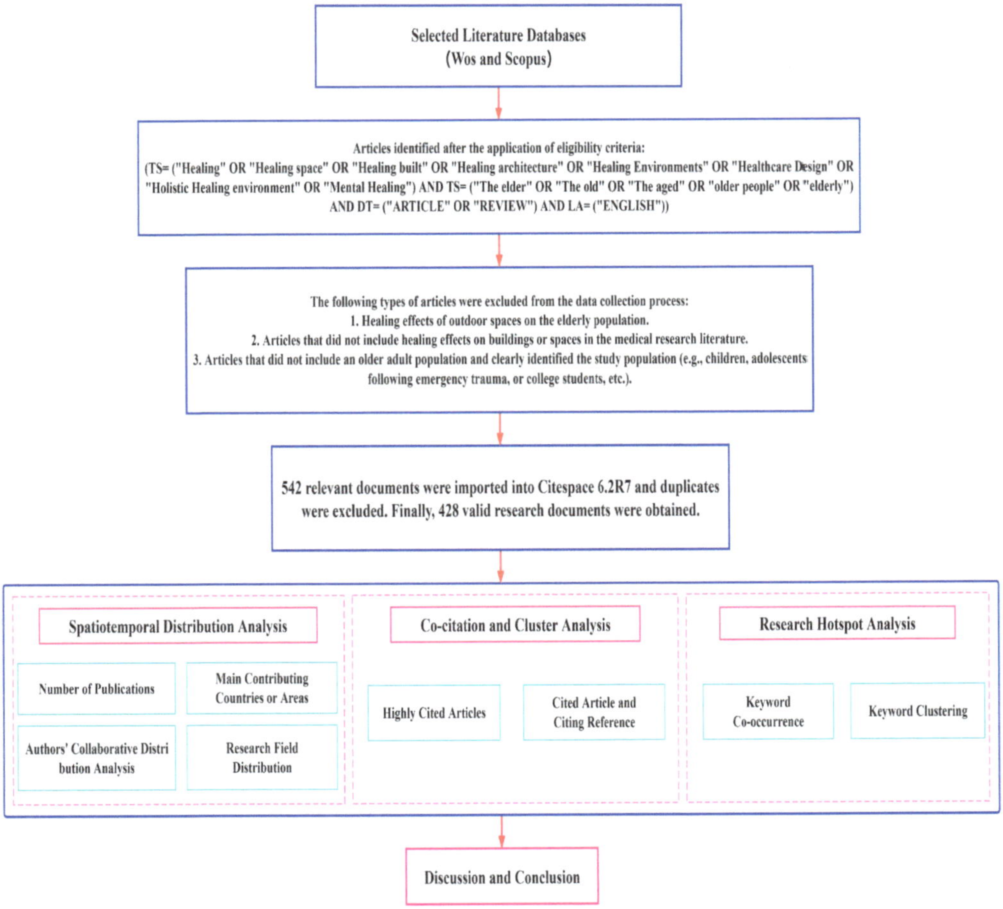

Figure 1. The study-design process.

2.2. Research Process

Bibliometric analysis offers a rigorous and widely adopted approach for exploring and analyzing large-scale scientific data [47], capable of revealing nuances and emerging trends within a domain [48]. Knowledge mapping, an advanced bibliometric and scientometric technique, directly represents quantitative research findings on specific topics [49]. Due to the extensive literature corpus, this review employed visualization tools to facilitate automatic data extraction, thereby validating the promoting effect of therapeutic spaces on elderly welfare. Visualization-analysis tools can efficiently extract key information from a large volume of literature and intuitively present data patterns and intrinsic connections [50], providing empirical support for the influence mechanism between therapeutic spaces and elderly welfare. Through a data-driven research approach, this paper aims to elucidate the intrinsic associations between the two and explore the application prospects of therapeutic spaces in improving the quality of life for the elderly. While software like HistCite [51], RefViz 1.0 [52], VOSviewer [53], and SATI [54] are available, CiteSpace 6.2R7 was selected as the primary tool for comprehensively analyzing the literature, particularly the subject's evolution. Its co-occurrence, co-citation analysis, and time-series capabilities enable targeted examinations [49,55]. CiteSpace 6.2R7, a Java-based information visualization software for knowledge mapping, presents scientific knowledge's structure, patterns, and distribution through visual analytics. After eliminating reviews, book chapters, news, and other non-research items, 428 valid publications were analyzed. CiteSpace facilitated exploring the role of healing spaces in enhancing elderly health and well-being through spatiotemporal distributions from 2001 to 2023, examining changes over time and across countries, institutions, and research topics. Collaborative networks were analyzed by setting node types as country and institution. The evolution of the research topic was visualized through a timeline view using the category node type. Keyword co-occurrence and cluster analyses identified frontier and hot issues at different stages. Integrating these analyses, this paper describes research trends on healing spaces' role in promoting elderly health and well-being, forecasting potential issues and breakthroughs meriting future attention. Metrics like "Centrality", indicating the likelihood of shortest paths through nodes [56], supplemented the interpretations. The study-design process is illustrated in Figure 1.

3. Results

3.1. Spatiotemporal Distribution Analysis

3.1.1. Number of Publications

The number of publications is the most tangible and objective metric for assessing a field's research status, representing a crucial indicator of its developmental trajectory. Figure 2 depicts the specific data on publication counts. Analyzing the changing number of publications offers vital insights into the impact of research on healing spaces for OA. After the screening, 428 articles published from 2001 to 2023 exploring the role of healing spaces in promoting elderly health and well-being were obtained for analysis. Taking 2004 and 2019 as the watershed years, although there were some relevant studies before 2004, they failed to form a systematic research domain. In 2004, the World Health Organization released the "International Plan of Action on Ageing: Implementation Report", providing guidance for countries to improve disease management and health promotion, leading to a steady increase in related publications. It was not until the COVID-19 pandemic outbreak in 2019 that specialized studies on therapeutic spaces for the elderly gained widespread attention and substantial publications in 2020. Therefore, the development of this research field can be broadly divided into three stages: slow start (2001–2004), steady development (2005–2019), and in-depth refinement (2020–2023). The publication trend reveals critical junctures and inflection points that demarcate these phases, shedding light on the field's evolution. By examining the quantitative changes in publication output over time, this study elucidates the overall development of healing space research for aging populations, its momentum shifts, and potential drivers underlying the observed patterns. These analyses lay the groundwork for a nuanced understanding of the field's research dynamics and intellectual trajectories.

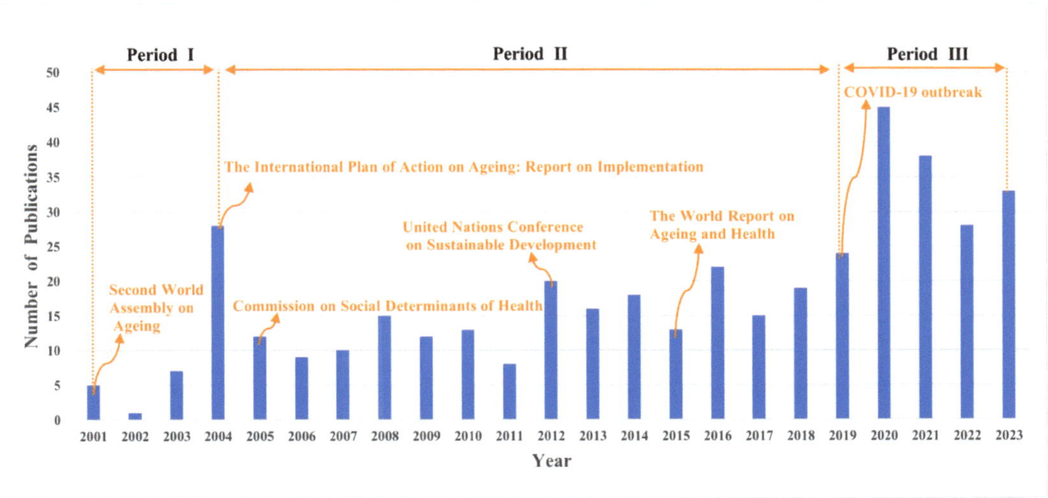

Figure 2. Number of publications per year.

The study's initial phase, from 2001 to 2004, exhibited slow and fluctuating growth. In 2001, the World Health Organization prepared the discussion paper Health and Ageing for the Second World Assembly on Ageing [57]. That same year, the United Nations held this assembly in Spain, approving WHO's active ageing policy framework. Subsequently, WHO published the Active Ageing Policy Framework [58], advancing global aging initiatives and stimulating extensive research. Consequently, studies on elderly well-being and health increased from 2002 onward. Additionally, a series of international aging conferences, such as the 16th and 17th Global Conferences of the International Federation on Ageing (held in Perth, Australia, in October 2002 and Singapore in September 2004), proposed the active aging concept and its three pillars of health, participation, and security. This gradually raised awareness and understanding of the importance of elderly healthcare. Furthermore, in 2004, WHO released the International Plan of Action on Ageing: Report on Implementation [59], guiding countries in improving disease management and health promotion. Consequently, relevant publications peaked during this initial stage in 2004 due to the increasing focus on aging and health.

The period from 2005 to 2019 witnessed a steady increase in publications. The World Health Organization established the Commission on Social Determinants of Health in 2005, operational until the end of 2008 [60], leading to a consistent rise in publication output during this time, with a brief decline after 2008. The 2007 World Health Report highlighted population aging as an increasing concern in developed nations and an emerging social issue in developing countries [60]. In 2012, the United Nations Conference on Sustainable Development proposed clear and practical steps for implementing sustainable development in the construction of architectural spaces [61]. Since 2012, targeted research has gained prominence in academia, and the number of publications has fluctuated upward. In 2015, the WHO released the World Report on Ageing and Health, emphasizing the urgent need for comprehensive public health actions to address population aging issues and the necessity of social, economic, and governmental support for the well-being of OA [62]. Consequently, academic research in this stage became more specialized, and the number of papers steadily increased. The role of healing spaces in improving the health and well-being of OA has been widely studied.

From 2020 to 2023, the number of publications remained relatively stable at around 25 per year. During this phase, research delved deeper into examining the role of healing spaces in enhancing elderly health and well-being with greater detail and nuance. A significant event impacting this field was the global spread of the COVID-19 pandemic [63]. After COVID-19, developing and investigating specialized, functional healing spaces tailored for elderly populations gained substantial momentum. This was driven by the increased vulnerability and long-term sequelae observed in aging demographics during the pandemic. Consequently, many publications emerged, focusing on designing targeted healing environments that cater to the unique needs of elderly individuals with lowered resistance and residual health impacts from COVID-19. This research thrust reflects the field's responsiveness to emerging public health challenges and the recognition of designing supportive spaces customized for this vulnerable population.

3.1.2. Analysis of Key Countries

Analyzing the regional distribution of collaborative publications sheds light on the significance and influence of significant cross-national research partnerships. It also elucidates the roles played by different countries in advancing knowledge on how healing spaces can improve elderly health and well-being. These insights can inform potential research directions and themes for future in-depth exploration in this field [64].

Using CiteSpace, the national collaboration landscape was examined, revealing 59 link strengths and 55 networks with a density of 0.0397 (Figure 3). Between 2001 and 2023, 138 publications resulted from international collaborations, marking the highest output. Researchers in New Zealand contributed the most, with 37 papers, followed by the United Kingdom, with 32 publications. This analysis uncovers the key nations driving cross-border research efforts and knowledge exchange in this domain. Understanding these collaboration networks is crucial for identifying influential country-level research hubs, facilitating future inter-institutional partnerships and aligning research agendas to tackle shared challenges in designing supportive healing environments tailored to aging populations globally.

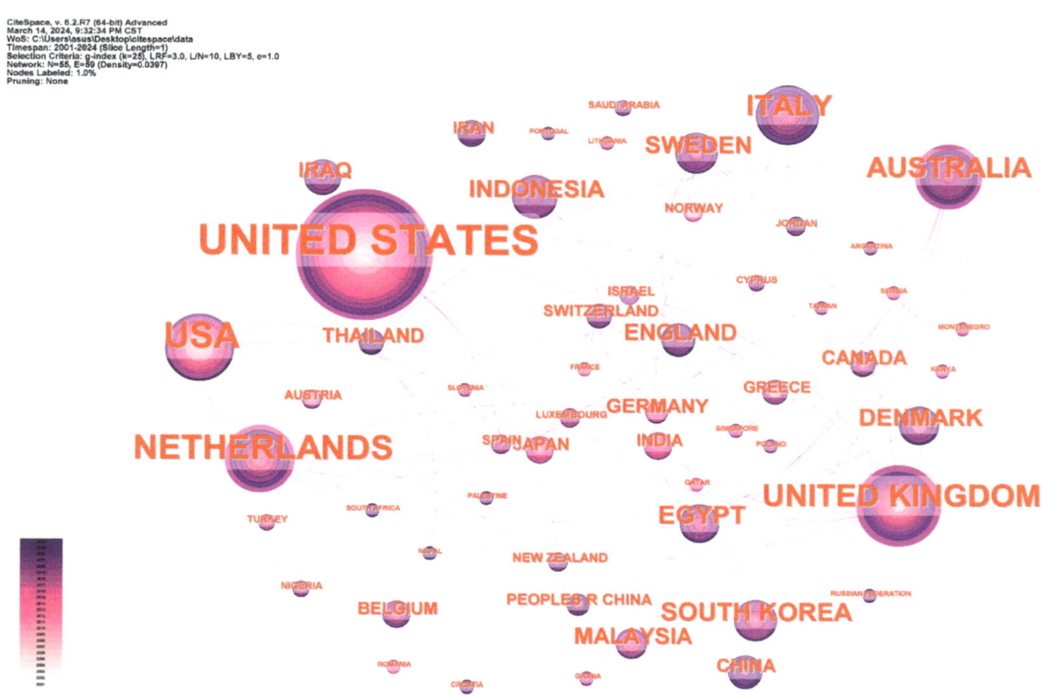

Figure 3. Knowledge map of cooperative countries and regions.

As depicted in Figure 3, the United States published the most papers, with a centrality of 0.18, indicating the most significant number of cross-country collaborative relationships and the broadest range of connections. Although ranking fourth in publication output, the United Kingdom exhibited the highest centrality value of 0.21 (Table 1), underscoring its pivotal role in information flow within this research network. The United States maintained close collaborative ties with New Zealand, the United Kingdom, Germany, Canada, and Japan. However, a lack of communication and cooperation was observed between Sweden, the United States, and Italy. Enhancing collaboration among these nations could potentially drive the globalization of research in this field, thereby fostering significant advancements. This centrality analysis highlights the influential country-level hubs shaping the collaborative landscape. The United States emerges as the primary producer of research output, while the United Kingdom acts as a critical bridge facilitating information exchange across national boundaries. Strengthening partnerships, especially among countries with limited current collaboration, can accelerate knowledge sharing, resource pooling, and cross-pollinating ideas vital for tackling the multifaceted challenges of designing healing spaces tailored to elderly populations worldwide.

Table 1. Top five countries in terms of centrality.

Rank	Centrality	Frequency	Country
1	0.21	25	United Kingdom
2	0.17	37	Netherlands
3	0.16	20	Australia
4	0.18	138	United States
5	0.08	4	Switzerland

3.1.3. Authors' Collaborative Distribution Analysis

Analyzing patterns of author co-authorship, cross-collaboration, and co-citation offers fundamental insights into research trends on the role of healing spaces in promoting elderly health and well-being. This analysis can inform new research directions driven by emerging interdisciplinary integration [65]. Leveraging CiteSpace, collaborative and cross-referencing relationships among critical researchers in this field were identified. Examining these research groups, their scholarly outputs were accessed to collect relevant data for analysis in CiteSpace. The author's collaboration network comprised 517 nodes and 521 connections with a density of 0.0039. Figure 4 details the leading authors and their affiliated institutions, while Table 2 presents the top five publications by citation count. This co-authorship analysis unveils the influential researchers shaping the intellectual base and driving collaborative efforts within this domain. Understanding these interconnected clusters of scholars is critical for mapping the knowledge landscape, identifying emerging interdisciplinary synergies and recognizing the potential for future cross-pollination of ideas and research integration. Highly collaborative author groups and their outputs can signal fertile areas for multidisciplinary inquiry into designing healing environments optimized for elderly populations.

Table 2. The top five primary authors of publications.

Rank	Frequency	Year	Author
1	13	2016	Capolongo, Stefano
2	11	2017	Gola, Marco
3	8	2017	Settimo, Gaetano
4	4	2021	Lundin, Stefan
5	4	2015	Macallister, Lorissa

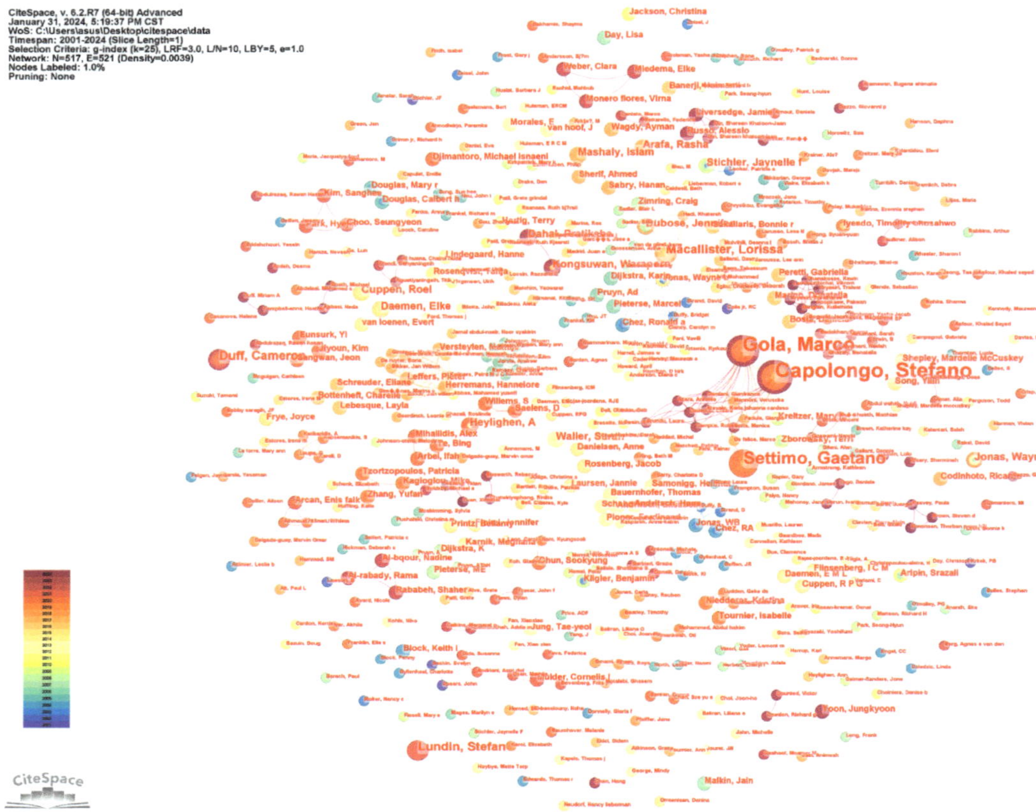

Figure 4. Knowledge map of co-authors of papers.

The author distribution exhibits a scattered landscape, albeit with several collaborative groups, which is indicative of a k-core cooperation structure. The largest cluster of co-authors is affiliated with Politecnico di Milano. Among the top five most prolific researchers, the top three are Italian: Capolongo, Stefano (13 publications), Gola, Marco (11), and Settimo, Gaetano (8), followed by Lundin, Stefan from Sweden (4), and Macallister, Lorissa from the US (4). Notably, five of the top ten authors are Italian scholars (Table 2), suggesting a concentration of discourse within the Italian research community on the role of healing spaces in promoting elderly health and well-being. The team, led by Settimo, Gaetano, and Gola, Marco, has focused on enhancing treatment efficacy through indoor air-quality interventions [66]. Collaborators Gola, Marco, and Capolongo, Stefano have investigated how healing spaces influence elderly well-being in the post-pandemic era [67]. Meanwhile, Lundin, Stefan, and Macallister, Lorissa have explored the psychological impacts of healing environments on patients [68,69]. While author collaborations exist, the distribution suggests weak cooperative ties among scholars in this research domain. Furthermore, a lack of communication and cooperation persists in specific sub-areas, indicating an absence of strongly interconnected, core academic teams with robust leadership driving advancements in understanding healing spaces' role for elderly populations. The academic community should prioritize strengthening interdisciplinary collaboration and in-depth exchanges among scholars from diverse fields. Concurrently, efforts should be directed toward establishing leading, cohesive research teams to spearhead progress in this vital area.

3.1.4. Research Field Distribution

By setting the node type to category and the time slice to 1 year, a collaboration network across disciplines is generated. This information offers insights into the transdisciplinary landscape of research on the role of healing spaces in promoting elderly health and well-being. The CiteSpace analysis yielded a knowledge map of 45 nodes and 60 link strengths. The most influential discipline, Health Care Sciences and Services, exhibited a centrality of 0.48. Public, Environmental, and Occupational Health emerged as the second-most central domain with a centrality of 0.30, followed by Medical Informatics at 0.16 centrality (Table 3). This interdisciplinary network highlights the core knowledge domains shaping the intellectual basis of healing space research for aging populations. The centrality metrics pinpoint the pivotal disciplines exerting the most significant influence, while the interconnected structure illuminates collaborative synergies driving conceptual and methodological cross-pollination. Visualizing this transdisciplinary ecosystem is crucial for identifying fertile areas for integrated research, potential knowledge gaps, and opportunities to bridge disparate perspectives through interdisciplinary dialogue and collaboration.

Table 3. The top 10 research fields ranked by centrality.

Rank	Frequency	Centrality	Year	Title
1	9	0.48	2007	Health Care Sciences and Services
2	28	0.30	2007	Public, Environmental, and Occupational Health
3	4	0.16	2003	Environmental Studies
4	5	0.16	2013	Medical Informatics
5	2	0.06	2011	Engineering, Manufacturing
6	2	0.06	2014	Computer Science, Artificial Intelligence
7	2	0.06	2017	Clinical Neurology
8	2	0.06	2017	Ergonomics
9	4	0.03	2012	Construction and Building Technology
10	3	0.03	2012	Engineering, Environmental

Within this research landscape, Health Care Sciences and Services and Clinical Neurology have significantly shaped the understanding of elderly healing spaces by integrating knowledge from medicine, nursing, psychology, and social sciences, facilitating the identifying and fulfilling of ageing populations' unique needs [45]. Public, Environmental, and Occupational Health, Engineering, Environmental Sciences, and Environmental Studies provide a critical theoretical foundation and practical guidance. These disciplines elucidate the intricate interactions between individuals and their environments, informing the design of healing spaces that promote physical, mental, and social well-being among OA Besterman–Dahan, et al. [70]. Moreover, Medical Informatics, Computer Science, Artificial Intelligence, and Ergonomics enable personalized healthcare delivery by collecting and analyzing large-scale health data. In geriatric healing environments, this data-driven approach allows medical teams to tailor treatment plans and care programs according to each elderly individual's health status, medical history, and preferences, enhancing treatment accuracy and efficacy [71]. Engineering, Manufacturing, Construction, and Building Technology further contribute by applying engineering principles to create ageing-friendly healing spaces that accommodate OAs' physiological and psychological requirements, supporting their health recovery, quality of life, independent living, and social engagement [72]. As research progresses historically, more profound and broader interdisciplinary communication and collaboration will be fostered. The academic community should remain attuned to pertinent societal issues and conduct comprehensive, cross-disciplinary investigations into the role of healing spaces in promoting elderly health and well-being. This will facilitate the updating or creation of novel spatial design systems tailored to geriatric healing. Consequently, the Social Sciences Citation Index is pivotal in advancing this multifaceted research domain.

3.2. Co-Citation and Cluster Analysis

3.2.1. High-Citation Articles

Citation analysis offers valuable insights into a field's knowledge base and research trends. Articles were retrieved from the WOS Core Collection and Scopus databases to identify the most frequently cited publications and imported into CiteSpace.

Settimo [73] was cited five times. The second-most frequently cited literature, each cited four times, are the works by Nartova–Bochaver and Mukhortova [74], Moscato, et al. [75], and Gola, et al. [76]. Details of the other cited articles are shown in Table 4. This citation analysis pinpoints the seminal and influential publications that have significantly shaped the intellectual foundations of research on healing spaces for elderly populations. The impact of these works, as quantified by their citation counts, underscores their conceptual or methodological contributions that have resonated across the field. Evaluating citation patterns illuminates the key ideas, theories, and empirical findings that have gained traction, guiding subsequent inquiries and driving the evolution of knowledge in this domain. These highly cited publications collectively represent scholarly outputs that have catalyzed discourse, spurred new lines of investigation, and advanced the understanding of designing supportive healing environments tailored to aging populations' needs.

Table 4. Top five most cited articles.

Rank	Frequency	Year	Reference
1	5	2017	Settimo [73]
2	4	2009	Nartova–Bochaver and Mukhortova [74]
3	4	2017	Moscato, Borghini, and Teleman [75]
4	4	2017	Gola, Mele, Tolino, and Capolongo [76]
5	3	2017	Settimo, et al. [77]

3.2.2. Cited Article and Citing Reference

Employing CiteSpace for cluster analysis reveals six thematically correlated clusters, as depicted in Figure 5 and detailed in Table 5. These clusters are labeled using indexing terms derived from the literature. These clusters were labeled as index terms, among which cluster #0 had the most significant number, which focused on the research on the physical environment of the healing environment, such as air quality, lighting, temperature, and humidity, which may affect the health condition of the patients. Environmental design focuses on the study of the physical environment of the healing environment, such as air quality, lighting, temperature, humidity, etc., which may impact the health status of patients. The clusters' biological environmental factors, mental health, and patients focus more on the elderly user group. Clusters are more concerned with the psychosocialization factors of the elderly recipient group, with the psychological feeling of the use of OAs as the base of the study. The biochemical environment and psychological state will affect the healing effect differently. The business case cluster focuses on the economic benefits of geriatric healing spaces, and the cost of using the space may also impact the degree of healing.

Table 5. Clustering of cited articles.

Cluster ID	Size	Silhouette	Year	Clusters
0	49	0	2014	Physical–chemical
1	40	0	2003	Patient
2	38	0	2017	Biological environmental factor
3	35	0	2006	Business case
4	34	0	2005	Environmental design
5	31	0	2009	Mental health

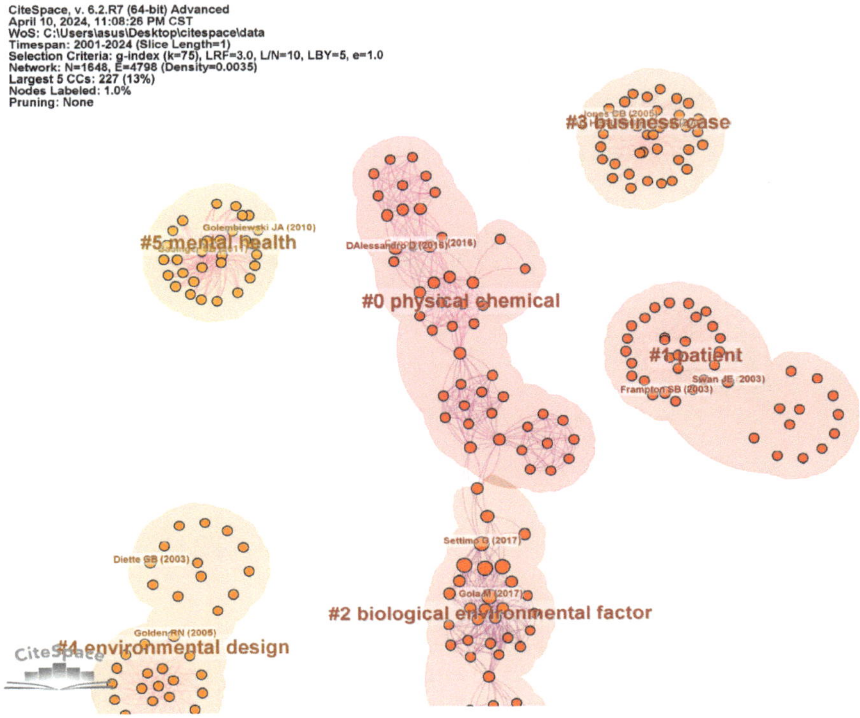

Figure 5. Literature cluster analysis.

The timeline graph (Figure 6) illustrates the temporal evolution of the identified clusters. Node size corresponds to citation counts, while interconnecting lines represent relationships between cited works. The initial clusters were #1 (2001–2006) and #4 (2001–2009), with #4's most extended duration spanning from 2001 to 2009. Cluster #5 is built upon the foundations of #4 and #0. Notably, Clusters #0 through #4 have maintained activity for over 5 years, underscoring their enduring representativeness within this research domain. Simultaneously, Clusters #0 and #2 demonstrate the highest levels of continuity, indicating their sustained status as research hotspots attracting consistent interest and citations over time. The thematic analysis revealed that Clusters #0 and #4 focused on the impacts of physical–chemical factors and environmental design on therapeutic spaces for the elderly, prompting reflections on commercial value in Cluster #3. From 2001 to 2010, the research foci of the patient, business case, and environmental design clusters converged on the human welfare and economic dimensions of elderly rehabilitation spaces. Since 2010, physical–chemical factors have emerged as a hotspot, peaking around 2015 and overlapping with research on biological environmental influences. This time-sequential analysis unveiled the knowledge trajectory and thematic shifts in the exploration of therapeutic spaces for the elderly population. Clusters #1 and #2 paid more attention to biological environmental impacts, fostering the rise of psychological research in Cluster #5. The evolving cluster landscape reflected the developmental trajectory of this field, transitioning from an initial emphasis on human-centeredness and economic considerations to the study of physical–chemical and biological–environmental influences. By mapping the temporal dynamics, the analysis elucidated how the constantly changing research foci, theoretical perspectives, and methodologies have collectively shaped the knowledge accumulation in this domain over the past two decades.

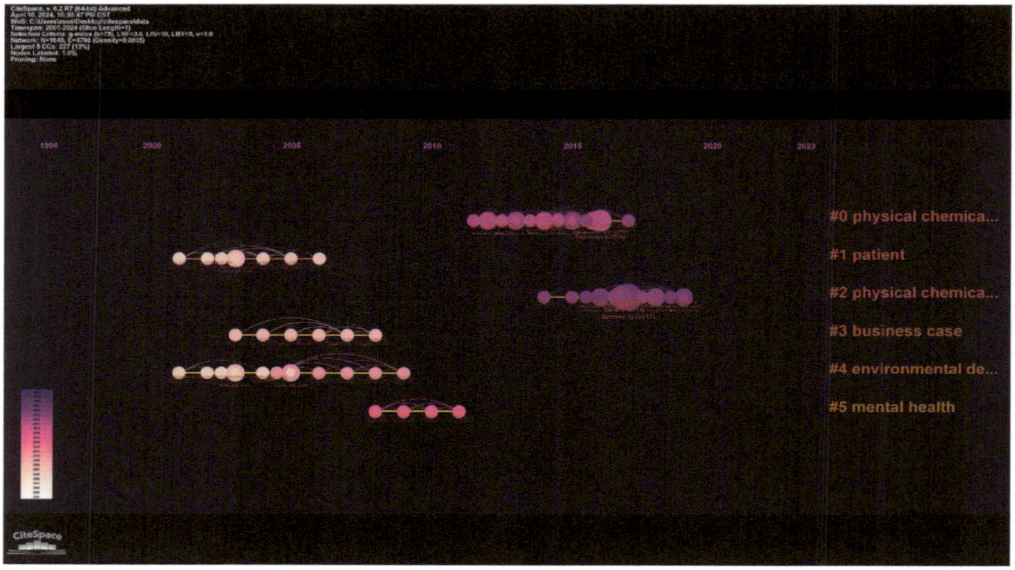

Figure 6. Literature clustering timeline analysis.

Among the cited article base and citing reference hotspots on the role of healing spaces in improving the health and well-being of OA, the clusters of physical, chemical, biological, environmental factors, environmental design, and mental health have been studied for a long time and have emerged as relatively new practices compared to the other clusters. The environmental design and mental health clusters that have been studied for a long time are relatively new to the emergence of practice compared to other clusters and have more significant implications for the role of healing spaces in improving health and well-being among OA. The topfour important clusters' citing articles and cited references information for each cluster are listed below (Cluster #0, Cluster #2, Cluster #4, Cluster #5):

- Cluster #0: physical–chemical

Cluster #0 centrally examines the impacts of biological, chemical, and physical risks present in therapeutic environments on the health and well-being of elderly users. These risks affect various aspects, such as physiological mechanisms [26], self-care capacity [78], and more. Specific examples include physical risks from inadequate microclimatic parameters causing thermal discomfort, chemical risks of toxic/harmful contamination, and biological risks from microbial exposure [79]. All are closely tied to the well-being and health of aging populations.

Among this research cluster, the cited references include Capolongo [80] and Nyhan, et al. [81], representing the foundational research in this field. The article references are Capolongo and Settimo [79] and Gola, et al. [82], presenting the cutting-edge research in this area with specific information provided in Table 6.

This cluster analysis pinpoints the core focus on the intricate relationships between healing environments' physical, chemical, and biological dimensions and their multifaceted impacts on elderly individuals' health and well-being. The cited works elucidate this area's fundamental concepts, theories, and empirical evidence. In contrast, the highly cited recent publications signal emerging perspectives, novel methodologies, and advanced findings, propelling the field forward.

Table 6. The top three articles cited in Cluster #0.

Cluster #0: Physical–Chemical			
Cited References		Citing Articles	
Cites	Author (Year)	Coverage %	Author (Year)
3	D'Alessandro, et al. [83]	14	Capolongo and Settimo [79]
2	Capolongo [80]	12	Gola, Settimo and Capolongo [82]
2	Nyhan, Sobolevsky, Kang, Robinson, Corti, Szell, Streets, Lu, Britter and Barrett [81]	-	-

- Cluster #2: biological–environmental factor

Cluster #2 is titled Biological–Environmental Factor. This cluster encompasses various factors closely related to the research topic, including biological–environmental health factors [84], social environmental factors of social engagement [85], and cognitive function and psychological health factors [86]. In this research direction, the studies by Gola and Settimo, et al. [87] can be considered foundational research. Simultaneously, the research by Settimo [73], Moscato, Borghini, and Teleman [75], and Settimo, Gola, Mannoni, De Felice, Padula, Mele, Tolino, and Capolongo [77] can be viewed as representing the cutting-edge perspectives of this field. Detailed information is provided in Table 7.

Table 7. The first three articles cited in Cluster #2.

Cluster #2: Biological–Environmental Factor			
Cited References		Citing Articles	
Cites	Author (Year)	Coverage %	Author (Year)
3	Settimo [73]	20	Gola, et al. [88]
2	Moscato, Borghini, and Teleman [75]	18	Settimo, Gola and Capolongo [87]
1	Settimo, Gola, Mannoni, De Felice, Padula, Mele, Tolino, and Capolongo [77]	-	-

- Cluster #4: environmental design

Cluster #4 is titled Environmental Design. In exploring the impact of healing spaces on the health and well-being of the elderly, environmental design plays a crucial role. Environmental design encompasses not only architectural spatial layout [89] but also encompasses multiple dimensions such as landscape design [38] and information design [90]. The comprehensive application of these design aspects is an indispensable factor in achieving the goals of health promotion and well-being enhancement. Environmental design's scientific and systematic nature is critical to supporting elderly health. In contrast, the multi-dimensional optimization of environmental design is crucial for realizing elderly health promotion and well-being improvement. Within Cluster #4, the research by Golden, et al. [91], Diette, Lechtzin, Haponik, Devrotes, and Rubin [31], and Kutash and Northrop [92] can be considered foundational. At the same time, Rashid [93] can be regarded as representing cutting-edge research. Detailed information is provided in Table 8.

Table 8. The first three articles cited in Cluster #4.

Cluster #4: Environmental Design			
Cited References		Citing Articles	
Cites	Author(Year)	Coverage %	Author (Year)
2	Golden, Gaynes, Ekstrom, Hamer, Jacobsen, Suppes, Wisner, and Nemeroff [91]	21	Rashid [93]
2	Diette, Lechtzin, Haponik, Devrotes, and Rubin [31]	-	-
1	Kutash and Northrop [92]	-	-

- Cluster #5: mental health

Cluster #5 is titled Mental Health. In exploring the impact of healing spaces on the health and well-being of the elderly, mental health plays a crucial role. For the mental healing of the elderly population, it is necessary to focus not only on the individual mental characteristics of OA but also on the psychological impact that the physical environment of the healing spaces has on them. Within healing spaces, the mental state of OA greatly influences the occurrence of healing. Many mental states can even alleviate physical suffering for OA. These applications of mental care for OA are indispensable factors in achieving the goals of health promotion and well-being enhancement. Furthermore, mental health healing presents a deep interdisciplinary fusion of medicine, environmental studies, and psychology. Therefore, attention to mental health is crucial for realizing elderly health promotion and well-being improvement. Within Cluster #5, the research by Cleary, et al. [94], Caspari, et al. [95], and Applebaum, et al. [96] can be considered foundational. In contrast, the research by Connellan, et al. [97] can be regarded as representing cutting-edge research. Detailed information is provided in Table 9.

Table 9. The first three articles cited in Cluster #5.

Cluster #5: Mental Health			
Cited References		Citing Articles	
Cites	Author(Year)	Coverage %	Author (Year)
2	Cleary, Hunt, O'Connor, and Snars [94]	31	Connellan, Gaardboe, Riggs, Due, Reinschmidt, and Mustillo [97]
1	Caspari, Eriksson, and Nåden [95]	-	-
1	Applebaum, Fowler, Fiedler, Osinubi, and Robson [96]	-	-

3.3. *Research Hotspot Analysis*

3.3.1. Keyword Co-Occurrence

Keyword co-occurrence analysis examines the co-occurrence patterns of selected keywords within a collected dataset. Compared to co-citation analysis, this approach more directly reveals the central content, methods, and core concepts explored in the literature on healing spaces' role in promoting elderly health and well-being. The dataset spanning from 2001 to 2024 was imported into CiteSpace to generate a yearly keyword co-occurrence network visualization, as shown in Figure 7. This network consists of 595 nodes and 2530 link strengths, with a density of 0.0143. Node size represents the frequency of occurrence, while link thickness corresponds to the co-occurrence count of the connected keyword pairs. Leveraging CiteSpace's word-frequency data, Table 10 presents the top 10 most frequently occurring keywords from 2001 to 2023, offering insights into the prominent themes and topics characterizing this research domain over the past two decades. This keyword co-occurrence analysis provides a robust overview of the intellectual landscape by illuminating the central concepts, theories, methods, and applications that have driven investigations into designing supportive healing environments tailored to ageing populations' needs. The visualization captures the interconnected nature of these critical topics, while the frequency data pinpoint the areas of sustained focus and emerging priorities within this multifaceted field of inquiry.

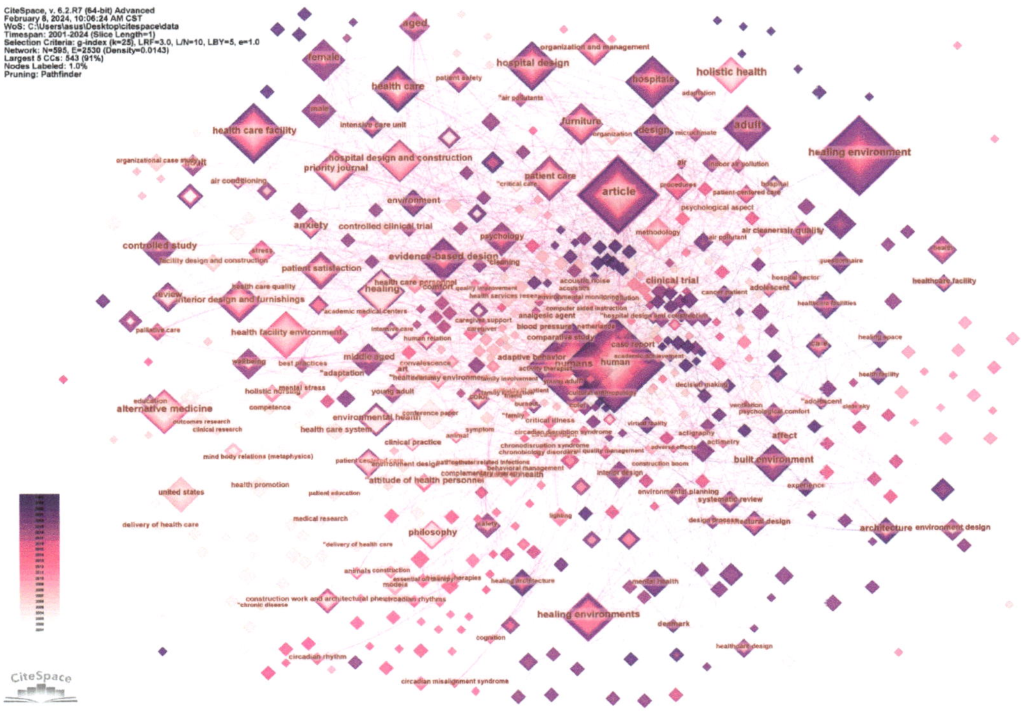

Figure 7. Keyword co-occurrence analysis chart.

Table 10. The top 10 keywords with co-occurrence frequency.

Rank	Frequency	Centrality	Year	Keywords
1	118	0.06	2001	Humans
2	87	0.16	2001	Healing environment
3	55	0.08	2001	Health care facility
4	37	0.09	2006	Hospital design
5	33	0.03	2001	Health facility environment
6	32	0.06	2003	Health care
7	29	0.06	2004	Patient care
8	28	0.19	2006	Evidence-based design
9	24	0.07	2003	Alternative medicine
10	23	0.05	2005	Hospital design and construction

The keyword analysis reveals that in research examining the role of healing spaces in promoting elderly health and well-being, the term "humans" is among the most frequently occurring, followed by healing environment, health care facility, hospital design, and health facility environment. Notably, the convergence of alternative medicine and patient care concepts within healthcare embodies an exploration of holistic patient care approaches, tending to incorporate alternative treatment modalities alongside traditional healing practices [98]. The keywords, including hospital design, health facility environment, hospital design and construction, and healthcare facility, emphasize healthcare facilities' physical structure and layout, utilizing research evidence to guide facility design and environmental construction to enhance support for patient rehabilitation [99]. Furthermore, health care and healing environments leverage health services' research methodologies to systematically evaluate evidence on how various design factors impact the health of OA, thereby providing a scientific foundation for constructing optimal healing environments [10].

3.3.2. Keyword Clustering

This study conducted a cluster analysis of keyword knowledge maps to explore the developmental trajectories of research on the role of healing spaces in promoting elderly health and well-being across different stages from 2001 to 2024. The 50 most cited articles were selected, and CiteSpace analysis yielded Figure 8. The cluster areas and intersections represent the cluster size and relationships, while node size indicates the frequency of occurrence, with larger nodes denoting higher frequencies. As depicted in Table 11, the role of healing spaces in enhancing elderly well-being formed 10 clusters around key nodes, representing 10 research directions: healing environment, healthcare facilities, healing architecture, gender factors, air quality, healthcare design, health services research, social dynamics, spiritual care, and optimal healing environments (Clusters 0 to 9 in descending order of keyword richness). The cluster modularity Q = 0.7097 (ranging from 0 to 1) and the mean silhouette value s = 0.835 (ranging from 0 to 1, with values above 0.7 indicating persuasive clustering) validate the cluster structure's robustness, detailed in Table 11. Three algorithms (LLR, LSI, and MI) were employed for cluster labeling, with LLR and MI emphasizing study characteristics. As the LLR algorithm proved more practical and less redundant through training, its results were selected as the primary cluster definitions.

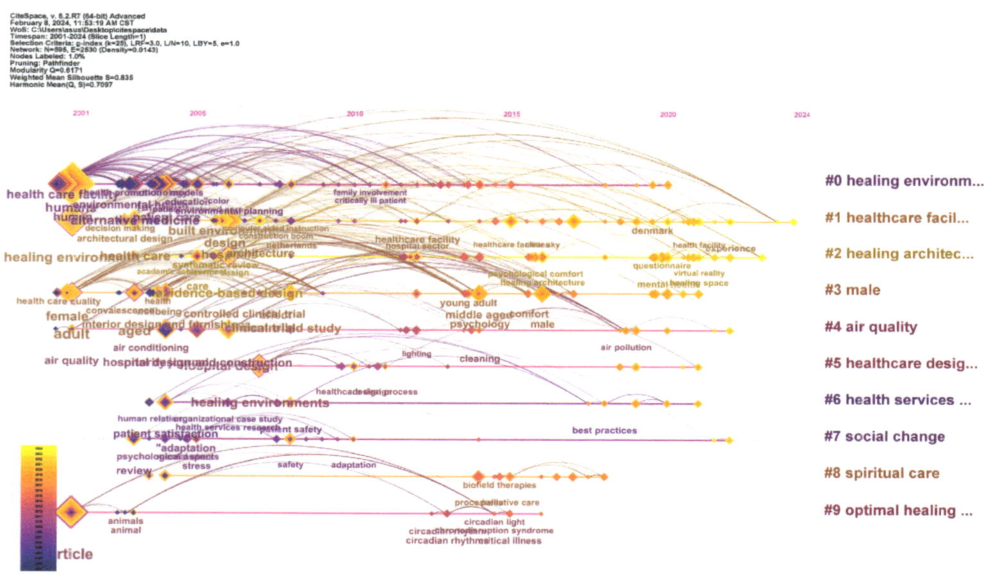

Figure 8. Keyword clustering timeline analysis chart.

Table 11. Clustering of the top 10 keywords.

Rank	Frequency	Centrality	Mean Year	LLR
1	92	0.718	2007	Healing Environment
2	78	0.744	2013	Healthcare Facilities
3	75	0.814	2017	Healing Architecture
4	67	0.812	2010	Male
5	43	0.854	2012	Air Quality
6	32	0.933	2012	Healthcare Design

Table 11. *Cont.*

Rank	Frequency	Centrality	Mean Year	LLR
7	27	0.863	2012	Health Services Research
8	26	0.918	2009	Social Change
9	24	0.980	2015	Spiritual Care
10	21	0.982	2011	Optimal Healing Environments

4. Analysis of Results

4.1. Spatiotemporal Analysis

The theme of the role of healing spaces in improving elderly health and well-being has undergone a transformative journey over the past two decades, marked by initiation, steady development, and further refinement. The breadth and depth of research in this domain have dramatically increased. Regarding active countries, the existing literature predominantly originates from developed nations. This trend can be attributed to the earlier onset and greater prevalence of population ageing in these regions. However, by 2024, developing countries such as China, Thailand, and Turkey have also witnessed rapid growth in their aging populations, consequently spurring an emergence of research investigating the impact of healing spaces on elderly health and well-being in these regions. While some countries like the United Kingdom, the United States, and Japan have maintained a focused research agenda on the role of healing spaces in promoting elderly well-being, it is crucial to recognize that this is a global issue, and different countries and regions may face distinct challenges and research needs. Therefore, transnational cooperation and knowledge-sharing remain pivotal aspects in achieving the overarching goal of leveraging therapeutic spaces to enhance elderly health and well-being. Through international collaboration, drawing upon the successful experiences of various nations, research and practice in this area can be collectively advanced, providing more targeted and effective solutions for the global challenge of population ageing.

4.2. Co-Citation and Cluster Analysis

4.2.1. Analysis of High-Citation Articles

Citation analysis was employed to rank this study's most frequently cited works (see Table 3 for ranking details). Among the highly cited publications, several focused on the impact of air quality (IAQ) on health and well-being. Settimo [73] underscored the critical importance of indoor air quality (IAQ) for health and advocated for developing a European IAQ strategy to optimize living and medical environments for OA. Moscato, Borghini, and Teleman [75] noted that in healthcare facilities, errors in the design, installation, and maintenance of air-conditioning systems can severely compromise IAQ, particularly for immunocompromised patients. Gola, Mele, Tolino, and Capolongo [76] analyzed IAQ monitoring in international hospitals, concentrating on chemical pollutants, and provided an overview of air biological pollution, aiming to review monitoring methods and key findings. Settimo, Gola, Mannoni, De Felice, Padula, Mele, Tolino, and Capolongo [77] investigated indoor air quality (IAQ) in medical institutions, assessing the effectiveness of improvement measures through monitoring formaldehyde, volatile organic compounds, and other parameters, reporting preliminary results on air quality in inpatient wards. Another highly cited work examined the potential of human–flora interactions for physical and mental health recovery, demonstrating the positive effects of interacting with the plant world on healing states [74]. These publications predominantly emerged during the second research phase, providing insights into this period's leading content and trends. Examining these highly cited works can facilitate a nuanced understanding of the second phase's research landscape.

4.2.2. Citation Cluster Analysis

To illustrate citation patterns, this research selected highly cited literature from the four most significant research areas identified by CiteSpace (#0, #2, #4, and #5). By conducting an in-depth analysis of these influential articles, characterized by high citation and co-citation rates, the research determined each research area's foundational theoretical basis and cutting-edge dynamics. This process involved a systematic organization of their knowledge structures.

- Cluster #0: physical–chemical

1. Research Basis

Many elements associated with biological, chemical, and physical risks in the therapeutic environment can impact the health and well-being of users. These factors affect physiological mechanisms [26], self-care capacity [78], and other aspects of the elderly population. For instance, physical risks may arise from changes in user well-being or thermal discomfort due to insufficient microclimate parameters, chemical risks from toxic, harmful, or carcinogenic contamination [81], and biological risks from microbial contamination or insufficient quantity [79]; each of these aspects is closely linked to the health and well-being of the users.

Architectural spaces serve as primary sites for healing. Building risks, such as managing significant external change, can increase physical, chemical, and biological risks that affect health and well-being. For example, construction activities are a known risk for the growth and spread of waterborne pathogens in building water systems [100], and environmental defects can negatively impact the building's water supply, leading to disease, injury, and death [101]. In recent decades, indoor air quality (IAQ) in buildings has received growing attention from the scientific community, political institutions, and environmental governance, as it impacts occupants' comfort, health, and well-being. Sources of indoor pollutants include outdoor infiltration (e.g., vehicle exhaust, industrial emissions) [81], as well as indoor sources (e.g., building materials, furniture, HVAC systems) [102]. These pollutants include a variety of harmful substances, such as carbon monoxide, ozone, radon, VOCs, particulate matter, fibers, bacteria, fungi, and pollen. For many people, the health risks associated with exposure to indoor air pollution may be more significant than those related to outdoor pollution. Notably, poor indoor air quality can be detrimental to vulnerable groups, such as children, young people, the elderly, or individuals with chronic respiratory and cardiovascular diseases. Therefore, for healing spaces or healthy buildings aimed at improving the health and well-being of OA, the design and construction process must address the above physical, chemical, and biological risk factors with even higher requirements. Not only should the functional spaces meet individual and collective needs, but the occupants' basic safety, health, and well-being must also be protected [81]. This requires integrating skills from various fields, including materials, engineering, technical systems, urban planning, and the medical domain, especially public health, which can be developed during design.

2. Research Frontiers

With the continuous improvement in the quality of life, the indoor breathing environment has become an important area of focus for researchers in the 21st century. Many studies have confirmed indoor air is more lethal than outdoor air [103]. Indoor air quality has received increasing attention in public health and has become one of the main concerns for governments. As a complex and dynamic issue, the physical factors, as well as biological and chemical pollutants generated in both outdoor and indoor environments, can affect users' health. This is because it is related to many major chronic diseases, such as asthma, hypersensitivity pneumonitis, and Legionnaires' disease, which are associated with the respiratory system, skin, mucous membranes, nervous system, and immune system [104]. In addition to these well-known diseases, there may also be some symptoms typically characterized by neurological sensory effects, which determine the state of discomfort, the decline in the health of occupants, and the negative perception of air quality [79]. All these ailments are caused by various aspects related to indoor air quality, which primarily impact vulnerable groups (OA, children, and those with pre-existing chronic conditions). In healing spaces, there have been increasing data analyses and research efforts to ensure and prevent the health of OA [79,82], with some studies even identifying indoor pollution as one of the most significant environmental risks to public health [88,105]. Particularly in the current pandemic-dominated era, more urgent actions are needed to address the burden of its impacts.

- Cluster #2: biological environmental factor

1. Research Basis

The indoor environment is a complex ecosystem that combines the user population, microorganisms, and the built environment [106]. Biological environmental control is a critical factor for the healing of elderly patients. In elderly surgery, cancer treatment, and others, controlling an appropriate biological environment can effectively curb highly infectious infections and improve the healing outcomes of elderly patients [73]. The indoor biological environment in healing spaces differs from other spaces' complexity. The continuous activities within the healing space bring about a constant influence of biological pollutants [107]. The concentration of biological pollutants is essential to the healing outcomes [108]. The biological pollutants that impact the healing effects in space are primarily airborne micro-pollutants, such as the influenza A virus (H1N1), the severe acute respiratory syndrome coronavirus 2 (SARS-CoV-2), and microorganisms like Aspergillus and Penicillium [109]. The use of HVAC systems can exacerbate the spread of biological pollutants within the indoor ecosystem. Environmental air quality, building materials, air circulation, appliances, cleaning products used, occupant habits (such as smoking), and building maintenance (such as energy-saving measures) can all influence the concentration changes of indoor biological pollutants [77]. Therefore, for specific population groups more sensitive to air quality, such as OA, it is necessary to continuously re-evaluate the importance of the healing space ecosystem and HVAC systems [75]. More efficient air-filtration systems, increased biological environment-quality monitoring points to precisely track microenvironmental changes, and regular air-treatment-system maintenance and upgrades can significantly improve OAs' well-being in healing spaces [73].

2. Research Frontiers

Biological environmental factors are the key to affecting the health and well-being of OA in healing spaces, with indoor air quality being the primary manifestation of these biological environmental factors [108]. High-quality indoor air quality can not only prevent the elderly from contracting hospital-acquired infections (HAIs) but also enhance their overall well-being [110]. Chemical pollution of indoor air is a focus that requires attention. Healing spaces are low-care environments easily exposed to chemical pollutants, exacerbating indoor air pollution (IAP) [108]. Therefore, factors such as the design and management of inpatient wards and management efficiency can influence indoor air quality [87]. The specific user needs, in-hospital infections, required technologies and ventilation systems, multi-departmental risk analysis, acceptable residual risk values, and related management procedures can all impact the air-quality outcomes in in-patient wards [111]. The construction and renovation of the biological environment for elderly healing must be adjusted based on the targeted impact factors (such as the light environment, humidity, thermal environment, and air environment), and the management and prevention system should be updated according to the health status of the users [88,112]. The dynamic monitoring of these impact factors and optimizing the management system are also the focus of future research [113]. Targeted design may directly affect the safety of the healing space, offsetting the probability of risk factors and harmful events, thereby improving the well-being and health of the elderly user population [88].

- Cluster #4: environmental design

1. Research Basis

Research on the design and planning of healing spaces has emphasized the substantive relationship between physical environmental factors and health [8]. The key to the environmental design of healing spaces is the indicator of indoor environmental quality (IEQ) [114]. Good indoor environmental quality (IEQ) can not only directly affect the psychological health of patients but also influence the physiological and psychological state of patients' families [115]. This is because the environmental design of healing spaces aims to promote the reduction of environmental stressors, facilitate the social, psychological, physical, spiritual, and behavioral components of healthcare support, and stimulate the innate self-healing ability of the body, playing an important role in alleviating pain, anxiety, and stress, and making patients feel comfortable and safe [116]. IEQ plays a crucial role in improving the well-being and comfort of occupants throughout the life cycle, especially for the elderly population [117]. The physical factors, such as the quality of architectural features, furniture, decorations, lighting, and ventilation, as well as the degree of daylighting, color temperature, and material temperature, can all influence the psychological factors of patients and affect the overall indoor environmental quality [118]. IEQ elements not only have a concentrated impact on the psychology of patients but also influence the physiological and psychological health status of patients' families [92]. The elderly population often requires the care and companionship of family members, and a sound healing environment can alleviate the stress of patient families and improve their quality of life. For example, comfortable seating, appropriate lighting, and good ventilation can create a relaxed and welcoming healthcare environment, enhancing patient families' sense of belonging and security [96]. The environmental design of healing spaces needs to keep up with the times and continuously optimize and upgrade. With the development of technology, some intelligent environmental control systems, such as automatic temperature and humidity adjustment, as well as lighting control, can also be incorporated into the design to enhance IEQ further and create a more comfortable and safe healthcare experience for OA [119].

2. Research Frontiers

Extensive patient-oriented research has demonstrated the close relationship between indoor environmental quality (IEQ) and the health and well-being of the healing population [120]. Ulrich proposed that the design of a healing environment should focus more on emotional and psychological responses, emphasizing that healing is based on the emotional feedback generated by the quality of the spatial environment [121]. There is a positive correlation between indoor environmental quality and the healing outcomes of the user group, which is influenced by sociological needs, psychological states, and individual differences. High-quality indoor environmental quality leads to favorable healing outcomes [122]. In recent years, research on the environmental quality of healing spaces has often been oriented towards high-energy-efficient buildings, where thermal optimization of insulation and airtight building envelopes, as well as high-efficiency heating, ventilation, and air conditioning (HVAC) systems, may lead to a deterioration of indoor environmental quality (IEQ) [123]. Most review studies have focused on the relationship between specific physical IEQ elements and patient psychological elements related to healing outcomes, lacking targeted research on the impact of staff activities on healing outcomes, such as the appearance, communication, and accessibility of healthcare professionals [124]. At the same time, the healing space's environmental design should be focused on the intangible dimensions of sustainable design, such as the semantics and regenerative properties of the healing space [120,125].

- Cluster #5: mental health

1. Research Basis

The design of healthcare environments has begun to be oriented towards rehabilitation and providing mental health services, including aesthetic enhancement, to alleviate stress and anxiety, improve patient satisfaction, and promote health and recovery [10,126]. In considering the psychological aspects of patients, the mainstream design approach for healing spaces has become to tailor the design based on the stage of the illness, considering factors such as the physical attributes of the space, the sensations evoked by the space, the familiarity of the space, the relationship with nature, and spiritual or religious meanings [94]. Users' ability to establish a spiritual connection in the healing environment, provide positive distractions, and enhance a sense of control can, to some extent, reduce the perception of pain [45]. For example, based on the problem of functional decline faced by the elderly population, their spiritual perception of autonomy, independence, and self-well-being in the healing space is reduced, which requires more emotional support services provided by healthcare professionals to support the healing process of OA [96].

2. Research Frontiers

With the intensification of population aging, mental health problems such as depression and anxiety are becoming increasingly prominent among the elderly population, which requires high attention [127]. The prevalence of depression among people aged 50 and above is 5.5–27.5%, and the prevalence of anxiety is 1.2–15%, far higher than other age groups [128]. In response to this issue, the specialized principle of rehabilitation-oriented mental healthcare, which includes therapeutic spaces, holistic care, and person-centered care, is considered a positive approach [126]. The patient-centered holistic care model can effectively improve OAs' treatment adherence and subjective well-being [129]. At the same time, creating a comfortable and safe therapeutic space, such as incorporating natural elements and an artistic atmosphere, can also help alleviate patients' anxiety and promote mental health [130]. This whole-person care based on the biopsychosocial model has become the mainstream direction of mental health services for OA [97]. Although the argument that the environmental elements within the healing space can positively impact the healing outcomes and the subjective well-being of patients has been established, this may be because these environmental elements can regulate the neuroendocrine system of patients, thereby producing positive psychological effects. However, the specific mechanisms of action still require further in-depth research, such as the impact of different sound characteristics on patients' emotional states [131] and even the establishment of clear standards and regulations to ensure service quality [132].

4.3. Keyword Analysis

In the keyword analysis, CiteSpace has clustered the keywords into 10 clusters closely related to the research topic, as shown in Table 11. In developing healing spaces for OA, the healing environment plays a crucial role, and appropriate environmental design can significantly impact OAs' physiological, psychological, and social health [133]. This is reflected in healthcare facilities and healthcare design and in optimizing architectural layouts and interior designs to provide OA with a safe, comfortable, and easy-to-navigate environment [36]. Healing architecture further emphasizes the importance of design elements, such as natural lighting, soothing color tones, and open views, to enhance OAs' well-being and recovery process. These design elements can effectively regulate the emotional states of OA, alleviating negative emotions such as anxiety and depression and thereby promoting their mental health. Particularly for OA who may face issues like heart disease or prostate health, the design needs to ensure that it can meet the specific health needs and psychological service needs of this population [134]. Moreover, the optimization of air quality is crucial for maintaining the health of the elderly, especially for those with respiratory system diseases [135]. Health services research provides scientific evidence for the effectiveness of healing spaces for OA, offering evidence on the impact of different elements on the health of OA [136]. Social change also plays a role in developing healing spaces for OA, as increasing attention is paid to the well-being of OA, and more resources are being invested in creating supportive and inclusive environments [137]. Research suggests that spiritual care can help create an environment that meets the spiritual needs of OA, assisting them in finding inner peace and a sense of meaning [138]. This holistic approach to the well-being of OA reflects the deepening recognition of the needs of an ageing society. The optimal healing environment is a design concept that comprehensively considers all the above factors, based on extensive health services research, providing a scientific basis for the best practices in space design [136]. This comprehensive and systematic design approach can improve the physical health of OA and meet their psychological, social, and spiritual needs, creating a truly healing environment for them. Developing healing spaces for OA is increasingly intertwined with perspectives such as healing architecture, healthcare design, health services research, spiritual care, and optimal healing environment, and they are mutually reinforcing [121]. This development trajectory reflects the deepening recognition of the aging society and its unique needs and the continuous pursuit of comprehensive and humanistic care in the healthcare field. As society pays increasing attention to the well-being of OA, more resources are being invested in creating supportive and inclusive environments. This is not only reflected in the optimization of hardware facilities but also requires efforts in soft power, such as enhancing healthcare professionals' service awareness and communication skills so that OA can feel more human care [96]. In the early stages, the concept of healing spaces for OA was relatively vague, focusing more on primary medical care and functional needs [139]. However, with in-depth health services research, especially on the physical and mental health needs of OA, people have begun to recognize the impact of the environment on health recovery. This has promoted innovations in the fields of healing architecture and healthcare design. Over time, the importance of spiritual care has gradually been incorporated, reflecting a comprehensive concern for the overall well-being of OA, including their emotional and spiritual needs. The integration of this concept has further driven the development of the optimal healing environment concept, aiming to comprehensively and systematically address the needs. Similar to the specialized research on male, female, and disabled populations, studies on specific groups like OA have also emerged and become increasingly popular. The development of healing spaces for OA is a multi-disciplinary and cross-domain integration process, reflecting society's increasing attention to the well-being of OA and the healthcare field's pursuit of humanistic care.

5. Discussion

5.1. Discussion of the Main Results

This paper analyzed the research on the impact of healing spaces on the elderly population using the WoS core database and Scopus database to determine the number of documents, journal classifications, research scope, institutional and author collaboration distribution, and the changes in research hotspots in this field. Using CiteSpace for domain analysis, the study has plotted the evolution of research hotspots and predicted their development trends. This study has obtained co-occurrence maps and timeline visualizations of the key terms by analyzing the hotspots in the research on the impact of healing spaces on OA.

In terms of geographic distribution, the research on healing spaces for OA is primarily concentrated in developed countries. However, as population ageing has become an important issue in developing countries [140], both developing and developed countries are taking proactive measures and exploring the construction of healing spaces for the ageing population, such as geriatric wards and integrated elderly care communities [141]. These initiatives are being incorporated into the overall future planning. Still, the specific actions taken vary [142,143] and must be tailored to the local context. Yet, most of the existing healing and wellness spaces targeting the elderly population are still experimental in practice and theoretical research. Updating or designing entirely new spaces that cater to the current and post-pandemic realities is a long-term process. At the same time, healing spaces for the elderly population are a crucial lever for future research and development. The uneven distribution of research and practical efforts across developed and developing countries highlights the need for a more globally inclusive approach to addressing the challenges and opportunities presented by population ageing. Collaborative efforts, knowledge sharing, and adapting best practices to local contexts will be essential in creating healing environments that meet the diverse needs of the elderly population worldwide.

5.2. The Basis of Research and the Future Development of Research

From the perspective of the research foundation, the main research themes have been the physical environment of healing spaces, mainly indoor air quality (IAQ), which has been shown to impact healing outcomes. While the research on IAQ is essential, it should not overlook other aspects of the physical environment, such as indoor thermal comfort, acoustic environment, and lighting conditions [144,145]. The elderly population is a unique user group in indoor environments [146]. Still, the development of healing spaces has not paid sufficient attention to this particular demographic's specific spatial environment requirements. This may be a promising direction for future research. This direction suggests that in future research on the health and well-being of the elderly population, the design and application of healing spaces will emphasize individualized needs, scientific evidence, and the interplay between the environment and patient well-being. Particularly for the elderly population, the development of these research directions will lead to the design and construction of healthcare facilities becoming more personalized and better able to meet their specific needs.

Furthermore, from the perspective of research hotspots, the focus has shifted more toward the psychological and spiritual aspects of the elderly user group, which is also a key area of in-depth development in current healing spaces [147]. Spiritual care is a subjective and dynamic concept, as it showcases the unique aspects of care and integrates all other elements [148]. However, the human spiritual healing effect is immeasurable by scientific means. Therefore, their exact effects and benefits are unpredictable [149]. More experimental research is needed to scientifically prove the spiritual healing effects that healing spaces can bring to OA. The subjective and intangible nature of spiritual well-being presents challenges in quantifying and validating its impact, which requires a nuanced, multi-faceted approach to research and evaluation.

Additionally, technological and equipment factors are important in improving the health and well-being of OA in healing spaces. The research focus has been on the technologically enhanced healing effects brought by the physical architectural environment and the improved healing effects on the psychological issues of the elderly population through changes in design elements. The elderly user group of healing spaces values the physical environmental context of the healing process and the changes in healing equipment during the healing process [10]. Most research exists within the theme of updating equipment comfort, as the increased ageing-friendliness of individual elements can lead to an overall increase in the healing effects of the environment, such as the updating of technologically advanced furniture, integrated ventilation systems, and safe coatings [19,150]. However, little attention has been paid to the psychological effects of equipment on the elderly user group [151]. In the 21st century, the rapid development of internet-connected devices has enabled the integration of consumer electronics and medical equipment, potentially allowing every space to become a healing space. However, some issues remain to be addressed, such as patient compliance, battery life, and concerns about safety and privacy [152]. Incorporating technology and equipment in healing spaces for OA represents an important frontier, as it can enhance the physical, psychological, and functional aspects of the healing experience. Future research should explore the optimal integration of these elements, balancing the technological capabilities with the unique needs and preferences of the elderly population to create healing spaces that are truly empowering and restorative.

As the global population continues to age, how to provide high-quality healthcare services and life security for the growing elderly population has become a significant challenge facing society. In this context, healing spaces, as an important platform connecting medical care, rehabilitation, and elderly care, play a crucial role in improving the health and well-being of OA and have become a hot research topic for the future. Future research on the impact of healing spaces on the health and well-being of OA should consider multiple aspects, including the physical, psychological, and social dimensions, to achieve a comprehensive improvement in the health and well-being of OA through integrated design. In the psychological and social domains, it is essential to pay attention to the loneliness faced by the elderly population, as social exclusion, lack of social connections, and weak ties with other users within the space can lead to a strong sense of rejection [153]. Current research tends to focus more on cognitively intact elderly individuals, overlooking the psychological issues faced by those with cognitive impairments and the lack of awareness among healing space staff regarding OAs' feelings of isolation [154]. In the physical environment, future research on healing spaces should also focus on the characteristics of the elderly population, emphasizing the integration of design and technology to meet the physiological healing needs of OA and optimizing the safety and comfort of the healing environment. For example, establishing support services and physical environments based on the highly diverse needs and preferences of OA [155] and utilizing electronic computing technologies (such as DHL technology) to manage the health and well-being of the elderly population [156]. However, research should also consider the cognitive decline, mental fatigue, and functional decline caused by electronic information technology among OA. Therefore, the impact of future healing spaces on the health and well-being of the elderly population will be far-reaching. This will improve the quality of life for OA and drive the healthcare field towards a more comprehensive, humanistic, and scientific direction, ultimately achieving better health management and quality of life for the elderly population.

5.3. Research Limitations

This study acknowledges several limitations in terms of data sources, analytical methods, and analytical perspectives that warrant further refinement and expansion. Regarding data sources, the study relied solely on the literature from the WoS and Scopus core databases, excluding other vital databases such as Google Scholar. This may restrict the representativeness and breadth of the research sample. Future research could consider integrating multiple databases to enhance the reliability of the findings. Additionally, the analysis was limited to English literature, overlooking publications in other languages, which could introduce bias. Regarding analytical methods, the study employed only the LLR algorithm for cluster analysis in constructing the scientific knowledge graph. Combining multiple association-analysis algorithms could potentially improve the accuracy of the analysis, an area for improvement in subsequent research. Furthermore, the analytical perspective of this study has spatial limitations, primarily focusing on the analysis and interpretation of core subject areas. In contrast, the exploration of non-core subject areas lacks depth and comprehensiveness. While bibliometric software analysis can aid in objectively identifying research frontiers and development trends, it still necessitates in-depth literature reading and meticulous examination to obtain more precise and comprehensive research conclusions. In summary, future research in this domain should consider the following refinements: (1) expanding data sources by integrating multiple databases, including non-English literature, (2) employing a combination of analytical methods to enhance the accuracy of the analysis, and (3) broadening the analytical perspective to include more comprehensive exploration of both core and non-core subject areas. These improvements will provide more reliable and nuanced insights into the research landscape and development trends.

6. Conclusions

This study systematically reviews the current status, knowledge base, and development trends of research on the role of healing spaces in promoting the health and well-being among the elderly population based on bibliometric analysis and visualization exploration. In summary, this study drew the following inferences:

- The temporal analysis reveals that research in this field has progressed through three stages: slow start, steady development, and continued growth. While developed countries initially dominated, relevant research has also recently emerged in developing countries. Different countries and regions should strengthen cooperative efforts and integrate global research to drive innovative progress.

- Analysis of highly cited literature and clustering shows that the research foundations cover multiple aspects of healing spaces, including physical and chemical environments (e.g., indoor air quality), biological factors, environmental design, and psychological health of OA. The research frontier is shifting towards a greater focus on addressing OAs' overall physical, mental, and social needs.
- Keyword analysis reflects a transition from a singular focus on physical environments towards a more interdisciplinary and integrated approach, with greater emphasis on the holistic psychological, social, and spiritual needs of OA. This transition reflects deepening societal awareness of ageing-related issues and a trend towards more humanized, individualized healthcare.
- Currently, the research foundation primarily concentrates on physical environmental factors, particularly the impact of indoor air quality on physiological health. However, future efforts need to place greater emphasis on individualized needs. Research hotspots are shifting towards the psychological and spiritual needs of OA, but a scientific evaluation system for assessing spiritual healing effects remains an area requiring further exploration. Technologies and equipment play a crucial role in optimizing healing spaces and enhancing the health and well-being of OA. However, their usage experience in this age group must also be considered. The integration of emerging technologies should adhere to a user-centered design philosophy to comprehensively meet the diverse needs of OA.

Overall, research in this field is evolving toward an interdisciplinary, individualized, and systematic approach. Future efforts should further emphasize the spiritual needs of OA and establish a scientific evaluation system for spiritual healing. By integrating emerging technologies like the Internet of Things and artificial intelligence, personalized and intelligent therapeutic environments can be optimized to holistically address OA' physiological, psychological, social, and spiritual health and well-being, driving sustainable innovation.

Author Contributions: All authors contributed to the study's conception and design. Material preparation, data collection, and analysis were performed by X.Y. and T.G. The first draft of the manuscript was written by X.Y. and all authors commented on previous versions of the manuscript; T.G. was responsible for the review and editing. All authors have read and agreed to the published version of the manuscript.

Funding: This research did not receive any specific grant from funding agencies in the public, commercial, or not-for-profit sectors.

Data Availability Statement: Data are contained within the article.

Conflicts of Interest: The authors have no relevant financial or non-financial interests to disclose. The authors declare no conflict of interest.

References

1. World Health Organization. Mental Health of Older Adults. Available online: https://www.who.int/news-room/fact-sheets/detail/mental-health-of-older-adults (accessed on 8 January 2024).
2. Nakano, M.M.; Otonari, T.S.; Takara, K.S.; Carmo, C.M.; Tanaka, C. Physical performance, balance, mobility, and muscle strength decline at different rates in elderly people. *J. Phys. Ther. Sci.* **2014**, *26*, 583–586. [CrossRef] [PubMed]
3. Kleinman, A.; Chen, H.; Levkoff, S.E.; Forsyth, A.; Bloom, D.E.; Yip, W.; Khanna, T.; Walsh, C.J.; Perry, D.; Seely, E.W. Social technology: An interdisciplinary approach to improving care for older adults. *Front. Public Health* **2021**, *9*, 729149. [CrossRef]
4. Sourdet, S.; Lafont, C.; Rolland, Y.; Nourhashemi, F.; Andrieu, S.; Vellas, B. Preventable iatrogenic disability in elderly patients during hospitalization. *J. Am. Med. Dir. Assoc.* **2015**, *16*, 674–681. [CrossRef]
5. Firth, K.; Smith, K.; Sakallaris, B.R.; Bellanti, D.M.; Crawford, C.; Avant, K.C. Healing, a concept analysis. *Glob. Adv. Health Med.* **2015**, *4*, 44–50. [CrossRef] [PubMed]
6. Liu, Y.; Chen, L.; Jiang, H. Influencing factor extraction of healing environment identifiability based on environmental psychoanalysis. *Psychiatr. Danub.* **2022**, *34*, 620.

7. Jaskiewicz, J.G., Jr. *Dignity in Palliative Care: The Hospice at Skogafoss Falls*; Virginia Tech: Blacksburg, VA, USA, 2011.
8. Day, C. *Places of the Soul: Architecture and Environmental Design as a Healing Art*; Routledge: London, UK, 2017.
9. Harris, P.B.; McBride, G.; Ross, C.; Curtis, L. A place to heal: Environmental sources of satisfaction among hospital patients 1. *J. Appl. Soc. Psychol.* **2002**, *32*, 1276–1299. [CrossRef]
10. Schweitzer, M.; Gilpin, L.; Frampton, S. Healing spaces: Elements of environmental design that make an impact on health. *J. Altern. Complement. Med.* **2004**, *10*, S-71–S-83. [CrossRef]
11. Whinfield, A.; Aitkenhead, I. The light revival: Does phototherapy promote wound healing? A review. *Foot* **2009**, *19*, 117–124. [CrossRef] [PubMed]
12. Goldman, J. *Healing Sounds: The Power of Harmonics*; Simon and Schuster: New York, NY, USA, 2022.
13. Gola, M.; Settimo, G.; Capolongo, S. Indoor air in healing environments: Monitoring chemical pollution in inpatient rooms. *Facilities* **2019**, *37*, 600–623. [CrossRef]
14. Cuthbertson, D.; Fell, G.; Smith, C.M.; Tilstone, W. Metabolism after injury. 1: Effects of severity, nutrition, and environmental temperature on protein potassium, zinc, and creatine. *Br. J. Surg.* **1972**, *59*, 925–931. [CrossRef]
15. Marquardt, G.; Bueter, K.; Motzek, T. Impact of the design of the built environment on people with dementia: An evidence-based review. *HERD Health Environ. Res. Des. J.* **2014**, *8*, 127–157. [CrossRef] [PubMed]
16. Lemprecht, B. The gap between design and healing. *Metropolis* **1996**, *77*, 123.
17. Norberg-Schulz, C.; Loci, G. *Towards a Phenomenology of Architecture*; Rizzoli: New York, NY, USA, 1980; p. 18.
18. Sabar, D.P.; Djimantoro, M.I. The application of healing space concept in holistic care facilities: A brief guideline for design. *Proc. IOP Conf. Ser. Earth Environ. Sci.* **2020**, *426*, 012068. [CrossRef]
19. Sternberg, E.M. *Healing Spaces: The Science of Place and Well-Being*; Harvard University Press: Cambridge, MA, USA, 2009.
20. Cant, S. Medical pluralism, mainstream marginality or subaltern therapeutics? Globalisation and the integration of 'Asian'medicines and biomedicine in the UK. *Soc. Cult. South Asia* **2020**, *6*, 31–51. [CrossRef]
21. Hampshire, K.R.; Owusu, S.A. Grandfathers, Google, and dreams: Medical pluralism, globalization, and new healing encounters in Ghana. *Med. Anthropol.* **2013**, *32*, 247–265. [CrossRef]
22. Gola, M.; Settimo, G.; Capolongo, S. Chemical pollution in healing spaces: The decalogue of the best practices for adequate indoor air quality in inpatient rooms. *Int. J. Environ. Res. Public Health* **2019**, *16*, 4388. [CrossRef] [PubMed]
23. Moore, D. Designed to heal modern healthcare. *Chicago* **2000**, *30*, 32.
24. Ozcan, H. Healing design: A holistic approach to social interaction in pediatric intensive care units in the United States and Turkey. In Proceedings of the 1st International CIB Endorsed METU Postgraduate Conference Built Environment & Information Technologies, Ankara, Turkey, 17–18 March 2006; pp. 244–245.
25. Aldridge, A.P.; Barbosa, C.; Barocas, J.A.; Bush, J.L.; Chhatwal, J.; Harlow, K.J.; Hyder, A.; Linas, B.P.; McCollister, K.E.; Morgan, J.R. Health economic design for cost, cost-effectiveness and simulation analyses in the HEALing Communities Study. *Drug Alcohol Depend.* **2020**, *217*, 108336. [CrossRef] [PubMed]
26. Kirmayer, L.J. The cultural diversity of healing: Meaning, metaphor and mechanism. *Br. Med. Bull.* **2004**, *69*, 33–48. [CrossRef]
27. Capolongo, S. Preface. Social aspects and well-being for improving healing processes' effectiveness. *Ann. Ist. Super. Sanita* **2016**, *52*, 11–14.
28. Ulrich, R.S. View through a window may influence recovery from surgery. *Science* **1984**, *224*, 420–421. [CrossRef] [PubMed]
29. Barlas, D.; Sama, A.E.; Ward, M.F.; Lesser, M.L. Comparison of the auditory and visual privacy of emergency department treatment areas with curtains versus those with solid walls. *Ann. Emerg. Med.* **2001**, *38*, 135–139. [CrossRef]
30. DuBose, J.; MacAllister, L.; Hadi, K.; Sakallaris, B. Exploring the concept of healing spaces. *HERD Health Environ. Res. Des. J.* **2018**, *11*, 43–56. [CrossRef] [PubMed]
31. Diette, G.B.; Lechtzin, N.; Haponik, E.; Devrotes, A.; Rubin, H.R. Distraction therapy with nature sights and sounds reduces pain during flexible bronchoscopy: A complementary approach to routine analgesia. *Chest* **2003**, *123*, 941–948. [CrossRef]
32. Innes, A.; Kelly, F.; Dincarslan, O. Care home design for people with dementia: What do people with dementia and their family carers value? *Aging Ment. Health* **2011**, *15*, 548–556. [CrossRef]
33. Moore, A.; Carter, B.; Hunt, A.; Sheikh, K. 'I am closer to this place'—Space, place and notions of home in lived experiences of hospice day care. *Health Place* **2013**, *19*, 151–158. [CrossRef]
34. Meng, L.; Zhu, C.; Wen, K.-H. Research on Constructing a Healing Environment for the Street Spaces of a High-Density City: Using Street Spaces in Macao's Old City Area. *Int. J. Environ. Res. Public Health* **2020**, *17*, 4767. [CrossRef] [PubMed]
35. Dijkstra, K.; Pieterse, M.; Pruyn, A. Physical environmental stimuli that turn healthcare facilities into healing environments through psychologically mediated effects: Systematic review. *J. Adv. Nurs.* **2006**, *56*, 166–181. [CrossRef]
36. DeMello, K. *Healing through Design: The Psychological Effects of Design on the Elderly*; University of Hawaii at Manoa: Honolulu, HI, USA, 2016.
37. Sakallaris, B.R.; Macallister, L.; Voss, M.; Smith, K.; Jonas, W.B. Optimal healing environments. *Glob. Adv. Health Med.* **2015**, *4*, 40–45. [CrossRef]
38. Marques, B.; McIntosh, J.; Kershaw, C. Healing spaces: Improving health and wellbeing for the elderly through therapeutic landscape design. *Int. J. Arts Humanit.* **2019**, *3*, 20–34.
39. Zhang, J.; Cenci, J.; Becue, V.; Koutra, S.; Ioakimidis, C.S. Recent evolution of research on industrial heritage in Western Europe and China based on bibliometric analysis. *Sustainability* **2020**, *12*, 5348. [CrossRef]

40. Zhu, Y.; Koutra, S.; Zhang, J. Zero-carbon communities: Research hotspots, evolution, and prospects. *Buildings* **2022**, *12*, 674. [CrossRef]
41. Falagas, M.E.; Pitsouni, E.I.; Malietzis, G.A.; Pappas, G. Comparison of PubMed, Scopus, web of science, and Google scholar: Strengths and weaknesses. *FASEB J.* **2008**, *22*, 338–342. [CrossRef] [PubMed]
42. Birkle, C.; Pendlebury, D.A.; Schnell, J.; Adams, J. Web of Science as a data source for research on scientific and scholarly activity. *Quant. Sci. Stud.* **2020**, *1*, 363–376. [CrossRef]
43. United Nations. *Madrid International Plan of Action on Ageing*; United Nations: New York, NY, USA, 2002.
44. Gaminiesfahani, H.; Lozanovska, M.; Tucker, R. A scoping review of the impact on children of the built environment design characteristics of healing spaces. *HERD Health Environ. Res. Des. J.* **2020**, *13*, 98–114. [CrossRef]
45. MacAllister, L.; Bellanti, D.; Sakallaris, B.R. Exploring inpatients' experiences of healing and healing spaces: A mixed methods study. *J. Patient Exp.* **2016**, *3*, 119–130. [CrossRef] [PubMed]
46. Schrank, S.; Ekici, D. *Healing Spaces, Modern Architecture, and the Body*; Routledge: London, UK, 2016.
47. Broadus, R.N. Toward a definition of "bibliometrics". *Scientometrics* **1987**, *12*, 373–379. [CrossRef]
48. Donthu, N.; Kumar, S.; Mukherjee, D.; Pandey, N.; Lim, W.M. How to conduct a bibliometric analysis: An overview and guidelines. *J. Bus. Res.* **2021**, *133*, 285–296. [CrossRef]
49. Chen, C. Science mapping: A systematic review of the literature. *J. Data Inf. Sci.* **2017**, *2*, 1–40. [CrossRef]
50. Chen, C. CiteSpace II: Detecting and visualizing emerging trends and transient patterns in scientific literature. *J. Am. Soc. Inf. Sci. Technol.* **2006**, *57*, 359–377. [CrossRef]
51. Garfield, E.; Pudovkin, A.I.; Istomin, V.S. Algorithmic citation-linked historiography—Mapping the literature of science. *Proc. Am. Soc. Inf. Sci. Technol.* **2002**, *39*, 14–24. [CrossRef]
52. Simboli, B.; Zhang, M. *Clustering Concepts*; American Association for the Advancement of Science: Washington, DC, USA, 2004.
53. Van Eck, N.; Waltman, L. Software survey: VOSviewer, a computer program for bibliometric mapping. *Scientometrics* **2010**, *84*, 523–538. [CrossRef] [PubMed]
54. Liu, Q.; Ye, Y. A study on mining bibliographic records by designed software SATI: Case study on library and information science. *J. Inf. Resour. Manag.* **2012**, *2*, 50–58.
55. Wu, C.; Cenci, J.; Wang, W.; Zhang, J. Resilient city: Characterization, challenges and outlooks. *Buildings* **2022**, *12*, 516. [CrossRef]
56. Chen, C. The citespace manual. *Coll. Comput. Inform.* **2014**, *1*, 1–84.
57. World Health Organization. *Health and Ageing: A Discussion Paper*, preliminary version ed.; World Health Organization: Geneva, Switzerland, 2001.
58. WHO. Active Ageing: A Policy Framework. *Aging Male* **2002**, *5*, 1–37. [CrossRef]
59. WHO. *International Plan of Action on Ageing: Report on Implementation*; WHO: Rome, Italy, 2004
60. He, W.; Goodkind, D.; Kowal, P.R. *An Aging World: 2015*; United States Census Bureau: Washington, DC, USA, 2016.
61. The United Nations. *United Nations Conference on Sustainable Development*; The United Nations: New York, NY, USA, 2012.
62. WHO. *World Report on Ageing and Health*; WHO: Rome, Italy, 2015; p. 267.
63. Fang, X.; Ming, C.; Cen, Y.; Lin, H.; Zhan, K.; Yang, S.; Li, L.; Cao, G.; Li, Q.; Ma, X. Post-sequelae one year after hospital discharge among older COVID-19 patients: A multi-center prospective cohort study. *J. Infect.* **2022**, *84*, 179–186. [CrossRef]
64. Wu, H.; Cheng, K.; Guo, Q.; Yang, W.; Tong, L.; Wang, Y.; Sun, Z. Mapping knowledge structure and themes trends of osteoporosis in rheumatoid arthritis: A bibliometric analysis. *Front. Med.* **2021**, *8*, 787228. [CrossRef]
65. Lu, H.; Feng, Y. A measure of authors' centrality in co-authorship networks based on the distribution of collaborative relationships. *Scientometrics* **2009**, *81*, 499–511. [CrossRef]
66. Settimo, G.; Manigrasso, M.; Avino, P. Indoor air quality: A focus on the European legislation and state-of-the-art research in Italy. *Atmosphere* **2020**, *11*, 370. [CrossRef]
67. Capolongo, S.; Rebecchi, A.; Buffoli, M.; Appolloni, L.; Signorelli, C.; Fara, G.M.; D'Alessandro, D. COVID-19 and cities: From urban health strategies to the pandemic challenge. A decalogue of public health opportunities. *Acta Bio Medica Atenei Parm.* **2020**, *91*, 13.
68. Nyrud, A.Q.; Lundin, S.; Blomdahl, C.; Caira, C. Wood in psychiatric in-patient rooms may reduce the length of stay for patients. In Proceedings of the 4th Conference on Architecture Research Care & Health (ARCH19), Trondheim, Norway, 12–13 June 2019.
69. MacAllister, L.; Zimring, C.; Ryherd, E. Environmental variables that influence patient satisfaction: A review of the literature. *HERD Health Environ. Res. Des. J.* **2016**, *10*, 155–169. [CrossRef] [PubMed]
70. Besterman-Dahan, K.; Bradley, S.E.; Arriola, N.B.; Downs, K. Blocking out the rabbit holes: Photovoice and community gardening as a healing space for veterans. *Wellbeing Space Soc.* **2021**, *2*, 100061. [CrossRef]
71. Lin, W.-H.; Juan, Y.-K. Examining the association between healing environments and work performance. *J. Build. Eng.* **2024**, *84*, 108624. [CrossRef]
72. Chahin, S.; Afify, A.; Mohsen, H.; Youssef, M. Role of 3D printed green walls in healing architecture. *BAU J.-Health Wellbeing* **2022**, *5*, 1. [CrossRef]
73. Settimo, G. Existing guidelines for indoor air quality: The case study of hospital environments. In *Indoor Air Quality in Healthcare Facilities*; Springer: Berlin/Heidelberg, Germany, 2017; pp. 13–26.
74. Nartova-Bochaver, S.K.; Mukhortova, E.A. Interaction with the plant world as a source of positive human functioning. *Психотерапия* **2020**, *28*, 152.

75. Moscato, U.; Borghini, A.; Teleman, A.A. HVAC management in health facilities. In *Indoor Air Quality in Healthcare Facilities*; Springer: Berlin/Heidelberg, Germany, 2017; pp. 95–106.
76. Gola, M.; Mele, A.; Tolino, B.; Capolongo, S. Applications of IAQ monitoring in international healthcare systems. In *Indoor Air Quality in Healthcare Facilities*; Springer: Berlin/Heidelberg, Germany, 2017; pp. 27–39.
77. Settimo, G.; Gola, M.; Mannoni, V.; De Felice, M.; Padula, G.; Mele, A.; Tolino, B.; Capolongo, S. Assessment of indoor air quality in inpatient wards. In *Indoor Air Quality in Healthcare Facilities*; Springer: Berlin/Heidelberg, Germany, 2017; pp. 107–118.
78. Rodin, J. Aging and health: Effects of the sense of control. *Science* **1986**, *233*, 1271–1276. [CrossRef]
79. Capolongo, S.; Settimo, G. Indoor Air Quality in Healing Environments: Impacts of Physical, Chemical, and Biological Environmental Factors on Users. In *Indoor Air Quality in Healthcare Facilities*; Capolongo, S., Settimo, G., Gola, M., Eds.; Springer International Publishing: Cham, Switzerland, 2017; pp. 1–11. [CrossRef]
80. Capolongo, S. *Architecture as a Generator of Health and Well-Being*; SAGE Publications: London, UK, 2014; Volume 3. [CrossRef]
81. Nyhan, M.; Sobolevsky, S.; Kang, C.; Robinson, P.; Corti, A.; Szell, M.; Streets, D.; Lu, Z.; Britter, R.; Barrett, S.R.; et al. Predicting vehicular emissions in high spatial resolution using pervasively measured transportation data and microscopic emissions model. *Atmos. Environ.* **2016**, *140*, 352–363. [CrossRef]
82. Gola, M.; Settimo, G.; Capolongo, S. Indoor air quality in inpatient environments: A systematic review on factors that influence chemical pollution in inpatient wards. *J. Healthc. Eng.* **2019**, *2019*, 8358306. [CrossRef]
83. D'Alessandro, D.; Tedesco, P.; Rebecchi, A.; Capolongo, S. Water use and water saving in Italian hospitals. A preliminary investigation. *Ann. Ist. Super. Sanita* **2016**, *52*, 56–62.
84. Salingaros, N.A. *Biophilia and Healing Environments*; Terrapin Bright Green: New York, NY, USA, 2015; pp. 1–44.
85. Dziechciaz, M.; Filip, R. Biological psychological and social determinants of old age: Bio-psycho-social aspects of human aging. *Ann. Agric. Environ. Med.* **2014**, *21*, 835–838. [CrossRef] [PubMed]
86. Kumar, J.; Patel, T.; Sugandh, F.; Dev, J.; Kumar, U.; Adeeb, M.; Kachhadia, M.P.; Puri, P.; Prachi, F.; Zaman, M.U. Innovative approaches and therapies to enhance neuroplasticity and promote recovery in patients with neurological disorders: A narrative review. *Cureus* **2023**, *15*, e41914. [CrossRef]
87. Settimo, G.; Gola, M.; Capolongo, S. The relevance of indoor air quality in hospital settings: From an exclusively biological issue to a global approach in the Italian context. *Atmosphere* **2020**, *11*, 361. [CrossRef]
88. Gola, M.; Settimo, G.; Capolongo, S. How Can Design Features and Other Factors Affect the Indoor Air Quality in Inpatient Rooms? Check-Lists for the Design Phase, Daily Procedures and Maintenance Activities for Reducing the Air Concentrations of Chemical Pollution. *Int. J. Environ. Res. Public Health* **2020**, *17*, 4280. [CrossRef]
89. Jaapar, S.A.B. Spatial Quality in Space Design for Promoting Self-Healing in Healthcare Design. Master's Thesis, Universiti Teknologi Malaysia, Skudai, Malaysia, 2017.
90. Woo, J.-Y. A Study on the Emotional design approach and the Value of the Space for Healing the Trauma-Focus on the periodical characteristics and changes. *Korean Inst. Inter. Des. J.* **2010**, *19*, 206–215.
91. Golden, R.N.; Gaynes, B.N.; Ekstrom, R.D.; Hamer, R.M.; Jacobsen, F.M.; Suppes, T.; Wisner, K.L.; Nemeroff, C.B. The efficacy of light therapy in the treatment of mood disorders: A review and meta-analysis of the evidence. *Am. J. Psychiatry* **2005**, *162*, 656–662. [CrossRef] [PubMed]
92. Kutash, M.; Northrop, L. Family members' experiences of the intensive care unit waiting room. *J. Adv. Nurs.* **2007**, *60*, 384–388. [CrossRef] [PubMed]
93. Rashid, M. Environmental design for patient families in intensive care units. *J. Healthc. Eng.* **2010**, *1*, 367–398. [CrossRef]
94. Cleary, M.; Hunt, G.; O'Connor, N.; Snars, J. The Concord Centre for Mental Health's new 'phase of illness' model of care: Are we on the right track? *Australas. Psychiatry* **2010**, *18*, 246–249. [CrossRef]
95. Caspari, S.; Eriksson, K.; Nåden, D. The importance of aesthetic surroundings: A study interviewing experts within different aesthetic fields. *Scand. J. Caring Sci.* **2011**, *25*, 134–142. [CrossRef]
96. Applebaum, D.; Fowler, S.; Fiedler, N.; Osinubi, O.; Robson, M. The impact of environmental factors on nursing stress, job satisfaction, and turnover intention. *JONA J. Nurs. Adm.* **2010**, *40*, 323–328. [CrossRef] [PubMed]
97. Connellan, K.; Gaardboe, M.; Riggs, D.; Due, C.; Reinschmidt, A.; Mustillo, L. Stressed spaces: Mental health and architecture. *HERD Health Environ. Res. Des. J.* **2013**, *6*, 127–168. [CrossRef] [PubMed]
98. Ulrich, R.S. Effects of interior design on wellness: Theory and recent scientific research. *J. Health Care Inter. Des.* **1991**, *3*, 97–109. [PubMed]
99. Ulrich, R.S.; Zimring, C.; Zhu, X.; DuBose, J.; Seo, H.-B.; Choi, Y.-S.; Quan, X.; Joseph, A. A review of the research literature on evidence-based healthcare design. *HERD Health Environ. Res. Des. J.* **2008**, *1*, 61–125. [CrossRef]
100. Gamage, S.D.; Ambrose, M.; Kralovic, S.M.; Roselle, G.A. Water safety and Legionella in health care: Priorities, policy, and practice. *Infect. Dis. Clin.* **2016**, *30*, 689–712.
101. Garrison, L.E.; Kunz, J.M.; Cooley, L.A.; Moore, M.R.; Lucas, C.; Schrag, S.; Sarisky, J.; Whitney, C.G. *Vital Signs: Deficiencies in Environmental Control Identified in Outbreaks of Legionnaires' Disease—North America, 2000–2014*; Wiley Online Library: Hoboken, NJ, USA, 2016; Volume 16, pp. 3049–3058.
102. Khoshnava, S.M.; Rostami, R.; Mohamad Zin, R.; Štreimikienė, D.; Mardani, A.; Ismail, M. The Role of Green Building Materials in Reducing Environmental and Human Health Impacts. *Int. J. Environ. Res. Public Health* **2020**, *17*, 2589. [CrossRef]
103. Cincinelli, A.; Martellini, T. Indoor Air Quality and Health. *Int. J. Environ. Res. Public Health* **2017**, *14*, 1286. [CrossRef]

104. Montagna, M.T.; De Giglio, O.; Cristina, M.L.; Albertini, R.; Pasquarella, C. Legionella Indoor Air Contamination in Healthcare Environments. In *Indoor Air Quality in Healthcare Facilities*; Capolongo, S., Settimo, G., Gola, M., Eds.; Springer International Publishing: Cham, Switzerland, 2017; pp. 63–71. [CrossRef]
105. Andriani, A.D.; Mulyana, D.; Dida, S.; Wahyudin, U. The role of healing environment in reducing the stress of patients with non-communicable disease. *J. Nat. Sci. Biol. Med.* **2021**, *12*, 300–306.
106. Adams, R.I.; Bhangar, S.; Dannemiller, K.C.; Eisen, J.A.; Fierer, N.; Gilbert, J.A.; Green, J.L.; Marr, L.C.; Miller, S.L.; Siegel, J.A. Ten questions concerning the microbiomes of buildings. *Build. Environ.* **2016**, *109*, 224–234. [CrossRef]
107. Ghanizadeh, F.; Godini, H. A review of the chemical and biological pollutants in indoor air in hospitals and assessing their effects on the health of patients, staff and visitors. *Rev. Environ. Health* **2018**, *33*, 231–245. [CrossRef]
108. Chawla, H.; Anand, P.; Garg, K.; Bhagat, N.; Varmani, S.G.; Bansal, T.; McBain, A.J.; Marwah, R.G. A comprehensive review of microbial contamination in the indoor environment: Sources, sampling, health risks, and mitigation strategies. *Front. Public Health* **2023**, *11*, 1285393. [CrossRef] [PubMed]
109. Lei, H.; Li, Y.; Xiao, S.; Lin, C.H.; Norris, S.L.; Wei, D.; Hu, Z.; Ji, S. Routes of transmission of influenza A H1N1, SARS CoV, and norovirus in air cabin: Comparative analyses. *Indoor Air* **2018**, *28*, 394–403. [CrossRef] [PubMed]
110. D'Alessandro, D.; Fara, G.M. Hospital environments and epidemiology of healthcare-associated infections. In *Indoor Air Quality in Healthcare Facilities*; Springer: Berlin/Heidelberg, Germany, 2017; pp. 41–52.
111. D'Amico, A.; Fara, G. The need to develop a multidisciplinary expertise for the microbiological safety of operating theatres. *Ann. Ig. Med. Prev. Comunità* **2016**, *28*, 379–380.
112. Śmiełowska, M.; Marć, M.; Zabiegała, B. Indoor air quality in public utility environments—A review. *Environ. Sci. Pollut. Res.* **2017**, *24*, 11166–11176. [CrossRef]
113. Signorelli, C.; Fara, G.M.; Odone, A.; Zangrandi, A. The reform of the Italian Constitution and its possible impact on public health and the National Health Service. *Health Policy* **2017**, *121*, 90–91. [CrossRef]
114. Ackley, A.; Olanrewaju, O.I.; Oyefusi, O.N.; Enegbuma, W.I.; Olaoye, T.S.; Ehimatie, A.E.; Ukpong, E.; Akpan-Idiok, P. Indoor environmental quality (IEQ) in healthcare facilities: A systematic literature review and gap analysis. *J. Build. Eng.* **2024**, *86*, 108787. [CrossRef]
115. Rangga, F.; Shaari, N.; Ismail, S.; Yuli, N.G.; Salmarisha, D. Hospital interior design with healing environment approach. *AIP Conf. Proc.* **2024**, *3001*, 040001.
116. Iyendo, T.O.; Uwajeh, P.C.; Ikenna, E.S. The therapeutic impacts of environmental design interventions on wellness in clinical settings: A narrative review. *Complement. Ther. Clin. Pract.* **2016**, *24*, 174–188. [CrossRef] [PubMed]
117. Alzoubi, H.; Bataineh, R.F. Pre-versus post-occupancy evaluation of daylight quality in hospitals. *Build. Environ.* **2010**, *45*, 2652–2665. [CrossRef]
118. Manca, S.; Bonaiuto, M.; Fornara, F. Perceived hospital environment quality indicators: The case of healthcare places for terminal patients. *Buildings* **2022**, *13*, 57. [CrossRef]
119. Moslehian, A.S.; Roös, P.B.; Gaekwad, J.S.; Van Galen, L. Potential risks and beneficial impacts of using indoor plants in the biophilic design of healthcare facilities: A scoping review. *Build. Environ.* **2023**, *233*, 110057. [CrossRef]
120. Kujundzic, K.; Stamatovic Vuckovic, S.; Radivojević, A. Toward Regenerative Sustainability: A Passive Design Comfort Assessment Method of Indoor Environment. *Sustainability* **2023**, *15*, 840. [CrossRef]
121. Ulrich, R.S. Effects of healthcare environmental design on medical outcomes. In *Design and Health: Proceedings of the Second International Conference on Health and Design*; Svensk Byggtjanst: Stockholm, Sweden, 2001; p. 59.
122. Gashoot, M.M. Holistic Healing Framework: Impact of the Physical Surrounding Design on Patient Healing and Wellbeing. *Art Des. Rev.* **2021**, *10*, 18–28. [CrossRef]
123. Dovjak, M.; Shukuya, M.; Krainer, A. User-centred healing-oriented conditions in the design of hospital environments. *Int. J. Environ. Res. Public Health* **2018**, *15*, 2140. [CrossRef]
124. Li, H.; Mahyuddin, N. The impact of physical environment on health-care workers' well-being in Chinese hospitals during COVID-19 pandemic. *Facilities* **2023**, *41*, 675–690. [CrossRef]
125. Cao, Y.; Huang, L. Research on the Healing Effect Evaluation of Campus' Small-Scale Courtyard Based on the Method of Semantic Differential and the Perceived Restorative Scale. *Sustainability* **2023**, *15*, 8369. [CrossRef]
126. Lorien, L.; Blunden, S.; Madsen, W. Implementation of recovery-oriented practice in hospital-based mental health services: A systematic review. *Int. J. Ment. Health Nurs.* **2020**, *29*, 1035–1048. [CrossRef] [PubMed]
127. Bhalla, R.; Chowdhary, N.; Ranjan, A. Spiritual tourism for psychotherapeutic healing post COVID-19. *J. Travel Tour. Mark.* **2021**, *38*, 769–781. [CrossRef]
128. Lutz, W.; Sanderson, W.; Scherbov, S. The coming acceleration of global population ageing. *Nature* **2008**, *451*, 716–719. [CrossRef]
129. Fu, E.; Ren, Y.; Li, X.; Zhang, L. Research on the Healing Potential of Rural Community Streets from the Perspective of Audiovisual Integration: A Case Study of Four Rural Communities in China. *Front. Public Health* **2022**, *10*, 861072. [CrossRef]
130. Wampold, B.E.; Flückiger, C. The alliance in mental health care: Conceptualization, evidence and clinical applications. *World Psychiatry* **2023**, *22*, 25–41. [CrossRef]
131. Andrews, G.J.; Duff, C. Matter beginning to matter: On posthumanist understandings of the vital emergence of health. *Soc. Sci. Med.* **2019**, *226*, 123–134. [CrossRef] [PubMed]

132. Solomon, B.; Sutton, D.; McKenna, B. The experience and meaning of recovery-oriented practice for nurses working in acute mental health services. *Int. J. Ment. Health Nurs.* **2021**, *30*, 963–974. [CrossRef] [PubMed]
133. Yoo, I.-Y. A study on healing environmental factors in the elderly patients with long-term care. *Int. J. Bio-Sci. Bio-Technol.* **2015**, *7*, 267–276. [CrossRef]
134. Naylor, M.D.; Brooten, D.A.; Campbell, R.L.; Maislin, G.; McCauley, K.M.; Schwartz, J.S. Transitional care of older adults hospitalized with heart failure: A randomized, controlled trial. *J. Am. Geriatr. Soc.* **2004**, *52*, 675–684. [CrossRef]
135. Gola, M.; Settimo, G.; Capolongo, S. Reaching Sustainability in Healthcare: Strategies for a Healthy Indoor Air Quality in Healing Environments. In Proceedings of the New Metropolitan Perspectives: Knowledge Dynamics, Innovation-Driven Policies Towards the Territories' Attractiveness, Reggio Calabria, Italy, 18–23 May 2020; Volume 1, pp. 166–177.
136. Jonas, W.B.; Chez, R.A. Toward optimal healing environments in health care. *J. Altern. Complement. Med.* **2004**, *10*, S-1–S-6. [CrossRef]
137. Nagda, B.A.; López, B. Intergroup Dialogue for Social Healing: Creating Spaces of Collective Hope and Transformation. In *Global Perspectives on Dialogue in the Classroom: Cultivating Inclusive, Intersectional, and Authentic Conversations*; Springer: Berlin/Heidelberg, Germany, 2022; pp. 101–114.
138. Tan, S.-Y.; Dong, N.J. Spiritual interventions in healing and wholeness. In *Faith and Health: Psychological Perspectives*; The Guilford Press: New York, NY, USA, 2001; pp. 291–310.
139. Wahl, H.-W.; Oswald, F. Environmental perspectives on ageing. *SAGE Handb. Soc. Gerontol.* **2010**, *124*, 111.
140. Sterns, H.L.; Miklos, S.M. The aging worker in a changing environment: Organizational and individual issues. *J. Vocat. Behav.* **1995**, *47*, 248–268. [CrossRef]
141. Wilson, K.; Rosenberg, M.W.; Abonyi, S. Aboriginal peoples, health and healing approaches: The effects of age and place on health. *Soc. Sci. Med.* **2011**, *72*, 355–364. [CrossRef]
142. Lyberaki, A.; Tinios, P.; Mimis, A.; Georgiadis, T. Mapping population aging in Europe: How are similar needs in different countries met by different family structures? *J. Maps* **2013**, *9*, 4–9. [CrossRef]
143. Li, L.; Loo, B.P. Mobility impairment, social engagement, and life satisfaction among the older population in China: A structural equation modeling analysis. *Qual. Life Res.* **2017**, *26*, 1273–1282. [CrossRef] [PubMed]
144. Khodakarami, J.; Nasrollahi, N. Thermal comfort in hospitals–A literature review. *Renew. Sustain. Energy Rev.* **2012**, *16*, 4071–4077. [CrossRef]
145. Rasheed, H. Therapeutic Architecture Illumination and Environments Within Healing Space. *Preprints* **2024**, 2024021305. [CrossRef]
146. Almeida-Silva, M.; Wolterbeek, H.T.; Almeida, S. Elderly exposure to indoor air pollutants. *Atmos. Environ.* **2014**, *85*, 54–63. [CrossRef]
147. Perriam, G. Sacred spaces, healing places: Therapeutic landscapes of spiritual significance. *J. Med. Humanit.* **2015**, *36*, 19–33. [CrossRef]
148. Ramezani, M.; Ahmadi, F.; Mohammadi, E.; Kazemnejad, A. Spiritual care in nursing: A concept analysis. *Int. Nurs. Rev.* **2014**, *61*, 211–219. [CrossRef]
149. Culliford, L. Spiritual care and psychiatric treatment: An introduction. *Adv. Psychiatr. Treat.* **2002**, *8*, 249–258. [CrossRef]
150. Zhuan, S. Research on Aging-friendly Home Space Design Based on Humanization. *Acad. J. Archit. Geotech. Eng.* **2023**, *5*, 46–50.
151. Huisman, E.R.; Morales, E.; Van Hoof, J.; Kort, H.S. Healing environment: A review of the impact of physical environmental factors on users. *Build. Environ.* **2012**, *58*, 70–80. [CrossRef]
152. Engineer, A.; Sternberg, E.M.; Najafi, B. Designing interiors to mitigate physical and cognitive deficits related to aging and to promote longevity in older adults: A review. *Gerontology* **2018**, *64*, 612–622. [CrossRef]
153. Jansson, A.H.; Karisto, A.; Pitkälä, K.H. Listening to the voice of older people: Dimensions of loneliness in long-term care facilities. *Ageing Soc.* **2023**, *43*, 2894–2911. [CrossRef]
154. Paque, K.; Bastiaens, H.; Van Bogaert, P.; Dilles, T. Living in a nursing home: A phenomenological study exploring residents' loneliness and other feelings. *Scand. J. Caring Sci.* **2018**, *32*, 1477–1484. [CrossRef] [PubMed]
155. Guzman, V.; Doyle, F.; Foley, R.; Craven, P.; Crowe, N.; Wilson, P.; Smith, A.; Hegarty, G.; Pertl, M.M. Socio-ecological determinants of older people's mental health and well-being during COVID-19: A qualitative analysis within the Irish context. *Front. Public Health* **2023**, *11*, 1148758. [CrossRef] [PubMed]
156. Dong, Q.; Liu, T.; Liu, R.; Yang, H.; Liu, C. Effectiveness of digital health literacy interventions in older adults: Single-arm meta-analysis. *J. Med. Internet Res.* **2023**, *25*, e48166. [CrossRef]

buildings

MDPI

Article

Age-Friendly Environment Design of High-Speed Railway Stations from a Healthy Ageing Perspective: A Case Implementation in Nanjing, China

Ying Cao *, Danming Li, Xuewen Gao, Shibo Bi, Kexin Yu and Di Zhou

School of Design Art & Media, Nanjing University of Science and Technology, Nanjing 210094, China; lidanming@njust.edu.cn (D.L.); gaoxuewen@njust.edu.cn (X.G.); bishibo@njust.edu.cn (S.B.); yukexin@njust.edu.cn (K.Y.); zhoudi@njust.edu.cn (D.Z.)
* Correspondence: caoying@njust.edu.cn; Tel.: +86-13912966176

Abstract: China stands out as one of the countries with the largest and fastest-growing elderly populations. Elderly requirements for the quality of high-speed railway stations environment are increasing. The relationship between the built environment of high-speed railway stations and elderly health has become an urgent problem to be solved. The challenge necessitates a systematic approach to age-friendly design. Based on the practical experience of the authors in an age-friendly design of an indoor environment at Nanjing South Railway Station, questionnaire surveys and user interviews were employed to profile four types of typical elderly passengers. Data collection methods were used to create behavior maps. (1) This study identifies the age-friendly needs of elderly passengers, pinpointing the key environment issues. (2) It proposes age-friendly design strategies for various functional areas of Nanjing South Railway Station. (3) It introduces an age-friendly module combination design system, which including basic type I, enhanced type II, and optimized type III, in order to meet the demands for six levels of high-speed railway stations in China. This study promotes the age-friendly transformation of environments at a high-speed railway station, offering theoretical foundations and methodological references for constructing an age-friendly environment from a healthy ageing perspective.

Keywords: healthy ageing; high-speed railway station; age-friendly design; elderly passengers

Citation: Cao, Y.; Li, D.; Gao, X.; Bi, S.; Yu, K.; Zhou, D. Age-Friendly Environment Design of High-Speed Railway Stations from a Healthy Ageing Perspective: A Case Implementation in Nanjing, China. *Buildings* **2024**, *14*, 3280. https://doi.org/10.3390/buildings14103280

Academic Editor: Theodore Stathopoulos

Received: 21 August 2024
Revised: 19 September 2024
Accepted: 15 October 2024
Published: 16 October 2024

1. Introduction

1.1. Research Background

China has now entered a moderately aged society, becoming one of the countries with the largest and fastest-growing age populations globally. According to data from the National Bureau of Statistics of China in 2023, the population aged 60 and above has reached 290 million, accounting for 21.1% of the total population [1]. The age process of the Chinese population is accelerating, and it has become a fundamental issue affecting the national economy and Chinese livelihood. As the proportion of elderly individuals in urban populations continues to grow, the construction of age-friendly indoor environments has become an important part of the current development strategy in China. The 14th Five-Year Plan for Healthy Age issued by the National Health Commission of China emphasized the need to strengthen the construction of age-friendly indoor environments to meet the health needs of the elderly [2].

China is currently experiencing a rapid development in smart transportation and urbanization. As key nodes in the transportation hub system, high-speed railway stations serve as transfer points for various modes of transportation and are given significant attention by the state in their development and construction. However, the indoor environments of high-speed railway stations in China are primarily designed for healthy young and middle-aged passengers, without considering the rapidly increasing elderly passenger population as an independent group and their specific needs within these environments. This oversight results in inconvenience and poor travel experiences for elderly passengers. For example, during station entry and ticket inspection, young passengers and middle-aged passengers can easily use their mobile phones to scan quick response codes or documents at electronic ticketing gates. In contrast, elderly passengers, who are more accustomed to paper tickets, may encounter difficulties with electronic ticketing. Therefore, incorporating age-friendly designs into transportation hub environments has become an inevitable trend. China urgently needs to age-friendly design of indoor environments for high-speed railway stations to quickly address the increasingly severe age traffic situation. Moreover, the rapid development of the internet and the widespread application of big data provide opportunities for age-friendly environment design at high-speed railway stations. The Guiding Opinions of the State Council on Actively Promoting the Internet Plus Initiative explicitly proposes promoting the development of smart indoor environments for the elderly [3].

This paper takes a forward-looking approach to the core strategy of age-friendly construction at high-speed railway stations in the era of intelligent transportation, tightly integrating age-friendly design theory with practical engineering. Based on the authors' practical experience with age-friendly design of an indoor environment at Nanjing South Railway Station, it shifts from fragmentary facility adaptations to systematic age-friendly design research for functional areas, providing case references for age-friendly design at different levels of high-speed railway stations in China. Additionally, it offers multiple theoretical bases and methodological references for planning and constructing age-friendly indoor environments and uses the demonstration effect of practical design projects to promote the age-friendly transformation of urban transportation hubs, facilitating healthy travel for the elderly and advancing the construction and development of age-friendly environments.

1.2. High-Speed Railway Stations in China

The high-speed railway construction of China has created an extensive transportation network, making intercity connections closer. According to data from the National Railway Administration of China in 2023, the railway passenger volume of China has reached 3.68 billion, with a total high-speed railway mileage of 45,000 km. Currently, high-speed railway construction remains a key focus on the development of urban railway transit in China. According to the 14th Five-Year Plan for Railway Safety Development, the operating mileage of high-speed railway in China will reach 50,000 km, covering over 95% of cities with populations of more than 500,000 people [4]. As high-speed railway construction rapidly advances, high-speed railway stations have also become a crucial part of national infrastructure development.

High-speed railway stations integrate subway stations, bus stops, parking lots, and taxi stands into a large comprehensive building, forming a hub station that combines various modes of transportation, facilitating both internal and external urban transportation functions. The primary purpose of high-speed railway stations is to handle the passenger flow and transfers of intercity, high-speed, and conventional railways. The stations mainly serve passengers while providing connections to other modes of transportation. Currently, the indoor environment of high-speed railway stations in China is characterized by comprehensive transportation, three-dimensional layouts, diversified functions, and intelligent operations. According to the "National Railway Station Grading Method", Chinese high-speed railway stations are divided into six grades [5]. Stations with a daily passenger flow of over 60,000 people are classified as special-class stations, those with over 15,000 people are first-class stations, over 5000 people are second-class stations, over 2000 people are third-class stations, fourth-class stations do not meet the comprehensive business requirements of third-class stations, and fifth-class stations are passing and overtaking stations. Different grades of high-speed railway stations vary in size, functional services, and the scale and level of facilities. Nanjing South Railway Station is a special-class station, Ma'anshan East Station is a second-class station, Liyang Railway Station is a third-class station, and Qidong Railway Station is fourth-class station (Table 1).

Table 1. Different levels of high-speed railway stations.

Special-Class Station	Second-Class Station	Third-Class Station	Fourth-Class Station
Nanjing South Railway Station	Ma'anshan East Station	Liyang Railway Station	Qidong Railway Station

1.3. *Age-Friendly Design of Indoor Environments at High-Speed Railway Stations*

As elderly passengers experience a decline in physical functions, they have specific indoor environment needs at high-speed railway stations. These needs include barrier-free access, appropriately scaled spaces, and comprehensive age-friendly facilities. Additionally, to prevent feelings of loneliness and loss in elderly passengers, age-friendly design of indoor environments at high-speed railway stations must consider their environmental needs for a safe, comfortable, private, and social interaction. For example, in terms of safety, emergency call buttons need to be installed in the restrooms at the high-speed railway station to allow elderly passengers to quickly call for help in case of emergencies. In terms of comfort, special seating with armrests and soft cushions needs to be provided in the waiting areas to offer a more comfortable waiting experience for elderly passengers. Therefore, the age-friendly design must be based on the physiological and psychological needs and behavioral characteristics of elderly passengers. It should integrate the spatial scale and functional needs of their activities, ensuring safety and convenience in functional areas within the station while adhering to accessibility design standards. Moreover, the design should consider the relationship between elderly passengers and other passengers, such as those with disabilities, young adults, and children. An inclusive design should be employed to ensure that the age-friendly environment does not inconvenience other passengers.

The age-friendly design of indoor environments at high-speed railway stations encompasses five main aspects: functional areas, traffic flow, environmental facilities, information guidance, and environmental ambiance.

(1) Age-friendly functional areas: This involves the rational planning of areas utilized by elderly passengers and enhancing the age-friendly features of these areas. It includes the establishment of dedicated security check areas for elderly passengers, enhancing the functionality of waiting areas, adding age-friendly service areas, improving the age-friendliness of commercial areas, and designing exhibition areas that cater to elderly aesthetic preferences.

(2) Age-friendly traffic flow: This includes ensuring the smooth connection of barrier-free routes, ensuring the rationality of traffic flow for elderly passengers, and minimizing their walking distances. For example, the main entrance should be conveniently located, with no steps or obstacles, a wide opening, ample space in front of the door, and clear signage and good lighting [6].

(3) Age-friendly environment facilities: This primarily involves improving barrier-free facilities, enhancing the age-friendly aspects of facilities, adding service facilities for

elderly passengers, and assisting elderly passengers in using smart facilities. For example, toilets and sanitation facilities should have clear signage, sufficient operational space, appropriate transfer options, and well-positioned equipment [6].

(4) Age-friendly information guidance: This includes setting up clear and comprehensible directional signs to avoid cognitive burdens for elderly passengers, employing multimedia guidance that combines text and images, and helping elderly passengers overcome the digital divide. Considering that some elderly passengers may have cognitive impairments, the information signage system should be designed with simplified interfaces, ensuring that all types of signage are clear and easy to understand, to better meet their wayfinding needs.

(5) Age-friendly environment ambiance: This involves adopting indoor design styles that are comfortable for elderly passengers, creating a safe and comfortable travel environment at high-speed railway stations, and incorporating natural ecological elements to enhance the therapeutic atmosphere.

1.4. Literature Review

Currently, most literature on age-friendly environment design focuses on senior residential buildings and urban public landscapes, emphasizing barrier-free systems, indoor environmental equipment, and outdoor facilities planning. Lee and Nam [7], based on universal design guidelines of South Korea and a green building certification system, proposed an eco-friendly residential design guideline. Li and Zhou [8] reviewed the academic research and design practices related to livable environments for the elderly in China. Liu and Yang [9] analyzed the impact of the built neighborhood environments on the outdoor leisure activities of the elderly. Hu and Chang [10] used space syntax to retrofit the planning layout of recreational facilities in old residential areas for age-friendliness. Marston et al. [11] elaborated on the concept and case studies of age-friendly smart environment systems in the post-pandemic era. Van et al. [12] explored the key issues in the building environment for elderly friendly community activities in urban areas. Van and Marston [13] also discussed the latest advancements in innovative design methods and technologies for age-friendly community housing.

Existing literature on the age-friendly design of indoor environments at high-speed railway stations often starts from the physiological conditions and behavioral characteristics of elderly passengers, studying the design of barrier-free systems, age-friendly facilities, and transfer areas within high-speed railway stations. Tan [14] proposed design principles and barrier-free design strategies for public transportation facilities aimed at the elderly. Kuo and Tang [15] conducted age-friendly research on the environmental facilities within subway stations based on the needs of elderly passengers. Zhao et al. [16] explored the feasibility of various transfer schemes for elderly passengers at high-speed railway stations and suggested improvements for the transfer environment. Zhang [17] proposed optimization paths for barrier-free design and humanized facilities at high-speed railway stations in China. Liu et al. [18] analyzed the travel characteristics and transfer methods of elderly passengers in urban railway transit in Denver, Colorado. Zhu and Shen [19] analyzed the conflicts between the sequence of railway transit transfer areas and the physiological characteristics of elderly passengers, proposing age-friendly design strategies for these transfer areas. Xie et al. [20] examined the issues of elderly passengers when using services at high-speed railway stations in China and offered suggestions to improve their travel experience.

In summary, existing literature primarily focuses on the age-friendly design details of high-speed railway stations, particularly in terms of environmental facility design and barrier-free system design. While these studies have made some progress and provide a foundational basis for this paper, the overall research is lacking in systematic studies on age-friendly design for the entire environment of high-speed railway stations. Additionally, existing literature tends to focus on theoretical research related to age-friendly design for high-speed railway stations, with a disconnect between theoretical concepts and practical design applications. This paper aims to address these research gaps by categorizing the functional areas of indoor environments at high-speed railway stations. Based on the practical experience of the authors in the age-friendly design of the indoor environment at Nanjing South Railway Station, the Shanghai Railway Bureau commissioned our design team to optimize the interior environment of Nanjing South Railway Station. The aim is to transform the station into a high-quality railway station with a more rational age-friendly functional space, a more distinct urban and station cultural character, and improved age-friendly facilities. This paper identifies the age-friendly needs of four typical elderly passengers, explores age-friendly design pathways for various functional areas at Nanjing South Railway Station, and proposes a modular combination design system for age-friendly functional areas.

2. Study Area and Methodology

2.1. Study Area

2.1.1. Overview of the Indoor Environment at Nanjing South Railway Station

As the largest high-speed railway station in Asia, Nanjing South Railway Station is a landmark transportation hub in China, embodying the local historical heritage of ancient Jinling city. Situated at the intersection of the Yangtze River and the Beijing–Shanghai Railway, and located along the urban development axis of Nanjing, Nanjing South Railway Station is not only a crucial high-speed railway station but also a key node in urban public transportation, connecting multiple high-speed railway lines and urban railway transit routes. Nanjing South Railway Station, which began operation in 2011, has a building area of 458,000 square meters and handles a daily passenger flow of over 100,000 people. The station comprises three interconnected floors. The second floor above ground is the waiting layer for passengers, spacious enough to accommodate a large number of passengers during peak hours. This floor also includes commercial areas on its mezzanine layer, and it is connected to the platforms via elevators, escalators, and stairs. The first floor above ground is the arrival layer, the area for passengers to enter and exit the station, with designated pick-up and drop-off areas for taxis and buses. The underground floor is the transfer layer, providing access to urban transit, including transfer passages to Lines 1 and 3 of the Nanjing South Subway Station (Table 2).

Table 2. Indoor environment of Nanjing South Railway Station.

Location	Three Floors of Nanjing South Railway Station	
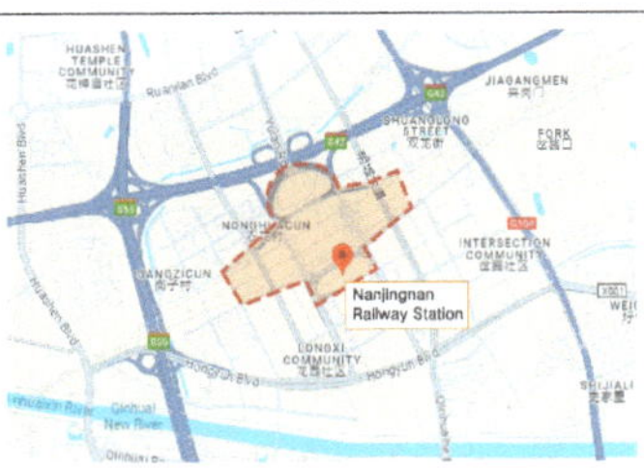	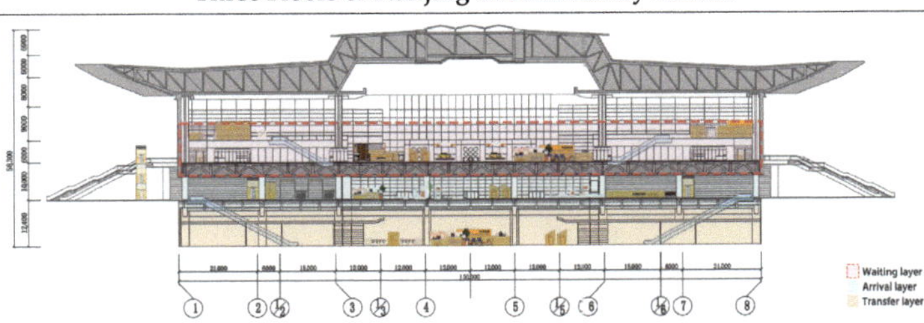	
−1F Transfer Floor	1F Arrival Floor	2F Waiting Floor

2.1.2. Functional Areas of the Indoor Environment at Nanjing South Railway Station

This design project, commissioned by the Shanghai Railway Bureau, focuses on the age-friendly design of the indoor environment at Nanjing South Railway Station. The functional areas consist of security check areas, waiting areas, service areas, and commercial areas. The security check areas, A1 and A2, are located at the north and south entrances, each with four security checkpoints. The waiting areas are divided into areas for regular passengers, VIP, and special passengers. The regular passenger waiting area B1 is symmetrically arranged on the east and west sides of the central axis. The VIP and special passenger waiting area B2 is located on the east and west sides of the central part, near the enquiry area. The service areas include a self-service area and an information service area. The self-service area C1 is located on the east side of the south entrance, which is relatively remote, while the information service area C2 is centrally located in the waiting area. The commercial areas are situated on the mezzanine layer, comprising dining shops D1 at the four corners and specialty product retail shops D2 along the connecting corridors (Figure 1).

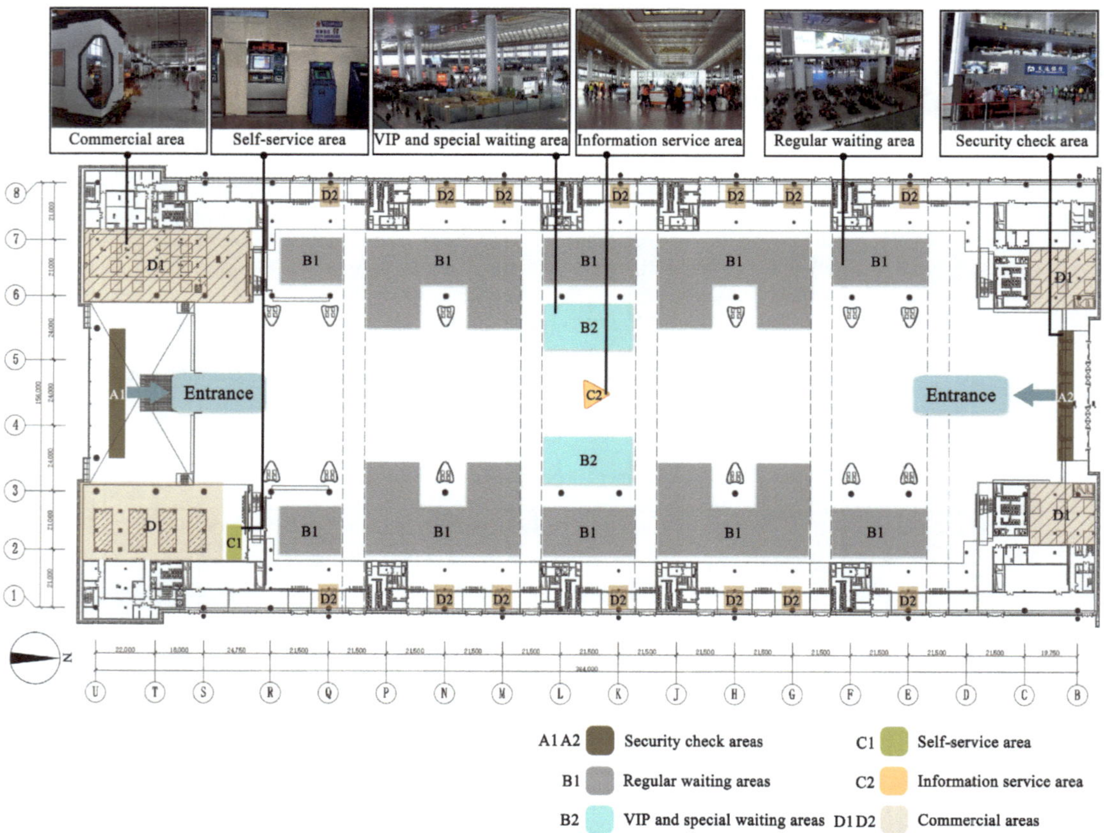

Figure 1. Previous function areas at Nanjing South Railway Station.

2.2. Methodology

2.2.1. Research Framework

This paper focuses on the age-friendly design of functional areas at high-speed railway stations. It begins by analyzing the theories and current age-friendly design of indoor environments at high-speed railway stations in China, clarifying the classification of high-speed railway stations and the current status of their environmental construction, as well as the concepts and contents of age-friendly design. Next, it outlines the age-friendly status of the indoor environment at Nanjing South Railway Station, analyzing the current functional areas. Then, it studies the age-friendly needs of elderly passengers in various functional areas at Nanjing South Railway Station by analyzing data on elderly passengers, establishing portraits of four typical elderly passengers, mapping their behavior, identifying age-friendly environment issues, and determining their needs for age-friendly design in different functional areas. The paper proceeds to design age-friendly strategies for each functional area at Nanjing South Railway Station, including functional zoning, security check areas, waiting areas, service areas, exhibition spaces, and commercial areas. Finally, it constructs a modular design system for age-friendly functional areas in high-speed railway stations, developing three models—basic type I, enhanced type II, and optimized type III—to meet the age-friendly design needs of six different levels of indoor environments at high-speed railway stations in China (Figure 2).

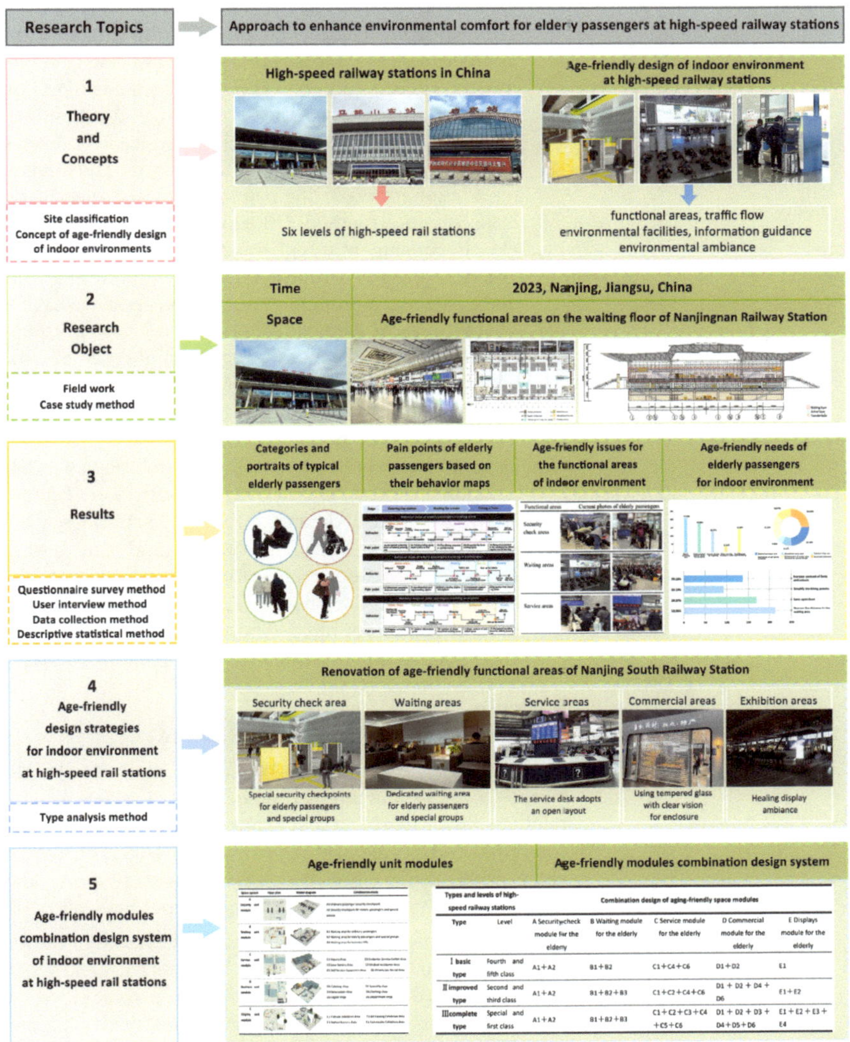

Figure 2. Research framework.

2.2.2. Research Methods

(1) Questionnaire survey method: This method is used to create a survey containing a series of questions to investigate the behaviors and attitudes of elderly passengers. From August to December 2023, this study targeted elderly passengers and their family members at high-speed railway stations as research objectives, distributing 324 questionnaires via an online survey platform and collecting 296 valid responses. The questionnaire content primarily included four aspects of elderly passengers: basic information, behavioral states, age-friendly pain points, and their demands for an age-friendly environment. The objective of the questionnaire survey was to clarify the types of elderly passengers at high-speed railway stations and establish portraits, identify their demands for age-friendly adaptations in various functional areas, and determine the main age-friendly issues, thus providing an objective basis for age-friendly design.
(2) User interview method: This qualitative analysis method involves interviewing and conversing with elderly passengers to supplement and verify the results of the questionnaire survey. This round of interviews used open-ended questions to delve deeply into the age-friendly issues and demands of elderly passengers in various functional areas at the high-speed railway station, understanding the underlying reasons. A total of 132 elderly passengers participated in the interviews, consisting of 68 males and 64 females aged between 60 and 90 years. The interview results provided a research foundation for exploring pathways for age-friendly design.
(3) Data collection method: Utilizing big data technology, this method collects GPS data of elderly passengers at high-speed railway stations, as well as data from smart devices and ticketing and station service software. By analyzing the GPS data provided by elderly passengers from their navigation apps, we have mapped their behavior patterns. This intelligent data collection method does not require manual data collection by researchers, significantly reducing labor costs. The data were characterized by large volume, quantifiability, and high credibility. It is used to analyze the different behavioral patterns of elderly passengers and establish behavior maps, further corroborating their age-friendly needs.
(4) Descriptive statistical method: This method uses tables, charts, and summary calculations to provide a statistical description of data characteristics. The statistical analysis of survey data mainly includes frequency analysis, central tendency analysis, dispersion analysis, distribution, and some basic statistical graphics. This study primarily uses descriptive statistics to categorize typical elderly passengers and their demands for age-friendly needs in various functional areas, specifically analyzing their demands for age-friendly adaptations in security check areas, waiting areas, service areas, commercial areas, and exhibition areas.

3. Results

3.1. Categories and Portraits of Typical Elderly Passengers

The elderly passengers at Nanjing South Railway Station are primarily aged between 60 and 69 years old, with their main travel purposes being business trips, tourism, visiting relatives and friends, and seeking medical treatment (Table 3). Based on their health conditions and travel purposes, elderly passengers are categorized into four types: traveling alone, traveling in a wheelchair, traveling with offspring, and traveling with peers. Portraits for these four typical elderly passenger types have been established.

(1) Elderly passengers traveling alone: Generally healthy, mobile, and highly educated. Their main travel purposes are business trips and visiting relatives and friends, with a travel frequency of about 13–16 times per year.
(2) Elderly passengers traveling in a wheelchair: Older, with health issues and mobility challenges, requiring assistance from others to travel by high-speed railway. Their main travel purpose is seeking medical treatment in other cities, with a travel frequency of about 4–6 times per year.

(3) Elderly passengers traveling with peers: Usually recently retired, healthy, and mobile. Their main travel purposes are group travel and visiting relatives and friends, with a travel frequency of about 8–12 times per year.

(4) Elderly passengers traveling with offspring: Usually in good health, with their main travel purposes being visiting relatives and friends. Their travel frequency is about 4–7 times per year (Figures 3 and 4).

Table 3. Individual characteristics of four typical elderly passengers (n = 132).

Demographic Data	Traveling Alone		Traveling by Wheelchair		Traveling with Children		Traveling with Peers		Total	
	n	%	n	%	n	%	n	%	n	%
Age (Years):										
60–69	41	45.05	1	1.10	26	28.57	23	25.27	91	68.93
70–79	12	42.86	2	7.14	3	10.71	11	39.29	28	21.21
80–89	0	0	0	0	9	90.00	1	10.00	10	7.57
90 years of age or older	0	0	1	33.33	2	66.67	0	0	3	2.27
Gender:										
Male	31	45.59	1	1.47	19	27.94	17	25.00	68	51.51
Female	22	34.38	3	4.69	21	32.81	18	28.13	64	48.48
Educational level:										
Elementary school and below	16	41.03	3	7.69	3	7.69	17	43.59	39	29.55
Junior high school	21	37.50	1	1.79	20	35.71	14	25.00	56	42.42
Senior high school	7	26.92	0	0	16	61.54	3	11.54	26	19.70
College and university	9	81.82	0	0	1	9.09	1	9.09	11	8.33
Travel purposes:										
Business trip	19	86.36	0	0	0	0	3	13.64	22	16.66
Tourism	2	5.88	0	0	17	50.00	15	44.12	34	25.75
Seeking medical treatment	8	25.81	3	9.68	18	58.06	2	6.45	31	23.48
Visiting relatives and friends	23	53.49	1	2.33	5	11.63	14	32.56	43	32.57
Other	1	50.00	0	0	0	0	1	50.00	2	1.51

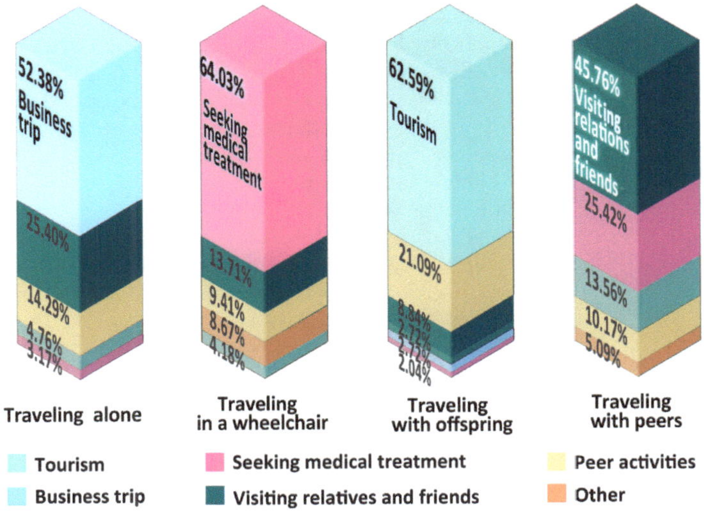

Figure 3. Types of typical elderly passengers.

Figure 4. Portraits of typical elderly passengers.

3.2. Pain Points of Elderly Passengers Based on Their Behavior Maps

To identify the pain points of elderly passengers, by analyzing the travel processes and behavioral experiences of 132 interviewed elderly passengers across five stages, which included the security check stage, waiting stage, commercial stage, service stage, and display stage, we have identified the main pain points faced by elderly passengers.

(1) Elderly passengers traveling alone: Their behavior map primarily includes moving from the waiting area to the tea room, restrooms, and shops. Their pain points include the distance of shops from the waiting area, difficulty in finding self-service ticketing machines, noisy waiting environments, and a lack of local cultural displays.

(2) Elderly passengers traveling in a wheelchair: These passengers need to use barrier-free elevators, accessible restrooms, and medical assistance services. Their pain points include difficulty finding the wheelchair service area, inconvenience during security checks while in a wheelchair, long overall distances to waiting area, incomplete support facilities, and a need for a therapeutic atmosphere.

(3) Elderly passengers traveling with peers: In addition to security checks and waiting, these passengers have business, service, and exhibition needs. Their pain points include the effort required for luggage security checks, unclear information signage, the distance of shops from the waiting area, and the monotonous content of exhibition areas.

(4) Elderly passengers traveling with offspring: Their behavior map, combined with the actions of their accompanying family members, focuses on security checks and waiting. Their pain points include the lack of dedicated security lanes for the elderly, noisy waiting environments, insufficient number of inquiry counters, and a lack of detailed service functions (Figure 5).

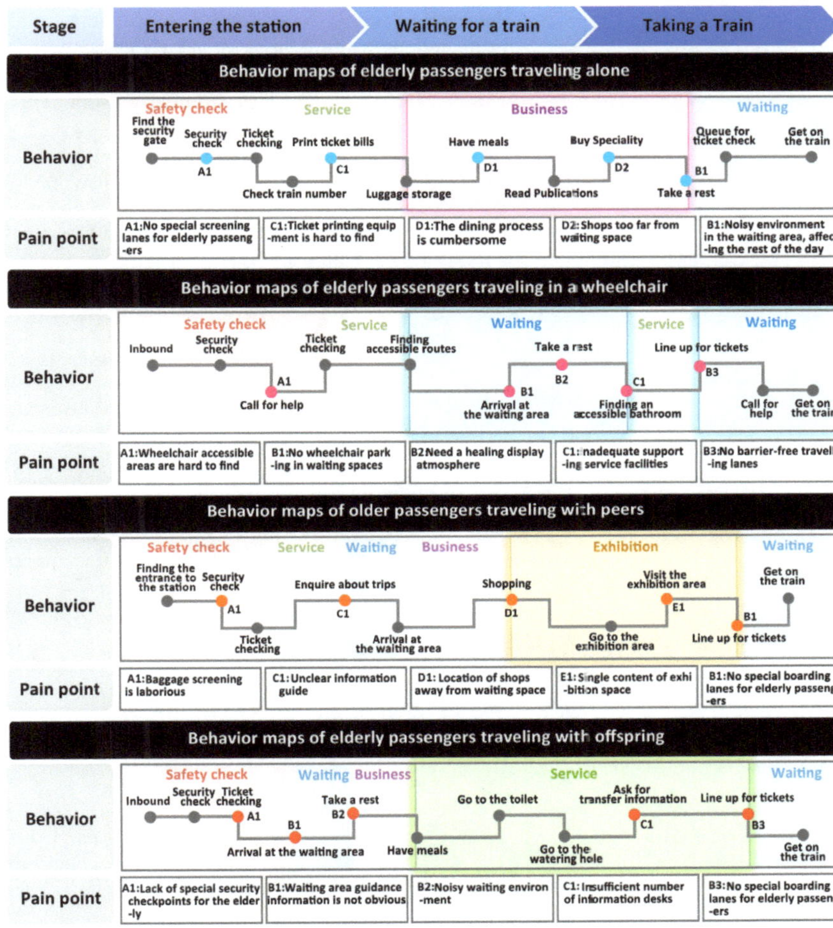

Figure 5. Behavioral maps of typical elderly passengers.

3.3. *Age-Friendly Issues for the Functional Areas of the Indoor Environment*

After investigating the functional areas of the indoor environment at Nanjing South Railway Station, we found that the current functional areas primarily focus on the safety and convenience of travel for elderly passengers with mobility issues, categorizing them under the special needs group alongside disabled individuals. The specific issues are as follows:

(1) Security check area: There is a lack of dedicated security lanes for elderly passengers, making security checks inconvenient for those in wheelchairs. Elderly passengers find it laborious to handle luggage during security checks, and the waiting times for security checks are long.
(2) Waiting area: There is no dedicated waiting areas for elderly passengers, and the waiting areas lack various supportive services for elderly passengers. Additionally, the overall distance to the waiting areas is long for elderly passengers.
(3) Service area: There are insufficient inquiry counters, and their functions are not well-defined. Currently, there is only one inquiry counter in the center of the waiting hall that handles inquiries, customer services, and medical assistance, among other functions. The facilities for elderly passengers within the service areas are inadequate.

There are not enough self-service information terminals, and the lack of guidance results in long waiting times for elderly passengers.

(4) Commercial area: The shops on the mezzanine layer are far from the waiting areas, scattered in layout, and lack a unified design. This results in low foot traffic and consequently low sales revenue.

(5) Exhibition area: There are no cultural display areas that appeal to elderly passengers, and there is a lack of a therapeutic atmosphere (Table 4).

Table 4. Age-friendly issues for functional areas of the indoor environment.

Functional Areas	Current Photos of Elderly Passengers	Age-Friendly Issues
Security Check Area	 	Lack of security channels for the elderly Difficulty in security checks for elderly passengers in wheelchairs Difficulty in luggage security for elderly passengers
Waiting Area	 	No waiting area for senior citizens Lack of supporting elderly service functions Far from the service areas
Service Area	 	Not enough inquiry desks Service functions are not detailed Inadequate facilities for the elderly
Commercial Area	 	Long distance from waiting areas Scattered store layout Lack of unified design
Exhibition Area	 	Lack of a regional cultural display area No psychological healing atmosphere Lack of ornamental value

3.4. *Age-Friendly Needs of Elderly Passengers for Functional Areas of the Indoor Environment*

The needs of elderly passengers are based on a sample of 296 valid survey responses. Security check areas are where elderly passengers undergo safety checks and baggage inspections. The age-friendly needs of elderly passengers for these areas include the following:

- 77.32% of elderly passengers wish for dedicated security check lanes for elderly and special passengers to ensure convenient passage and avoid delays caused by slower movement.
- 64.68% of elderly passengers look forward to the implementation of document-free security checks. By using facial recognition technology, elderly passengers can pass through without presenting tickets and identification documents, thereby skipping the ticket verification step.
- 46.47% of elderly passengers find it necessary to eliminate height differences in baggage inspection machines to facilitate the handling of large luggage.

- 51.68% of elderly passengers hope for the establishment of fast lanes for small luggage to increase the efficiency of the security check process.
- 12.64% of elderly passengers who use wheelchairs need wider security check lanes to accommodate their needs.

The waiting areas are where elderly passengers rest after obtaining ticket information, making it the area where they spend the most time. The age-friendly needs of elderly passengers for the waiting area include:

- 87.36% of elderly passengers hope for dedicated waiting areas for elderly and other special passengers, providing a quiet and comfortable environment.
- 68.03% of elderly passengers wish to shorten the distance between the waiting area and other functional areas, to reduce intersecting and confusing traffic flows.
- 53.53% of elderly passengers believe that medical assistance should be an essential facility within the waiting areas to provide emergency help in case of sudden illness.
- 34.94% of elderly passengers hope for the installation of accessible passageways at the ticket barriers within the waiting area.
- 9.29% of elderly passengers who use wheelchairs look forward to having wheelchair parking areas next to the waiting seats (Table 5).

Table 5. Age-friendly needs for waiting areas.

Age-Friendly Demand	Frequency	Percentage
Provision of waiting area for elderly passengers	235	87.36
Reducing distance between waiting area and functional areas	183	68.03
Medical assistance	144	53.53
Barrier-free access	94	34.94
Wheelchair parking area	25	9.29

The service areas provide various services for elderly passengers, including an inquiry area and customer service counters, a medical assistance area, an accessible restroom, a wheelchair rental area, a luggage storage area, and a ticket printing area. The age-friendly needs of elderly passengers for the service area include:

- 66.17% of elderly passengers hope for the establishment of a medical assistance area to address medical emergencies and be equipped with emergency medication and facilities.
- 63.57% of elderly passengers believe that the number of accessible restrooms should be increased.
- 21.56% of elderly passengers expect the establishment of a wheelchair rental area, with staff providing assistance.
- 58.36% of elderly passengers think that the functions of the customer service center should be refined, separating the functions of inquiry counters from the customer service center to improve service efficiency.
- 36.80% of elderly passengers believe that the age-friendly design of signage system in the service area should be strengthened to reduce cognitive load for elderly passengers.
- 13.75% of elderly passengers think a ticket printing area should be set up to meet the needs of self-service ticket printing for reimbursement purposes, with the equipment designed considering the height of elderly passengers (Figure 6).

Exhibition areas are important for creating emotional connections with elderly passengers through displayed content. They reflect the spirit of the high-speed railway station and convey the regional culture of the city. The age-friendly needs of elderly passengers for exhibition areas include:

- 43.49% of elderly passengers are interested in historical and cultural exhibition spaces and wish to learn about the local history and culture of Nanjing.
- 63.57% of elderly passengers expect the establishment of viewing exhibition areas to create a psychologically therapeutic atmosphere.

- 46.10% of elderly passengers hope for the setup of station culture exhibition areas to understand the spirit and character of Nanjing South Railway Station.

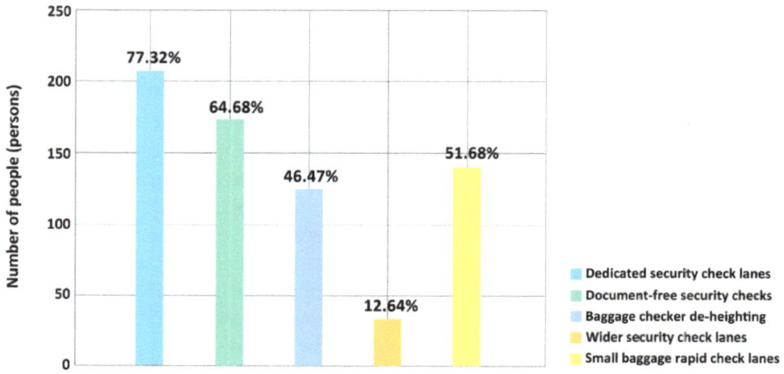

Figure 6. Age-friendly needs for service areas.

Commercial areas serve the dining and shopping needs of elderly passengers. Currently, the commercial areas at Nanjing South Railway Station are located on the mezzanine layer, far from the waiting areas, with some commercial areas having a high degree of enclosure. The age-friendly needs of elderly passengers for commercial areas include:

- 79.18% of elderly passengers believe that the distance between commercial areas and the waiting areas should be shortened so that they can shop while waiting.
- 63.20% of elderly passengers think that commercial areas should be semi-open, allowing them to stay informed about train schedules while shopping.
- 34.57% of elderly passengers believe that the dining service process should be simplified to make the dining experience more efficient.
- 50.93% of elderly passengers think that the font size in shops should be increased and the color contrast enhanced to convey shop information more effectively (Figure 7).

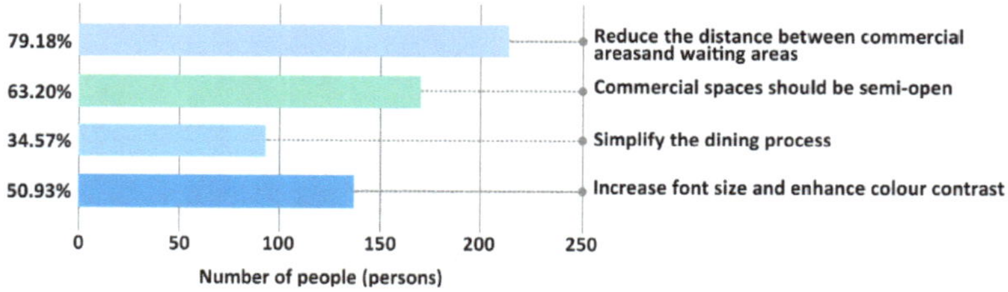

Figure 7. Age-friendly needs for commercial areas.

3.5. Age-Friendly Design Strategies for Indoor Environments at High-Speed Railway Stations

The renovated age-friendly functional areas of Nanjing South Railway Station include:

(1) Security check areas: Adding security check area for elderly and special passengers A3 at the north and south entrances to enhance the security check experience.

(2) Waiting areas: Adjusting waiting areas for elderly and special passengers B2 to the southeast corner of the south entrance and the northeast corner of the north entrance to shorten the travel distance for elderly passengers and avoid overlapping traffic flows with regular passenger waiting areas B1.

(3) Service areas: Adding south inquiry counter C1, customer service center C2, and north inquiry counter C3 to refine customer service functions. Equipping medical assistance area C4 and self-service equipment area C5 near the elderly and special passenger waiting area B2, and adding a C6 wheelchair rental area at the entrance.

(4) Commercial areas: Concentrating dining shops D1 on the mezzanine layer and setting up four rows of shops along the central axis. Unifying specialty products shops D2, newspapers shops D3, clothing shops D4, electronics shops D5, and general merchandise shops D6 near regular passenger waiting areas B1.

(5) Exhibition areas: Arranging cultural exhibition area E1, scenic exhibition area E2, station culture area E3, and product exhibition area E4 along the central axis at intervals. Creating a visual central corridor with multiple exhibition areas nodes combined with the display wall of the inquiry counter (Figure 8).

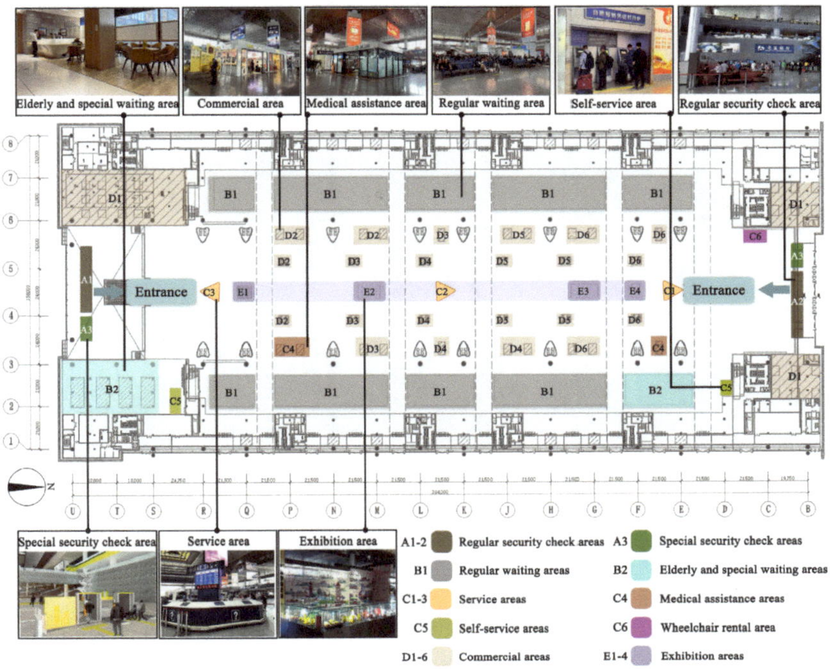

Figure 8. The renovated age-friendly function areas at Nanjing South Railway Station.

Upon arriving at the security check area, elderly passengers will benefit from the addition of dedicated security lanes for elderly and special passengers A3 to avoid prolonged waiting times. The width of the security lanes at the east and west entrances should be widened to accommodate elderly passengers in wheelchairs or those with special needs. Facial recognition technology should be implemented to allow elderly passengers to pass without showing tickets and IDs. Elderly passengers carrying small packages or traveling light can directly pass through the security gates to speed up the process. Lowering the height of luggage inspection machines will facilitate elderly passengers carrying large luggage to undergo inspections more conveniently (Table 6).

Dedicated waiting areas for elderly and special passengers B2 should be set up in the southeast and northeast corners of Nanjing South Railway Station. The navigation features of smart devices (such as smartphones and smartwatches) should be utilized to guide elderly passengers and other special groups from the security checkpoint to designated waiting areas. These areas should be equipped with additional service spaces such as medical assistance, inquiry counters, and a wheelchair rental area to reduce the distance that elderly passengers need to travel to access services. To enhance comfort, waiting areas for elderly passengers should include a wheelchair seating area to provide a stable environment for interaction. The seating height should be adjusted to alleviate difficulties in standing up after prolonged sitting. The seat width should be appropriately increased, and the seat depth should not be too deep to ensure that elderly passengers can rest their backs against the seat while keeping their feet flat on the ground. Service and medical buttons should be installed next to the seats to facilitate help requests. Additionally, ensuring unobstructed accessible pathways and consistent ground height will reduce the risk of movement for elderly passengers. The waiting area should ensure the smooth flow of accessible pathways, be free of steps or obstacles, provide ample maneuvering space, have wide openings and easily operable doors, and ensure good signage, lighting, and visual clarity [6]. Passenger information boards should account for cognitive impairments in elderly passengers by using a simple layout, displaying key information such as train numbers, platforms, and times in large fonts with high-contrast colors. Intuitive icons should be used to aid understanding, reducing the need for complex text descriptions.

Table 6. Age-friendly design of the security check area, waiting area, and service area.

	A Security Check Area	B Waiting Area	C Service Area			
Before Renovation						---
	A3 Security Check Areas for Elderly	B2 Waiting Area for Elderly	C2 Customer Service Center	C4 Medical Aid Area	C5 Self-service Equipment Area	C6 Wheelchair Rental Area
After Renovation						

Incorporating the traditional partition walls with the circular gates of Nanjing, the south inquiry desk C1 and north inquiry desk C3 will feature wall designs—frosted glass inscribed with poems celebrating the traditional culture of Nanjing. The counter of customer service center C2 should have an open layout, facilitating face-to-face communication between elderly passengers and service personnel. Surrounding the area are touch-screen electronic inquiry devices, allowing elderly passengers to independently access information. The medical aid area C4 should be equipped with emergency facilities, medicines, emergency call linkage devices, and a remote video consultation system, maintaining contact with the nearest hospital emergency center. The self-service equipment area C5 will be furnished with self-service devices, enabling elderly passengers to conveniently print tickets and other documents. The wheelchair rental area C6 will be located at the entrance and offers both on-site and online reservation options, providing wheelchair assistance to elderly passengers as needed. Navigation between service areas should be implemented by using the Internet of Things (IoT) to connect and coordinate devices across different zones. While navigation information is sent to the smart devices of elderly passengers, station staff can remotely operate the equipment in the target service area via the IoT system, such as preparing a wheelchair in advance at the C6 wheelchair rental station.

The dining shops D1 will be concentrated on the mezzanine layer, with escalators connecting the upper and lower layers, complemented by appropriate commercial signage and accessibility features. To guide elderly passengers to the dining shops, signage will feature larger fonts and enhanced color contrast. The dining shops will adopt a design style that feels friendly to elderly passengers and use lightweight tables and chairs to facilitate easy movement. The seat height should be set between 45–50 cm, with the seat surface made of soft, moderately elastic materials. The seat width should accommodate the body of elderly passengers, and armrests should be installed on both sides of the seat. The armrest height should be 20–25 cm from the seat surface to facilitate gripping and standing up. The armrests should have a rounded shape and be made of non-slip materials. To shorten lengthy traffic routes and increase shop revenues, specialty shops D2, newspaper shops D3, clothing shops D4, digital shops D5, and department stores D6 will be concentrated near the waiting areas for both elderly and regular passengers, arranged in four rows along the central axis, with a uniform design for shop sizes and facades. The shops will feature transparent tempered glass enclosures or semi-open designs to enhance the recognition of products and prevent elderly passengers from missing important vehicle information. Additionally, the shops will have island-style free shopping routes to attract both elderly and regular passengers (Table 7).

Table 7. Age-friendly design of commercial areas and exhibition areas.

	Before Renovation	After Renovation			
D Commercial Area	D2 Specialty Shop 	D1 Dining Shop 	D2 Specialty Shop 	D4 Clothing Shop 	D6 Department Shop 
E Exhibition Area	E2 Scenery Exhibition 	E1 Cultural Exhibition 	E2 AR Exhibition 	E3 Station Culture Area 	E4 Product Exhibition 

Cultural exhibition area E1 will be established to help elderly passengers understand the regional culture of Nanjing. Landmark architectures from the Ming and Qing dynasties will be depicted on Gaochun ceramic murals, displayed as scenic walls. Near the elderly passenger waiting room, AR viewing area E2 will combine physical and virtual landscapes to create a therapeutic exhibition atmosphere. Along the central axis, station culture area E3 will convey the humanitarian care of Nanjing South Railway Station, enhancing the travel experience for elderly passengers. Product exhibition area E4 will be set up to display various branded products from Jiangsu, including Yun brocade, Guoyuan liquor, and Hailan Home clothing.

3.6. Usability Testing

To evaluate the effectiveness of the age-friendly design modifications at Nanjing South Railway Station, our design team conducted a usability testing activity. We randomly selected 20 elderly passengers as test subjects and used a questionnaire to assess their satisfaction with the redesigned security check area, waiting area, service area, commercial area, and display area. The questionnaire consisted of five questions, each scored from 0 to 5, with higher scores indicating better evaluations. The rating criteria were as follows: A: 23–25 points, B: 20–22 points, C: 17–19 points, D: 14–16 points, and E: below 14 points.

The test results indicated that elderly passengers rated the overall renovation design of Nanjing South Railway Station as B, suggesting a high level of satisfaction with the age-friendly design. Among the areas evaluated, elderly passengers expressed the highest satisfaction with the waiting area, followed by the commercial area, service area, and display area, with the security check area receiving the lowest satisfaction rating. Further interviews revealed that some elderly passengers had difficulty operating the smart devices in the service areas, the display areas needed to create a more therapeutic atmosphere, and the security check area was crowded and noisy during peak times. There is room for improvement in the age-friendly environment and service efficiency of the security check area.

4. Discussion

4.1. Age-Friendly Unit Modules of Indoor Environments at High-Speed Railway Stations

The functional areas of high-speed railway stations are divided into five age-friendly modular units: security check unit module A, waiting unit module B, service unit module C, commercial unit module D, and exhibition unit module E. The security check unit module A includes regular passenger security check area A1 and security check area for elderly passengers and special groups A2. The waiting area unit module B includes regular passenger waiting area B1, waiting area for elderly passengers and special groups B2, and VIP waiting area B3. The service unit module C includes inquiry area C1, customer service center C2, care service area C3, medical assistance area C4, self-service equipment area C5, and wheelchair rental area C6. The commercial unit module D includes dining shops D1, specialty shops D2, newspaper shops D3, clothing shops D4, digital shops D5, and department stores D6. The exhibition unit module E includes cultural exhibition area E1, AR scenic exhibition area E2, station culture area E3, and product exhibition area E4 (Table 8).

Table 8. Age-friendly unit modules of indoor environments at high-speed railway stations.

Unit Modules	Modules Plan	Unit Model	Combination Mode
A Security Check Unit Module	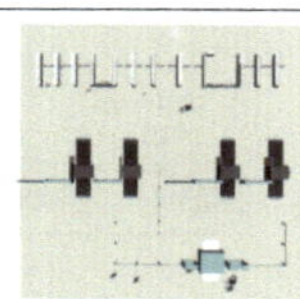	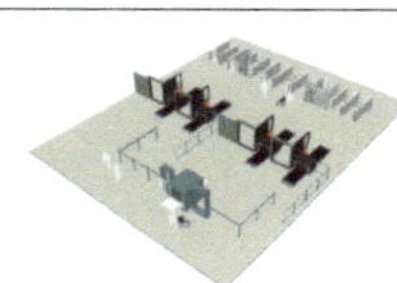	A1 Regular passenger security check area A2 Security check area for elderly passengers and special groups
B Waiting Unit Module			B1 Regular passenger waiting area B2 Waiting area for elderly passengers and special groups B3 VIP waiting area
C Service Unit Module	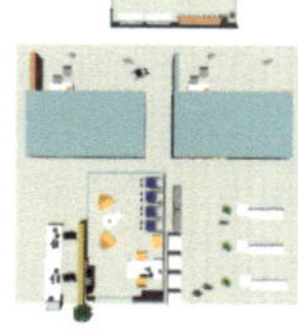		C1 Inquiry area C2 Customer service center C3 Care service area C4 Medical assistance area C5 Self-service equipment C6 Wheelchair rent
D Commercial Unit Module	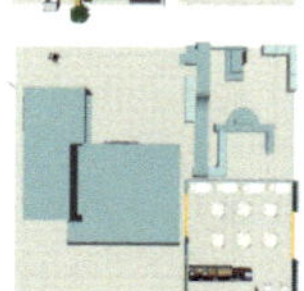		D1 Dining shop D2 Specialty shop D3 Newspaper shop D4 Clothing shop D5 Digital shop D6 Department shop
E Exhibition Unit Module			E1 Cultural exhibition area E2 AR scenic exhibition area E3 Station culture area E4 Product exhibition area

4.2. Age-Friendly Unit Modules Combination Design System of Indoor Environments at High-Speed Railway Stations

To meet age-friendly demands for six levels of high-speed railway stations in China, a modules combination design system of basic type I, enhanced type II, and optimized type III was constructed. This system allows for menu combinations of the five modules according to the specific conditions of different high-speed railway stations, ensuring the maximum universality of the age-friendly module combination design system (Table 9).

(1) The age-friendly modules combination design of basic type I is suitable for fourth- and fifth-class high-speed railway stations, meeting the basic needs of elderly passengers. The design modules mainly include the regular passenger security check area A1, security check area for elderly passengers and special groups A2, regular passenger waiting area B1, waiting area for elderly passengers and special groups B2, inquiry area C1, medical assistance area C4, wheelchair rental area C6, dining shops D1, specialty shops D2, and cultural exhibition area E1.
(2) The age-friendly modules combination design of enhanced type II is suitable for second- and third-class high-speed railway stations, focusing on the comfort experience of elderly passengers. The design modules mainly include the regular passenger security check area A1 and security check area for elderly passengers and special groups A2, regular passenger waiting area B1, waiting area for elderly passengers and special groups B2, and VIP waiting area B3, inquiry area C1, customer service center C2, medical assistance area C4, wheelchair rental area C6, dining shops D1, specialty shops D2, clothing shops D4, department stores D6, cultural exhibition area E1, and AR scenic exhibition area E2.
(3) The age-friendly modules combination design of optimized type III is suitable for first-class and special-class high-speed railway stations, focusing on providing comprehensive services for elderly passengers. The design modules mainly include the regular passenger security check area A1 and security check area for elderly passengers and special groups A2, regular passenger waiting area B1, waiting area for elderly passengers and special groups B2, VIP waiting area B3, inquiry area C1, customer service center C2, care service area C3, medical assistance area C4, self-service equipment area C5, wheelchair rental area C6, dining shops D1, specialty shops D2, newspaper shops D3, clothing shops D4, digital shops D5, department stores shop D6, cultural exhibition area E1, AR scenic exhibition area E2, station culture area E3, and product exhibition area E4.

Table 9. Age-friendly modules combination design system of basic type I, enhanced type II, and optimized type III.

Types and Levels		**Age-Friendly Modules Combination Design System**				
Types	Levels	A Security Check Unit Module	B Waiting Unit Module	C Service Unit Module	D Commercial Unit Module	E Exhibition Unit Module
Basic Type I	Fourth and Fifth Class	A1 + A2	B1 + B2	C1 + C4 + C6	D1 + D2	E1
Enhanced Type II	Second and Third Class	A1 + A2	B1 + B2 + B3	C1 + C2 + C4 + C6	D1 + D2 + D4 + D6	E1 + E2
Optimized Type III	Special and First Class	A1 + A2	B1 + B2 + B3	C1 + C2 + C3 + C4 + C5 + C6	D1 + D2 + D3 + D4 + D5 + D6	E1 + E2 + E3 + E4

5. Conclusions

To address the increasingly severe traffic challenges, design practice projects are being used to promote the transformation of indoor environments at high-speed railway station to be more age-friendly, providing theoretical and methodological references for an age-friendly indoor environment. This paper analyzes the current age-friendly indoor environments at high-speed railway stations in China. Based on the authors' design practice for the age-friendly design of the indoor environment at Nanjing South Railway Station, the paper explores the design strategies for age-friendly design by investigating the needs of typical elderly passengers for various functional areas at Nanjing South Railway Station. The paper proposes an age-friendly modules combination design system of basic type I, enhanced type II, and optimized type III. In subsequent studies, we will use cross-validation to further ensure the accuracy of data. The following conclusions were drawn.

The portraits for four types of typical elderly passengers were created. In the survey, the proportions of the four typical elderly passenger categories are as follows: 40.15% travel alone, 3.03% use wheelchairs, 30.30% travel with their offspring, and 26.52% travel with their peers. The specific needs for each category of elderly passengers are as follows:

- Elderly passengers traveling alone need a quiet and comfortable waiting environment, regional cultural display areas offering cultural experiences, and clear signage and navigation systems.
- Elderly passengers using wheelchairs require accessible elevators, barrier-free restrooms, and medical assistance service points.
- Elderly passengers traveling with peers need convenient luggage security check processes and a clear information signage system.
- Elderly passengers traveling with offspring need dedicated elderly security check lanes and additional inquiry counters.

Their age-friendly needs for functional areas were defined as follows:

(1) Security check areas need to add dedicated security check channels for elderly and special passengers, implement paperless security checks, eliminate height differences in luggage screening machines, and widen security check channel widths.
(2) Waiting areas need to set up exclusive waiting areas for elderly and other special passengers, provide wheelchair parking areas, and nearby medical aid areas, and shorten the distance between waiting areas and other functional areas.
(3) Service areas should be including a medical aid area, accessible restroom, wheelchair rental area, and ticket printing area, refining the functions of a customer service center.
(4) Exhibition areas need to set up cultural exhibition areas, viewing areas, and station-style exhibition areas.
(5) Commercial areas need to shorten the distance between commercial areas and waiting areas, adopt semi-open forms, simplify dining service processes, and enlarge shop fonts and enhance color contrast.

This paper outlines the key strategies for the age-friendly design of the indoor environment at Nanjing South Railway Station as follows:

(1) Adding elderly and special passenger security check areas at the south and north entrances.
(2) Adjusting the waiting areas for elderly and special passengers to the southeast corner of the south entrance and the northeast corner of the north entrance.
(3) Adding south and north inquiry counters and customer service centers, refining customer service functions, equipping medical aid areas and self-service equipment areas near the waiting areas for elderly and special passengers, and adding wheelchair rental areas at the entrances.
(4) Concentrating dining shops on the mezzanine layer and unifying the layout of specialty, newspaper, clothing, digital, and general merchandise shops near the waiting area for regular passengers.
(5) Arranging the cultural exhibition area, AR scenic exhibition area, station culture area, and product exhibition area at intervals along the central axis.

This paper introduces an age-friendly modules combination design system, categorized into basic type I, enhanced type II, and optimized type III, to address the age-friendly design needs of indoor environments at six levels of high-speed railway stations in China. These three types of age-friendly design module combination systems are derived from the categorized data on needs of elderly passengers as follows: Basic type I primarily addresses fundamental physiological and safety needs of elderly passengers, enhanced type II focuses on improving social and comfort needs of elderly passengers, and optimized type III aims to fulfill psychological needs of elderly passengers related to culture and a sense of belonging. The age-friendly modules are designed in a menu mode, allowing for menu combination and adjustment based on the age-friendly needs of different levels high-speed railway stations to ensure the maximum universality. This study shows that the requirements for security check unit modules and waiting unit modules are similar across different levels of stations, but the age-friendly needs for a service unit module, commercial unit module, and exhibition unit module vary significantly. Higher-level stations require a higher degree of completeness in the age-friendly provision of the latter three modules.

Author Contributions: Conceptualization, Y.C.; methodology, D.L.; software, X.G.; validation, S.B.; formal analysis, X.G.; investigation, K.Y.; resources, Y.C.; data curation, D.Z.; writing—original draft preparation, Y.C. and D.L.; writing—review and editing, Y.C.; visualization, X.G.; supervision, S.B.; project administration, Y.C.; funding acquisition, Y.C. All authors have read and agreed to the published version of the manuscript.

Funding: This research was funded by the Jiangsu Provincial Social Science Foundation Project (grant number 21YSB008).

Institutional Review Board Statement: This study was conducted in accordance with the Declaration of Helsinki, and approved by the Ethical Review Board of Eindhoven University of Technology (protocol code ERB2023ID414 and date of approval 18 July 2023).

Informed Consent Statement: Informed consent was obtained from all subjects involved in the study.

Data Availability Statement: The original contributions presented in this study are included in the article, further inquiries can be directed to the corresponding authors.

Conflicts of Interest: The authors declare no conflicts of interest.

References

1. Public Data from the National Bureau of Statistics. 17 January 2024. Available online: https://m.gmw.cn/2024-01/17/content_1303633858.htm (accessed on 1 May 2024).
2. The State Council of People's Republic of China. *The Fourteenth Five-Year Plan for the Healthy Age Plan*; The Bulletin of the State Council of the People's Republic of China: Beijing, China, 2022.
3. The State Council of People's Republic of China. *Guiding Opinions of the State Council on Actively Promoting the Action of "Internet Plus"*; The Bulletin of the State Council of the People's Republic of China: Beijing, China, 2015.
4. The State Council of People's Republic of China. *The Fourteenth Five-Year Plan for Modern Comprehensive Transportation System Development Plan*; The Bulletin of the State Council of the People's Republic of China: Beijing, China, 2021.
5. Ministry of Railways of the People's Republic of China. *Method for Determining the Classification of Railway Stations*; Ministry of Railways of the People's Republic of China: Beijing, 1980.
6. *ISO 21542:2021*; Building Construction—Accessibility and Usability of the Built Environment. International Organization for Standardization: Geneva, Switzerland, 2021.
7. Lee, J.W.; Nam, K.S. An Eco-Friendly Residential Housing Design Guideline as an Elderly Housing Alternative. *Korean Inst. Inter. Des. J.* **2015**, *24*, 111–123.
8. Li, X.; Xu, Y.; Zhou, D. The Study and Practice of Livable Environment for the Elderly in China. *Archit. J.* **2016**, *2*, 16–21.
9. Liu, Z.Y.; Yang, D.F. Exploring the Impact of the Built Environment on Outdoor Recreational Activities of the Elderly in the Neighborhood: A Comparative Study of Four Typical Neighborhoods in Dalian. *Archit. J.* **2016**, *6*, 25–29.
10. Hu, H.Q.; Chang, L. Planning Cultural Entertainment Facilities Adaptive to the Elderly in Old Residential Areas A Case Study of Hongmiaobeili Community in Beijing. *Archit. J.* **2016**, *2*, 22–27.
11. Marston, H.R.; Shore, L.; White, P.J. How does a (smart) age-friendly ecosystem look in a post-pandemic society. *Int. J. Environ. Res. Public Health* **2020**, *17*, 8276. [CrossRef] [PubMed]
12. Van Hoof, J.; Marston, H.R.; Kazak, J.K.; Buffel, T. Ten questions concerning age-friendly cities and communities and the built environment. *Build. Environ.* **2021**, *199*, 107922. [CrossRef]
13. Van Hoof, J.; Marston, H.R. Age-friendly cities and communities: State of the art and future perspectives. *Int. J. Environ. Res. Public Health* **2021**, *18*, 1644. [CrossRef] [PubMed]
14. Tan, D. Research on Design on Nubile Transport Facilities Based on Cities with City Population Age. Master's Thesis, Hubei University of Technology, Wuhan, China, 2013.
15. Kuo, C.W.; Tang, M.L. Relationships among service quality, corporate image, customer satisfaction, and behavioral intention for the elderly in high-speed rail services. *J. Adv. Transp.* **2013**, *47*, 512–525. [CrossRef]
16. Zhao, W.; Xu, L.; Dong, Z.S.; Qi, B.; Qin, L.; Ran, B. Improving transfer feasibility for older travelers inside high-speed train station. *Transp. Res. Part A Policy Pract.* **2018**, *113*, 302–317. [CrossRef]
17. Zhang, J.R. Research on the Inspiration of Barrier free and Humanized Design for Japanese Railway Transport Passenger Stations. *2020 Wanzhi Sci. Dev. Forum Proc. (Smart Eng. I)* **2020**, 876–886. [CrossRef]
18. Liu, C.; Bardaka, E.; Palakurthy, R.; Tung, L.-W. Analysis of travel characteristics and access mode choice of elderly urban rail riders in Denver, Colorado. *Travel Behav. Soc.* **2020**, *19*, 194–206. [CrossRef]

19. Zhu, D.Z.; Shen, F.Y. Analysis of spatial aging strategies for urban rail transit transfer sequences. *South. Archit.* **2021**, *4*, 39–46.
20. Xie, J.; Zhan, S.; Wong, S.; Wen, K.; Qiang, L.; Lo, S. High-speed rail services for elderly passengers: Ticket-booking patterns and policy implications. *Transp. Policy* **2022**, *125*, 96–106. [CrossRef]

buildings

MDPI

Article

Exploring the Link Between Social Media Engagement, Social Anxiety, and Environmental Preferences in University Students

Lan Wang [1], Xia Zhang [1,*] and Aolin Xie [2]

[1] School of Urban Design, Wuhan University, Wuhan 430072, China; lanwang@whu.edu.cn
[2] School of Architecture and Urban Planning, Huazhong University of Science and Technology, Wuhan 430074, China; xieaolin@hust.edu.cn
* Correspondence: xiazhang@whu.edu.cn

Abstract: As mental health issues become increasingly prominent, social anxiety is prevalent among university students and significantly affects their social behaviors and environmental preferences. With the increasing reliance on social media, understanding how these behaviors shape preferences for physical social environments is crucial for designing supportive campus spaces. This study investigates the relationship between social anxiety, social media usage, and environmental preferences in campus settings. We aim to identify the key environmental factors that can promote healthier social interactions for students with varying levels of social anxiety and social media use. Data were collected from university students through a structured questionnaire assessing demographic characteristics, online and offline behavior, social media usage intensity, social anxiety level, and environmental preferences. Using K-means clustering, participants were grouped into four clusters based on their social media usage and social anxiety levels. ANOVA and Multivariate Multiple Regression (MMR) were conducted to examine differences in environmental preferences across these clusters. Significant differences in environmental preferences were observed across the four clusters. Gender and grade influenced social anxiety levels, with females and undergraduate students reporting greater anxiety. Individuals with high social media use, regardless of their anxiety levels, generally exhibited stronger preferences for various environmental dimensions, including openness, convenience, facility provision, safety/comfort, and cultural landscape. However, individuals with higher social anxiety tended to prefer enclosed, facility-rich, and more controlled environments. The relationship between social anxiety and preferences for convenience, safety/comfort, and cultural landscape was minimal or non-significant. This study highlights the importance of considering both behavioral and psychological factors when designing university social environments. By identifying specific environmental preferences of students with high social anxiety, the findings offer evidence-based recommendations for creating more inclusive and supportive campus spaces that promote positive social interactions and well-being.

Keywords: social anxiety; social media; social behavior; environmental preferences; supportive campus spaces

Citation: Wang, L.; Zhang, X.; Xie, A. Exploring the Link Between Social Media Engagement, Social Anxiety, and Environmental Preferences in University Students. *Buildings* **2024**, *14*, 3284. https://doi.org/10.3390/buildings14103284

Academic Editor: Yung Yau

Received: 20 August 2024
Revised: 12 October 2024
Accepted: 14 October 2024
Published: 17 October 2024

1. Introduction

The World Health Organization defines health as "a state of complete physical, mental, and social well-being", encompassing physical, mental, and social dimensions [1]. Social health specifically involves possessing good social adaptability, which entails an individual's ability to adapt to social life, play social roles, and fulfill social functions positively [2]. The college stage is a critical period for individual socialization, where the myriad of complex social activities often induces feelings of anxiety and unease among students [3]. As a vulnerable group to social anxiety, college students often exhibit heightened interpersonal sensitivity during this stage, frequently experiencing elevated levels of social anxiety [4]. Social anxiety among college students adversely affects their health and hinders societal development.

With the prevalence of smartphones and the advancements in internet technology, the use of social networks has transformed traditional face-to-face social interaction patterns [5]. As the primary users of social networks, college students benefit from the convenience and immediacy of online social interaction, leading to the gradual digitalization of social interactions [6]. However, excessive use of social networks may negatively affect college students, leading to psychological issues [7–9]. These issues reduce the initiative for real-life social interactions and impair their social interaction abilities. Social interaction ability is essential for college students to adapt to social development. Therefore, creating a supportive campus environment and guiding college students to engage in healthy social behaviors are crucial for cultivating their social interaction skills and alleviating social anxiety.

This study investigates the relationship between social anxiety, social media use, and environmental preferences in campus settings, understanding online and offline social behavior among college students. We aim to identify the key environmental factors that can promote healthier social interactions for students with varying levels of social anxiety and social media use. The findings offer evidence-based recommendations for creating more inclusive and supportive campus spaces that enhance occupants' health and well-being.

1.1. Social Anxiety and Cognitive Behavioral Therapy (CBT)

Social anxiety (SA) is a prevalent psychological condition characterized by an intense fear of social interactions, often driven by concerns about negative evaluation and judgment by others [10]. Individuals with SA tend to avoid social situations, which reinforces their anxiety and leads to a cycle of avoidance and isolation [11–13]. Social anxiety frequently emerges during adolescence and early adulthood, particularly during transitional periods such as the college years, when students face heightened social and academic pressures [14]. These pressures can make college students especially vulnerable to social anxiety, affecting their ability to form relationships, participate in academic activities, and adapt to new environments [15,16].

Research into social anxiety has identified several key factors that contribute to its development and persistence. Genetic predisposition and temperamental traits such as behavioral inhibition are well-documented predictors of social anxiety [17,18]. However, environmental factors play an equally critical role, especially school environment [19]. Studies show that early negative social experiences, peer rejection, and bullying during adolescence are strongly associated with the development of social anxiety [20–22]. In particular, experiences that undermine an individual's self-esteem or lead to feelings of exclusion tend to heighten sensitivity to social scrutiny [23].

Cognitive Behavioral Therapy (CBT) is widely recognized as one of the most effective treatments for social anxiety [24,25]. The core principle of CBT is that thoughts, emotions, behaviors, and physical sensations are interrelated. Exposure therapy, a key component of CBT, involves gradually confronting feared social situations in a safe and controlled environment [26]. By progressively exposing individuals to these situations, CBT helps reduce the overwhelming anxiety associated with them, allowing the individual to develop more adaptive social behaviors [27]. Therefore, supportive campus environments—those that provide opportunities for peer support, collaboration, and inclusive social activities—can mitigate social anxiety by promoting positive social experiences.

1.2. The Impact of the Built Environment on Mental Health and Well-Being

Environmental psychology explores the dynamic relationship between humans and their physical environments [28], focusing on how perception, cognition, and behavior are influenced by environmental factors, and how individuals in turn affect their surroundings (Figure 1). There is an increasing body of research examining the role of the built environment in promoting mental health and well-being. Much of this research has focused on how environments can alleviate mental health issues such as stress and depression [29,30]. Access to natural environments and green spaces has been shown to reduce stress and enhance emotional well-being [31,32]. Studies have also highlighted the role of spatial design in reducing symptoms of depression by promoting light exposure, creating open areas for movement, and encouraging social interaction [33]. To our knowledge, there is little empirical investigation into connections between environments and social anxiety, which is crucial for promoting healthier social engagement among socially anxious individuals. Given the pervasive impact of social anxiety, particularly among college students, understanding the environmental preferences of socially anxious individuals represents a critical gap.

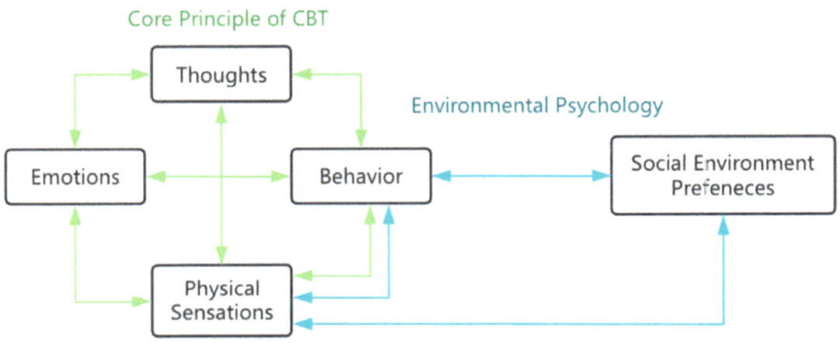

Figure 1. The correlation between CBT and environmental psychology.

1.3. The Role of Social Media in Shaping Social Anxiety and Environmental Preferences

The rise of social media has not only transformed the way people communicate but has also influenced how individuals interact with physical environments [5]. Social media allows users, especially those with social anxiety, to engage in social interactions without the immediate pressures of face-to-face encounters, offering an alternative platform for communication. While this digital interaction provides convenience [6], it also has implications for how individuals perceive and prefer their real-world social environments. Research indicates that social network use can maintain group relationships, promote group identity [34], and deepen offline social interactions [35]. This use fulfills personalized psychological needs and positively impacts college students' psychological well-being and socialization processes [36]. However, over-reliance on social media may lead to psychological issues such as loneliness, low self-esteem, and depression [7–9]. Additionally, Studies indicate that excessive use of social media—especially behaviors like social comparison, passive consumption, and the fear of missing out—can exacerbate social anxiety [37,38]. In this sense, a thorough comprehension of college students' social behaviors, particularly their engagement with social media, is essential for elucidating their social anxiety and preferences regarding social environments.

1.4. Campus Third Places and Environmental Preferences

The concept of third places, introduced by sociologist Ray Oldenburg, refers to informal public spaces beyond work and home, such as streets, street parks, retail, dining, leisure, and entertainment facilities [39]. These spaces are essential for promoting social interaction and enhancing community cohesion [40,41]. Third places are known to support the development of weak ties—social connections with acquaintances or neighbors—which play a significant role in individuals' self-rated health and well-being [42–44].

In the context of universities, campus third places serve as crucial venues for student interaction and social engagement. Defined as informal public spaces on university campuses, separate from classrooms, self-study areas, and residential spaces, these environments facilitate social interaction, fostering a vibrant and dynamic campus atmosphere. Current research suggests that high-quality third places contribute to the psychological well-being of students, offering restorative benefits and opportunities for social participation [45]. Despite their importance, the formal classification of campus third places in the literature remains limited, with most studies focusing on their function and the types of activities they support [46–48]. This study categorizes campus third places into five types based on their purpose of use and management:

(1) Public Open Spaces: accessible to all students and the public, including campus squares and green areas, providing open areas for relaxation and informal social activities.
(2) Campus Service Spaces: managed by the university to deliver services such as cafeterias, student activity centers, and library atriums; these spaces support administrative, educational, and recreational functions.
(3) Dining Commercial Spaces: operated by businesses offering dining options such as cafes and restaurants; these spaces provide essential services while fostering social interactions.
(4) Creative, Athletic, and Entertainment Spaces: these venues cater to students' cultural, athletic, and leisure pursuits, including gyms, game rooms, and art centers, promoting both physical and social activities.
(5) Personal Service Spaces: providing convenience services such as barbershops and convenience stores; these spaces address students' personal needs while offering additional opportunities for social engagement.

Research on public spaces emphasizes their role as central venues for social interaction, with key influencing factors including accessibility, the volume and variety of people and activities, and the duration of social engagements [46,49]. The quality of these spaces is often assessed based on inclusiveness, meaningfulness, safety, comfort, and pleasurability [50].

Demographic factors shape individuals' spatial needs and perceptions. For instance, males tend to prioritize environmental features and design, while females value accessibility and safety [3,51]. Enhanced accessibility is believed to foster positive lifestyles, while attributes such as safety, aesthetics, convenient facilities, cleanliness, and comfort impact behaviors and contribute to the health and well-being of users, including adolescents and older individuals [52].

Studies on public space utilization show that factors like shading [53], thermal comfort [54], spatial openness, material appearance [55], and cultural elements [56] play significant roles in influencing users' behaviors. The richness of greenery, its layout, and the quality of sports, lifestyle, and convenience facilities [57–60] also determine how well these spaces are utilized, particularly by adolescents, who are drawn to areas that offer recreational opportunities [61].

On university campuses, public spaces serve as crucial venues for student interaction, contributing to the overall vibrancy of the campus environment [62,63]. Research in this domain focuses on aspects such as thermal comfort [64–66], spatial configuration [2,67], health benefits [68–71], and user behavior [72]. The environmental quality of these spaces is generally assessed based on physical/ecological, behavioral/functional, and aesthetic/visual components [73], with important design principles like openness, accessibility, spatial hierarchy, comfort, and participation influencing social interactions [74–76]

This literature provides a solid foundation for understanding the environmental factors that influence social interactions in public spaces, particularly within university campuses. The selection of these indicators is grounded in the established research on public space quality and user behavior, ensuring that the study is based on well-documented factors that are essential for fostering healthy social environments.

1.5. Current Study

This study integrates the core principle of Cognitive Behavioral Therapy (CBT) and Environmental Psychology as its theoretical framework, emphasizing a comprehensive approach that considers both behavioral and psychological factors (Figure 2). By focusing on how social media use and social anxiety interact with social environment preferences, this research aims to understand the differences in environmental preferences among individuals with varying levels of social media use and social anxiety. The study addresses how these factors shape preferences for physical environments and offers insights into how campus environments can be designed to promote healthy social interactions, particularly for students experiencing social anxiety. The study focuses on the following research questions:

(1) What is the current state of social behavior and social anxiety among college students?
(2) What are the differences in social environment preferences among individuals with varying levels of social media use and social anxiety?
(3) How does social media use influence social environment preferences?
(4) How does social anxiety affect social environment preferences?

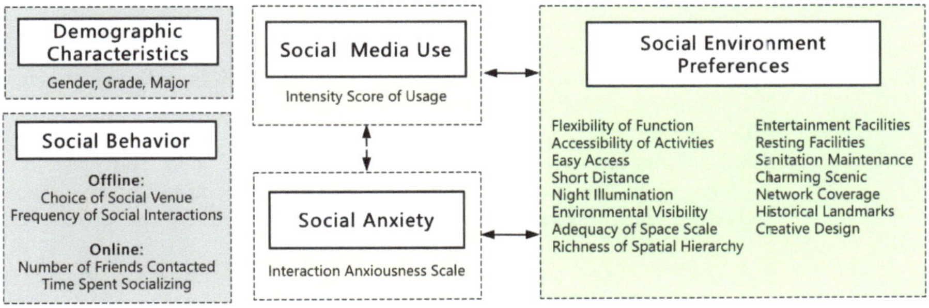

Figure 2. Study variable selection and hypothesis.

2. Materials and Methods

2.1. Research Design

This study employs a quantitative research approach to investigate the relationship between social media usage, social anxiety, and social environmental preferences (Figure 3, see Supplementary File for details). The questionnaire survey method was selected as the primary data collection technique because it provides a standardized means of gathering data from a large sample. This method is widely used in studies that aim to quantify psychological and behavioral traits like social anxiety and environmental preferences. Furthermore, surveys are effective in capturing self-reported data, which is essential for understanding individual perceptions of social spaces and personal social media usage.

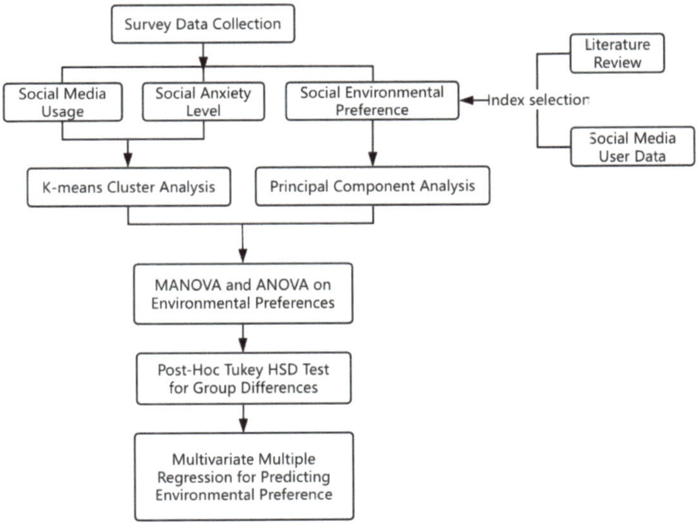

Figure 3. Research design diagram.

2.2. Participants

This study was conducted at Wuhan University, involving undergraduate, master's, and doctoral students. As a top-tier comprehensive university in China, Wuhan University boasts outstanding faculty, excellent academic resources, and a rich campus culture, providing students with a favorable learning and living environment. The student body is diverse, representing various backgrounds from across the country. Additionally, Wuhan University's abundant campus facilities and beautiful landscapes offer a diverse social environment. Therefore, selecting students from Wuhan University as research subjects enables a comprehensive understanding of social behavior and social anxiety among Chinese university students.

Participants in this study were recruited through an online survey from October to December 2023. Informed consent was obtained from all participants before they filled out the questionnaire, and all responses were anonymized to ensure confidentiality. A total of 385 questionnaires were distributed, with 359 deemed valid. The questionnaire consisted of five parts: demographic characteristics, usage of campus third places, usage of social media, preferences for social environments, and degree of social anxiety.

2.3. Measures

2.3.1. Interaction Anxiousness Scale

To accurately measure the level of social anxiety among university students, this study employed the Interaction Anxiousness Scale, revised and validated by Leary and subsequently by multiple scholars [77–79]. This scale has been demonstrated to effectively assess the subjective feelings of interpersonal anxiety among Chinese university students [80]. The scale comprises 13 items (Table 1), including 2 reverse-scored items. Each item is rated on a 5-point Likert scale (1 = "strongly disagree" to 5 = "strongly agree"). After reverse scoring the appropriate items, the average score across all items is calculated, with higher scores indicating higher levels of social anxiety.

Table 1. Measure of social anxiety level.

Items	Options
You often feel nervous, even at informal gatherings.	1 = Strongly Disagree, 2 = Disagree, 3 = Neutral, 4 = Agree, 5 = Strongly Agree; Including two reverse questions, after converting the reverse questions, the average score of all items is calculated. The higher the score, the higher the level of social anxiety.
You usually feel uncomfortable around a group of people you do not know.	
You usually feel uncomfortable talking to someone of the opposite sex.	
You feel nervous when you have to talk to your teacher or supervisor.	
Parties often make you feel anxious and uncomfortable.	
You often feel nervous when talking to someone of the same sex you do not know very well.	
You feel nervous during a job interview.	
In social situations, you rarely feel anxious.	
Generally speaking, you are a shy person.	
You often feel nervous when talking to an attractive person of the opposite sex.	
You are usually nervous when calling someone you do not know very well.	
You feel nervous when talking to authority figures.	
Even if you are in a group of people who are very different from you, you still feel relaxed.	

2.3.2. Social Media Usage Intensity Scale

To measure online social behavior among university students, this study utilized the Facebook Intensity Scale [35], which is widely used in studies examining university students' social media behaviors and has been proven to be reliable in assessing intensity of use [81]. Additionally, the scale was translated and revised by other researchers [82] to ensure it is well-suited for Chinese students. The scale comprises 8 items (Table 2), with the first 2 measuring the number of friends and time spent on social media platforms, and the remaining 6 items assessing the emotional connection with social media and its integration into individuals' lives. Responses are scored on a 5-point Likert scale (1 = "strongly disagree" to 5 = "strongly agree"), with higher scores indicating greater intensity of social media usage. In this study, social media refers to online platforms in China that allow users to create, share, and communicate information, including text, images, audio, and video, such as WeChat, QQ, Weibo, and LittleRedBook.

Table 2. Measure of social media usage intensity.

Items		Options
Number of Friends Contacted Per Day		One or two Three to five Six to ten More than ten
Time Spent Socializing Per Day		Less than 1 h 1–2 h 3–4 h More than 5 h
Intensity Score of Usage	Using social media is part of your life.	1 = Strongly Disagree, 2 = Disagree, 3 = Neutral, 4 = Agree, 5 = Strongly Agree; calculate the average score of all items. The higher the score, the higher the intensity of social media usage.
	You are proud of using social media.	
	You spend most of your time on social media.	
	When you are not on social media, you feel lonely.	
	You feel a strong sense of identification with social media communities.	
	Turning off social media can make you feel bad and sorry.	

2.3.3. Social Environmental Preferences

The environmental preference data were collected via a 15-item questionnaire developed through a review of literature and analysis of user social media data. Responses were recorded on a 5-point Likert scale (1 = very unimportant, 5 = very important).

To tailor the indicators to the unique culture and circumstances of Wuhan University, relevant social media data were crawled and analyzed. Using the keyword "Wuhan University", notes from LittleRedBook (a social media platform) were collected. After removing advertisements and promotional content, 995 valid data points were obtained. ROST CM6 (5.8.0.603) software was employed to segment the text and manually input specific terms related to Wuhan University, such as "Plum Garden" and "Sakura Castle". After merging synonymous terms, a word cloud was generated to visualize high-frequency words (Figure 4). High-frequency words such as "canteen", "playground", "sunset", and "delicacies" reflect popular activity locations and types on campus. The appearance of high-frequency words like "Old Dormitory", "Sakura Castle", "history", and "art" indicates the unique historical and cultural heritage of Wuhan University, suggesting that the cultural aspects of the campus environment may be significant for students' social interactions. The analysis of social media data served as validation and supplementation of existing environmental factor indicators. Ultimately, the environmental preferences were comprised of 15 items, which were later reduced to 5 dimensions using Principal Component Analysis (Table 3).

2.4. Statistical Analyses

2.4.1. Principal Component Analysis (PCA)

To ensure the internal consistency of the questionnaire items, reliability tests were conducted (see Table A1 in Appendix A). The obtained Cronbach's alpha values were all greater than 0.7, indicating good reliability of the questionnaire.

PCA was conducted to reduce the 15 environmental indicators into 5 key dimensions (see Table A2 in Appendix A), identifying composite variables that reflect the original data and making the data more manageable for further analysis. The KMO (Kaiser–Meyer–Olkin) value suggested the suitability of the data for factor analysis. The cumulative variance contribution of the factors indicated that the new variables could effectively reflect the information of the original variables. Using principal component analysis and varimax rotation to generate factor loadings, the results showed that all factor loadings were greater than 0.5, meeting the requirements [83]. The extracted components were used in subsequent analyses to examine their relationship with social media use and social anxiety.

2.4.2. K-Means Clustering

Based on social media usage and social anxiety levels, participants were grouped into four distinct clusters using K-means clustering. The clustering process grouped students into four quadrants reflecting high/low social media usage and high/low social anxiety levels. The optimal number of clusters was determined by evaluating the elbow plot and silhouette score. The clustering revealed meaningful subgroups, which were used for comparison in later analyses.

2.4.3. MANOVA and ANOVA

Multivariate Analysis of Variance (MANOVA) was conducted to examine differences in environmental preferences across the identified clusters. In addition, a series of one-way ANOVA tests were performed for each environmental preference dimension to explore significant differences among the groups. This approach enabled us to identify which environmental factors varied significantly among individuals with different social media use and social anxiety profiles.

2.4.4. Post-Hoc Tukey HSD Test

Following the ANOVA results, Post-hoc Tukey HSD tests were applied to further investigate group differences in environmental preferences. This post-hoc test helped clarify the specific group-level variations in each environmental dimension and ensured that the differences were statistically significant.

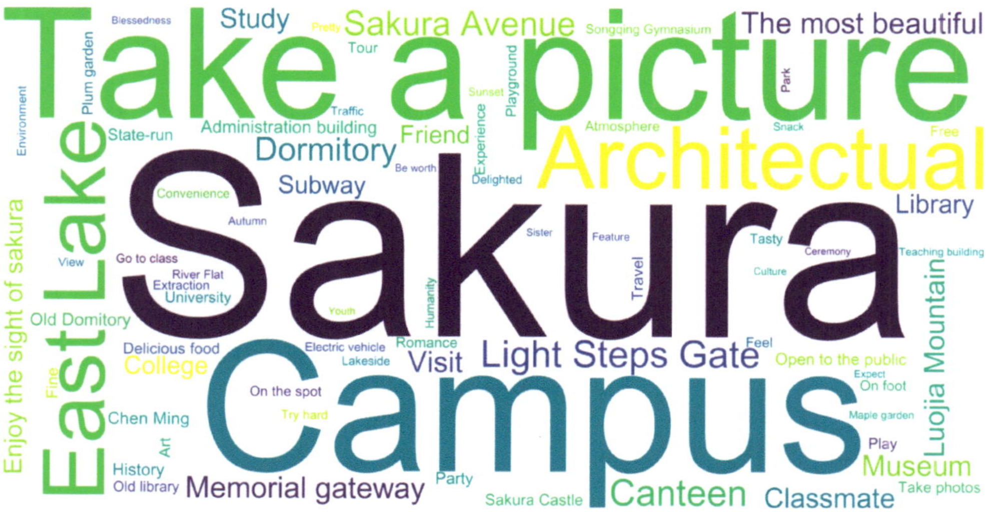

Figure 4. Word cloud of high-frequency words.

Table 3. Measure of social environmental preferences.

Items	Explanation	Source	Options
Flexibility of Function	Adaptability of a space to accommodate various activities.	Literature Review	
Accessibility of Activities	How easily individuals can engage in diverse activities within the environment.	Literature Review, Social Media	
Easy Access	Convenience of reaching and entering the space from various points.	Literature Review, Social Media	
Short Distance	Proximity between different areas or facilities within the environment.	Literature Review, Social Media	
Night Illumination	Adequacy and quality of lighting in the environment during nighttime, affecting safety and usability.	Literature Review, Social Media	
Environmental Visibility	Clarity with which the surroundings can be seen and perceived from different vantage points.	Literature Review	
Adequacy of Space Scale	Dimensions and layout of the space are sufficient for its intended functions and user capacity.	Literature Review	1 = Very Unimportant, 2 = Not Important, 3 = Neutral 4 = Important, 5 = Very Important
Richness of Spatial Hierarchy	Variety and complexity of spatial arrangements and levels within the environment, contributing to a dynamic experience.	Literature Review	
Entertainment Facilities	Availability and quality of recreational and leisure facilities.	Literature Review, Social Media	
Resting Facilities	Presence and quality of areas designed for relaxation and repose.	Literature Review, Social Media	
Sanitation Maintenance	Cleanliness and upkeep of the environment, contributing to overall hygiene and user comfort.	Literature Review, Social Media	
Charming Scenic	Aesthetic appeal and visual attractiveness of the environment, enhancing its pleasantness and allure.	Literature Review, Social Media	
Network Coverage	Extent and reliability of wireless network connectivity available.	Social Media	
Historical Landmarks	Presence and significance of culturally or historically important features.	Literature Review, Social Media	
Creative Design	Originality and innovativeness in the design elements and aesthetics of the environment.	Social Media	

2.4.5. Multivariate Multiple Regression (MMR)

To predict environmental preferences, a Multivariate Multiple Regression analysis was conducted, with the five PCA-derived environmental dimensions (openness, convenience, safety and comfort, facilities, and cultural landscape) as the dependent variables. Social media usage and social anxiety levels were included as independent variables. This allowed us to examine how these psychological and behavioral factors influence different environmental preferences.

2.4.6. Statistical Software

All statistical analyses were conducted using Python 3.9 and the following libraries: Pandas, Scikit-learn, Statsmodels, and Matplotlib.

3. Results

3.1. Descriptive Analyses

The survey yielded a total of 359 valid responses, with 166 (46.2%) from male participants, and 193 (53.8%) from female participants. The academic level distribution included 148 (41.2%) undergraduate students, 152 (42.3%) master's students, and 59(16.4%) doctoral students. These proportions closely align with the overall student distribution at the university.

3.1.1. Usage of Campus Third Places

Regarding the usage of third places (Table 4), campus service facilities were the most popular social venues, with a selection rate of 81.6% (n = 293). Public open spaces followed closely with a selection rate of 65.7% (n = 236). Dining commercial spaces had a selection rate of 42.1% (n = 151), ranking third. In terms of social frequency, the majority of students socialized weekly: 31.5% (n = 113) socialized 3–5 times per week, and 36.2% (n = 130) socialized 1–2 times per week. However, 13.3% (n = 48) of students socialized only 1–2 times per month or less.

Table 4. Descriptive statistics of usage of third places.

Variables		n	%
Choice of Social Venue	Public open spaces	236	65.7
	Campus service facilities	293	81.6
	Dining and commercial establishments	151	42.1
	Creative sports and entertainment venues	74	20.6
	Personal service establishments	59	16.4
Frequency of Social Interactions	Almost everyday	68	18.9
	3–5 times per week	113	31.5
	1–2 times per week	130	36.2
	1–2 times per month or even less	48	13.3

3.1.2. Usage of Social Media

Regarding social media usage (Table 5), 53.8% (n = 193) of students contact 3–5 friends daily through social media, while 30.6% (n = 110) contact 1–2 friends daily. A smaller proportion (n = 56, 15.6%) contacts more than 6 friends daily. In terms of daily usage time on social media, 30.1% (n = 108) of students spend 1–2 h, 38.4% (n = 138) spend 3–4 h, and 24.3% (n = 91) spend over 5 h. The average intensity of social media usage is 20.3 (Table 6), with no significant gender difference. However, there were significant differences by academic grade (F-statistic = 4.723, $p < 0.001$), with freshmen reporting the highest mean use at 21.9. Additionally, a significant difference was found by major (t-statistic = 3.551, $p < 0.001$), with students in Humanities and Social Sciences using social media more intensively (mean = 21.4) compared to those in Science and Engineering (mean = 19.8).

Table 5. Descriptive statistics of usage of social media.

Variables		n	%
Number of Friends Contacted Per Day	One or two	110	30.6
	Three to five	193	53.8
	Six to ten	32	8.9
	More than ten	24	6.7
Time Spent Socializing Per Day	Less than 1 h	22	6.1
	1–2 h	108	30.1
	3–4 h	138	38.4
	More than 5 h	91	24.3

Table 6. Descriptive statistics of usage intensity score.

Variables			Mean	Chi-Square Test
Intensity of Social Media Use	Total		20.3	—
	Gender	Male Female	20.2 20.5	t-statistic = −0.691 *p*-value = 0.490
	Grade	Freshman Sophomore Junior Senior Graduate Doctoral	21.9 19.8 21.1 18.6 20.9 18.9	F-statistic = 4.723 *p*-valuec < 0.001
	Major	Humanities and Social Science Science and Engineering	21.4 19.8	t-statistic = 3.551 *p*-value < 0.001

3.1.3. Social Anxiety

The analysis of social anxiety levels across demographic variables revealed notable differences (Table 7). The mean social anxiety level for males was 39.6, compared to 43.4 for females. A *t*-test revealed a significant difference, with females exhibiting higher levels of social anxiety (t = −3.376, $p < 0.001$). Social anxiety levels varied significantly by grade level (F = 2.609, $p = 0.025$). Undergraduate students have higher average scores than master's (41.6) and doctoral students (39.4). No significant difference was found between majors. Students in Humanities and Social Sciences had an average social anxiety level of 40.6, while those in Science and Engineering reported an average of 42.1 (t = −1.173, $p = 0.242$).

Table 7. Descriptive statistics of social anxiety level.

Variables			Mean	Chi-Square Test
Social Anxiety Level	Total		41.6	—
	Gender	Male	39.6	t-statistic = −3.376 *p*-value <0.001
		Female	43.4	
	Grade	Freshman Sophomore Junior Senior Graduate Doctoral	41.5 38.8 44.5 45.2 41.6 39.4	F-statistic = 2.609 *p*-valuec = 0.025
	Major	Humanities and Social Science Science and Engineering	40.6 42.1	t-statistic = −1.173 *p*-value = 0.242

3.2. K-Means Clustering

K-means clustering analysis was conducted to classify participants into distinct groups based on their social media usage and social anxiety levels. The optimal number of clusters was determined using the elbow method (Figure 5), where the within-cluster sum of squares (WCSS) was plotted against the number of clusters, revealing a noticeable "elbow" at k = 4 clusters. This indicated that the optimal number of clusters for segmentation was 4:

Group 0: low social media use/high social anxiety (n = 103).
Group 1: high social media use/low social anxiety (n = 107).
Group 2: low social media use/low social anxiety (n = 88).
Group 3: high social media use/high social anxiety (n = 61).

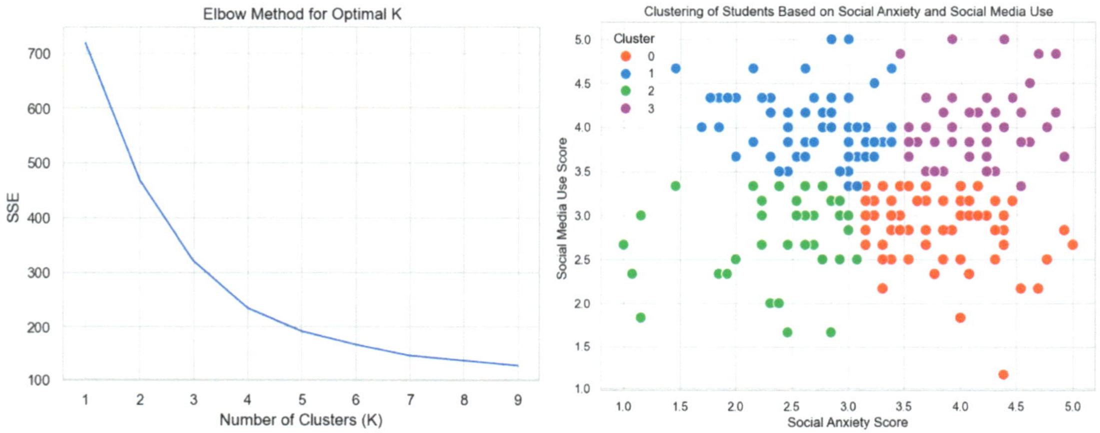

Figure 5. Results of K-means clustering.

3.3. MANOVA and ANOVA

The MANOVA test was conducted to assess whether there were significant differences in the environmental preferences (openness, convenience, safety and comfort, facility provision, and cultural landscape) across the four groups. The results indicated a significant multivariate effect of group membership on environmental preferences, Wilks' $\lambda = 0.9679$, $F(5, 353) = 2.3443$, $p = 0.0410$, suggesting that the four groups differ in their combined environmental preferences.

To further investigate the differences for each environmental preference dimension, individual one-way ANOVAs were conducted for each of the five dimensions (Table 8). These ANOVA results suggest significant differences between clusters in their preferences for openness, convenience, safety, and facility provision. However, for cultural landscape, while significant differences were found, the F-value was lower, indicating a weaker effect.

Table 8. ANOVA results.

Environmental Preference	Group 0: Low SMU, High SA	Group 1: High SMU, Low SA	Group 2: Low SMU, Low SA	Group 3: High SMU, High SA	ANOVA F-Value	p-Value
Openness	3.835	4.224	3.955	4.279	8.162	0.000
Convenience	4.165	4.371	4.008	4.388	6.174	0.000
Safety and Comfort	4.070	4.275	3.839	4.292	6.174	0.000
Facility Provision	3.956	3.981	3.341	4.254	16.501	0.000
Cultural Landscape	3.366	3.514	3.250	3.601	3.486	0.160

3.4. Post-Hoc Test

A Tukey HSD post-hoc test was conducted to further explore the specific group differences for each environmental preference (Table 9).

Table 9. Post-hoc test results.

Environmental Preference	Group A	Group B	Mean Difference	p-adj	95% CI (Lower)	95% CI (Upper)	Significance (Reject)
Openness	0	1	0.3893	0.0004	0.1400	0.6387	TRUE
	0	3	0.4437	0.0006	0.1519	0.7356	TRUE
	1	2	−0.2698	0.0386	−0.5297	−0.0098	TRUE
	2	3	0.3241	0.0291	0.0232	0.6251	TRUE
Convenience	1	2	−0.3631	0.0012	−0.6136	−0.1127	TRUE
	2	3	0.3800	0.0044	0.0904	0.6704	TRUE
Safety and Comfort	1	2	−0.4361	0.0000	−0.6804	−0.1919	TRUE
	2	3	0.4533	0.0003	0.1704	0.7359	TRUE
Facility Provision	0	2	−0.6154	0.0000	−0.9337	−0.2971	TRUE
	2	3	0.9132	0.0000	0.5479	1.2785	TRUE
Cultural Landscape	2	3	0.3511	0.0251	0.031	0.6712	TRUE

For openness, Group 0 (Low SMU/High SA) significantly preferred more openness compared to Group 1 (High SMU/Low SA) and Group 3 (High SMU/High SA), with meandiffs of 0.3893 ($p = 0.0004$) and 0.4437 ($p = 0.0006$), respectively.

In terms of convenience, Group 1 significantly differed from Group 2 (Low SMU/Low SA) with a meandiff of −0.3631 ($p = 0.0012$), and Group 2 differed significantly from Group 3 with a meandiff of 0.3804 ($p = 0.0044$).

Similarly, in the safety and comfort dimension, Group 1 significantly differed from Group 2 with a meandiff of −0.4361 ($p < 0.001$), and Group 2 differed significantly from Group 3 with a meandiff of 0.4532 ($p = 0.0003$).

For facility provision, Group 0 (Low SMU-High SA) significantly differed from Group 2 (Low SMU/Low SA) with a meandiff of −0.6154 ($p < 0.001$), and Group 2 differed significantly from Group 3 with a meandiff of 0.9132 ($p < 0.001$).

Lastly, in cultural landscape, Group 2 and Group 3 significantly differed with a meandiff of 0.3511 ($p = 0.0251$).

These post-hoc results help clarify where the significant differences exist among the clusters for each environmental preference dimension, providing a clearer understanding of how different levels of social media use and social anxiety relate to environmental preferences.

3.5. Multivariate Multiple Regression

The multivariate multiple regression (MMR) analysis was conducted to examine the simultaneous effects of social media use and social anxiety on the five environmental preference dimensions: openness, convenience, safety and comfort, facility, and cultural landscape (Table 10). The results highlight the distinct relationships between these psychological factors and various aspects of environmental preferences.

Table 10. Multivariate multiple regression results.

Dependent Variables	Independent Variables	Coefficient	Standard Error	t-Value	*p*-Value	R^2
Openness	Social media use	0.2912	0.053	5.544	0.000	0.091
	Social anxiety	−0.1156	0.044	−2.614	0.009	
Convenience	Social media use	0.2023	0.051	3.951	0.000	0.055
	Social anxiety	0.0843	0.043	1.955	0.051	
Safety and Comfort	Social media use	0.2185	0.050	4.338	0.000	0.059
	Social anxiety	0.0670	0.042	1.580	0.115	
Facility Provision	Social media use	0.2525	0.064	3.919	0.000	0.129
	Social anxiety	0.3164	0.054	5.832	0.000	
Cultural Landscape	Social media use	0.2601	0.056	4.665	0.000	0.060
	Social anxiety	0.0282	0.047	0.600	0.549	

3.5.1. Openness

The MMR results for openness indicated a statistically significant model (F (2, 356) = 17.88, $p < 0.001$, $R^2 = 0.091$), meaning that 9.1% of the variance in openness could be explained by the independent variables. Social media use had a significant positive effect ($\beta = 0.2912$, $p < 0.001$), suggesting that higher social media engagement is associated with a greater preference for open environments. In contrast, social anxiety showed a significant negative effect ($\beta = -0.1156$, $p = 0.009$), indicating that individuals with higher levels of anxiety tend to prefer less open spaces.

3.5.2. Convenience

For convenience, the model was also significant (F (2, 356) = 10.29, $p < 0.001$, $R^2 = 0.055$), explaining 5.5% of the variance in preferences. The results showed a positive relationship between social media use and preference for convenient environments ($\beta = 0.2023$, $p < 0.001$). The effect of social anxiety was marginally significant ($\beta = -0.0843$, $p = 0.051$), suggesting that higher social anxiety might be associated with a preference for convenience, although this effect was not as strong.

3.5.3. Safety and Comfort

The regression model for safety and comfort was statistically significant (F (2, 356) = 11.18, $p < 0.001$, $R^2 = 0.059$), explaining 5.9% of the variance. Social media use had a significant positive effect on safety and comfort preferences ($\beta = 0.2185$, $p < 0.001$), indicating that individuals who engage more with social media are likely to prefer safer and more comfortable spaces. However, social anxiety did not significantly influence preferences for safety and comfort ($\beta = 0.0670$, $p = 0.115$).

3.5.4. Facility Provision

The MMR model for facility provision showed the highest explanatory power (F (2, 356) = 26.37, $p < 0.001$, $R^2 = 0.129$), with 12.9% of the variance explained. Both social media use ($\beta = 0.2525$, $p < 0.001$) and social anxiety ($\beta = 0.3164$, $p < 0.001$) had significant positive effects on facility preferences, suggesting that individuals who use social media frequently and experience higher social anxiety are more likely to value environments with comprehensive facilities.

3.5.5. Cultural Landscape

The regression model for cultural landscape was also significant (F (2, 356) = 11.30, $p < 0.001$, $R^2 = 0.060$), explaining 6% of the variance. Social media use was positively associated with a preference for cultural landscapes ($\beta = 0.2601$, $p < 0.001$), indicating that individuals who are active on social media prefer environments with rich cultural features. However, social anxiety had no significant effect on this dimension ($\beta = 0.0282$, $p = 0.549$).

4. Discussion

4.1. Offline and Online Social Behavior

The findings show that university students favor campus service facilities and public open spaces for offline interactions, with many engaging in social activities several times a week. In contrast, online behaviors reveal a moderate to high level of social interaction, with most students contacting 3–5 friends daily and spending 3–4 h on socializing.

Regarding social media use, students show a moderate level on average. There was no significant gender difference in social media use, aligning with the findings of prior research [84]. However, grade played a significant role in determining usage intensity. Freshmen reported the highest level of social media use, which may be due to their need to find campus events and establish social networks in the early stages of university life [85]. Significant differences in social media usage were also found by major. Students in Humanities and Social Sciences were more intensive users of social media compared to those in Science and Engineering. This difference may reflect the different academic cultures and communication needs of these disciplines, with humanities students potentially relying more on social platforms for networking and collaboration [86,87].

When examining social anxiety, the results reveal significant demographic variations. Females reported higher levels of social anxiety compared to males. This finding aligns with existing research [88,89], which can potentially be explained from a "vulnerability-stress perspective" [90]. Exposure to variable psychosocial stressors and an increased biological and psychological vulnerability towards anxiety in females may explain the sex differences in anxiety [91]. Significant differences were also found based on grade level. Undergraduate students exhibited higher social anxiety levels than master's and doctoral students, possibly due to the accumulation of social experience and skills as students progress through their academic years [92]. Interestingly, no significant differences were observed between students from different majors.

4.2. Environmental Preferences of Different Groups

The cluster analysis identified four distinct groups based on their levels of social media use and social anxiety. Each group displayed unique environmental preferences, reflecting the complex interactions between their psychological characteristics and spatial needs.

4.2.1. Group 0: Low Social Media Use/High Social Anxiety

Individuals in Group 0 showed a marked preference for more enclosed and private environments. Their high levels of social anxiety may explain their inclination towards spaces that provide a sense of safety and comfort, as these environments can minimize social interaction and reduce stress. This finding aligns with previous studies that link social anxiety to avoidance behaviors in public settings [93].

4.2.2. Group 1: High Social Media/Low Social Anxiety

In contrast, Group 1 exhibited a strong preference for open and interactive spaces. Their low levels of social anxiety, coupled with high social media engagement, may contribute to a greater comfort level in environments that encourage social interactions. These individuals likely perceive open spaces as opportunities for socialization and exploration, aligning with their communication capacity and higher self-confidence [94]. Additionally, Group 1's preference for convenience-oriented spaces could be attributed to their need for accessible and efficient environments that support an active lifestyle.

4.2.3. Group 2: Low Social Media Use/Low Social Anxiety

Group 2, characterized by both low social media use and low social anxiety, shows a moderate to lower preference for most environmental factors compared to the other groups. This group seems less concerned with convenience, safety, comfort, and facility provision, suggesting that they may prefer more minimalistic or functional environments that do not require many amenities or high levels of comfort.

4.2.4. Group 3: High Social Media Use/High Social Anxiety

Group 3 exhibits strong preferences across all environmental dimensions. They prioritize openness, convenience, safety and comfort, facility provision, and cultural landscapes. Their heightened preference for these features suggests that they seek environments that not only meet practical needs but also provide emotional comfort and relief from the pressures of social anxiety. This group may benefit most from environments designed to reduce stress and promote well-being through functional, accessible, and aesthetically pleasing spaces.

4.3. Impact of Social Media Use and Social Anxiety on Environmental Preferences

4.3.1. Openness

Individuals who are more active on social media tend to favor open, flexible environments that encourage interaction and social engagement. This association suggests that frequent social media users, accustomed to open and accessible virtual spaces, seek similar qualities in their physical surroundings [95]. In contrast, social anxiety was negatively associated with openness, indicating that individuals with higher levels of social anxiety prefer more enclosed, private environments that limit exposure to uncomfortable social situations [96,97]. This aligns with previous research showing that anxious individuals often engage in avoidance behaviors when faced with large, unbounded spaces where they feel vulnerable to social interactions they cannot control [93].

Altman defines privacy as an active process of boundary regulation used to control interactions [98], which allows individuals to manage their exposure to others in social settings [99,100]. In the context of campus third places, maintaining a balance between openness and privacy is essential for creating environments that meet the diverse needs of individuals with different levels of social comfort. To address these needs, public open spaces like campus squares and green areas can incorporate vegetation arrangements and landscape elements to establish spatial hierarchies and provide a sense of territoriality and privacy [101]. This would allow students with higher social anxiety to engage in social interactions while maintaining control over their personal space. Similarly, in campus service facilities such as cafeterias and library courtyards, flexible partitions and spatial enclosures can be used [102], allowing for a range of social interactions while also ensuring that individuals can regulate their exposure to others.

4.3.2. Facility Provision

Both social media use and social anxiety had significant positive effects on preferences for facility-rich environments. High social media users may value spaces with diverse facilities due to their need for versatile, adaptable settings that support their social and recreational activities. Likewise, individuals with higher social anxiety may also favor such environments as they offer a range of options for engagement while potentially providing opportunities for avoidance or withdrawal if needed. Therefore, enhancing facility amenities tailored to different types of places is essential for promoting social interactions among university students. For public open spaces like campus squares and green areas, additional amenities such as seating, shade structures, and artistic installations can increase attractiveness and comfort [103,104]. For campus service facilities such as cafeterias, playgrounds, and library courtyards, incorporating recreational facilities like art corners and communal areas can diversify functionalities and boost utilization rates [73,105,106]. For dining and commercial establishments, as well as personal service establishments, enhancing the provision of office and study discussion facilities can create more comfortable and convenient spaces for students to share and communicate [107,108].

4.3.3. Convenience: Safety/Comfort and Cultural Landscape

Social media use consistently exhibited a positive relationship with preferences for convenience, safety and comfort, and cultural landscape, while the influence of social anxiety on these environmental preferences was either minimal or non-significant.

Overall, individuals with high social media use, regardless of their anxiety levels, generally exhibited stronger preferences for various environmental dimensions, including openness, convenience, and facility provision. This suggests that higher social media users may have greater expectations for their physical environments, seeking spaces that are adaptable, engaging, and conducive to both social and personal activities. The demands of their digital lifestyles may translate into a heightened need for environments that support flexibility, connectivity, and aesthetic appeal, reflecting their online behaviors in the real world.

In contrast, social anxiety appears to shape environmental preferences in a more protective manner. Individuals with higher social anxiety tended to prefer enclosed, safe, and more controlled environments, which likely serve as a refuge from potentially overwhelming or anxiety-inducing social encounters. Their lower preference for openness and social interaction indicates that their primary concern is minimizing social discomfort in public spaces.

4.4. Limitations and Future Directions

This study relied on self-reported survey data, which may introduce subjective bias. The lack of a significant correlation between self-reported social anxiety and preferences for safety/comfort, convenience, and cultural landscape may be due to the subjective nature of these concepts, which are influenced by individual differences. However, as safety is a fundamental need in Maslow's hierarchy of needs, it remains essential for campus environmental design [109]. Future research should employ more precise measurement methods to better assess the relationship between social anxiety and perceptions of environmental safety and comfort.

Additionally, this study focused on demographic characteristics, online and offline social behaviors, social media usage, environmental preferences, and social anxiety levels. However, research suggests that biological, psychological, and family social environmental factors also play a role in shaping social anxiety [18,110,111]. Future studies should consider incorporating these additional variables and exploring their potential mediating effects on the relationship between environmental preferences and social anxiety.

5. Conclusions

The findings of this study highlight the significant influence of both social media use and social anxiety on environmental preferences among university students, revealing distinct patterns across different user groups. Individuals with high social media use, regardless of their anxiety levels, generally exhibited stronger preferences for various environmental dimensions, including openness, convenience, facility provision, safety/comfort, and cultural landscape. Conversely, individuals with higher social anxiety tended to prefer enclosed, facility-rich, and more controlled environments. The relationship between social anxiety and preferences for convenience, safety/comfort, and cultural landscape was minimal or non-significant, highlighting the need for further investigation.

The study underscores the importance of considering both modern behaviors and psychological traits when designing campus environments. By addressing these diverse psychological needs, we can design more inclusive and adaptive public spaces that cater to both the connectedness of social media users and the psychological well-being of socially anxious individuals.

Supplementary Materials: The following supporting information can be downloaded at: https://www.mdpi.com/article/10.3390/buildings14103284/s1.

Author Contributions: Conceptualization, L.W., X.Z. and A.X.; methodology, L.W., X.Z. and A.X.; data curation, L.W., X.Z. and A.X.; formal analysis, L.W.; investigation, L.W. and X.Z.; writing—original draft, L.W.; writing—review&editing, L.W.; supervision, X.Z. All authors have read and agreed to the published version of the manuscript.

Funding: This research received no external funding.

Institutional Review Board Statement: According to Article 32 of the Measures for Ethical Review of Research in the Life Sciences and Medicine Involving Human Beings issued by the National Health Commission of the People's Republic of China, this study, which involves observation data that does not interfere with public behavior and anonymized data, is exempt from ethical review.

Data Availability Statement: The raw data supporting the conclusions of this article will be made available by the authors on request.

Conflicts of Interest: The authors declare no conflict of interest.

Appendix A

Table A1. Reliability analysis of constructs.

Variables	Number of Questions	Cronbach's Alpha
Social Media Usage Intensity	6	0.754
Social Anxiety Level	15	0.854
Social Environmental Preferences	13	0.924
Overall Reliability	34	0.879

Table A2. Principal component analysis of social environmental preference variables.

Common Factor	Category	Factor (Loading)
X1: Facility Availability	Entertainment facilities	0.797
	Resting facilities	0.755
X2: Openness	Flexibility of function	0.861
	Accessibility of activities	0.780
X3: Convenience	Easy access	0.778
	Short distance	0.686
	Network coverage	0.731
X4: Safety and Comfort	Night illumination	0.815
	Environmental visibility	0.624
	Adequacy of space scale	0.704
	Richness of spatial hierarchy	0.615
	Sanitation maintenance	0.579
X5: Cultural Landscape	Historical landmarks	0.87
	Creative design	0.667
	Charming scenic	0.635
KMO (Kaiser–Meyer–Olkin) Measure		0.832
Cumulative Explained Variance (%)		68.033

References

1. World Health Organization. Health is a state of complete physical, mental and social well-being and not merely the absence of disease or infirmity. In Proceedings of the International Health Conference, New York, NY, USA, 19–22 June 1946.
2. Thoits, P.A. Stress, coping, and social support processes: Where are we? What next? *J. Health Soc. Behav.* **1995**, 53–79. [CrossRef]
3. Knapstad, M.; Sivertsen, B.; Knudsen, A.K.; Smith, O.R.F.; Aarø, L.E.; Lønning, K.J.; Skogen, J.C. Trends in self-reported psychological distress among college and university students from 2010 to 2018. *Psychol. Med.* **2021**, *51*, 470–478. [CrossRef]
4. Miers, A.; Blöte, A.; De Rooij, M.; Bokhorst, C.; Westenberg, P. Trajectories of social anxiety during adolescence and relations with cognition, social competence, and temperament. *J. Abnorm. Child Psychol.* **2013**, *41*, 97–110. [CrossRef]
5. Dwyer, R.J.; Kushlev, K.; Dunn, E.W. Smartphone use undermines enjoyment of face-to-face social interactions. *J. Exp. Soc. Psychol.* **2018**, *78*, 233–239. [CrossRef]
6. Urista, M.A.; Dong, Q.; Day, K.D. Explaining why young adults use MySpace and Facebook through uses and gratifications theory. *Hum. Commun.* **2009**, *12*, 215–229.
7. Hou, Y.; Xiong, D.; Jiang, T.; Song, L.; Wang, Q. Social media addiction: Its impact, mediation, and intervention. *Cyberpsychol. J. Psychosoc. Res. Cyberspace* **2019**, *13*, 4. [CrossRef]
8. Müller, K.W.; Dreier, M.; Beutel, M.E.; Duven, E.; Giralt, S.; Wölfling, K. A hidden type of internet addiction? Intense and addictive use of social networking sites in adolescents. *Comput. Hum. Behav.* **2016**, *55*, 172–177. [CrossRef]
9. Wright, K.B.; Rosenberg, J.; Egbert, N.; Ploeger, N.A.; Bernard, D.R.; King, S. Communication competence, social support, and depression among college students: A model of Facebook and face-to-face support network influence. *J. Health Commun.* **2013**, *18*, 41–57. [CrossRef]
10. Schlenker, B.R.; Leary, M.R. Social anxiety and self-presentation: A conceptualization model. *Psychol. Bull.* **1982**, *92*, 641. [CrossRef]

11. Moitra, E.; Herbert, J.D.; Forman, E.M. Behavioral avoidance mediates the relationship between anxiety and depressive symptoms among social anxiety disorder patients. *J. Anxiety Disord.* **2008**, *22*, 1205–1213. [CrossRef]
12. Morrissette, M. School closures and social anxiety during the COVID-19 pandemic. *J. Am. Acad. Child Adolesc. Psychiatry* **2021**, *60*, 6. [CrossRef]
13. Teo, A.R.; Lerrigo, R.; Rogers, M.A. The role of social isolation in social anxiety disorder: A systematic review and meta-analysis. *J. Anxiety Disord.* **2013**, *27*, 353–364. [CrossRef]
14. Leigh, E.; Clark, D.M. Understanding social anxiety disorder in adolescents and improving treatment outcomes: Applying the cognitive model of Clark and Wells (1995). *Clin. Child Fam. Psychol. Rev.* **2018**, *21*, 388–414. [CrossRef]
15. Strahan, E.Y. The effects of social anxiety and social skills on academic performance. *Personal. Individ. Differ.* **2003**, *34*, 347–366. [CrossRef]
16. Vilaplana-Pérez, A.; Pérez-Vigil, A.; Sidorchuk, A.; Brander, G.; Isomura, K.; Hesselmark, E.; Kuja-Halkola, R.; Larsson, H.; Mataix-Cols, D.; de la Cruz, L.F. Much more than just shyness: The impact of social anxiety disorder on educational performance across the lifespan. *Psychol. Med.* **2021**, *51*, 861–869. [CrossRef]
17. Banerjee, R.; Henderson, L. Social-cognitive factors in childhood social anxiety: A preliminary investigation. *Soc. Dev.* **2001**, *10*, 558–572. [CrossRef]
18. Brook, C.A.; Schmidt, L.A. Social anxiety disorder: A review of environmental risk factors. *Neuropsychiatr. Dis. Treat.* **2008**, *4*, 123–143.
19. Rapee, R.M.; Creswell, C.; Kendall, P.C.; Pine, D.S.; Waters, A.M. Anxiety disorders in children and adolescents: A summary and overview of the literature. *Behav. Res. Ther.* **2023**, *168*, 104376. [CrossRef]
20. Coyle, S.; Malecki, C.K.; Emmons, J. Keep your friends close: Exploring the associations of bullying, peer social support, and social anxiety. *Contemp. Sch. Psychol.* **2021**, *25*, 230–242. [CrossRef]
21. Pörhölä, M.; Almonkari, M.; Kunttu, K. Bullying and social anxiety experiences in university learning situations. *Soc. Psychol. Educ.* **2019**, *22*, 723–742. [CrossRef]
22. Wu, X.; Kou, Z.; Oldfield, P.; Heath, T.; Borsi, K. Informal learning spaces in higher education: Student preferences and activities. *Buildings* **2021**, *11*, 252. [CrossRef]
23. Smart Richman, L.; Leary, M.R. Reactions to discrimination, stigmatization, ostracism, and other forms of interpersonal rejection: A multimotive model. *Psychol. Rev.* **2009**, *116*, 365. [CrossRef]
24. Hofmann, S.G. Cognitive factors that maintain social anxiety disorder: A comprehensive model and its treatment implications. *Cogn. Behav. Ther.* **2007**, *36*, 193–209. [CrossRef]
25. Lucock, M.P.; Salkovskis, P.M. Cognitive factors in social anxiety and its treatment. *Behav. Res. Ther.* **1988**, *26*, 297–302. [CrossRef]
26. Hofmann, S.G.; Asnaani, A.; Vonk, I.J.; Sawyer, A.T.; Fang, A. The efficacy of cognitive behavioral therapy: A review of meta-analyses. *Cogn. Ther. Res.* **2012**, *36*, 427–440. [CrossRef]
27. Wenzel, A. Basic strategies of cognitive behavioral therapy. *Psychiatr. Clin.* **2017**, *40*, 597–609. [CrossRef]
28. Stokols, D. Environmental psychology. *Ann. Rev. Psychol.* **1978**, *29*, 253–295. [CrossRef]
29. Melis, G.; Gelormino, E.; Marra, G.; Ferracin, E.; Costa, G. The effects of the urban built environment on mental health: A cohort study in a large northern Italian city. *Int. J. Environ. Res. Public Health* **2015**, *12*, 14898–14915. [CrossRef]
30. Moore, T.H.; Kesten, J.M.; López-López, J.A.; Ijaz, S.; McAleenan, A.; Richards, A.; Gray, S.; Savović, J.; Audrey, S. The effects of changes to the built environment on the mental health and well-being of adults: Systematic review. *Health Place* **2018**, *53*, 237–257. [CrossRef]
31. Jabbar, M.; Yusoff, M.M.; Shafie, A. Assessing the role of urban green spaces for human well-being: A systematic review. *GeoJournal* **2022**, *87*, 4405–4423. [CrossRef]
32. Wood, L.; Hooper, P.; Foster, S.; Bull, F. Public green spaces and positive mental health–investigating the relationship between access, quantity and types of parks and mental wellbeing. *Health Place* **2017**, *48*, 63–71. [CrossRef]
33. Schweitzer, M.; Gilpin, L.; Frampton, S. Healing spaces: Elements of environmental design that make an impact on health. *J. Altern. Complement. Med.* **2004**, *10*, S-71–S-83. [CrossRef]
34. Zhao, S.; Grasmuck, S.; Martin, J. Identity construction on Facebook: Digital empowerment in anchored relationships. *Comput. Hum. Behav.* **2008**, *24*, 1816–1836. [CrossRef]
35. Ellison, N.B.; Steinfield, C.; Lampe, C. The benefits of Facebook "friends:" Social capital and college students' use of online social network sites. *J. Comput.-Mediat. Commun.* **2007**, *12*, 1143–1168. [CrossRef]
36. Gentile, B.; Twenge, J.M.; Freeman, E.C.; Campbell, W.K. The effect of social networking websites on positive self-views: An experimental investigation. *Comput. Hum. Behav.* **2012**, *28*, 1929–1933. [CrossRef]
37. Boer, M.; Stevens, G.W.; Finkenauer, C.; de Looze, M.E.; van den Eijnden, R.J. Social media use intensity, social media use problems, and mental health among adolescents: Investigating directionality and mediating processes. *Comput. Hum. Behav.* **2021**, *116*, 106645. [CrossRef]
38. Vogel, E.A.; Rose, J.P.; Roberts, L.R.; Eckles, K. Social comparison, social media, and self-esteem. *Psychol. Pop. Media Cult.* **2014**, *3*, 206. [CrossRef]
39. Oldenburg, R. *The Great Good Place: Cafés, Coffee Shops, Community Centers, Beauty Parlors, General Stores, Bars, Hangouts, and How They Get You Through the Day*; Paragon House Pulishers: New York, NY, USA, 1989.

40. Klinenberg, E. *Palaces for the People: How Social Infrastructure Can Help Fight Inequality, Polarization, and the Decline of Civic Life*; Crown: New York, NY, USA, 2018.
41. Latham, A.; Layton, J. Social infrastructure and the public life of cities: Studying urban sociality and public spaces. *Geogr. Compass* **2019**, *13*, e12444. [CrossRef]
42. Cabras, I.; Mount, M.P. How third places foster and shape community cohesion, economic development and social capital: The case of pubs in rural Ireland. *J. Rural Stud.* **2017**, *55*, 71–82. [CrossRef]
43. Granovetter, M.S. The strength of weak ties. *Am. J. Sociol.* **1973**, *78*, 1360–1380. [CrossRef]
44. Verhaeghe, P.-P.; Pattyn, E.; Bracke, P.; Verhaeghe, M.; Van De Putte, B. The association between network social capital and self-rated health: Pouring old wine in new bottles? *Health Place* **2012**, *18*, 358–365. [CrossRef]
45. Lee, N. Third place and psychological well-being: The psychological benefits of eating and drinking places for university students in Southern California, USA. *Cities* **2022**, *131*, 104049. [CrossRef]
46. Gehl, J. *Life Between Buildings: Using Public Space*; The Danish Architectual Press: Copenhagen, Denmark, 2001.
47. Rhubart, D.; Sun, Y.; Pendergrast, C.; Monnat, S. Sociospatial disparities in "third place" availability in the United States. *Socius* **2022**, *8*, 23780231221090301. [CrossRef]
48. Sugiyama, M.; Chau, H.-W.; Abe, T.; Kato, Y.; Jamei, E.; Veeroja, P.; Mori, K.; Sugiyama, T. Third places for older adults' social engagement: A scoping review and research agenda. *Gerontol.* **2023**, *63*, 1149–1161. [CrossRef]
49. Jacobs, J. *The Death and Life of Great American Cities*; Random: New York, NY, USA, 1961.
50. Mehta, V. Evaluating public space. *J. Urban Des.* **2014**, *19*, 53–88. [CrossRef]
51. Akpınar, A. Investigating the barriers preventing adolescents from physical activities in urban green spaces. *Urban For. Urban Green.* **2020**, *53*, 126724. [CrossRef]
52. Cerin, E.; Zhang, C.J.; Barnett, A.; Sit, C.H.; Cheung, M.M.; Johnston, J.M.; Lai, P.-c.; Lee, R.S. Associations of objectively-assessed neighborhood characteristics with older adults' total physical activity and sedentary time in an ultra-dense urban environment: Findings from the ALECS study. *Health Place* **2016**, *42*, 1–10. [CrossRef]
53. Huang, K.-T.; Lin, T.-P.; Lien, H.-C. Investigating thermal comfort and user behaviors in outdoor spaces: A seasonal and spatial perspective. *Adv. Meteorol.* **2015**, *2015*, 423508. [CrossRef]
54. Paukaeva, A.A.; Setoguchi, T.; Luchkova, V.I.; Watanabe, N.; Sato, H. Impacts of the temporary urban design on the people's behavior-The case study on the winter city Khabarovsk, Russia. *Cities* **2021**, *117*, 103303. [CrossRef]
55. Lenzholzer, S.; van der Wulp, N.Y. Thermal experience and perception of the built environment in Dutch urban squares. *J. Urban Des.* **2010**, *15*, 375–401. [CrossRef]
56. Knez, I.; Thorsson, S. Influences of culture and environmental attitude on thermal, emotional and perceptual evaluations of a public square. *Int. J. Biometeorol.* **2006**, *50*, 258–268. [CrossRef]
57. Adinolfi, C.; Suárez-Cáceres, G.P.; Carinanos, P. Relation between visitors' behaviour and characteristics of green spaces in the city of Granada, south-eastern Spain. *Urban For. Urban Green.* **2014**, *13*, 534–542. [CrossRef]
58. De Jong, K.; Albin, M.; Skärbäck, E.; Grahn, P.; Björk, J. Perceived green qualities were associated with neighborhood satisfaction, physical activity, and general health: Results from a cross-sectional study in suburban and rural Scania, southern Sweden. *Health Place* **2012**, *18*, 1374–1380.
59. Grilli, G.; Mohan, G.; Curtis, J. Public park attributes, park visits, and associated health status. *Landsc. Urban Plan.* **2020**, *199*, 103814. [CrossRef]
60. Zhang, W.; Yang, J.; Ma, L.; Huang, C. Factors affecting the use of urban green spaces for physical activities: Views of young urban residents in Beijing. *Urban For. Urban Green.* **2015**, *14*, 851–857. [CrossRef]
61. Kabisch, N.; Kraemer, R. Physical activity patterns in two differently characterised urban parks under conditions of summer heat. *Environ. Sci. Policy* **2020**, *107*, 56–65. [CrossRef]
62. Peker, E.; Ataöv, A. Exploring the ways in which campus open space design influences students' learning experiences. *Landsc. Res.* **2020**, *45*, 310–326.
63. Soares, I.; Yamu, C.; Weitkamp, G. The relationship between the spatial configuration and the fourth sustainable dimension creativity in university campuses: The case study of Zernike campus, Groningen, the Netherlands. *Sustainability* **2020**, *12*, 9263. [CrossRef]
64. Ghaffarianhoseini, A.; Berardi, U.; Ghaffarianhoseini, A.; Al-Obaidi, K. Analyzing the thermal comfort conditions of outdoor spaces in a university campus in Kuala Lumpur, Malaysia. *Sci. Total Environ.* **2019**, *666*, 1327–1345.
65. He, X.; An, L.; Hong, B.; Huang, B.; Cui, X. Cross-cultural differences in thermal comfort in campus open spaces: A longitudinal field survey in China's cold region. *Build. Environ.* **2020**, *172*, 106739.
66. Wang, Y.; de Groot, R.; Bakker, F.; Wörtche, H.; Leemans, R. Thermal comfort in urban green spaces: A survey on a Dutch university campus. *Int. J. Biometeorol.* **2017**, *61*, 87–101. [CrossRef]
67. Alnusairat, S.; Ayyad, Y.; Al-Shatnawi, Z. Towards meaningful university space: Perceptions of the quality of open spaces for students. *Buildings* **2021**, *11*, 556. [CrossRef]
68. Loder, A.K.; Schwerdtfeger, A.; Van Poppel, M.N. Perceived greenness at home and at university are independently associated with mental health. *BMC Public Health* **2020**, *20*, 802. [CrossRef]
69. Lu, M.; Fu, J. Attention restoration space on a university campus: Exploring restorative campus design based on environmental preferences of students. *Int. J. Environ. Res. Public Health* **2019**, *16*, 2629. [CrossRef]

70. Malekinezhad, F.; Courtney, P.; bin Lamit, H.; Vigani, M. Investigating the mental health impacts of university campus green space through perceived sensory dimensions and the mediation effects of perceived restorativeness on restoration experience. *Front. Public Health* **2020**, *8*, 578241. [CrossRef]
71. McFarland, A.; Waliczek, T.; Zajicek, J.M. The relationship between student use of campus green spaces and perceptions of quality of life. *HortTechnology* **2008**, *18*, 232–238. [CrossRef]
72. Lee, J.; Shepley, M.M. College campuses and student walkability: Assessing the impact of smartphone use on student perception and evaluation of urban campus routes. *Sustainability* **2020**, *12*, 9986. [CrossRef]
73. Abu-Ghazzeh, T.M. Communicating behavioral research to campus design: Factors affecting the perception and use of outdoor spaces at the University of Jordan. *Environ. Behav.* **1999**, *31*, 764–804. [CrossRef]
74. O'Rourke, V.; Baldwin, C. Student engagement in placemaking at an Australian university campus. *Aust. Plan.* **2016**, *53*, 103–116. [CrossRef]
75. Timm-Bottos, J.; Reilly, R.C. Learning in third spaces: Community art studio as storefront university classroom. *Am. J. Community Psychol.* **2015**, *55*, 102–114. [CrossRef]
76. Wu, X.; Qi, J.; Zhen, R. Bullying victimization and adolescents' social anxiety: Roles of shame and self-esteem. *Child Indic. Res.* **2021**, *14*, 769–781. [CrossRef]
77. Leary, M.R. Social anxiousness: The construct and its measurement. *J. Personal. Assess.* **1983**, *47*, 66–75. [CrossRef]
78. Peng, C.; Gong, Y.; Zhu, X. The Applicability of Interaction Anxiousness Scale in Chinese Undergraduate Students. *Chin. Ment. Health J.* **2004**, *18*, 39–41.
79. Yang, X.; Bai, Y.; Yu, Y.; Wang, X.; Lv, J.; Cao, J. The relationship among loneliness, social anxiety and mobile phone dependence in college students: A cross-lag analysis. *Chin. J. Clin. Psychol.* **2022**, *1*, 64–67.
80. Li, D.; Li, Z.; Wang, Z. Validation of the Chinese Version of the Interaction Anxiety Scale in College Students. *J. Chin. J. Appl. Psychol.* **2023**, *30*, 251–259.
81. Kalpidou, M.; Costin, D.; Morris, J. The relationship between Facebook and the well-being of undergraduate college students. *Cyberpsychol. Behav. Soc. Netw.* **2011**, *14*, 183–189. [CrossRef]
82. Sun, X.; Lian S-l Niu, G.; Yan, J.; Tong, Y.; Zhou, Z. Social network site use and depression in adolescents: Mediating of upward social comparison. *Chin. J. Clin. Psychol.* **2016**, *24*, 32–35.
83. Lubben, J.; Blozik, E.; Gillmann, G.; Iliffe, S.; von Renteln Kruse, W.; Beck, J.C.; Stuck, A.E. Performance of an abbreviated version of the Lubben Social Network Scale among three European community-dwelling older adult populations. *Gerontologist* **2006**, *46*, 503–513. [CrossRef]
84. Alshalawi, A.S. Social Media Usage Intensity and Academic Performance among Undergraduate Students in Saudi Arabia. *Contemp. Educ. Technol.* **2022**, *14*, ep361. [CrossRef]
85. Dennen, V.P.; Bagdy, L.M.; Ayan Karabatman, H. Navigating the high school to university transition with social media: Intensity of use, sense of belonging, and meaningful change. *Inf. Learn. Sci.* **2024**. [CrossRef]
86. Bashir, I.; Malik, A.; Mahmood, K. Social media use and information-sharing behaviour of university students. *IFLA J.* **2021**, *47*, 481–492. [CrossRef]
87. Ortega, J.L. Disciplinary differences in the use of academic social networking sites. *Online Inf. Rev.* **2015**, *39*, 520–536. [CrossRef]
88. Furmark, T.; Tillfors, M.; Everz, P.-O.; Marteinsdottir, I.; Gefvert, O.; Fredrikson, M. Social phobia in the general population: Prevalence and sociodemographic profile. *Soc. Psychiatry Psychiatr. Epidemiol.* **1999**, *34*, 416–424. [CrossRef]
89. Stein, D.J.; Lim, C.C.; Roest, A.M.; De Jonge, P.; Aguilar-Gaxiola, S.; Al-Hamzawi, A.; Alonso, J.; Benjet, C.; Bromet, E.J.; Bruffaerts, R. The cross-national epidemiology of social anxiety disorder: Data from the World Mental Health Survey Initiative. *BMC Med.* **2017**, *15*, 143. [CrossRef]
90. Al-Johani, W.M.; AlShamlan, N.A.; AlAmer, N.A.; Shawkhan, R.A.; Almayyad, A.H.; Alghamdi, L.M.; Alqahtani, H.A.; Al-Shammari, M.A.; Gari, D.M.K.; AlOmar, R.S. Social anxiety disorder and its associated factors: A cross-sectional study among medical students, Saudi Arabia. *BMC Psychiatry* **2022**, *22*, 505. [CrossRef]
91. Hallers-Haalboom, E.T.; Maas, J.; Kunst, L.E.; Bekker, M.H. The role of sex and gender in anxiety disorders: Being scared "like a girl"? *Handb. Clin. Neurol.* **2020**, *175*, 359–368.
92. Givertz, C.S.M. Methods of social skills training and development. In *Handbook of Communication and Social Interaction Skills*; Routledge: London, UK, 2003; pp. 153–194.
93. Kashdan, T.B.; Farmer, A.S.; Adams, L.M.; Ferssizidis, P.; McKnight, P.E.; Nezlek, J.B. Distinguishing healthy adults from people with social anxiety disorder: Evidence for the value of experiential avoidance and positive emotions in everyday social interactions. *J. Abnorm. Psychol.* **2013**, *122*, 645. [CrossRef]
94. Lai, F.; Wang, L.; Zhang, J.; Shan, S.; Chen, J.; Tian, L. Relationship between social media use and social anxiety in college students: Mediation effect of communication capacity. *Int. J. Environ. Res. Public Health* **2023**, *20*, 3657. [CrossRef]
95. Ahmed, M.; Zeile, P. Places representation on social media–A study to analyze the differences between the virtual communities and the offline environment. SHAPING URBAN CHANGE–Livable City Regions for the 21st Century. In Proceedings of the REAL CORP 2020, 25th International Conference on Urban Development, Regional Planning and Information Society, Virtual, 15–18 September 2020.

96. Kashdan, T.B.; Goodman, F.R.; Machell, K.A.; Kleiman, E.M.; Monfort, S.S.; Ciarrochi, J.; Nezlek, J.B. A contextual approach to experiential avoidance and social anxiety: Evidence from an experimental interaction and daily interactions of people with social anxiety disorder. *Emotion* **2014**, *14*, 769. [CrossRef]
97. Khan, A.N.; Bilek, E.; Tomlinson, R.C.; Becker-Haimes, E.M. Treating social anxiety in an era of social distancing: Adapting exposure therapy for youth during COVID-19. *Cogn. Behav. Pract.* **2021**, *28*, 669–678. [CrossRef]
98. Altman, I. *The Environment and Social Behavior: Privacy, Personal Space, Territory, and Crowding*; Brooks/Cole Publishing Company: Pacific Grove, CA, USA, 1975.
99. Cattell, V.; Dines, N.; Gesler, W.; Curtis, S. Mingling, observing, and lingering: Everyday public spaces and their implications for well-being and social relations. *Health Place* **2008**, *14*, 544–561. [CrossRef]
100. Friedman, B.; Kahn, P.H., Jr.; Hagman, J.; Severson, R.L.; Gill, B. The watcher and the watched: Social judgments about privacy in a public place. *Hum.-Comput. Interact.* **2006**, *21*, 235–272. [CrossRef]
101. Loidl, H.; Bernard, S. *Open(ing) Spaces: Design as Landscape Architecture*; Birkhäuser: Basel, Switzerland, 2022.
102. Mostafa, M. The Autism Friendly University Design Guide. Dublin City University. 2021. Available online: https://issuu.com/magdamostafa/docs/the_autism_friendly_design_guide (accessed on 9 May 2021).
103. Działek, J.; Homiński, B.; Miśkowiec, M.; Świgost-Kapocsi, A.; Gwosdz, K. The assessment of the quality of campus public spaces as key parts of the learning landscape: Experience from a crowdsensing study on the Third Campus of Jagiellonian University, Krakow, Poland. *Urban Des. Int.* **2024**, *29*, 77–92. [CrossRef]
104. El-Darwish, I.I. Enhancing outdoor campus design by utilizing space syntax theory for social interaction locations. *Ain Shams Eng. J.* **2022**, *13*, 101524. [CrossRef]
105. Fang, C. University Library: The "Third Place" for Students. In Proceedings of the IFLA 2008 Conference, Québec, QC, Canada, 10–14 August 2008.
106. Harrington, K.D. Community on Campus: The Role of Physical Space. Ph.D. Thesis, Georgia State University, Atlanta, GA, USA, 2014.
107. Campbell, N. There's no place like third place: Starting to generalize the qualities and value of third places. *Hous. Soc.* **2017**, *44*, 157–172. [CrossRef]
108. Farahani, L.M.; Beynon, D. Third places and their contribution to the street life. In *Rethinking Third Places*; Edward Elgar Publishing: Cheltenham, UK, 2019; pp. 196–218.
109. Abbas, J. Service quality in higher education institutions: Qualitative evidence from the students' perspectives using Maslow hierarchy of needs. *Int. J. Qual. Serv. Sci.* **2020**, *12*, 371–384. [CrossRef]
110. Al-Naggar, R.A. Prevalence and associated factors of phobia and social anxiety among university students. *ASEAN J. Psychiatry* **2012**, *13*, 112–121.
111. Spence, S.H.; Rapee, R.M. The etiology of social anxiety disorder: An evidence-based model. *Behav. Res. Ther.* **2016**, *86*, 50–67. [CrossRef]

buildings

MDPI

Article

Residential Care Facilities for Users with Alzheimer's Disease: Characterisation of Their Architectural Typology

Santiago Quesada-García *, Pablo Valero-Flores and María Lozano-Gómez

Department of Architectural Design, University of Seville, Av. Reina Mercedes 2, 41012 Sevilla, Spain; pvalero@us.es (P.V.-F.); mlozano@us.es (M.L.-G.)
* Correspondence: sqg@us.es

Abstract: The design and construction of residences for persons with Alzheimer's disease (AD) have been based on the recommendations of design guides, the results of empirical tests with samples of the population, and the experience of architects and planners. The reiteration of certain patterns, criteria, and guidelines has given rise to a new type of building that has not yet been explicitly described. The aim of this paper is to determine the main characteristics of this typology. This research is based on a critical review methodology, analysing 30 care homes built over the last four decades across various global contexts. Detailed surveys of plans, projects, and buildings were carried out, allowing a comparative analysis of the architectural attributes to determine the most influential parameters for these buildings. The results indicate that environments designed with safety, accessibility, and opportunities for social interaction in mind—and, above all, those that are personalised to the needs of this collective—significantly enhance the behaviour, emotional state, and cognitive state of their residents. The main theoretical contributions include identifying and stating the key features of this type, such as small scale, basic cell housing, comprehensible organisation, and sensory stimulation of spaces, among others. The breakthrough of this study that differentiates it from other works in this field is that it provides concrete guidelines to approach the planning, design, and construction of these kinds of residences. The significance of this research lies in the definition of this unique typology, which is not characterised by its morphology, shape, or formal composition but rather focused on promoting an adequate cognitive and physiological reception of the space by the users. This building concept has important management implications, as its construction must provide for and integrate specific care services in a residential setting for people with AD.

Keywords: architecture; Alzheimer's disease; planning of residential care homes; organisation of geriatric buildings; active assisted living; nursing homes; environmental interventions

Citation: Quesada-García, S.; Valero-Flores, P.; Lozano-Gómez, M. Residential Care Facilities for Users with Alzheimer's Disease: Characterisation of Their Architectural Typology. *Buildings* **2024**, *14*, 3307. https://doi.org/10.3390/buildings14103307

Academic Editor: Haifeng Liao

Received: 20 September 2024
Revised: 11 October 2024
Accepted: 15 October 2024
Published: 19 October 2024

1. Introduction

Today's society faces the ambitious challenge of responding to the problems caused by the progressive increase in the ageing of the population [1]. Ageing is a risk factor directly linked to dementia [2], and Alzheimer's disease (AD) accounts for between 60% and 70% of diagnosed cases [3]. This disease is a multifactorial syndrome characterised by a progressive alteration of brain function that increases significantly from the age of 65 onwards. It also has a high incidence and prevalence in developed countries, becoming not only a medical but also a social, economic, and architectural problem [4].

Traditional models of care, often characterised by institutional settings that prioritise efficiency over residents' well-being, have proven inadequate in addressing the complex needs of those living with dementia [5]. This inadequacy has sparked a growing interest in the role of architectural design in enhancing the quality of life for individuals with cognitive impairments. As with tuberculosis and cholera at the beginning of the 20th century, architecture is now also being used as an important aid to alleviate some of the symptoms of a disease for which there is still no cure [6]. Although there are great methodological difficulties, which constitute some of the greatest problems in this field, there is a growing body of evidence supporting the positive effects caused by the appropriate adaptation of a space or environment to the needs of Alzheimer's sufferers (see Table A1) [7,8].

The importance of AD has influenced some architectural practices that, for the last forty years, have been used to improve the environmental conditions of people with Alzheimer's dementia, focusing their efforts on providing qualities and services. During this period, there has been a shift from an inpatient hospital model to a new paradigm based on caring for these patients in a residential environment with care facilities adapted to their needs [9]. This change of model has generated a new residential architectural typology specialised for people with AD, which has been the embryo and genesis of the current models of geriatric residences. It is possible to maintain the hypothesis that this unique type of residential-care building, dedicated to users with AD dementia, has ended up being applied and extended to all nursing homes for the elderly, with or without dementia. This has been due to the undoubted advantages and benefits that the implementation of this innovative residential model has brought to all groups of people over 65 years of age.

Research has shown that the built environment in hospitals, residences, day care centres, and homes can profoundly influence the cognitive and emotional experiences of people with dementia [5]. Factors such as spatial orientation, sensory stimulation, small scale, and social interaction play critical roles in shaping their daily lives. However, the existing literature often lacks a complete exploration of how specific architectural features can evoke positive feelings and support cognitive function. On the other hand, some of the solutions applied in residences incorporate criteria and guidelines expressed in specialised design manuals and guides. Other times, these solutions are based on results obtained in empirical research intended to find out how certain spatial and environmental variables influence people with dementia [10,11]. However, most of the buildings constructed in recent decades are based on the experience of the architects' or designers' teams, who have been testing various practical measures as part of a trial-and-error process.

The milestones achieved with these buildings have not been analysed and compared in depth. There has been no methodical review of this type of architecture, nor has there been a rigorous evaluation, from an architectural point of view, of the main contributions that have contributed to the genesis of this typology of residential facilities for AD. This gap in knowledge highlights the need for a systematic approach to the design of residential facilities that cater specifically to the needs of users with Alzheimer's disease.

This paper aims to address these gaps by defining a new architectural typology tailored for residential care facilities serving users with AD. By synthesising insights from various disciplines—including architecture, psychology, gerontology, and health sciences—this study advocates for a multidisciplinary approach that considers the diverse factors influencing residents' well-being. The proposed typology emphasises the importance of creating environments that not only meet the functional requirements of care but also enhance residents' cognitive and emotional experiences.

In addition to establishing a theoretical framework, this research employs a robust methodological approach that combines qualitative analysis with observational data. This research has been carried out by performing a critical review of a significant sample of 30 care homes for AD sufferers in various parts of the world. The method involves studying the facilities built in the last 40 years that have made some kind of progress in improving these residents' quality of life. This research has found that, in these buildings, certain patterns are repeated that apply architectural attributes or parameters and employ similar organisational and functional aspects. This reiteration of elements has produced the decantation of a new type of building, the characteristics of which have not yet been explicitly stated. By identifying key characteristics, this study provides empirical validation for the proposed architectural attributes. This empirical foundation strengthens the theoretical framework and demonstrates how specific design features can lead to improved outcomes for individuals with Alzheimer's disease.

The novelty of this work lies in the analysis and review of a set of residential architectures intended for people with AD, with the aim of establishing their main patterns and common elements. The main contributions of this research consist of defining the features of an architectural typology for a residential building for people with AD, establishing the functional programme that these types of residences must have, empirically validating design guidelines, and providing evidence supporting specific design principles that enhance the quality of life for this collective. This research highlights the importance of user experience in architectural design, focusing on how the physical environment can enhance residents' cognitive and emotional well-being. It explores the interplay between design and the lived experiences of residents, carers, and family members, recognising that a holistic approach to design must consider the needs of all stakeholders involved in the care ecosystem.

This knowledge is essential for investors, developers, and managers of these facilities as a prior step to their planning, and it is very useful for architects, interior designers, and technicians who undertake the design, construction, and environmental renovation of these residences. This study is intended to contribute to a paradigm shift in architectural practice, emphasising the necessity of creating residential environments that foster dignity, autonomy, and a sense of belonging for individuals living with dementia. By laying the groundwork for future research and practice, this study invites ongoing dialogue about the impact of design on health and well-being, encouraging architects and planners to embrace innovative solutions that address the evolving needs of ageing populations. Through this comprehensive exploration of architectural design for Alzheimer's care facilities, this paper aims to enhance our understanding of how they should be built.

2. Background and State of the Art

In 1945, 45% of admissions to mental hospitals had a diagnosis of neurodegenerative disorder associated with dementia [12]. Until the late 1950s, people with dementia were usually admitted to institutional psychiatric hospitals. The first examples of architectures for users with dementia begin to emerge in the United States. The approval of the social action programmes 'Medicare' and 'Medicaid' in 1965 were key to the changes introduced to the social and healthcare models aimed at these people. These programmes provided the financial support necessary for many people with cognitive disabilities to be eligible to reside in specialised facilities with more developmental and stage-based care [9].

This change in the American healthcare system led to the development of new initiatives for this group of people between 1955 and 1980, with hospitals losing 75% of dementia patients to what were initially known as nursing homes. From that moment on, a model dedicated to care for the elderly called the Green House Project proliferated (Figure 1). These were residential homes located in peripheral or rural environments that, with a controlled size and scale, grouped residents into small groups of people who could thus have constant and personalised care. The main aim was to provoke memories of home and stimulate the memory of the residents. Thus, the first dementia care units emerged, accommodating a limited number of users, from six to fifteen people, and introducing places for social interaction with family and others [13].

Alongside this, a model of dementia care known as Gentle Care emerged in Canada. This programme was based on the assumption that Alzheimer's disease should be approached with the aim of treating its symptoms in order to improve the patient's quality of life, maintaining residual abilities and supporting those that are deteriorating. It was thought that it could be useful to design an environment dedicated to the specific needs of these users, trying to make up for their cognitive deficits and facilitate the implementation of therapeutic rehabilitation programmes. The need to build an external prosthesis to support the patient in relation to their environment arose, which would be proportionally more complex the more severe the cognitive loss. The definition of prosthetic environment thus appeared to describe a spatial context of support for the person with Alzheimer's dementia.

Figure 1. Green House Project. Photo of The Green House Homes at Green Hill. Bottom left, the general plan of the Green House Project for the 2008 National Design Competition. Bottom right, the group-living unit of the same Green House. (Source: Prepared by the authors and adapted from NKArchitects).

Moyra Jones, programme sponsor and director of Moyra Jones Resources, an international consultancy specialising in Alzheimer's dementia care, maintains that there is a need for designs or projects aimed at adapting the living environment, as dementia patients are among those most affected by the quality of the place in which they live [14]. This approach led to the idea that the patient should be accommodated in an environment free of obstacles and restrictions, with intimate and familiar features, much more like a home than a hospital, and in which technological aids, necessary for control and safety, are transparent and invisible to the person in order not to restrict their freedom.

Senior living communities progressed rapidly in North America from the early 1980s onwards, as they provided optimal and specialised care, especially for people with cognitive impairment. Their intentions were focused on creating friendly environments that would bring spatial quality to the living conditions of this group. This type of accommodation differed substantially from the typology of aseptic sanitary or hospital facilities. The main novelty was that they had domestic characteristics with communal living areas, thus facilitating the maintenance of the user's social ties. In the places where these units were created, the positive results of this person-centred care became evident.

Lawton stated in 1984 that a building for these people had to meet the requirements of safety, orientation, functionality, integration, and personalisation. These concepts were reflected in the construction of The Abram and Helen Weiss Institute [15,16]. This institute reduced the number of people indoors and incorporated outdoor areas, with the aim of reducing disorientation and memory loss in residents [17]. In this building, it became clear that many of the problems experienced by people with dementia and their caregivers are directly related to the planning and design of the environment [18].

2.1. Design Guidelines, Manuals, and Principles

As a result of these early experiences, various manuals for the design of buildings for people with dementia began to be published in the mid-1980s. The design guidelines offer recommendations for planning, spatial organisation and guidelines focused on safety, accessibility, homelikeness and, in general, on improving quality of life in dementia care facilities (Figure 2). Managers and developers of these facilities are starting to see the importance of the architectural project as an essential resource to promote, in the long-term, well-being and functionality among people with dementia, beyond applying solutions based on purely decorative aspects.

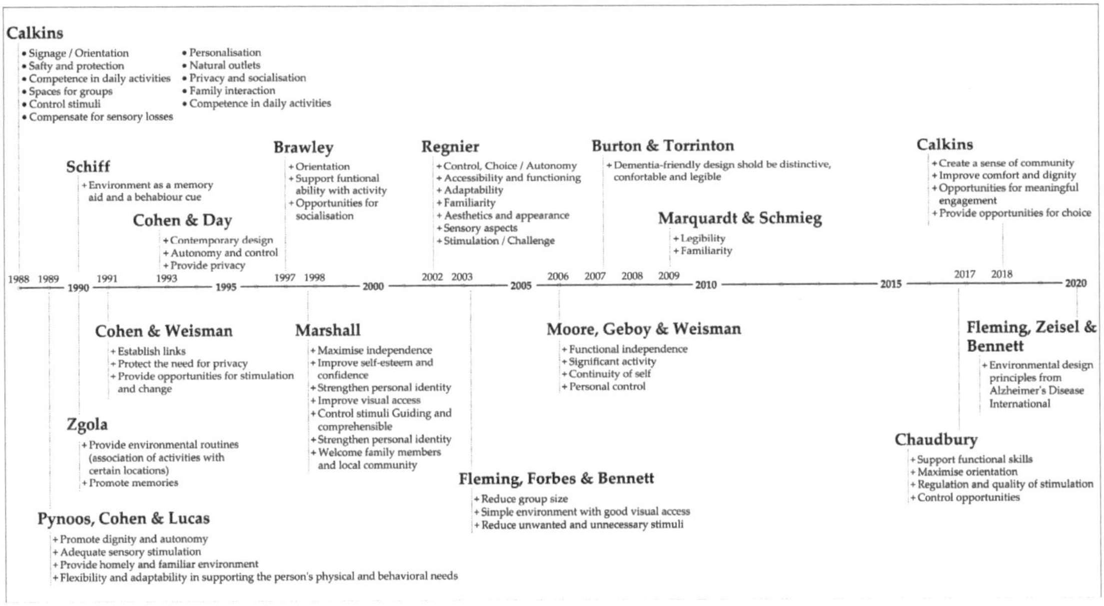

Figure 2. Main contributions of various authors over the last forty years to design principles for environments for people with dementia. (Source: Prepared by the authors).

In 1987, Fleming and Bowles [19] published some initial design principles, later expanded in 2003, which were incorporated in the first specific building for people with dementia, built in the Australian city of Sydney, called The Meadows. This work was followed by other manuals such as that of M. Calkins [12], which establishes the following as basic criteria for these buildings: attention to safety, orientation, privacy, and personalisation of space.

Shortly afterwards, Cohen and Weisman [18] developed new criteria, highlighting the influence and importance of environmental design to maintain the functional capacities of users with dementia. These authors established the following as central elements for designing this type of building: safety, orientation, support for functional abilities, stimulation, autonomy, privacy, and the family social link. They also proposed a classification in which it was established that the design of these buildings should take into account certain planning principles, the attributes or qualities of the building, the organisation and layout of the different spaces, and, finally, the provision of specific rooms and spaces for activities that preserve the dignity and privacy of the residents. Shortly afterwards, two other manuals appeared, that of Cohen and Day [20] and that of E.C. Brawley [21], which also emphasised the need to support functional abilities and provide opportunities for stimulation and socialisation.

During the 1990s, The Dementia Services Development Centre was set up at the University of Stirling (UK) to research designs aimed at people with dementia. Its director, Professor M. Marshall, published the guide Working with Dementia: Guidelines for Professionals, which is still a benchmark for the design of buildings aimed at this group [22]. In 1993, R. Pollock applied some of these design principles in a pilot example of a housing project called St. Leonard's Initiative, built for the Places for People Housing Association in Edinburgh. Gradually, certain design guidelines and criteria were consolidated at facilities for people with dementia, producing residential buildings equipped with care areas, which provided direct and personalised assistance to the user. The results obtained in some of these experiences were evaluated and assessed by M. Kelly in 1995. One of the main conclusions of his study was the confirmation that it was necessary to promote a safe and accessible design that would facilitate the autonomy of the residents [23].

Another important contribution was made by J. Zeisel, director of Alzheimer's in the United States, together with J. Hyde and S. Levkoff, geriatricians from Boston [24]. In a paper published in 1994, these authors propose a building organisation model called Environmental-Behaviour (E-B). Its aim is to identify the special needs of these patients, motivated by the behavioural deficits caused by the disease. The results pointed out some of the characteristics that these facilities should have in order to be congruent with the requirements of the user frequenting them. A series of critical performance criteria were identified, that is, the specific requirements that a space intended for this group must meet in order to respond to the user's behavioural model and, on this basis, to be able to apply certain intervention strategies. The criteria of the E-B model are summarised into eight environmental characteristics, each with two parameters that define them quantitatively. These characteristics are:

1. Exit control; its parameters are immediacy of control and not seeing the exits.
2. Wandering paths; their parameters are continuity and localisation/orientation.
3. User rooms: their parameters are privacy and personalisation.
4. Communal spaces: their parameters are quantity and variability.
5. External freedom: its parameters are availability and assistance.
6. Residential: its parameters are familiarity and size.
7. Autonomy support: its parameters are safety and external prostheses.
8. Sensory comprehension: its parameters are noise management and comprehensibility.

Later, J. Zeisel and his team investigated the correlation between optimal environmental conditions and the health of people with Alzheimer's dementia. In a new article, published in 2003, they set out several design principles, emphasising aspects such as safety control, enhancing autonomy, sensory stimulation, spatial comprehension, and open-air spaces [25].

In the early years of this century, these concepts on the importance of adequately designing spaces and environments inhabited by this group of people spread throughout Europe. Various guides and manuals began to appear in different places. In Spain, the *Guía de arquitectura. Adaptación y habilitación de la vivienda para personas con Alzheimer y deficiencias de movilidad* was published in 2002, which contains a series of recommendations aimed at organising, planning and adapting a home inhabited by a person with Alzheimer's disease [26]. Shortly afterwards, in 2010, J. van Hoof's team reviewed different studies and works related to domestic interventions for the elderly [11].

In Italy, the book *Architettura per l'Alzheimer. Il malato di Alzheimer e l'ambiente* was published, which maintains that the environments experienced and lived in by these patients must be conceived, designed, structured, and built taking into account their reduced capabilities and their daily discomfort; this is an essential prerequisite to avoid reactive and depressive states that greatly compromise therapeutic treatments [27]. The physical space, its morphological dimension, its functions, and its environmental connotations can contain fear, reactivate memory, facilitate orientation, and reduce the feeling of frustration through the generation of a security that the patient must perceive. To this end, during the process of designing the space, the perceptual-sensory component must take precedence over functionality.

The design principles culminate in the World Alzheimer Report 2020 [28], which contains a set of recommendations to be observed when designing spaces for people with dementia. This series of criteria are summarised in the following points:

- Reduce risk. Prioritise safety and avoid elements that could lead to increased agitation, anger, or apathy.
- Provide a human scale, through three key factors: the number of people, the physical size of the space, and the size of the elements in it.
- Allow the user to see and be seen. An environment that is easy to understand and recognise helps to minimise confusion. Visual relationships give people the opportunity and confidence to explore their environment.
- Reduce unnecessary stimulation, both visual and auditory. The environment must be designed to minimise exposure to stimuli that are not useful to the person with dementia.
- Optimise useful stimulation. Allow the person living with dementia to see, hear, touch, or smell elements present in the space that give them clues as to where they are and what they can do.
- Support movement and participation. Design and propose well-defined, unobstructed internal and external pathways and avoid complex decision points. Create points of interest and opportunities for activity and social interaction.
- Create a friendly place by using familiar and personal elements that are recognisable in the living environment.
- Provide the opportunity to be alone or in company, which requires a variety of spaces, some for tranquillity, others for personal interaction, activity, or relaxation for users.
- Allow interaction with the community, as well as friends and family, to maintain a sense of identity; to this end, the building should provide meeting spaces.
- Facilitate various life options, allowing the focus to be on ordinary or instrumental activities of daily living, leisure, and recreational lifestyles, healthy lifestyles with exercise and walks, etc.

A summary of the main authors and theoretical contributions which have been made over the last thirty-six years, is presented in Table 1.

All the criteria set out in the different guides, manuals, and design principles consist of a series of very useful recommendations. The problem lies in the fact that they are somewhat generic and do not specify, with specific ranges and values, how they should be applied during the process of planning, designing, and building these facilities. Likewise, none of the aforementioned works define the architectural characteristics that a residential typology intended for these users should contemplate. This information is very useful for tackling, practically and with guarantees, the challenge posed by the planning of this type of building.

Table 1. Design principles for people with dementia according to different authors.

Author(s)	Main Contributions	
Calkins, 1988 [12]	Signage/orientation Safety and protection Competence in daily activities Spaces for groups Control stimuli Compensate for sensory losses	Personalisation Natural outlets Privacy and socialisation Family interaction Competence in daily activities
Pynoos et al., 1989 [29]	Adequate sensory stimulation Promote dignity and autonomy Provide security Provide a homely and familiar environment Generate an appropriate level of activity/task Provide individual control and privacy	Create opportunities for socialising Emphasise well-being and maintain a connection to health and family Flexibility and adaptability in supporting the person's physical and behavioural needs
Zgola, 1990 [30]	Stable and structured environment Provide environmental routines (associating activities with certain locations)	Serve as a memory cue Promote memories Facilitate orientation Provide security
Schiff, 1990 [31]	Stable and familiar environment Clear and well-structured environment	Serve as a memory aid Serves as a behavioural cue Support orientation to reality
Cohen and Weisman, 1991 [18]	Ensure safety and security Maximise autonomy and control Adapt to needs	Establish links Protect the need for privacy Provide opportunities for stimulation and change
Cohen and Day, 1993 [20]	Contemporary design of environments Safety and protection Autonomy and control	Opportunities for stimulation Endow and provide privacy
Brawley, 1997 [21]	Ensure safety Maximise autonomy and control Adapt to needs Establish links Protect privacy	Orientation Support functional ability with activity Opportunities for stimulation. Opportunities for socialisation
Marshall, 1998 [32]	Maximise independence Improve self-esteem and confidence Strengthen personal identity Improve visual access	Control stimuli Guiding and comprehensible Strengthen personal identity Welcome family members and local community

Table 1. *Cont.*

Author(s)	Main Contributions	
Regnier, 2002 [33]	Control, choice/autonomy Safety Accessibility and functioning Adaptability Familiarity Aesthetics and appearance	Orientation Sensory aspects Stimulation/challenge Personalisation Privacy Social interaction
Fleming et al., 2003 [34]	Ensure safety Reduce group size Highlight useful stimuli Reduce unwanted and unnecessary stimuli Simple environment with good visual access Provide wandering paths	Make the environment as familiar as possible and the atmosphere homelike Provide spaces for visitors to interact with the community Provide both privacy and community opportunities
Moore et al., 2006 [35]	Safety and protection Functional independence Significant activity Continuity of self Orientation	Sensory stimulation Personal control Privacy Social interaction
Burton and Torrington, 2007 [36]	Familiar Distinctive Legible	Secure Accessible Comfortable
Marquardt and Schmieg, 2009 [37]	Autonomy Legibility Familiarity	Sensory stimulation Social interaction
Chaudbury, 2017 [38]	Safety and protection Support functional skills Provision of privacy Maximise orientation	Regulation and quality of stimulation Control opportunities Facilitation of social contact
Calkins, 2018 [39]	Support security Create a sense of community Improve comfort and dignity	Opportunities for meaningful engagement Provide opportunities for choice
Fleming et al., 2020 [28]	Environmental design principles from Alzheimer's Disease International	

2.2. Empirical and Experimental Tests

There are essential values and attributes involving security, accessibility, privacy, autonomy, etc. that are inalienable and therefore do not require specific research for their validation [40]. Therefore, not all design guidelines are required to incorporate findings from patient trials to justify their recommendations for implementation.

Design guidelines often offer hypotheses on how a building can improve the quality of life and well-being of people with Alzheimer's disease. Some of the solutions proposed, both in these manuals and in the construction of buildings, are the result of research findings and studies with people which assess and verify how certain environmental and spatial variables influence people with dementia. These tests are necessary to properly understand and know the impact that certain design decisions may have on people. Table A2, in Appendix A, is a summary of some of the main studies carried out to determine the impact of certain environmental variables and architectural factors on people with dementia.

These validations are also necessary when the proposed solutions are of unknown effectiveness, when they involve controversial and important impacts, or when situations that provoke certain conflicts in the behaviour of these people need to be resolved. Another value of these research studies is that they serve to discriminate and find out which architectural attributes or parameters have the greatest impact on these people.

The disadvantage of incorporating these tests on design and dementia into the projects of these buildings is that, as they are experimental, they usually address a very specific problem and do so punctually and partially. Furthermore, they generally fail to assess the inevitable interactions that the variables under investigation have with the other spatial components comprising a building. Another limitation is that most of these research studies are conducted on a population sample that tends to be small and variable in size, which raises problems about the validity and generalisability of the results obtained [8]. Another important aspect to consider is that these trials tend to focus on patients in the mild or intermediate stages of the disease, but do not address the unavoidable impact of living spaces as the disease progresses.

Due to the ageing of the population, there has been a proliferation of therapeutic models aimed at the retired population and the elderly, with the aim of improving their quality of life, preventing their cognitive deterioration, and reducing their possible dependence. In recent years, knowledge of cognitive functioning in the elderly has improved enormously through specific techniques for training their cognitive skills [41,42]. The starting point for interventions aimed at promoting therapeutic action in the field of brain ageing has a proven scientific basis in the plastic capacity of the brain. Neuroplasticity is the brain's ability to restore disturbed equilibrium and adapt to new situations. Neuronal plasticity is regulated by two types of factors: intrinsic factors, which refer to the physiological characteristics of the human being, and extrinsic stimuli, which include environmental and socio-economic influences, but also the physical environment, such as the domestic or living space in which these users live. These factors can be addressed with pharmacological and non-pharmacological treatments, both with their respective limitations [43].

Furthermore, two causes of cognitive deficit coexist: organic and functional. The first is caused by the onset of the degenerative pathological process; in the case of Alzheimer's disease, this is neuronal death, loss of synapses, and alterations in neurotransmitters. The second cause would result from the continued absence of the exercise of a certain function, be it cognitive, physical, or social. This decline in function is often brought about by the sufferer themselves and sometimes by their relatives. This withdrawal from daily activities can be prevented by psychosocial interventions, but also by cognitive, physical, and sensory stimulation through an appropriately designed environment [43]. The latest trends have a holistic, person-centred approach and address the need to support the person's permanence in their own environment, with a special emphasis on maintaining the autonomy of people with dementia [44].

These fundamentals support the concept on which the particularised designs of environments and physical spaces for people affected by Alzheimer's disease are based. It is a matter of designing and building environments that facilitate effective psychostimulation and that are also as accessible, safe, and comfortable as possible. Sensory stimulation is understood as a set of stimuli generated by the physical environment, with the aim of preventing functional losses, as well as maintaining the skills and activities of daily living [45]. With this approach, the spaces inhabited and used by those affected by this disease can be used as effective tools to improve their cognitive aspects, optimising and improving their quality of life and that of their family environment.

2.3. Residential Homes for People with Alzheimer's Disease

Alongside the publication of the different design guides and the conducting of studies and tests on the environmental variables that influence people with dementia, buildings specifically designed for this group have also been constructed over the last forty years. One of the main objectives of this work is to review and evaluate the experiences obtained from these residential facilities, in order to know what results have been obtained with the practical application of the recommendations and design principles previously established theoretically and experimentally.

The first institution designed with specific criteria for people with Alzheimer's dementia was the Corinne Dolan Alzheimer Center, designed by Taliesin Associated Architects and built in 1985 in Heather Hill (Cleveland) (Figure 3).

Figure 3. Corinne Dolan Alzheimer Center, designed by Taliesin Associated Architects and built in 1985 in Heather Hill (Cleveland). (Source: Prepared by the authors and adapted from photography by Pete Guerrero and Daniel Ruark).

This is a nursing home with medical assistance, which arose as an architectural experiment with the aim of finding out which environmental design features could keep Alzheimer's patients relaxed, safe, independent, and active for as long as possible [46]. The building proposes a spatial organisation model whose main contribution is the elimination of long paths, the creation of communal relationship areas, the design of open spaces with visuals to and from the residents' bedrooms, and the use of warm materials to create specific atmospheres.

In 1991, the Woodside Place residential complex was built in Oakmont, Pennsylvania, USA, designed by the architectural firm Perkins Eastman (Figure 4). This building was an important milestone in residential homes for this group of people. With this facility, designed for 36 residents, the development of a new residential model for Alzheimer's patients began, which is still in force today and has been extended to all homes for the elderly. The main innovation introduced by this centre is its spatial organisation, based on the grouping of rooms into cohabitation units or dwellings, each with a kitchen, living room, and eight bedrooms. These housing units are distributed into three modules, with a floor plan in the shape of a comb or inverted E, in which each of the arms is a housing unit. The design of each of these modules allows the residents to maintain close social relations, as well as permanent contact with the outside spaces, thus changing the secrecy and control that had hitherto existed over the patient. The user with dementia can go out independently to the garden areas, encouraging independence in some daily activities, and thus using the environment as a therapeutic mechanism for the resident. The architectural language of the building proposes space as an important factor of reminiscence, using archetypal forms linked to the concept of the house, such as gable roofs. Moreover, reduced proportions and dimensions are used in the volumes, or adornments and furnishings are incorporated into the rooms, allowing residents to reminisce about their homes.

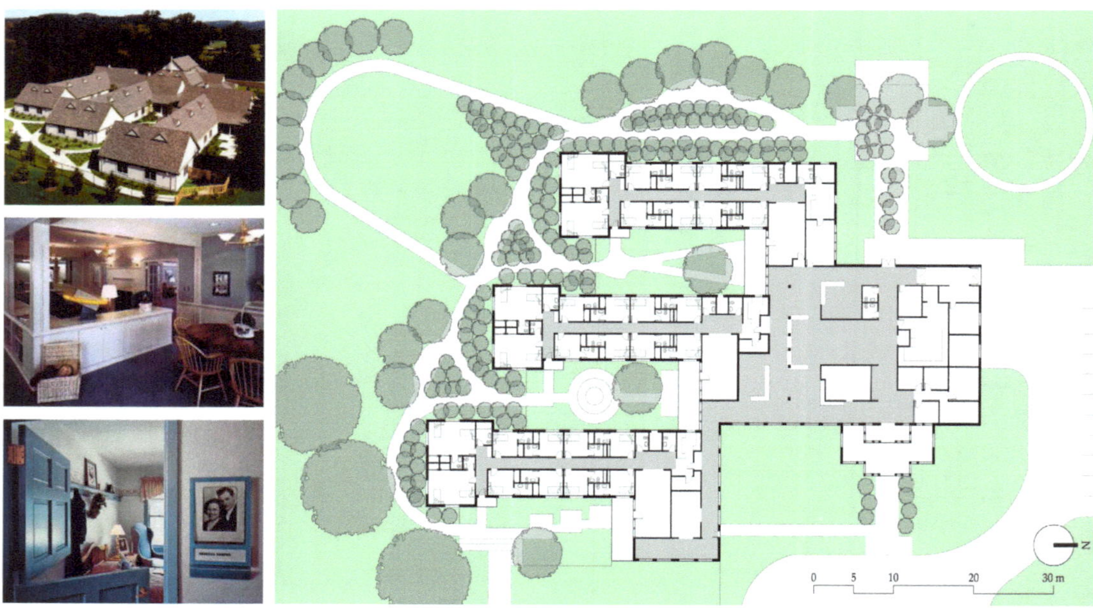

Figure 4. Woodside Place, designed by the architectural firm Perkins Eastman, built in 1991 in Oakmont (Pennsylvania, USA). (Source: prepared by the authors and adapted from photography by Robert Ruschak).

The residential models, tested in the United States, in which the housing unit, similar to a dwelling, is the generating and structuring element of the residence, are also starting to be introduced in Asia and Europe. One example is the Kompetenzzentrum Demenz in Nuremberg (Germany), designed by Feddersen Architeckten in 2006. The contribution of this building was that it developed the extensive single-storey residential model, common in the USA, over several floors. Another interesting case is the Hogeweyk Dementia Village, a residence built in the city of Wees (Holland), which opened in 2009. A one-hectare enclosed complex for 150 residents with twenty-three homes for six to eight inhabitants each, which recreates a kind of urban centre of a small town. This building, with a closed block morphology, develops a type of residence in which a traditional urban environment is recreated with restaurants, a supermarket, a post office, shops. and another urban services. The idea is for its inhabitants to develop a daily life that maintains their daily routines prior to the disease as far as possible.

An example of an extensive residence located in a peri-urban environment is the interesting experience of Boswijk, built in 2010 by EGM Architecten in the Dutch city of Vught. Another case is the Village Landais Alzheimer built in 2020 in the French town of Dax, designed by the French-Danish team of architects Champagnat and Grégoire (Landes) and Nord Architects (Copenhagen). This extensive complex is prepared to accommodate 120 inhabitants affected by this disease and other dementias. The complex recreates a population centre, whose life is organised around a garden with a central lake, around which the buildings are located. The buildings are small in scale, with shapes and materials that favour and guarantee a functioning in accordance with the needs of these patients.

3. Materials and Methods: Case Studies and Parameter Analysis

In order to draw solid and solvent conclusions about the contributions of this typology and its genesis and evolution, this paper is focused on the review, analysis, and comparison of 30 of these residences built over the last four decades. Due to their configuration and characteristics, these residences have contributed significant advances in the design of spaces and environments for patients with AD and have formed a specific architectural typology (Figure 5).

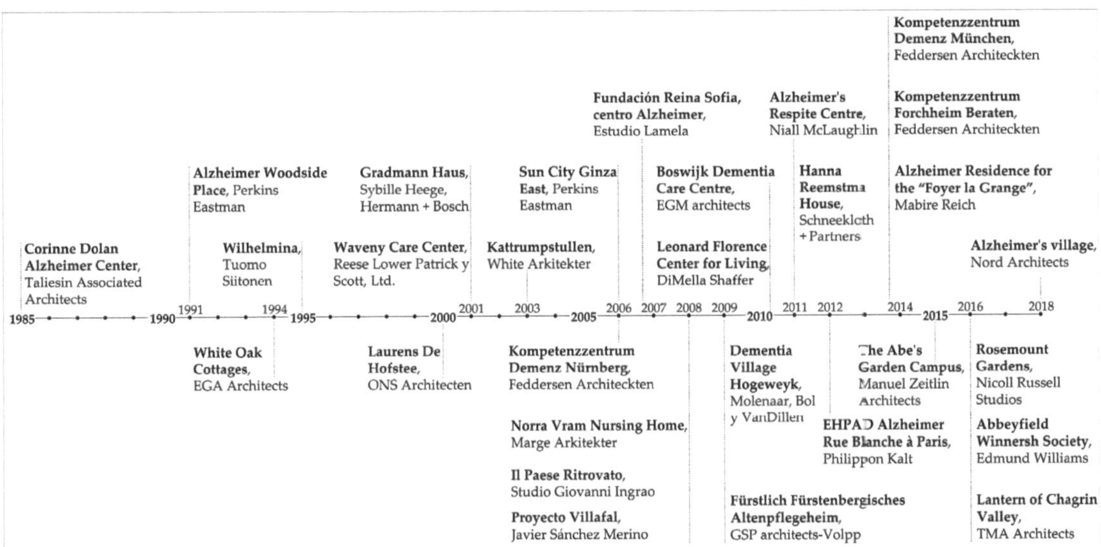

Figure 5. Chronology of the Alzheimer's residential facilities reviewed and analysed in this research. (Source: Prepared by the authors).

This research adopts the methodology of critical review, using an analytical framework with a search, assessment, synthesis, and analysis process. The methodology used in the review of the residential homes for individuals with Alzheimer's disease included an exhaustive review, selection of facilities, comparative observational process, data collection, and identification of patterns. These steps are described in the following:

1. Exhaustive review:

An extensive review of design principles proposed in various specialized guides and the scientific literature was conducted to provide a complete and objective view of what has been published to date on this subject. Moreover, a compilation of the existing information about facilities for people with AD from the first initial examples, during the last quarter of the last century, and in the present day provided knowledge and understanding of the chronological evolution of the solutions that have been adopted in the construction of these facilities.

To define the sample for this critical review, the main database literature used comprised books published in 2018 and 2019 by the Healthy Architecture & City research group, representing 28 residences designed for users with AD [47,48]. Additionally, the World Alzheimer Report 2020, Volume 2, presents a compilation of 84 buildings for people with AD in 27 countries [28]: twenty day care centres, fifty-eight residential homes, five public facilities and gardens, and one hospital. This list of buildings is based on self-reports sent to Alzheimer's Disease International following an initial survey carried out by this institution. The document does not provide any specific critical assessment of these buildings, so no conclusions can be drawn from this compilation as to what value they add to knowledge on improving the quality of life of AD patients.

2. Selection of residential facilities for people with AD:

The criteria for the initial selection included buildings primarily intended for this collective; a representative global sample incorporating countries with a significant incidence of dementia was sought. No geographical limitation was envisaged, which is why the initial selection of buildings was broad. It also needed to be a diverse and varied sample in terms of the buildings' surroundings, covering those found in urban, peri-urban, and rural environments.

Buildings were selected through a screening process. Initially, the duplicate buildings from different databases were excluded. The criterion for inclusion was centres with long-term care units exclusively for the use of Alzheimer's patients or, failing this, with a separate area in their formal and functional structure. Moreover, these buildings needed to have facilities, care services, and uses complementary to residential ones, which provide users various types of services focused on socio-health or care aspects. Buildings that, despite being care and assistance centres for people with dementia, did not have residential areas for this population of patients in their spatial organisation or formal configuration were excluded. For the final selection, quality and architectural uniqueness criteria were established, considering the projects carried out by reputable architectural firms or by those with experience in developing facilities for people with dementia. Buildings that have received quality awards or recognitions were also included, as well as buildings that serve as a reference in their closest geographical setting due to their social impact and/or significance in their contributions, as they have become a landmark in this type of construction.

3. Comparative observational process:

This step involved analysing several aspects of the selected buildings, including:

(a) Locations of the residences;
(b) Organisation and functional programme of the building;
(c) Presence of architectural attributes or parameters;
(d) Integration of new technologies.

This process made it possible to obtain useful quantifiable data to compare and identify the patterns that are repeated in them, and thus determine the structure and organisation of these buildings. The results of this study make it possible to establish the main characteristics forming this typology, which has been shaped and decanted over time.

By applying the above criteria, an initial selection of facilities was obtained, the global distribution of which corresponds to the following percentages: America (35%), Europe (21%), Oceania (6%), and Asia (3%). Based on this initial compilation of buildings, a significant group of buildings was selected to analyse the ways in which people with AD live in residences in various parts of the world, distributed across 11 countries. The final sample comprised 30 buildings which, due to their quality, contributions, and architectural characteristics, form a reference for determining the common patterns and rules in this typology (Table A2). The process is shown in Figure 6.

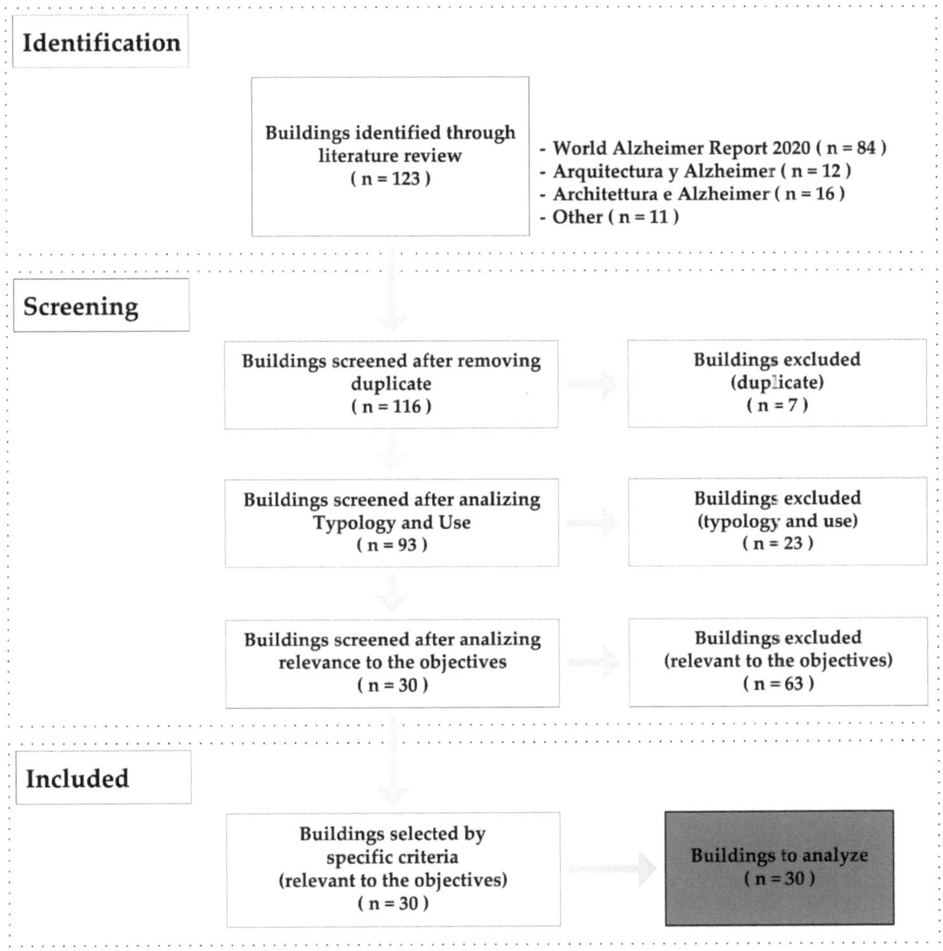

Figure 6. Sample selection flow diagram of the Alzheimer's residential facilities reviewed and analysed in this research. (Source: Prepared by the authors).

4. Data collection:

After obtaining the sample, we first proceeded to carry out a detailed and rigorous survey of all the floors of these buildings at a scale of 1:200 and 1:300 so as to properly compare them with each other. With the detailed analysis of these buildings, results were extracted that provided the main architectural characteristics comprising these residential facilities. From here, we determined which common spatial conditions were repeated and the architectural attributes or parameters that influence the well-being and quality of life of this group. With the aim of collecting data in each building, a data sheet (Figure A1) was drawn up that included the various aspects to be analysed and evaluated (Table A3). To prepare this file, a review of the existing empirical tests from design guides and the scientific literature was considered, making it possible to evaluate and compare the various solutions. It was also analysed whether the buildings have specific architectural measures aimed at alleviating some of the specific symptoms of AD, such as memory loss, difficulty in performing daily tasks, disorientation in time and place, mood swings, problems with abstract thinking, misplacement of items, anxiety, and stress.

5. Identification of patterns:

The methodology allowed for the identification of common patterns and characteristics that have come to define the architectural typology of these residential facilities over time.

4. Results

The results obtained in the research are presented in this section in four different parts: (a) Planning related to the location of the facility, (b) organisation of the building and its functional programme, (c) built-in architectural attributes or parameters, and (d) integration of information and communication technologies.

4.1. Location of the Residence and Its Relationship with the Surroundings

It should be noted that familiarity with the place in which these patients live is key to enhancing their daily activities, maintaining a sense of autonomous living and maintaining their social and spatial relationships [49]. The relationship with the surroundings is directly linked to the location of the building, its contact with the close space and its relationship with urban centres. Direct connection with outdoor spaces full of vegetation is a crucial aspect, which has become one of the priority elements to be taken into account in the planning of these buildings.

The first facilities, built for this group, are usually located in rural or peri-urban environments, proposing occupancy solutions on extensive plots, with low density and low height. One example is White Oak Cottages in Colorado, built in 2006, a building located in contact with large green and wooded spaces. Another similar example in Europe is the Boswijk complex, built in 2010 in the city of Vught (Holland), an interesting example that combines a central area with collective services to which a series of housing units are attached, directly connected to the outdoor green space (Figure 7). In other places, determined by the availability of land and the urban situation of the facilities, it is necessary for a residential development to be a high-rise. In this case, the facilities also have a functional organisation based on housing as the basic cell but incorporate green spaces on terraces or rooftops. An example would be the Kompetenzzentrum Demenz in Nuremberg (Germany), built in 2006, designed by Feddersen Architeckten (Figure 7).

Figure 7. On the left, the Boswijk Residence, built in 2010 in Vught (Holland), an extensive single-storey building designed by EGM Architecten. On the right, the Kompetenzzentrum Demenz Nürnberg residence, designed by Feddersen Architeckten, built in 2006 in Nuremberg (Germany), an example of a high-rise development of this residential model. (Source: Prepared by the authors).

The proximity or closeness to services, considered as everyday or basic necessities outside the home, depends on the environment in which the building is located. In the buildings analysed, 47.6% of them are located in a rural or peri-urban environment. The distance to urban areas with services such as shops, cultural, and other facilities is 1.1 km on average. In these residences, the relationship established with the natural environment and open spaces allows users to maintain direct contact with landscaped outdoor areas. Table 2 shows the average distances between the residences in the sample and service facilities, as well as the average distances to outdoor green or natural spaces.

Table 2. Average distances between residences to urban centres and to green or natural spaces.

Parameter	Units	Range of Optimal Values
Relationship with the surroundings	Distance (km)	Average distance to services
		Large cities 1.5 ± 0.5 km Small cities 1 ± 0.25 km Rural environment 0.5 ± 0.25 km
		Average distance to green/natural spaces
		Large cities 1 ± 0.25 km Small cities 0.5 ± 0.25 km

4.2. Organisation of the Building

4.2.1. Size, Scale, and Number of Residents

The results of an investigation of fifty-three special respite units in four US states indicated that units with few residents reduce the overstimulation of people with dementia, mainly due to noise control [50]. Therefore, the reduction of the number of residents to be served and the provision of specialised care has been a constant since the first buildings were constructed in the 1980s.

A total of 73% of the buildings analysed in this research have an average number of users that is between 80 and 120 residents per residence, distributed in independent group-living units or households, the number of which in a residence varies between nine and twelve units (Table A4). Within each household, also referred to as a special care units (SCUs), the spaces shared by its inhabitants occupy 50–55% of the total surface area of the unit. These are areas where the user spends most of the time outside their room and are usually equipped with a living room, kitchen, dining room, or rooms for other uses shared with other patients or family members (Figure 8). In 87% of the buildings in the sample, the rooms are usually single, with an average size of 17.68 m^2. The average number of residents per household ranges from eight to twelve residents (Table 3).

Figure 8. Norra Vram Nursing Home, designed by Marge Architeckten, built in 2008, in Billesholm (Sweden), with homely environment and domestic scale. (Source: Prepared by the authors and adapted from photography by Johan Fowelin).

Table 3. Number of residents and households or group-living unit in a residence or nursing home.

Number of total residents:	80–120 people
Number of residents per household:	8–12 people
Average number of households:	9–12 units per residence
Number of rooms per unit:	5–12 rooms/dwelling
Average surface areas of a 12 group-living units with 9 persons per unit	
Average floor area of the residence	7160 m^2 (average value of floor area)
Average living area of the residence	5812 m^2 (average value of usable area)
Surface area of outdoor spaces	9760 m^2
Plot occupancy	43%
Building height	One floor

The above characteristics have a direct and notable influence on the design of this type of building, since the concept of a housing or living unit is inescapably associated with the way in which residents are grouped in these modules and their relationship with the common use or care spaces. This requirement entails a particular determining factor in the composition of these facilities, because it is necessary to design them not as hospital accommodation with rooms in a row, but as a grouping of large dwellings equipped with care services. Moreover, closely linked to the number of residents are the size and scale of the buildings that, in 86% of the cases analysed, tend to have a very controlled size, both in terms of height and floor plan.

4.2.2. Distribution of the Functional Programme

Another decisive element in these residential facilities is their functional programme. The organisational structure of these buildings has evolved from the first buildings to the present day. Nowadays, services, facilities, and rooms are incorporated that complement the residential area and favour the daily activities of these users. Currently, the spaces that users use and require in these facilities are not limited to the private space of their bedrooms, but also include other places of coexistence existing in the centre.

From the analysis of the buildings in the sample, it can be seen that the spatial organisation of these facilities corresponds to three types of areas, ranging from a more public use to the private rooms of the patient (Table 4). The area that brings together the collective or common uses aims to provide the necessary support so that users can develop activities during the day, providing social interaction, psychomotor skills, cognitive stimulation, etc. These functions are generally located in the central areas of the buildings, where the user can interact with the other residents. This area also incorporates the management and administration areas of the centre and is complemented by places for interaction with family members, as well as rooms for carers or healthcare staff.

The semi-private and private areas correspond to those existing in the households or group-living units, in which their functional and organisational programme corresponds to the layout of an autonomous or independent house. The shared or semi-private uses incorporated in these units are the usual ones in a large dwelling, such as kitchens, dining rooms, living rooms, and cleaning or laundry areas, among others (Figure 9).

Table 4. Zoning, functions, and uses of a residence for people with AD.

Areas	Classification		Uses
1. Public area: Collective area dedicated to care activities and uses (22–25% of the total usable area of the residence)	1.1	Main areas	Access/reception—waiting room
			Workshops/multipurpose activity rooms
			Evaluation rooms and other services
			Auxiliary room for carers—staff
			Medical—nursing assistance room
			Administration—offices
			Living area—cafeteria
			Outdoor areas—gardens—terraces
	1.2.	Complementary areas	Relaxation room—library
			Physiotherapy room—gym/swimming pool
			Psychomotor stimulation room
			Geriatric bathrooms for collective use
			Resting area—lounge area for care staff
2. Semi-private areas: Residential area of the dwelling or household (30–35% of the total useful area of the residence)	2.1	Main rooms	Kitchen
			Dining hall
			Living room—communal rooms
			Multipurpose room—small groups
	2.2	Complementary rooms	Laundry—cleaning room
			Facilities—storage
3. Private areas: Residential area of the dwelling or household. (42–45% of the total usable area of the residence)	3.1	Main rooms	Rooms for users—patients
			Bathrooms for private use
	3.2	Secondary areas	Private outdoor spaces—terraces—gardens
			Living area

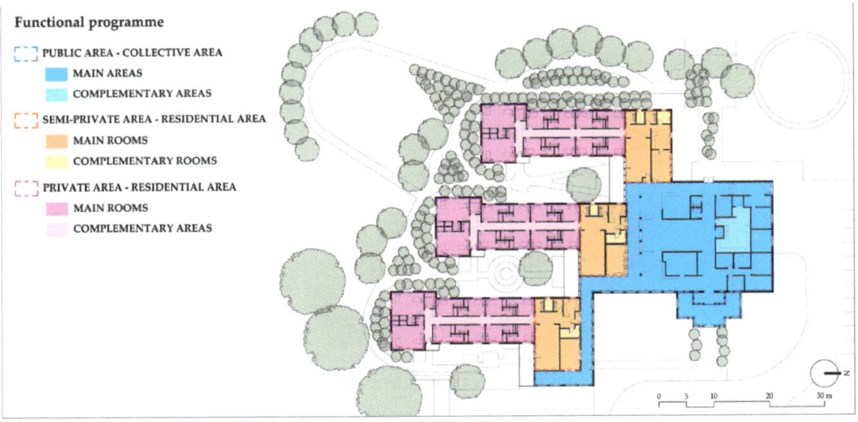

Figure 9. Functional programme and zoning of the Woodside Place residence. (Source: Prepared by the authors).

- Residential area

As mentioned in the previous section, the number of residents per household is between nine and twelve people. A dwelling or group-living unit of nine users usually has a surface area of approximately 425 m^2. Below is a summary table with the average surface areas of the different rooms obtained from the analysis of the sample of buildings (Table 5).

Table 5. Functional programme of semi-private and private areas with the calculated usable areas for a nine-user group-living unit or household.

	Classification	Length of Stay	Surface Area m^2	Surface Area (m^2/User)
Semi-private areas				
2.1	Main rooms	Kitchen	22.86	2.54
		Dining hall	36.27	4.03
		Living room—communal rooms	31.32	3.48
		Multipurpose room—small group	35.30	3.92
		Laundry—cleaning room	16.35	1.82
		Facilities—storage	23.75	2.64
Total surface area of semi-private areas: 1 group-living unit with 9 users			165.87 m^2	18.43 m^2/user
Total surface area of semi-private areas: 12 group-living units with 108 users			1990.44 m^2	18.43 m^2/user
Private areas				
3.1	Main rooms	9 single bedrooms	159.12	17.68
		Bathrooms for private use	50.49	5.61
		Bedrooms (Single)	16–18 m^2	
		Bedrooms (Double)	26–28 m^2	
		Bathroom size	4–6 m^2	
		Outdoor areas associated with the bedrooms (terraces, gardens, courtyard) per user	3–4 m^2 (min)	
		Average clear height of rooms	2.70 m/3.23 m	
Total surface area of private areas 1 group-living unit with 9 users			209.61 m^2	23.29 m^2/user
Total surface area of private areas 12 group-living units with 108 users			2515.32 m^2	23.29 m^2/user

- Care area with communal uses and daytime activities

In recent decades, the spatial model of care for users with dementia has generated various profound changes in the organisational structure of these facilities, especially determined by the need to provide the building not only with a residential structure but also with complementary services that attend to the symptoms and evolution of the disease. According to A. Casares, the hospitalisation area corresponding to the surface area used by the patient accounts for 21.54% of the total surface area in a hospital, compared to 86.3% of the surface area used by a person with dementia in this new residential typology [51]. Thus, there is a move towards a mixed model that includes care services in residential accommodation.

This paradigm shift is reflected in the typology of these architectural structures, which now have controlled areas that regulate all the patient's daily activities, while also integrating services such as medical care areas and special care or specific healthcare. In 76% of the cases studied in the sample, residential use is associated with a functional programme dedicated to services and facilities for social, therapeutic or medical care (Table 6).

Table 6. Functional programme of the public area, with usable areas calculated for a residence with 12 households or group-living units and a total of 108 users.

	Classification	Length of Stay	Surface Area m^2	Surface Area (m^2/user)
Public area				
1.1	Main areas	Access/reception—waiting room	21.60	0.20
		Multipurpose activity workshops (4 units) [1]	179.28	1.66
		Evaluation rooms (2 units) [2]	32.40	0.30
		Geriatric bathrooms (2 units)	23.76	0.22
		Administration, management and offices	45.36	0.42
		Cafeteria area	108.00	1.00
		Outdoor area for physical activities	324.00	3.00
		Psychological care room [2]	16.20	0.15
		Occupational therapy care room [2]	16.20	0.15
		Family interaction room (4 units)	54.00	0.50
		Group activity room	108.00	1.00
1.2	Secondary areas	Group relaxation room (2 units) [3]	41.04	0.38
		Physiotherapy room—gym	108.00	1.00
		Psychomotor stimulation room [3]	41.04	0.38
		Swimming pool (sheet of water) [4]	79.92	0.74
		Changing rooms and toilets	54.00	0.50
		Rest area—lounge area for staff	54.00	0.50
Total (public area)			1306.80 m^2	12.10 m^2/user

[1] Workshops of 45 m^2/unit, for small groups of 15 users (3 m^2/user). [2] Room for individualised use for user + assistance/healthcare staff. [3] Rooms of 21 m^2/unit, for small groups of seven users (3 m^2/user). [4] Swimming pool with capacity for 40 users (2 m^2/user).

A paradigmatic example of this new mixed typology is the Reina Sofía Foundation centre in Madrid (Spin), built in 2007 (Figure 10), a complex for 156 residents subdivided into several modules, six of which are residential, which have care areas according to the different degrees of their disease. It also has research areas, a day care centre, and a respite unit. The building is located in a congested urban environment, which leads to the construction of the group-living units with interior courtyards that allow an adequate relationship with the light and the outside air, as well as a suitable orientation of the dwellings.

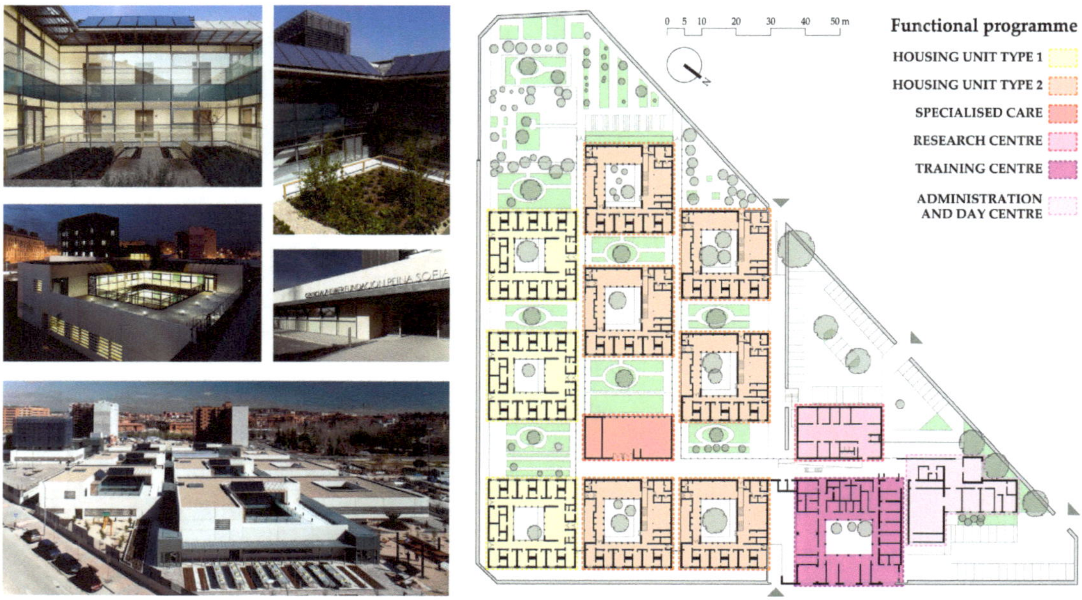

Figure 10. Reina Sofía Foundation, Alzheimer's Centre, designed by Estudio Lamela, built in 2007 in Madrid (Spain). (Source: Prepared by the authors and adapted from Estudio Lamela Arquitectos).

Other facilities resolve the incorporation of care services by recreating a traditional urban environment inside with supermarkets, pharmacies, shops, etc., similar to the original neighbourhood or town, with the aim of making the residents feel like they are living in an environment they recognise, as in the case of Dementia Village in Hogeweyk (Holland), built in 2009.

4.2.3. Paths and Wayfinding

Spatial disorientation in these patients is generated by the existence of an unrecognisable space, by confusion regarding the reference of the time in which they live, aggravated by the loss of personal identity [18]. Paths and wayfinding support movement and participation, and are also closely associated with the size of the building. A simple and clear layout, with explicit environmental information in the building, generates a good orientation for the resident [52]. It is important to design and propose well-defined, unobstructed internal and external pathways, avoiding complex decision points. This means providing points of interest, clear visuals, and opportunities for activities and social interaction [53].

In the facilities analysed, as they are predominantly residential and small in scale, the paths tend to be short, with low to medium distances between rooms and other essential spaces, excluding long corridors where there are no visual references or landmarks that allow the user to orientate themselves in the best possible way. In 42% of the cases in the sample, the paths are arranged around communal areas or rooms, allowing the user to quickly identify the services of the dwelling that are independent of their room. A total of 68% of the buildings have circular or concentric circuits. The average distance travelled by a person from their room to the communal services of the residence is 18.10 m. The average distance travelled by the user in the outdoor areas is around 142 m (Table 7).

Table 7. Average values of the paths in the buildings studied.

Parameter	Units	Range of Optimal Values
Paths	Distance (lm)	Interior paths
		Bedroom—service area: 10 ± 2 lm (max) Bedroom—secondary rooms: 15 ± 3 lm
		Paths of private outdoor spaces.
		Outdoor areas 142 ± 10 lm

Some of the creative practices recently used by architecture allow the patient to recognise certain spaces and orient themselves. At Woodside Place Pennsylvania, built in 1995 or at the Alzheimer's Respite Centre, built in 2011 in Dublin (Figure 11), memory boxes, photographs of rooms, or the chromatic identification on the doors of the rooms are used, among other resources. In 81% of the projects analysed in this research, furniture or ornamental aspects of the house in which the residents lived are incorporated into the rooms, with the aim of generating a reminiscence of their own home in the inhabitant. Some buildings, such as Abe's Garden, built 2015 in Nashville, create a fictitious outdoor environment that reproduces the environment in which the users have lived, so that they maintain a link with environments that are familiar to them.

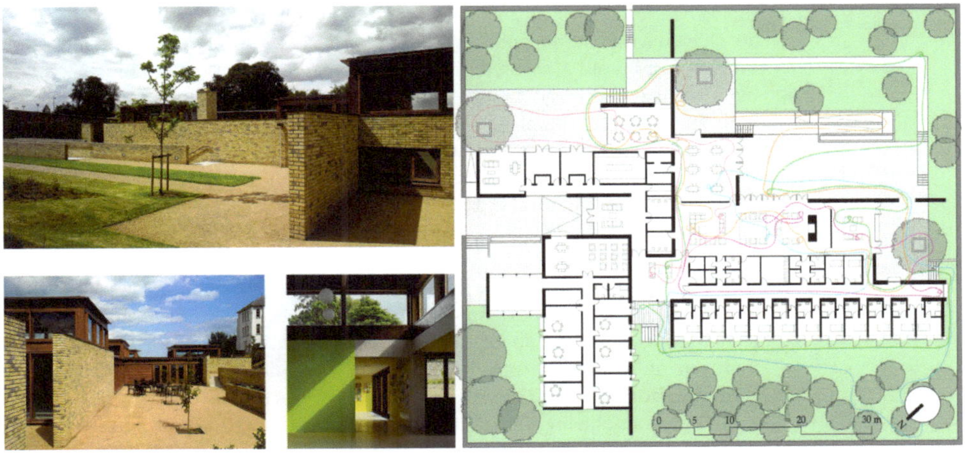

Figure 11. Alzheimer's Respite Centre, designed by Niall McLaughlin, built in 2011, in Dublin (Ireland). (Source: Prepared by the authors and adapted from Niall McLaughlin Architects and photography by Nick Kane).

4.3. Architectural and Environmental Attributes or Parameters

4.3.1. Lighting and Visual Contrast

People with AD often have sleep disorders, so exposure to light during the day may improve their night-time sleep and reduce daytime sleepiness. Lighting is an essential attribute in the care of people with Alzheimer's dementia, as it has a particular impact on circadian rhythm, reduction of agitation and anxiety, and improved safety, orientation, and cognitive stimulation, all of which are critical to the well-being and quality of life of people with dementia. It is beneficial to use natural lighting whenever possible, as its influence on people with AD has been demonstrated by different scientific evidence, as shown in Table A2. Artificial lighting systems should avoid glare, shadows, and dark areas, as these can cause disorientation or anxiety. Adequate lighting helps people with AD to move around more confidently and safely, as well as preventing falls and accidents. Moreover, good lighting, which makes it possible to perceive changes in the intensity and type of light throughout the day, helps residents to orient themselves in time and space (Table 8).

Table 8. Average lighting values.

Parameters	Units	Range of Optimal Values
Lighting	Illuminance (lux)	Predominant light source
		Natural light, in day + night rooms.
		Illuminance, by areas
		Communal areas: 300–500 lux Activities and therapy areas: 500–1000 lux Outdoor areas: 100–200 lux Bedrooms: 300–500 lux Bathrooms: 300–500 lux

It is therefore important to design and maintain a well-lit environment, adapted to the specific needs of these patients. Designing and building interiors with adequate light, both natural and artificial, has a very positive influence on people's moods. In this sense, the orientation of the building and, in particular, of the rooms comprising it, is closely linked to this parameter.

Providing adequate levels of light in interior environments is also achieved through the design of façades, as well as windows, atriums, skylights, or courtyards, all of which are elements used in the buildings in the sample to incorporate daylight into specific spaces. The interior layout and organisation of the space also influences the exposure to natural light that users receive. The use of significant contrasts, both with the correct lighting and with the use of colours, generates identification benefits in the memory of these users.

Generally, the buildings analysed in this study are located in the northern hemisphere, so their rooms are usually oriented towards the south-east, meaning that their users are in direct contact with natural light during most of their daytime activities. The importance of natural light is evident in 92% of the buildings studied.

4.3.2. Thermal and Climatic Comfort

Thermal comfort is essential for general well-being. A well-designed and maintained climate control system can significantly improve the quality of life of residents, facilitating their care and contributing to their physical and emotional well-being. However, climate control in buildings for people with Alzheimer's is not only a matter of comfort, but also an integral component of a safe and healthy environment. It contributes to their safety, for example, in the prevention of falls, as falls may increase in Alzheimer's patients if they feel cold or if there is condensation on surfaces due to poor ventilation. It is also important to control air quality, in aspects such as eliminating dust, pollen, or other allergens or maintaining the relevant humidity, preventing environments that are too dry or too humid, which can favour the proliferation of mould and bacteria [54].

On the other hand, people with Alzheimer's dementia may have difficulty regulating their body temperature due to symptoms of the disease that affect thermoregulation, and sudden changes in temperature can cause stress and confusion. A constant and appropriate room temperature environment contributes to a more stable daily routine, reducing anxiety, but also helps to avoid problems such as hypothermia or hyperthermia, which increase the risk of dehydration and heat stroke. Equally, good climate control can help to regulate the circadian rhythm of residents, maintaining appropriate temperatures at night to improve the quality of sleep.

Thermal conditioning in the buildings analysed is influenced by a number of elements including building orientation, local location, organisational structure, occupant density, ventilation strategy, and mode of operation. These combined variables influence and interact in the comfort and thermal sensation of the users, which are basically achieved with an adequate combination of outdoor temperature and humidity, both of which influence people's well-being.

When the air temperature is within a comfortable range for the occupants, the effects of humidity do not play a role. However, in warm temperature environments, humidity can influence the human body's ability to release heat through evaporation and the level of discomfort from excessive sweating. Conversely, cold and dry spaces can cause discomfort and irritation of the respiratory tract, skin, eyes, and throat. This is why the operating temperature is useful for the assessment of thermal comfort, as it reliably represents the temperature felt by a person in an indoor environment. This temperature is the mean value between the air temperature and the average radiant temperature and is variable depending on the geography and location of the building.

In this research, buildings located in the northern temperate climate zone have been studied, where the appropriate humidity level should be between 50% and 60%. Maintaining humidity within this range helps to ensure a comfortable and healthy environment, reducing respiratory problems and ensuring the general well-being of residents [36] (Table 9).

Table 9. Optimal residential temperature values in the northern temperate climate zone, where most of the case studies in the sample are located.

Parameter	Units	Range of Optimal Values
		Operating temperature
Thermal comfort	Temperature (°C/°F)	Daytime temperature: 20–25 °C; 68–77 °F Night-time temperature: 18–21 °C; 65–70 °F Bedrooms in winter: 21 ± 2 °C; 70 °F Bedrooms in summer: 22 ± 1 °C; 71.6 °F Daytime areas in winter: 22 ± 2 °C; 71.6 °F Daytime areas in summer: 24 ± 2 °C; 75.2 °F

4.3.3. Sensory or Environmental Stimulation

One of the innovations introduced since the first experiences of these architectures has been the personalisation of the surroundings or environment as a mechanism of spatial stimulation for the resident. The parameter of sensory or environmental stimulation is what differentiates this typology from other types of residences or facilities for the elderly. It is also one of the most important in providing quality of life for people with dementia. The use of this attribute consists of managing and balancing a series of environmental stimuli in such a way as to have a significant impact on the perceptions, feelings, and behaviours of users.

The use of certain materials and their corresponding textures, the use of colours, the control of noise, or the smells themselves are stimuli that have a direct impact on the well-being of the person with AD, as they report to a certain extent a reduction in their stress, anxiety, or agitation. Over time, certain measures that influence patients' senses have been incorporated into the different nursing homes. Elements that favour sensory or environmental stimulation have been introduced, such as those described below (Table 10):

- Materials: A total of 78% of the buildings studied use warm materials such as wood or similar in their interiors. As opposed to colder materials such as marble or porcelain elements, 69% of the buildings use warm flooring such as floorboards/parquet/carpets; these are materials that also cushion the effects of possible falls.

- Textures: There is a tendency to use smooth textures or surfaces, without excessively greyed drawings, so that they do not generate uneasiness, fear, stress, or situations of apathy on the part of the user when they see or touch them. In 32% of the case studies, the use of textures related to materials that allow the user to use them comfortably is identified, as opposed to cold metal elements that normally cause rejection.
- Colours: A total of 78% of the case studies analysed use light colours in both vertical and horizontal surfaces. A total of 22% use differentiating colour elements to increase contrast and identify a colour with a use or room. Contrasts are used between different rooms and places, as well as between different walls, in order to identify different zones or areas. As a general rule, warm colours predominate in night-time spaces or in spaces associated with tranquillity and relaxation.
- Smells: The presence of smells inside spaces is produced with the incorporation of certain plant elements, among other resources. Some studies show that certain fragrances can have a favourable effect on factors such as sleep. These smells can be achieved through the presence of jasmine, lavender, or other pleasant scents. A total of 72% of the buildings analysed have a direct relationship with outdoor spaces that incorporate natural vegetation with its associated smells, thus allowing the presence of certain smells at different times of the year.
- Climate control: Today's ability to adjust the climate control differently in different areas of the building allows the creation of personalised environments to the different needs of people, such as warmer areas for residents with greater sensitivity to cold or rooms with more ventilation for physical activities.

Table 10. Adequate values for sensory stimulation.

Parameter	Categories	Optimal Characteristics
Sensory stimulation	Touch	Warm, naturally sourced, medium-elasticity, high-strength materials
		Smooth, soft, and uniform textures
		Customised climatic conditioning
	Appearance	Colours with light tones, high saturation to generate contrasts and medium brightness
	Smell	Aromas of plant origin, preferably from plants or vegetation
	Noise	Noise levels < 80 dB, natural sounds from animals, such as birds singing

The proposals for the most recent buildings do not forget proposing contemporary architectural languages. This is the case at the *Etablissement d'hébergement pour les personnes âgées dépendantes*, built in 2012, in Paris and the Foyer la Grange Alzheimer's residence, built in 2014, in Nantes, which include exposed concrete or prefabricated panels on their façade so as to sift the light according to the time of day and the needs of their residents.

4.4. Integration of New Emerging Technologies

Alongside the development of this type of building, in recent decades, contemporary society has been immersed in a digital revolution that is also transforming living spaces and environments. Architecture designed for people with AD has been gradually integrating the new emerging information and communication technologies. Its objective has been, first and foremost, to provide these buildings with maximum safety, comfort, and well-being for their inhabitants. Recently, the new paradigm being established by new technologies is that the habitat and environments can interact proactively with the Alzheimer's sufferer.

With the architectural integration of new technologies, it is possible to build a prosthesis, which becomes proportionally more complex the more complex the cognitive loss. In this way, the living space can become a kind of 'exo-brain' that actively supports the affected person in their daily relationship with the physical environment. The current development of technology makes it possible to create assisted environments and spaces that interact with the person, making up for and complementing some of their cognitive or sensory deficiencies, and also helping them to preserve their cognitive capacity in the early stages of the disease. A degree of technological development has been reached in which it is possible to build a prosthetic environment where, after assessing the patient's condition, deficits, and capabilities, the environment can provide a prosthetic response, providing the possibility of maintaining their capabilities and slowing down those that are deteriorating.

In the design of these new spaces, the emergence of new paradigms such as Active and Assisted Living (AAL) plays a fundamental role, an innovative assistance model that, using the Internet of Things (IoT), Ambient Intelligence (AmI), and generative Artificial Intelligence (AI), introduces intelligent assistance systems embedded in furniture, objects, and domestic spaces. These systems are made up of different devices that communicate with each other and interact invisibly and proactively with the inhabitant, making it possible to generate assisted environments that facilitate a healthier, safer, and more comfortable life for people with physical, sensory, or cognitive functional diversity, as well as greater personal autonomy [55].

Through the use of technologies such as facial or biometric readers, smart mirrors, voice control, contact or presence locks, location, movement and presence sensors, and alarm systems in kitchens and bathrooms, it is possible to improve the instrumental activities of daily living of people with cognitive impairment, improving their orientation, motivation, or mood. It is possible, for example, that a person may not remember where the bathroom is but can be shown the way to it by automatically activating LED lights to lead them to that room. Likewise, position sensors are used linked to the switching on and off of directed paths, the control of air-conditioning units that provide comfortable aspects to the domestic environment. These are techniques that make up the new residential care space for these inhabitants and have a significant influence on aspects such as stress and anxiety control.

5. Definition of a New Architectural Typology, Characteristics

What is common to all the examples analysed in this research is that they repeat patterns that frequently use the elements described in the previous section. The application and repetition of these architectural measures gives rise to the main characteristics that define the typology of residential care facilities for patients with Alzheimer's dementia. The peculiarities comprising this type of architecture are adapted to the circumstances and determining factors of the different places where the buildings are located, as well as to the specific requirements of the developers and planners.

One of the main conclusions drawn from this paper is that, unlike traditional residences or hospital architecture where the space used by the patient is confined solely to the room and bed, the innovative nature of this new typology is its functional programme and organisational structure. In this new type, the residential component has been complemented by care services and areas that complement the living area and seek to favour the daily activities of these users. The building has to respond not only to the geographical, climatic, and cultural characteristics of the place, but also, to a certain extent, to propose a continuation of the customs and ways of life of the people who live there.

The analysis conducted during the research has identified a series of eight characteristics (Figure 12) that define the main aspects of this new typology, which are set out below:

1. Close and familiar location with large outdoor areas with vegetation.
   The surface area available for gardens, terraces or walkways in the residences should be between 55–60% of the total plot. This means that the occupation of the built-up area should be between 40% and 45%, always considering it as a single-storey

development. In order to maintain the person's sense of identity, it is necessary to maintain interaction with the territory, the climate, and the community with which they have lived. The location must be linked to the ways of life of the resident user, so that the location of the residence facilitates maintaining the social and spatial relationship with the environment they have known. In other words, it is important that the locations of these residential centres ensure, as far as possible, that their residents do not abandon the traditional habitat where they have lived their lives. Sites should provide a welcoming, safe, and stimulating environment with meeting places. The location needs to provide a variety of living options, allowing a focus on ordinary activities or leisure and recreational modes of living, healthy lifestyles with exercise and walking, encouraging social interaction and providing constant emotional support, and better managing the symptoms of the disease.

2. Residential scale with non-institutional character, reduced size, and dimension.
The appropriate maximum number of residents is usually between 80 and 120 people, grouped into several group-living units, household, or dwellings. A building for these patients needs to be on a human scale, with a small number of residents, small spaces and controlled environments. As far as possible, a building with few storeys in height and extensive development is preferable, so that the building allows the user to visually and immediately recognise the volumes comprising it. The size of the interior spaces should have characteristics and dimensions similar to those of a home. Instead of proposing institutional or hospital environments, it is necessary to design environments that are reminiscent of the home through the use of familiar elements and that allow the user to easily recognise the environment in which they live. The patient must be given the opportunity to be alone or accompanied, which requires a variety of spaces. To this end, it is necessary to provide welcoming spaces in which to be at ease and others for shared use, with communal living areas and kitchens that encourage and facilitate interaction and coexistence. These communal areas are large, accessible spaces with uses that are not restricted to resident users.
3. The dwelling as a basic cell in the building distribution.
The grouping of the residents' rooms in these residences is by dwellings or households, with a variable number of rooms that varies between five and twelve rooms per group-living unit. The number of dwellings per residence varies from nine to twelve units. These units are equipped with the usual spaces in a dwelling such as kitchens, living rooms, launderettes, etc. and which, as far as possible, have a relationship with outdoor spaces such as terraces, gardens, courtyards, etc. The layout should be orderly with short paths, avoiding right angles and blind spots; the design of the routes should make it easy for residents to walk without encountering barriers or dead ends. Elements that could lead to increased agitation, anger, or apathy should be avoided.
4. Mixed functional programme: residential and care services, with comprehensible organization.
The programme of uses should incorporate a mixed organisation chart that integrates both residential and care services. It must also take into account the temporary needs of the user and the state or phase of the disease. The spatial organisation of these facilities consists of different areas with private, semi-private, and public uses. Semi-private areas favour socialisation and generally have an open design so that they are visible to the patients themselves and also to carers. Public care spaces incorporate services and facilities with areas for medical care, special care, specific healthcare, visiting spaces, or multifunctional activity rooms. The orientation of the building should be clear and understandable, with cross visuals and a unitary perception of space, and clear and understandable visual signs are used to help residents find their way around. These visual relationships give people the opportunity and confidence to explore their environment. An environment that is easy to understand and recognise helps to minimise confusion, so it is important to establish visual landmarks or direct connections to outdoor spaces through windows or access to gardens.

5. Safety and mobility adapted to cognitive accessibility and universal design.
As far as possible, compliance with the seven principles of universal design or design for all should be taken into account, so that the environment always facilitates the movement, well-being, and comfort of the residents. On the other hand, one of the priorities in this type of building is to minimise risk. A safe building is more accessible and promotes greater freedom in decision-making for users. To achieve this, it is necessary to implement cognitive accessibility measures in buildings, complemented by technical solutions for access control, with security systems that prevent residents from wandering outside designated areas. Ramps, handrails, and adapted toilets should be provided, as well as avoiding open stairways, balconies without secure handrails, or any other element that could pose a risk of falls or accidents.
6. Personalisation of spaces through sensory stimulation.
Spaces should be designed to minimise exposure to stimuli that are not useful to the person with dementia, so unnecessary stimulation, both visual and auditory, should be reduced. On the other hand, it is necessary to optimise useful stimulation and allow the person living with dementia to see, hear, touch, or smell elements present in the space that give them clues as to where they are and what they can do. To do this, materials, colours, and textures that are stimulating to the senses should be used in a controlled way, creating a calm environment that does not cause anxiety, confusion or counterproductive effects due to a high presence of stimuli. It is important to implement construction techniques and solutions that minimise noise to create a calm and relaxing environment. In private rooms, which are the places where residents can maintain some autonomy, independence, and privacy, it is important to allow them to be personalised with furniture, personal objects, and decorations that are familiar and comforting to them. In public areas of the building, furnishings should be comfortable and safe for people with reduced mobility. It is important that outdoor areas have areas with greenery or water features arranged to stimulate the senses and provide a tranquil environment that provides relaxation and well-being.
7. Conditioning for well-being: lighting and climatic comfort.
Proper environmental conditioning is essential for the well-being and comfort of Alzheimer's patients. Two parameters are important in this typological feature: natural lighting and climate control. Adequate lighting improves sleep cycles, reduces agitation, improves safety and supports temporal and spatial orientation. On the other hand, well-managed climate control ensures a comfortable temperature, reduces heat stress, controls humidity, and maintains good ventilation. Both factors combined create a safer, more comfortable, and more therapeutic environment, which is crucial for the quality of life of people with AD.
8. Integration of emerging information and communication technologies.
The integration of technologies, as one of the main characteristics of this new architectural typology, is a fait accompli in new residential buildings. The application of new information and telecommunication technologies makes it possible to ensure the well-being of residents by monitoring them in a non-invasive way. Care management systems, consisting of software and digital tools for daily care management and monitoring of the health status of residents, must be integrated. On the other hand, it is possible to design and build an intelligent or exo-brain environment that, through the use of various types of assistance devices, sensors, and monitoring systems, can make up for some of the shortcomings of these people with cognitive impairment. Internal communication systems can also be implemented to facilitate staff coordination and immediate attention to the needs of the residents.

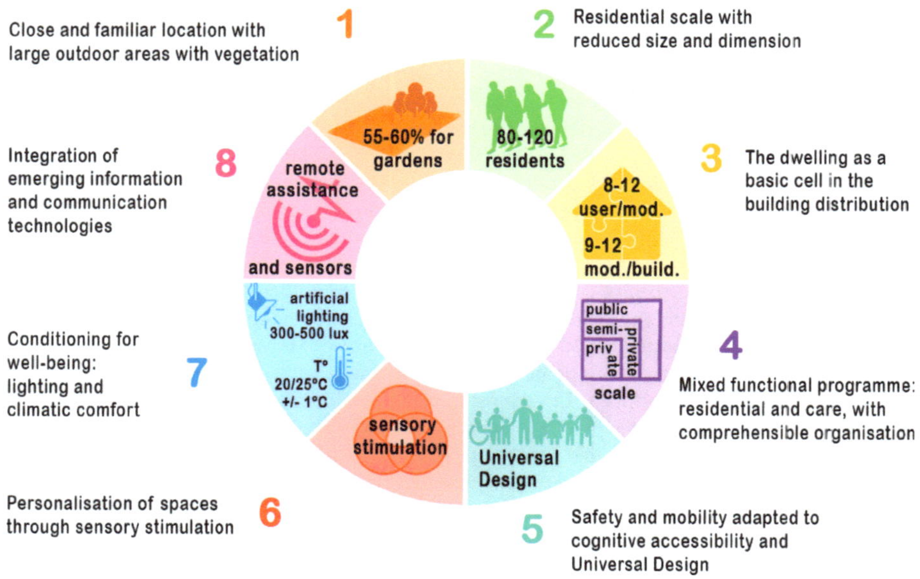

Figure 12. Characteristics of the typology of residential facilities for Alzheimer's patients. (Source: Prepared by the authors).

6. Discussion

The findings of this research underscore the importance of creating environments that are not only functional but also supportive of the cognitive and emotional needs of residents. This discussion explores the implications of this study's theoretical contributions, the weaknesses of the research, a comparative analysis between the results obtained in this work and the findings of the existing literature, and the potential for future research and practice in this area.

The establishment of a concrete architectural typology for Alzheimer's care facilities is a significant advancement in the field. By defining this typology, this study challenges traditional notions of architectural design, which often prioritise composition, aesthetics, or functionality over the cognitive, emotional, or memory needs of this collective of persons. The emphasis on the individual's cognitive and emotional well-being as central to design principles represents a paradigm shift that aligns with contemporary understandings of health and wellness in built environments. This research not only enriches the theoretical discourse but also provides a practical framework for architects, planners, and investors to follow. The results demonstrate that effective design for Alzheimer's inhabitants requires a comprehensive understanding of how environmental factors influence behaviour and well-being. This approach encourages architects to engage with psychologists, gerontologists, and healthcare professionals, fostering a holistic understanding of the challenges faced by persons with AD. Such collaboration can lead to more innovative and effective design solutions that address the complex needs of this population.

One of the main limitations of this work has been the lack of a previously established methodology, a circumstance that constituted a major barrier at the beginning of the work. However, this lack of methodology made it necessary to design an innovative method that has allowed us to tackle the problem under study with certainty. This new methodology can serve as a basis for future research and studies on this subject, and it is one of the main contributions of this work.

The results of this study present several advances compared to existing literature findings regarding the design of residential facilities for persons with AD. The paper defines a new architectural typology specifically for residential facilities catering to AD patients, detailing characteristics that have not been explicitly stated in the previous literature. Additionally, while the existing literature proposes various general design principles, this paper synthesises these into eight concrete architectural features that emphasise cognitive and physiological receptions of space for these people. Previous studies may not have systematically reviewed such a wide range of residential facilities for people with AD, limiting their applicability to current architectural practices.

The existing literature may address user experience, but it frequently lacks a detailed exploration of how specific and architectural design features can evoke positive feelings in residents. This research highlights the importance of user experience in architectural design, focusing on how architectural attributes can enhance residents' cognitive and emotional well-being. Many previous studies have relied on less systematic approaches, which can limit the generalisability of their findings. This paper employs a robust methodological framework that combines qualitative analysis with observational data, allowing for a more comprehensive understanding of design impacts.

In summary, the research advances the field by providing a well-defined architectural typology, a contemporary review of facilities, a holistic design approach, a focus on user experience, and methodological rigour, all of which enhance the understanding of effective residential environments for individuals with Alzheimer's disease.

This study opens several avenues for future research that can further improve the architectural design in the context of Alzheimer's care. Longitudinal studies examining the impact of specific design features on residents' cognitive and emotional outcomes would provide valuable empirical data to support the characteristics of this typology. Additionally, exploring the role of technology in enhancing the design of these facilities could yield insights into how smart environments can further support residents' needs. Furthermore, future research could investigate the experiences of carers and family members within these environments, as their well-being is also crucial to the overall care ecosystem. Understanding how design can facilitate family engagement and support carer needs could lead to more comprehensive care models that benefit all stakeholders involved.

This research represents a significant step forward in the theoretical understanding of architectural design for residential facilities serving persons with AD. By establishing a new typology, the research contributes to a more nuanced understanding of how built environments can enhance residents' quality of life. However, challenges in implementation and the need for further research highlight the complexities of translating theory into practice. As the population of users with Alzheimer's continues to grow, the importance of evidence-based design in residential care settings will only become more critical.

7. Conclusions

The main contribution of this work is the definition of the main architectural characteristics that a typology of residential care facilities for people with Alzheimer's dementia should have. Residences for people with AD dementia must attend to aspects such as spatial and temporal orientation, control of scale, relationships with exterior spaces and interior paths, participation in daily activities, the use of forgotten skills, and improvement in behavioural changes or in the well-being of the people affected. Additionally, these aspects need to be integrated with consideration for the necessary provision and adaptation of spaces for the family and the carer. The theoretical contributions of this study are significant and contribute to the broader discourse in architectural design, particularly in the context of residential care for individuals with AD. Key theoretical implications include the following:

1. Establishment of an architectural typology: This research defines a unique architectural typology specifically tailored for Alzheimer's care facilities. By focusing on the interplay between design and residents' cognitive and emotional needs, it challenges existing paradigms that often treat architectural design as a purely aesthetic or functional endeavour. This typology emphasises the importance of creating environments

that actively support cognitive function and emotional well-being, thus expanding the theoretical framework for designing spaces for vulnerable populations.

2. Integration of multidisciplinary perspectives: This study synthesises insights from architecture, psychology, gerontology, and health sciences, promoting a multidisciplinary approach to design. This integration underscores the necessity of considering various factors—such as safety, accessibility, and social interaction—when designing for persons with cognitive impairments. It encourages future research to adopt similar interdisciplinary methodologies, fostering collaboration across fields to enhance the understanding of how built environments impact health and well-being through thoughtful design. This contribution broadens the theoretical landscape by promoting collaboration across disciplines, emphasising that effective design for Alzheimer's care requires a comprehensive understanding of the interplay between environment and human behaviour.
3. Empirical validation of design guidelines: Through a critical review of 30 residential homes, this study provides evidence that supports the proposed design principles. This empirical foundation strengthens the theoretical framework by demonstrating how specific architectural attributes can lead to improved outcomes for residents with AD. It encourages future researches to adopt similar empirical approaches, thereby enhancing the applicability of architectural guidelines in dementia care.
4. Framework for future research and practice: This study lays the groundwork for future research by identifying key characteristics and design principles that can be further explored and refined. This framework not only guides architects and planners in their practice but also invites ongoing investigation into the evolving needs of ageing populations and the role of architecture in addressing these needs. It encourages a continuous dialogue in the field about the influence of design on health and well-being.

In summary, the theoretical contributions of this paper significantly enhance the understanding of how architectural design can serve to meet the specific needs of persons with AD, promoting a paradigm shift in the design of residential care facilities. This work not only informs current practices but also sets the stage for future innovations in the field.

We can conclude that designing a residence for Alzheimer's patients requires a deep understanding of the disease and designing with a holistic and comprehensive approach that includes aspects such as safety, accessibility, and emotional well-being, as well as social, family, and community interaction. One of the fundamental novelties of this typology is that it is configured by means of an organisation made up of households or dwellings, which recreate the atmosphere of a home, provoking reminiscences in their inhabitants. These residential facilities also incorporate care services for users and places for carers, and in some cases, they are complemented by day care centres and/or respite units. The peculiarity of this architectural typology is the sensorial personalisation of the spaces for the users for whom it is intended.

This typology is not defined by the morphology or composition of the building but focuses on achieving an adequate cognitive and physiological reception of the space by the people who inhabit it, which is why it is essential to use certain architectural parameters that stimulate all the senses. Its purpose is to stimulate the physical–neurological relations between the human body and the environment to naturally improve residents' comfort and well-being. The understanding and inclusion of the physical–chemical mechanisms that govern spaces for people with dementia implies a change in the way of thinking and understanding these environments, consequently influencing the ways of designing to make them more habitable, healthy, and inclusive for this group.

In this new type of architecture, conventional forms of representation (analogical, aesthetic, or rhetorical) are overcome in favour of other design mechanisms that encourage the relationship between environment and organism. These residences are designed using not only semantic, cultural, or plastic means, but using alternative architectural attributes that promote and favour the existing stimuli between space and body. As a result, these spaces and environments become an effective element to physically, sensorially, and emotionally stimulate people with dementia, improving the conditions of comfort and habitability, and increasing their well-being and quality of life.

Author Contributions: Conceptualization, S.Q.-G.; methodology, S.Q.-G. and P.V.-F.; validation, S.Q.-G.; formal analysis, M.L.-G.; investigation, S.Q.-G. and P.V.-F.; data curation, P.V.-F. and M.L.-G.; writing—original draft preparation, S.Q.-G.; writing—review and editing, M.L.-G.; visualization, S.Q.-G. and M.L.-G.; supervision, S.Q.-G.; funding acquisition, S.Q.-G. All authors have read and agreed to the published version of the manuscript.

Funding: Grant PID2020-115790RB-I00 funded by MICIU/AEI/10.13039/501100011033.

Data Availability Statement: The datasets below are part of the analysis process carried out at 30 residential facilities for people with Alzheimer's disease. They are presented in several data sheets, one per building. Each dataset provides data on the buildings analysed, including location, functional programme, organisation, characteristics, photographs, drawings and diagrams, among other aspects. All of them are open access data and, in accordance with FAIR principles, are stored in the institutional repository of the University of Seville (iDUS). The list of data relating to the research covered by this publication is as follows: Characteristics and data of residential facility for people with Alzheimer's disease. 01. Waveny Care Centre. 1975 [56]. Characteristics and data of residential facility for people with Alzheimer's disease. 02. Corinne Dolan Alzheimer Center. 1985 [57]. Characteristics and data of residential facility for people with Alzheimer's disease. 03. Alzheimer Woodside Place. 1991 [58]. Characteristics and data of residential facility for people with Alzheimer's disease. 04. White Oak Cottages. 1994 [59]. Characteristics and data of residential facility for people with Alzheimer's disease. 05. Wilhelmiina. 1995 [60]. Characteristics and data of residential facility for people with Alzheimer's disease. 06. De Naber. 1995 [61]. Characteristics and data of residential facility for people with Alzheimer's disease. 07. Laurens De Hofstee. 2000 [62]. Characteristics and data of residential facility for people with Alzheimer's disease. 08. Gradmann Haus. 2001 [63]. Characteristics and data of residential facility for people with Alzheimer's disease. 09. Kattrumpstullen. 2003 [64]. Characteristics and data of residential facility for people with Alzheimer's disease. 10. Sun City Ginza East. 2006 [65]. Characteristics and data of residential facility for people with Alzheimer's disease. 11. Kompetenzzentrum Demenz Nürnberg. 2006 [66]. Characteristics and data of residential facility for people with Alzheimer's disease. 12. Fundación Reina Sofía, Centro de Alzheimer. 2007 [67]. Characteristics and data of residential facility for people with Alzheimer's disease. 13. Norra Vram Nursing Home. 2008 [68]. Characteristics and data of residential facility for people with Alzheimer's disease. 14. Il Paese Ritrovato. 2008 [69]. Characteristics and data of residential facility for people with Alzheimer's disease. 15. Proyecto Villafal. 2008 [70]. Characteristics and data of residential facility for people with Alzheimer's disease. 16. Hogeweyk. 2009 [71]. Characteristics and data of residential facility for people with Alzheimer's disease. 17. Fürstlich Fürstenbergisches Altenpflegeheim. 2009 [72]. Characteristics and data of residential facility for people with Alzheimer's disease. 18. Alzheimer's Respite Centre. 2009 [73]. Characteristics and data of residential facility for people with Alzheimer's disease. 19. Leonard Florence Center for Living. 2010 [74]. Characteristics and data of residential facility for people with Alzheimer's disease. 20. Boswijk. 2010 [75]. Characteristics and data of residential facility for people with Alzheimer's disease. 21. Hanna Reemstma House. 2011 [76]. Characteristics and data of residential facility for people with Alzheimer's disease. 22. EHPAD Etablissement d'Hébergement pour les Personnes Âgées Dépendantes. 2012 [77]. Characteristics and data of residential facility for people with Alzheimer's disease. 23. Alzheimer Residence for the "Foyer la Grange". 2014 [78]. Characteristics and data of residential facility for people with Alzheimer's disease. 24. Kompetenzzentrum Demenz München. 2014 [79]. Characteristics and data of residential facility for people with Alzheimer's disease. 25. Kompetenzzentrum Beraten Wohnen Pflegen. 2014 [80]. Characteristics and data of residential facility for people with Alzheimer's disease. 26. The Abe's Garden Campus. 2015 [81]. Characteristics and data of residential facility for people with Alzheimer's disease. 27. Abbeyfield Winnersh Society. 2016 [82]. Characteristics and data of residential facility for people with Alzheimer's disease. 28. Lantern of Chagrin Valley. 2016 [83]. Characteristics and data of residential facility for people with Alzheimer's disease. 29. Rosemount Gardens. 2016 [84]. Characteristics and data of residential facility for people with Alzheimer's disease. 30. Village Landais Alzheimer. 2018 [85].

Acknowledgments: This paper has been possible thanks to funding obtained competitively from grant PID2020-115790RB-I00 funded by MICIU/AEI/10.13039/501100011033.

Conflicts of Interest: The authors declare no conflicts of interest.

Appendix A

Table A1. Studies on the influence on people with AD of environmental and spatial variables that affect planning, organisation, or architectural attributes/parameters.

Variables Related to Planning (P)/Organisation (O)/Attributes (A)	Author(s)/Year	Results of Studies and Trials
(P) Non-institutional	Annersted, 1994 [86]; Kihlgren et al., 1992 [87]; Slone et al., 1998 [50]; Cohen-Mansfield and Werner, 1998 [88]	Home environments are associated with improved emotional well-being, reduced agitation, increased social interaction, and improved functioning.
	Cohen-Mansfield et al., 1990 [89]	In non-institutional settings, compared to institutional settings such as nursing homes and hospitals, residents are less aggressive and have less anxiety
(P) Provision of outdoor areas	Mooney and Nicell, 1992 [90]	Violent episodes between residents decrease at facilities with outdoor environments compared to those without such outdoor spaces.
(P) Dimension and scale	Sloane et al., 1998 [50]	Larger sized spaces are associated with higher levels of agitation and emotional disturbance.
	Morgan and Stewart, 1998 [91]	Larger units exhibit more frequent territorial conflicts, spatial invasions and aggression towards other users.
	McAllister and Silverman, 1999 [92], Moore, 1999 [93]; Netten, 1993 [94]	Small units experience less anxiety and depression and more mobility. Small groups are positively associated with increased supervision and interaction among users.
	Annerstedt, 1993 [95]	Residents living in groups show higher motor functions and slightly improved performance in activities of daily living.
	Skea and Lindesay, 1996 [96]; Annerstedt, 1997 [97]	People with dementia living in small units experience less depression and anxiety, and are more mobile.
(O) Paths and wayfinding	Netten, 1993 [94]	Higher levels of orientation were associated with calm environments.
	Elmsthal et al., 1997 [98]	Corridor design is associated with higher degrees of restlessness and dyspraxia. Residents show more orientation in places with shorter corridors, with more space and with L, H, or square shapes.
	Passini et al., 1998 [52]	The orientation depends on the configuration of the building.
	Marquardt and Schmieg, 2009 [37]	Paths with interconnected visuals and avoiding longitudinal and back-and-forth corridors were conducive to greater orientation.
	Chaudbury et al., 2017 [38] Orfield, 2015 [99]	The environment should maximise clarity of perception and reduce optical perceptual noise such as visual clutter.

Table A1. *Cont.*

Variables Related to Planning (P)/Organisation (O)/Attributes (A)	Author(s)/Year	Results of Studies and Trials
(O) Private and semi-private rooms and activity spaces: bathrooms, dining rooms, kitchens, resident bedrooms	Pynoos and Ohta, 1991 [100]; Sloane et al., 1995 [101]; Kovach and Meyer-Arnold, 1996 [102]; Namazi and Johnson, 1996 [103]	The bathroom is one of the spaces associated with the greatest degree of stress in people with dementia. With aspects of design involving lighting, mirrors, bathtubs, etc. Natural elements such as water or animal sounds have a calming effect.
	Namazi and Johnson, 1991 [104]	The accessible location and visualisation of the toilet is important to avoid accidents and incontinence problems. It is important that they are visible through glass or windows preserving privacy when occupied.
	Gotestam and Melin, 1987 [105]	Need to locate dining and kitchen activity areas within each housing unit so that meals are at small tables and "non-institutional".
	Lawton et al., 1970 [106]	The number of residents and the design of rooms can affect levels of social interaction. In single rooms grouped around a communal space, residents spent less time indoors and increased their interactions.
(A) Lighting and visual contrast	Brawley, 1997 [21]	Increase overall light levels and reduce glare. Increase contrast to minimise confusion over depth perception.
	Mishima et al., 1994 [107]	Bright light treatment consistently regulates circadian rhythms and improves sleep patterns among people with dementia.
	Sloane et al., 1998 [50]	Residents in facilities with low overall lighting showed higher levels of agitation.
(A) Thermal and climatic comfort	Linares et al., 2017 [108]	The AD population is a group at risk to thermal extremes and temperature variations, especially heat waves.
	Jung et al., 2022 [109]	Environmental temperature is linked to the body temperature of patients with AD being significantly higher than that of other elderly people.
(A) Sensory stimulation	Calkins, 1988 [12]	Certain levels of sensory stimulation may be necessary to elicit participation in activities and social interaction.
	Brawley, 1997 [21]; Cohen and Weisman, 1991 [18]; Evans, 1989 [110]	Sensory overstimulation must be reduced as it can increase confusion and reduce social interaction and self-esteem. Therefore, a balance must be maintained between overstimulation and that needed to motivate participation.
	Evans 1989 [81]; Hall et al., 1986 [111]	Provide tactile stimulation on surfaces and walls and eliminate overstimulation of dispensable elements
	Heller, 2004 [112]; Reginald, [113] 2008	The use and classification of colours based on the psychology of colour according to what each colour conveys and the influence they can cause.
	Cohen-Mansfield et al., 1990 [89] Negley and Manley, 1990 [114]	High stimulation, as measured by agitation levels, generally occurs in lifts, corridors, bathrooms, and other residents' bedrooms or shared bedrooms.

Table A1. *Cont.*

Variables Related to Planning (P)/Organisation (O)/Attributes (A)	Author(s)/Year	Results of Studies and Trials
(A) Safety	Mayer and Darby, 1991 [115]	Placement of a full-length mirror on the exit door reduces exit attempts by half.
	Morgan and Steward, 1999 [91]; Pynoos and Ohta, 1991 [100]	Increase vigilance and supervision by facility staff, as well as prevention of falls for residents.
	Dickinson et al., 1996 [116]; Namazi et al., 1989 [117]	Create optical illusions on doors by disguising door handles and knobs.
	Hussian and Brown, 1987 [118]	Due to depth perception problems, people with AD may interpret certain elements placed on the floor as three-dimensional. The use of grids or two-dimensional elements reduced exit attempts.

Table A2. Buildings selected as case studies in the sample.

Cases	Building Name	Year	Architect(s) Author(s)	Location
C01	Corinne Dolan Alzheimer Center [56]	1985	Taliesin Associated Architects	Ohio, U.S.A.
C02	Alzheimer Woodside Place [57]	1991	Perkins Eastman	Pennsylvania, U.S.A.
C03	White Oak Cottages [58]	1994	EGA Architects	Massachusetts, U.S.A.
C04	Wilhelmina [59]	1995	Tuomo Siitonen	Helsinki, Finland
C05	De Naber [60]	1995	Massa Bureau	Rotterdam, Netherlands
C06	Laurens De Hofstee [61]	2000	ONS Architecten	Rotterdam, Netherlands
C07	Waveny Care Center [62]	2001	Reese Lower Patrick, Scott, Ltd.	Connecticut, U.S.A.
C08	Gradmann Haus [63]	2001	Sybille Heeg, Hermann + Bosch	Stuttgart, Germany
C09	Kattrumpstullen [64]	2003	White Arkitekter	Stockholm, Sweden
C10	Sun City Ginza East [65]	2006	Perkins Eastman	Tokyo, Japan
C11	Kompetenzzentrum Demenz Nürnberg [66]	2006	Feddersen Architeckten	Nuremberg, Germany
C12	Fundación Reina Sofía, centro Alzheimer [67]	2007	Estudio Lamela	Madrid, Spain
C13	Norra Vram Nursing Home [68]	2008	Marge Arkitekter	Billesholm, Sweden
C14	Il Paese Ritrovato [69]	2008	Studio Giovanni Ingrao	Monza, Italy
C15	Proyecto Villafal [70]	2008	Javier Sánchez Merino	Soria, Spain
C16	Dementia Village Hogeweyk/ [71]	2009	Molenaar, Bol y VanDillen	Wess, Netherlands
C17	Fürstlich Fürstenbergisches Altenpflegeheim [72]	2009	GSP architects-Volpp	Hüfingen, Germany
C18	Leonard Florence Center for Living [73]	2010	DiMella Shaffer	Massachusetts, U.S.A.
C19	Boswijk Dementia Care Centre [74]	2010	EGM architects	Vught, Netherlands
C20	Alzheimer's Respite Centre [75]	2011	Niall McLaughlin	Dublin, Ireland
C21	Hanna Reemstma House [76]	2011	Schneekloth + Partners	Hamburg, Germany
C22	EHPAD. Alzheimer Rue Blanche à Paris [77]	2012	Philippon Kalt	Paris, France

Table A2. *Cont.*

Cases	Building Name	Year	Architect(s) Author(s)	Location
C23	Alzheimer Residence for the "Foyer la Grange" [78]	2014	Mabire Reich	Nantes, France
C24	Kompetenzzentrum Demenz München [79]	2014	Feddersen Architeckten	Munich, Germany
C25	Kompetenzzentrum Forchheim Beraten [80]	2014	Feddersen Architeckten	Beraten, Germany
C26	The Abe's Garden Campus [81]	2015	Manuel Zeitlin Architects	Tennessee, U.S.A.
C27	Abbeyfield Winnersh Society [82]	2016	Edmund Williams	Winnersh, U.K.
C28	Lantern of Chagrin Valley [83]	2016	TMA Architects	Ohio, U.S.A.
C29	Rosemount Gardens [84]	2016	Nicoll Russell Studios	Bathgate, U.K.
C30	Alzheimer's village [85]	2018	Nord Architects	Dax, France

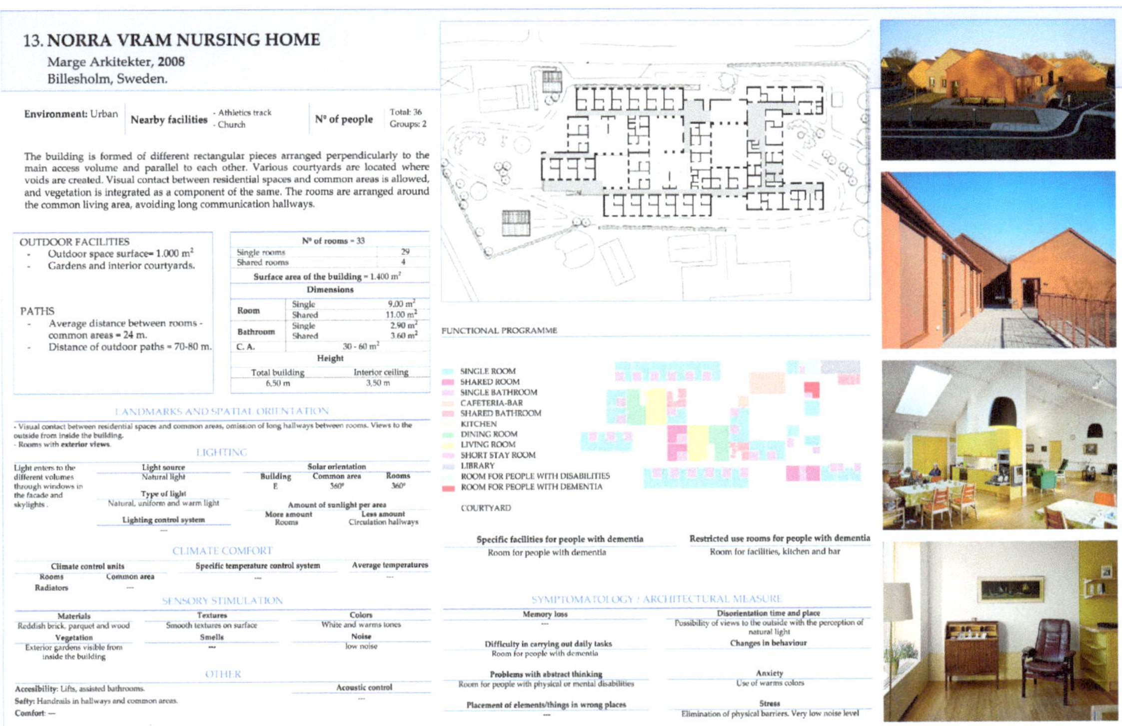

Figure A1. Data collection sheet for the selected buildings in the sample (Source: Prepared by the authors).

Table A3. Analysis parameters and measurers in residential facilities for people with Alzheimer's disease. Quantitative and qualitative variables.

		Name		
Identifying information	Architects			Year of construction
	Location	Site	City and Country	Context in the territory
		Environment	Type (rural, peri-urban, urban)	Nearby facilities
Categories	Subcategories	Quantitative analysis parameters		Unit of measurement
Functional programme	No. of people	Residents—groups	Caregivers/service staff	No.
	Outdoor facilities	Type (garden, courtyard, paths…)		No.
	Paths	Distance of paths between rooms—common areas		lm
		Distance of outdoor paths		
	No. of rooms	Number of single rooms/shared rooms		No.
		Number of communal rooms and contact with other people		
	Size of surface area	Plot surface area	Floor area of building	m^2
		Surface area of outdoor facilities	Usable area of communal areas	
		Usable area of single room/double room	Usable area of single bathroom/shared bathroom	
	Height dimension	Number of floors	Total building—interior ceiling	lm
	Other	Specific facilities for people with dementia (type)		No.
		Restricted use rooms for people with dementia (type)		No.
Parameters	Lighting	Light source	Predominant lighting	Lux N-S-E-W
		Type of light	Lighting control system	
		Solar orientation	Amount of sunlight per area	
	Climatic comfort	Average temperatures	Climate control units	°C g/m^3
		Specific temperature control system		
	Acoustic control	Systems or elements for sound control or noise cancellation		dB
	New technologies	Existence, location, and use		No.

Table A3. *Cont.*

Categories	Qualitative analysis parameters		
Sensory stimulation	Touch: materials—type/location	Textures—soft/delicate/smooth/rough/dry	
	Smell: Smells—type/source	Vegetation—type/location	
	Appearance: Colours—cool (blue/green/purple)/warm (red/orange/yellow)		
	Noise: Quiet environment (0 to 20 dB)/low noise (40 to 80 dB)/very noisy (80 to 100 dB)		
Safety	Accessibility—ramps/lifts/obstacle-free paths/adapted and assisted bathrooms		
	Mobility—handrails/short paths/articulated beds/night safety light		
	Access control—wide and sliding doors/remote assistance		
Landmarks and spatial orientation	Exterior views	Cross visuals	Yes/No
	Recreation of environments	Homely environment	
	Signs and markings on paths	Signs indicating spaces	
	Change of wall textures	Colour-coded floors	
	Personal furniture	Circular paths	
	Display cabinets with souvenirs and photographs at the entrance to rooms (memory box)		

Table A4. Average values in buildings for users with AD.

Cases	Number of Residents	Residential Group-Living Units by Building	Residents per Group-Living Unit
C01	24	2	12
C02	36	3	12
C03	24	2	12
C04	60	12	5
C05	71	1	71
C06	107	17	8 + 7 + 6 [1]
C07	52	4	13
C08	24	2	12
C09	97	9	8 + 13 [2]
C10	276	23	12
C11	80	8	10
C12	156	9	22 + 8 [3]
C13	36	2	18
C14	64	8	8
C15	28	1	28
C16	174	29	6
C17	36	3	12
C18	100	10	10
C19	120	12	10
C20	11	1	11
C21	100	5	20
C22	70	5	14
C23	118	4 + 11	20 + 3 [4]
C24	14	1	14
C25	91	8	11 + 12 [5]
C26	42	3	17 + 12 + 13
C27	63	6	10 + 11 [6]
C28	48	2	24
C29	48	2	24
C30	108	16	4 + 7 + 8 [7]

[1] 7 units of 8 persons + 5 units of 7 persons + 5 units of 6 persons. [2] 4 units of 8 persons + 5 units of 13 persons. [3] 6 units of 22 persons + 3 units of 8 persons. [4] 4 units of 20 persons + 11 apartments for up to 3 persons. [5] 5 units of 11 persons + 3 units of 12 persons. [6] 3 units of 10 persons + 3 units of 11 persons. [7] 4 units of 4 persons + 4 units of 7 persons + 8 units of 8 persons.

References

1. United Nations Department of Economic and Social Affairs, Population Division. *World Population Prospects 2022: Summary of Results*; UN DESA/POP/2022/TR/NO.3; United Nations: New York, NY, USA, 2022.
2. Livingston, G.; Huntley, J.; Liu, K.Y.; Costafreda, S.G.; Selbæk, G.; Alladi, S.; Ames, D.; Banerjee, S.; Burns, A.; Brayne, C.; et al. Dementia prevention, intervention, and care: 2024 report of the Lancet standing Commission. *Lancet* **2024**, *404*, 572–628. [CrossRef] [PubMed]
3. World Health Organization. Dementia. Key Factors. 2023. Available online: https://www.who.int/news-room/fact-sheets/detail/dementia (accessed on 23 June 2024).

4. Alzheimer's Disease International. *World Alzheimer Report 2024: Global Changes in Attitude to Dementia*; Alzheimer's Disease International: London, UK, 2024.
5. Kirch, J.; Marquardt, G. Towards human-centred general hospitals: The potential of dementia-friendly design. *Archit. Sci. Rev.* **2021**, *66*, 382–390. [CrossRef]
6. Quesada-García, S.; Valero-Flores, P.; Lozano-Gómez, M. Towards a Healthy Architecture: A New Paradigm in the Design and Construction of Buildings. *Buildings* **2023**, *13*, 2001. [CrossRef]
7. Karol, E.; Smith, D. Impact of design on emotional, psychological, or social well-being for people with cognitive impairment. *Health Environ. Res. Des. J.* **2019**, *12*, 220–232. [CrossRef]
8. Golembiewski, J. Architectural design gives hope for dementia. *Archit. Sci. Rev.* **2022**, *66*, 355–357. [CrossRef]
9. National Academies of Sciences, Engineering, and Medicine. *Reducing the Impact of Dementia in America: A Decadal Survey of the Behavioral and Social Sciences*; The National Academies Press: Washington, DC, USA, 2021; p. 182.
10. Day, K.; Carreon, D.; Stump, C. The Therapeutic Design of Environments for People with Dementia: A Review of the Empirical Research. *Gerontologist* **2000**, *40*, 397–416. [CrossRef]
11. van Hoof, J.; Kort, H.S.M.; van Waarde, H.; Blom, M.M. Environmental Interventions and the Design of Homes for Older Adults with Dementia: An Overview. *Am. J. Alzheimer's Dis. Other Dement.* **2010**, *25*, 202–232. [CrossRef]
12. Calkins, M.P. *Design for Dementia: Planning Environments for the Elderly and the Confused*; National Health Publishing: Owing Mills, MD, USA, 1988.
13. Cohen, L.W.; Zimmerman, S.; Reed, D.; Brown, P.; Bowers, B.J.; Nolet, K.; Hudak, S.; Horn, S. The Green House Model of Nursing Home Care in Design and Implementation. *Health Serv. Res.* **2016**, *51*, 352–377. [CrossRef]
14. Jones, M. A sensitive solution to caring for the mentally impaired. *Archit. J.* **1993**, *197*, 18–20.
15. Lawton, M.P.; Fulcomer, M.; Kleban, M.H. Architecture for the mentally impaired elderly. *Environ. Behav.* **1984**, *16*, 730–757. [CrossRef]
16. Liebowitz, B.; Lawton, M.P.; Waldman, A. Evaluation: Designing for confused elderly people. *Am. Inst. Archit. J.* **1979**, *68*, 59–61.
17. Devlin, A.S.; Arneill, A.B. Health Care Environments and Patient Outcomes: A Review of the Literature. *Environ. Behav.* **2003**, *35*, 665–694. [CrossRef]
18. Cohen, U.; Weisman, G.D. *Holding on to Home: Designing Environments for People with Dementia*; Johns Hopkins University Press: Baltimore, MD, USA, 1991.
19. Fleming, R.; Bowles, J. Units for the confused and disturbed elderly: Development, design, programming and evaluation. *Aust. J. Ageing* **1987**, *6*, 25–28. [CrossRef]
20. Cohen, U.; Day, K. *Contemporary Environments for People with Dementia*; Johns Hopkins University Press: Baltimore, MD, USA, 1993.
21. Brawley, E.C. *Designing for Alzheimer's Disease Strategies for Creating Better Care Environments*; John Wiley & Sons: Hoboken, NJ, USA, 1997.
22. Marshall, M. *Working with Dementia: Guidelines for Professional*; Venture: State College, PA, USA, 1990.
23. Kelly, M.; Carr, J.S. *An Evaluation of the Design of Specialist Residential Care Units for People with Dementia*; Stirling Dementia Services Development Centre: Stirling, UK, 1995.
24. Zeisel, J.; Hude, J.; Levkoff, S.E. Best practices: An Environment Behavior (EB) model for Alzheimer special care units. *Am. J. Alzheimer's Dis. Other Dement.* **1994**, *9*, 4–21. [CrossRef]
25. Zeisel, J.; Silverstein, N.M.; Hyde, J.; Levkoff, S.; Lawton, M.P.; Holmes, W. Environmental correlates to behavioural health outcomes in Alzheimer's special care units. *Gerontologist* **2003**, *43*, 697–711. [CrossRef]
26. Trías Salas, G.; Soler, P. *Guía de Arquitectura. Adaptación y Habilitación de la Vivienda para Personas con Alzheimer y Deficiencias de movilidad*; Fundación "La Caixa": Barcelona, Spain, 2002.
27. Del Nord, R. *Architettura per l'Alzheimer. Il Malato di Alzheimer e L'ambiente*; Edizione Regione Toscana: Firenze, Italy, 2002.
28. Fleming, R.; Zeisel, J.; Bennett, K. *World Alzheimer Report 2020. Design, Dignity, Dementia: Dementia-Related Design and the Built Environment (Vol. 1 and Vol. 2)*; Alzheimer's Disease International: London, UK, 2020.
29. Pynoos, J.; Cohen, E.; Lucas, C. Environmental coping strategies for Alzheimer's caregivers. *Am. J. Alzheimer's Care Relat. Disord. Res.* **1989**, *4*, 4–8. [CrossRef]
30. Zgola, J. Alzheimer's disease and the home: Issues in environmental design. *Am. J. Alzheimer's Care Relat. Disord. Res.* **1990**, *5*, 15–22. [CrossRef]
31. Schiff, M.R. Designing environments for individuals with Alzheimer's disease: Some general principles. *Am. J. Alzheimer's Care Relat. Disord. Res.* **1990**, *5*, 4–8. [CrossRef]
32. Marshall, M. Therapeutic buildings for people with dementia. In *Design for Dementia. Journal of Dementia*; Judd, S., Phippen, P., Marshall, M., Eds.; Hawker Publications: London, UK, 1998.
33. Regnier, V. *Design for Assisted Living Guidelines for Housing the Physically and Mentally Frail*; John Wiley & Sons: New York, NY, USA, 2002.
34. Fleming, R.; Forbes, I.; Bennett, K. *Adapting the Ward—For People with Dementia*; NSW Department of Health: Sydney, Australia, 2003.
35. Moore, K.; Geboy, L.; Weisman, G. *Designing a Better Day: Guidelines for Adult and Dementia Day Services Centers*; The Johns Hopkins University Press: Baltimore, MD, USA, 2006.
36. Burton, E.; Torrington, J. Designing environments suitable for older people. *CME J. Geriatr. Med.* **2007**, *9*, 39–45.

37. Marquardt, G.; Schmieg, P. Dementia-friendly architecture: Environments that facilitate wayfinding in nursing homes. *Am. J. Alzheimer's Dis. Other Dement.* **2009**, *24*, 333–340. [CrossRef]
38. Chaudbury, H.; Cooke, H.; Cowie, H.; Razaghi, L. The influence of the physical environment on residents with dementia in long-term care settings: A review of the empirical literature. *Gerontologist* **2017**, *58*, 325–337. [CrossRef] [PubMed]
39. Calkins, M.P. From Research to Application: Supportive and Therapeutic Environments for People Living with Dementia. *Gerontologist* **2018**, *58*, 114128. [CrossRef] [PubMed]
40. Lawton, M.P. Sensory deprivation and the effect of the environment on management of the senile dementia patient. In *Clinical Studies of Alzheimer's Disease and Senile Dementia*; Miller, N., Cohen, G., Eds.; Raven Press: New York, NY, USA, 1981; pp. 227–251.
41. Grandmaison, E.; Simard, M. A critical review of memory stimulation programs in Alzheimer's disease. *J. Neuropsychiatry Clin. Neurosci.* **2003**, *15*, 130–144. [CrossRef] [PubMed]
42. Francés, I.; Barandian, M.; Marcellán, T.; Moreno, L. Estimulación psicocognoscitiva en las demencias. In *Anales del Sistema Sanitario de Navarra*; Gobierno de Navarra. Departamento de Salud: Navarra, España, 2003; Volume 26, pp. 405–422.
43. Cullel Gómez, N.; Vendrell Brucet, J.M. Aplicación de las nuevas tecnologías en los programas de intervención cognitiva para personas con deterioro. *Alzheimer. Real. Investig. Demenc.* **2008**, *38*, 14–20.
44. Tárrega, L.; Boada, M. Terapias blandas: Programa de psicoestimulación integral. Alternativa terapéutica para personas con enfermedad de Alzheimer. *Neurología* **1999**, *27*, 51–61. [CrossRef]
45. Uzzel, B.P.; Gross, Y. (Eds.) *Clinical Neuropsychology of Intervention*; Martinus Nijhoff: Boston, MA, USA, 1986. [CrossRef]
46. Lewin, T. Strategies to Let Elderly Keep Some Control. *The New York Times*, 28 March 1990. Available online: https://link.gale.com/apps/doc/A175428953/HRCA?u=anon~7ba5f68&sid=googleScholar&xid=a56d11a0 (accessed on 21 May 2024).
47. Quesada-García, S.; Valero-Flores, P.; Romero Vergara, G. (Eds.) *Arquitectura y Alzhéimer. 12 Nuevos Edificios para Necesidades Emergentes*; HAC University Books: Sevilla, España, 2018. Available online: https://hdl.handle.net/11441/70199 (accessed on 12 June 2024).
48. Quesada-García, S.; Valero-Flores, P.; d'Ospina, A. (Eds.) *Architettura y Alzheimer (II). 16 Nuovi Edifici per Necessità Emergenti*; HAC University Books: Sevilla, España, 2019. Available online: https://hdl.handle.net/11441/88994 (accessed on 15 June 2024).
49. Førsund, L.H.; Grov, E.K.; Helvik, A.S.; Juvet, L.; Skovdahl, K.; Eriksen, S. The experience of lived space in persons with dementia: A systematic meta-synthesis. *BMC Geriatr.* **2018**, *18*, 33. [CrossRef]
50. Sloane, P.D.; Mitchell, C.M.; Preisser, J.S.; Phillips, C.; Commander, C.; Burker, E. Environmental correlates of resident agitation in Alzheimer's disease special care units. *J. Am. Geriatr. Soc.* **1998**, *46*, 862–869. [CrossRef]
51. Casares, A. Arquitectura Sanitaria y Hospitalaria. Escuela Nacional de Sanidad. Tema 12.1. 2012. Available online: https://www.docsity.com/es/arquitectura-sanitaria/5657107/ (accessed on 24 May 2024).
52. Passini, R.; Rainville, C.; Marchand, N.; Joanette, Y. Wayfinding and dementia: Some research findings and a new look at design. *J. Archit. Plan. Res.* **1998**, *15*, 133–151.
53. van Buuren, L.P.G.; Mohammadi, M. Dementia-Friendly Design: A Set of Design Criteria and Design Typologies Supporting Wayfinding. *Health Environ. Res. Des. J.* **2022**, *15*, 150–172. [CrossRef]
54. Castellani, B.; Bartington, S.; Wistow, J.; Heckels, N.; Ellison, A.; Van Tongeren, M.; Arnold, S.R.; Barbrook-Johnson, P.; Bicket, M.; Pope, F.D.; et al. Mitigating the impact of air pollution on dementia and brain health: Setting the policy agenda. *Environ. Res.* **2022**, *215*, 114362. [CrossRef]
55. Quesada-García, S.; Valero-Flores, P.; Lozano-Gómez, M. Active and Assisted Living, a Practice for the Ageing Population and People with Cognitive Disabilities: An Architectural Perspective. *Int. J. Environ. Res. Public Health* **2023**, *20*, 5886. [CrossRef] [PubMed]
56. Valero-Flores, P.; Lozano-Gómez, M.; Quesada-García, S. *Characteristics and Data of Residential Facility for People with Alzheimer's Disease. 01. Waveny Care Centre. 1975. [Dataset]*; idUS Depósito de Investigación de la Universidad de Sevilla: Sevilla, Spain, 2023. [CrossRef]
57. Valero-Flores, P.; Lozano-Gómez, M.; Quesada-García, S. *Characteristics and Data of Residential Facility for People with Alzheimer's Disease. 02. Corinne Dolan Alzheimer Center. 1985. [Dataset]*; idUS Depósito de Investigación de la Universidad de Sevilla: Sevilla, Spain, 2023. [CrossRef]
58. Valero-Flores, P.; Lozano-Gómez, M.; Quesada-García, S. *Characteristics and Data of Residential Facility for People with Alzheimer's Disease. 03. Alzheimer Woodside Place. 1991. [Dataset]*; idUS Depósito de Investigación de la Universidad de Sevilla: Sevilla, Spain, 2023. [CrossRef]
59. Valero-Flores, P.; Lozano-Gómez, M.; Quesada-García, S. *Characteristics and Data of Residential Facility for People with Alzheimer's Disease. 04. White Oak Cottages. 1994. [Dataset]*; idUS Depósito de Investigación de la Universidad de Sevilla: Sevilla, Spain, 2023. [CrossRef]
60. Valero-Flores, P.; Lozano-Gómez, M.; Quesada-García, S. *Characteristics and Data of Residential Facility for People with Alzheimer's Disease. 05. Wilhelmiina. 1995. [Dataset]*; idUS Depósito de Investigación de la Universidad de Sevilla: Sevilla, Spain, 2023. [CrossRef]
61. Valero-Flores, P.; Lozano-Gómez, M.; Quesada-García, S. *Characteristics and Data of Residential Facility for People with Alzheimer's Disease. 06. De Naber. 1995. [Dataset]*; idUS Depósito de Investigación de la Universidad de Sevilla: Sevilla, Spain, 2023. [CrossRef]

62. Valero-Flores, P.; Lozano-Gómez, M.; Quesada-García, S. *Characteristics and Data of Residential Facility for People with Alzheimer's Disease. 07. Laurens De Hofstee. 2000. [Dataset]*; idUS Depósito de Investigación de la Universidad de Sevilla: Sevilla, Spain, 2023. [CrossRef]
63. Valero-Flores, P.; Lozano-Gómez, M.; Quesada-García, S. *Characteristics and Data of Residential Facility for People with Alzheimer's Disease. 08. Gradmann Haus. 2001. [Dataset]*; idUS Depósito de Investigación de la Universidad de Sevilla: Sevilla, Spain, 2023. [CrossRef]
64. Valero-Flores, P.; Lozano-Gómez, M.; Quesada-García, S. *Characteristics and Data of Residential Facility for People with Alzheimer's Disease. 09. Kattrumpstullen. 2003. [Dataset]*; idUS Depósito de Investigación de la Universidad de Sevilla: Sevilla, Spain, 2023. [CrossRef]
65. Valero-Flores, P.; Lozano-Gómez, M.; Quesada-García, S. *Characteristics and Data of Residential Facility for People with Alzheimer's Disease. 10. Sun City Ginza East. 2006. [Dataset]*; idUS Depósito de Investigación de la Universidad de Sevilla: Sevilla, Spain, 2023. [CrossRef]
66. Valero-Flores, P.; Lozano-Gómez, M.; Quesada-García, S. *Characteristics and Data of Residential Facility for People with Alzheimer's Disease. 11. Kompetenzzentrum Demenz Nürnberg. 2006. [Dataset]*; idUS Depósito de Investigación de la Universidad de Sevilla: Sevilla, Spain, 2023. [CrossRef]
67. Valero-Flores, P.; Lozano-Gómez, M.; Quesada-García, S. *Characteristics and Data of Residential Facility for People with Alzheimer's Disease. 12. Fundación Reina Sofía, Centro de Alzheimer. 2007. [Dataset]*; idUS Depósito de Investigación de la Universidad de Sevilla: Sevilla, Spain, 2023. [CrossRef]
68. Valero-Flores, P.; Lozano-Gómez, M.; Quesada-García, S. *Characteristics and Data of Residential Facility for People with Alzheimer's Disease. 13. Norra Vram Nursing Home. 2008. [Dataset]*; idUS Depósito de Investigación de la Universidad de Sevilla: Sevilla, Spain, 2023. [CrossRef]
69. Valero-Flores, P.; Lozano-Gómez, M.; Quesada-García, S. *Characteristics and Data of Residential Facility for People with Alzheimer's Disease. 14. Il Paese Ritrovato. 2008. [Dataset]*; idUS Depósito de Investigación de la Universidad de Sevilla: Sevilla, Spain, 2023. [CrossRef]
70. Valero-Flores, P.; Lozano-Gómez, M.; Quesada-García, S. *Characteristics and Data of Residential Facility for People with Alzheimer's Disease. 15. Proyecto Villafal. 2008. [Dataset]*; idUS Depósito de Investigación de la Universidad de Sevilla: Sevilla, Spain, 2023. [CrossRef]
71. Valero-Flores, P.; Lozano-Gómez, M.; Quesada-García, S. *Characteristics and Data of Residential Facility for People with Alzheimer's Disease. 16. Hogeweyk. 2009. [Dataset]*; idUS Depósito de Investigación de la Universidad de Sevilla: Sevilla, Spain, 2023. [CrossRef]
72. Valero-Flores, P.; Lozano-Gómez, M.; Quesada-García, S. *Characteristics and Data of Residential Facility for People with Alzheimer's Disease. 17. Fürstlich Fürstenbergisches Altenpflegeheim. 2009. [Dataset]*; idUS Depósito de Investigación de la Universidad de Sevilla: Sevilla, Spain, 2023. [CrossRef]
73. Valero-Flores, P.; Lozano-Gómez, M.; Quesada-García, S. *Characteristics and Data of Residential Facility for People with Alzheimer's Disease. 18. Alzheimer's Respite Centre. 2009. [Dataset]*; idUS Depósito de Investigación de la Universidad de Sevilla: Sevilla, Spain, 2023. [CrossRef]
74. Valero-Flores, P.; Lozano-Gómez, M.; Quesada-García, S. *Characteristics and Data of Residential Facility for People with Alzheimer's Disease. 19. Leonard Florence Center for Living. 2010. [Dataset]*; idUS Depósito de Investigación de la Universidad de Sevilla: Sevilla, Spain, 2023. [CrossRef]
75. Valero-Flores, P.; Lozano-Gómez, M.; Quesada-García, S. *Characteristics and Data of Residential Facility for People with Alzheimer's Disease. 20. Boswijk. 2010. [Dataset]*; idUS Depósito de Investigación de la Universidad de Sevilla: Sevilla, Spain, 2023. [CrossRef]
76. Valero-Flores, P.; Lozano-Gómez, M.; Quesada-García, S. *Characteristics and Data of Residential Facility for People with Alzheimer's Disease. 21. Hanna Reemstma House. 2011. [Dataset]*; idUS Depósito de Investigación de la Universidad de Sevilla: Sevilla, Spain, 2023. [CrossRef]
77. Valero-Flores, P.; Lozano-Gómez, M.; Quesada-García, S. *Characteristics and Data of Residential Facility for People with Alzheimer's Disease. 22. EHPAD Etablissement d'Hébergement pour les Personnes Âgées Dépendantes. 2012. [Dataset]*; idUS Depósito de Investigación de la Universidad de Sevilla: Sevilla, Spain, 2023. [CrossRef]
78. Valero-Flores, P.; Lozano-Gómez, M.; Quesada-García, S. *Characteristics and Data of Residential Facility for People with Alzheimer's Disease. 23. Alzheimer Residence for the "Foyer la Grange". 2014. [Dataset]*; idUS Depósito de Investigación de la Universidad de Sevilla: Sevilla, Spain, 2023. [CrossRef]
79. Valero-Flores, P.; Lozano-Gómez, M.; Quesada-García, S. *Characteristics and Data of Residential Facility for People with Alzheimer's Disease. 24. Kompetenzzentrum Demenz München. 2014. [Dataset]*; idUS Depósito de Investigación de la Universidad de Sevilla: Sevilla, Spain, 2023. [CrossRef]
80. Valero-Flores, P.; Lozano-Gómez, M.; Quesada-García, S. *Characteristics and Data of Residential Facility for People with Alzheimer's Disease. 25. Kompetenzzentrum Beraten Wohnen Pflegen. 2014. [Dataset]*; idUS Depósito de Investigación de la Universidad de Sevilla: Sevilla, Spain, 2023. [CrossRef]
81. Valero-Flores, P.; Lozano-Gómez, M.; Quesada-García, S. *Characteristics and Data of Residential Facility for People with Alzheimer's Disease. 26. The Abe's Garden Campus. 2015. [Dataset]*; idUS Depósito de Investigación de la Universidad de Sevilla: Sevilla, Spain, 2023. [CrossRef]

82. Valero-Flores, P.; Lozano-Gómez, M.; Quesada-García, S. *Characteristics and Data of Residential Facility for People with Alzheimer's Disease. 27. Abbeyfield Winnersh Society. 2016. [Dataset]*; idUS Depósito de Investigación de la Universidad de Sevilla: Sevilla, Spain, 2023. [CrossRef]
83. Valero-Flores, P.; Lozano-Gómez, M.; Quesada-García, S. *Characteristics and Data of Residential Facility for People with Alzheimer's Disease. 28. Lantern of Chagrin Valley. 2016. [Dataset]*; idUS Depósito de Investigación de la Universidad de Sevilla: Sevilla, Spain, 2023. [CrossRef]
84. Valero-Flores, P.; Lozano-Gómez, M.; Quesada-García, S. *Characteristics and Data of Residential Facility for People with Alzheimer's Disease. 29. Rosemount Gardens. 2016. [Dataset]*; idUS Depósito de Investigación de la Universidad de Sevilla: Sevilla, Spain, 2023. [CrossRef]
85. Valero-Flores, P.; Lozano-Gómez, M.; Quesada-García, S. *Characteristics and Data of Residential Facility for People with Alzheimer's Disease. 30. Village Landais Alzheimer. 2018. [Dataset]*; idUS Depósito de Investigación de la Universidad de Sevilla: Sevilla, Spain, 2023. [CrossRef]
86. Annerstedt, L. An attempt to determine the impact of group living care in comparison to traditional long-term care on demented elderly patients. *Aging Clin. Exp. Res.* **1994**, *6*, 372–380. [CrossRef] [PubMed]
87. Kihlgren, M.; Bråne, G.; Karlsson, I.; Kuremyr, D.; Leissner, P.; Norberg, A. Long-term influences on demented patients in different caring mileaus, a collective living unit and a nursing home: A descriptive study. *Dementia* **1992**, *3*, 342–349. [CrossRef]
88. Cohen-Mansfield, J.; Werner, P. The effects of an enhanced environment on nursing home residents who pace. *Gerontologist* **1998**, *38*, 199–208. [CrossRef]
89. Cohen-Mansfield, J.; Werner, P.; Marx, M.S. The spatial distribution of agitation in agitated nursing home residents. *Environ. Behav.* **1990**, *22*, 408–419. [CrossRef]
90. Mooney, P.; Nicell, P.L. The importance of exterior environment for Alzheimer residents: Effective care and risk management. *Healthc. Manag. Forum* **1992**, *5*, 23–29. [CrossRef]
91. Morgan, D.G.; Stewart, N.J. High versus low density special care units: Impact on the behavior of elderly residents with dementia. *Can. J. Aging* **1998**, *17*, 143–165. [CrossRef]
92. McAllister, C.L.; Silverman, M.A. Community formation and community roles among persons with Alzheimer's disease: A comparative study of experiences in a residential Alzheimer's facility and a traditional nursing home. *Qual. Health Res.* **1999**, *9*, 65–85. [CrossRef]
93. Moore, K.D. Dissonance in the dining room: A study of social interaction in a special care unit. *Qual. Health Res.* **1999**, *9*, 133–155. [CrossRef]
94. Netten, A. *A Positive Environment? Physical and Social Influences on People with Senile Dementia in Residential Care*; Ashgate: Aldershot, UK, 1993.
95. Annerstedt, L. Development and consequences of group living in Sweden. *Soc. Sci. Med.* **1993**, *37*, 1529–1538. [CrossRef]
96. Skea, D.; Lindesay, J. An evaluation of two models of long-term residential care for elderly people with dementia. *Int. J. Geriatr. Psychiatry* **1996**, *11*, 233–241. [CrossRef]
97. Annerstedt, L. Group-living care: An alternative for the demented elderly. *Dement. Geriatr. Cogn. Disord.* **1997**, *8*, 136–142. [CrossRef] [PubMed]
98. Elmståhl, S.; Annerstedt, L.; Åhlund, O. How should a group living unit for demented elderly be designed to decrease psychiatric symptoms? *Alzheimer Dis. Assoc. Disord.* **1997**, *11*, 47–52. [CrossRef] [PubMed]
99. Orfield, S. Dementia environment design in seniors housing: Optimizing resident perception and cognition. *Sr. Hous. Care J.* **2015**, *23*, 58–69.
100. Pynoos, J.; Ohta, R.J. In-home interventions for persons with Alzheimer's disease and their caregivers. *Phys. Occup. Ther.* **1991**, *9*, 83–92. [CrossRef]
101. Sloane, P.D.; Honn, V.J.; Dwyer, S.A.R.; Wieselquist, J.; Cain, C.; Myers, S. Bathing the Alzheimer's patient in long term care: Results and recommendations from three studies. *Am. J. Alzheimer's Dis.* **1995**, *10*, 3–11. [CrossRef]
102. Kovach, C.R.; Meyer-Arnold, E.A. Coping with conflicting agendas: The bathing experience of cognitively impaired older adults. *Sch. Inq. Nurs. Pract. Int. J.* **1996**, *10*, 23–36.
103. Namazi, K.H.; Johnson, B.D. Issues related to behavior and the physical environment: Bathing cognitively impaired patients. *Geriatr. Nurs.* **1996**, *17*, 234–239. [CrossRef]
104. Namazi, K.H.; Johnson, B.D. Physical environmental cues to reduce the problems of incontinence in Alzheimer's disease units. *Am. J. Alzheimer's Care Relat. Disord. Res.* **1991**, *6*, 22–29. [CrossRef]
105. Götestam, K.G.; Melin, L. Improving well-being for patients with senile dementia by minor changes in the ward environment. In *Society, Stress, and Disease*; Levi, L., Ed.; Oxford University Press: Oxford, UK, 1987; pp. 295–297.
106. Lawton, M.P.; Liebowitz, B.; Charon, H. Physical structure and the behavior of senile patients following ward remodeling. *Aging Hum. Dev.* **1970**, *1*, 231–239. [CrossRef]
107. Mishima, K.; Okawa, M.; Hishikawa, Y.; Hozumi, S.; Hori, H.; Takahashi, K. Morning bright light therapy for sleep and behaviour disorders in elderly patients with dementia. *Acta Psychiatry Scand.* **1994**, *89*, 1–7. [CrossRef] [PubMed]
108. Linares, C.; Carmona, R.; Ortiz, C.; Díaz, J. *Temperatura Extremas y Salud. Cómo nos Afectan las Olas de Calor y de Frío*; Catarata: Madrid, Spain, 2017.

109. Jung, C.G.; Kato, R.; Zhou, C.; Abdelhamid, M.; Shaaban, E.; Yamashita, H.; Michikawa, M. Sustained high body temperature exacerbates cognitive function and Alzheimer's disease-related pathologies. *Sci. Rep.* **2022**, *12*, 12273. [CrossRef] [PubMed]
110. Evans, B. *Managing from Day to Day: Creating a Safe and Workable Environment*; Department of Veterans Affairs Medical Center: Minneapolis, MN, USA, 1989.
111. Hall, G.; Kirschling, M.V.; Todd, S. Sheltered freedom—An Alzheimer's unit in an ICF. *Geriatr. Nurs.* **1986**, *7*, 132–137. [CrossRef] [PubMed]
112. Heller, E. *Psicología del Color, Cómo Actúan los Colores Sobre los Sentimientos y la Razón*; Gustavo Gili: Barcelona, Spain, 2004.
113. Reginald, R. *Psicología del Color, Curso de Formación en Psicología y Terapia del Color*; Yug: Mexico City, México, 2008.
114. Negley, E.N.; Manley, J.T. Environmental interventions in assaultive behavior. *J. Gerontol. Nurs.* **1990**, *16*, 29–33. [CrossRef]
115. Mayer, R.; Darby, S.J. Does a mirror deter wandering in demented older people? *Int. J. Geriatr. Psychiatry* **1991**, *6*, 607–609. [CrossRef]
116. Dickinson, J.I.; McLain-Kark, J.; Marshall-Baker, A. The effects of visual barriers on exiting behavior in a demented care unit. *Gerontologist* **1995**, *35*, 127–130. [CrossRef]
117. Namazi, K.H.; Rosner, T.T.; Calkins, M.P. Visual barriers to prevent ambulatory Alzheimer's patients from exiting through an emergency door. *Gerontologist* **1989**, *29*, 699–702. [CrossRef]
118. Hussian, R.A.; Brown, D.C. Use of two-dimensional grid to limit hazardous ambulation in demented patients. *J. Gerontol.* **1987**, *42*, 558–560. [CrossRef]

buildings

MDPI

Article

Is the Children's 15-Minute City an Effective Framework for Enhancing Children's Health and Well-Being? An Empirical Analysis from Western China

Xiao Wang [1,2], Jingping Huang [1], Zhengyou Qin [3], Wei Gan [4], Zhen He [4] and Xiang Li [5,*]

1 College of Tourism & Landscape Architecture, Guilin University of Technology, Guilin 541004, China; wangxiaoglut@glut.edu.cn (X.W.); 1020220708@glut.edu.cn (J.H.)
2 Institute of Guangxi Tourism Industry, Guilin University of Technology, Guilin 541004, China
3 China Academy of Urban Planning & Design, Beijing 100835, China; 6613042@glut.edu.cn
4 Department of Urban Planning, Tongji University, Shanghai 200092, China; ganw@tongji.edu.cn (W.G.); 2210292@tongji.edu.cn (Z.H.)
5 School of Design and Architecture, Zhejiang University of Technology, Hangzhou 310023, China
* Correspondence: lixiang@zjut.edu.cn

Citation: Wang, X.; Huang, J.; Qin, Z.; Gan, W.; He, Z.; Li, X. Is the Children's 15-Minute City an Effective Framework for Enhancing Children's Health and Well-Being? An Empirical Analysis from Western China. *Buildings* **2025**, *15*, 248. https://doi.org/10.3390/buildings15020248

Academic Editor: Geun Young Yun

Received: 8 December 2024
Revised: 7 January 2025
Accepted: 13 January 2025
Published: 16 January 2025

Citation: Wang, X.; Huang, J.; Qin, Z.; Gan, W.; He, Z.; Li, X. Is the Children's 15-Minute City an Effective Framework for Enhancing Children's Health and Well-Being? An Empirical Analysis from Western China. *Buildings* **2025**, *15*, 248. https://doi.org/10.3390/buildings15020248

Correction Statement: This article has been republished with a minor change. The change does not affect the scientific content of the article and further details are available within the backmatter of the website version of this article.

Abstract: Increasing attention is being directed toward understanding the impact of urban built environments on children's health. The concept of the 15-minute city, or "15-minute life circle", in China is significant; however, the needs of the pediatric population are often overlooked. Walkability and accessibility to urban service facilities are crucial for promoting children's well-being. This paper focuses on the old city of Guilin in western China as a case study. It investigates the service facilities within a 15-minute life circle that affect children's physical health, mental health, and social engagement. Utilizing the Walking Index method, this study assesses the walkability and rationality of these facilities in 621 neighborhoods. Data were gathered from 200 questionnaires and interviews with children, concentrating on their daily use of these facilities, supplemented by a multi-source analysis. The findings reveal that the Walking Index in the study area is generally low, with an uneven distribution, primarily influenced by road network density, block size, and urban landscape patterns. This study proposes the development of a "Children's Healthy 15-minute Life Circle", which would increase child-friendly service facilities, pedestrian infrastructure, and urban road network density. It also advocates for constructing service facilities that enhance children's health and the optimization of the spatial arrangement of these facilities. Furthermore, it suggests the effective utilization of urban landscapes to create a diverse system of leisure spaces for children. This study offers valuable insights for enhancing the walkability of urban life circle service facilities and establishing a child-friendly city.

Keywords: 15-minute life circle; child health; walkability; service facilities; walking index; child-friendly city; western China

1. Introduction

The construction of a healthy China is a national strategy. The "Outline of the Healthy China 2030 Plan [1]" emphasizes the health of people, with special attention to vulnerable groups such as older people and children. At the city and community levels, the vision of building healthy cities and realizing the 15-minute healthy life circle of cities is emphasized. As the main walking distance area of children, the spatial distribution of service facilities in the Neighborhood Life Circle ① will affect children's willingness to walk and thus affect the frequency and types of children's use of service facilities [2–5]. Relevant studies have shown that improving the walking accessibility of service facilities will have a positive impact on children's health [6–9]. However, at present, there are problems such as poor walking accessibility and insufficient child-friendliness [10] in the Neighborhood Life Circle [11], inevitably forcing children to choose less walking and outdoor activities [12]. Neighborhood Life Circle refers to the basic unit that meets the work and life needs of urban and rural residents in the whole life cycle within a suitable daily walking range. It usually takes a 15-minute walk as the standard to build a comprehensive community service center to ensure that residents can meet their basic living needs within walking distance. Resulting children's health problems, such as anxiety, obesity, and weakened immunity, are on the rise [13,14]. Existing studies mainly focus on the service radius and accessibility of service facilities from the perspective of the adult population and rarely consider the walkability of service facilities and the needs of children's health from the children's perspective [12]. Therefore, based on children's health needs, this paper selects service facilities that could affect children's physical health, mental health, and social adaptation, analyzes the walkability of service facilities, and provides scientific reference for establishing a healthy and friendly Neighborhood Life Circle for children.

2. Research Review

2.1. Children and Cities

The international community's attention to children originated from the publication of the "Convention on the Rights of the Child" in 1989 [15,16], which put forward the principle of "The best interests of the child" as the theoretical basis for formulating public policies and made it clear that children enjoy the rights to survival, development, protection, and participation. In 1996, the United Nations International Children's Emergency Fund (UNICEF) and the United Nations Human Settlements Programme (UN-Habitat) jointly put forward the "Child-Friendly Cities Initiative (CFCI)", suggesting that children are entitled to a nurturing environment that ensures safety, access to vital services, and clean resources. They should have spaces for play, education, and personal growth, with their opinions valued. The CFCI works towards these goals by engaging local authorities, NGOs, businesses, academia, media, and children across more than 40 countries [17,18]. China has also made "persisting in encouraging children's participation" one of the basic principles in the Program for the Development of Chinese Children (2021–2030) [19].

In 2001, Habitat II included children in environmental decision-making systems. In 2016, at the United Nations Conference on Housing and Sustainable Urban Development (Habitat III), UNICEF released "Your Voice in Habitat", stating that children need an overall urban environment that is more convenient, safe, inclusive, and resilient. In May 2018, UNICEF published *Shaping Urbanization for Children—A Handbook on Child-Responsive Urban Planning*, which integrates planning, urban design, social behavior, economics, public management, and other disciplines to explore how cities can better serve the real needs of children [20].

2.2. 15-Minute City

The 15-minute city is an urban planning concept that aims to focus residents' basic needs, such as living, working, and playing, within a 15-minute walk or bike ride by reducing their dependence on cars. The concept emphasizes density, proximity, diversity, and digitalization to promote sustainable urban development. Its core is to improve the spatial efficiency and quality of life of cities while reducing traffic congestion and carbon emissions. The origins of the 15-minute city idea can be traced back to urban planning movements of the early 20th century, such as neighborhood unit plans and new urbanism.

These early ideas, which emphasized community self-sufficiency and walkability, laid the foundation for the development of the 15-minute city. In recent years, especially in the wake of the COVID-19 pandemic, 15-minute cities have received a lot of attention for their emphasis on local communities.

The 15-minute city concept has already been implemented in several cities around the world. Paris, for example, took the lead in promoting the idea in 2019, achieving its goals by increasing green space and improving public transport. Other cities such as Portland, Melbourne, and Stockholm are also actively exploring related practices. However, many urban practices are still in the planning stage and have yet to form a comprehensive implementation strategy.

In the future, the development of the 15-minute city will require further research and practice to address its adaptability and implementation strategies in different urban environments. At the same time, combined with the development of digital technology and smart cities, 15-minute cities are expected to play a greater role in improving the efficiency of urban governance and the quality of life of residents.

2.3. Review of Walkability Research

In urban settings, there are complex links between human health and urban planning and transportation design. The impact of urban environment on human health is multifaceted. Studies have shown that air pollution, noise, and traffic congestion in cities have a negative impact on the health of residents. In addition, the increase of green and natural spaces in cities is believed to improve the mental health and overall well-being of residents.

The discussion on the travel mode of urban residents has been going on for a long time. From the earliest urban critics, including Jane Jacobs, to the later scholars of New Urbanism, all advocated the establishment of pedestrian-friendly neighborhoods and communities to improve the status of pedestrian transportation [21,22]. In the 1990s, Cervero, a scholar in the field of transportation planning, proposed three built environment factors that affect travel demand—density, diversity, and design—known as "3Ds" [23], which have become important theoretical sources for the study of walkability. Also in the 1990s, researchers in the field of public health found that the reduction of physical activity represented by walking would lead to a series of health problems [24,25], and actively sought interdisciplinary cooperation to enhance residents' physical activity and improve residents' health.

In 2003, Sealens and other scholars put forward the theory that built environment factors at the community scale affect residents' walking behavior and used the term "walkability" for the first time [26]. From 2003 to 2010, large-scale ecological empirical studies in the Americas, Australia, and Europe proved that environmental factors have a significant impact on walking behavior [27–29]. Since then, walkability research has received more attention, and a large number of articles have summarized it.

From both subjective and objective perspectives, some scholars expanded walkability to "the degree of support for walking in the built environment and pedestrians' evaluation of walking experience in the environment" [30]. In more specific research practices, some scholars have summarized walkability as "the degree of support and attraction of urban environments for walking, including providing a comfortable and safe environment for pedestrians, enabling people to reach various destinations within a reasonable time and physical capacity, and providing visual attraction in the walking network" [31,32]. Walkability is affected by many environmental factors, including the distribution of daily service facilities, road network density, population density, and other objectively measurable urban spatial attributes [26,33–35], as well as subjective usage feelings such as pedestrians' perception of space safety when walking and the attractiveness of streetscapes to pedestrians [36,37]. Therefore, there are still many discussions on how to measure the factors affecting walkability more accurately. For example, Krizek created a community walking environment evaluation table, which constructs an indicator system from the dimensions of community living type, accessibility of shops and service facilities, street continuity, walking space, community beauty, and so on [38]. According to objective environmental factors, such as the diversity of destination types and spatial layout of users' travel needs as measurement indicators, the most widely used method, such as Walk Score, is used.

2.4. Links Between Child Health and Service Facilities

This topic focuses on the environmental factors that influence children's walking behavior. Children are active, and the demand and intensity of outdoor games are higher than other groups. As a socially vulnerable group, the guardian's judgment on the safety of the walking environment will greatly affect the travel mode of children, so children have special environmental needs for walking [39–41]. However, as the improvement of physical activity will have a profound impact on the health level and quality of life throughout life [42,43], it has attracted particular attention from the government and researchers and has formed a relatively independent long-term discussion topic around it.

The concept of child health includes three aspects: physical health, mental health, and social adaptation [44–46]. The physiological characteristics of children are as follows: children's immunity is relatively low, and they are susceptible to physiological diseases caused by various factors, such as obesity and respiratory system infection, while the psychological characteristics of children are as follows. As a special group, children have multiple needs for curiosity, growth, and participation. If the urban service system and social support system are weakened, children's negative emotions, such as anxiety and panic, may occur, leading to psychological problems. In terms of social adaptation, if children get along harmoniously with the surrounding environment, maintain good psychological quality, and moderately participate in social activities, it will have a positive impact on children's healthy growth [47]. Studies have shown that children's health is closely related to the service facilities of Neighborhood Life Circle. For example, the obesity rate of children is negatively correlated with the accessibility and quantity of outdoor sports fields [48], e.g., small parks, playgrounds, and other activity facilities in their neighborhoods. Green spaces, such as parks and pedestrian greenways, can promote children's psychological feelings of security, comfort, and pleasure. Sports facilities such as gymnasiums and science and technology museums can not only provide children with education, culture, and sports needs but also provide places for children's social interaction.

To sum up, there is room for further discussion on the walkability of Neighborhood Life Circle. In the construction of the index system, the existing index system studies mainly the dimensions of safety, comfort, aesthetics, and accessibility, and less considers the health dimension [49,50]. In terms of research content, most of the existing studies focus on public service facilities, commercial spaces, traffic stations, or urban streets from a single perspective, and few studies consider the public and commercial service facilities of the Neighborhood Life Circle comprehensively. In terms of research objects, less attention has been paid to special groups such as children and older people. Due to physiological reasons, the walking needs of this special group are different from those of ordinary adults, and the environmental factors affecting their walkability need to be discussed separately. Based on this, this study takes the old urban area of Guilin, China, as an example [51]. The 15-minute city is an urban planning concept that aims to create a community where residents can reach all basic amenities and daily needs within 15 min by walking, biking, or public transportation. Based on the theory of the 15-minute life circle, this study uses the method of the Walking Index to conduct a quantitative study on the walkability of the service facilities related to children's health, to create a pedestrian-friendly and healthy growing neighborhood for children, and thus provide planning suggestions.

3. Research Framework and Data Sources

3.1. Research Framework

Based on the 15-minute life circle and the theory of children's health, we put forward the concept of the "Children's Healthy 15-minute Life Circle". In the Children's Healthy 15-minute Life Circle, children can safely and quickly access public and commercial service facilities to meet children's physical [52], psychological, and social health needs. This paper aims to study the walkability of service facilities in Children's Healthy 15-minute Life Circle, which is divided into the following three parts: First, we select all kinds of service facilities related to children's health needs, that is, select public and commercial service facilities that can meet their needs based on children's physical, mental, and social health needs, and improve the indicators of these service facilities. Second, we measure the walkability of service facilities and calculate the Single Point Walking Index (SPWI) and Area Walking Index (AWI) through ArcGIS spatial analysis and correlation analysis, in which the Area Walking Index (AWI) includes the Sub-district Walking Index (SWI) and Overall Walking Index (OWI). Third, we evaluate the walkability of service facilities in each Children's Healthy 15-minute Life Circle, analyze the configuration of service facilities in depth according to the scores of the walkability of service facilities in each life circle, and then propose improvement strategies and suggestions (Figure 1, Table 1).

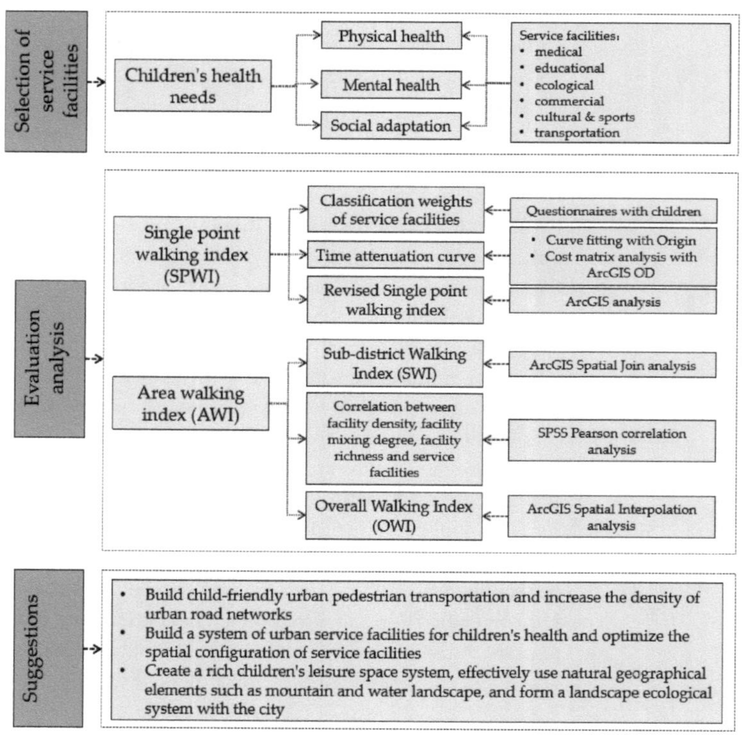

Figure 1. Research framework.

Table 1. Walking Index methods in this research.

Walking Index		Methods
Single Point Walking Index (SPWI)		First, OD Cost Matrix Analysis in ArcGIS software 10.8 was used to calculate the time taken by children to reach various service facilities within 15 minutes, taking POI of 621 urban neighborhoods in the old urban area of Guilin as the starting point of walking. POI data (Point of Interest) are from the Amap open API. Then, the time attenuation rate is obtained by substituting the corresponding time attenuation function. Finally, the initial SPWI is obtained by summarizing the weights of all service facilities after attenuation.
Sub-district Walking Index (SWI)	Area Walking Index (AWI)	The SPWI of an urban neighborhood was spatially linked through the ArcGIS Spatial Join function to obtain the Walking Index at the sub-district scale, which was named the Sub-district Walking Index (SWI), to observe the changing trend of the Walking Index from the sub-district scale.
Overall Walking Index (OWI)		In order to obtain the values of other data points except urban neighborhood points, Ordinary Kriging in ArcGIS was applied to perform Spatial Interpolation on the Walking Index of each neighborhood in the old urban area of Guilin City, and the OWI prediction of the old urban area was obtained.

3.2. Data Sources

We use multi-source data, including the data of Guilin's old urban area. The main sources are as follows:

(1) Road network data are derived from the Open Street Map website, and the obtained road network data are topologically processed by ArcGIS software (Figure 2).
(2) POI data (Point of Interest) are from the Amap open API.
(3) Twelve sub-district administrative areas data are from the website of the China National Bureau of Statistics.
(4) Data on the daily use of service facilities relating to 621 neighborhoods are from 200 questionnaires and interviews conducted by authors in the study area.

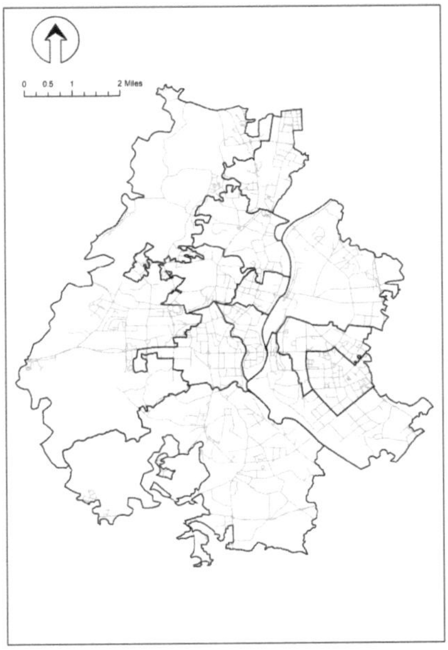

Figure 2. Road network of the research area.

3.3. Research Area and Object

Guilin, renowned as a preeminent tourist destination within China, has garnered international acclaim for its distinctive karst landscape, which is emblematic of the diverse topography found across the country. Situated within the Guangxi Zhuang Autonomous Region, Guilin serves as a microcosm of the expansive western frontier of China, a region that embodies the broader challenges and characteristics of China's less developed areas. Given the sheer geographical scope of China, there exists a pronounced disparity in economic advancement between its eastern and western territories [53].

The paper focuses on the old city of Guilin in western China as a case study. The city of Guilin, known for its unique karst landforms, represents the vast frontier areas of China's western regions, which face the challenge of uneven economic development with China's eastern regions. The special situation of Guilin in urban development, including its status as a tourist city and its representation in the development of the western region. According to the data of the seventh national population census, the population between 0 and 14 years old in Guilin is 968064, accounting for 19.63% of the total population. Compared with 2010, the proportion of the population in this age group increased by 3.48 percentage points. This indicates that the population of children plays an important role in the population structure of Guilin.

The eastern regions, which are more prosperous, have been the focal point for numerous scholarly inquiries into the assessment of walkability within the context of healthy living environments and the creation of child-friendly urban spaces [54]. However, such research is notably scarce in the western regions, which are comparatively less developed and thus present unique challenges and opportunities for urban planning and public health.

This study seeks to bridge this gap by conducting an in-depth investigation and analyzing a multitude of data sources. It focuses on the old urban district of Guilin, a representative area within the underdeveloped western region, to assess the walkability of neighborhoods and its implications for the health and well-being of children in the area. Through this research, we aim to contribute to the broader discourse on urban design and public health by providing insights that are grounded in the realities of China's western regions (Figure 3).

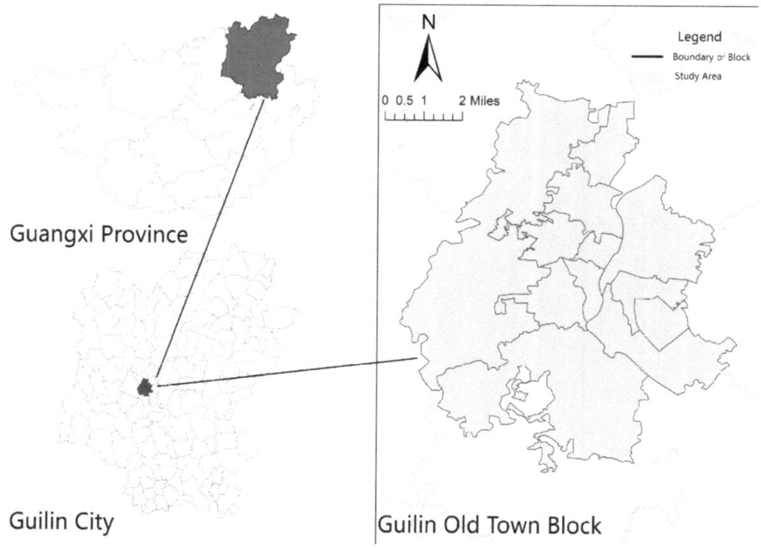

Figure 3. Location map of research area.

The scope of this research is selected from the old urban area of Guilin, Guangxi, China, including 12 sub-districts that belong to 5 urban districts (Table 2, Figure 4).

Figure 4. Research area and sub-districts administrative map.

Table 2. Research area sub-districts and their belonging districts.

	Sub-Districts	District
1	Beimen	Diecai District
2	Diecai	
3	Lijun	Xiufeng District
4	Xiufeng	
5	Jiashan	
6	Xiangshan	Xiangshan District
7	Nanmen	

Table 2. *Cont.*

	Sub-Districts	District
8	Dongjiang	Qixing District
9	Qixing	
10	Lidong	
11	Chuanshan	
12	Pingshan	Xiangshan District

According to the "United Nations Convention on the Rights of the Child" and other relevant provisions, children generally refer to minors between the ages of 0 and 18 [55]. Combined with the research theme, the selected research objects are limited to children between the ages of 4 and 18 who can walk independently.

3.4. Selection of Service Facilities

Because children are not mature physically and psychologically, they have special needs for the physical space and social environment that are different from adults. Children's health needs are reflected in three aspects: physical health, mental health, and social adaptation. In terms of physical health, we select service facilities that could impact children's physical health such as fitness activities, myopia, obesity, and other diseases; these service facilities include comprehensive gymnasiums, Internet cafes, general hospitals, etc. In terms of mental health, service facilities related to children's anxiety, panic, depression, and other mental health problems were selected, such as parks and plazas, amusement parks, schools, etc. In terms of social adaptation, we choose service facilities that can provide places for children to carry out social interaction and activities, such as museums, science and technology museums, community cultural centers, etc. Therefore, based on the needs of children's health, this paper selects relevant service facilities according to Chinese national guidelines and standards regarding children's health built environment design, such as "Urban Residential Area Planning and Design Standards" and "Urban Child-Friendly Space Construction Guidelines", combined with the Amap POI classification, and constructs a Children's Healthy 15-minute Life Circle service facilities index system, which contains 6 categories of first-level indicators. It includes types of medical service facilities, educational service facilities [51], ecological service facilities, commercial service facilities, cultural and sports service facilities, and transportation service facilities, as well as 22 categories of second-level indicators (Figure 5, Table 3).

Table 3. Selection of service facilities related to children's health and weights of service facilities.

First-Level Indicators	Second-Level Indicators	Frequency (%)	Weights (%)	Effects on Children's Health
Medical	Community health service center	0.23	0.87	The scope and accessibility of medical service facilities have an important impact on the first aid time of children and the convenience of obtaining drugs for children with chronic diseases.
	General hospital	0.24	0.93	
	Pharmacy	0.29	1.11	
	Clinic	0.18	0.70	
Educational	School (kindergarten, primary school, secondary school)	8.05	31.09	Functional destinations such as kindergartens, primary and secondary schools will encourage students to walk or ride to school to promote daily physical activities for people of different ages.

Table 3. *Cont.*

First-Level Indicators	Second-Level Indicators	Frequency (%)	Weights (%)	Effects on Children's Health
Ecological	Parks and squares	1.75	6.76	Open spaces such as parks and squares can promote children's physical and mental health, and their accessibility determines children's exposure to the natural environment, which in turn affects children's environmental perception ability and physical activity level.
	Public toilet	0.61	2.38	Sanitary toilet conditions directly affect children's health and experience, such as slips, bacterial infections and odors.
Commercial	Supermarket	1.81	6.98	The accessibility of food facilities such as supermarkets, convenience stores and restaurants has an impact on the balance of children's food choices and nutritional intake, thus affecting children's exposure to health risks such as overweight and obesity. The absence of such services might be correlated to malnutrition.
	Restaurants	1.15	4.44	
	Convenience store	2.52	9.72	
	Tobacco and alcohol retail	0.07	0.26	The accessibility and visibility of tobacco and alcohol retail facilities will affect the exposure probability of children and adolescents, increase tobacco smoke exposure, and affect the neurological and respiratory health of children.
	Playground	0.15	0.58	Playground space can increase children's physical activity, and can also help children reduce anxiety and release pressure.
	Amusement park	0.67	2.60	
	Internet bar	0.05	0.19	The change of electronic media to game ecology also changes children's game conditions and indirectly affects children's growth.
Cultural and sports	Sports complex	0.80	3.10	Physical activity can meet and promote children's needs for healthy growth, including the development of their cognitive abilities, emotional expression, social skills, and independence.
	Other sports venues	0.90	3.49	
	Training institution	1.36	5.24	Cultural facilities provide children with a comprehensive development environment by providing educational, cultural, sporting and social opportunities to promote their physical and mental health and personal growth.
	Library	1.11	4.27	
	Museum	0.41	1.58	
	Science and technology museum	0.33	1.28	
	Community cultural activity center	1.20	4.65	
Transportation	Bus station	2.01	7.77	Bus stops are transportation hubs for children commuting to school or outside activities, and are essential for them to develop healthy travel habits.

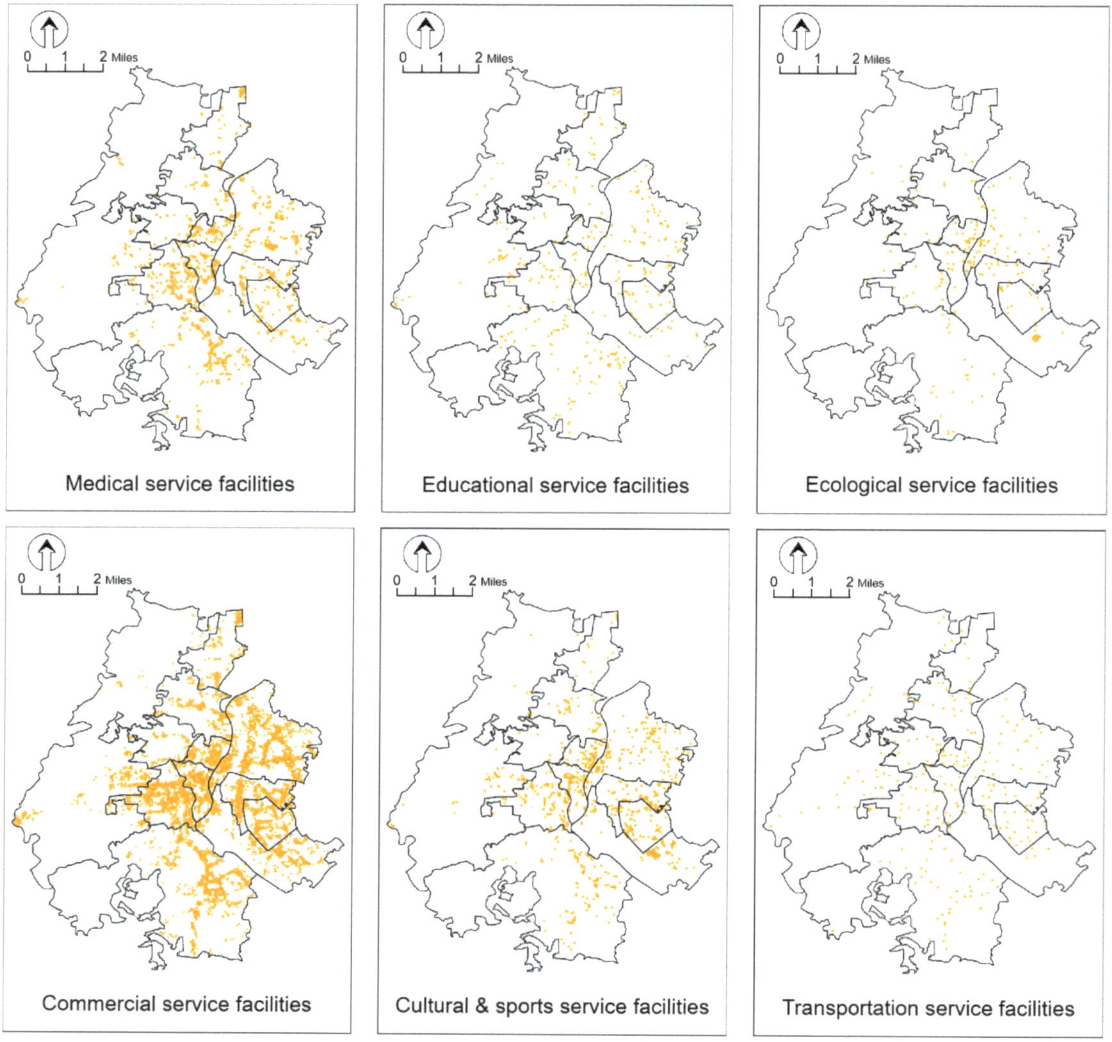

Figure 5. POI distribution of research area.

4. Children's Healthy 15-Minute Life Circle

4.1. Single Point Walking Index (SPWI)

4.1.1. Classification Weights of Service Facilities

This research investigated the use of service facilities by children in the form of questionnaires and interviews. A total of 200 questionnaires were distributed in the study area, and 188 valid questionnaires were collected. According to the results of the questionnaire, the weights of various service facilities were calculated, and the weight table of service facilities related to children's health was developed (Table 3).

4.1.2. Time Attenuation Curve

The time attenuation curve reflects a walking pattern, that is, the possibility of walking to a service facility decreases with the increase of time. Similarly, the shorter the time to reach the facility, the more likely people are to walk. Based on the survey results of the longest walking time acceptable to children for various service facilities in the questionnaire, the curve fitting is carried out in Origin software 2021 to obtain the time attenuation curves of different service facilities (Figure 6). The horizontal axis is the maximum walking time that children can accept to reach the facility, divided into six levels: 5 min, 10 min, 15 min, 20 min, 25 min, and 30 min. The vertical axis is the value of the time attenuation coefficient, which reflects children's willingness to travel on foot.

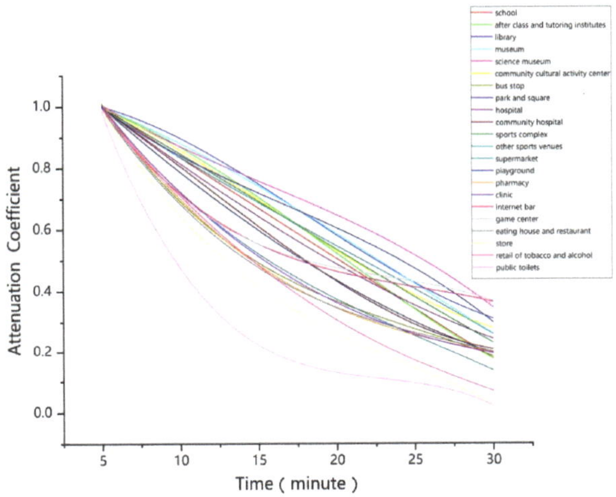

Figure 6. Time attenuation curve of different service facilities.

In general, when the time exceeds 10 min, the time attenuation curve of all kinds of service facilities has an obvious downward trend, and each facility presents different time attenuation laws. The attenuation curve of service facilities that are required for daily life and are highly substitutable decreases rapidly, such as public toilets and convenience stores. The time decline of service facilities with high scarcity and sense of value is slower, such as science and technology museums, museums, libraries, and other cultural and sports facilities.

4.1.3. Calculation and Correction of SPWI

When calculating the SPWI, we consider that children walk slower than adults and select 70 m/min as the walking speed of children by referring to relevant literature ②. Firstly, OD Cost Matrix Analysis in ArcGIS software was used to calculate the time taken by children to reach various service facilities within 15 min, taking POI of 621 urban neighborhoods in the old urban area of Guilin as the starting point of walking. Secondly, the time attenuation rate is obtained by substituting the corresponding time attenuation function. Finally, the initial Walking Index of a single point is obtained by summarizing the weights of all service facilities after attenuation.

Since the SPWI represents the linear distance between the urban neighborhood and the service facilities, to consider that people walking along the street in real life will be affected by road intersections and other factors, it is necessary to modify the SPWI by using urban neighborhood length and intersection density. The attenuation laws of urban neighborhood length and intersection density are based on the existing research walk score (Table 4).

Table 4. Attenuation rate reference for neighborhood length and intersection density based on the walk score.

Intersection Density (per km^2)	Attenuation Rate (%)	Neighborhood Length/100 m	Attenuation Rate (%)
>200	0	≤120	0
150–200	1	120–150	1
120–150	2	150–165	2
90–120	3	165–180	3
60–90	4	180–195	4
<60	5	>195	5

Source: https://www.walkscore.com (accessed on 2 July 2024).

4.1.4. Result Analysis

The revised SPWI of the old urban area of Guilin is shown below (Figure 7). We normalized the revised SPWI to make its value range between 0 and 100. Therefore, the higher the score of the neighborhood, the more suitable it is for children to walk. To facilitate the study, we divided the SPWI into 5 levels (Table 5).

Table 5. Grade classification of Single Point Walking Index (SPWI).

SPWI	Walkability Rating
90–100	Walker's Paradise
70–89	Very Walkable
50–69	Somewhat Walkable
25–49	Poorly walkable
0–24	Very poorly walkable

Quantity distribution: The number of neighborhoods with an SPWI above 70 in the old urban area of Guilin is small, accounting for only 3.06% of the total. The service facilities in these neighborhoods are more reasonable within the Children's Healthy 15-minute Life Circle, and children can meet most of the needs of a healthy life by walking. In neighborhoods with an SPWI of 50–69, some service facilities can be reached within walking distance, and walkability is average, accounting for 11.76%. There are more neighborhoods with poor SPWI; 24.80% are poor with SPWI between 25 and 49, and 60.39% are very poor, with SPWI between 0 and 24. The number of service facilities within the walking range of such neighborhoods is small, and most trips need to rely on other means of transportation such as motorcycles and cars (Figure 8).

Spatial distribution: residential areas with high SPWI are concentrated in the central area of the old city, among which Binjiang Road, the greenway around the city water system of "Two rivers and four lakes" and the characteristic pedestrian streets such as Zhengyang Pedestrian Street and Dongxi Alley provide rich walking space for children. Yijia Huating, Guiming Commercial Building, and other residential areas surrounded by Ximen Bridge–Xinyi Bridge–Wenming Road–South Ring Road have a more comfortable walking experience. The residential areas with poor walking experience are mainly distributed in the urban edge with low road network density, industrial and mining areas, or areas with more natural mountains and rivers. For example, there are many mountains and waters in the areas on both sides of the North Huancheng Road, and the distribution of service facilities is divided by mountains and rivers, resulting in a poor walking experience for children in nearby neighborhoods such as Hanlinju and Huguangshanse.

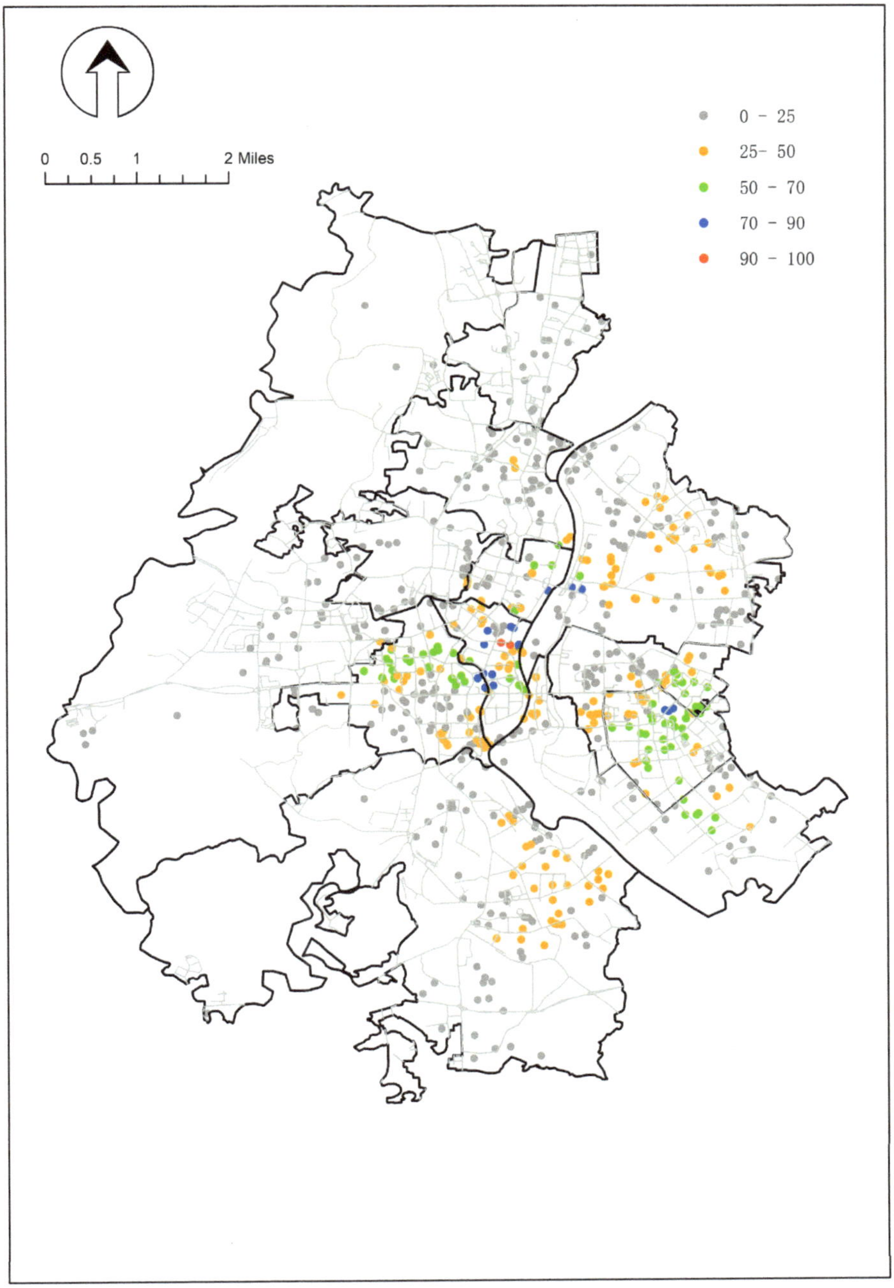

Figure 7. Visualized results of SPWI.

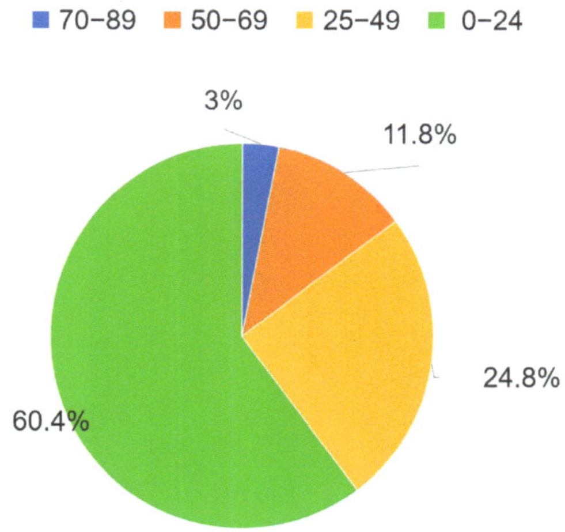

Figure 8. The classification proportion of SPWI of neighborhoods.

4.2. Area Walking Index (AWI)

4.2.1. Sub-District Walking Index (SWI)

The SPWI of an urban neighborhood was spatially linked through the ArcGIS Spatial Join function to obtain the Walking Index at the sub-district scale, which was named the Sub-district Walking Index (SWI) (Figure 9), to observe the changing trend of the Walking Index from the sub-district scale.

(1) Overall, the distribution trend of SWI is higher in central, eastern, and southern areas, and lower in northwest areas. The sub-districts with higher SWI mainly include Nanmen, Dongjiang, and Lidong sub-districts, etc., while those with lower SWI mainly include Beimen, Lijun, and Jiashan sub-districts. The SWI is not only affected by road network density and natural geographic factors but also by urban neighborhood scale. For example, the Jiashan sub-district is located on the edge of the old urban area due to its large urban neighborhood size and is separated by more natural mountains and water bodies, resulting in a relatively dispersed distribution of service facilities, so the SWI weakens quickly, thus affecting the SWI of the sub-district.

(2) From the perspective of service facility types, areas with higher overall SWI generally have higher walkability among their service facilities in various types (Figure 10). First of all, transportation service facilities dominated by bus stations have a higher SWI than other service facilities. Public transportation in Guilin is mainly based on buses, so its distribution quantity is large with high coverage. Secondly, in the actual measurement, it is found that most of the service facilities do not contain all types of health service facilities. The SWI of ecological service facilities such as parks and squares is not high in all sub-districts, which may be because parks and squares mostly rely on municipal investment, which costs a lot of capital investment but has less income, so the overall number is small. Thirdly, the survey finds that the sub-districts with high SWI of educational, cultural and sports, commercial, and medical service facilities are mainly located in the central area of the old city. For example, Pingshan has better supporting facilities for education, Lidong has high scores for cultural and sports facilities, and Nanmen has outstanding supporting services for commercial and medical services, which can be attributed to the influence of urban development history, namely, older community service facilities have a complete range of infrastructure. However, the sub-districts located at the edge of the urban development area, such as Jiashan and Lijun, have poor performance in the walkability of many

types of service facilities, indicating that the level of supporting facilities and construction is lagging, with poor child-friendliness.

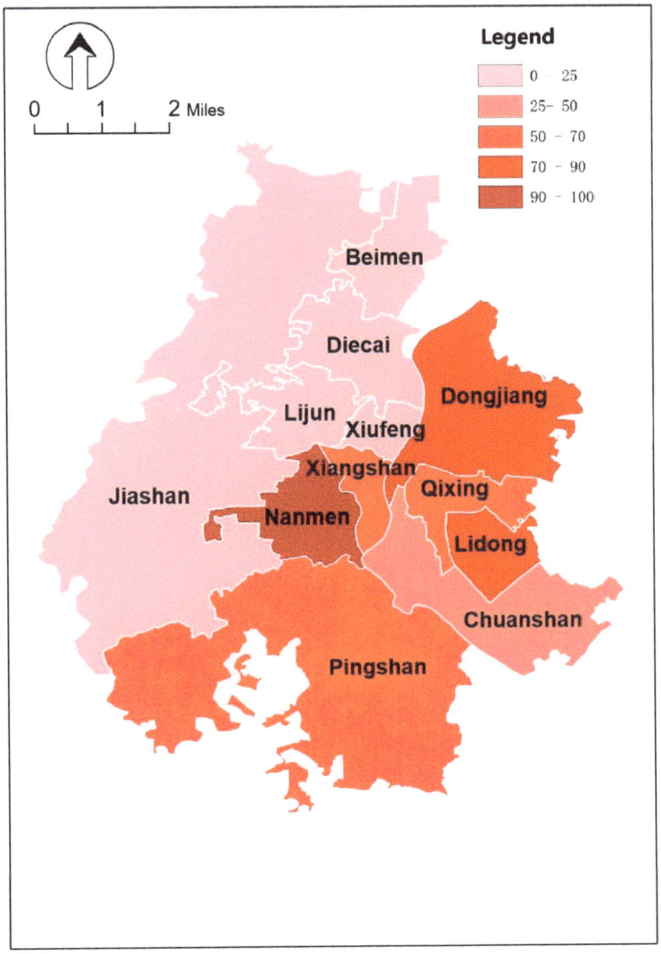

Figure 9. SWI visualization.

(3) Calculate the correlation between facility density, facility mixing degree, facility richness, and service facilities of each sub-district, and find the internal relationship. Facility density refers to the number of service facilities per hectare in the sub-district. Facility mixing degree refers to the ratio of the number of various types of service facilities in a sub-district to the total number of all service facilities, which is used to describe the degree of diversity of various service facilities. Facility richness refers to the mathematical ratio between the number of facility types in the sub-district and all 22 facility types and describes the richness of facility types (Table 6). SPSS software 27.0 was used to conduct a Pearson Correlation Analysis between the SWI and the above three indicators. The results are shown below (Figure 11). The correlation coefficient ranges from −1 to 1. When $r > 0$, there is a positive correlation between the two factors; when $r < 0$, there is a negative correlation between the two factors. Among them, the correlation coefficients of SWI with facility density, facility mixing degree, and facility richness are 0.486, 0.110, and 0.577, respectively. It can be seen that there is no significant positive correlation between facility

mixing degree and SWI, but there is a large positive correlation with facility richness, that is, the more types of service facilities, the more walkable. The correlation between facility density and SWI was close to moderate, indicating that the density of service facilities helps to promote children's walking experience.

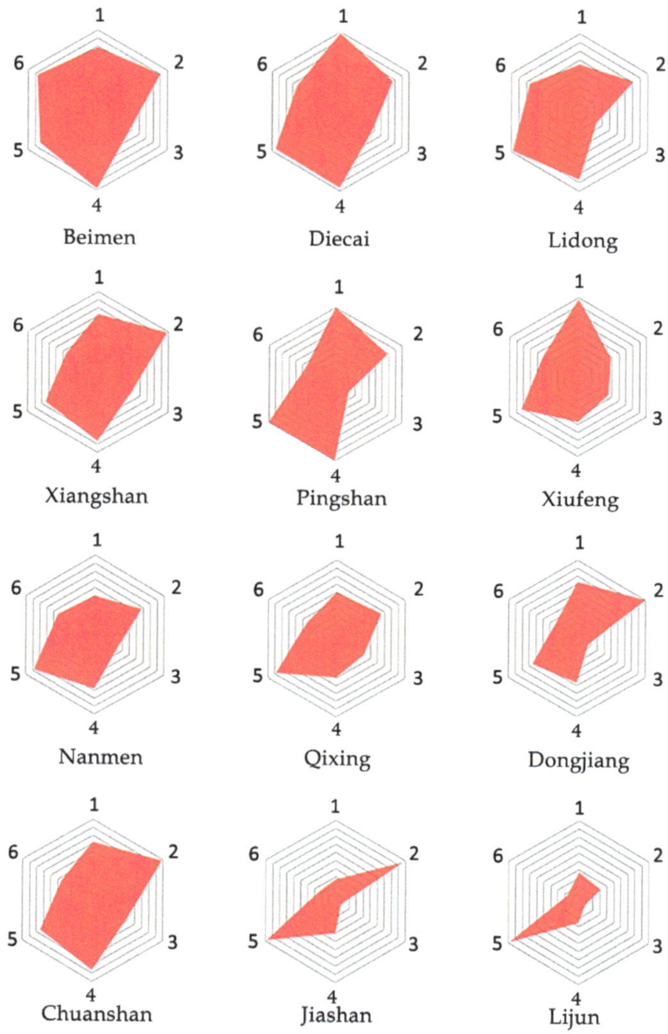

Figure 10. Radar map of SWI in classified facilities (1 is medical, 2 is educational, 3 is environmental, 4 is commercial, 5 is transportation, and 6 is cultural and sports).

Table 6. The correlation between facility density, facility mixing degree, facility richness, and service facilities of each sub-district.

Indicators	Calculation Formula	Method Description
Facility Density	X = Sn/ha	The number of service facilities per hectare in the sub-district is S
Facility Mixing Degree	$X = -\sum_{i=1}^{n} (pi \times \ln pi)$	*pi* is the ratio of Class *i* facilities to the number of all facilities in the sub-district
Facility Richness	X = C/N	C represents the number of service facility types in the sub-district, and N represents all 22 types of service facilities

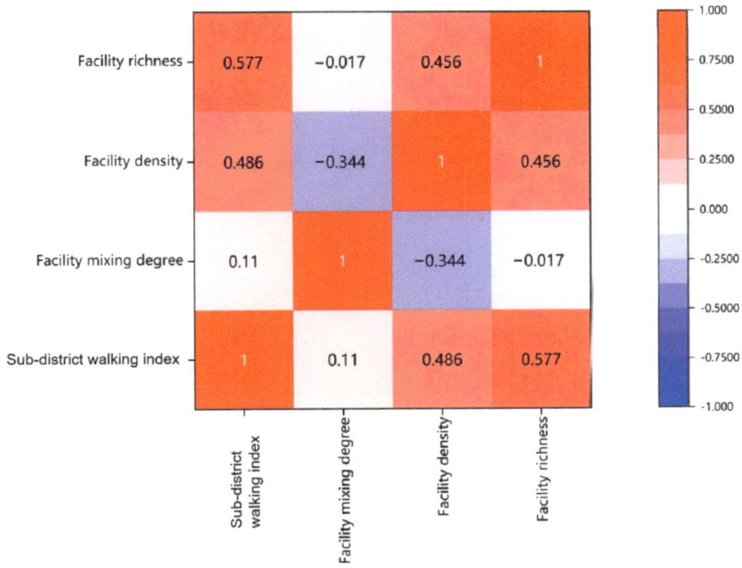

Figure 11. Pearson correlation analysis heat map.

4.2.2. Overall Walking Index (OWI)

The Kriging method is a method to predict or simulate the value of unknown data points by calculating the weighted average of the surrounding known data points. To obtain the values of other data points except urban neighborhood points, Ordinary Kriging in ArcGIS was applied to perform Spatial Interpolation on the Walking Index of each neighborhood in the old urban area of Guilin City, and the OWI prediction of the old urban area was obtained (Figure 12).

(1) From the perspective of overall spatial distribution, the spatial distribution characteristics of the OWI and the SWI are similar. The central area of the old city forms a pedestrian-friendly area for children, and the second-level pedestrian-friendly area is mainly located in the southern and eastern areas of the old city. It is worth noting that the division of natural geographical elements affects the spatial distribution of service facilities to a certain extent, and Guilin, as a famous city with mountains and rivers, this situation is particularly obvious. For example, due to the many mountains within the scope of Pingshan, the forest park in the west covers a large area, and the eastern area is close to the Lijiang River valley, which has a certain impact on the walkability of the area.

(2) From the spatial distribution of classified service facilities (Figure 13), the spatial distribution of the six categories of service facilities, namely medical service facilities, educational service facilities, ecological service facilities, commercial service facilities, cultural and sports service facilities, and transportation service facilities, are clustered in the southeast region. Compared with the ecological, transportation, and cultural and sports service facilities, the clustering tendency of educational, medical, and commercial service facilities is more obvious. To a certain extent, this reflects the imbalance of resource allocation. Children living in communities with a higher OWI have a higher chance of walking to obtain better educational, cultural, medical, and other resources.

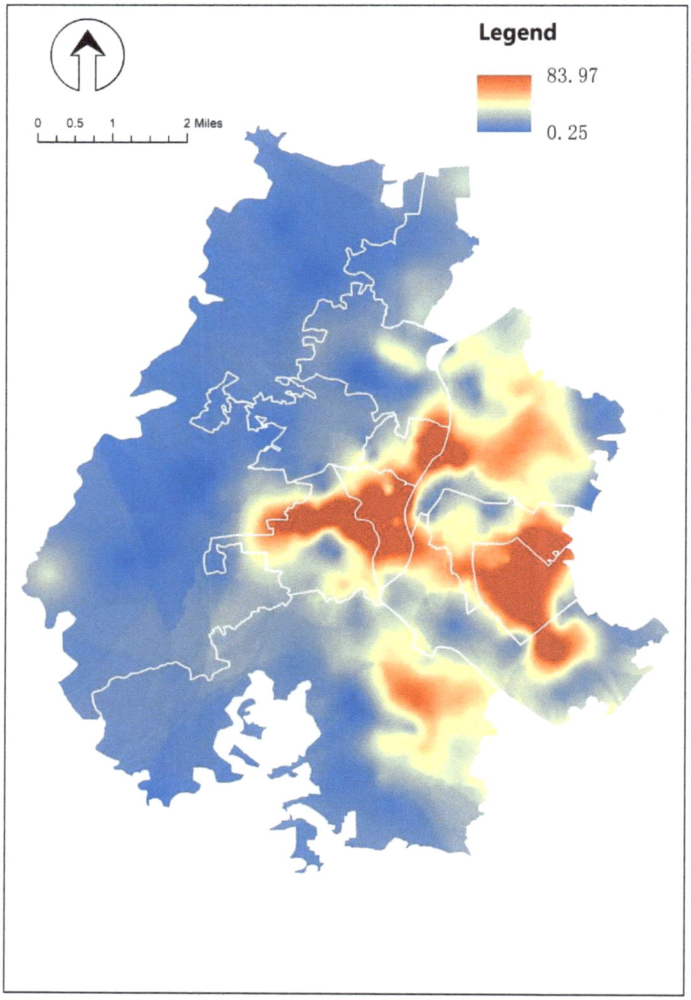

Figure 12. OWI visualization.

4.3. The Relationship Between Walking Accessibility and Children's Social and Mental Health

Walkability is a key factor in promoting social engagement in an urban environment. According to the present study, the Walking Index shows that walking accessibility is generally low and unevenly distributed in the study area. This has a direct impact on children's accessibility to services, including social and leisure spaces. In places with high walking accessibility, children are more likely to be out and about frequently, interacting with peers and community members, thereby enhancing social skills and social networks. For example, sports facilities and science and technology museums not only meet the educational, cultural, and sporting needs of children, but also provide a place for social interaction.

Walkability promotes mental health. The link between walking accessibility and mental health can be understood in several ways. First, walking is itself a physical activity, and physical activity has been linked to a reduced risk of mental health problems such as anxiety and depression. Research shows that services within walking distance can positively affect children's health, including mental health. Second, walkable green spaces, such as parks and pedestrian greenways, can promote children's sense of psychological security, comfort, and pleasure, which are important components of mental health In addition, the effective use of the urban landscape can create a diverse leisure space system for children, which helps to enhance not only their physical health, but also their mental health.

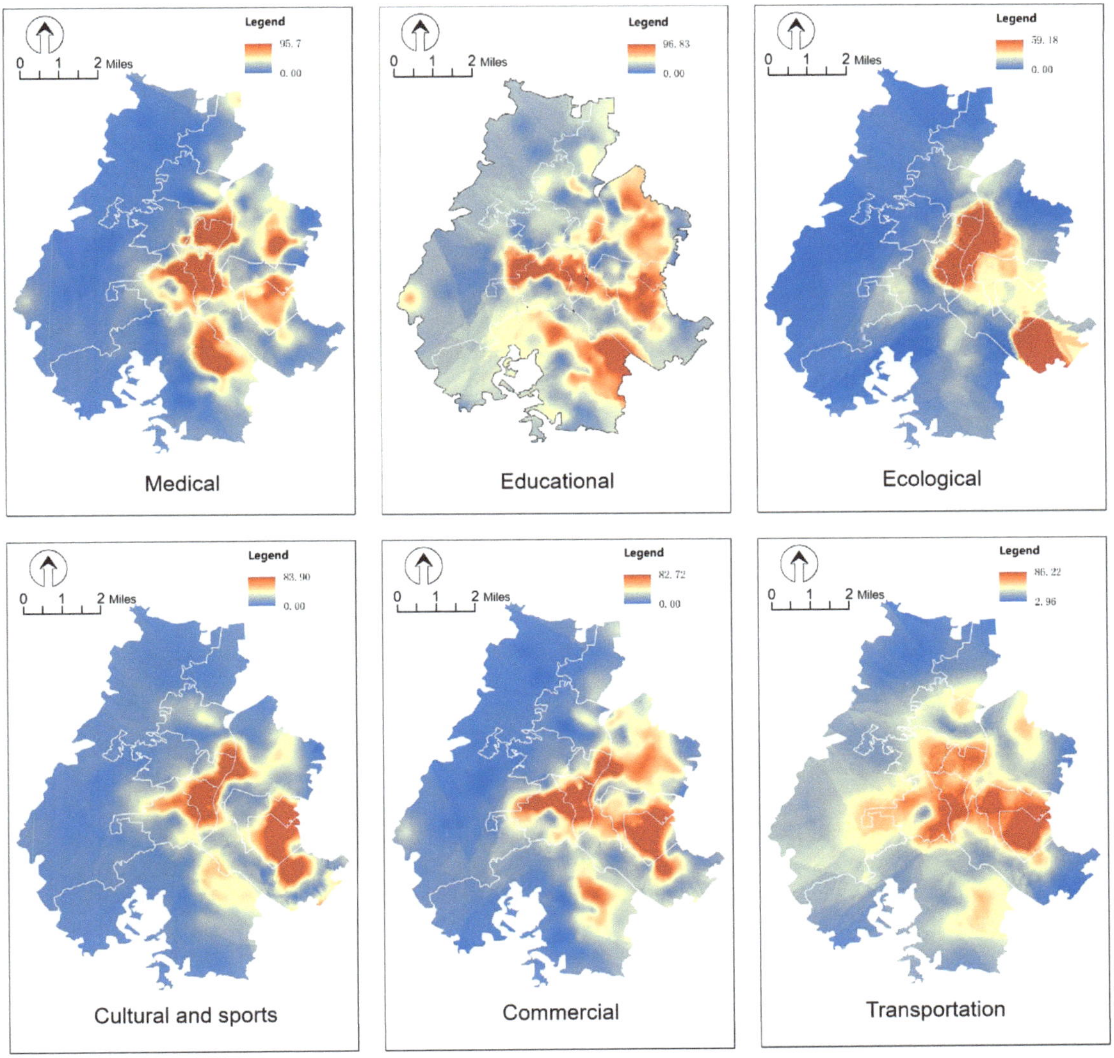

Figure 13. Visualization of OWI for classification facilities.

Based on the above analysis, we propose specific policy and planning recommendations in the paper to improve children's walking environments, thereby promoting their social and psychological well-being. For example, increasing walking paths, improving street safety, creating more green and public spaces, and optimizing the spatial layout of services to make them more pedestrian-friendly are all strategies that can improve children's social and mental health.

4.4. The Relationship Between Urban Planning and Child Health

Urban environments, urban communities, and urban planning have a profound impact on children's physical and mental health. Studies have shown that urban living is more likely than rural living to lead to serious mental health problems, such as depression and anxiety, This poses a major challenge to child Health worldwide [56]. Physical design of the urban environment, including factors such as green space, traffic exposure, and housing density, directly impacts children's physical development and mental health [57]. In addition, rapid urbanization has brought many health problems, such as environmental pollution, malnutrition, and infectious diseases, which not only affect urban residents; it also has implications for children in rural and remote areas [58].

At a theoretical and practical level, there is a growing global focus on child-friendly cities. The study highlights that urban planning should consider the needs of children, such as providing safe outdoor spaces and rich natural environments, to promote the physical and mental development of children [59]. Research in environmental psychology also points out that children's attachment to the urban environment is crucial to the development of their personal identity, and cities should provide children with rich opportunities for exploration and participation in order to promote their social and cognitive development [60]. These studies provide important guidance for policy makers and urban planners to design more inclusive and sustainable urban spaces that support the healthy growth of children.

This study used a walkability index method to assess children's walkability, which is used in several cities around the world [61], which suggests that the walking index is an effective tool that can be scaled up globally to assess and improve children's walking environments. The research can draw on global evaluation indicators such as safety, comfort, fun, diversity, and accessibility, which have been used to evaluate child-friendly community public spaces in several cities around the world. From the perspective of children's philosophy, the study incorporates children's rights and needs into the core elements of urban planning, which is in line with the global trend of child-friendly city construction.

Through the above analysis, we can see that in the process of implementing the "15-minute life circle for children's health" in Guilin City, its research findings have a wide range of links and mutual learning with global experience. The city of Guilin can use these global experiences to optimize and upgrade its urban planning to better meet children's health and well-being needs.

5. Conclusions

In the current discourse surrounding the development of China's 15-minute life circles, there is a discernible oversight regarding the specific needs of vulnerable demographics, such as older people, children, and individuals with disabilities. This paper posits that to foster a healthy 15-minute life circle for children, it is imperative to integrate their voices into the urban planning process to address their multifaceted needs, encompassing physical well-being, mental health, and social adaptability. The following recommendations are proposed to enhance the child-centric urban fabric:

Enhancing Urban Pedestrian Infrastructure for Children: To foster a child-friendly urban environment, it is essential to enhance the pedestrian infrastructure and increase the density of urban road networks. This strategy not only boosts the likelihood of children walking but also promotes their physical health by providing low-intensity physical activity that strengthens heart and lung function. A dense road network facilitates social interactions, nurturing children's social skills as they encounter peers and neighbors while walking. The contrast between the high-density eastern part and the low-density western part of the old city highlights the impact of network density on walkability. Improving the western region's branch roads, as Jane Jacobs noted, can enhance children's access to services and create a healthier, safer walking environment [22], aligning with the findings of Carver [50]. In essence, targeted improvements to road network density, especially in less connected areas, are crucial for creating urban spaces that support children's physical activity and social development.

Establishing a Comprehensive System of Urban Service Facilities for Children's Health: The paper emphasizes the need to optimize the spatial configuration of service facilities to ensure they are child-friendly. It points out the uneven distribution of ecological and cultural-sports facilities in certain areas of Guilin, suggesting that urban planning should consider the health needs of children. The paper recommends prioritizing the planning and quality improvement of service facilities frequently used by children, such as schools, parks, squares, and bus stations. It also calls for government measures to limit the visibility and accessibility of facilities that may pose risks to children's health, such as tobacco and alcohol retail outlets and internet cafes within the Children's Healthy 15-minute Life Circle.

Creating a Rich System of Children's Leisure Spaces: The paper advocates for the effective use of natural geographical elements like mountains and water to create a landscape ecological system that blends with the urban environment. For a city like Guilin, known for its superior natural landscapes, it is crucial to protect these resources and integrate them with the urban fabric to provide children with diverse outdoor spaces for healthy growth. This approach can enhance children's physical and mental health and promote social development. The paper also suggests using landscape resources to develop recreational and ecological facilities that reflect local culture, creating a child-friendly pedestrian environment (Table 7).

Table 7. Recommendations for child-centric urban fabric.

Suggestions	Detailed Lists	Health Benefits
1. Enhancing Urban Pedestrian Infrastructure for Children	a. To foster a child-friendly walking environment. b. To increase the density of urban road networks, especially in less connected areas.	Physical activity Social development
2. Establishing a Comprehensive System of Urban Service Facilities for Children's Health	a. To optimize the spatial configuration of service facilities and ensure they are child-friendly. b. To prioritize the planning and quality improvement of service facilities frequently used by children. c. To limit the visibility and accessibility of facilities that may pose risks to children's health.	Physical activity Social development Keeping away from deleterious environments
3. Creating a Rich System of Children's Leisure Spaces	a. To create a landscape ecological system that blends with the urban environment. b. To develop recreational and ecological facilities.	Physical activity Physical and mental health Social development

At the neighborhood level of Children's Healthy 15-minute Life Circles, the paper evaluates the walkability of health-related service facilities from a child-centric perspective. It aims to provide insights into the development of children's health service facilities in the context of building a child-friendly city. The paper acknowledges its limitation in focusing on quantitative research and suggests that future studies should consider children's subjective experiences and perceptions, increase the sample size, and thus make the research findings more relevant, practical, and comprehensive.

In promoting the transformation of urban public space, Guilin needs to increase capital investment, optimize urban planning, and encourage community and public participation in order to improve the quality of urban space and service efficiency. At the same time, Guilin will vigorously develop universal childcare and preschool education and improve the quality and equity of public services such as education and medical care. In the protection of children's rights, Guilin will strengthen the classified protection and care measures for children in difficulties and improve the child welfare system. In addition, Guilin will make use of its characteristic cultural tourism resources to expand children's growth space, develop outdoor education and recreational activities suitable for children, and strengthen the construction of urban green space and public space to provide more opportunities for natural experience. In terms of community governance, Guilin will improve the level of community service by establishing convenient facilities and promoting the construction of community elderly care and care service facilities and create a 15-minute life circle.

Through the above measures, the city of Guilin can gradually overcome the challenges and obstacles encountered in the implementation of the "Children's healthy 15-minute life circle" and create a healthier and more friendly living environment for children.

This academic paper contributes to the discourse on urban planning by advocating for a more inclusive approach that prioritizes the needs of children, ensuring that the 15-minute life circle is not only functional but also conducive to the well-being of the youngest members of society.

Notes

① Neighborhood, or community, including residential buildings, residential units, public housing, and so on. The "15-minute city" in China refers to the concept of the "15-minute life circle". It usually means that within a 15-minute walk from home, residents can enjoy the public services they need every day.

② Children's walking speed varies according to age and individual differences. According to clinical experience, children who have just learned to walk may walk 10–20 steps per minute, while children aged 6–12 can walk about 100 steps per minute, and can reach 120 steps when walking fast. The Physical Activity Guide for Children and Adolescents in China (2018 Edition) [62] provides the average walking speed of primary school students in different age groups: aged 6 to 7 walk 55–65 m per minute, aged 8 to 10 walk 65–75 m per minute, and aged 11–13 walk 75–85 m per minute. Average values were used in this study, with 70 m/min selected as the walking speed of children.

Author Contributions: Conceptualization, X.W. and X.L.; methodology, X.W. and X.L.; software, J.H. and W.G.; validation, X.L., Z.H. and X.W.; formal analysis, J.H. and Z.Q.; investigation, J.H. and Z.Q.; resources, X.W.; data curation, X.W.; writing—original draft preparation, X.W.; writing—review and editing, X.L.; visualization, Z.H.; supervision, X.L.; project administration, X.W.; funding acquisition, X.W. All authors have read and agreed to the published version of the manuscript.

Funding: This research was funded by the Guangxi Key Laboratory of Green Building Materials and Construction Industrialization, grant number GKN18-J-21-3; the National Natural Science Foundation of China, grant number 52308082; the Guangxi Philosophy and Social Sciences Planning Office, grant number 23FSH018; the International Knowledge Centre for Engineering Sciences and Technology (IKCEST) under the auspices of UNESCO, grant number CKCEST-2021-2-19; and the Guangxi Zhuang Autonomous Region Department of Education: 2024KY0248.

Institutional Review Board Statement: The study was conducted in accordance with the Declaration of Helsinki, and the protocol was approved by the Ethics Committee of College of Tourism & Landscape Architecture, Guilin University of Technology (CTLA-LL20231003) on [29 October 2023].

Informed Consent Statement: Informed consent was obtained from all subjects involved in the study.

Data Availability Statement: The raw data supporting the conclusions of this article will be made available by the authors on request.

Acknowledgments: Special thanks go to the 200 children and their guardians who took part in our study, offering invaluable data on the health needs and perspectives of urban youth. Their participation has been crucial in informing our policy recommendations. We appreciate their time and the insights they have shared.

Conflicts of Interest: The authors declare no conflicts of interest.

References

1. Liu, J.; Li, W.; Yao, H.; Liu, J. Proactive Health: An Imperative to Achieve the Goal of Healthy China. *China CDC Wkly.* **2022**, *4*, 799–801. [PubMed]
2. Ma, W.; Wang, N.; Li, Y.; Sun, D. 15-min pedestrian distance life circle and sustainable community governance in Chinese metropolitan cities: A diagnosis. *Humanit. Soc. Sci. Commun.* **2023**, *10*, 1–14. [CrossRef]
3. Molina-García, J.; Queralt, A. Neighborhood Built Environment and Socioeconomic Status in Relation to Active Commuting to School in Children. *J. Phys. Act. Health* **2017**, *14*, 761–765. [CrossRef]
4. King, K.E.; Clarke, P.J. A Disadvantaged Advantage in Walkability: Findings From Socioeconomic and Geographical Analysis of National Built Environment Data in the United States. *Am. J. Epidemiol.* **2014**, *181*, 17–25. [CrossRef] [PubMed]
5. E Laxer, R.; Janssen, I. The proportion of youths' physical inactivity attributable to neighbourhood built environment features. *Int. J. Heal. Geogr.* **2013**, *12*, 31. [CrossRef]
6. Chen, S.; Knöll, M. Environmental Justice in the Context of Access to Urban Green Spaces for Refugee Children. *Land* **2024**, *13*, 716. [CrossRef]
7. Molina-García, J.; Menescardi, C.; Estevan, I.; Queralt, A. Associations between Park and Playground Availability and Proximity and Children's Physical Activity and Body Mass Index: The BEACH Study. *Int. J. Environ. Res. Public Health* **2021**, *19*, 250. [CrossRef]
8. An, R.; Shen, J.; Yang, Q.; Yang, Y. Impact of built environment on physical activity and obesity among children and adolescents in China: A narrative systematic review. *J. Sport Health Sci.* **2018**, *8*, 153–169. [CrossRef]
9. An, R.; Yang, Y.; Li, K. Residential Neighborhood Amenities and Physical Activity Among U.S. Children with Special Health Care Needs. *Matern. Child Health J.* **2016**, *21*, 1026–1036. [CrossRef]
10. Liao, Y.; Furuya, K. A Bibliometric Analysis of Child-Friendly Cities: A Cross-Database Analysis from 2000 to 2022. *Land* **2023**, *12*, 1919. [CrossRef]
11. Rakhimova, N. Child-Friendly Cities and Neighborhoods: An Evaluation Framework for Planners. Master's Thesis, Arizona State University, Tempe, AZ, USA, 2011.
12. Weng, M.; Ding, N.; Li, J.; Jin, X.F.; Xiao, H.; He, Z.M.; Su, S.L. The 15-minute walkable neighborhoods: Measurement, social inequalities and implications for building healthy communities in urban China. *J. Transp. Health* **2019**, *13*, 259–273. [CrossRef]
13. Duncan, D.T.; Sharifi, M.; Melly, S.J.; Marshall, R.; Sequist, T.D.; Rifas-Shiman, S.L.; Taveras, E.M. Characteristics of Walkable Built Environments and BMI *z* -Scores in Children: Evidence from a Large Electronic Health Record Database. *Environ. Health Perspect.* **2014**, *122*, 1359–1365. [CrossRef]
14. Spence, J.C.; Cutumisu, N.; Edwards, J.; Evans, J. Influence of neighbourhood design and access to facilities on overweight among preschool children. *Pediatr. Obes.* **2008**, *3*, 109–116. [CrossRef]
15. Grugel, J. Children's rights and children's welfare after the Convention on the Rights of the Child. *Prog. Dev. Stud.* **2012**, *13*, 19–30. [CrossRef]
16. Svevo-Cianci, K.; Lee, Y. Twenty years of the Convention on the Rights of the Child: Achievements in and challenges for child protection implementation, measurement and evaluation around the world. *Child Abus. Negl.* **2010**, *34*, 1–4. [CrossRef] [PubMed]
17. UNICEF. *The State of the World's Children 2012: Children in an Urban World*; UNICEF: New York, NY, USA, 2012.
18. Riggio, E.; Kilbane, T. The international secretariat for child-friendly cities: A global network for urban children. *Environ. Urban.* **2000**, *12*, 201–205. [CrossRef]
19. State Council of China. Program for the Development of Chinese Children (2021–2030). 2021. Available online: https://www.gov.cn/lianbo/2023-04/20/content_5752337.htm (accessed on 2 July 2024).
20. UNICEF. *Shaping Urbanization for Children: A Handbook on Child-Responsive Urban Planning*; UNICEF: New York, NY, USA, 2018.
21. Jacobs, A.B. *Great Streets*; MIT Press: Cambridge, MA, USA, 1993; p. 331.
22. Jacobs, J. *The Death and Life of Great American Cities*; Penguin: Harmondsworth, UK, 1972; p. 474.
23. Cervero, R.; Kockelman, K. Travel demand and the 3Ds: Density, diversity, and design. *Transp. Res. Part D-Transp. Environ.* **1997**, *2*, 199–219. [CrossRef]
24. Booth, M.L.; Bauman, A.; Owen, N.; Gore, C.J. Physical activity preferences, preferred sources of assistance, and perceived barriers to increased activity among physically inactive Australians. *Prev. Med.* **1997**, *26*, 131–137. [CrossRef]

25. Schmid, T.L.; Pratt, M.; Howze, E. Policy as intervention: Environmental and policy approaches to the prevention of cardiovascular disease. *Am. J. Public Health* **1995**, *85*, 1207–1211. [CrossRef] [PubMed]
26. Saelens, B.E.; Sallis, J.F.; Frank, L.D. Environmental correlates of walking and cycling: Findings from the transportation, urban design, and planning literatures. *Ann. Behav. Med.* **2003**, *25*, 80–91. [CrossRef] [PubMed]
27. Van Dyck, D.; Cardon, G.; Deforche, B.; Sallis, J.F.; Owen, N.; De Bourdeaudhuij, I. Neighborhood SES and walkability are related to physical activity behavior in Belgian adults. *Prev. Med.* **2010**, *50*, S74–S79. [CrossRef] [PubMed]
28. Owen, N.; Cerin, E.; Leslie, E.; Dutoit, L.; Coffee, N.; Frank, L.D.; Bauman, A.E.; Hugo, G.; Saelens, B.E.; Sallis, J.F. Neighborhood walkability and the walking behavior of Australian adults. *Am. J. Prev. Med.* **2007**, *33*, 387–395. [CrossRef]
29. Frank, L.D.; Schmid, T.L.; Sallis, J.F.; Chapman, J.; Saelens, B.E. Linking objectively measured physical activity with objectively measured urban form: Findings from SMARTRAQ. *Am. J. Prev. Med.* **2005**, *28*, 117–125. [CrossRef] [PubMed]
30. Wang, H.; Yang, Y. Neighbourhood walkability: A review and bibliometric analysis. *Cities* **2019**, *93*, 43–61. [CrossRef]
31. Southworth, M. *Walkable Suburbs? An Evaluation of Neotraditional Communities at the Urban Edge*; UC Berkeley, Fisher Center for Real Estate and Urban Economics: Berkeley, CA, USA, 1995.
32. Lynch, K. A Walk Around the Block. In *City Sense and City Design: Writings and Projects of Kevin Lynch*; Banerjee, M.S.T., Banerjee, M.S.T., Eds.; MIT Press: Cambridge, MA, USA, 1959.
33. Ewing, R.; Schmid, T.; Killingsworth, R.; Zlot, A.; Raudenbush, S. Relationship between urban sprawl and physical activity, obesity, and morbidity. *Am. J. Health Promot.* **2003**, *18*, 47–57. [CrossRef]
34. Humpel, N.; Owen, N.; Leslie, E. Environmental factors associated with adults' participation in physical activity—A review. *Am. J. Prev. Med.* **2002**, *22*, 188–199. [CrossRef]
35. Handy, S.L.; Boarnet, M.G.; Ewing, R.; Killingsworth, R.E. How the built environment affects physical activity: Views from urban planning. *Am. J. Prev. Med.* **2002**, *23*, 64–73. [CrossRef]
36. Ewing, R.; Handy, S.; Brownson, R.C.; Clemente, O.; Winston, E. Identifying and Measuring Urban Design Qualities Related to Walkability. *J. Phys. Act. Health* **2006**, *3*, S223–S240. [CrossRef]
37. Pikora, T.; Giles-Corti, B.; Bull, F.; Jamrozik, K.; Donovan, R. Developing a framework for assessment of the environmental determinants of walking and cycling. *Ocial Sci. Med.* **2002**, *56*, 1693–1703. [CrossRef]
38. Krizek, K. Residential Relocation and Changes in Urban Travel: Does Neighborhood-Scale Urban Form Matter? *J. Am. Plan. Assoc.* **2003**, *69*, 265–281. [CrossRef]
39. Ding, D.; Sallis, J.F.; Kerr, J.; Lee, S.; Rosenberg, D.E. Neighborhood Environment and Physical Activity Among Youth A Review. *Am. J. Prev. Med.* **2011**, *41*, 442–455. [CrossRef] [PubMed]
40. Kerr, J.; Rosenberg, D.; Sallis, J.F.; Saelens, B.E.; Frank, L.D.; Conway, T.L. Active commuting to school: Associations with environment and parental concerns. *Med. Sci. Sports Exerc.* **2006**, *38*, 787–794. [CrossRef] [PubMed]
41. Timperio, A.; Crawford, D.; Telford, A.; Salmon, J. Perceptions about the local neighborhood and walking and cycling among children. *Prev. Med.* **2004**, *38*, 39–47. [CrossRef] [PubMed]
42. McDonald, N.C.; Brown, A.L.; Marchetti, L.M.; Pedroso, M.S. U.S. School Travel, 2009: An Assessment of Trends. *Am. J. Prev. Med.* **2011**, *41*, 146–151. [CrossRef] [PubMed]
43. Gordon-Larsen, P.; Nelson, M.C.; Page, P.; Popkin, B.M. Inequality in the built environment underlies key health disparities in physical activity and obesity. *Pediatrics* **2006**, *117*, 417–424. [CrossRef] [PubMed]
44. National Research Council (US); Institute of Medicine (US). *Children's Health, The Nation's Wealth: Assessing and Improving Child Health*; National Academies Press: Washington, DC, USA, 2004.
45. Green, M. *Bright Futures: Guidelines for Health Supervision of Infants, Children and Adolescents*; National Center for Education in Maternal and Child Health: Arlington, VA, USA, 2002.
46. Anonymous. Ottawa Charter for Health Promotion. *Bull. Pan Am. Health Organ.* **1987**, *21*, 200–204.
47. Lewis, M.; Feiring, C. The child's social network: Social object, social functions, and their relationship. In *The Child and Its Family*; Lewis, M., Rosenblum, L.A., Eds.; Springer: Boston, MA, USA, 1979.
48. Jia, P.; Xue, H.; Cheng, X.; Wang, Y.; Wang, Y. Association of neighborhood built environments with childhood obesity: Evidence from a 9-year longitudinal, nationally representative survey in the US. *Environ. Int.* **2019**, *128*, 158–164. [CrossRef]
49. Carver, A.; Timperio, A.; Crawford, D. Playing it safe: The influence of neighbourhood safety on children's physical activity—A review. *Health Place* **2008**, *14*, 217–227. [CrossRef] [PubMed]
50. Horelli, L. Constructing a Theoretical Framework for Environmental Child-Friendliness. *Child. Youth Environ.* **2007**, *17*, 267–292. [CrossRef]
51. Wang, X.; Zhao, J.; Lu, Y.; Li, X. Spatial Pattern, Quality Evaluation, and Implications of Preschool Education Facilities in New Urban Areas Using Multi-Source Data: A Case Study from Lingui New District in West China. *Buildings* **2024**, *14*, 1718. [CrossRef]
52. Liangwa, C. The Research on Safety Children's Travel Route on Child-Friendly City of Netherlands. *Int. J. Environ. Prot. Policy* **2017**, *5*, 94. [CrossRef]

53. Zhang, Y.; Shao, C.; Shi, Y. Guilin Sustainable Development City Construction Countermeasure and Idea Design for SDGs in 2030. *IOP Conf. Ser. Earth Environ. Sci.* **2019**, *310*, 052005. [CrossRef]
54. Nan, F. Policy innovation on building child friendly cities in China: Evidence from four Chinese cities. *Child. Youth Serv. Rev.* **2020**, *118*, 105491. [CrossRef]
55. UNICEF. *The Convention on the Rights of the Child: The Children's Version*; UNICEF: New York, NY, USA, 1989.
56. Buttazzoni, A.; Doherty, S.; Minaker, L. How Do Urban Environments Affect Young People's Mental Health? A Novel Conceptual Framework to Bridge Public Health, Planning, and Neurourbanism. *Public Health Rep.* **2022**, *137*, 48–61. [CrossRef] [PubMed]
57. Villanueva, K.; Badland, H.; Kvalsvig, A.; O'Connor, M.; Christian, H.; Woolcock, G.; Giles-Corti, B.; Goldfeld, S. Can the Neighborhood Built Environment Make a Difference in Children's Development? Building the Research Agenda to Create Evidence for Place-Based Children's Policy. *Acad. Pediatr.* **2016**, *16*, 10–19. [CrossRef] [PubMed]
58. Gracey, M. Child health implications of worldwide urbanization. *Rev. Environ. Health* **2003**, *18*, 51–64. [CrossRef]
59. Kodali, H.P.; Hitch, L.; Dunlap, A.F.; Starvaggi, M.; E Wyka, K.; Huang, T.T. A systematic review on the relationship between the built environment and children's quality of life. *BMC Public Health* **2023**, *23*, 2472. [CrossRef] [PubMed]
60. Spencer, C.; Woolley, H. Children and the city: A summary of recent environmental psychology research. *Child Care Health Dev.* **2000**, *26*, 181–198. [CrossRef]
61. Ubiali, A.; Gori, D.; Rochira, A.; Raguzzoni, G.; Fantini, M.P. Measures of walkability in the pediatric population: A qualitative review of the literature. *Ann. Ig. Med. Prev. Comunità* **2021**, *33*, 67–85. [CrossRef]
62. General Administration of Sport of China. The Physical Activity Guide for Children and Adolescents in China. 2018. Available online: https://www.sport.gov.cn/n20001280/n20745751/n20767277/c21359075/content.html (accessed on 8 July 2024).

buildings

MDPI

Article

Spatio-Temporal Analysis of Beijing Residents' Lifestyles: Data-Driven Insights into Apartment Interior Design

Feifei Liu, Yuzhe Wang and Qi An *

Academic of Arts and Design, Beijing City University, Beijing 101309, China; feifei_79625@sina.com (F.L.); wangyuzhe041004@163.com (Y.W.)
* Correspondence: anqi@bcu.edu.cn; Tel.: +86-132-6910-5617

Abstract: The urbanization of Beijing has precipitated a shift in the design of living spaces, with the focus transitioning from the design of new houses to existing residential properties. The concept of the living environment is inherently multifaceted, evolving in response to societal and lifestyle shifts. The employment of computer-assisted observation in acquiring lifestyle data about residential behavior circumvents the subjectivity inherent in questionnaires, thereby providing a novel approach to considering user behavior within the design process. This observational study utilizes video data collected by the Oriental Culture & Design Center, offering a comprehensive depiction of the daily lives of Beijing residents. The Noldus Observer XT program was utilized to encode and analyze the data, thereby facilitating the acquisition of insights into urban dwelling patterns in China. Over 14 days, 53,550 behavioral codes were recorded for six households, meticulously organized axially based on a 24 h cycle to capture the behavioral facts of living spaces. Through the synthesis of quantitative data analysis and qualitative observations, this study aims to provide a comprehensive overview of the general lifestyle patterns exhibited by these urban residents. In addition, based on the insights gained, we propose four directions for the future design of living spaces. This comprehensive temporal dataset on living behaviors offers significant data support for design practitioners and researchers developing residential spaces. This study's findings can optimize living environments for mental health and well-being by providing empirical data and design recommendations grounded in real-life observations.

Keywords: user observation; lifestyle; living space; spatio-temporal analysis; behavior visualization

Citation: Liu, F.; Wang, Y.; An, Q. Spatio-Temporal Analysis of Beijing Residents' Lifestyles: Data-Driven Insights into Apartment Interior Design. *Buildings* **2025**, *15*, 576. https://doi.org/10.3390/buildings15040576

Academic Editor: Adrian Pitts

Received: 30 December 2024
Revised: 22 January 2025
Accepted: 11 February 2025
Published: 13 February 2025

Citation: Liu, F.; Wang, Y.; An, Q. Spatio-Temporal Analysis of Beijing Residents' Lifestyles: Data-Driven Insights into Apartment Interior Design. *Buildings* **2025**, *15*, 576. https://doi.org/10.3390/buildings15040576

1. Introduction

In the context of Beijing's rapid urbanization, there has been a shift in the residential design paradigm, with an increasing emphasis on the renovation of existing housing, particularly in established districts. While this transition has addressed numerous pressing concerns, the intricacies inherent in the process necessitate a nuanced approach. The interplay between housing ownership patterns and the heterogeneity of residents' demands poses substantial challenges, impeding the efficacy of government-funded renovations in addressing all issues concurrently. Consequently, a considerable gap persists in the systematic enhancement in housing quality. Concurrently, families with existing children express a heightened demand for social housing [1]. The design and remodeling of existing housing by families on their own initiative continue to be the mainstay of residential design.

Domestic spaces reflect people's physical and spiritual needs through various human activities and are where people spend most of their lives [2]. The built environment influences people's productivity and well-being while satisfying safety [3–5]. User-centered design (UCD) is increasingly recognized as necessary to provide occupants with a comfortable living experience and achieve the design of living spaces [6,7]. China abolished its long-standing "one-child policy" and, starting in 2016, adopted the "two-child policy" to address declining birth rates [8]. The relationship between parents and children is much closer in Chinese families, and family space is an important entry point for studying living space [9].

While emphasizing dwelling dimensions holds significance, the cultural relativist paradigm underscores the imperative of investigating need fulfillment within the framework of culturally specific norms [10]. Consequently, user-centered design transforms an approach encompassing both external factors and social and cultural facets centered around the user [11]. Nevertheless, establishing a direct correlation between culture and the examination of living spaces proves implausible; instead, it necessitates scrutiny through distinct variables [12]. The daily activity patterns of residents are very regular and are closely related to land use and urban built environments [13]. Given that the amalgamation of various variables configures the living milieu, a profound understanding of user culture, the fulfillment of user requisites, and the extraction of user attributes becomes indispensable in residential space design [14]. From a consumer standpoint, delving into user culture constitutes a pivotal aspect to contemplate in design [15]. Hence, exploring the nexus between culture and the design of living spaces should scrutinize user-centricity and living behaviors as the nexus connecting the investigation of housing and its associated design domains.

In contrast, for the design of living spaces, the personal experience of the user's perspective and the information received from the user may not describe their actual needs [16]. User observation, a core method of user research, consists of observing phenomena that directly reflect the behavior of the observed, and researchers use observation to derive a general understanding of specific activities and environments to help develop designs [17].

Owing to the difficulty of obtaining observational data about living spaces, a lot of data are needed to analyze the overall picture of life. Currently, only some existing user observations are carried out to study the generalization of living spaces. Inadequate attention to users in living spaces may lead to incorrect design decisions and development assumptions [18]. Accordingly, furniture design for living spaces needs to systematically consider the user's needs and incorporate relevant design criteria to meet the user's needs.

In this study, the non-participatory observation of stable families with children in Beijing was conducted, and the data were coded and transcribed through the Noldus Observer XT. Lifestyle data were then visualized to obtain an analyzable lifestyle map. By examining the lifestyle data from different perspectives, this study proposes design insights and suggests paths for future home remodeling that families and design firms can explore and consider.

2. Materials and Methods

2.1. Ethnographic User Observation

User-centered research methods can be divided into methods for communicating with the user, methods for investigating what users do, and methods for analyzing what users do, including communicating with the user [19]. User observation is the first step in driving innovation [20]. Widely used in all phases of design [21], user observation can be broadly classified into three categories: Type A, matter extraction (hypothesis search); Type B, hypothesis validation; and Type C, effect confirmation [22]. Type A is an effective method for users that need research in furniture and living space design derivation. This study uses an ethnographic user observation method to derive users' hidden needs and provide a database from which designers and researchers can obtain hypotheses.

Ethnographic-based user observation studies have been widely used in the study of living space [23–25], dating back to the study of kitchens by the Viennese architect Margarete Schutte Lihotzky in 1926 [26]. In such an observation process, the collaboration-based user observation process is optimized through memos and the researcher's notes [27]. Moreover, in ethnographic observations, the considerations of use and culture influence the conduct of research [28,29].

Figure 1 represents the understanding of the user's lifestyle in the living space based on the framework of the "user–space–object" relationship of the living culture, and by observing these three factors, we can understand the ideology of the living culture and people's lifestyles to export the direction of interior design better.

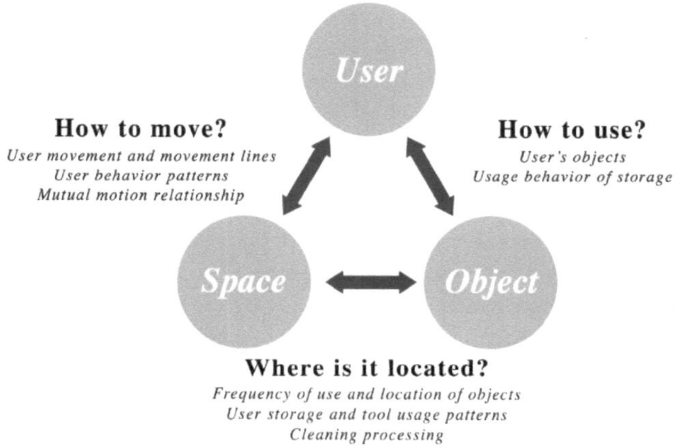

Figure 1. User–space–object factors in user observation.

2.2. *Noldus Observer XT*

To enhance the efficacy of the extant user observation methodology within residential environments, this paper improves a user observation framework grounded in user culture, supplemented by the application of Observer XT 16 software. Leveraging computer-aided user observation facilitates the expeditious acquisition of profound insights into user behavior and their interactions with computing systems during the developmental phase of exporting design direction [30]. This study systematically categorizes coding schemas, observational data, independent variables, data filters, and analytical outcomes from video data sourced from the OCDC. Filters can be intricately devised based on an amalgamation of independent variables, behavioral parameters, and temporal criteria, offering a versatile toolset for designers and researchers.

The Observer XT is a behavioral coding and analysis software that supports all research project steps, including coding scheme development, data entry, data management, data analysis, and researcher data sharing [31]. It has been applied in a range of user research processes, such as the tracking of the behaviors of children on the autism spectrum [32], classroom acoustical design and repetitive behaviors in children with autism [33], pilot behavior [34], and the driver–vehicle interface [35]. There is also a large volume of research on living spaces for furniture design and space planning [36,37].

2.3. *Sample Selection*

The spatial distribution reveals that the hotspots of the residents' daily activities are mainly concentrated in the central city, which is located on Fifth Ring Road [38]. To understand the actual usage of living space for a nuclear family in Beijing and to obtain more stable residential lifestyle data, considering the problem of population flow in Beijing, this research surveyed the state of affairs of families with children living in two- and three-bedroom apartments. The survey chose 25 typical families living around Fifth Ring in Beijing. The sample households were divided into four groups according to the children's age. Tables 1–4 describe the household code, household size, and the age of the children for the selected households. At the same time, semi-structured interviews were conducted with the users of each household to understand their perceptions of the space.

Table 1. Research participant information—Group A (0–3 years old).

Family Code	Plan	Information
a2.01		88 m^2—Girl, 3rd
a2.02		89 m^2—Boy, 3rd
a2.03		90 m^2—Girl, 2nd
a2.04		100 m^2—Boy, 3rd
a3.01		102 m^2—Boy, 3rd
ac3.01		70 m^2—Boy, 9th; Girl, 1st
ac3.02		110 m^2—Boy, 10th; Girl, 1st

Table 2. Research participant information—Group B (4–7 years old).

Family Code	Plan	Information
ab2.01	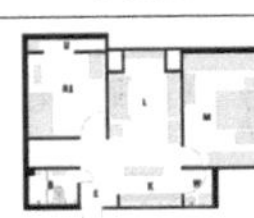	80 m^2—Boy, 6th; Boy, 3m
ab3.01		150 m^2—Boy, 7th; Boy, 7m
b3.01	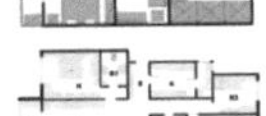	141 m^2—Boy, 7th
b2.01		89 m^2—Girl, 5th
b2.02	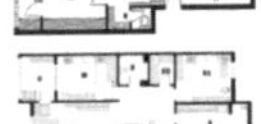	141 m^2—Boy, 7th

Table 3. Research participant information—Group C (9–12 years old).

Family Code	Plan	Information
c3.01		138 m^2—Girl, 9th
c3.02	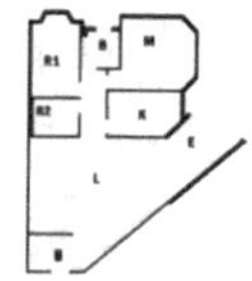	143 m^2—Boy, 9th
c3.03		80 m^2—Girl, 8th
c3.04		76 m^2—Boy, 7th
c2.01		120 m^2—Girl, 10th
c2.02	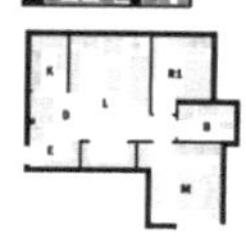	89 m^2—Boy, 8th
c2.03	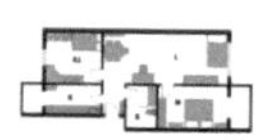	65 m^2—Boy, 10th

Table 4. Research participant information—Group D (13–17 years old).

Family Code	Plan	Information
d2.01		94 m^2—Boy, 13th
d2.02		70 m^2—Boy, 13th
d2.03		99 m^2—Boy, 16th
d2.04		150 m^2—Boy, 13th
d3.01		135.5 m^2—Boy, 16th
d3.02		75 m^2—Boy, 16th

During the interviews, it was ascertained that in Group D families, the age of the children, the children's residence in school, and the change in their learning situation had led to a decrease in parent–child interactions. Consequently, the present study was videotaped, and stratified random sampling was used based on three non-overlapping groups of children's age strata that had previously been delineated. Simple random sampling was conducted within each stratum in Groups A, B, and C.

Regarding the protection of participants' privacy, a series of rigorous measures was implemented. The process of obtaining informed consent from participants was elucidated in detail, ensuring their voluntary participation in this study after comprehending the research purposes, data usage, storage, and potential sharing methods. To mitigate the privacy concerns associated with the 24 h video-recording in private household environments, we strictly limited the scope and duration of recording, collecting only the necessary data for this study. This study endeavored to minimize observer effects when video-recording family activities, and unnatural behaviors were reduced by familiarizing participants with the video equipment before recording.

Problematic parts were removed by filming from fixed cameras in households and recording the residential life of Beijing residents over 24 h for 21 days to screen the image data. To obtain more regular lifestyle videos, the shooting time was chosen to be March-April, avoiding children's vacations. Finally, 14 days of video from each household was selected as the information used for the following analysis step. The camera arrangement is shown in Table 5.

Table 5. Recorded family information.

Family Code	Recorded Plan
ab2.01	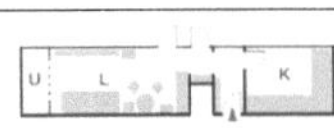
c3.04	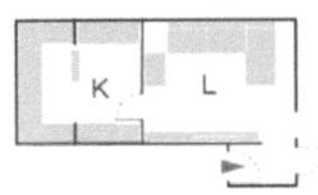
a2.02	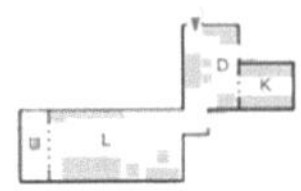
c2.02	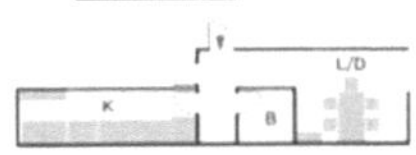
b2.01	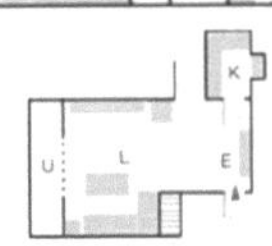
c3.02	

In this research, participants were protected by blurring video data, removing facial recognition information, and deleting or coding any information that could identify the participants.

2.4. Setting Code Schema

According to Figure 2a, studies on lifestyle types were divided by OCDC in 2011, and the behavioral classification was updated based on the observed life videos. A thorough review of the videos was conducted, and the various behaviors exhibited by the subjects were meticulously documented using a card-based system. Following this, card clustering was used to systematically group the behaviors into 11 main behavior categories and 40 sub-behavior codes. Most of the behavioral activities in the living space were covered. When behaviors did not fit neatly into these categories, they were recorded in the INSIGHT column. The behavior card classification method generated 11 main behaviors.

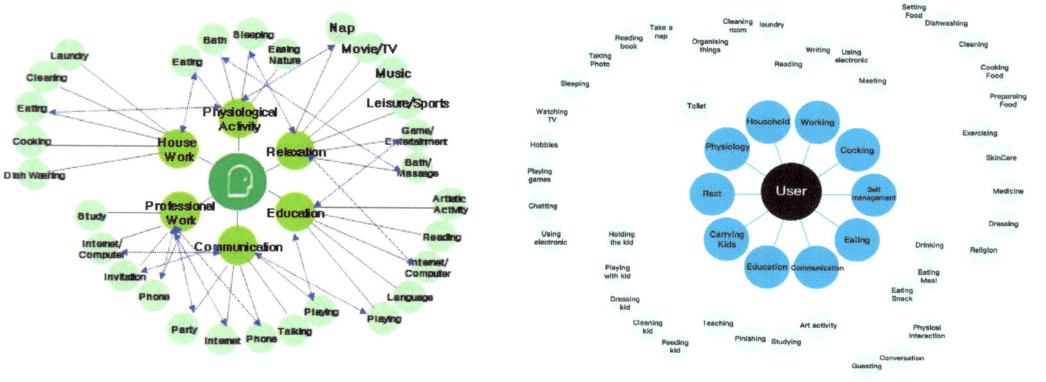

(a) (b)

Figure 2. (**a**) Description of existing living behavior classification; (**b**) behavioral codes for composite lifestyle were reorganized through behavioral card classification method.

Table 6 shows the coding settings for data input, mainly considering the recorded codes as user, behavior, space, and object codes.

Table 6. Coding categories.

User Code	Behavior Code	Space Code	Object Code
Mother Father Kid1 Kid2 Grandmother Grandfather Guest Nanny	Cooking (CO) Doing housework (DH) Working (WO) Caring for kid (CK) Self-management (SM) Eating (EA) Communicating (CM) Education (ED) Resting (RE) Physiological activities (PAs) Using object (UO)	Living room Bathroom Kitchen Entrance Utility Dining room Main room Room1 Room2	Cooking Object Housework Object Office Object Bedding Personal Object Food Furniture Electronics Kid Object

After obtaining licenses from Hanssem and obtaining approval from OCDC, the videos were coded by 11 researchers affiliated with the Graduate School of Techno Design at Kookmin University in 2022. Trained researchers were assigned to code the video materials using the Noldus Observer XT. Before initiating the initial round of video observation, a card sorting method was employed to convene all the video observation team members for a collective discussion and establish a behavioral code, as shown in Table 7, through the typing of the behaviors. This approach was adopted to mitigate potential observer bias, ensuring the clear delineation of the behaviors to be observed and the norms for their documentation. Table 7 describes the behavioral codes, sub-behavioral codes, and descriptions of each behavior that help the researcher better understand this study when recording and reviewing.

Table 7. Coding categories.

Behavior Coding	Sub-Behavior Coding	Description
Cooking (CO) Act of cooking in all processes, specifically refers to cooking	Preparing food	Process of preparing food, including washing vegetables, panning rice, cutting vegetables, and other acts in preparation process of cooking
	Dishwashing	Washing dishes and cooking utensils
	Cooking food	Main process of cooking, frying, steaming, and other acts in process of cooking
	Cleaning	Cleanup of cooking areas and other related cleanup actions
Doing housework (DH) Housework-related behaviors	Organizing things	Behaviors of organizing stuff
	Cleaning room	Act of cleaning up room
	Laundry	Act of washing clothes and other fabrics
Working (WO) Work-related behaviors	Reading	Specifically refers to act of reading at work
	Writing	Acts of writing and other writing-related actions
	Using electronics	Use of electric appliances at work
Caring for kid (CK) Parenting behaviors that are specific to parents and relate to their children	Dressing kid	Behaviors of helping kids get dressed
	Cleaning kid	Behaviors of helping kids keep proper hygiene, such as wiping child's mouth
	Feeding kid	Feeding children, such as food and snacks; helping them drink water, take medicine, etc.
	Holding kid	Act of caring that refers specifically to holding child
	Playing with kid	Recreational activities with children
Self-management (SM) Behaviors of taking care of one's body and beauty	Exercising	-
	Skin care	-
	Taking medicine	Act of taking medicine and using health care products
	Dressing	Behaviors such as putting on clothes and shoes
Eating (EA) Behaviors related to eating	Eating meal	Act of eating regular meals such as breakfast, lunch, and dinner
	Eating snack	-
	Drinking	-
Communicating (CM) Communication-related behaviors	Conversation	Communicative act of dialog, specifically verbal communication
	Physical interaction	Physical interaction, specifically act of physical contact
	Receiving guests	-
Education (ED) Child-related educational behaviors	Teaching	Specifically referring to parents teaching children
	Punishing	Chastising child, including verbal punitive behaviors
	Art activity	Art-related activities (e.g., singing, dancing, painting, playing instruments, etc.)
	Studying	Here, exclusively refers to children's learning behaviors, and parents' learning behaviors are categorized as "working"

Table 7. *Cont.*

Behavior Coding	Sub-Behavior Coding	Description
Resting (RE) Leisure-related behaviors	Taking nap	Naps, brief acts of rest
	Reading book	Recreational reading (does not refer to learning through reading)
	Taking photo	Take photos and videos and perform other actions
	Sleeping	Refers specifically to sleeping
	Watching TV	-
	Doing hobbies	-
	Playing games	Recreational playing behavior, game-related behaviors
	Non-in-person chatting	Talking on phone, chatting online, and other related behaviors
Physiological activities (PAs)	Physiological activities	Physiological behaviors, such as urination and defecation
Using object (UO)	Using object	Refers specifically to use of objects that cannot be recorded at same time as other acts, including appliances and furniture

2.5. Observation Process

Entering the code of observation can be conducted quickly via shortcut keys. It is also possible to adjust the pace of observation recording, slowing down or speeding up according to different behavioral characteristics. The position of the image can be adjusted via the progress bar to repeat or confirm the observation. It is also possible to click on the code of each entry to find the impact at that time. Take the example of the content shown in Figure 3, which shows the mother finishing eating, feeding her child, and leaving the living room.

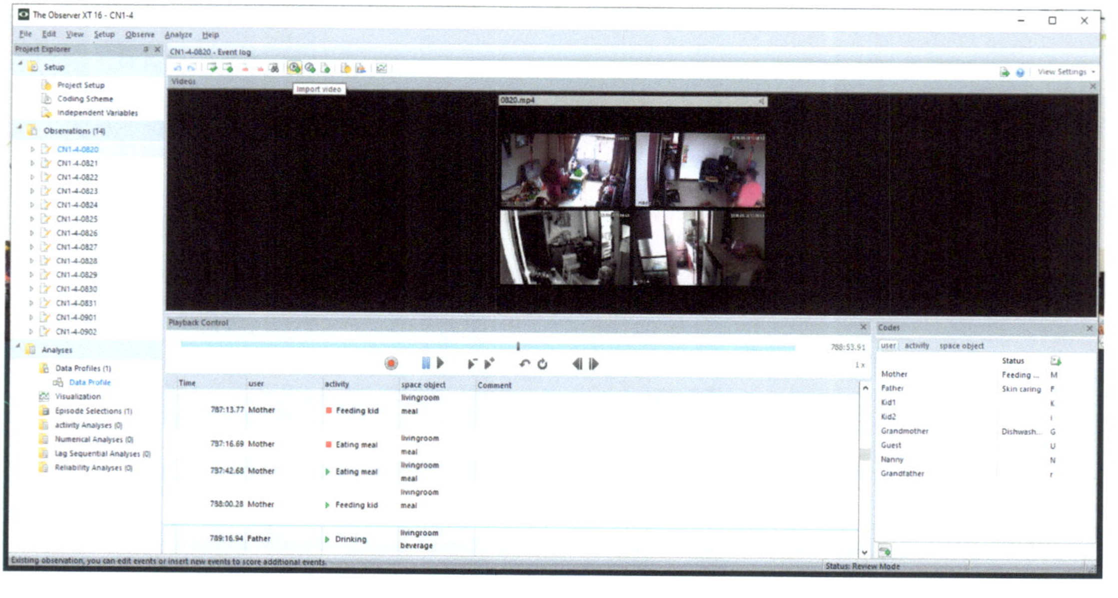

Figure 3. Code input interface.

Researchers were trained to initiate the observation coding process. Utilizing the designated behavioral coding schema, the researchers meticulously documented each transition of the participants. The behavioral codes were recorded in the following sequence for each behavior in a single observation sequence: user–behavior–sub-behavior–space–object–insight. Researchers are instructed to record the facts of the behavior through observation rather than making judgments about the behavior. Insights are only added as additional data in later data analysis.

3. Results

3.1. Living Behavior Dataset

A total of 53,550 coded records were recorded in this Beijing resident life study. The research objects could be identified based on the different data models for the subsequent, more focused, qualitative analysis. The data were exported into Excel (Table 8), and the data were filtered and counted after fixing the variables according to the different studies.

Table 8. Data output.

Case Code	Time (h.m.s)	Duration (s)	User	Behavior	Object	Space	Start–Stop	Insights
ab2.01-1018	08:10:51	196.404	Mother	Preparing food	Food ingredient	Kitchen	State starts	
ab2.01-1018	08:14:08	0	Mother	Preparing food	Food ingredient	Kitchen	State stops	
ab2.01-1018	08:14:08	203.954	Mother	Cooking food	Cooker	Kitchen	State starts	
ab2.01-1018	08:17:32	0	Mother	Cooking food	Cooker	Kitchen	State stops	
ab2.01-1018	08:22:16	53.0112	Father	Toileting		Bathroom	State starts	
ab2.01-1018	08:23:09	0	Father	Toileting		Bathroom	State stops	
ab2.01-1018	08:23:48	62.8544	Kid1	Toileting		Bathroom	State starts	
ab2.01-1018	08:24:51	0	Kid1	Toileting		Bathroom	State stops	m: wear socks

- The Excel file contains information about the behavior of the users in the observation, as well as the space and the objects used.
- The "Case Code" column shows the households' coding and the recording date, with a total of six household codes: ab2.01; c3.04; a2.02; c2.02; b2.01; and c3.02. Each household contains 14 days of data in the format xx months xx day.
- The "Time" column indicates when the behavior occurred in the hour: minute: second format in 24 h format.
- The "Duration" column records the duration of the behavior; the duration is only present in the code of the start of the behavior, and the data are blank in the stop code.
- The "User" column records the emitter of the behavior.
- The "Behavior" column indicates the specific behavior performed, which is divided into 41 types and grouped into 11 main behavior categories (specific codes are detailed in Table 6).
- The "Space" column indicates where the behavior occurred.
- The "Start–Stop" column expresses the start attribute of the behavior, divided into start and stop.
- The "Insight" column records the researcher's perception of the behavior during observation.

3.2. Behavior Duration and Frequency Data

Data filtered by workday–weekend in the living room show the total duration of each family member's behavior in the living room (in minutes) and its proportion of a member's total usage time of the space during the same period. The top four behaviors with the longest total duration are included in Table 9. The behaviors in the living room on weekdays are mainly concentrated on RE (c3.04; a2.02; b2.01), CM (a2.02; c2.02), UO (ab2.01; b2.01), and EA (c3.04; a2.02; b2.01). The living room activities on weekends are diverse. However, the proportion of RE behaviors increases overall. The children's living room use behaviors are more similar: in both cases, they are dominated by resting behaviors (ab2.01–0.74; c3.04–0.64; a2.02–0.67; b2.01–0.61; c3.02–0.29).

Table 9. Behavior duration data in living room.

		Workday								Weekend							
Code	**User**	**1st**		**2nd**		**3rd**		**4th**		**1st**		**2nd**		**3rd**		**4th**	
ab2.01	F	UO	50.0 0.41	EA	24.1 0.20	CM	19.8 0.16	RE	11.6 0.10	RE	75.3 0.44	EA	50.5 0.30	UO	37.6 0.22	CM	29.4 0.17
	M	EA	25.8 0.26	UO	22.4 0.22	CM	21.7 0.22	CK	11.8 0.12	EA	67.8 0.34	UO	55.5 0.28	CM	49.6 0.25	DH	27.9 0.14
	K1	RE	146.9 0.74	CM	21.3 0.11	EA	19.6 0.10	ED	9.7 0.05	RE	51.8 0.24	CM	15.4 0.07	EA	11.2 0.05	ED	9.7 0.04
	K2	RE	156.0 0.75	CM	26.3 0.13	EA	25.5 0.12	UO	1.3 0.01	RE	188.9 0.73	EA	35.3 0.14	CM	32.0 0.12	UO	1.1 0.00
c3.04	F	RE	82.9 0.54	EA	23.7 0.15	UO	21.5 0.14	CM	8.5 0.06	RE	313.3 0.75	UO	64.7 0.16	EA	18.9 0.05	SM	9.0 0.02
	M	RE	30.4 0.30	UO	22.9 0.23	EA	17.7 0.18	CM	11.0 0.11	UO	24.1 0.26	RE	21.2 0.23	ED	15.4 0.16	SM	12.4 0.13
	K	RE	111.0 0.64	EA	42.2 0.24	UO	11.3 0.07	CM	3.9 0.02	RE	24.2 0.17	UO	16.7 0.12	EA	7.5 0.05	ED	3.5 0.02
a2.02	F	RE	112.2 0.55	CO	26.9 0.13	EA	25.7 0.13	WO	23.0 0.11	RE	291.3 0.70	EA	74.7 0.18	WO	26.2 0.06	CM	8.1 0.02
	M	RE	292.9 0.44	EA	128.5 0.19	CK	94.0 0.14	WO	68.3 0.10	RE	275.7 0.42	EA	159.1 0.24	WO	91.8 0.14	DH	20.9 0.03
	K	RE	233.4 0.67	EA	59.6 0.17	UO	23.8 0.07	CM	21.3 0.06	RE	130.3 0.31	EA	17.3 0.04	UO	14.6 0.03	CM	5.8 0.01
	G	CM	55.7 0.37	RE	23.4 0.16	CK	23.2 0.15	DH	13.6 0.09	CM	22.4 0.26	RE	22.3 0.26	CK	14.8 0.17	CO	10.7 0.13
c2.02	F	EA	24.1 0.40	CM	23.8 0.39	UO	6.3 0.10	DH	2.4 0.04	EA	14.7 0.30	CM	12.8 0.26	UO	7.8 0.16	WO	7.2 0.15
	M	DH	21.4 0.25	CM	20.9 0.24	WO	12.2 0.14	ED	6.2 0.07	WO	36.9 0.56	DH	11.1 0.17	CM	9.8 0.15	EA	3.6 0.06
	K	ED	37.4 0.39	CM	27.4 0.28	EA	23.4 0.24	UO	3.6 0.04	RE	7.0 0.11	ED	7.0 0.11	CM	4.3 0.07	UO	2.3 0.04
b2.01	F	UO	58.7 0.39	RE	40.7 0.27	EA	21.4 0.14	CM	13.5 0.09	RE	109.6 0.27	UO	105.5 0.26	CM	68.3 0.17	WO	64.1 0.16
	M	UO	15.8 0.20	EA	15.0 0.19	CM	11.3 0.14	CK	10.1 0.13	UO	31.5 0.28	CM	24.5 0.22	CK	20.5 0.18	EA	21.3 0.19
	K	RE	122.2 0.61	EA	30.6 0.15	CM	28.6 0.14	SM	5.1 0.03	RE	66.9 0.23	EA	21.6 0.08	CM	17.5 0.06	SM	2.7 0.01
	G	RE	123.4 0.34	WO	73.2 0.20	DH	71.0 0.20	EA	51.2 0.14	UO	86.5 0.26	RE	81.9 0.24	EA	58.5 0.17	DH	37.2 0.11
c3.02	F	WO	83.6 0.67	WO	15.1 0.12	CM	8.1 0.06	EA	6.3 0.05	WO	195.1 0.75	CM	30.9 0.12	UO	12.1 0.05	EA	6.6 0.03
	M	UO	46.0 0.41	WO	21.5 0.19	EA	18.5 0.17	CM	15.0 0.14	UO	73.0 0.48	CM	23.7 0.16	EA	21.8 0.14	SM	13.4 0.09
	K	RE	54.5	ED	45.8	UO	39.1	EA	15.3	ED	33.0	RE	15.3	CM	14.8	EA	8.6

3.3. User Behavior Timeline Data

As illustrated in Figure 4, the Observer XT program enables the visualization of data results and the acquisition of a timeline of user behavior. The primary intervals of the father's family activities are visible, and the associated behaviors are color-coded. This arrangement provides a convenient foundation for the analysis of future research on different types of behaviors. For instance, composite behavioral research can concentrate on intervals characterized by poor spatial repetition and repeated behavioral crossover, facilitating more profound observations to derive insights for analyzing user problems and characteristics. Visual data analysis facilitates designers' research and enables communication between designers and users.

As shown in Figure 5, the timeline provides a comprehensive overview of pertinent information, which can be expeditiously retrieved using the user code, the date, and the designated space for the act and behavior.

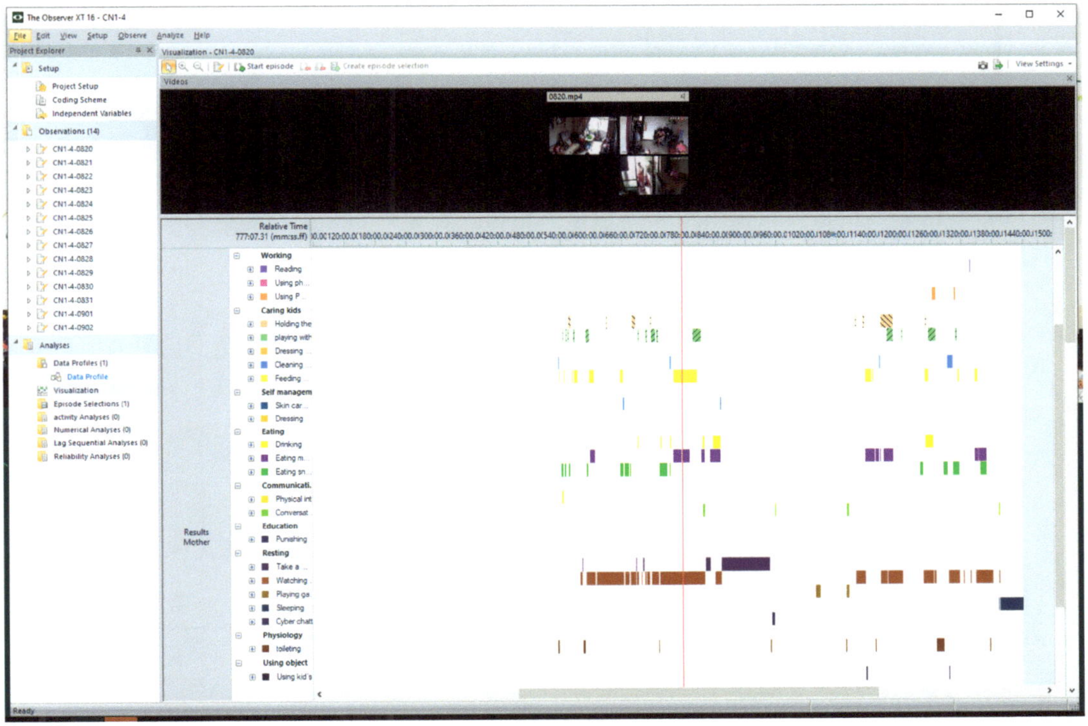

Figure 4. Visualization interface in the Noldus Observer XT.

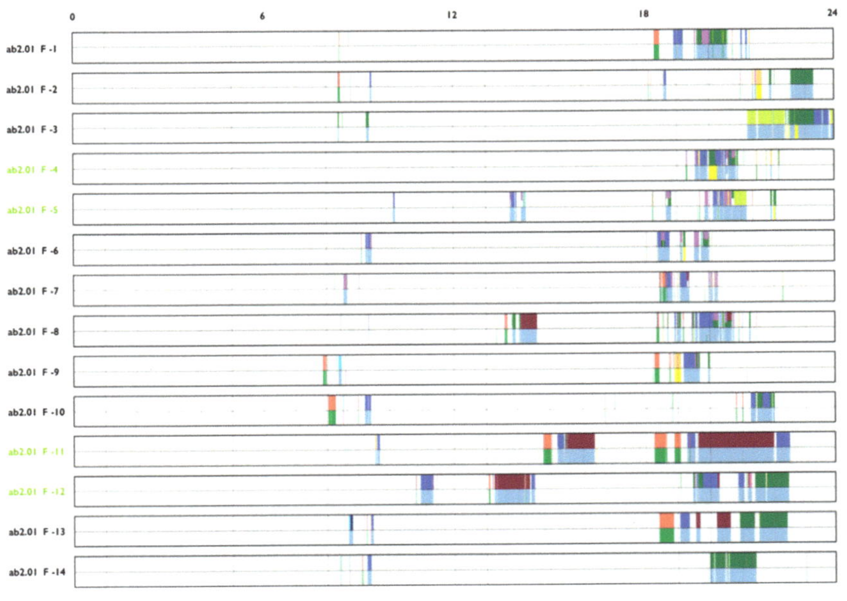

Figure 5. Timeline of 14-day behaviors of ab2.01 father case.

Figure 6 is a graphical representation of the spatio-temporal data of residential behavior. On the left side of the figure, the coding of the families (six families) and the users (F—father; M—mother; G—grandmother; K—kid) is presented. The figure also presents the day on which the data were collected (1–14). The upper portion of the axial data depicts behavioral patterns, while the lower portion focuses on spatial elements. Notably, the data for each day were recorded in a 24 h format. Color coding facilitates the identification of specific behaviors and spatial locations, thereby providing a visual representation that can be subjected to subsequent analysis. Table 10 shows the results of visualizing the behavioral and spatial–temporal data of 6 families and 21 persons for 14 days each. A subsequent analysis of the obtained data revealed a certain regularity of recurrence in the spatio-temporal data of the behaviors, showing a midweek weekend pattern. This finding aligns with the expectations formulated before this research.

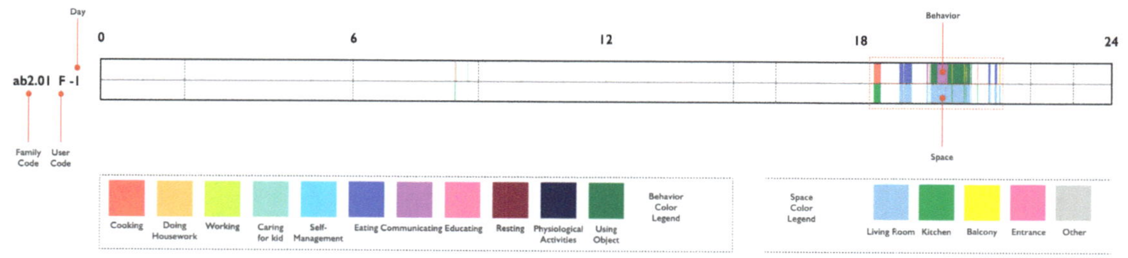

Figure 6. Timeline data legend.

Table 10. Behavior duration data in kitchen.

Family Code	Father	Mother	Kid1	Kid2/Grandmother
ab2.01	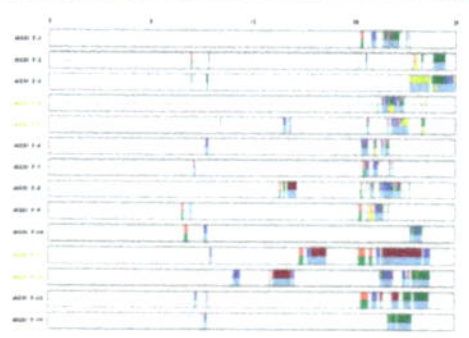	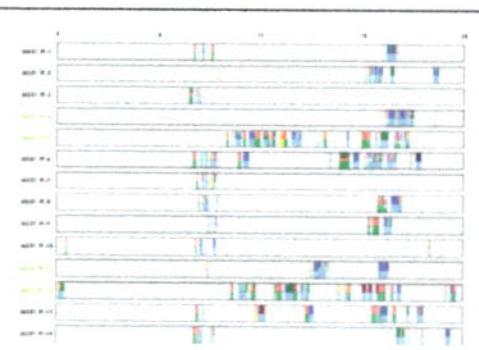	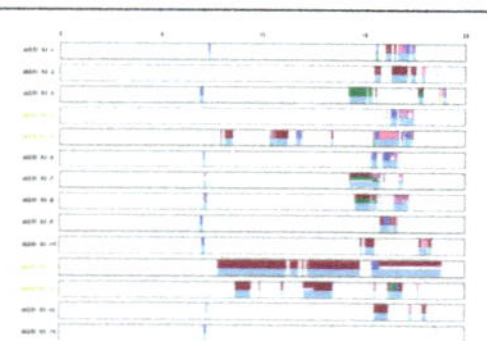	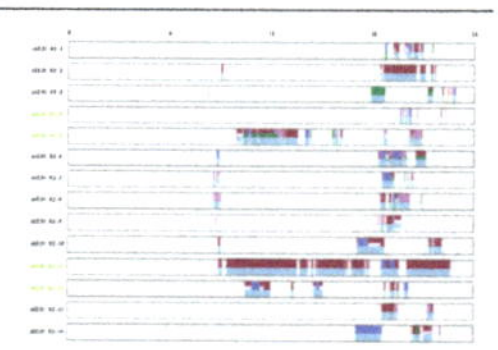
c3.04	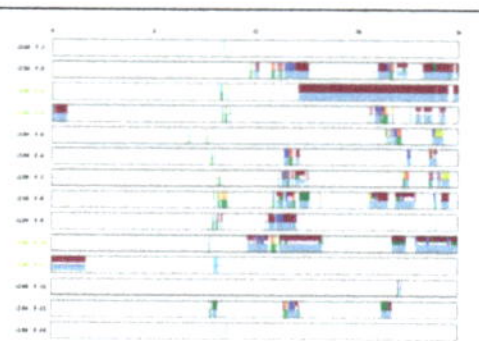	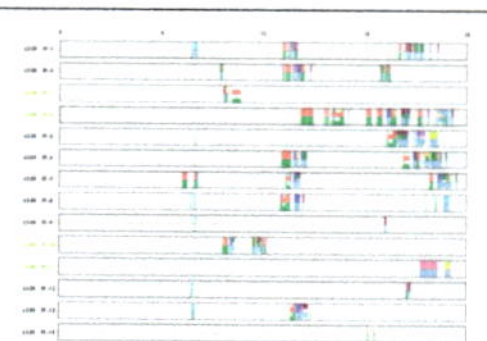	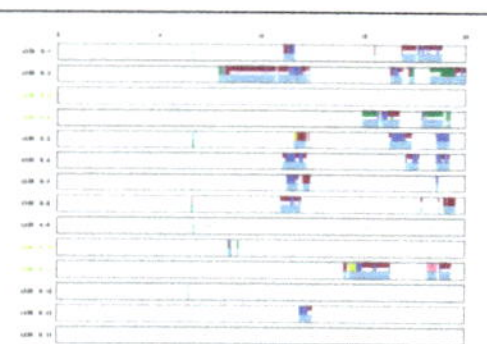	-
a2.02	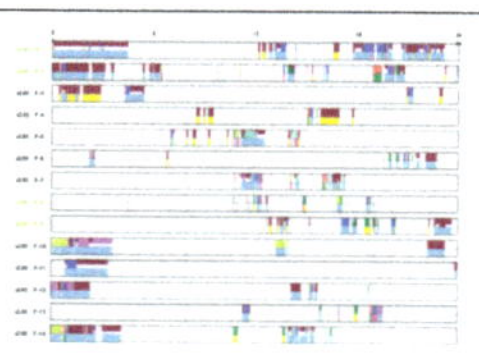	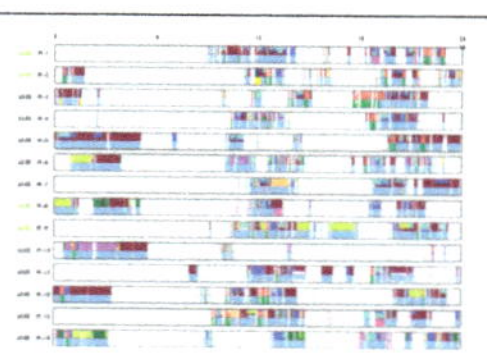	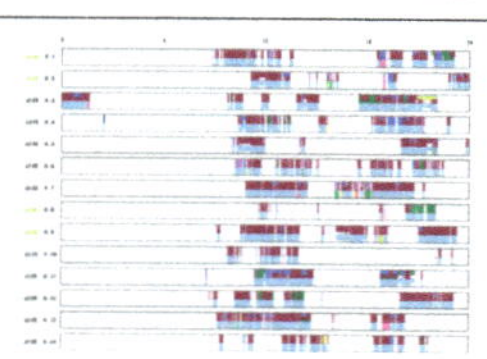	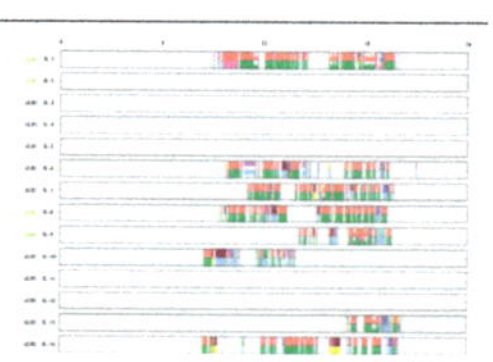
c2.02	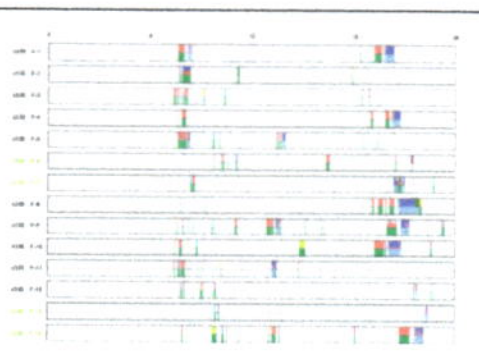	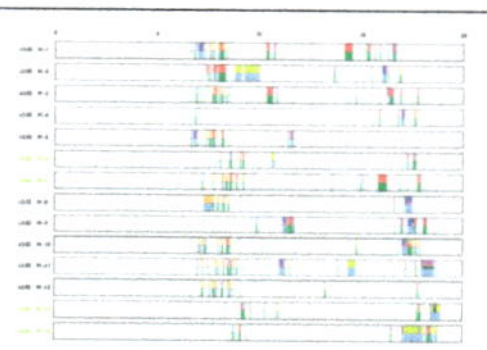	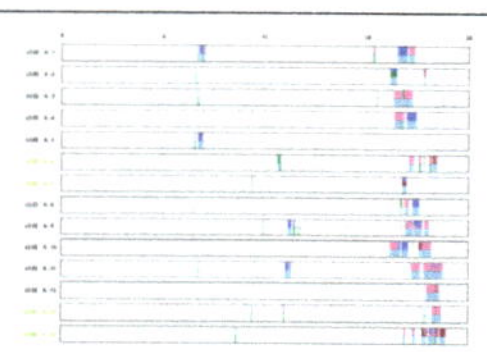	-

Table 10. *Cont.*

Family Code	Father	Mother	Kid1	Kid2/Grandmother
b2.01	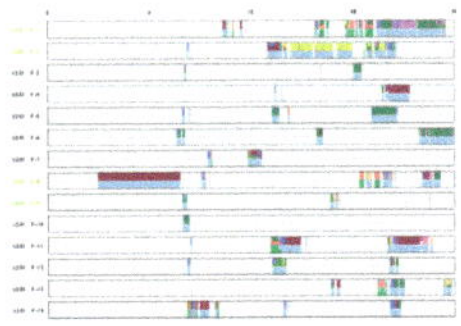	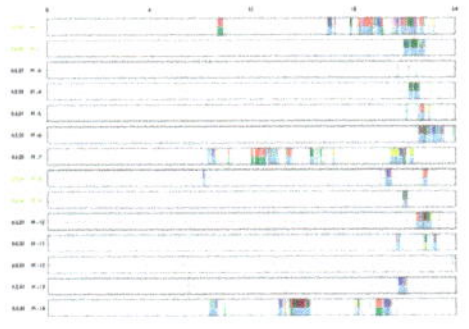	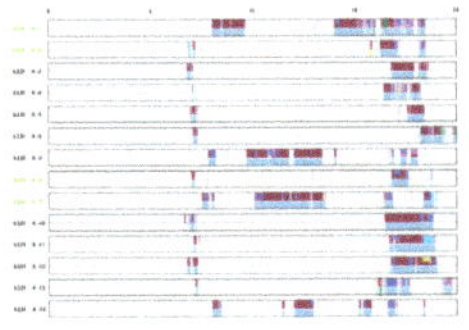	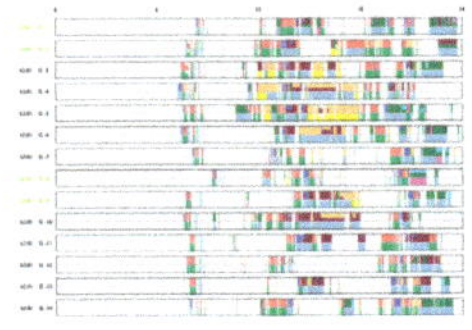
c3.02	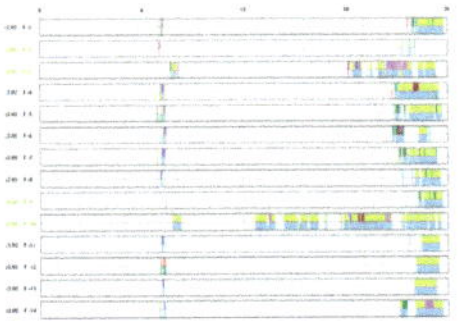	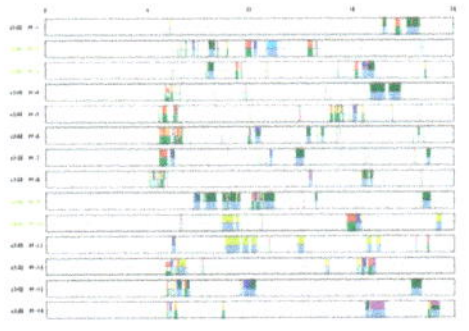	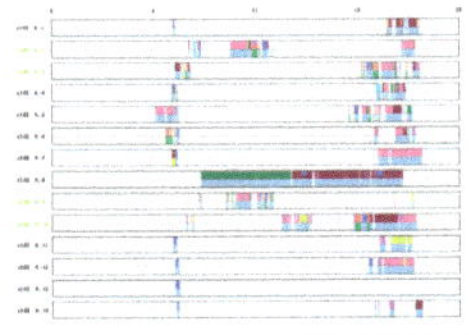	-

4. Discussion

4.1. General Analysis of Lifestyles

The data can be filtered by "user–object," "user–space," and "user–behavior" benchmarks to obtain data from different perspectives. The utilization of "user–space" frequency and interval, "user–behavior" weight disparity, and "user–object" usage frequency as filters for the data underscores the versatility of this method.

The combined statistics of the duration and frequency of the users' behaviors are shown in Figure 7. The left side of Figure 7 illustrates a breakdown of the blue area in the right-hand graph. The grandmother has the highest index of most behaviors, including DH (more than 1500 times, 100,000 s); CO (more than 100 times, 90,000 s), EA (more than 500 times, 50,000 s), and RE (more than 200 times, 100,000 s), followed by the child's RE (more than 250 times 100,000 s), CM, and EA behaviors. Next is the father's RE and CM behaviors. Overall, resting behavior is the primary life behavior among family members. While grandmothers do not constitute the primary occupants of the domestic environment, they nevertheless allocate a more significant proportion of their time and behavior to the household. This also confirms that the three-generation family with the parent–grandparent co-parenting of children is a prevalent phenomenon in Chinese families, especially at the preschool stage [39].

Figure 8a was obtained by collating the mean values of the duration of users' 11 types of behavior in the different spaces. A cursory examination of the images reveals that the behaviors exhibited by the father, mother, and grandmother in the entrance and balcony areas bear a striking resemblance. A further analysis reveals that the behaviors the grandmother and mother exhibited in each space are analogous. A notable distinction emerges between the behaviors the children and the parents exhibited, particularly concerning inter-behavioral proportions.

According to the three types of behavioral division of living and housing science, the data from 11 behaviors were summarized to obtain a ratio graph of the three types of living behaviors, as shown in Figure 8b. According to the graph, the following is evident. 1. Fathers have a relatively high proportion of the third style of life behaviors in each space compared to mothers. 2. Mothers and grandmothers have similar behavior patterns, but grandmothers have a high proportion of the second style of life behaviors. 3. The father's behavior on the balcony is like that of the mother and grandmother. However, his actions are more like those of the third style of life behaviors. The father's living room behavior is like that seen in the interactive model on the balcony. The proportions of the three types of behaviors demonstrate that within the domestic space, gender differences are evident, particularly in the leisure and hedonic behaviors and domestic behaviors of fathers and mothers.

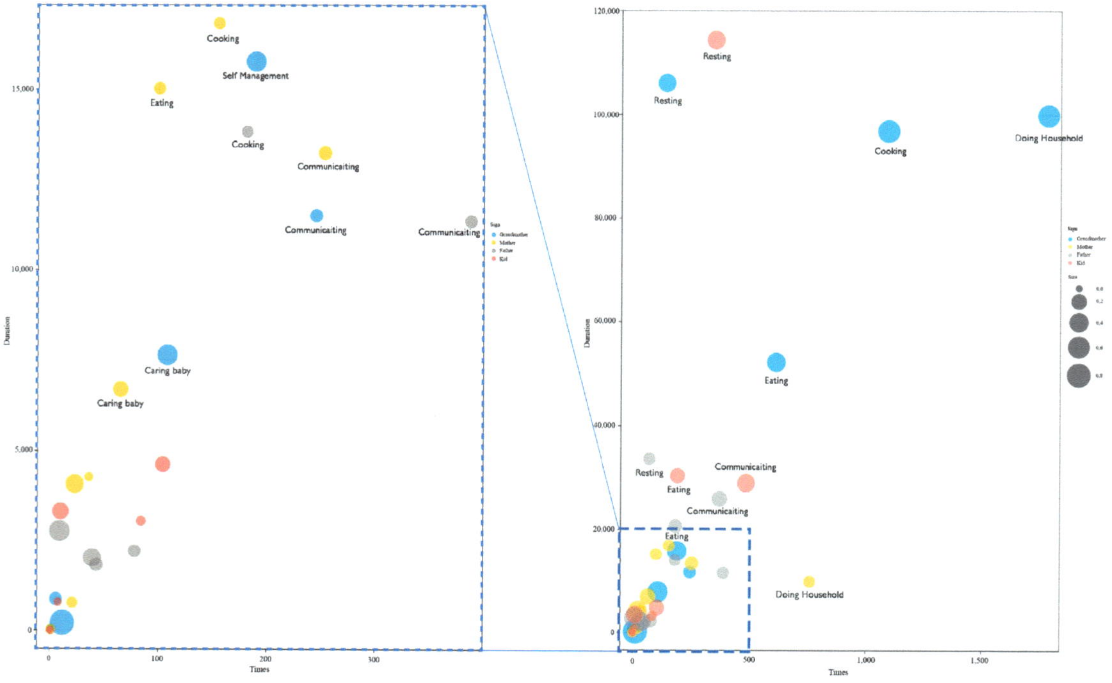

Figure 7. Distribution of frequency and duration of behaviors.

Figure 8. (**a**) Behavior ratio of 11 main behaviors; (**b**) behavior ratio of 3 styles of life behaviors.

The data for 288 days from six families were integrated to generate a life behavior timeline characteristic graph. In Figure 8, yellow represents the father's behavior, red represents the mother's, and blue represents the child's behavior. Figures 9–12 is divided into 24 h, and the thickness of the behavior line segment indicates the frequency of the behavior occurring in that current period. By conceptualizing behavioral data within the living space, its multifaceted nature can be deconstructed through diverse lenses, establishing a foundational framework for subsequent analysis.

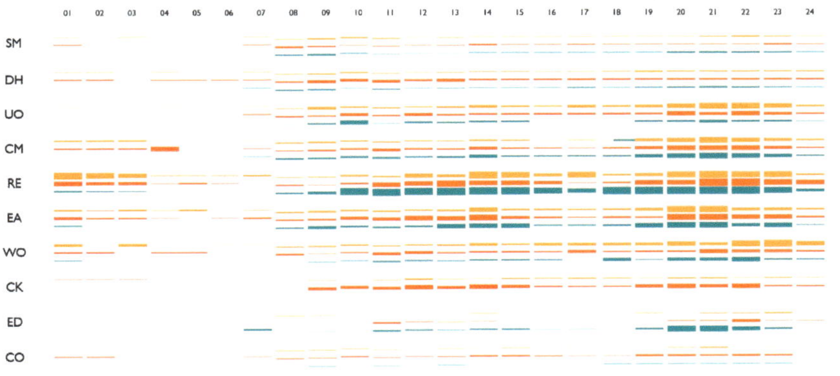

Figure 9. Spatio-temporal data on residential behavior.

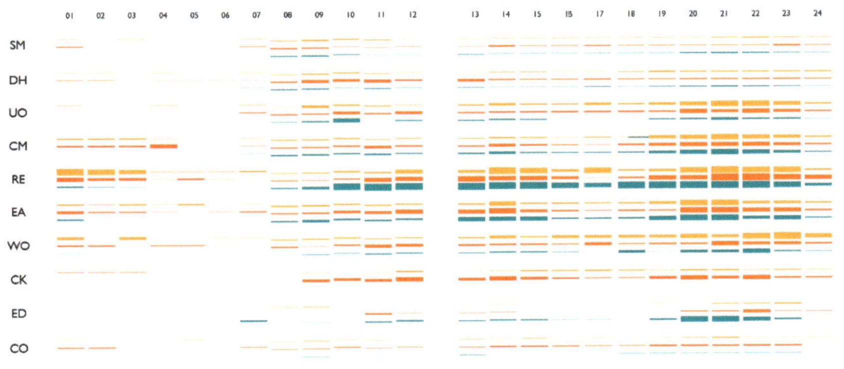

Figure 10. Temporal differences in behavior.

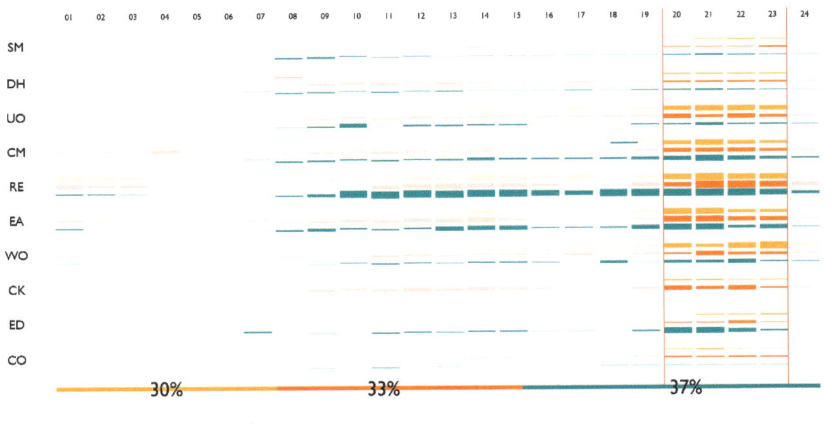

Figure 11. Child's behavior in living room.

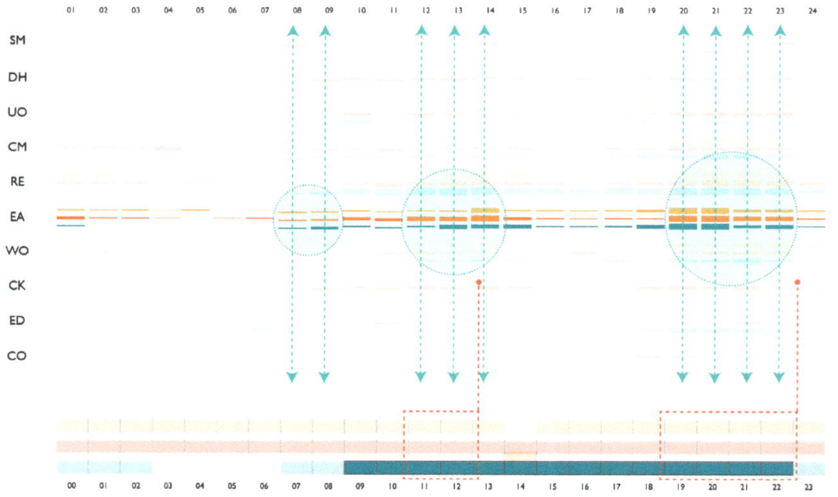

Figure 12. Eating-centered behavior.

4.2. The Design Direction of the Interior Through Lifestyles

Through the organization of quantitative data, users can be categorized. Subsequently, a qualitative analysis was undertaken to delineate distinct user archetypes and succinctly compile attributes associated with each user category. Consequently, derived from the overarching analysis, interior design strategies for diverse user typologies for Beijing residents can be delineated along four distinct trajectories.

1. Design that considers temporal differences in behavior.

Figure 10 shows that in dual-income families, there is less behavior in the morning than in the afternoon; the whole family eats breakfast at home, which is not always the case because parents go to work, and children go to school; and breakfast takes less time than lunch or dinner. The child is young and falls asleep at dawn, so the parents use the living room; the child's behaviors are average from 9 a.m. onwards. The peak of family activity is between 20 and 21 h; the child's primary learning time is between 19 and 22 h. Resting is the main family activity event in the living room; for all family members, the living room is a place for all-weather relaxation; fathers use the living room more extensively, and children use the living room for more extended periods. Between 10 and 14 h and between 18 and 22 h are periods of high diversity and long durations of family behaviors. Interior design should consider the intersection of the diverse behaviors of family members. Resting is the most critical behavior in the living room; relaxation in the living room occurs almost all day. The two main mealtimes and mealtime activities are evenly distributed throughout the day. Interactive activities may occur between different behaviors, especially during meals; between the study and work behaviors in the living room; and between most behaviors in relative proximity throughout the home. Consequently, when undertaking spatial planning and design, it is imperative to consider the sequence of behaviors observed during the morning hours and the high-frequency crossover of behaviors in the afternoon. The configuration of space and furniture must be designed to accommodate the temporal variability in behaviors in their totality, thereby facilitating the transition between periods of high and low density of behaviors.

2. Living room design centered on child behavior.

Figure 11 shows that the child's main activities in the living room are resting, eating, communicating, and educating; parental activities increase as the child's activities increase;

learning work in the living room may conflict with the parent's recreational activities. On weekdays, children should go to school, but the child's living room hours starting at 9 o'clock are clearly shown in this chart. Therefore, most weekends are estimated to be spent by children using the living room during the day. The child's resting behavior is the longest; although the child spends significantly less time using electricity than the parent, electricity use is still the child's main activity. In many households, furniture does not fit the child's size. The child's eating and entertainment activities are estimated to be difficult to focus on. Therefore, the child element must be considered during the planning and design of the space. It has been observed that the needs of children are frequently overlooked in the design of family spaces. However, children are the most significant users of these spaces, particularly in learning, eating, entertainment, and other behavioral aspects of their lives. Inadequate furniture dimensions and impractical home layouts can potentially inflict physiological harm on children. Furthermore, the absence of designated independent space for children in Beijing family apartments hinders the cultivation of their autonomy. Interior design should consider the intersection of the diverse behaviors of family members. Resting is the most critical behavior in the living room; relaxation in the living room occurs almost all day. The two main mealtimes and mealtime activities are evenly distributed throughout the day. Interactive activities may occur between different behaviors, especially during meals; between the study and work behaviors in the living room; and between most behaviors in relative proximity throughout the home.

3. Common family behavior linked to eating behavior.

Figure 12 shows that behavior changes in the living room are based on eating behavior. The diversity of different behaviors increases when eating. It can be assumed that people move according to different eating behaviors. Resting behavior, electricity use behavior, and communication behavior are highly related to meals; during the eating period, these acts increase noticeably. Children eat lunch later than their parents and dinner earlier than their parents; their behaviors are average from 9 a.m onwards. The activities at dinner are the most frequent and diverse. There is the most family interaction at dinner and after dinner. In addition to dinner, families eat snacks in the living room. Other behaviors can also occur when eating snacks. Families often drink water in the living room, which is related to having a hot water bucket in the living room. Family members' mealtimes are relatively concentrated, but meals can be eaten at different times. Different family members have different behavioral characteristics at mealtimes. The use location of the living room space when the family eats is also different, especially when using the sofa; the father usually sits on it and uses it, but the child has a small stool because of their height.

4. The consideration of the home office and the massive use of electronic devices in the digital age.

Two-income Chinese families usually have two or three bedrooms, so there is no study room; therefore, the living room becomes the place to work at home and for the children to study. The most used things in the living room are TVs, cell phones, and computers, so electronics are the main items used in the living room. The area where the children study is separate from where the parents work. All actions in the family, such as eating, resting, and studying, are accompanied by using electrical appliances; mobile phone charging problems also occur. The office furniture may not meet the needs of office workers because there is no dedicated office space at home, and other furniture needs to be moved frequently to meet these needs.

5. Conclusions

This study summarizes the user research process related to interior design under the design framework system that combines user culture-form and discusses the utility of the Observer XT in this process. The framework system evaluates specific knowledge that can be used to continuously assess the quality of design alternatives before providing design practitioners with relevant decision-making information. This enhances the efficiency, quality, and consistency of user-centered design (UCD) decision-making, optimizing the decision process for user-centered interior and furniture design.

Based on the lifestyle dimension, four design directions for residential space were postulated and substantiated through an investigation into Beijing residents' living morphologies to validate this methodology's efficacy. This study highlighted several key aspects of family behavior in living spaces, particularly in the context of dual-income families in China. Firstly, the design of living spaces should account for temporal differences in behavior, acknowledging that family activities vary throughout the day, with peaks in activity and a high diversity of behaviors occurring during specific times. The configuration of space and furniture must adapt to these temporal variations, facilitating smooth transitions between periods of high and low behavior density. Secondly, the living room design should be centered around child behavior, recognizing that children are significant users of this space for learning, eating, entertainment, and rest. Inadequate furniture dimensions and impractical home layouts can pose physiological risks to children, emphasizing the need for child-friendly designs. Thirdly, common family behaviors are closely linked to eating behavior, with mealtimes as focal points for increased diversity and interaction among family members. Lastly, this study underscores the importance of considering the home office and the prevalent use of electronic devices in the digital age, as the living room often doubles as a workspace and study area due to the limited number of bedrooms in Chinese homes. These findings emphasize the need for holistic and adaptive living room designs that cater to modern families' diverse and dynamic needs. In the process of redesigning existing residential properties, the demolition of existing space is often more challenging than altering the layout. However, the redesign of residential spaces should prioritize the needs of children and be adaptable to behavioral changes. Consequently, the variability, mobility, and accessibility of furniture become the primary focus, superseding the emphasis on space remodeling.

Considering this research's industrial and academic contributions, observing residential space behaviors is labor-intensive, requiring substantial effort. The user observation process based on the USO framework system and Observer XT software can significantly reduce the recording and analysis of redundant data. Sharing observation results among observers can shorten the time required for observations while systematically extracting a substantial amount of data generated during long-term observations, thereby reducing data analysis challenges. Additionally, this study provides a logical and rational basis for subsequent qualitative analyses, enabling design practitioners to conduct user observation research theoretically and systematically. It is imperative to acknowledge that the process of coding and recording behaviors necessitates the establishment of clear standards and comprehensive training to mitigate potential biases in the observation process.

From an academic perspective, this study transforms traditional qualitative user observation methods into more scientifically and rigorously engineered problems, addressing concerns about overly subjective judgments in observation methods lacking reasonable logic. Furthermore, this research explores the framework built using the user observation database proposed in the USO framework, facilitating knowledge transfer among individuals within an organization during the design process in different fields. This study offers a general analysis grounded in the 11 main categories of spatio-temporal data. Future research has the potential to expand on and delve deeper into existing data. For example, the broad classification of "resting" encompasses a wide range of actions that could be further subdivided into more specific categories, thereby facilitating a more nuanced examination of the phenomena. Alternatively, this database could catalyze subsequent research, prompting the formulation of additional sub-questions to guide more in-depth investigation.

Inevitably, this study has some limitations despite some advances. The first is that it is an observational study. In the framework of this observational study, supplementing the data to compensate for the disadvantages of the survey is overly subjective, but inevitably, the sample size cannot be guaranteed. The second is that the composition of families and family cycles are diverse, and this study cannot comprehensively cover and categorize families in different situations, which is also the future direction of our team's research. Another constraint of the present study is that due to the implementation of stratified random sampling for the selection of subgroups across diverse age groups, establishing a control group for direct comparison was not viable. This limitation imposes a restriction on this study's capacity to make specific comparisons and draw conclusions that would be bolstered by the incorporation of a control group in subsequent research endeavors.

Author Contributions: F.L.: conceptualization, methodology, and writing—review; Y.W.: data curation; Q.A.: writing—original draft, writing—review and editing, and visualization. All authors have read and agreed to the published version of the manuscript.

Funding: This work was supported by the China National Arts & Crafts Society, CNACS2024-I-1, and Research on the Inheritance and Innovation Path of Jingzuo Furniture under the Change of Beijing Urban Lifestyle, Beijing City University, KYQH202449. This work was also supported by the Oriental Cultural & Design Center.

Institutional Review Board Statement: All subjects gave their informed consent for inclusion before they participated in this study. This study was conducted in accordance with the Declaration of Helsinki, and the protocol was approved by the Ethics Committee of Kookmin University (A2022-0501).

Informed Consent Statement: Informed consent was obtained from all subjects involved in this study.

Data Availability Statement: The data presented in this study are openly available in Mendeley Data: https://data.mendeley.com/datasets/b9bpyw7grt; https://data.mendeley.com/datasets/5874fmcfn3.

Acknowledgments: We thank all the participants in the recording and all the researchers for coding.

Conflicts of Interest: The authors declare no conflicts of interest.

References

1. Jin, C.; Chen, Z. Housing Inclusion Perspective: The Causal Structure of Housing Cost Burden, Fertility Intentions, and Social Housing in Urban China. *Sustainability* **2024**, *16*, 9352. [CrossRef]
2. Andargie, M.S.; Touchie, M.; O'Brien, W. A review of factors affecting occupant comfort in multi-unit residential buildings. *Build. Environ.* **2019**, *160*, 106182. [CrossRef]
3. Ellsworth-Krebs, K.; Reid, L.; Hunter, C.J. Integrated framework of home comfort: Relaxation, companionship and control. *Build. Res. Inf.* **2019**, *47*, 202–218. [CrossRef]
4. Mitton, M.; Nystuen, C. *Residential Interior Design: A Guide to Planning Spaces*; John Wiley & Sons: Hoboken, NJ, USA, 2021.
5. Yang, H.; Lee, W.; Lee, H. IoT smart home adoption: The importance of proper level automation. *J. Sens.* **2018**, *2018*, 6464036. [CrossRef]
6. Heydarian, A.; Pantazis, E.; Wang, A.; Gerber, D.; Becerik-Gerber, B. Towards user-centered building design: Identifying end-user lighting preferences via immersive virtual environments. *Autom. Constr.* **2017**, *81*, 56–66. [CrossRef]
7. Choi, W. Design Approach to Support Prospective Memory for Independent Living among Aging People. *Arch. Des. Res.* **2016**, *29*, 79–96. [CrossRef]

8. Liu, J.; Liu, T.X. Two-child policy, gender income and fertility choice in China. *Int. Rev. Econ. Financ.* **2020**, *69*, 1071–1081. [CrossRef]
9. Xu, X.; Hanafi, Z.; Razak, N.A. Exploring Sociodemographic Correlates of Fathers' and Mothers' Behavioral Control. *Behav. Sci.* **2024**, *14*, 1203. [CrossRef] [PubMed]
10. Ytrehus, S. Interpretation of housing needs? A critical discussion. *Hous. Theory Soc.* **2000**, *17*, 166–174. [CrossRef]
11. Choi, K.R.; Choi, J.M.; Sato, K. Socio-Cultural Factors for New Product Acceptance in Home Environment Design. *J. Hum.-Environ. Syst.* **2008**, *11*, 65–71. [CrossRef]
12. Rapoport, A. Theory, culture and housing. *Hous. Theory Soc.* **2000**, *17*, 145–165. [CrossRef]
13. Ettema, D.; van der Lippe, T. Weekly rhythms in task and time allocation of households. *Transportation* **2009**, *36*, 113–129. [CrossRef]
14. Gao, C.; Wang, Y. Study on the Coupling Association between Human and House under the Influence of the Evaluation Index of Urban Residents' Behavior. *J. Phys. Conf. Ser.* **2021**, *1838*, 012056. [CrossRef]
15. Ganvir, L.; Kalita, C.P. Adoption of Socio-Cultural Aspects in PSS Design for Smart Home Products: An Integrative Review. *Arch. Des. Res.* **2022**, *35*, 7–29. [CrossRef]
16. Kuo, T.C.; Wu, H.H.; Shieh, J.I. Integration of environmental considerations in quality function deployment by using fuzzy logic. *Expert Syst. Appl.* **2009**, *36*, 7148–7156. [CrossRef]
17. Kim, H.J.; Choi, K.R.; Sung, Y.J. Multi-Functional Furniture Design in Small Living Space. *J. Korea Furnit. Soc.* **2011**, *22*, 190–198.
18. Pea, R.D. User centered system design: New perspectives on human-computer interaction. *J. Educ. Comput. Res.* **1987**, *3*, 129–134.
19. Daae, J.; Boks, C. A classification of user research methods for design for sustainable behaviour. *J. Clean. Prod.* **2015**, *106*, 680–689. [CrossRef]
20. Rosted, J. *User-Driven Innovation. Results and Recommendations*; Fora: Copenhagen, Denmark, 2005.
21. Won, H. Designs for sensing radiation: Deployment of a tangible interface and a visual projection interface for user interaction. *Arch. Des. Res.* **2020**, *33*, 57–71. [CrossRef]
22. Tsukada, A.; Maeda, T.; Yanagisawa, H.; Kaneko, S.; Mirakami, T.; Tozaki, M.; Tamaru, E.; Hasuike, K. Proposal for User-observation Design Assistant System Focused on Field Work. *IFAC Proc. Vol.* **2010**, *43*, 245–250. [CrossRef]
23. Jaglarz, A. Color as a Key Factor in Creating Sustainable Living Spaces for Seniors. *Sustainability* **2024**, *16*, 10251. [CrossRef]
24. Zhou, M.; Li, M.; Ono, K.; Watanabe, M. A Comparative Study of the User Interaction Behavior and Experience in a Home-Oriented Multi-User Interface (MUI) During Family Collaborative Cooking. *Future Internet* **2024**, *16*, 478. [CrossRef]
25. Nippert-Eng, C. *Watching Closely: A Guide to Ethnographic Observation*; Oxford University Press: Oxford, UK, 2015.
26. Lee, R.P. A Study on the early modernist woman architect, Margarete Schutte-Lihotzky-A Revolution of woman's space or an assignment to the new industrialization? *Korean Inst. Inter. Des. J.* **2005**, *14*, 20–27.
27. Yanagisawa, H.; Kozuka, Y.; Matsunaga, M.; Murakami, T. Observation support system for recording, reviewing and sharing observed design problems. In Proceedings of the International Association of Societies of Design Research (IASDR 2009), Seoul, Republic of Korea, 18–22 October 2009; pp. 1–10.
28. Janesick, V.J. Ethnographic inquiry: Understanding culture and experience. In *Forms of Curriculum Inquiry*; State University of New York Press: New York, NY, USA, 1991; pp. 101–119.
29. Lopez-Dicastillo, O.; Belintxon, M. The challenges of participant observations of cultural encounters within an ethnographic study. *Procedia-Soc. Behav. Sci.* **2014**, *132*, 522–526. [CrossRef]
30. Martin, J.L.; Clark, D.J.; Morgan, S.P.; Crowe, J.A.; Murphy, E. A user-centred approach to requirements elicitation in medical device development: A case study from an industry perspective. *Appl. Ergon.* **2012**, *43*, 184–190. [CrossRef] [PubMed]
31. Ducharme, D.A.; Arcand, I. Using Noldus Observer XT for research on deaf signers learning to read: An innovative methodology. *Behav. Res. Methods* **2009**, *41*, 833–840. [CrossRef] [PubMed]
32. Cohen, I.L.; Gardner, J.M.; Karmel, B.Z.; Kim, S.Y. Rating scale measures are associated with Noldus EthoVision-XT video tracking of behaviors of children on the autism spectrum. *Mol. Autism* **2014**, *5*, 15. [CrossRef] [PubMed]
33. Kanakri, S.M.; Shepley, M.; Tassinary, L.G.; Varni, J.W.; Fawaz, H.M. An observational study of classroom acoustical design and repetitive behaviors in children with autism. *Environ. Behav.* **2017**, *49*, 847–873. [CrossRef]
34. Sun, R.; Zhang, G.; Yuan, Z. The Preliminary Application of Observer XT (12.0) in a Pilot-Behavior Study. In Proceedings of the Engineering Psychology and Cognitive Ergonomics: 15th International Conference, EPCE 2018, Held as Part of HCI International 2018, Las Vegas, NV, USA, 15–20 July 2018; Proceedings 15. Springer International Publishing: Berlin/Heidelberg, Germany, 2018; pp. 686–700.
35. Noldus LP, J.J.; Spink, A.J.; Bollen, R.; Heffelaar, T. Smart mobility: Driver state estimation and advanced driver-vehicle interfaces. In Mobility Engineering, Proceedings of the CAETS 2015 Convocation on Pathways to Sustainability, New Delhi, India, 13–15 October 2015; Springer: Singapore, 2017; pp. 11–18.
36. Huang, Y. *The Analysis of User Research Method Based on User Cultural Factors-Focus on Behavior in Children's Roomactors*; Kookmin University: Seoul, Republic of Korea, 2008. Available online: http://www.riss.kr/link?id=T11475238 (accessed on 1 January 2022).

37. Kim, M.; Choi, K.R. A Study on Kitchen Use Behaviors through User Observation-Focused on Small-sized Apartments in Shanghai, China. *Korean Inst. Inter. Des. J.* **2015**, *24*, 121–128.
38. Liu, J.; Meng, B.; Wang, J.; Chen, S.; Tian, B.; Zhi, G. Exploring the Spatiotemporal Patterns of Residents' Daily Activities Using Text-Based Social Media Data: A Case Study of Beijing, China. *ISPRS Int. J. Geo-Inf.* **2021**, *10*, 389. [CrossRef]
39. Guan, C.Q.; Wang, Y.; Wang, Y. Grandparenting Role on Math Online Learning in Chinese Multigenerational Households. *Sustainability* **2022**, *14*, 11551. [CrossRef]

MDPI AG
Grosspeteranlage 5
4052 Basel
Switzerland
Tel.: +41 61 683 77 34

Buildings Editorial Office
E-mail: buildings@mdpi.com
www.mdpi.com/journal/buildings

www.ingramcontent.com/pod-product-compliance
Lightning Source LLC
Chambersburg PA
CBHW041429210926
51850CB00013B/3388